Plants as Bioreactors for Industrial Molecules

Plants as Bioreactors for Industrial Molecules

Edited by

Santosh Kumar Upadhyay
Department of Botany, Panjab University
Chandigarh, India

Sudhir P. Singh
Center of Innovative and Applied Bioprocessing (CIAB)
Mohali, Punjab, India

WILEY

About the Editors

Dr. Santosh Kumar Upadhyay is currently working as an Assistant Professor at the Department of Botany, Panjab University, Chandigarh, India. Prior to this, Dr. Upadhyay was DST-INSPIRE faculty at the National Agri-Food Biotechnology Institute, Mohali, Punjab, India. He did his doctoral work at the CSIR-National Botanical Research Institute, Lucknow and received his Ph.D. in Biotechnology from UP Technical University, Lucknow, India. He has been working in the field of Plant Biotechnology for more than 14 years. His present research focuses in the area of functional genomics. He is involved in the bio-prospecting and characterization of various insect toxic proteins from plant biodiversity and defence and stress signaling genes in bread wheat. His research group at PU has characterized numerous important gene families and long noncoding RNAs related to the abiotic and biotic stress tolerance and signaling in bread wheat. He has also established the method for genome editing in bread wheat using CRISPR-Cas system and developed a tool SSinder for CRISPR target site prediction. His research contribution led to the publication of more than 70 research papers in leading journals of international repute. Further, there are more than five national and international patents, 40 book chapters, and 11 books in his credit. In recognition of his substantial research record, he has been awarded NAAS Young scientist award (2017–2018) and NAAS-Associate (2018) from the National Academy of Agricultural Sciences, India, INSA Medal for Young Scientist (2013) from the Indian National Science Academy, India, NASI-Young Scientist Platinum Jubilee Award (2012) from the National Academy of Sciences, India, and Altech Young Scientist Award (2011). He has also been the recipient of the prestigious DST-INSPIRE Faculty Fellowship (2012), and SERB-Early Career Research Award (2016) from the Ministry of Science and Technology, Government of India. Dr. Upadhyay also serves as a member of the editorial board and reviewer of several peer-reviewed international journals.

Dr. Sudhir P. Singh is currently Scientist-D at the Center of Innovative and Applied Bioprocessing (CIAB), Mohali, India. He has been working in the area of molecular biology and biotechnology for more than a decade. Currently, his primary focus of research is gene mining, biocatalyst engineering, and characterization of novel enzyme variants for the production of high-value functional biomolecules from low-cost feedstock. He has generated metagenomic data resources from diverse ecological niches and ethnic food samples and has characterized novel genes encoding enzyme variants with desirable catalytic properties for the biosynthesis of functional sugar molecules such as D-allulose, turanose, trehalose,

Kricelle Mosquera Deamici
Laboratory of Biochemical Engineering, College of Chemistry and Food Engineering, Federal University of Rio Grande, Rio Grande, Rio Grande do Sul, Brazil

Zeuko'o Menkem Elisabeth
Department of Biomedical Sciences, University of Buea, Buea, Cameroon; Antimicrobial and Biocontrol Agents Unit, Laboratory for Phytobiochemistry and Medicinal Plant Studies, Department of Biochemistry, University of Yaoundé I, Yaoundé, Cameroon

Dhara Gandhi
Department of Botany, The Maharaja Sayajirao University of Baroda, Vadodara, Gujarat, India

Rajeswaree Gohel
Department of Microbiology, Marwadi University, Rajkot, Gujarat, India

Juliane Karine Ishida
Botany Department, Institute of Biological Sciences, Federal University of Minas Gerais, Belo Horizonte, Brazil

Kirthikah Kadiresen
Division of Applied Biomedical Sciences and Biotechnology, School of Health Sciences, International Medical University, Kuala Lumpur, Malaysia

Madhu Kamle
Applied Microbiology Laboratory, Department of Forestry, North Eastern Regional Institute of Science and Technology, Nirjuli, Arunachal Pradesh, India

Elif Karlik
Department of Molecular Biology and Genetics, University of Istinye, Istanbul, Turkey

Amandeep Kaur
Department of Botany, Panjab University, Chandigarh, Punjab, India

Simardeeep Kaur
Division of Biochemistry, ICAR-Indian Agricultural Research Institute, New Delhi, India

Pradeep Kumar
Applied Microbiology Laboratory, Department of Forestry, North Eastern Regional Institute of Science and Technology, Nirjuli, Arunachal Pradesh, India; Department of Botany, University of Lucknow, Lucknow, Uttar Pradesh, India

Suelen Goettems Kuntzler
Laboratory of Microbiology and Biochemistry, College of Chemistry and Food Engineering, Federal University of Rio Grande, Rio Grande, Rio Grande do Sul, Brazil

Céline Laroche
Institut Pascal, Université Clermont Auvergne, Clermont-Ferrand, France

Jia Choo
Division of Applied Biomedical Sciences and Biotechnology, School of Health Sciences, International Medical University, Kuala Lumpur, Malaysia

Anna Pick Kiong Ling
Division of Applied Biomedical Sciences and Biotechnology, School of Health Sciences, International Medical University, Kuala Lumpur, Malaysia

Madhu
Department of Botany, Panjab University, Chandigarh, Punjab, India

A. Méndez-Zavala
Department of Chemical Engineering, Autonomous University of Coahuila, Saltillo, México

Carolina Merlo
Área de Bioplaguicidas, Instituto Multidisciplinario de Biología Vegetal (IMBIV-CONICET), Córdoba, Argentina; Área Aromas y Pigmentos, Instituto de Ciencia y Tecnología de los Alimentos (ICTA), Universidad Nacional de Córdoba (UNC), Córdoba, Argentina; Departamento de Recursos Naturales, Cátedra de Microbiología Agrícola, Universidad Nacional de Córdoba, Córdoba, Argentina

Amit Kumar Mishra
Department of Botany, School of Life Sciences, Mizoram University, Aizawl, Mizoram, India

Pragati Misra
Centre for Tissue Culture Technology, Jacob Institute of Biotechnology and Bioengineering, Sam Higginbottom University of Agriculture Technology and Sciences, Prayagraj, Uttar Pradesh, India

J.C. Montañez
Department of Chemical Engineering, Autonomous University of Coahuila, Saltillo, México

Michele Greque de Morais
Laboratory of Microbiology and Biochemistry, College of Chemistry and Food Engineering, Federal University of Rio Grande, Rio Grande, Rio Grande do Sul, Brazil

L. Morales-Oyervides
Department of Chemical Engineering, Autonomous University of Coahuila, Saltillo, México

Juliana Botelho Moreira
Laboratory of Microbiology and Biochemistry, College of Chemistry and Food Engineering, Federal University of Rio Grande, Rio Grande, Rio Grande do Sul, Brazil

Rupasree Mukhopadhyay
Department of Genetics and Biotechnology, University College for Women, Hyderabad, Telangana, India

Alejandra B. Omarini
Laboratorio de Biotecnología Fúngica y de los Alimentos, Asociación para el Desarrollo de Villa Elisa y Zona (ADVEZ), Consejo Nacional de Investigaciones Científicas y Técnicas (CONICET), Entre Ríos, Argentina

Elif Aylin Ozudogru
Department of Molecular Biology and Genetics, University of Istinye, Istanbul, Turkey

Ghanshyam Pandey
Department of Plant Protection and Plant Pathology, Naini Agricultural Institute, Sam Higginbottom University of Agriculture, Technology and Sciences, Prayagraj, Uttar Pradesh, India

María L. Peschiutta
Área de Bioplaguicidas, Instituto Multidisciplinario de Biología Vegetal (IMBIV-CONICET), Córdoba, Argentina; Área Aromas y Pigmentos, Instituto de Ciencia y Tecnología de los Alimentos (ICTA), Universidad Nacional de Córdoba (UNC), Córdoba, Argentina

Laura Oliveira Pires
Botany Department, Institute of Biological Sciences, Federal University of Minas Gerais, Belo Horizonte, Brazil

Romina P. Pizzolitto
Área de Bioplaguicidas, Instituto Multidisciplinario de Biología Vegetal (IMBIV-CONICET), Córdoba, Argentina; Área Aromas y Pigmentos, Instituto de Ciencia y Tecnología de los Alimentos (ICTA), Universidad Nacional de Córdoba (UNC), Córdoba, Argentina; Departamento de Recursos Naturales, Cátedra de Microbiología Agrícola, Universidad Nacional de Córdoba, Córdoba, Argentina

D. Rodríguez-Zuñiga
Department of Chemical Engineering, Autonomous University of Coahuila, Saltillo, México

Isabela Sandy Rosa
Botany Department, Institute of Biological Sciences, Federal University of Minas Gerais, Belo Horizonte, Brazil

Aryadeep Roychoudhury
Post Graduate Department of Biotechnology, St. Xavier's College (Autonomous), Kolkata, West Bengal, India

Gaurav Sanghvi
Department of Microbiology, Marwadi University, Rajkot, Gujarat, India

Karishma Seem
Division of Biochemistry, ICAR-Indian Agricultural Research Institute, New Delhi, India

Alok Sharma
Department of Botany, Panjab University, Chandigarh, Punjab, India

Chikanshi Sharma
Applied Microbiology Laboratory, Department of Forestry, North Eastern Regional Institute of Science and Technology, Nirjuli, Arunachal Pradesh, India

Archana Shukla
Department of Biological Sciences, Sam Higginbottom University of Agriculture, Technology and Sciences, Prayagraj, Uttar Pradesh, India

Pradeep Kumar Shukla
Department of Biological Sciences, Sam Higginbottom University of Agriculture, Technology and Sciences, Prayagraj, Uttar Pradesh, India

Prashant Kumar Singh
Department of Biotechnology, Pachhunga University College Campus, Mizoram University, Aizawl, Mizoram, India

Sudhir P. Singh
Center of Innovative and Applied Bioprocessing (CIAB), Mohali, Punjab, India

Vipin Kumar Singh
Department of Botany, K.S. Saket P.G. College, Ayodhya, Uttar Pradesh, India

O. Solís-Quiroz
Department of Chemical Engineering, Autonomous University of Coahuila, Saltillo, México

Nidhi Sonkar
Department of Molecular and Cellular Engineering, Jacob Institute of Biotechnology and Bioengineering, Sam Higginbottom University of Agriculture Technology and Sciences, Prayagraj, Uttar Pradesh, India

Himani Thakkar
Department of Biochemistry, All India Institute of Medical Sciences, New Delhi, India

Rufin Marie Kouipou Toghueo
Antimicrobial & Biocontrol Agents Unit (AmBcAU), Laboratory for Phytobiochemistry and Medicinal Plants Studies, Department of Biochemistry, University of Yaoundé I, Yaoundé, Cameroon

Santosh Kumar Upadhyay
Department of Botany, Panjab University, Chandigarh, Punjab, India

Vinnyfred Vincent
Department of Biochemistry, All India Institute of Medical Sciences, New Delhi, India

Ying Pei Wong
Division of Applied Biomedical Sciences and Biotechnology, School of Health Sciences, International Medical University, Kuala Lumpur, Malaysia

Megan Min Tse Yew
Division of Applied Biomedical Sciences and Biotechnology, School of Health Sciences, International Medical University, Kuala Lumpur, Malaysia

Jia Qi Yip
Division of Applied Biomedical Sciences and Biotechnology, School of Health Sciences, International Medical University, Kuala Lumpur, Malaysia

María P. Zunino
Área de Bioplaguicidas, Instituto Multidisciplinario de Biología Vegetal (IMBIV-CONICET), Córdoba, Argentina; Área Aromas y Pigmentos, Instituto de Ciencia y Tecnología de los Alimentos (ICTA), Universidad Nacional de Córdoba (UNC), Córdoba, Argentina; Departamento de Química, Cátedra de Química Orgánica, Universidad Nacional de Córdoba, Córdoba, Argentina

Julio A. Zygadlo
Área de Bioplaguicidas, Instituto Multidisciplinario de Biología Vegetal (IMBIV-CONICET), Córdoba, Argentina; Área Aromas y Pigmentos, Instituto de Ciencia y Tecnología de los Alimentos (ICTA), Universidad Nacional de Córdoba (UNC), Córdoba, Argentina; Departamento de Química, Cátedra de Química Orgánica, Universidad Nacional de Córdoba, Córdoba, Argentina

Preface

In the past two decades, the application of plants as bioreactors for the production of various industrial molecules has evolved into an important research area with numerous new opportunities. Low production cost, easy to scale up, good-quality produce, and easy downstream processing have attracted the rapid growth of plants as bioreactors in recent years. Genetic and gene engineering methods are helpful in further improving the yield and quality of the product in various plants. Several industrial products have been produced in plant bioreactors, such as pharmaceutical proteins, vaccines, medical diagnostics proteins, industrial proteins, antibodies, attenuated viral particles, and nutritional supplements, including carbohydrates, minerals, vitamins, etc. Since the product quality, concentration, and yield, etc., are essential for commercial products, various strategies such as tissue-specific expression, enhanced transcript stability, translation optimization, and subcellular accumulation are developed and further improved to increase proteins and other products yields in transgenic plants. Several plant-derived products have also been reached in the market.

The high-value biomolecules in the biosphere happen to be an excellent attraction for research and development. The scientific community's primary objectives are the technological developments to exploit biotic and abiotic components of the ecosystem for societal benefits in a sustainable manner. It is desirable to develop cost-effective biological systems to produce biomolecules vital in various sectors. Advancement in biotechnological research has enabled the engineering of various plants to produce biomolecules such as proteins, carbohydrates, and lipids, with crucial effects on health and agriculture. These biological systems have been proved to be economical devices for expressing natural molecules of pharmaceutical and nutraceutical significance and, therefore, called bioreactors. Recent biosynthetic technologies have paved the way to develop expression platforms for pilot scale biosynthesis of the metabolites of medical and agricultural importance. The leading advantages of plant bioreactors have emerged the opportunities for the development of edible vaccines and molecular farming of pharmaceutical proteins, insecticidal proteins, antioxidant molecules, secondary metabolites, bioavailable micronutrients, functional food products, etc.

The present book covers the holistic knowledge about plants as bioreactors from a general introduction to the applications in numerous fields, including pharmaceuticals, nutraceuticals, secondary metabolites, carotenoids, flavonoids, biopesticides, biofuels, etc. This book will act as a repository to get comprehensive information on the application of plants as a bioreactor in one place. This information would be significantly valuable for graduate students, academicians, researchers, and the general public.

Santosh Kumar Upadhyay
Department of Botany, Panjab University, Chandigarh, India

Sudhir P. Singh
Center of Innovative and Applied Bioprocessing (CIAB), Mohali, Punjab, India

Acknowledgments

We are thankful to the Panjab University, Chandigarh and the Department of Biotechnology, Government of India, for providing the facility to complete this book. We are grateful to all the esteemed authors for their exceptional contributions and reviewers for their critical evaluation and suggestions for the quality improvement of the book.

We would like to thank Miss Rebecca Ralf (Commissioning Editor), Kelly Labrum (Managing Editor) and Mahalakshmi Pitchai (Production Editor) from John Wiley & Sons Ltd for their excellent management of this project and anonymous reviewers for their positive recommendations about the book.

We also appreciate the support of our friends and research students, whose discussion and comments were beneficial to shaping this book. We thank our numerous colleagues for direct or indirect help in shaping this project.

SKU wish to express gratitude to his parents, wife, and daughter for their endless support, patience, and inspiration. SPS is grateful to his parents and family for consistent moral support. SPS acknowledges the support from CIAB and the Department of Biotechnology.

We would like to warmly thank the department and university faculties and staff for providing a great working environment.

1

Plants as Bioreactors: An Overview

Madhu[1], *Alok Sharma*[1], *Amandeep Kaur*[1], *Deepika Antil*[1], *Sudhir P. Singh*[2], *and Santosh Kumar Upadhyay*[1]

[1] *Department of Botany, Panjab University, Chandigarh, Punjab, India*
[2] *Center of Innovative and Applied Bioprocessing (CIAB), Mohali, Punjab, India*

1.1 Introduction

Plants are the primary producers of the ecosystem on which all other living organisms rely (Malmstrom 2010). They provide essential human diet components, such as carbohydrates, proteins, vitamins, and minerals, and are also a significant source of various phytochemicals (Kumar et al. 2021; Upadhyay and Singh 2021). Phytochemicals are naturally produced by plants as a result of their metabolic pathways and are also referred to as secondary metabolites (Anulika et al. 2016). For a long time, several pharmaceutical companies have used secondary metabolites to treat a variety of diseases. Furthermore, these chemicals are used in various industries, including cosmetics, herbicides, insecticides, flavors, and perfumes (Singh et al. 2017; Dixit et al. 2021a). On the other hand, traditional farming results in variation in the quality and quantity of raw materials due to different environmental conditions in different geographical areas (Seufert et al. 2012). Furthermore, many plants that produce commercially useful metabolites are difficult to cultivate outside their native environments, and as a result, they are overexploited, leading to their extinction (D'Amelia et al. 2021). A survey revealed surprising results that one-fifth of the 50 000 medicinal plants used today belong to threatened species (Pan et al. 2014). To safeguard these plants, several technologies such as plant-cell bioreactors and tissue culture have been used (Hussain et al. 2012). But with the discovery of recombinant-DNA (rDNA) technology, the entire research attention shifted to the development of transgenic crops. Now rDNA technology is mainly focused on improving and modifying proteins of commercial importance (Khan et al. 2016). Transgenic crops are considered bioreactors for the development of commercially important proteins. To develop transgenic crops, we can employ a variety of strategies. The most frequent and widely used strategy is Agrobacterium-mediated transformation, which includes the transformation of the desired gene into Agrobacterium cells, and then these cells are used to infect plants to transfer the essential genes to the plant

Plants as Bioreactors for Industrial Molecules, First Edition.
Edited by Santosh Kumar Upadhyay and Sudhir P. Singh.

genome. Electroporation, particle bombardment, and polyethylene glycol-mediated gene absorption are some more ways of direct gene transfer (Basso et al. 2020). The recombinant proteins produced in transgenic plants are then directed to different organelles for regular eukaryotic post-translational modifications (Hofbauer and Stoger 2013). A strong promoter sequence that boosts the expression of the desired product is necessary for excessive production of foreign protein (Gopal and Kumar 2013). Transgenic plants used as a bioreactor to create necessary plant products are a simpler, more appealing, and less expensive technique than microbiological systems (Yao et al. 2015). One big advantage is that plants frequently bloom throughout the year, making it simple to deliver the product to market on time. Plants that produce more biomass have the potential to produce more genetically modified commodities. We can also keep modified proteins in seeds for an extended period without losing their biological capabilities (Gunn et al. 2012). Furthermore, due to their natural and higher biochemical potential, plants are regarded as the best source of alkaloids, carbohydrates, fatty acids, proteins, phenolic compounds, and steroids (Sharma and Sharma 2009). However, for transgenics to be successful, contamination of food crops and their products as a result of gene integration and expression must be addressed (Bawa and Anilakumar 2013). As a result, strict regulatory issues must be addressed to establish transgenics for the development of recombinant proteins (Bawa and Anilakumar 2013). Plant bioreactors have recently achieved major advances in several fields. There is a myriad of products that are produced by plant bioreactors, such as medical, nutritional, industrial, and biodegradable plastics (Sharma and Sharma 2009). The main focus, however, is on the development of therapeutic proteins that can be taken orally or applied topically following extraction and purification (Daniell et al. 2009). The fermentors and bioreactors can be replaced with greenhouses with a proper plant growth chamber to reduce the upstream facility. In addition, plant tissues can be processed for oral delivery of food proteins, which would also reduce downstream processing (Fischer and Buyel 2020). Despite more than 20 years of research and reports about the efficacy of plant-produced products, very few products have gone all the way through the production and regulatory hurdles (Whitelam et al. 1993). In this chapter, we will discuss the factors that are essential for the optimization of the production of recombinant proteins in plants and different products developed by this approach.

1.2 Factors Controlling the Production of Recombinant Protein

1.2.1 Choice of the Host Species

The host plant species must be carefully chosen to ensure the efficient production of recombinant protein. Before selecting the host, we must consider several factors such as host species, biomass yield, and cost of the scale-up process (Sharma and Sharma 2009). To achieve great success, we must first investigate the species- or tissue-specific factors that influence recombinant protein accumulation and quality (Whitelam et al. 1993). There are several model plant species, such as alfa–alfa, Arabidopsis, banana, lettuce, pea, potato, purple false brome, maize, rice, soybean, tobacco, and tomato, which we can choose for recombinant protein production (Richter et al. 2000). Plants can produce a variety of products, but each has its own set of requirements for production. As a result, no single species can be ideal for producing all these products (Sharma and Sharma 2009). Plants from the wild are ideal for molecular farming because they do not pose a risk of cross-contamination (Bawa and Anilakumar 2013). On the other hand, scientists favor domesticated species due to their commercialization and ability to adapt to a wide range of environmental conditions (Sharma and Sharma 2009). Furthermore, self-pollinating crops should be considered over cross-pollinating crops to reduce the risk of transgene spread (Daniell 2002). The production sites of recombinant proteins in different plants are different, it may be leaves, cereal grains, oil, etc. (Xu et al. 2018). Major leafy crops used as bioreactors are alfalfa, tobacco, clover, and lettuce. However, leafy crops have a limited shelf life and need immediate processing (Sharma and Sharma 2009). On the other hand, in cereal crops like barley, maize, rice, and wheat, the storage site of protein is grain, where the protein may remain stable for more extended periods without any loss of its proteolytic activity (Garg et al. 2018). Moreover, avidin, which is the first plant-derived product, also came from a cereal crop (Hood et al. 1997). The recombinant proteins are also produced from oil crops due to their low-priced downstream processing (Zhu et al. 2018). Some proteins are targeted to the oil bodies, for instance, the protein oleosin. SemBioSys Genetics Inc. has developed this recombinant oleosin protein in safflower seeds (http://www.sembiosys.com) (Fischer et al. 2004). By using the same strategy, thrombin inhibitor hirudin was also produced in transgenic seeds of *Brassica napus*. It is estimated that more than 1 kg of protein is recovered from 1 ton of seeds (Parmenter et al. 1995). In the field of biopharmaceuticals, bryophytes are considered the best option due to their cost-effective purposes and also there are no risks associated with the release of the transgenic (Reski et al. 2015). A company called Greenovation has adopted moss as a bioreactor system for the manufacturing of various medicines (http://www.greenovation.com/). Other aquatic plants and green algae can be used to produce recombinant proteins (Fabris et al. 2020). Another essential factor that should be considered before selecting the plant is that the plant genome should be sequenced, annotated, and available on public databases.

1.2.2 Optimization of Expression of Recombinant Protein

In the era of biotechnology, protein production is now accomplished by expression systems derived from different lineages (Ullrich et al. 2015). In recent years, we have had great success in increasing higher protein production in transgenic plants. Transgenic plants are now a viable alternative to traditional expression systems for the production of recombinant proteins (Ullrich et al. 2015). The critical question that arises here is how we can optimize the expression of recombinant protein to commercialize it. If we have to maximize the expression of heterologous protein, we have to optimize it. For optimization, we should consider the adjustable parameters like the structure of mRNA and protein, vector system, and culture conditions (Hayat et al. 2018; Upadhyay et al. 2010a,b, 2011). Each of the parameters should be evaluated individually for the experiments. Moreover, to control the expression of the transgene, other factors like transcription, translation, and posttranslational modifications (PTMs) should also be considered. Here we will discuss how these factors are responsible for the optimization of recombinant protein expression.

1.2.2.1 Transcription

Transcription is a fundamental process that transcribes DNA into RNA. It is a significant determinant of the recombinant protein expression level in transgenic plants. The majority of strategies aimed at increasing recombinant protein expression are based on increasing transgene transcription. Therefore, we should consider several factors to improve transcription levels.

1.2.2.1.1 Promoters Promoters are the most important regulatory components in the transcription process. Promoters are the nucleotide sequences and crucial binding sites for RNA polymerase to initiate the transcription. In addition, the promoter sequences also act as regulatory sequences (Haberle and Stark 2018). They are usually present adjacent to the gene that is about to be transcribed. Moreover, understanding how a promoter works have opened up a whole new world of possibilities for gene expression regulation. A strong promoter is required to enhance the gene expression in transgenic plants. The most commonly used promoter is the cauliflower mosaic virus 35S (CaMV) promoter, which works constitutively and can induce the expression regardless of the environment or development factors. The CaMV promoter is mainly more effective for dicot plant species than monocot plant species (Bak and Emerson 2020). To improve and strengthen the CaMV 35S promoter, some elements of the promoter have been duplicated, which has ultimately been found to enhance the expression of the reporter transgene in the tobacco plant (Kay et al. 1987). Moreover, when the upstream region of the promoter was repeated, it had no significant impact on the expression of the *gusA* reporter gene in both tobacco and tomato transformants (Comai et al. 1990). There are several antigenic proteins, which have been produced in plants by using CaMV 35S promoter, for instance, protective antigen, hemagglutinin, cholera toxin subunit B (CTB), severe acute respiratory syndrome coronavirus (SARS-CoV) vaccine, and rabies virus glycoprotein (Aziz et al. 2005; Huang et al. 2001; Jani et al. 2002; Li et al. 2006; McGarvey et al. 1995).

Another important type of promoter is inducible promoters, which are required for inducible expression vectors (Corrado and Karali 2009). These inducible systems are mainly designed for fundamental and applied research and now they are becoming popular in plant molecular biology. These promoters are usually triggered by wounding and pathogen attacks. For example, the expression of some proteins is only required during any pathogen attack, and their constitutive expression may cause a detrimental effect on the growth and development of the plant. These inducible promoters are mainly regulated by chemical compounds like alcohol, steroids, and tetracycline. Moreover, to make the inducible promoters to be highly efficient, these chemical compounds are known as inducers, should be highly specific to the promoter, have a fast response to induction, and immediately turn off upon withdrawal, nonpoisonous to host, and should be easily available. These inducers provide high inducibility and specificity for the regulation of gene expression. There are many examples where inducible promoters are used to increase the expression of the different heterologous proteins. Human interferon-gamma and α1-antitrypsin are expressed by a sucrose starvation-inducible promoter of *the alpha-amylase* gene in rice plants (Chen et al. 2004; Terashima et al. 1999). The Ramy3D promoter, which is induced to express recombinant human granulocyte-macrophage colony-stimulating factor (hGM-CSF) by sucrose deprivation, is another important example of an inducible promoter. Moreover, when compared to tobacco cell-suspension cultures, the production of hGM-CSF in transgenic rice seedlings was found to be higher (Shin et al. 2003). Although inducible systems are widely accepted in plant biology, their application is still very limited. As a result, more research is needed to increase the number of inducers that are useful in the field of plant biotechnology.

Some recombinant proteins are expressed in specific tissue or organ, so to control the expression of such proteins, there is a need for tissue-specific promoters. These promoters are very useful to restrict the final product in specific organs like seeds, tubers, or fruit. For instance, fruits are used to store vaccine antigens, a fruit-specific promoter E8, is primarily used for various vaccine antigens, and this promoter was identified in the tomato (Deikman et al. 1992; Jiang et al. 2007; Ramirez et al. 2007; Sandhu et al. 2000). There are some other promoters that concentrate the expression of foreign bio-molecules in seeds, such as porcelain promoter, maize globulin-1, maize zein, 7S globulin, rice glutelin, soybean P-conglycinin a′-subunit (Belanger and Kriz 1991; Chen et al. 1986; Fogher 2000; Lau and Sun 2009; Marks et al. 1985; Osborn et al. 1988; Russell and Fromm 1997; Wu et al. 2005). Some promoters are chloroplast specific, for instance, the 16S ribosomal RNA promoter, which is used for the transgenic products like CTB and labile toxin B (LTB) (Daniell et al. 2001; Kang et al. 2004a; Koya et al. 2005; Ruhlman et al. 2007; Staub et al. 2000). Moreover, a plant-based pharmaceutical company "Chlorogen Inc." has patented a technology named chloroplast transformation technology (CTT™) to produce recombinant proteins in the chloroplast. Some recombinant proteins produced by this technology are interferon, cholera vaccine, insulin, and polymers. Another group of researchers has developed a smallpox vaccine and xylanase or E1 endoglucanase using a leaf-specific *rbcS* gene promoter in the plant leaves (Dai et al. 2000; Golovkin et al. 2007; Hyunjong et al. 2006).

1.2.2.1.2 DNA Methylation DNA methylation is an epigenetic modification of the DNA sequence primarily caused by DNA methyltransferases (DNMTs). DNA methylation is found to be involved in the regulation of the expression of various endogenous genes. It has been found that in both plants and mammals, DNA methylation is conserved and critical for growth and development (Zhang et al. 2018). Similarly, DNA methylation also causes suppression of transgenes in transgenic plants, which reduces the expression of chimeric transgenes. For example, in the tobacco plant, two transfer DNA (T-DNAs) were transformed, and after transformation, it has been found that the expression of T-DNAs was reduced due to the methylation in the promoters of these genes (Matzke et al. 1989). Moreover, it has been found that in *Agrobacterium tumefaciens* itself, 0.5% of cytosines are methylated. Thus, it may be possible that transgenes are methylated and get inactivated before transformation (Palmgren et al. 1993). Therefore, to prevent the transgene from methylation, some demethylating agents can be used before and after the transformation. For instance, Palmgren and colleagues used demethylating agent 5-azacytidine to increase the expression of the *β-glucuronidase* gene, which was transferred to tobacco leaf disks by *A. tumefaciens* (Palmgren et al. 1993).

1.2.2.1.3 Transacting Factors Transacting factors are the proteins that bind to the cis-regulatory elements for the regulation of gene expression. These factors are the positive regulators of gene expression. So, these factors are considered important in controlling transgene expression in the plant system (Carrier et al. 2020). To regulate transgene expression, these factors either directly bind to the promoter or interact with other factors and promote their recruitment to the promoter (Carrier et al. 2020). Yang and colleagues have engineered a transcription factor rice endosperm bZIP (REB) in rice plants. It has been found that the engineered REB binds with the rice globulin promoter and enhances the expression level of the *lysozyme* gene (Yang et al. 2001). In an experiment, the function of a transcription factor on transgene expression was studied by fusing the steroid-binding domain of the glucocorticoid receptor to a plant transcription factor. This glucocorticoid-responsive GAL4-VP16 fusion protein mainly acts as an inducer for the activation of a *luciferase* reporter gene in transgenic Arabidopsis and tobacco plants. In this study, it has been found that in the absence of a ligand, transcription factor functioning is suppressed. But after induction, the suppression is relieved and the transcription factor resumes its functioning in transgenic plants (Aoyama and Chua 1997). In another study, Lloyd and colleagues have seen that transparent *testa glabra* (*ttg*) mutant Arabidopsis plants were not able to produce trichomes, anthocyanins, and seed coat pigments; however, they were generating excess root hairs. So, they have experimented and found that the trichomes and anthocyanin production can be restored by overexpressing the transcription factor R of maize in the constitutive and inducible manner (Lloyd et al. 1994). In one more study, which was conducted by Kermode et al. in 2007, it is mentioned that the recombinant human α-L-iduronidase protein production in transgenic tobacco plants could be enhanced by the expression of a conifer abscisic acid insensitive 3 (ABI3) transcription factor (Kermode et al. 2007).

1.2.2.2 Post-Transcription Modifications

Following the transcription process, the pre-mRNA undergoes certain modifications known as post-transcriptional modifications and after these modifications, pre-mRNA becomes mature mRNA. These modifications include splicing, capping, and polyadenylation. All these modifications occur in the nucleus; mature transcripts are then translocated into the cytoplasm for protein synthesis. These posttranscriptional modifications play a critical role in the gene expression and thus influence the yield of recombinant proteins in transgenic plants (Bandopadhyay et al. 2010). Splicing is the most important modification of pre-mRNA involving the removal of introns and joining of exons. Introns are the non-coding intragenic sequences containing acceptor and donor sites required for splicing (Wang et al. 2015). The presence of introns has been found to affect transcription levels, but the mechanism is not fully understood. Enhancer sequences in introns are thought to increase expression levels, but the efficiency of splicing and other mRNA stability processing events could also be involved. In a number of cases, introns have been discovered to aid transgene expression in transgenic plants. For example, in Arabidopsis, a profilins (PRF2) promoter and complete expression of β-glucuronidase require an intron PRF2 (Jeong et al. 2006). In addition, the Arabidopsis *agamous* gene has an intron with enhancer sequences that are found to be involved in the expression of a reporter from a minimal promoter. Other Arabidopsis genes, such as *STK* and *FLC*, and wheat *VRN-1* have enhancer sequences in their introns (Rose 2008). Likewise, in maize plants, the expression of the *alcohol dehydrogenase 1* (*ADH1*) gene is dependent upon the introns (Callis et al. 1987). Moreover, the introns of the *ADH1* gene are also responsible for the expression of heterologous gene *chloramphenicol acetyltransferase* of bacteria in maize protoplasts (Mascarenhas et al. 1990). There are some other examples where introns can influence the expression of genes, for instance, introns of *OsTua2*, *OsTua3*, *OsTub4*, and *OsTub6* (Gianì et al. 2009).

1.2.2.3 Translation

The translation is the process where the genetic codons are translated from mRNA to protein by ribosome translocation. To enhance the yield of recombinant protein in a heterologous system, the translation of the transgenes should be regularized. There are several important parameters to increase the translational efficiency and some are discussed later.

1.2.2.3.1 Position and Context of the Initiation Codon In eukaryotes, thousands of proteins are encoded on which a substantial amount of cellular energy is utilized. If there is any error occurs in the process of translation, it will lead to the production of inactive proteins that will disrupt cellular fitness (Harper and Bennett 2016). So, mRNA must be rightly translated into protein. For this, the ribosome must initiate the translation process with an appropriate codon, incorporate the suitable amino acids, and terminate only at the stop codon (Lutcke et al. 1987). The translation in eukaryotes is mainly initiated with a start codon, i.e. AUG. Furthermore, it has been noticed that the sequences near the initiation codon have an important role in initiating the translation process. In plants, sequences such as AACAAUGGC, GCCAUGGCG, and UAAACAAUGGCU have been identified near the initiation codon (Lutcke et al. 1987). Furthermore, Guerineau et al. 1992 studied the effect of these sequences near the initiation codon on the expression of a *GUS* reporter gene in tobacco mesophyll protoplast (Guerineau et al. 1992). They have formed an expression cassette by fusing these sequences with a duplicated CaMV 35S promoter. After that, the GUS activity was found to be three times more in translational fusions as compared to transcriptional fusions (Guerineau et al. 1992). Furthermore, Sawant and colleagues studied the effect of sequences downstream of the initiator codon on the gene expression and protein stability in tobacco leaves. They have inserted a sequence GCT TCC TCC after the initiator codon of two reporter genes, *uidA* and *gfp*, which further augment the expression of both the genes (Sawant et al. 2001). It has been found that to optimize transgene expression in plants, nucleotide sequence ACC or ACA insertion near the initiation codon will be preferred. In tobacco plants, increasing the expression of the *ctxB* gene as a C-terminal fusion protein with native plant protein "ubiquitinin" fragment has been achieved by codon optimization (Mishra et al. 2006).

1.2.2.3.2 Control of Gene Expression by 5′ and 3′ Untranslated Region To optimize the expression of the transgene, translation efficiency and RNA stability should be there. In some reports, it has been established that 5′ and 3′ untranslated regions (UTRs) of mRNAs play an important role in enhancing translation efficiency (Zeyenko et al. 1994). For instance, in tobacco plants, it has been reported that 5′ and 3′ UTRs of tobacco mosaic virus and alfalfa mosaic virus enhanced the gene expression (Zeyenko et al. 1994). There is another report where 5′ and 3′ UTRs of the *ADH1* gene in maize are found to be involved in increasing the translation efficiency in hypoxic protoplasts (Bailey-Serres and Dawe 1996; Hulzink et al. 2002). To enhance the translational efficiency, the structural features of 5′ leader sequences were found to be very helpful. A report published where it has been demonstrated that for the enhanced expression of pollen *ACT1* gene, 5′ UTRs are required in *Arabidopsis thaliana* (Vitale et al. 2003). Lu et al. and Samadder et al. also showed the role of 5′UTRs of polyubiquitin gene *RUBI3* along with its promoter to enhance the expression of *GUS*, which also enhances the translational efficiency in rice (Lu et al. 2008;

Samadder et al. 2008). Improving the translational initiation efficiency-related to increasing gene expression mainly depends upon the specific structure and optimization of 5′ UTR leader sequences. Moreover, the expression of genes can also be augmented by 5′ UTR sequences from heat shock genes or plant viruses (Datla et al. 1993). Likewise, 3′ UTR sequences are involved in enhancing gene expression, regulating polyadenylation, nuclear export, subcellular targeting, translational efficiency, and protecting mRNA from RNases. Haffani et al. characterized the AAUAAA sequence element in the coding part of the *Bacillus thuringiensis cry3Ca1* gene and found that it causes premature polyadenylation and reduced expression of the transgene in transgenic potato plants (Haffani et al. 2000). Consequently, for transgene expression in heterogeneous systems, such sequence elements should be avoided or modified (Kang et al. 2004b; Mishra et al. 2006; Richter et al. 2000).

1.2.2.4 Posttranslational Modifications (PTMs) of Recombinant Proteins

Nowadays, transgenic plants are a very convenient platform for the production of various important commercial and pharmaceutical recombinant proteins. Plants are gaining more acceptance for recombinant protein production than traditional systems like bacteria, viruses, yeast, and mammalian cells, especially where PTMs are needed (Burnett and Burnett 2020). The most important and studied PTM of plant-based recombinant protein is glycosylation. But for the production of high-quality proteins, other modifications are also required like deamidation, glycation, hydroxylation, lipidation, methylation, and phosphorylation (Friso and van Wijk 2015). However, during the production of recombinant proteins in transgenic plants, PTMs are found to be challenging. Although plant-specific PTMs are very important to improving the stability of recombinant proteins, which are produced for biopharmaceutical purposes (Bosch and Schots 2010; Singh et al. 2009; Xu et al. 2010). Glycosylation is the most studied PTM of plant-made recombinant proteins. It has been reported that the physicochemical properties like folding, self-association, resistance to proteolytic degradation, solubility, and protein sorting of recombinant proteins are affected by the attachment of glycans (Jacobs and Callewaert 2009; Walsh 2010). For glycosylation of recombinant proteins, the plant expression system is very suitable as in a prokaryotic system, glycosylation is not possible. It mainly occurs in the endoplasmic reticulum and golgi apparatus. Plants can add the N-linked glycans to the recombinant proteins with the core substituted by two *N*-acetylglucosamine residues. Moreover, there is a need to explore the PTMs except for N-glycosylation in plant expression systems required for the confirmation and functionality of pharmaceutical proteins. For instance, O-glycosylation modifications of recombinant proteins in transgenic plants, and there is a need to improve the production strategies of pharmaceuticals. However, there are some disadvantages to the plant expression system, like the inability to add galactose and terminal salic acids by plants to the plant-made proteins.

1.2.3 Downstream Processing

There should be an optimization for the full recovery of recombinant proteins in the plant-based production system. To get the purified form of recombinant protein, the undesired molecules like oils and phenolic compounds should be removed (Whitelam et al. 1993). The whole process for harvesting recombinant proteins from plants includes fractionation, extraction, and purification. Different strategies are there for the downstream processing of plant-produced recombinant proteins. Consequently, the economic factors should also be considered during the harvesting. According to the calculations about the cost of manufacturing the recombinant proteins in transgenic plants, it has been found that a plant-based production system is the most convenient system for large-scale production of recombinant proteins (Whitelam et al. 1993). It has been demonstrated that to extract the proteins from plant tissue, we should use a secretory system as in this system there is no need to disrupt the plant tissue (Schillberg et al. 1999). There is another method where proteins are targeted to oil bodies. As protein bodies are insoluble and oil bodies are light in weight, so, recombinant proteins can be separated by simple floatation centrifugation (Parmenter et al. 1995). Additionally, we can also use affinity tags for protein recovery, these tags should be removed after the recovery of proteins. Several tags are available, for instance, histidine- ricin toxin binding subunit B (HIS-RTB) was tagged with N-terminal hexahistidine, which was expressed in tobacco. To purify the recombinant protein lactose, an affinity column was used (Reed et al. 2005). Valdez-Ortiz et al. used metal-ion affinity chromatography to purify recombinant His-tagged amarantin (Valdez-Ortiz et al. 2005). Moreover, it was found that elastin-like polypeptide tags also enhance our target proteins expression. Similarly, other affinity tags like histamine, tryptamine, and phenylamine have also been evaluated for the purification of recombinant proteins. We can also employ protein splicing elements like inteins in protein purification. For instance, in tobacco plants, a mammalian antimicrobial protein, SMAP 29, was expressed with an intein fusion tag and purified by using a chitin column (Morassutti et al. 2005). Lacorte and colleagues used the nucleoprotein of tomato spotted wilt virus as an affinity tag for the purification of recombinant proteins from plant tissues (Lacorte et al. 2007). It showed that for the improvement of downstream processing of recombinant proteins from plants, affinity fusion tag methods are found to be very successful. It has been demonstrated that fusion proteins increase the stability of recombinant proteins and also protect these proteins from degradation (Whitelam et al. 1993).

1.3 Recombinant Proteins in Plants

1.3.1 Pharmaceutical Proteins

Proteins are highly complex substances present in all living organisms. They are used in several areas, including research and pharmacy. However, the extraction of the protein from natural sources is difficult and expensive, and sometimes they can also pose risks (Gupta et al. 2017). So, there is a need to develop a simple, inexpensive system that also allows the production of recombinant proteins at a larger scale. Many studies reported that plant systems have many economic, practical, and safety advantages for the large-scale production of proteins compared to other traditional systems (Gupta et al. 2017). Now plant-based production of pharmaceutical proteins has developed into a new viable industry (Spok 2007). Several pharmaceutically important mammalian proteins have been produced in transgenic plants, such as serum albumin, erythropoietin, human α-interferon, and growth hormone. Glucocerebrosidase (GC) is an important enzyme that develops in plants and catalyzes the degradation of glycosylceramide lipids. The deficiency of GC causes Gaucher disease in humans; it is an inherited disease that leads to the expansion of bone marrow, deterioration of the bones, and also causes visceromegaly. To treat this disease, enzyme replacement therapy is required, where modified GC is regularly administrated intravenously. This enzyme is the most expensive in the world; it costs about \$70 000–\$300 000 for a 50 kg person. To make it cost-effective, scientists have produced this enzyme in the tobacco plant. For this, first they have inserted the gene of human GC (hGC) in the tobacco plants by agrobacterium-mediated transformation. This was the first active human enzyme produced in plants. It has been found that one plant of tobacco can yield a sufficient amount of enzyme for one standard therapeutic dose (Cramer et al. 1996). Another important example of therapeutic protein produced in plants is human serum albumin (HSA). HSA is a globular, soluble, unglycosylated monomeric protein, which functions as a carrier protein for various fatty acids, steroids, and thyroid hormones (Peters 1995). It is used for the treatment of diseases like lethal erythroblastosis, hemorrhagic shock, hypoproteinemia, and ascites caused by liver cirrhosis (Hastings and Wolf 1992). It has been estimated that the market demand for HSA is about 500 tons per year globally (Xinhua News Agency 2007). The commercial production of HSA is mainly dependent upon the human plasma presently. But its supply is very limited and demand is very high, which has led to the rise in the price of HSA also, there are many cases of fake albumin in the market. Moreover, it has been reported that plasma-derived HSA has a risk for transmission of blood-derived infectious diseases like hepatitis and human immunodeficiency virus (HIV). Therefore, the demand for the development of recombinant HSA (rHSA) was raised, which should be economical and derived from nonanimal sources (Chamberland et al. 2001; Erstad 1996). Sijmons et al. developed the first plant-derived rHSA in transgenic tobacco and potato plants and have achieved 0.02% expression of soluble protein (Sijmons et al. 1990). Furthermore, in 2003, Millán and colleagues reported an 11.1% expression of total soluble protein in tobacco leaf chloroplasts (Fernández-San Millan et al. 2003). Huang et al. in 2005 achieved an 11.5% expression level of rHSA in rice cell cultures by sugar starvation-induced promoter (Huang et al. 2005). But the main disadvantage of these systems is they are not cost effective. So, in 2011, He et al. published a research article where they described the cost-effective large-scale production of serum albumin in rice seeds (He et al. 2011). Similarly, various other important pharmaceutical proteins are produced in different transgenic plants. Erythropoietin, a glycoprotein encoding complementary DNA (cDNA), is transformed into cultured tobacco BY2 cells. However, this recombinant erythropoietin can induce the differentiation and proliferation of erythroid cells only *in vitro* (Matsumoto et al. 1995). Furthermore, in 2010, Xu et al described a new method to increase the serum half-life of recombinant human growth hormone when it is expressed as an arabinogalactan protein in BY-2 cells of tobacco (Xu et al. 2010). A recombinant protein, the basic fibroblast growth factor (FGF-2) has been produced in rice grains. It is an important factor to stimulate cell proliferation during the wound healing process. It has been found that the expression of recombinant protein was 185.66 mg/kg in brown rice (An et al. 2013). A fibrinolytic enzyme lumbrokinase (LK) has been extracted from the earthworm. However, the extraction of the enzyme is a very laborious and time-consuming process. So, scientists have transformed the *LK* gene in *Helianthus annus*. To confirm the expression of recombinant LK, sodium dodecyl-sulfate polyacrylamide gel electrophoresis (SDS-PAGE) and western blot analysis were performed. The total yield of protein was found to be 5.1 g/kg of flower seeds. This finding provides a method to develop low-cost and seed-kernel edible delivery of therapeutic proteins (Guan et al. 2014). In 2017, Rattanapisit et al. have produced recombinant human osteopontin in *Nicotiana benthamiana*, and its expression was confirmed by western blot and enzyme-linked immunosorbent assay (ELISA) and purified by Nickel affinity chromatography. This report also described that the enhanced expression of recombinant osteopontin also stimulates the expression of osteogenesis-related genes *OSX*, *DMP1*, and *Wnt3a*, which was later confirmed by quantitative real time-polymerase chain reaction (qRT-PCR) (Rattanapisit et al. 2017). In 2019, another dentin protein, dentin matrix protein-1 (DMP-1), a noncollagenous protein in the matrix of bone was expressed in *N. benthamiana* and was studied that it can induce osteogenesis in human periodontal ligament stem cells (hPDLSCs). The findings demonstrated that this protein successfully induces osteogenic differentiation in hPDLSCs and that it could be a potential candidate for bone inducer in bone tissue engineering (Table 1.1) (Ahmad et al. 2019). A detailed study of pharmaceutical and therapeutic proteins is provided in Chapters 2, 4, and 5.

Table 1.1 List of recombinant proteins produced in various plant species.

S.No.	Recombinant protein	Plant species for expression	Function	References
Pharmaceutical proteins				
1.	Human glucocerebrosidase	*Nicotiana benthamiana*	Degradation of glycosylceramide lipids	Cramer et al. (1996)
2.	Human serum albumin	*Oryza sativa*	Carrier protein	He et al. (2011)
3.	Erythropoietin	*Nicotiana tabacum*	Erythrocytes formation	Matsumoto et al. (1995)
4.	Human growth hormone	*Nicotiana tabacum*	Growth	Xu et al. (2010)
5.	Fibroblast growth factor	*Oryza sativa*	Stimulate cell proliferation	An et al. (2013)
6.	Lumbrokinase	*Helianthus annus*	Antithrombosis protein	Guan et al. (2014)
7.	Human osteopontin	*Nicotiana benthamiana*	Dental bone regeneration	Rattanapisit et al. (2017)
8.	Dentin matrix protein-1	*Nicotiana benthamiana*	Induce osteogenesis	Ahmad et al. (2019)
Vaccines				
9.	Hepatitis B surface antigen	*Nicotiana tabacum*	Protection from hepatitis B virus infection	Mason et al. (1992)
10.	Structural protein	*Solanum tuberosum*	Immunogen of rabbit hemorrhagic disease virus	Castanon et al. (1999)
11.	Spike protein of TGEV	*Nicotiana tabacum*	Mucosal immunity	Tuboly et al. (2000)
12.	16 L1 major capsid protein	*Nicotiana tabacum*	Vaccine antigens of human papillomavirus	Varsani et al. (2003)
13.	Hepatitis B surface antigen	*Solanum tuberosum*	Edible vaccine for hepatitis B virus infection	Thanavala et al. (2005)

(*Continued*)

Table 1.1 (Continued)

S.No.	Recombinant protein	Plant species for expression	Function	References
14.	Haemagglutinin protein	*Nicotiana benthamiana*	Immunogenicity from H1N1 and H5N1	D'Aoust et al. (2008)
15.	Hemagglutinin-based VLP	*Nicotiana benthamiana*	Immunogenicity from H6N2 virus	Smith et al. (2020)
Antibodies				
16.	Single-chain Fv antibodies	*Oryza sativa* and *Triticum aestivum*	Immunity for carcinoembryonic antigen	Stöger et al. (2000)
17.	IgG monoclonal antibodies	*Nicotiana benthamiana*	Immunity for ebola virus	Huang et al. (2010)
18.	mAb CO17-1A	*Arabidopsis thaliana*	To treat anti-colorectal cancer	Song et al. (2013)
19.	mAb, E60	*Nicotiana benthamiana*	Immunity for dengue virus	Dent et al. (2016)
20.	mAb, 2G12	*Oryza sativa*	HIV-neutralizing antibody	Vamvaka et al. (2016)
21.	mAb D5 anti-EV71	*Nicotiana benthamiana*	Hand–foot–mouth disease	Rattanapisit et al. (2019)
22.	mAb anti-CHIKV E1	*Nicotiana benthamiana*	Immunity for Chikungunya virus	Hurtado et al. (2020)
Nutritional molecules				
23.	Lactoferrin	*Oryza sativa*	Increased iron content	Suzuki et al. (2001)
24.	Ferritin	*Oryza sativa*	Increased iron content	Goto and Yoshihara (2001)
25.	Nicotianamine synthase	*Oryza sativa*	Increased iron content	Wirth et al. (2009)
26.	Phytoene desaturase, phytoene synthase, and lycopene-cyclase	*Oryza sativa*	Increased vitamin A	Ye et al. (2000)
27.	PDX1 and PDX2	*Arabidopsis thaliana*	Increased vitamin B6	Raschke et al. (2011)
28.	GTP cyclohydrlase I	*Lycopersicon esculentum*	Increased folate content	de la Garza et al. (2004)
29.	Pterin and para-aminobenzoate	*Oryza sativa*	Increased vitamin B9	Storozhenko et al. (2007)
30.	GDP-1 galactose phosphorylase	*Solanum lycopersicum* and *Fragaria ananassa*	Increased vitamin C	Bulley et al. (2012)
31.	DHAR	*Oryza sativa*	Increased vitamin C	Gallie (2013)

1.3.2 Vaccine Antigens

Vaccination aims to protect people and animals from infectious diseases before the infection. Edward Jenner was the pioneer in the field of vaccinology. He discovered the vaccine for smallpox more than 200 years ago (Riedel 2005). With time there is a need to develop the improvement of vaccines. In the new era of biotechnology, scientists started to use intact or inactivated strains of pathogens or sometimes also use subunit vaccines, which are commercially produced in bacteria, yeast, or mammalian cells. First in 1990, Curtiss and Cardineau used the tobacco plant for the expression of *Streptococcus mutans* surface protein antigen A (Curtiss and Cardineau 1990). It has been found that plant-based expression systems have been used to develop vaccine antigens and antibodies for therapeutic applications in human and animal medicine over the past two decades. Several monocot and dicot species have been genetically engineered to produce edible vaccines and antibodies. After the first vaccine antigen production in plants, in 1992, hepatitis B surface antigen was also produced in transgenic tobacco (Mason et al. 1992). In transgenic potato plants, rabbit hemorrhagic disease virus major protein VP60 was expressed under the control of CaMV 35S promoter (Castanon et al. 1999). In one more study, it was found that the transmissible disease gastroenteritis is very severe and fatal for newborn pigs. The presence of transmissible gastroenteritis virus (TGEV)-specific antibody secretory immunoglobulin A (sIgA) develops mucosal immunity in young pigs. Another protein, i.e. spike protein of TGEV, reported being majorly involved in providing immunity. Tuboly and contemporaries generated the spike protein in transgenic tobacco plants (Tuboly et al. 2000).

The 16 L1 major capsid protein of human papillomavirus was also expressed in transgenic tobacco plants. To check the antigenicity of the virus, proteins in concentrated leaf extracts were tested using monoclonal antibodies (mAb) (Varsani et al. 2003). An edible vaccine of hepatitis B surface antigen was also created in potato plants. To check the immunogenicity, the clinical trials were organized, and the results showed that this edible vaccine could be considered a viable source of global immunization of hepatitis B (Thanavala et al. 2005). Another very well-known example of the plant-based production of vaccines is the influenza vaccine. Scientists expressed the hemagglutinin protein from two strains of influenza virus (H1N1 and H5N1) in *N. benthamiana* plants. To test the immunogenicity, two doses of 0.1 μg of purified influenza H5-virus-like particles (VLPs) were injected into the mice, and results showed that mice developed a strong immune response against the virus. It demonstrated that plants are a potential source of a bioreactor for vaccine antigens (D'Aoust et al. 2008). In 2009, Scotti et al. performed the agrobacterium-mediated transformation of human immunodeficiency syndrome virus Gag structural protein precursors encoding gene targeted to plastids. It was found that the protein expressed in plastids is similar to the proteins that are produced in the bacterial and viral systems. It showed that plastid transformation is a promising tool to develop HIV antigen in plants (Scotti et al. 2009). Recently, in 2020, antigenic molecule H6 subtype VLP of H6N2 virus is transiently expressed in *N. benthamina*. It was found that a single dose of this plant-based vaccine generates a strong immune response (Table 1.1) (Smith et al. 2020). The complete information about the vaccine antigens is given in Chapter 3.

triphosphate (GTP) cyclohydrolase I was expressed, which enhanced the folate content in the fruits (de la Garza et al. 2004). Transgenic rice was developed to increase the content of vitamin B9 by transforming Arabidopsis *pterin* and *para-aminobenzoate* genes. It was discovered that the content of vitamin B9 in transgenic rice is 100 times higher than in wild types (Storozhenko et al. 2007). Similarly, transgenic crops with increased vitamin C levels are available. *GDP-1 galactose phosphorylase* gene was used to increase vitamin C content in tomatoes, potatoes, and strawberries (Bulley et al. 2012). To increase ascorbate content in rice, the *dehydroascorbate reductase* gene was expressed under the control of the barley D-hordein promoter, and results showed that the transgenic rice had six times more ascorbate content than the wild type (Gallie 2013) (Table 1.1). Moreover, vitamin E content was also increased in transgenic tobacco, corn, and Arabidopsis (Dolde and Wang 2011; Tanaka et al. 2015). Likewise, various other nutraceuticals are also available that are produced in plants; their information are given in different chapters of the book.

1.3.5 Other Valuable Products

Various other products are also produced in plants like bio-pesticides, lipids/fatty acids, and nanoparticles. For instance, *Camelina sativa*, a member of the Brassicaceae family, can synthesize and store large amounts of fatty acids and oils in its seeds. Scientists studied the lipid metabolic pathways in *C. sativa* and genetically modified them to accumulate high levels of fatty acids and oils in seeds. They redesign the acetyl triacylglycerol synthesis in order to produce superior biodiesel and lubricant oils (Yuan and Li 2020). For the higher production of resistance oils, biosynthesis of hydroxylated fatty acids has been engineered. Furthermore, for the production of high-quality jet oils, medium-chain fatty acids are assembled (Yuan and Li 2020). Another very promising example of metabolic engineering is sandalwood oil, which is a high-demanding raw source for perfume industries. The main active principles of sandalwood oil are santolols, bergamotols, and other sesquiterpene compounds. A new subfamily, CYP76F, was identified in sandalwood plants via transcriptomic studies (Banerjee and Roychoudhury 2014). It has been reported that this subfamily, along with santalene synthase and nicotinamide adenine dinucleotide phosphate (NADPH)-dependent cytochrome P450 reductase (CPR), was found to be involved in the synthesis of α, β, and epi β-santalols and bergamotols. Moreover, the cloning of santalene synthases, CYP76, and CPR cDNAs is considered an important tool to increase the sandalwood oil content (Banerjee and Roychoudhury 2014). Metabolic engineering is beneficial for the overproduction of waxes. For instance, *fatty acyl reductase* (*FAR*) and *wax ester synthase* (*WS*) from *Mus musculus* were transformed in Arabidopsis to produce long-chained monounsaturated wax esters (Heilmann et al. 2012). Additionally, the production of bioplastics in plants has been also reported. Polyhydroxyalkanoates are natural polymers with thermoplastic properties. These are biodegradable plastic and have been found in a diverse range of bacteria under excess carbon and limited nitrogen conditions (Bhuwal et al. 2013). The bacterial *polyhydroxyalkanoic acids (PHA)* genes have been cloned in the Arabidopsis plant for their higher production. Nowadays, environmental protection is a great concern; and in order to aid it, scientists have developed biopesticides (Bhuwal et al. 2013). *Azadirachta indica* is a rich source of bioactive secondary metabolites. Azadirachtin is the most active constituent explored in the sector of eco-friendly and biodegradable biopesticides, which is characterized by low toxicity toward nontarget organisms (Thakore and Srivastava 2017). The biopesticide azadirachtin has been produced by the plant cell and hairy root cultures (Thakore and Srivastava 2017). Currently, phyto-nanotechnology is an emerging field for the production of plant-based nanostructures. The metallic nanoparticles prepared in plant seed extracts have been reported (Magnabosco et al. 2020). It showed that in the era of biotechnology, we can utilize plant systems for the production of different molecules, which will be cost-effective, stable, and eco-friendly. Similarly, other valuable products are produced in plants, and their detailed study is given in different chapters of the book.

1.4 Conclusions

Transgenic plants are of great importance for the manufacturing of different proteins, chemicals, oils, etc. They are now an important resource for multinational companies due to their numerous advantages over other traditional technologies. In transgenic plants, we can scale up the production of desired products and can also store the desired products in grains for longer periods without the loss of their functionality. A significant advantage provided by transgenic plants is the ability of glycosylation, which is not possible in microbial systems. However, there are some challenges in plants that need to be addressed. For instance, there is a need to pay attention to increasing the yield, bio-safety, and acceptability issues; manufacturing of mAb and other therapeutic proteins; and technical problems related to molecular farming. Currently, plant-derived commercialized products have a lower cost than native products. In the future, it is expected that a higher number of transgenic plant-based products will get commercialized. Moreover, attempts have already been made in this direction. Moreover, these products also required a favorable regulatory climate and public acceptance for their release. In the near future, probably all the major crops will become fully accessible to transgenic expertise for the overall production of plant-derived products.

References

Ahmad, A.R., Kaewpungsup, P., Khorattanakulchai, N. et al. (2019). Recombinant human dentin matrix protein 1 (hDMP1) expressed in *Nicotiana benthamiana* potentially induces osteogenic differentiation. *Plants* 8 (12): 566.

An, N., Ou, J., Jiang, D. et al. (2013). Expression of a functional recombinant human basic fibroblast growth factor from transgenic rice seeds. *International Journal of Molecular Sciences* 14 (2): 3556–3567.

Anulika, N.P., Ignatius, E.O., Raymond, E.S. et al. (2016). The chemistry of natural product: plant secondary metabolites. *International Journal of Technology Enhancements and Emerging Engineering Research* 4 (8): 1–9.

Aoyama, T. and Chua, N.H. (1997). A glucocorticoid-mediated transcriptional induction system in transgenic plants. *The Plant Journal* 11 (3): 605–612.

Aziz, M.A., Sikriwal, D., Singh, S. et al. (2005). Transformation of an edible crop with the *pagA* gene of *Bacillus anthracis*. *The FASEB Journal* 19 (11): 1501–1503.

Dolde, D. and Wang, T. (2011). Oxidation of crude corn oil with and without elevated tocotrienols. *Journal of the American Oil Chemists' Society* 88 (9): 1367–1372.

Düring, K., Hippe, S., Kreuzaler, F., and Schell, J. (1990). Synthesis and self-assembly of a functional monoclonal antibody in transgenic *Nicotiana tabacum*. *Plant Molecular Biology* 15 (2): 281–293.

Erstad, B.L. (1996). Viral infectivity of albumin and plasma protein fraction. *Pharmacotherapy* 16: 996–1001.

Fabris, M., Abbriano, R.M., Pernice, M. et al. (2020). Emerging technologies in algal biotechnology: toward the establishment of a sustainable, algae-based bioeconomy. *Frontiers in Plant Science* 11: 279.

Fernández-San Millán, A., Mingo-Castel, A., Miller, M., and Daniell, H. (2003). A chloroplast transgenic approach to hyper-express and purify human serum albumin, a protein highly susceptible to proteolytic degradation. *Plant Biotechnology Journal* 1 (2): 71–79.

Fischer, R., Stoger, E., Schillberg, S. et al. (2004). Plant-based production of biopharmaceuticals. *Current Opinion in Plant Biology* 7 (2): 152–158.

Fischer, R. and Buyel, J.F. (2020). Molecular farming–the slope of enlightenment. *Biotechnology Advances* 40: 107519.

Fogher, C.A. (2000). Synthetic polynucleotide coding for human lactoferrin, vectors, cells and transgenic plants containing it. Gene bank acc no AX006477, Patent WO0004146.

Friso, G. and van Wijk, K.J. (2015). Posttranslational protein modifications in plant metabolism. *Plant Physiology* 169 (3): 1469–1487.

Gallie, D.R. (2013). L-ascorbic acid: a multifunctional molecule supporting plant growth and development. *Scientifica* 2013: 795964.

Garg, M., Sharma, N., Sharma, S. et al. (2018). Biofortified crops generated by breeding, agronomy, and transgenic approaches are improving lives of millions of people around the world. *Frontiers in Nutrition* 5: 12.

Gianì, S., Altana, A., Camanoni, P. et al. (2009). In transgenic rice, α- and β-tubulin regulatory sequences control GUS amount and distribution through intron mediated enhancement and intron dependent spatial expression. *Transgenic Research* 18: 151–162.

Golovkin, M., Spitsin, S., Andrianov, V. et al. (2007). Smallpox subunit vaccine produced in planta confers protection in mice. *Proceedings of the National Academy of Sciences of the United States of America* 104: 6864–6869.

Gopal, G.J. and Kumar, A. (2013). Strategies for the production of recombinant protein in *Escherichia coli*. *The Protein Journal* 32 (6): 419–425.

Goto, F. and Yoshihara, T. (2001). Improvement of micronutrient contents by genetic engineering development of high iron content crops. *Plant Biotechnology* 18: 7–15.

Guan, C., Du, X., Wang, G. et al. (2014). Expression of biologically active anti-thrombosis protein lumbrokinase in edible sunflower seed kernel. *Journal of Plant Biochemistry and Biotechnology* 23 (3): 257–265.

Guerineau, F., Lucy, A., and Mullineaux, P. (1992). Effect of two consensus sequences preceding the translation initiator codon on gene expression in plant protoplasts. *Plant Molecular Biology* 18 (4): 815–818.

Gunn, K.S., Singh, N., Giambrone, J., and Wu, H. (2012). Using transgenic plants as bioreactors to produce edible vaccines. *Journal of Biotech Research* 4 (1): 92–99.

Gupta, V., Sengupta, M., Prakash, J., and Tripathy, B.C. (2017). Production of recombinant pharmaceutical proteins. In: *Basic and Applied Aspects of Biotechnology*, 77–101. Singapore: Springer.

Haberle, V. and Stark, A. (2018). Eukaryotic core promoters and the functional basis of transcription initiation. *Nature Reviews Molecular Cell Biology* 19 (10): 621–637.

Haffani, Y.Z., Overney, S., Yelle, S. et al. (2000). Premature polyadenylation contributes to the poor expression of the *Bacillus thuringiensis cry3Ca1* gene in transgenic potato plants. *Molecular & General Genetics* 264: 82–88.

Harper, J.W. and Bennett, E.J. (2016). Proteome complexity and the forces that drive proteome imbalance. *Nature* 537 (7620): 328–338.

Hastings, G.E. and Wolf, P.G. (1992). The therapeutic use of albumin. *Archives of Family Medicine* 1: 281–287.

Hayat, S.M., Farahani, N., Golichenari, B., and Sahebkar, A. (2018). Recombinant protein expression in *Escherichia coli* (*E. coli*): what we need to know. *Current Pharmaceutical Design* 24 (6): 718–725.

He, Y., Ning, T., Xie, T. et al. (2011). Large-scale production of functional human serum albumin from transgenic rice seeds. *Proceedings of the National Academy of Sciences* 108 (47): 19078–19083.

Hefferon, K.L. (2015). Nutritionally enhanced food crops; progress and perspectives. *International Journal of Molecular Sciences* 16 (2): 3895–3914.

Heilmann, M., Iven, T., Ahmann, K. et al. (2012). Production of wax esters in plant seed oils by oleosomal cotargeting of biosynthetic enzymes. *Journal of Lipid Research* 53 (10): 2153–2161.

Hiatt, A., Caffferkey, R., and Bowdish, K. (1989). Production of antibodies in transgenic plants. *Nature* 342 (6245): 76–78.

Hofbauer, A. and Stoger, E. (2013). Subcellular accumulation and modification of pharmaceutical proteins in different plant tissues. *Current Pharmaceutical Design* 19 (31): 5495–5502.

Hood, E.E., Witcher, D.R., Maddock, S. et al. (1997). Commercial production of avidin from transgenic maize: characterization of transformant, production, processing, extraction and purification. *Molecular Breeding* 3 (4): 291–306.

Huang, L.F., Liu, Y.K., Lu, C.A. et al. (2005). Production of human serum albumin by sugar starvation induced promoter and rice cell culture. *Transgenic Research* 14 (5): 569–581.

Huang, Z., Dry, I., Webster, D. et al. (2001). Plant-derived measles virus hemagglutinin protein induces neutralizing antibodies in mice. *Vaccine* 19 (15–16): 2163–2171.

Huang, Z., Phoolcharoen, W., Lai, H. et al. (2010). High-level rapid production of full-size monoclonal antibodies in plants by a single-vector DNA replicon system. *Biotechnology and Bioengineering* 106 (1): 9–17.

Hulzink, R.J.M., de Groot, P.F.M., Croes, A.F. et al. (2002). The 50-untranslated region of the *ntp303* gene strongly enhances translation during pollen tube growth, but not during pollen maturation. *Plant Physiology* 129: 342–353.

Hurtado, J., Acharya, D., Lai, H. et al. (2020). In vitro and in vivo efficacy of anti-chikungunya virus monoclonal antibodies produced in wild-type and glycoengineered *Nicotiana benthamiana* plants. *Plant Biotechnology Journal* 18 (1): 266–273.

Hussain, A., Qarshi, I.A., Nazir, H., and Ullah, I. (2012). Plant tissue culture: current status and opportunities. *Recent Advances in Plant In Vitro Culture* 6 (10): 1–28.

Hyunjong, B., Lee, D.S., and Hwang, I. (2006). Dual targeting of xylanase to chloroplasts and peroxisomes as a means to increase protein accumulation in plant cells. *Journal of Experimental Botany* 57: 161–169.

Jacobs, P.P. and Callewaert, N. (2009). N-glycosylation engineering of biopharmaceutical expression systems. *Current Molecular Medicine* 9: 774–800.

Jani, D., Meena, L.S., Rizwan-ul-Haq, Q.M. et al. (2002). Expression of cholera toxin B subunit in transgenic tomato plants. *Transgenic Research* 11 (5): 447–454.

Jeong, Y.M., Mun, J.H., Lee, I. et al. (2006). Distinct roles of the first introns on the expression of *Arabidopsis profilin* gene family members. *Plant Physiology* 140: 196–209.

Jiang, X.L., He, Z.M., Peng, Z.Q. et al. (2007). Cholera toxin B protein in transgenic tomato fruit induces systemic immune response in mice. *Transgenic Research* 16: 169–175.

Kang, T.J., Han, S.C., Kim, M.Y. et al. (2004a). Expression of non-toxic mutant of *Escherichia coli* heat-labile enterotoxin in tobacco chloroplasts. *Protein Expression and Purification* 38: 123–128.

Kang, T.J., Loc, N., Jang, M., and Yang, M. (2004b). Modification of the cholera toxin B subunit coding sequence to enhance expression in plants. *Molecular Breeding* 13: 143–153.

Kay, R., Chan, A.M.Y., Daly, M., and McPherson, J. (1987). Duplication of CaMV 35 S promoter sequences creates a strong enhancer for plant genes. *Science* 236 (4806): 1299–1302.

Kermode, A.R., Zeng, Y., Hu, X. et al. (2007). Ectopic expression of a conifer abscisic acid insensitive3 transcription factor induces high-level synthesis of recombinant human α-L-iduronidase in transgenic tobacco leaves. *Plant Molecular Biology* 63 (6): 763–776.

Khan, S., Ullah, M.W., Siddique, R. et al. (2016). Role of recombinant DNA technology to improve life. *International Journal of Genomics* 2016: 2405954.

Kim, S., Takahashi, M., Higuchi, K. et al. (2005). Increased nicotianamine biosynthesis confers enhanced tolerance of high levels of metals, in particular nickel, to plants. *Plant and Cell Physiology* 46 (11): 1809–1818.

Koya, V., Moayeri, M., Leppla, S.H., and Daniell, H. (2005). Plant-based vaccine: mice immunized with chloroplast-derived anthrax protective antigen survive anthrax lethal toxin challenge. *Infection and Immunity* 73 (12): 8266–8274.

Kumar, M., Puri, S., Pundir, A. et al. (2021). Evaluation of nutritional, phytochemical, and mineral composition of selected medicinal plants for therapeutic uses from cold desert of Western Himalaya. *Plants* 10 (7): 1429.

Lai, H., He, J., Hurtado, J. et al. (2014). Structural and functional characterization of an anti-West Nile virus monoclonal antibody and its single-chain variant produced in glycoengineered plants. *Plant Biotechnology Journal* 12 (8): 1098–1107.

Lau, O.S. and Sun, S.S. (2009). Plant seeds as bioreactors for recombinant protein production. *Biotechnology Advances* 27 (6): 1015–1022.

Lacorte, C., Ribeiro, S.G., Lohuis, D. et al. (2007). The nucleoprotein of Tomato spotted wilt virus as protein tag for easy purification and enhanced production of recombinant proteins in plants. *Protein Expression and Purification* 55 (1): 17–22.

Li, H.Y., Ramalingam, S., and Chye, M.L. (2006). Accumulation of recombinant SARS-CoV spike protein in plant cytosol and chloroplasts indicate potential for development of plant-derived oral vaccines. *Experimental Biology and Medicine* 231 (8): 1346–1352.

Lloyd, A.M., Schena, M., Walbot, V., and Davis, R.W. (1994). Epidermal cell fate determination in Arabidopsis: patterns defined by a steroid-inducible regulator. *Science* 266 (5184): 436–439.

Lu, J., Sivamani, E., Azhakanandam, K. et al. (2008). Gene expression enhancement mediated by the 5′ UTR intron of the rice *rubi3* gene varied remarkably among tissues in transgenic rice plants. *Molecular Genetics and Genomics* 279: 563–572.

Lutcke, H.A., Chow, K.C., Mickel, F.S. et al. (1987). Selection of AUG initiation codons differs in plants and animals. *The EMBO Journal* 6: 43–48.

Magnabosco, G., Pantano, M.F., Rapino, S. et al. (2020). A plant bioreactor for the synthesis of carbon nanotube bionic nanocomposites. *Frontiers in Bioengineering and Biotechnology* 8: 560349.

Malmstrom, C. (2010). Ecologists study the interactions of organisms and their environment. *Nature Education Knowledge* 3 (10): 88.

Marks, M.D., Lindell, J.S., and Larkins, B.A. (1985). Quantitative analysis of the accumulation of Zein mRNA during maize endosperm development. *The Journal of Biological Chemistry* 260: 16445–16450.

Mascarenhas, D., Mettler, I.J., Pierce, D.A., and Lowe, H.W. (1990). Intron-mediated enhancement of heterologous gene expression in maize. *Plant Molecular Biology* 15 (6): 913–920.

Mason, H.S., Lam, D.M.K., and Arntzen, C.J. (1992). Expression of hepatitis B surface antigen in transgenic plants. *Proceedings of the National Academy of Sciences of the United States of America* 89: 11745–11749.

Matsumoto, S., Ikura, K., Ueda, M., and Sasaki, R. (1995). Characterization of a human glycoprotein (erythropoietin) produced in cultured tobacco cells. *Plant Molecular Biology* 27 (6): 1163–1172.

Matzke, M.A., Primig, M., Trnovsky, J., and Matzke, A.J.M. (1989). Reversible methylation and inactivation of marker genes in sequentially transformed tobacco plants. *The EMBO Journal* 8 (3): 643–649.

McGarvey, P.B., Hammond, J., Dienelt, M.M. et al. (1995). Expression of the rabies virus glycoprotein in transgenic tomatoes. *Bio/Technology* 13 (12): 1484–1487.

Mishra, S., Yadav, D.K., and Tuli, R. (2006). Ubiquitin fusion enhances cholera toxin B subunit expression in transgenic plants and the plant-expressed protein binds GM1 receptors more efficiently. *Journal of Biotechnology* 127 (1): 95–108.

Morassutti, C., De Amicis, F., Bandiera, A., and Marchetti, S. (2005). Expression of SMAP-29 cathelicidin-like peptide in bacterial cells by intein-mediated system. *Protein Expression and Purification* 39 (2): 160–168.

Naqvi, S., Zhu, C., Farre, G. et al. (2009). Transgenic multivitamin corn through biofortification of endosperm with three vitamins representing three distinct metabolic pathways. *Proceedings of the National Academy of Sciences* 106 (19): 7762–7767.

Osborn, T.C., Burow, M., and Bliss, F.A. (1988). Purification and characterization of arcelin seed protein from common bean. *Plant Physiology* 86: 399–405.

Palmgren, G., Mattson, O., and Okkels, F.T. (1993). Treatment of agrobacterium or leaf disks with 5-azacytidine increases transgene expression in tobacco. *Plant Molecular Biology* 21 (3): 429–435.

Pan, S.Y., Litscher, G., Gao, S.H. et al. (2014). Historical perspective of traditional indigenous medical practices: the current renaissance and conservation of herbal resources. *Evidence-based Complementary and Alternative Medicine* 2014: 525340.

Parmenter, D.L., Boothe, J.V., Van Rooijen, G.J.H. et al. (1995). Production of biologically active hirudin in plant seeds using oleosin partitioning. *Plant Molecular Biology* 29 (6): 1167–1180.

Peters, T. Jr. (1995). *All About Albumin: Biochemistry, Genetics, and Medical Applications.* Academic Press.

Ramirez, Y.J., Tasciotti, E., Gutierrez-Ortega, A. et al. (2007). Fruit-specific expression of the human immunodeficiency virus type 1 tat gene in tomato plants and its immunogenic potential in mice. *Clinical and Vaccine Immunology* 14: 685–692.

Rattanapisit, K., Abdulheem, S., Chaikeawkaew, D. et al. (2017). Recombinant human osteopontin expressed in *Nicotiana benthamiana* stimulates osteogenesis related genes in human periodontal ligament cells. *Scientific Reports* 7 (1): 1–9.

Rattanapisit, K., Chao, Z., Siriwattananon, K. et al. (2019). Plant-produced anti-enterovirus 71 (EV71) monoclonal antibody efficiently protects mice against EV71 infection. *Plants* 8 (12): 560.

Raschke, M., Boycheva, S., Crèvecoeur, M. et al. (2011). Enhanced levels of vitamin B6 increase aerial organ size and positively affect stress tolerance in Arabidopsis. *The Plant Journal* 66 (3): 414–432.

Reed, D.G., Nopo-Olazabal, L.H., Funk, V. et al. (2005). Expression of functional hexahistidine-tagged ricin B in tobacco. *Plant Cell Reports* 24 (1): 15–24.

Reski, R., Parsons, J., and Decker, E.L. (2015). Moss-made pharmaceuticals: from bench to bedside. *Plant Biotechnology Journal* 13 (8): 1191–1198.

Richter, L.J., Thanavala, Y., Arntzen, C.J., and Mason, H.S. (2000). Production of hepatitis B surface antigen in transgenic plants for oral immunization. *Nature Biotechnology* 18 (11): 1167–1171.

Riedel, S. (2005). Edward Jenner and the history of smallpox and vaccination. In: *Baylor University Medical Center Proceedings*, vol. 18, 21–25. Taylor & Francis.

Rose, A.B. (2008). Intron-mediated regulation of gene expression. In: *Nuclear Pre-mRNA Processing in Plants: Current Topics in Microbiology and Immunology*, vol. 326 (ed. A.S.N. Reddy and M. Golovkin), 277–290. Berlin: Springer.

Ruhlman, T., Ahangari, R., Devine, A. et al. (2007). Expression of cholera toxin B–proinsulin fusion protein in lettuce and tobacco chloroplasts–oral administration protects against development of insulitis in non-obese diabetic mice. *Plant Biotechnology Journal* 5 (4): 495–510.

Russell, D.A. and Fromm, M.E. (1997). Tissue-specific expression in transgenic maize of four endosperm promoters from maize and rice. *Transgenic Research* 6: 157–168.

Samadder, P., Sivamani, E., Lu, J. et al. (2008). Transcriptional and post-transcriptional enhancement of gene expression by the 5′ UTR intron of rice *rubi3* gene in transgenic rice cells. *Molecular Genetics and Genomics* 279: 429–439.

Sandhu, J.S., Krasnyanski, S.F., Domier, L.L. et al. (2000). Oral immunization of mice with transgenic tomato fruit expressing respiratory syncytial virus-F protein induces a systemic immune response. *Transgenic Research* 9: 127–135.

Sawant, S.V., Kiran, K., Singh, P.K., and Tuli, R. (2001). Sequence architecture downstream of the initiator codon enhances gene expression and protein stability in plants. *Plant Physiology* 126 (4): 1630–1636.

Schillberg, S., Zimmermann, S., Voss, A., and Fischer, R. (1999). Apoplastic and cytosolic expression of full-size antibodies and antibody fragments in Nicotiana tabacum. *Transgenic Research* 8 (4): 255–263.

Scotti, N., Alagna, F., Ferraiolo, E. et al. (2009). High-level expression of the HIV-1 Pr55gag polyprotein in transgenic tobacco chloroplasts. *Planta* 229 (5): 1109–1122.

Seufert, V., Ramankutty, N., and Foley, J.A. (2012). Comparing the yields of organic and conventional agriculture. *Nature* 485 (7397): 229–232.

Sharma, A.K. and Sharma, M.K. (2009). Plants as bioreactors: recent developments and emerging opportunities. *Biotechnology Advances* 27 (6): 811–832.

Shin, Y.J., Hong, S.Y., Kwon, T.H. et al. (2003). High level of expression of recombinant human granulocyte-macrophage colony stimulating factor in transgenic rice cell suspension culture. *Biotechnology and Bioengineering* 82 (7): 778–783.

Sijmons, P.C., Dekker, B.M., Schrammeijer, B. et al. (1990). Production of correctly processed human serum albumin in transgenic plants. *Bio/technology* 8 (3): 217–221.

Singh, S.K., Stephani, J., Schaefer, M. et al. (2009). Targeting glycan modified OVA to murine DC-SIGN transgenic dendritic cells enhances MHC class I and II presentation. *Molecular Immunology* 47: 164–174.

Singh, R., Kumar, M., Mittal, A., and Mehta, P.K. (2017). Microbial metabolites in nutrition, healthcare and agriculture. *3 Biotech* 7 (1): 1–14.

Slavin, J.L. and Lloyd, B. (2012). Health benefits of fruits and vegetables. *Advances in Nutrition* 3 (4): 506–516.

Smith, T., O'Kennedy, M.M., Wandrag, D.B. et al. (2020). Efficacy of a plant-produced virus-like particle vaccine in chickens challenged with Influenza A H6N2 virus. *Plant Biotechnology Journal* 18 (2): 502–512.

So, Y., Lee, K.J., Kim, D.S. et al. (2013). Glycomodification and characterization of anti-colorectal cancer immunotherapeutic monoclonal antibodies in transgenic tobacco. *PCTOC* 113 (1): 41–49.

Song, I., Kang, Y.J., Kim, D.H. et al. (2020). Expression and in vitro function of anti-cancer mAbs in transgenic Arabidopsis thaliana. *BMB Reports* 53 (4): 229.

Spök, A. (2007). Molecular farming on the rise–GMO regulators still walking a tightrope. *TRENDS in Biotechnology* 25 (2): 74–82.

Staub, J.M., Garcia, B., Graves, J. et al. (2000). High-yield production of a human therapeutic protein in tobacco chloroplasts. *Nature Biotechnology* 18 (3): 333–338.

Stöger, E., Vaquero, C., Torres, E. et al. (2000). Cereal crops as viable production and storage systems for pharmaceutical scFv antibodies. *Plant Molecular Biology* 42 (4): 583–590.

Storozhenko, S., De Brouwer, V., Volckaert, M. et al. (2007). Folate fortification of rice by metabolic engineering. *Nature Biotechnology* 25 (11): 1277–1279.

Suzuki, Y.A., Shin, K., and Lönnerdal, B. (2001). Molecular cloning and functional expression of a human intestinal lactoferrin receptor. *Biochemistry* 40: 15771–15779.

Tanaka, H., Yabuta, Y., Tamoi, M. et al. (2015). Generation of transgenic tobacco plants with enhanced tocotrienol levels through the ectopic expression of rice homogentisate geranylgeranyl transferase. *Plant Biotechnology* 32: 233–238.

Terashima, M., Murai, Y., Kawamura, M. et al. (1999). Production of functional human alpha 1-antitrypsin by plant cell culture. *Applied Microbiology and Biotechnology* 52: 516–523.

Thakore, D. and Srivastava, A.K. (2017). Production of biopesticide azadirachtin using plant cell and hairy root cultures. *Engineering in Life Sciences* 17 (9): 997–1005.

Thanavala, Y., Mahoney, M., Pal, S. et al. (2005). Immunogenicity in humans of an edible vaccine for hepatitis B. *Proceedings of the National Academy of Sciences of the United States of America* 102 (9): 3378–3382.

Tuboly, T., Yu, W., Bailey, A. et al. (2000). Immunogenicity of porcine transmissible gastroenteritis virus spike protein expressed in plants. *Vaccine* 18 (19): 2023–2028.

Ullrich, K.K., Hiss, M., and Rensing, S.A. (2015). Means to optimize protein expression in transgenic plants. *Current Opinion in Biotechnology* 32: 61–67.

Upadhyay, S.K., Saurabh, S., Rai, P. et al. (2010a). SUMO fusion facilitates expression and purification of garlic lectin but modifies some of its properties. *Journal of Biotechnology* 146: 1–8. IF 2.9.

Upadhyay, S.K., Mishra, M., Singh, H. et al. (2010b). Interaction of *Allium sativum* leaf agglutinin (ASAL) with midgut BBMV proteins and its stability in *Helicoverpa armigera*. *Proteomics* 10: 4431–4440.

Upadhyay, S.K., Saurabh, S., Singh, R. et al. (2011). Purification and characterization of a lectin with high hemagglutination property isolated from *Allium altaicum*. *The Protein Journal* 30: 374–383.

Upadhyay, S.K. (2021). *Genome Engineering for Crop Improvement*. Wiley https://doi.org/10.1002/9781119672425.

Upadhyay, S.K. and Singh, S.P. (2021). *Bioprospecting of Plant Biodiversity for Industrial Molecules*. Wiley https://doi.org/10.1002/9781119718017.

Valdez-Ortiz, A., Rascón-Cruz, Q., Medina-Godoy, S. et al. (2005). One-step purification and structural characterization of a recombinant His-tag 11S globulin expressed in transgenic tobacco. *Journal of Biotechnology* 115 (4): 413–423.

Vamvaka, E., Twyman, R.M., Murad, A.M. et al. (2016). Rice endosperm produces an underglycosylated and potent form of the HIV-neutralizing monoclonal antibody 2G12. *Plant Biotechnology Journal* 14 (1): 97–108.

Vaquero, C., Sack, M., Chandler, J. et al. (1999). Transient expression of a tumor-specific single-chain fragment and a chimeric antibody in tobacco leaves. *Proceedings of the National Academy of Sciences of the United States of America* 96 (20): 11128–11133.

Varsani, A., Williamson, A.L., Rose, R.C. et al. (2003). Expression of Human papillomavirus type 16 major capsid protein in transgenic *Nicotiana tabacum* cv. Xanthi. *Archives of Virology* 148 (9): 1771–1786.

Vitale, A., Wu, R.J., Cheng, Z., and Meagher, R.B. (2003). Multiple conserved 5′ elements are required for high-level pollen expression of the Arabidopsis reproductive actin ACT1. *Plant Molecular Biology* 52: 1135–1151.

Walsh, G. (2010). Post-translational modifications of protein biopharmaceuticals. *Drug Discovery Today* 15: 773–780.

Wang, Y., Liu, J., Huang, B.O. et al. (2015). Mechanism of alternative splicing and its regulation. *Biomedical Reports* 3 (2): 152–158.

Whitelam, G.C., Cockburn, B., Gandecha, A.R., and Owen, M.R. (1993). Heterologous protein production in transgenic plants. *Biotechnology and Genetic Engineering Reviews* 11 (1): 1–32.

Wirth, J., Poletti, S., Aeschlimann, B. et al. (2009). Rice endosperm iron biofortification by targeted and synergistic action of nicotianamine synthase and ferritin. *Plant Biotechnology Journal* 7: 631–644.

Wu, G., Truksa, M., Datla, N. et al. (2005). Stepwise engineering to produce high yields of very long-chain polyunsaturated fatty acids in plants. *Nature Biotechnology* 23: 1013–1017.

Xinhua News Agency. (2007). Drug watchdog tightens supervision of albumin medicine. Beijing Review.

Xu, J., Okada, S., Tan, L. et al. (2010). Human growth hormone expressed in tobacco cells as an arabinogalactan-protein fusion glycoprotein has a prolonged serum life. *Transgenic Research* 19 (5): 849–867.

Xu, J., Towler, M., and Weathers, P.J. (2018). Platforms for plant-based protein production. In: *Bioprocessing of Plant In Vitro Systems* (ed. A. Pavlov and T. Bley), 509. Cham: Springer.

Yang, D., Wu, L., Hwang, Y.S. et al. (2001). Expression of the REB transcriptional activator in rice grains improves the yield of recombinant proteins whose genes are controlled by a Reb-responsive promoter. *Proceedings of the National Academy of Sciences of the United States of America* 98: 11438–11443.

Yao, J., Weng, Y., Dickey, A., and Wang, K.Y. (2015). Plants as factories for human pharmaceuticals: applications and challenges. *International Journal of Molecular Sciences* 16 (12): 28549–28565.

Ye, X., Al-Babili, S., Klöti, A. et al. (2000). Engineering the provitamin A (β-carotene) biosynthetic pathway into (carotenoid-free) rice endosperm. *Science* 287: 303–305.

Yuan, L. and Li, R. (2020). Metabolic engineering a model oilseed *Camelina sativa* for the sustainable production of high-value designed oils. *Frontiers in Plant Science* 11: 11.

Zeyenko, V.V., Ryabova, L.A., Gallie, D.R., and Spirin, A.S. (1994). Enhancing effect of the 30′-untranslated region of tobacco mosaic virus RNA on protein synthesis in vitro. *FEBS Letters* 354: 271–273.

Zhang, H., Lang, Z., and Zhu, J.K. (2018). Dynamics and function of DNA methylation in plants. *Nature Reviews Molecular Cell Biology* 19 (8): 489–506.

Zhu, Y., Xie, L., Chen, G.Q. et al. (2018). A transgene design for enhancing oil content in Arabidopsis and Camelina seeds. *Biotechnology for Biofuels* 11 (1): 1–11.

2

Molecular Farming for the Production of Pharmaceutical Proteins in Plants

Gaurav Augustine[1], Pragati Misra[2], Archana Shukla[3], Ghanshyam Pandey[4], and Pradeep Kumar Shukla[3]

[1] *Department of Agricultural Biotechnology, National Agri-food Biotechnology Institute (An Autonomous Institute of the Department of Biotechnology), Mohali, Punjab, India*
[2] *Centre for Tissue Culture Technology, Jacob Institute of Biotechnology and Bioengineering, Sam Higginbottom University of Agriculture, Technology and Sciences, Prayagraj, Uttar Pradesh, India*
[3] *Department of Biological Sciences, Sam Higginbottom University of Agriculture, Technology and Sciences, Prayagraj, Uttar Pradesh, India*
[4] *Department of Plant Protection and Plant Pathology, Naini Agricultural Institute, Sam Higginbottom University of Agriculture, Technology and Sciences, Prayagraj, Uttar Pradesh, India*

2.1 Introduction

Wide-reaching mandate for quantitative, qualitative, and miscellany production of chimeric proteins is augmenting gradually. Thereby drawing global considerations for the advancement of new technologies for synthesis of novel chimeric proteins and engineering of microorganism or mammalian cell-culture based advanced expression systems. Recombinant proteins are intricate peregrine proteins synthesized in hosts and employed in human healthcare as vaccinations, monoclonal antibodies (mAbs), or medicines (Burnett and Burnett 2020). Because of the inimitable portrayal and intensifying market challenges for high-value chimeric proteins in novel drug revelation, it is now possible to construct a variety of proteins expression hosts to synthesize unusual proteins while adhering to the contemporary stringent ethics for veterinary and human pertinences. Existing ventures are mainly focusing on already familiar protein production platforms through a diverse array of expression systems from prokaryotic and eukaryotic host cells comprising *Escherichia coli* cultures, yeast cell cultures, insect cell cultures, and mammalian cell cultures, as they exhibit widely known compliant processes along with optimized manufacturing procedures (Schillberg et al. 2019). Supplementally, regulatory approbation for industrially engendered mammalian and other cell cultures is stringent, making it arduous for the incipient generation or manufacturing technology to gain widespread adoption. Bacterial expression methods synthesize high yields expeditiously, whereas *Saccharomyces cerevisiae* and *Pichia pastoris* (yeast) offer post-translational modifications (PTMs), obligatory for recombinant protein activity (Vieira Gomes et al. 2018).

Plants as Bioreactors for Industrial Molecules, First Edition.
Edited by Santosh Kumar Upadhyay and Sudhir P. Singh.

Plant molecular farming (PMF) has progressed in the recent decennium, making it a winsome minuscule bio-factories with prodigious economic viability along with paramount production stages in a short duration. Plant expression systems' pertinence and possibilities for cost-efficacious recombinant protein synthesis are paramountcy and prospects in this ever-growing domain industrial enzymes and mAbs, produced in plants are all examples of potential vaccine candidates.

Molecular farming is the utilization of whole plants or plant cells/tissues cultured *in vitro* for the synthesis of valuable recombinant proteins, and it has been proven to be a cost-efficacious alternative to traditional production systems such as microbes and mammalian cells grown in immensely colossal-scale bioreactors.

Plants supplementally sanction non-native proteins to be congruously synthesized, packed, and glycosylated, as well as provide adaptable scalability with minimal peril of human pathogen contamination (Yusibov et al. 2011). Human growth hormone (hGH) expressed in transgenic tobacco and sunflower callus was the first pharmaceutically manufactured protein in plants (Barta et al. 1986). Recombinant human serum albumin (rHSA) was genetically engineered and incited initially in transgenic tobacco and potato a few years later (Sijmons et al. 1990). Many more recombinant proteins have now been synthetized in genetically modified plants (Figure 2.1). With an ascending number of specialized startup and biotechnology firms throughout the world, PMF is becoming a valuable biotechnology industry. An incipient ecumenical organization for PMF has been composed, with its first meeting set to take place in Germany (Paul 2015). Incipient health concerns and perplexed illnesses such as cancer and degenerative disorders will necessitate intricate treatment solutions in the future years (Andrianantoandro 2014).

2.2 Plant as an Expression Platform

Utilization of plant cells as miniaturized bio-factories for producing high-quality chimeric proteins limiting from pharmaceutical therapeutics to nonpharmaceutical products (including enzymes, growth factors, antibodies, vaccine antigens, reagents for research/diagnostic, and propriety constituents) has advanced during the period of time and amended pointedly in recent years, subsequently resulting in a key paradigm shift in the pharmaceutical industry. These applied sciences engineering have been upgraded competently, and the potential hindrances related to PMF in their early phases of development, such as the essentiality for a high caliber of protein expression and efficacious downstream processing, have already been addressed. Figure 2.1 demonstrates multistep process in production of pharmaceutical/nonpharmaceutical protein using plant as a bioreactor.

Following the efficacious production of therapeutic proteins in plants for the first time (Hiatt et al. 1989), it became indubitable that plants have a number of advantages over several well-established platforms including mammalian cell cultures (Figure 2.2a). Some of the advantages are listed below (Buyel 2019):

i) Plants are inherently safe due their avirulent nature for human ailments, resulting in a low pathogen load and negligible hazard of contagion through the procedure (Commandeur et al. 2003).

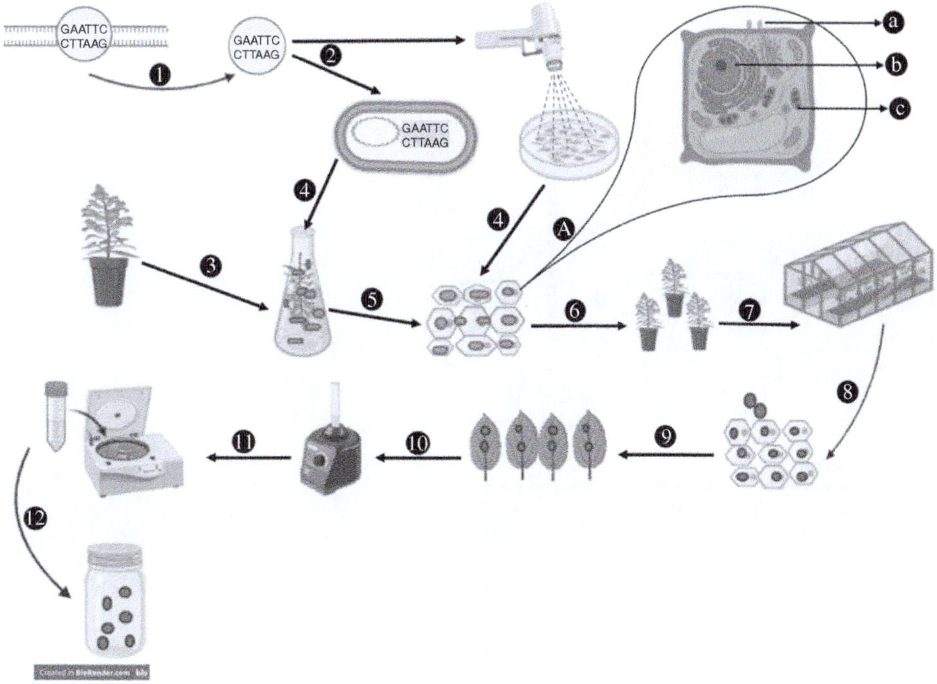

Figure 2.1 An overview of production of pharmaceutical/non-pharmaceutical protein using plant as a bioreactor. (1. Identification and extraction of gene of the protein of interest; 2. insertion of the gene into *Agrobacterium tumefaciens*/gene transfer through gene gun; 3. selection of a disease-free plant; 4. coinoculation of *Agrobacterium tumefaciens* with leaves of the plant/bombardment of plant tissue with DNA-coated particles; 5. isolation of plant cells with the inserted gene of the interest; 6. regeneration of the plants; 7. transfer of the plant in green house for controlled environment; 8. screening of plants with efficient production of protein of interest; 9. harvesting of leaves from the plants; 10. blending/vortexing of plant leaves into appropriate solution for protein extraction; 11. centrifugation of the solution for purified protein; and 12. extraction of protein and final product can be used for pharmaceutical use.) A. Enlarged view of the cell (a. transient expression offers high yield and limited scalability; b. nuclear expression offers multiple genetic manipulation tools; c. plastid expression offers high production due to multiple plastid production and easy manipulation with resistance to gene silencing).

ii) Plant cultivation is straightforward since no sterile environment is required. Ergo, plants that are salubrious can enact their inherent immune system to keep infections at bay (Sack et al. 2015). Transgenic plant cultivation, in particular, may theoretically be scaled up to the agricultural scale, i.e. several thousand hectares.

iii) Few weeks (~8) after incorporation of appropriate DNA sequence, plants can express/synthesize recombinant proteins (Shoji et al. 2012), through transient/unstable expression driven by *Agrobacterium tumefaciens* and/or viral vectors (Peyret and Lomonossoff 2013; Gleba et al. 2014).

iv) As witnessed by the manufacturing of an anti-Ebola antibody cocktail, fast production enables for swift responses to epidemic or pandemic threats (Qiu et al. 2014).

Further, plants can also perform the same PTMs as mammalian cells (Strasser 2016), making them preferable for prokaryotic systems, viz. the expression of intricate proteins,

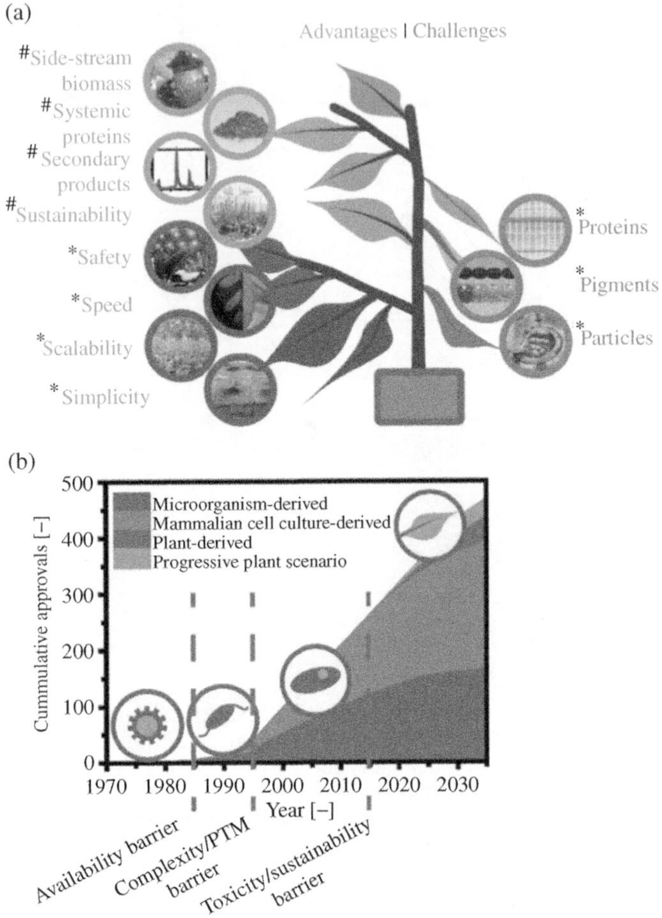

Figure 2.2 Overview of plant-based expression systems. (a) Advantages and challenges typically encountered during plant molecular farming. This study expands the previously reported aspects (*) with new potential benefits (#). (b) Approvals of recombinant-protein-based biopharmaceuticals. Starting in 1985, the approved drugs expressed in microorganisms (prokaryotic and eukaryotic), mammalian cells, or plant cells are plotted in five-year intervals based on data published up to 2014 (Walsh 2014). The forecast up to 2035 is based on reported product pipelines (Gleba et al. 2014). Source: Adapted from Buyel (2019).

e.g. mAbs or membrane proteins (He et al. 2014). Additionally, plants can serve as an excellent assembly line to synthesize nature-abundant innately disarrayed proteins (Gengenbach et al. 2018b; Covarrubias et al. 2017) or to produce anticancer lectins, e.g. viscumin (Gengenbach et al. 2018a; Zwierzina et al. 2011).

Furthermore, in comparison to expensive media necessary for mammalian cell cultures, affordable, and specified fertilizer solutions are sufficient for plant cultivation (Buyel and Fischer 2012; Xu et al. 2017). Multi-tonne scale production may be conceivable even in highly controlled environment agriculture like greenhouses (Ma et al. 2015; Sack et al. 2015) or agriculture in vertically stacked layers (Wirz et al. 2012; Holtz et al. 2015; Buyel et al. 2017).

Despite the above-mentioned promising discoveries, large biopharmaceutical companies have yet to adopt plants as biofactories to express novel proteins in defiance of validation of first plant-predicated therapeutic protein for use of mankind (Mor 2015). The consummate depreciation of established fermentation capacities, the steadily incrementing product titers reported for mammalian cell cultures (Rader and Langer 2015; Xu et al. 2017), and the introduction of increasingly sophisticated single-use technologies that sanction flexible process layouts could all contribute to this inertia (Gottschalk 2009).

There will be a voluminous shift toward plant-predicated expression systems once an incipient type of biopharmaceutical product hits the market and pushes conventional manufacturing techniques to their circumscriptions. This has transpired afore: proteins were obtained from natural sources afore the recombinant DNA era, pertaining to limited accessibility of novel protein products. This was addressed with the development of recombinant DNA technologies, which led to the development of microbial fermentation methods that made it possible to produce divergent synthetic proteins. Microbes, on the other hand, are unable to create more complex proteins, hence mammalian cell cultures have been conventionally recognized for the synthesis of complex proteins including mAbs (Figure 2.2b) (Buyel 2019).

The effects of the plant on the natural atmosphere, human life, food security, and accessibility should all be considered while avowing plant as potential host. Pharmaceutical proteins can be instigated in a variety of plant systems, as mentioned in Figure 2.3.

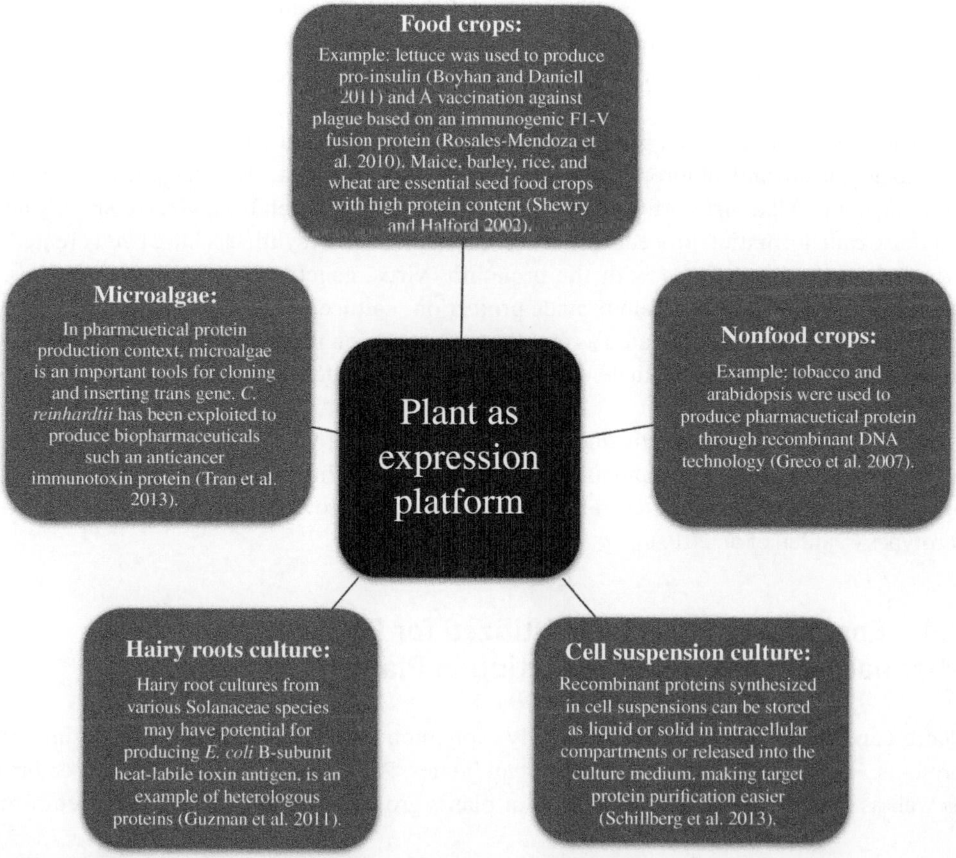

Figure 2.3 Overview of plant expression platform for pharmaceuticals protein production.

All plant-predicated systems have several benefits, including ease of culture, low-cost inputs, minimum or zero pathogen load, prompt voluminous recombinant protein synthesis, and plants' capacity to intermix involute proteins similar to eukaryotic PTMs (Chen and Davis 2016).

The *Nicotiana* genus is customarily utilized for genetic transformation studies because of its exemplary growth and facile genetic manipulation. Tobacco, which is considered as a functional biology or molecular biology engine of the plant kingdom, produces the majority of recombinant proteins, including medicines, vaccines, hormones, cytokines, growth regulators, and industrial items. *Nicotiana benthamiana* and *Nicotiana tabacum* are two popular species utilized for recombinant protein synthesis, both steady and transient. Furthermore, depending on the protein target and application, many cereal crops, fruits, and vegetables were studied for their utility in PMF, including rice (*Oryza sativa*), maize (*Zea mays*), lettuce (*Lactuca sativa*), tomato (*Solanum lycopersicum*), potato (*Solanum tuberosum*), and alfalfa (*Medicago sativa*), etc. (Burnett and Burnett 2020).

2.3 Plant-Derived Recombinant Proteins

The concept of employing plants to engender novel proteins, from pharmacological and nonpharmaceutical sectors, has been intensively investigated and documented. Many researches have shown that plant systems can develop vaccine candidates *in vivo* and *in vitro* for animal and mankind uses, as well as the fact that plant-derived antigens trigger possible immune replications in animal models or even confer safety in animal-based studies.

Tobacco genome has been genetically engineered to construct a variety of antigens in the nucleoplasm and chloroplast together with chikungunya, dengue pyrexia, Ebola, influenza, and Zika virus. However, in case of fruits and vegetables, viz. tomatoes and potatoes, transformation procedures for recombinant protein synthesis have been created. When chickens are infected with the bronchitis virus, genetically engineered potatoes, constructing the S1 glycoprotein provide protection. Lettuce, alfalfa, and clover and other leafy crops have been investigated as excellent construction platforms for molecular farming to acquire oral vaccine antigen dispersion without the desideratum for purification or infusion.

In mice, primed with inactivated poliovirus vaccine (IPV), utilization of lyophilized plant cells expressing the poliovirus capsid protein in the lettuce chloroplast-derived booster vaccine synthesized neutralizing antibodies and provided shield against all polio serotypes (Daniell et al. 2019).

2.4 Engineering Strategies Utilized for Recombinant Pharmaceutical Protein Production in Plants

Plants can utilize either transient or steady expression strategies to construct recombinant proteins. For the development of vaccine candidates, PMF uses plant-specific viral vectors, as well as stable nuclear transformation of plants grown through hydroponic agriculture used for extraction of recombinant proteins from media, stable chloroplast transformation, or transient expression (Figures 2.1 and 2.5).

Plant-predicated recombinant protein construction is a multistep process. It commences with the optimization of the foreign gene's coding sequence afore to its introduction into the host. Following that, the expression construct containing the gene of interest is constructed and integrated into the plant genome (stable or transient). Plants expressing the desired protein are propagated, and the desired protein is purified (Egelkrout et al. 2012). External DNA can be introduced into the nuclear or plastid genomes of plant cells to ensue stable transformants. Efficient vectors such as *A. tumefaciens* (Leuzinger et al. 2013) or particle bombardment are habituated to accomplish such transformations (Chen and Lai 2014). Transgenes may be meticulously targeted to a categorical locus owing to stable transformation into the chloroplast genome, preventative gene knockouts and ensuring that foreign DNA does not spread into the environment via pollen (Bock 2015). Plant viruses that replicate themselves are utilized in the other technique for heterologous protein synthesis (Hefferon 2013). Viral genomes are an intriguing alternative to stable transgenic systems for expeditious transient expression because of their minute size, simplicity of modification, and infection mechanism (Gleba et al. 2014) (Figure 2.4).

2.4.1 Nuclear Transformation

A. tumefaciens-mediated transformation or particle bombardment can be used to introduce the transgene into the plant expression vector, resulting in *in vitro* produced stable transgenic lines. After that, the pool of transgenic strains may be tested for a high-quality transgenic line for protein synthesis. Because the transgene has been successfully incorporated into the genome of the plant, recombinant proteins may be synthesized in consecutive generations utilizing this method. Model plants like *Arabidopsis thaliana* and tobacco were

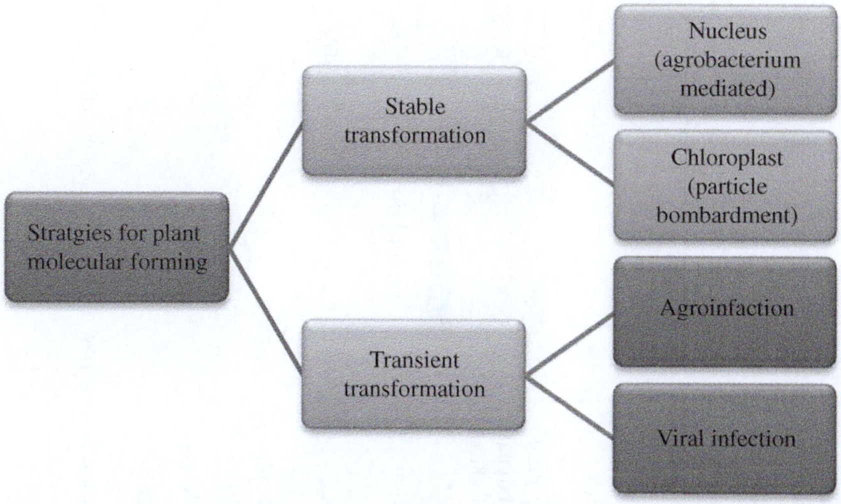

Figure 2.4 An overview of plant transformation strategies used to make recombinant pharmaceutical and nonpharmaceutical proteins in plants.

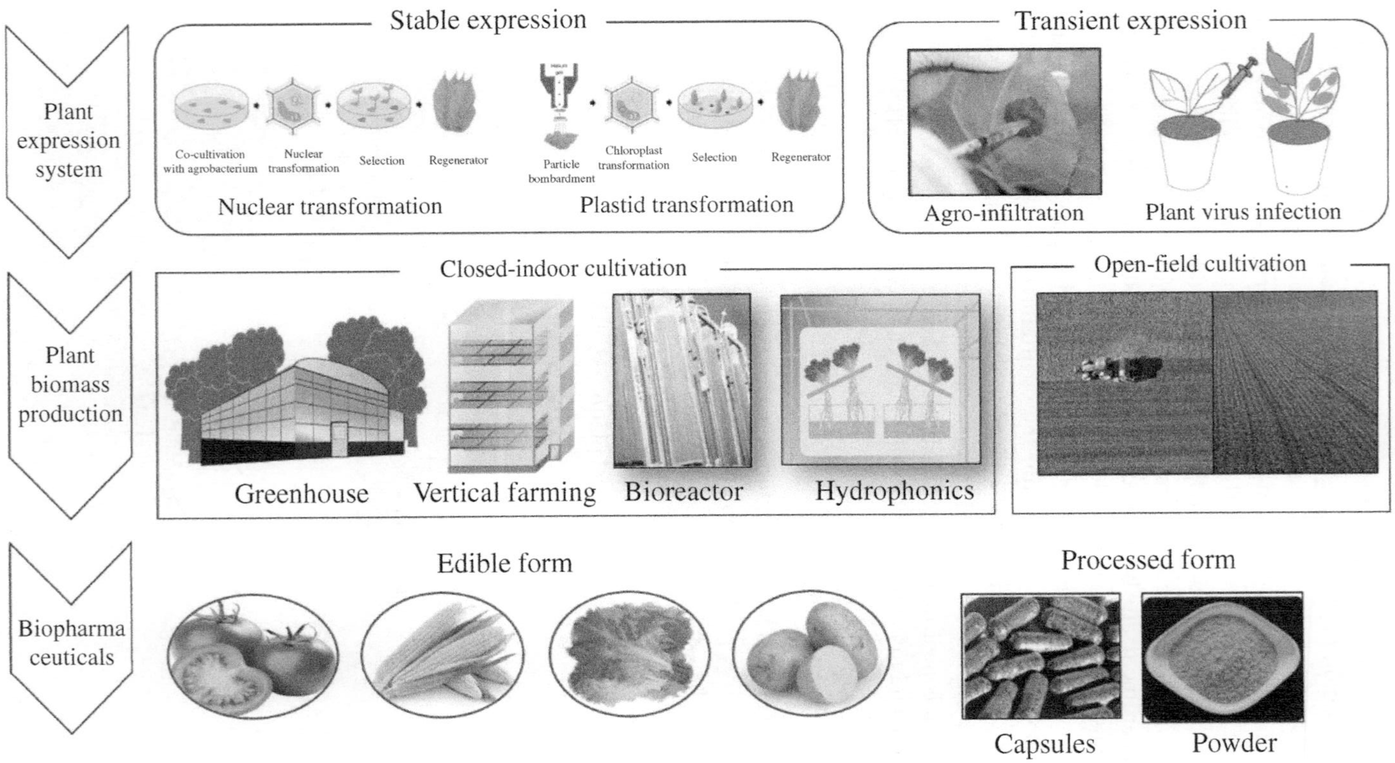

Figure 2.5 Overview of production system of plant-based biopharmaceuticals and plant biomass. Source: Adapted from Moon et al. (2019).

more commonly used in the early stages of genetic transformation to produce stable transformants. Transient expression predicated on agroinfiltration or virus-predicated vectors, on the other hand, has been constructed to complement transgenic plants that provide expeditious and high-level protein synthesis in a matter of days.

2.4.2 Chloroplast Transformation

Foreign gene expression in chloroplast is opportune to result in an astronomically immense number of proteins. In a cell, each enlarged plant leaf has roughly 100 chloroplast organelles, each with 100 replica of the circular chloroplast genome. As a result, foreign gene inserted into the chloroplast genome has a significant and extensive gene dosage impact. When the gene for green fluorescent protein was inserted into the chloroplast genome, tobacco leaves accumulated 30% of total soluble proteins (Yabuta et al. 2008). Chloroplast transformation is achieved by bombarding green organelles with gold particles coated with vectors. Regeneration of transformed plants is opulent in various plants, including tobacco, potato, and lettuce (Boyhan and Daniell 2011). Another advantage of transmuting the chloroplast genome is that it makes it more facile to grow transgenic plants in the field. Further, pollen does not contain chloroplasts, introducing a foreign gene into the chloroplast genome is far less hazardous than introducing it into the nuclear genome.

Stable or transient expression methods are employed to create target proteins, and they are dependent on applications that use whole or slightly transformed plants or plant components (Moon et al. 2019; Upadhyay 2021). Each type of method can be further subdivided utilizing (i) Agrobacterium-mediated transformation and (ii) particle bombardment enabled stable nucleoplast or chloroplast transformation; nevertheless, transient expression can be produced by using a plant virus or by agroinfiltration (Fischer et al. 2004; Stoger et al. 2005). Currently, there are two main methods for producing plant biomass that expresses the target of interest: open-field farming and closed indoor systems. The term "open-field" simply refers to a plantation that is located outside. It is simple to scale-up and its economic benefits may be highlighted because it employs current agricultural facilities. Closed-indoor systems include greenhouses, vertical farming units, cell bioreactors, and other smart agriculture regimes (Figure 2.5).

2.5 Pharmaceutical Protein Developed Using Plant Expression Platform

The conception of employing plants to produce novel proteins, for both medicinal and non-pharmaceutical utilizations, has been well researched and documented. Many studies have demonstrated that plant-reliant antigens trigger plausible immunological replications in animal models and even provide safeguard in animal-based trails and researches, revealing the capabilities of *in vivo* and *in vitro* plant systems to develop vaccine candidates for veterinary and human use. Tables 2.1 and 2.2 demonstrate the expression of a variety of pharmaceutical and nonpharmaceutical proteins in plant-based models.

Table 2.1 List of various pharmaceutical proteins produced in plants.

Recombinant protein	Pathogen/disease	Expression system	Transformation method	Expression level	Reference
Hepatitis B surface antigen	Hepatitis B virus	Tobacco (*Nicotiana tabacum*)	Agrobacterium mediated	66 ng/mg of soluble protein	Mason et al. (1992)
Structural protein VP60	Rabbit hemorrhagic disease virus	Potato (*Solanum tuberosum*)	Agrobacterium mediated	0.3% of total soluble protein	Castanon et al. (1999)
Spike (S) protein of transmissible gastroenteritis virus	Transmissible gastroenteritis virus (TGEV)	Tobacco (*Nicotiana tabacum*)	Agrobacterium mediated	0.1–0.2% of total soluble protein	Tuboly et al. (2000)
Hemagglutinin protein of rinderpest virus	Rinderpest virus (RPV)	Peanut (*Arachis hypogea* L.)	Agrobacterium mediated	0.2–1.3% of total soluble protein	Khandelwal et al. (2003)
Glycoprotein D (gD) of bovine herpes virus	Bovine herpes virus	Tobacco (*Nicotiana benthamiana*)	Mechanical inoculation	20 μg/g fresh weight (FW)	Pérez Filgueria et al. (2003)
L1 major capsid protein	Human papillomavirus	Tobacco (*Nicotiana tabacum*)	Agrobacterium mediated	2–4 μg/kg FW	Varsani et al. (2003)
Spike (S) protein of transmissible gastroenteritis virus	Transmissible gastroenteritis virus (TGEV)	Corn (*Zea mays*)	Agrobacterium mediated	2–4 μg/kg FW	Lamphear et al. (2004)
Spike (S) protein of infectious bronchitis virus	Infectious bronchitis virus (IBV)	Potato (*Solanum tuberosum*)	Agrobacterium mediated	2.39–2.53 μg/g FW	Zhou et al. (2004)
Anthrax protective antigen (PA)	Anthrax	Tobacco (*Nicotiana tabacum*)	Biolistic method (stable expression/chloroplast)	14.2% of total soluble protein	Koya et al. (2005)
Hepatitis B virus surface antigen	Hepatitis B virus (HBV)	Potato (*Solanum tuberosum*)	Agrobacterium mediated	8.5 μg/g FW	Thanavala et al. (2005)
Fusion (F) protein of Newcastle disease virus	Newcastle disease virus (NDV)	Corn (*Zea mays* L.)	Biolistic method (stable expression/chloroplast)	3.0% of total soluble protein	Guerrero-Andrade et al. (2006)
F4 fimbrial adhesin FaeG	Enterotoxigenic *E. coli*	Alfalfa (*Medicago sativa* L.)	Agrobacterium mediated	1.0% of total soluble protein	Joensuu et al. (2006)

L1 capsid protein gene	Cottontail rabbit papillomavirus	Tobacco (*Nicotiana tabacum*)	Agrobacterium mediated	0.4–1 mg/kg of total leaf mass	Kohl et al. (2006)
Structural protein VP2	Infectious bursal disease virus (IBDV)	Rice	Agrobacterium mediated	40.21 μg/g FW	Wu et al. (2007)
Hepatitis B virus surface antigen	Hepatitis B virus (HBV)	Tobacco (*Nicotiana benthamiana*)	Agrobacterium mediated	295 μg/g FW	Huang et al. (2008)
Hemagglutinin (HA)	H5N1 (avian influenza virus) and H1N1 (human) influenza strains	Tobacco (*Nicotiana benthamiana*)	Agrobacterium mediated	50 mg/kg FW	D'Aoust et al. (2008)
Heat-labile toxin B subunit (LTB)	Enterotoxigenic *E. coli*	Carrot (*Daucus carota*)	Agrobacterium mediated	0.3% of total soluble protein	Rosales-Mendoza et al. (2008)
Norwalk virus capsid protein	Norwalk virus	Tobacco (*Nicotiana benthamiana*)	Agrobacterium mediated (transient expression)	0.8 mg/g FW	Santi et al. (2008)
Structural protein VP1	Foot-and-mouth disease virus (FMDV)	Legume (*Stylosanthes guianensis*)	Agrobacterium mediated	0.1–0.5% of total soluble protein	Wang et al. (2008)
HIV-1 Pr55gag polyprotein	Human immunodeficiency virus type 1 (HIV)	Tobacco (*Nicotiana tabacum*)	Biolistic method	312–363 mg/kg FW	Scotti et al. (2009)
Japanese encephalitis virus (JEV) envelope protein (E)	Japanese encephalitis virus	Japonica rice (*Nipponbare*)	Agrobacterium mediated	1.1–1.9 μg/mg of total soluble protein	Wang et al. (2009)
Hemagglutinin (HA)	Avian influenza (H5N1)	Tobacco (*Nicotiana benthamiana*)	Agrobacterium mediated	0.3 g/kg FW	Kalthoff et al. (2010)
Hemagglutinin (HA)	Influenza virus	Tobacco (*Nicotiana benthamiana*)	Agrobacterium mediated	400–1300 mg/kg FW	Shoji et al. (2012)
Hemagglutinin (HA)	Avian influenza A (H7N7)	Tobacco (*Nicotiana benthamiana*)	Agrobacterium mediated	0.2 g/kg FW	Kanagarajan et al. (2012)
Structural protein VP2	Infectious bursal disease virus (IBDV)	Tobacco (*Nicotiana benthamiana*)	Agrobacterium mediated	1.0% of total soluble protein	Gomez et al. (2013)

(*Continued*)

Table 2.1 (Continued)

Recombinant protein	Pathogen/disease	Expression system	Transformation method	Expression level	Reference
Structural protein E2	Bovine viral diarrhea virus (BVDV)	Alfalfa (*Medicago sativa* L.)	Agrobacterium mediated	1 µg/g FW	Perez Aguirreburualde et al. (2013)
Bluetongue virus-like particles	Bluetongue virus	Tobacco (*Nicotiana benthamiana*)	Agrobacterium mediated	70 mg/kg FW	Thuenemann et al. (2013)
Narita 104 virus virus-like particles	Narita 104 virus	Tobacco (*Nicotiana benthamiana*)	Agrobacterium mediated	0.3 mg/g FW	Mathew et al. (2014)
Glycoprotein (GP) of PRRSV	Porcine reproductive and respiratory syndrome virus (PRRSV)	*Arabidopsis thaliana*	Agrobacterium mediated	2.74% of total soluble protein	Piron et al. (2014)
Matrix protein 2 ectodomain (M2e)	Avian influenza (H5N1)	Duckweed (*Lemna minor*)	Agrobacterium mediated	90–970 mg/kg FW	Firsov et al. (2015)
Matrix protein 2 ectodomain (M2e)	Avian influenza (H5N1)	Tobacco (*Nicotiana benthamiana*)	Agrobacterium mediated	125–205 mg/kg FW	Mbewana et al. (2015)
Consensus domain III of dengue virus E glycoprotein (cEDIII)	Dengue virus	Tobacco (*Nicotiana benthamiana*)	Agrobacterium mediated	5.2 mg/g FW	Kim et al. (2015)
Dengue envelop protein domain III (EDIII)	Dengue virus	Tobacco (*Nicotiana tabacum*)	Biolistic method	0.8–1.6% of total soluble protein	Gottschamel et al. (2016)
PV3 VLPs	Poliovirus (PV)	Tobacco (*Nicotiana benthamiana*)	Agrobacterium mediated	60 mg/kg FW	Marsian et al. (2017)
Eimeria maxima gametocyte antigen (Gam82)	*Eimeria maxima*	Tobacco (*Nicotiana tabacum*)	Agrobacterium mediated	20 mg/kg FW	Kota et al. (2017)
CHIKV E1 and E2	Chikungunya virus	Tobacco (*Nicotiana benthamian*)	Agrobacterium mediated	8–13 mg/kg of fresh leaf weight	Iyappan et al. (2018)

ZIKV envelope (E) protein	Zika virus (ZIKV)	Tobacco (*Nicotiana benthamiana*)	Agrobacterium mediated	160 µg/g FW	Yang et al. (2018)
Porcine circovirus type 2 (PCV-2) capsid protein	Porcine circovirus type 2	Tobacco (*Nicotiana benthamiana*)	Agrobacterium mediated	6.5 mg/kg leaf wet weight	Gunter et al. (2019)
HIV Env gp140	Human immunodeficiency virus (HIV)	Tobacco (*Nicotiana benthamiana*)	Agrobacterium mediated	5–6 mg/kg FW	Margolin et al. (2019)
H6 subtype hemagglutinin (HA)	Influenza A virus (H6N2)	Tobacco (*Nicotiana benthamiana*)	Agrobacterium mediated	95 mg/kg FW	Smith et al. (2020)
		Antibodies			
cT84.66	Cancer (tumor marker)	Tobacco (*Nicotiana tabacum*)	Agrobacterium mediated	1 mg/kg FW	Vaquero et al. (1999)
scFvT84.66	Cancer (tumor marker)	Tobacco (*Nicotiana tabacum*)	Agrobacterium mediated	5 mg/kg FW	Vaquero et al. (1999)
scFvT84.66	Cancer (tumor marker)	Rice (*Oryza sativa*)	Biolistic method	3.8 µg/g FW	Torres et al. (1999)
scFvT84.66	Cancer (tumor marker)	Cereal crops (wheat and rice)	Biolistic method	30 µg/g FW	Stoger et al. (2000)
BR55-2	Human colorectal cancer	Tobacco (*Nicotiana tabacum*)	Agrobacterium mediated	30 mg/kg FW	Brodzik et al. (2006)
2F5	HIV	Tobacco (*Nicotiana tabacum*)	Agrobacterium mediated	0.01% of total soluble protein	Floss et al. (2008)
2G12	HIV	Tobacco (*Nicotiana benthamiana*)	Agrobacterium mediated	0.3 g/kg FW	Sainsbury and Lomonossoff (2008)
2G12	HIV	Tobacco (*Nicotiana benthamiana*)	Agrobacterium mediated	8 mg/l culture medium	Holland et al. (2010)
6D8	Ebola virus	Tobacco (*Nicotiana benthamiana*)	Agrobacterium mediated	0.5 mg/g FW	Huang et al. (2010)

(*Continued*)

Table 2.1 (Continued)

Recombinant protein	Pathogen/disease	Expression system	Transformation method	Expression level	Reference
6D8	Ebola virus	Lettuce (*L. sativa*)	Agrobacterium mediated	0.23–0.27 mg/g FW	Lai et al. (2012)
CO17-1AK	Human colorectal cancer	Tobacco (*Nicotiana tabacum*)	Agrobacterium mediated	0.25 mg/kg FW	So et al. (2013)
Palivizumab-N	Respiratory syncytial virus	Tobacco (*Nicotiana benthamiana*)	Agrobacterium mediated	180 mg/kg FW	Zeitlin et al. (2013)
E559	Rabies	Tobacco (*Nicotiana tabacum*)	Agrobacterium mediated	1.8 mg/kg FW (0.04% of total soluble protein)	Van Dolleweerd et al. (2014)
pE16	West Nile virus	Tobacco (*Nicotiana benthamiana* Δ*XF*)	Agrobacterium mediated	0.74 mg/g FW	Lai et al. (2014)
pE16scFv-CH	West Nile virus	Tobacco (*Nicotiana benthamiana* Δ*XF*)	Agrobacterium mediated	0.77 mg/g FW	Lai et al. (2014)
E60	Dengue virus	Tobacco (*Nicotiana benthamiana*)	Agrobacterium mediated	120 μg/g FW	Dent et al. (2016)
2G12	HIV	Rice (*Oryza sativa*)	Biolistic method	46.4 μg/g dry seed weight	Vamvaka et al. (2016)
8B10 and 5F10	Chikungunya virus	Tobacco (*Nicotiana benthamiana*)	Agrobacterium mediated	20–30 mg/kg FW	Iyappan et al. (2018)
SO57	Rabies virus	Tobacco (*Nicotiana tabacum*)	Agrobacterium mediated	0.014–0.019% of total soluble protein	Shafaghi et al. (2018)
cD5	Enterovirus 71	Tobacco (*Nicotiana benthamiana*)	Agrobacterium mediated	50 μg/g FW	Rattanapisit et al. (2019a)
PD1	Cancer	Tobacco (*Nicotiana benthamiana*)	Agrobacterium mediated	140 μg/g FW	Rattanapisit et al. (2019b)

c2A10G6	Zika virus	Tobacco (*Nicotiana benthamiana*)	Agrobacterium mediated	1.47 mg/g FW	Diamos et al. (2019)
6D8	Ebola	Tobacco (*Nicotiana benthamiana*)	Agrobacterium mediated	1.21 mg/g FW	Diamos et al. (2019)
HSV8	Herpes simplex virus	Tobacco (*Nicotiana benthamiana*)	Agrobacterium mediated	1.42 mg/g FW	Diamos et al. (2019)
CHKV mab	Chikungunya virus	Tobacco (*Nicotiana benthamiana*)	Agrobacterium mediated	100 μg/g FW	Hurtado et al. (2020)

Source: Adapted from Shanmugaraj et al. (2020).

Table 2.2 List of various nonpharmaceutical proteins produced in plants.

Recombinant protein	Expression system	Transformation method	Expression level	Reference
Human serum albumin	Potato (*Solanum tuberosum*)	Agrobacterium mediated	0.25 μg/mg (0.02% of total soluble protein)	Sijmons et al. (1999)
Erythropoietin	Tobacco (*Nicotiana tabacum*)	Agrobacterium mediated	26 pg/mg total protein	Matsumoto et al. (1995)
α1-antitrypsin	Rice (*Japonica rice*)	Biolistic method	4.6–5.7 mg/g dry cell	Terashima et al. (1999)
Aprotinin	Corn	Biolistic method	0.069% of total extractable seed protein	Zhong et al. (1999)
Human-secreted alkaline phosphatase	Tobacco (*Nicotiana tabacum*)	Agrobacterium mediated	1.1 μg/g FW (3% of total soluble protein)	Komarnytsky et al. (2000)
Collagen	Tobacco (*Nicotiana tabacum*)	Agrobacterium mediated	0.03 g/kg powdered plants	Ruggiero et al. (2000)
Human somatotropin (hST)	Tobacco	Biolistic method	>7% of total soluble protein	Staub et al. (2000)
Bacillus thuringiensis (Bt) cry2Aa2	Tobacco	Biolistic method	5 mg/g FW (45.3–46.1% of total soluble protein)	De Cosa et al. (2001)
Human serum albumin	Tobacco (*Nicotiana tabacum*)	Biolistic method	11.1% of total protein	Fernandez-San Millan et al. (2003)
Human epidermal growth factor (hEGF)	Tobacco (*Nicotiana tabacum*)	Agrobacterium mediated	34.2 μg/g FW	Wirth et al. (2004)
Human basic fibroblast growth factor (bFGF)	Soybean (*Glycine max*)	Cotyledonary node explant method (stable expression/nucleus)	2.3% of total soluble protein	Ding et al. (2006)
Type I interferon (IFNα2b)	Tobacco (*Nicotiana tabacum*)	Biolistic method	3 mg/g FW (20% of total soluble protein)	Arlen et al. (2007)
Human growth hormone (hGH)	Rice (*Oryza sativa*)	Biolistic method	57 mg/l culture medium	Kim et al. (2008)
PlyGBS lysin	Tobacco (*Nicotiana tabacum*)	Biolistic method	>70% of the total soluble protein	Oey et al. (2009)
Human growth hormone (hGH)	Tobacco BY-2 cells	Agrobacterium mediated	35 mg/l culture medium (2–4% of total soluble protein)	Xu et al. (2010)

Table 2.2 (Continued)

Recombinant protein	Expression system	Transformation method	Expression level	Reference
Human basic fibroblast growth factor (bFGF)	Rice (*Oryza sativa*)	Agrobacterium mediated	185.66 mg/kg	An et al. (2013)
Lumbrokinase	Sunflower (*Helianthus annuus* L.)	Agrobacterium mediated	5.1 g/kg seeds	Guan et al. (2014)
Human acidic fibroblast growth factor 1 (FGF-1)	*Salvia miltiorrhiza*	Agrobacterium mediated	272 ng/g FW	Tan et al. (2014)
Glucocerebrosidase (GCase)	Tobacco (*Nicotiana benthamiana*)	Agrobacterium mediated	68 μg/g FW (1.45% of total soluble protein)	Limkul et al. (2015)
Human acid alpha glucosidase	Tobacco (*Nicotiana tabacum*)	Biolistic method	6.38 μg/g FW	Su et al. (2015)
Human basic fibroblast growth factor (bFGF)	Tobacco (*Nicotiana tabacum*)	Biolistic method	0.1% of total soluble protein	Wang et al. (2016)
Endo-β-1,4-xylanase	Tobacco (*Nicotiana tabacum*)	Biolistic method	35.7% of total soluble protein	Castiglia et al. (2016)
β-Glucosidase	Tobacco (*Nicotiana tabacum*)	Biolistic method	>75% of total soluble protein	Castiglia et al. (2016)
Osteopontin (OPN)	Tobacco (*Nicotiana benthamiana*)	Agrobacterium mediated	100 ng/g FW	Rattanapisit et al. (2017)
Dentin matrix protein-1 (DMP1)	Tobacco (*Nicotiana benthamiana*)	Agrobacterium mediated	0.3 μg/g FW	Ahmad et al. (2019)

Source: Adapted from Shanmugaraj et al. (2020).

The construction of pharmaceutically substantial and economically accessible proteins in plants is avowed as molecular farming. Plant-predicated molecular farming exhibits the competency to synthesize essentially illimitable quantities of chimeric proteins to be utilized as diagnostic and therapeutic purposes in medicine and the life sciences (Fischer and Emans 2000). Plants are peculiarly ideal for the production of animal vaccinations as crude plant extracts can be administered/injected into animals directly (Topp et al. 2016). A vast variety of proteins of medicinal importance, viz. antibodies, vaccines, hormones, and enzymes, as well as proteins for analytic, research, and cosmetic uses, have been prosperously synthesized in plants. The medley of plant species and contemporary manufacturing technologies employed in plant-predicated protein expression is also a distinctive trait (Schillberg and Finnern 2021; Upadhyay and Singh 2021). Plants have gifted humans with functional chemicals since long back, but in the past 20 years, it has only become possible to employ plants to produce explicit heterologous proteins. Many mammalian-derived medicinal proteins have been developed in plants. These include everything from blood components like human serum albumin, which has a 500-tonne annual requisite, to cytokines and other signaling molecules, which are needed in much more minute quantities (Fischer and Emans 2000).

Therapeutic protein manufactured through plant system has an abundance of potential for boosting and ameliorating protein synthesis in humans and other animals for disease prophylaxis and therapy purposes. This approach has an abundance of paramount advantages for scientists and society (Burnett and Burnett 2020; Dixit et al. 2021). Antibody synthesis in plants is categorically burdensome since the molecules must fold and assemble precisely in order to identify their cognate antigens. Tetramers comprising two identical enormously ponderous chains and two identical light chains make up most serum antibodies (Ma et al. 1995). Molecular farming requires simply a virus-infected or transgenic plant, supplied with appropriate sunlight, mineral salts, and water in order to produce utmost quality of safe, efficient chimeric proteins on an agricultural scale (Whitelam et al. 1993, 1994; Whitelam and Garry 1996). Recombinant antibodies (rAbs) (Ma and Hein 1995a,b), enzymes (Hogue et al. 1990; Verwoerd et al. 1995), hormones, interleukins (Magnuson et al. 1998), plasma proteins (Sijmons et al. 1990), and vaccines are among the current applications of plant-based expression systems in biotechnology (Mason and Arntzen 1995; Walmsley and Arntzen 2000).

Chimeric plant viruses, which are created in plants, can also be utilized to provide vaccinations (Johnson et al. 1997). Plants have a significant advantage over microbial technologies, mammalian cell culture, and transgenic animal technology in terms of their ability to be genetically modified and grown in single cell suspension culture or scaled up for field scale production. It is possible to create COVID-19 plant-based vaccines by producing the antigenic component of severe acute respiratory syndrome coronavirus 2 (SARS-CoV-2) to induce active immunity or by expressing the virus's antibodies to give passive protection (Laere et al. 2016; Naderi and Fakheri 2015; Dhama et al. 2020). Vaccines developed from plants are relegated as third-generation vaccines. A plant-based vaccine is developed by cloning the vaccine candidate into a plant expression system that can increase the candidate gene's expression in the plant, which subsequently synthesizes the antigenic or protective protein. Utilizing plants as bioreactors and cultivating them for numerous generations, this method sanctions for vaccine manufacture, ensuring continuous production and availability (Streatfield et al. 2001; Liew and Hair-Bejo 2015; Phan and Conrad 2016).

2.6 Perspectives

Plants are alluring because they provide distinct benefits; however, they may not be able to vie with today's well-known and prominent microbial and mammalian systems, particularly in terms of (GMP) development and regulatory ratification in a commercial context. Despite years of extensive research, demonstrating the feasibility of expressing several therapeutic proteins in plants, the leap from the lab-bench to commercialization has proven arduous. Therefore, in order to move forward, veterinary vaccines, non-pharmaceutical diagnostics, cosmetics, and commercial enzymes, which have a lower regulatory load than therapeutic proteins, must be developed in plants. Several scientific and technological obstacles connected to the plant platform have been overcome in recent years. Nonetheless, the regulatory problem related to therapeutic protein synthesis is a consequential road-block to the full potential of plant system. Because of the lower costs and better scalability of plant manufacturing systems, nonpharmaceutical protein commercialization is more facile and more expeditious owing to fewer regulatory hurdles. As a result, the licit framework and laws applied to ecumenical plant-derived goods might significantly boost the technology's appeal. The urge for pharmaceutically or industrially related recombinant proteins, along with proven production potential and economic viability of plant reliant systems, all indicate to an optimistic future for plant-based molecular farming.

2.7 Conclusion

One of the most exciting problems in molecular pharming is the manufacturing of recombinant proteins in transgenic plants for medicinal reasons. Plants have several benefits over other protein manufacturing systems, including quick scaling, low cost, and a substantially lower risk of human contamination. While both plant and cell culture technologies suffer identical extraction and purification costs, plant-made medicines can be quickly scaled up to generate huge quantities with little production expenses.

References

Ahmad, A.R., Kaewpungsup, P., Khorattanakulchai, N. et al. (2019). Recombinant human dentin matrix protein 1 (hDMP1) expressed in *Nicotiana benthamiana* potentially induces osteogenic differentiation. *Plants* 8: 566.

An, N., Ou, J., Jiang, D. et al. (2013). Expression of a functional recombinant human basic fibroblast growth factor from transgenic rice seeds. *Int. J. Mol. Sci.* 14: 3556–3567.

Andrianantoandro, E. (2014). Next-generation therapeutics. *Trends Biotechnol.* 32 (9): 431–432.

Arlen, P.A., Falconer, R., Cherukumilli, S. et al. (2007). Field production and functional evaluation of chloroplast-derived interferon-alpha2b. *Plant Biotechnol. J.* 5: 511–525.

Barta, A., Sommergruber, K., Thompson, D. et al. (1986). The expression of a nopaline synthase - human growth hormone chimeric gene in transformed tobacco and sunflower callus tissue. *Plant Mol. Biol.* 6 (5): 347–357.

Bock, R. (2015). Engineering plastid genomes: methods, tools, and applications in basic research and biotechnology. *Annu. Rev. Plant Biol.* 66: 211–241.

Boyhan, D. and Daniell, H. (2011). Low-cost production of proinsulin in tobacco and lettuce chloroplasts for injectable or oral delivery of functional insulin and C-peptide. *Plant Biotechnol. J.* 9 (5): 585–598.

Brodzik, R., Glogowska, M., Bandurska, K. et al. (2006). Plant-derived anti-Lewis Y mAb exhibits biological activities for efficient immunotherapy against human cancer cells. *Proc. Natl. Acad. Sci. U. S. A.* 103: 8804–8809.

Burnett, M.J.B. and Burnett, A.C. (2020). Therapeutic recombinant protein production in plants: challenges and opportunities. *Plants People Planet* 2: 121–132.

Buyel, J.F. (2019). PMF–Integration and exploitation of side streams to achieve sustainable biomanufacturing. *Front. Plant Sci.* 9: 1893.

Buyel, J.F. and Fischer, R. (2012). Predictive models for transient protein expression in tobacco (*Nicotiana tabacum* L.) can optimize process time, yield, and downstream costs. *Biotechnol. Bioeng.* 109: 2575–2588. https://doi.org/10.1002/bit.24523.

Buyel, J.F., Twyman, R.M., and Fischer, R. (2017). Very-large-scale production of antibodies in plants: the biologization of manufacturing. *Biotechnol. Adv.* 35: 458–465. https://doi.org/10.1016/j.biotechadv.2017.03.011.

Castanon, S., Marin, M.S., Martin-Alonso, J.M. et al. (1999). Immunization with potato plants expressing VP60 protein protects against rabbit hemorrhagic disease virus. *J. Virol.* 73: 4452–4455.

Castiglia, D., Sannino, L., Marcolongo, L. et al. (2016). High-level expression of thermostable cellulolytic enzymes in tobacco transplastomic plants and their use in hydrolysis of an industrially pretreated *Arundo donax* L. biomass. *Biotechnol. Biofuels* 9: 154.

Chen, Q. and Lai, H. (2014). Gene delivery into plant cells for recombinant protein production. *Biomed. Res. Int.* 2014: 932161.

Chen, Q. and Davis, K.R. (2016). The potential of plants as a system for the development and production of human biologics. *F1000Res.* 5: F1000.

Commandeur, U., Twyman, R.M., and Fischer, R. (2003). The biosafety of molecular farming in plants. *AgBiotechNet* 5: 1–9.

Covarrubias, A.A., Cuevas-Velazquez, C.L., Romero-Pérez, P.S. et al. (2017). Structural disorder in plant proteins: where plasticity meets sessility. *Cell. Mol. Life Sci. 74* (17): 3119–3147.

D'Aoust, M.-A., Lavoie, P.-O., Couture, M.M.-J. et al. (2008). Influenza virus-like particles produced by transient expression in *Nicotiana benthamiana* induce a protective immune response against a lethal viral challenge in mice. *Plant Biotechnol. J.* 6: 930–940.

Daniell, H., Rai, V., and Xiao, Y. (2019). Cold chain and virus-free oral polio booster vaccine made in lettuce chloroplasts confers protection against all three poliovirus serotypes. *Plant Biotechnol. J.* 17: 1357–1368.

De Cosa, B., Moar, W., Lee, S.B. et al. (2001). Overexpression of the Bt cry2Aa2 operon in chloroplasts leads to formation of insecticidal crystals. *Nat. Biotechnol.* 19: 71–74.

Dent, M., Hurtado, J., Paul, A.M. et al. (2016). Plant-produced anti-dengue virus monoclonal antibodies exhibit reduced antibody-dependent enhancement of infection activity. *J. Gen. Virol.* 97: 3280–3290.

Dhama, K., Natesan, S., Yatoo, M.I. et al. (2020). Plant-based vaccines and antibodies to combat COVID-19: current status and prospects. *Hum. Vaccin. Immunother.* 16 (12): 2913–2920.

Diamos, A.G., Hunter, J.G.L., Pardhe, M.D. et al. (2019). High level production of monoclonal antibodies using an optimized plant expression system. *Front. Bioeng. Biotechnol.* 7: 472.

Ding, S.H., Huang, L.Y., Wang, Y.D. et al. (2006). High-level expression of basic fibroblast growth factor in transgenic soybean seeds and characterization of its biological activity. *Biotechnol. Lett.* 28: 869–875.

Dixit S., Shukla, A., Singh, V., Upadhyay, S.K. (2021) Bioprospecting of natural compounds for industrial and medical applications; current scenario and bottleneck. In Upadhyay SK, Singh SP (Ed.) *Bioprospecting of Plant Biodiversity for Industrial Molecules*. p. 53–57. Wiley. https://doi.org/10.1002/9781119718017.ch3

Egelkrout, E., Rajan, V., and Howard, J.A. (2012). Overproduction of recombinant proteins in plants. *Plant Sci.* 184: 83–101.

Fernandez-San Millan, A., Mingo-Castel, A., Miller, M., and Daniell, H. (2003). A chloroplast transgenic approach to hyper-express and purify human serum albumin, a protein highly susceptible to proteolytic degradation. *Plant Biotechnol. J.* 1: 71–79.

Firsov, A., Tarasenko, I., Mitiouchkina, T. et al. (2015). High-yield expression of M2e peptide of Avian Influenza virus H5N1 in transgenic duckweed plants. *Mol. Biotechnol.* 57: 653–661.

Fischer, R. and Emans, N. (2000). Molecular farming of pharmaceutical proteins. *Transgenic Res.* 9: 279–299.

Fischer, R., Stoger, E., Schillberg, S. et al. (2004). Plant-based production of biopharmaceuticals. *Curr. Opin. Plant Biol.* 7 (2): 152–158.

Floss, D.M., Sack, M., Stadlmann, J. et al. (2008). Biochemical and functional characterization of anti-HIV antibody–ELP fusion proteins from transgenic plants. *Plant Biotechnol. J.* 6: 379–391.

Gengenbach, B.B., Keil, L.L., Opdensteinen, P. et al. (2018a). Comparison of microbial and transient expression (tobacco plants and plant cell packs) for the production and purification of the anti-cancer mistletoelectin viscumin. *Plant Biotechnol. J.* 116 (9): 2236–2249.

Gengenbach, B.B., Müschen, C.R., and Buyel, J.F. (2018b). Expression and purification of human phosphatase and actin regulator 1 (PHACTR1) in plant based systems. *Protein Expr. Purif.* 151: 46–55. https://doi.org/10.1016/j.pep.2018.06.003.

Gleba, Y. Y., Tuse, D., and Giritch, A. (2014). Plant viral vectors for delivery by *Agrobacterium*. *Curr. Top. Microbiol. Immunol.* 375, 155–192. https://doi.org/10.1007/82_2013_352

Gomez, E., Lucero, M.S., Chimeno Zoth, S. et al. (2013). Transient expression of VP2 in *Nicotiana benthamiana* and its use as a plant-based vaccine against infectious bursal disease virus. *Vaccine* 31: 2623–2627.

Gottschalk, U. (2009). *Process Scale Purification of Antibodies*. Hoboken, NJ: Wiley https://doi.org/10.1002/9780470444894.

Gottschamel, J., Lössl, A., Ruf, S. et al. (2016). Production of dengue virus envelope protein domain III-based antigens in tobacco chloroplasts using inducible and constitutive expression systems. *Plant Mol. Biol.* 91: 497–512.

Guan, C., Du, X., Wang, G. et al. (2014). Expression of biologically active anti-thrombosis protein lumbrokinase in edible sunflower seed kernel. *J. Plant Biochem. Biotechnol.* 23: 257–265.

Guerrero-Andrade, O., Loza-Rubio, E., Olivera-Flores, T. et al. (2006). Expression of the Newcastle disease virus fusion protein in transgenic maize and immunological studies. *Transgenic Res.* 15: 455–463.

Gunter, C.J., Regnard, G.L., Rybicki, E.P., and Hitzeroth, I.I. (2019). Immunogenicity of plant-produced porcine circovirus-like particles in mice. *Plant Biotechnol. J.* 17: 1751–1759.

He, Y., Wang, K., and Yan, N. (2014). The recombinant expression systems for structure determination of eukaryotic membrane proteins. *Protein Cell* 5, 658–672. https://doi.org/10.1007/s13238-014-0086-4

Hefferon, K. (2013). Plant-derived pharmaceuticals for the developing world. *Biotechnol. J.* 8: 1193–1202.

Hiatt, A., Caffferkey, R., and Bowdish, K. (1989). Production of antibodies in transgenic plants. *Nature 342* (6245): 76–78.

Hogue, R.S., Lee, J.M., and An, G. (1990). Production of a foreign protein product with genetically modified plant cells. *Enzym. Microb. Technol.* 12: 533–538.

Holland, T., Sack, M., Rademacher, T. et al. (2010). Optimal nitrogen supply as a key to increased and sustained production of a monoclonal full-size antibody in BY-2 suspension culture. *Biotechnol. Bioeng.* 107: 278–289.

Holtz, B.R., Berquist, B.R., Bennett, L.D. et al. (2015). Commercial-scale biotherapeutics manufacturing facility for plant-made pharmaceuticals. *Plant Biotechnol. J.* 13: 1180–1190. https://doi.org/10.1111/pbi.12469.

Huang, Z., LePore, K., Elkin, G. et al. (2008). High-yield rapid production of hepatitis B surface antigen in plant leaf by a viral expression system. *Plant Biotechnol. J.* 6: 202–209.

Huang, Z., Phoolcharoen, W., Lai, H. et al. (2010). High-level rapid production of full-size monoclonal antibodies in plants by a single-vector DNA replicon system. *Biotechnol. Bioeng.* 106: 9–17.

Hurtado, J., Acharya, D., Lai, H. et al. (2020). In vitro and in vivo efficacy of anti-chikungunya virus monoclonal antibodies produced in wild-type and glycoengineered *Nicotiana benthamiana* plants. *Plant Biotechnol. J.* 18: 266–273.

Iyappan, G., Shanmugaraj, B.M., Inchakalody, V. et al. (2018). Potential of plant biologics to tackle the epidemic like situations—case studies involving viral and bacterial candidates. *Int. J. Infect. Dis.* 73: 363.

Joensuu, J.J., Verdonck, F., Ehrstrom, A. et al. (2006). F4 (K88) fimbrial adhesin FaeG expressed in alfalfa reduces F4+ enterotoxigenic *Escherichia coli* excretion in weaned piglets. *Vaccine* 24: 2387–2394.

Johnson, J., Lin, T., and Lomonossoff, G. (1997). Presentation of heterologous peptides on plant viruses- genetics, structure and function. *Annu. Rev. Phytopathol.* 35: 67–86.

Kalthoff, D., Giritch, A., Geisler, K. et al. (2010). Immunization with plant-expressed hemagglutinin protects chickens from lethal highly pathogenic avian influenza virus H5N1 challenge infection. *J. Virol.* 84: 12002–12010.

Kanagarajan, S., Tolf, C., Lundgren, A. et al. (2012). Transient expression of hemagglutinin antigen from low pathogenic avian influenza A (H7N7) in *Nicotiana benthamiana*. *PLoS One* 7: e33010.

Khandelwal, A., Lakshmi Sita, G., and Shaila, M.S. (2003). Oral immunization of cattle with hemagglutinin protein of rinderpest virus expressed in transgenic peanut induces specific immune responses. *Vaccine* 21: 3282–3289.

Kim, T.G., Baek, M.Y., Lee, E.K. et al. (2008). Expression of human growth hormone in transgenic rice cell suspension culture. *Plant Cell Rep.* 27: 885–891.

Kim, M.-Y., Jang, Y.-S., Yang, M.-S., and Kim, T.-G. (2015). High expression of consensus dengue virus envelope glycoprotein domain III using a viral expression system in tobacco. *Plant Cell Tissue Organ Cult.* 122: 445–451.

Kohl, T., Hitzeroth, I.I., Stewart, D. et al. (2006). Plant-produced cottontail rabbit papillomavirus L1 protein protects against tumor challenge: a proof-of-concept study. *Clin. Vaccine Immunol.* 13: 845–853.

Komarnytsky, S., Borisjuk, N.V., Borisjuk, L.G. et al. (2000). Production of recombinant proteins in tobacco guttation fluid. *Plant Physiol.* 124: 927–934.

Kota, S., Subramanian, M., Shanmugaraj, B. et al. (2017). Subunit vaccine based on plant expressed recombinant Eimeria gametocyte antigen Gam82 elicit protective immune response against chicken coccidiosis. *J. Vaccines Vaccin* 8: 6.

Koya, V., Moayeri, M., Leppla, S.H., and Daniell, H. (2005). Plant-based vaccine: mice immunized with chloroplast-derived anthrax protective antigen survive anthrax lethal toxin challenge. *Infect. Immun.* 73: 8266–8274.

Laere E, Ling APK, Wong YP, Koh RY, Mohd Lila MA, Hussein S. Plant-based vaccines: production and challenges. *J. Bot.* 2016;2016:1–11. https://doi.org/10.1155/2016/4928637.

Lai, H., He, J., Engle, M. et al. (2012). Robust production of virus-like particles and monoclonal antibodies with geminiviral replicon vectors in lettuce. *Plant Biotechnol. J.* 10: 95–104.

Lai, H., He, J., Hurtado, J. et al. (2014). Structural and functional characterization of an anti-West Nile virus monoclonal antibody and its single-chain variant produced in glycoengineered plants. *Plant Biotechnol. J.* 12: 1098–1107.

Lamphear, B.J., Jilka, J.M., Kesl, L. et al. (2004). A corn-based delivery system for animal vaccines: an oral transmissible gastroenteritis virus vaccine boosts lactogenic immunity in swine. *Vaccine* 22: 2420–2424.

Leuzinger, K., Dent, M., Hurtado, J. et al. (2013). Efficient agroinfiltration of plants for high-level transient expression of recombinant proteins. *J. Vis. Exp.* 77: 50–52.

Liew, P.S. and Hair-Bejo, M. (2015). Farming of plant-based veterinary vaccines and their applications for disease prevention in animals. *Adv. Virol.* 2015: 936940.

Limkul, J., Misaki, R., Kato, K., and Fujiyama, K. (2015). The combination of plant translational enhancers and terminator increase the expression of human glucocerebrosidase in *Nicotiana benthamiana* plants. *Plant Sci.* 240: 41–49.

Ma, J. and Hein, M. (1995a). Immunotherpeuic potential of antibodies produced in plants. *Trends Biotechnol.* 13: 522–527.

Ma, J. and Hein, M. (1995b). Plant antibodies for Immunotherapy. *Plant Physiol.* 109: 341–346.

Ma, J.K.C., Drossard, J., Lewis, D. et al. (2015). Regulatory approval and a first-in-human phase I clinical trial of a monoclonal antibody produced in transgenic tobacco plants. *Plant Biotechnol. J. 13* (8): 1106–1120.

Ma, J.K., Hiatt, A., Hein, M. et al. (1995). Generation and assembly of secretory antibodies in plants. *Science* 268: 716–719.

Magnuson, N.S., Linzmaier, P.M., Reeves, R. et al. (1998). Secretion of biologically active human interleukin-2 and interleukin-4 from genetically modified tobacco cells in suspension culture. *Protein Expr. Purif.* 13: 45–52.

Margolin, E., Chapman, R., Meyers, A.E. et al. (2019). Production and immunogenicity of soluble plant-produced HIV-1 subtype C envelope gp140 immunogens. *Front. Plant Sci.* 10: 1378.

Marsian, J., Fox, H., Bahar, M.W. et al. (2017). Plant-made polio type 3 stabilized VLPs-a candidate synthetic polio vaccine. *Nat. Commun.* 8: 245.

Mason, H.S., Lam, D.M., and Arntzen, C.J. (1992). Expression of hepatitis B surface antigen in transgenic plants. *Proc. Natl. Acad. Sci. U. S. A.* 89: 11745.

Mason, H.S. and Arntzen, C.J. (1995). Transgenic plants as vaccine production systems. *Trends Biotechnol.* 13: 388–392.

Mathew, L.G., Herbst-Kralovetz, M.M., and Mason, H.S. (2014). Norovirus Narita 104 virus-like particles expressed in *Nicotiana benthamiana* induce serum and mucosal immune responses. *Biomed. Res. Int.* 2014: 807539.

Matsumoto, S., Ikura, K., Ueda, M., and Sasaki, R. (1995). Characterization of a human glycoprotein (erythropoietin) produced in cultured tobacco cells. *Plant Mol. Biol.* 27: 1163–1172.

Mbewana, S., Mortimer, E., Pera, F.F. et al. (2015). Production of H5N1 influenza virus matrix protein 2 ectodomain protein bodies in tobacco plants and in insect cells as a candidate universal influenza vaccine. *Front. Bioeng. Biotechnol.* 3: 197.

Moon, K.B., Park, J.S., Park, Y.I. et al. (2019). Development of systems for the production of plant-derived biopharmaceuticals. *Plants* 9 (1): 30.

Mor, T.S. (2015). Molecular pharming's foot in the FDA's door: Protalix's trailblazing story. *Biotechnol. Lett.* 37: 2147–2150. https://doi.org/10.1007/s10529-015-1908-z.

Naderi, S. and Fakheri, B. (2015). Overview of plant-based vaccines. *Res. J. Fish. Hydrobiol.* 10: 275–289.

Oey, M., Lohse, M., Kreikemeyer, B., and Bock, R. (2009). Exhaustion of the chloroplast protein synthesis capacity by massive expression of a highly stable protein antibiotic. *Plant J.* 57: 436–445.

Paul, M. (2015). International Society for Plant Molecular Farming: 1st Conference, Berlin-Dahlem, Germany, 17th-19th June 2014. *Transgenic Res.* 24 (2): 375–380.

Perez Aguirreburualde, M.S., Gomez, M.C., Ostachuk, A. et al. (2013). Efficacy of a BVDV subunit vaccine produced in alfalfa transgenic plants. *Vet. Immunol. Immunopathol.* 151: 315–324.

Pérez Filgueira, D.M., Zamorano, P.I., Domínguez, M.G. et al. (2003). Bovine herpes virus gD protein produced in plants using a recombinant tobacco mosaic virus (TMV) vector possesses authentic antigenicity. *Vaccine* 21: 4201–4209.

Peyret, H. and Lomonossoff, G.P. (2013). The pEAQ vector series: the easy and quick way to produce recombinant proteins in plants. *Plant Mol. Biol.* 83: 51–58. https://doi.org/10.1007/s11103-013-0036-1.

Phan, H.T. and Conrad, U. (2016). Plant-based vaccine antigen production. *Methods Mol. Biol.* 1349: 35–47.

Piron, R., De Koker, S., De Paepe, A. et al. (2014). Boosting in planta production of antigens derived from the porcine reproductive and respiratory syndrome virus (PRRSV) and subsequent evaluation of their immunogenicity. *PLoS One* 9: e91386.

Qiu, X.G., Wong, G., Audet, J. et al. (2014). Reversion of advanced ebola virus disease in nonhuman primates with ZMapp. *Nature* 514: 47–53. https://doi.org/10.1038/nature13777.

Rader, R.A. and Langer, E.S. (2015). 30 years of upstream productivity improvements. *BioProcess Int.* 13: 10–14.

Rattanapisit, K., Abdulheem, S., Chaikeawkaew, D. et al. (2017). Recombinant human osteopontin expressed in *Nicotiana benthamiana* stimulates osteogenesis related genes in human periodontal ligament cells. *Sci. Rep.* 7: 17358.

Rattanapisit, K., Chao, Z., Siriwattananon, K. et al. (2019a). Plant-produced anti-enterovirus 71 (EV71) monoclonal antibody efficiently protects mice against EV71 infection. *Plants* 8: 560.

Rattanapisit, K., Phakham, T., Buranapraditkun, S. et al. (2019b). Structural and in vitro functional analyses of novel plant-produced anti-human PD1 antibody. *Sci. Rep.* 9: 15205.

Rosales-Mendoza, S., Soria-Guerra, R.E., Lopez-Revilla, R. et al. (2008). Ingestion of transgenic carrots expressing the *Escherichia coli* heat-labile enterotoxin B subunit protects mice against cholera toxin challenge. *Plant Cell Rep.* 27: 79–84.

Ruggiero, F., Exposito, J.Y., Bournat, P. et al. (2000). Triple helix assembly and processing of human collagen produced in transgenic tobacco plants. *FEBS Lett.* 469: 132–136.

Sack, M., Rademacher, T., Spiegel, H. et al. (2015). From gene to harvest: insights into upstream process development for the GMP production of a monoclonal antibody in transgenic tobacco plants. *Plant Biotechnol. J.* 13: 1094–1105. https://doi.org/10.1111/pbi.12438.

Sainsbury, F. and Lomonossoff, G.P. (2008). Extremely high-level and rapid transient protein production in plants without the use of viral replication. *Plant Physiol.* 148: 1212–1218.

Santi, L., Batchelor, L., Huang, Z. et al. (2008). An efficient plant viral expression system generating orally immunogenic Norwalk virus-like particles. *Vaccine* 26: 1846–1854.

Schillberg, S. and Finnern, R. (2021, 2021). PMF for the production of valuable proteins – critical evaluation of achievements and future challenges. *J. Plant Physiol.* 258–259: 153359.

Schillberg, S., Raven, N., Spiegel, H. et al. (2019). Critical analysis of the commercial potential of plants for the production of recombinant proteins. *Front. Plant Sci.* 10: 720.

Scotti, N., Alagna, F., Ferraiolo, E. et al. (2009). High-level expression of the HIV-1 Pr55gag polyprotein in transgenic tobacco chloroplasts. *Planta* 229: 1109–1122.

Shafaghi, M., Maktoobian, S., Rasouli, R. et al. (2018). Transient expression of biologically active anti-rabies virus monoclonal antibody in tobacco leaves. *Iran. J. Biotechnol.* 16: e1774.

Shanmugaraj, B., Bulaon, C.J.I., and Phoolcharoen, W. (2020). PMF: A viable platform for recombinant biopharmaceutical production. *Plants* 9 (7): 842.

Shoji, Y., Farrance, C.E., Bautista, J. et al. (2012). A plant-based system for rapid production of influenza vaccine antigens. *Influenza Other Respir. Viruses* 6: 204–210.

Sijmons, P.C., Dekker, B.M., Schrammeijer, B. et al. (1990). Production of correctly processed human serum albumin in transgenic plants. *Biotechnology (N Y)* 8 (3): 217–221.

Smith, T., O'Kennedy, M.M., Wandrag, D.B.R. et al. (2020). Efficacy of a plant-produced virus-like particle vaccine in chickens challenged with Influenza A H6N2 virus. *Plant Biotechnol. J.* 18: 502–512.

So, Y., Lee, K.-J., Kim, D.-S. et al. (2013). Glycomodification and characterization of anti-colorectal cancer immunotherapeutic monoclonal antibodies in transgenic tobacco. *Plant Cell Tissue Organ Cult.* 113: 41–49.

Staub, J.M., Garcia, B., Graves, J. et al. (2000). High-yield production of a human therapeutic protein in tobacco chloroplasts. *Nat. Biotechnol.* 18: 333–338.

Stoger, E., Ma, J.K., Fischer, R., and Christou, P. (2005). Sowing the seeds of success: pharmaceutical proteins from plants. *Curr. Opin. Biotechnol.* *16* (2): 167–173.

Stoger, E., Vaquero, C., Torres, E. et al. (2000). Cereal crops as viable production and storage systems for pharmaceutical scFv antibodies. *Plant Mol. Biol.* 42: 583–590.

Strasser, R. (2016). Plant protein glycosylation. *Glycobiology* 26: 926–939. https://doi.org/10.1093/glycob/cww023.

Streatfield, S.J., Jilka, J.M., Hood, E.E. et al. (2001). Plant-based vaccines: unique advantages. *Vaccine* 19 (17–19): 2742–2748. https://doi.org/10.1016/S0264-410X(00)00512-0.

Su, J., Sherman, A., Doerfler, P.A. et al. (2015). Oral delivery of Acid Alpha Glucosidase epitopes expressed in plant chloroplasts suppresses antibody formation in treatment of Pompe mice. *Plant Biotechnol. J.* 13: 1023–1032.

Tan, Y., Wang, K.Y., Wang, N. et al. (2014). Ectopic expression of human acidic fibroblast growth factor 1 in the medicinal plant, *Salvia miltiorrhiza*, accelerates the healing of burn wounds. *BMC Biotechnol.* 14: 74.

Terashima, M., Murai, Y., Kawamura, M. et al. (1999). Production of functional human α1-antitrypsin by plant cell culture. *Appl. Microbiol. Biotechnol.* 52: 516–523.

Thanavala, Y., Mahoney, M., Pal, S. et al. (2005). Immunogenicity in humans of an edible vaccine for hepatitis B. *Proc. Natl. Acad. Sci. U. S. A.* 102: 3378–3382.

Thuenemann, E.C., Meyers, A.E., Verwey, J. et al. (2013). A method for rapid production of heteromultimeric protein complexes in plants: assembly of protective bluetongue virus-like particles. *Plant Biotechnol. J.* 11: 839–846.

Topp, E., Irwin, R., McAllister, T. et al. (2016). The case for plant-made veterinary immunotherapeutics. *Biotechnol. Adv.* 34 (5): 597–604.

Torres, E., Vaquero, C., Nicholson, L. et al. (1999). Rice cell culture as an alternative production system for functional diagnostic and therapeutic antibodies. *Transgenic Res.* 8: 441–449.

Tuboly, T., Yu, W., Bailey, A. et al. (2000). Immunogenicity of porcine transmissible gastroenteritis virus spike protein expressed in plants. *Vaccine* 18: 2023–2028.

Upadhyay, S.K. (2021). *Genome Engineering for Crop Improvement*. Wiley https://doi.org/10.1002/9781119672425.

Upadhyay, S.K. and Singh, S.P. (2021). *Bioprospecting of Plant Biodiversity for Industrial Molecules*. Wiley https://doi.org/10.1002/9781119718017.

Vamvaka, E., Twyman, R.M., Murad, A.M. et al. (2016). Rice endosperm produces an underglycosylated and potent form of the HIV-neutralizing monoclonal antibody 2G12. *Plant Biotechnol. J.* 14: 97–108.

Van Dolleweerd, C.J., Teh, A.Y., Banyard, A.C. et al. (2014). Engineering, expression in transgenic plants and characterisation of E559, a rabies virus-neutralising monoclonal antibody. *J. Infect. Dis.* 210: 200–208.

Vaquero, C., Sack, M., Chandler, J. et al. (1999). Transient expression of a tumor-specific single-chain fragment and a chimeric antibody in tobacco leaves. *Proc. Natl. Acad. Sci. U. S. A.* 96: 11128–11133.

Varsani, A., Williamson, A.L., Rose, R.C. et al. (2003). Expression of human papillomavirus type 16 major capsid protein in transgenic *Nicotiana tabacum* cv. Xanthi. *Arch. Virol.* 148: 1771–1786.

Verwoerd, T.C., van Paridon, P.A., van Ooyen, A.J.J. et al. (1995). Stable accumulation of *Aspergillus niger* phytase in transgenic tobacco leaves. *Plant Physiol.* 109: 1199–1205.

Vieira Gomes, A.M., Souza Carmo, T., Silva Carvalho, L. et al. (2018). Comparison of yeasts as hosts for recombinant protein production. *Microorganisms* 6: 38.

Walmsley, A. and Arntzen, C. (2000). Plants for delivery of edible vaccines. *Curr. Opin. Biotechnol.* 11: 126–129.

Walsh, G. (2014). Biopharmaceutical benchmarks 2014. *Nat. Biotechnol. 32* (10): 992–1000.

Wang, D.M., Zhu, J.B., Peng, M., and Zhou, P. (2008). Induction of a protective antibody response to FMDV in mice following oral immunization with transgenic *Stylosanthes spp.* as a feedstuff additive. *Transgenic Res.* 17: 1163–1170.

Wang, Y., Deng, H., Zhang, X. et al. (2009). Generation and immunogenicity of Japanese encephalitis virus envelope protein expressed in transgenic rice. *Biochem. Biophys. Res. Commun.* 380: 292–297.

Wang, Y.P., Wei, Z.Y., Zhong, X.F. et al. (2016). Stable expression of basic fibroblast growth factor in chloroplasts of tobacco. *Int. J. Mol. Sci.* 17: 19.

Whitelam, G.C. and Garry, W. (1996). Antibody expression in transgenic plants. *Trends Plant Sci.* 1: 268–271.

Whitelam, G.C., Cockburn, B., Gandecha, A.R., and Owen, M.R. (1993). Heterologous protein production in transgenic plants. *Biotechnol. Genet. Eng. Rev.* 11: 1–29.

Whitelam, G.C., Cockburn, W., and Owen, M.R. (1994). Antibody production in transgenic plants. *Biochem. Soc. Trans.* 22: 940–944.

Wirth, S., Calamante, G., Mentaberry, A. et al. (2004). Expression of active human epidermal growth factor (hEGF) in tobacco plants by integrative and non-integrative systems. *Mol. Breed.* 13: 23–35.

Wirz, H., Sauer-Budge, A.F., Briggs, J. et al. (2012). Automated production of plant-based vaccines and pharmaceuticals. *J. Lab. Autom.* 17: 449–457. https://doi.org/10.1177/2211068212460037.

Wu, J., Yu, L., Li, L. et al. (2007). Oral immunization with transgenic rice seeds expressing VP2 protein of infectious bursal disease virus induces protective immune responses in chickens. *Plant Biotechnol. J.* 5: 570–578.

Xu, J., Okada, S., Tan, L. et al. (2010). Human growth hormone expressed in tobacco cells as an arabinogalactan-protein fusion glycoprotein has a prolonged serum life. *Transgenic Res.* 19: 849–867.

Xu, S., Gavin, J., Jiang, R., and Chen, H. (2017). Bioreactor productivity and media cost comparison for different intensified cell culture processes. *Biotechnol. Prog.* 33: 867–878. https://doi.org/10.1002/btpr.2415.

Yabuta, Y., Tamoi, M., Yamamoto, K. et al. (2008). Molecular design of photosynthesis-elevated chloroplasts for mass accumulation of a foreign protein. *Plant Cell Physiol.* 49: 375–385.

Yang, M., Sun, H., Lai, H. et al. (2018). Plant-produced Zika virus envelope protein elicits neutralizing immune responses that correlate with protective immunity against Zika virus in mice. *Plant Biotechnol. J.* 16: 572–580.

Yusibov, V., Streatfield, S.J., and Kushnir, N. (2011). Clinical development of plant-produced recombinant pharmaceuticals: vaccines, antibodies and beyond. *Hum. Vaccin.* 7 (3): 313–321.

Zeitlin, L., Bohorov, O., Bohorova, N. et al. (2013). Prophylactic and therapeutic testing of Nicotiana-derived RSV-neutralizing human monoclonal antibodies in the cotton rat model. *MAbs* 5: 263–269.

Zhong, G.-Y., Peterson, D., Delaney, D.E. et al. (1999). Commercial production of aprotinin in transgenic maize seeds. *Mol. Breed.* 5: 345–356.

Zhou, J.Y., Cheng, L.Q., Zheng, X.J. et al. (2004). Generation of the transgenic potato expressing full-length spike protein of infectious bronchitis virus. *J. Biotechnol.* 111: 121–130.

Zwierzina, H., Bergmann, L., Fiebig, H. et al. (2011). The preclinical and clinical activity of aviscumine: a potential anticancer drug. *Eur. J. Cancer* 47: 1450–1457. https://doi.org/10.1016/j.ejca.2011.02.022.

3

Plants as Edible Vaccine

Jia Qi Yip, Jia Choo, Kirthikah Kadiresen, Megan Min Tse Yew, Ying Pei Wong, and Anna Pick Kiong Ling

Division of Applied Biomedical Sciences and Biotechnology, School of Health Sciences, International Medical University, Kuala Lumpur, Malaysia

3.1 Introduction

Immunizations have been one of the most successful medical interventions that have saved millions of lives. Vaccination has prevented deaths from diseases including tetanus, pertussis, diphtheria, measles, and influenza each year (World Health Organisation (WHO) n.d.). Vaccines refer to biological preparations of weakened or dead microorganisms that stimulate an immune response against a particular disease. The history of vaccinology goes all the way back to the year 1796 when Edward Jenner noticed a dairymaid with a fresh cowpox lesion, which he used to inoculate James Phipps, an eight-year-old boy (Zakir et al. 2019). Although the boy developed mild symptoms, he recovered nine days later and when Jenner inoculated the eight-year old boy with matter from a fresh smallpox lesion two months later, no disease developed (Zakir et al. 2019). Hence, he was confirmed having protection from smallpox. This finding led to further studies and the development of vaccines.

Vaccine development is an extensive process involving several different stages, including the exploratory stage, preclinical stage, and clinical trial stages before obtaining regulatory approval. In the exploratory stage of this complex process, basic laboratory research is conducted with regard to conceptual ideas and to identify natural antigens or develop synthetic antigens against the disease (Dutta 2020; Sharma et al. 2020). This research-intensive exploratory stage takes 2–4 years (Dutta 2020). Vaccine candidates that show promising results from the exploratory stage would move on to the preclinical stage that involves both *in vitro* and *in vivo* studies. The preclinical stage serves as a safety and immunogenicity testing stage by utilizing cell or tissue culture systems and animal models (Rather et al. 2021). This stage is essential to provide an idea on the expected response in humans toward the candidate vaccine. Besides that, results from preclinical studies may suggest the safest administration method and safest starting dose for the next stage

Plants as Bioreactors for Industrial Molecules, First Edition.
Edited by Santosh Kumar Upadhyay and Sudhir P. Singh.

(Baylor 2016). The following clinical trial stage is further divided into three phases and phase IV, which is post-marketing. The Phases I, II, and III of clinical trials are aimed to assess the safety, dosage, and immune response; safety, dosage, immunogenicity, administration route, and immunization schedule; and immunogenicity, respectively (Rather et al. 2021). Candidate vaccines that have successfully passed these stages will be reviewed by regulatory bodies and given approval. However, phase IV of the clinical trial will continue with post-marketing where the vaccine is monitored for its long-term safety.

Some of the conventional vaccines that have been developed include live attenuated vaccines, killed whole organism vaccines, subunit vaccines, and toxoid vaccines. Apart from these other vaccine platforms, including protein–polysaccharide conjugate, outer membrane vesicle, viral vectored, and nuclei acid, vaccines have also been developed with a similar goal to provide immunity by stimulating an immune response and immunological memory (Pollard and Bijker 2021). Conventional vaccines possess several limitations. One of the major limitations is associated with its lengthy and costly process of production, hence making them less accessible to developing countries. The special requirements of these vaccines in terms of storage conditions, transportation conditions, administration route, which is intramuscular or subcutaneous for most vaccines, and the need for trained healthcare personnel to administer the vaccine also contributed to the limitations of conventional vaccines (Zhang et al. 2015; Kurup and Thomas 2020). In addition, another major problem is its safety concern as unsuccessful inactivation of the pathogen in live-attenuated vaccines may result in the reversion to its virulent form (Hanley 2011; Kurup and Thomas 2020). With these limitations associated with conventional vaccines, alternative vaccine development platforms that are safer and cheaper have arisen, leading to the development of plant-based edible vaccines.

The development of edible vaccines began with the introduction of the concept by Charles Arntzen in 1990 (Bhairy and Hirlekar 2017; Sahoo et al. 2020). The first successful edible vaccine was done on tobacco where a surface antigen of *Streptococcus mutans* was expressed in a tobacco plant (Sonia and Negi 2021). The mucosal immune response stimulated by the edible vaccine expressing *S. mutans* antigen was proposed to provide protection against tooth decay by inhibiting bacteria colonization in an individual's teeth (Bhairy and Hirlekar 2017). Edible vaccines are developed by introducing a gene of interest into a plant by stable or transient transformation. This transformation will produce a transgenic plant expressing the protein coded by the gene of interest in its tissues, which provides immunization when administered orally. Edible vaccines hold promises to a better vaccination strategy with a cost-effective and safer approach compared to traditional vaccines. Some of the many advantages of the edible vaccine include a simpler route of administration that does not require a trained healthcare personnel nor the addition of adjuvants to produce better immune responses, can be mass-produced easily, does not require refrigeration during transportation and storage, lack contamination risks, as well as easier expression separation and protein purification (Sohrab 2020). It is also known that traditional injectable vaccines possess the ability to stimulate systemic humoral response but does not elicit robust T cell and mucosal-mediated immunity (Criscuolo et al. 2019; Kurup and Thomas 2020). However, edible vaccines have been reported to be capable of eliciting systemic immunity as well as mucosal immunity when administered. These advantages hold great promises toward plant-based edible vaccines.

3.2 Mechanism of Action

A major difference between conventional injectable vaccines and edible plant vaccine is its route of administration. Hence, variation in the mechanism of action of the vaccines can be observed. The oral route of administration of edible plant vaccines enables the stimulation of the mucosal immune response system (Gunasekaran and Gothandam 2020). This mucosal immune response refers to immune responses that take place at the mucosal membranes of the respiratory tract, gastrointestinal tract, and urogenital tracts. This form of immune response is essential as the first line of defence against pathogens when invading the human body (Sahoo et al. 2020). When an edible vaccine is administered, the mucosal immune response is stimulated in the gut-associated lymphoid tissue (GALT) where Peyer's patches play an essential role. Peyer's patches refer to groupings of subepithelial lymphoid follicles that are located in the intestine (Shah and Misra 2011). Antigens released from the edible vaccines are captured primarily by microfold (M) cells, which are found in the follicle-associated epithelium that covers the Peyer's patches (Lamichhane et al. 2014; Kurup and Thomas 2020).

Upon capturing the antigens, M cells will present the antigens to antigen-presenting cells (APCs) such as dendritic cells and macrophages (Wosen et al. 2018; Diaz-Dinamarca et al. 2020; Gunasekaran and Gothandam 2020). Among many of the APCs, mature dendritic cells are known to be the strongest activator of naïve T cells (Janeway et al. 2001). Immature dendritic cells have low capacity of activating T cells due to decreased levels of MHC and costimulatory molecule expressions (Sabado et al. 2017). Mature dendritic cells possess the ability to stimulate regulatory T cell, helper T cell, or T follicular helper cell differentiations (Sabado et al. 2017; Tanne and Bhardwaj 2017). Alternative to obtaining antigen from the M cell, dendritic cells also possess the ability to penetrate into the epithelium of the gut by extending its membrane extensions, allowing it to capture the antigen directly from the gut (Diaz-Dinamarca et al. 2020). The dendritic cells will carry out endocytosis of the antigen presented, which it then processes. Then, the resulting processed antigens are presented to $CD4^+$ T cells via the T cell receptor present on the surface of naïve $CD4^+$ T cells by major histocompatibility complex (MHC) class II of the dendritic cells (Kiyono and Fukuyama 2004; Lamichhane et al. 2014). This stimulates the differentiation of naïve $CD4^+$ T cells into helper T cells. In the lymph nodes' T-cell areas, primed T cells are differentiated into T_H1, T_H2, T_H17, and Treg (Powell et al. 2010; Sahoo et al. 2020). T_H1, T_H2, and T_H17 are essential in providing immunity against bacteria and viruses, stimulating antibody production, and providing immunity against mucosal pathogens, respectively (Kim et al. 2012).

Secretion of IgA antibody plays an important role in mucosal immune responses. This process is stimulated by the activation of B cells. These B cells are activated by the interaction between MHC class II and T cell receptor and between receptor CD40 present on B cell surface and its CD40L ligand expressed on the surface of $CD4^+$ T cells, which lead to the development of germinal center in the B cell follicle (Kim et al. 2012; Lycke and Bemark 2017; Li et al. 2020). This event is essential for the process of affinity maturation and IgA class switch. Cytokines including interleukin (IL)-2, IL-4, IL-5, IL-6, IL-10, and transforming growth factor (TGF)-β play a crucial role in class switching of B cells to IgA producing B cells and for the process of differentiation into plasma cells producing IgA antibody (Lamichhane et al. 2014; Pantazica et al. 2021). Upon class switching and maturation of affinity, the IgA committed B cells migrate from the inductive site, which is the Peyer's patches to mesenteric lymph nodes, then to the effector sites via thoracic duct and blood circulation (Kiyono and Fukuyama 2004).

At the effector site, the IgA committed B cells are differentiated into plasma cells. This is enhanced by the presence of cytokines including IL-5 and IL-6, which are produced by T_H2 cells to produce IgA antibodies in the form of dimers and polymers (Marasini et al. 2014). The basolateral surface of the lumen epithelial cells expresses polymeric immunoglobulin receptors (pIgR), and the interaction between dimeric forms of IgA and pIgR allows the transport of IgA antibodies in the form of secretory IgA into the gut lumen (Kim et al. 2012; Lycke and Bemark 2017). Recent studies have also suggested that another mechanism of action involved in immunity elicited by edible vaccines, which are through the activity of goblet cells. Goblet cells are simple columnar cells found mostly in the mucosal epithelial cells of the intestine (Knoop and Newberry 2018). These goblet cells were found to be a suitable route of antigen delivery to APC as the goblet cells are able to directly capture antigens in the lumen and deliver the captured antigen to APCs through goblet cell-associated antigen passages (Knoop and Newberry 2018; Kurup and Thomas 2020).

3.3 Edible Plant Vaccines

The term "molecular farming" was coined based on the first usage of transgenic tobacco and potato plants to produce human serum albumin in the 1990s (Sijmons et al. 1990). Over the years, the implication of the term has evolved from its novel nature to a well-established method in which recombinant proteins can be produced safely, efficiently, and economically through a variety of protocols (Shakoor et al. 2019). Plants as a bioreactor for producing parental vaccines have become a promising subject of research since then.

3.3.1 Candidate Plants and Selection of Desired Gene

There are certain criteria to take note of when selecting a plant candidate to house the recombinant antigen of interest. Aside from considering the palatability of the plant, the ideal candidate for an edible vaccine should be able to grow quickly, be affordable, allow easy transformation, have a high shelf life, and be edible with minimal prior processing (Gunasekaran and Gothandam 2020; Kurup and Thomas 2020). Tobacco plants were popular as a model to develop edible vaccines due to their short growth time and versatile transformation ability (Shakoor et al. 2019).

- **Fast-growing** plants have an advantage in commercial settings since they can be harvested more frequently, generating a higher yield of the vaccine.
- Plants that are a staple food anywhere in the world or specifically in the region of high demand for edible vaccines would be **affordable** to produce with certain techniques on the market.
- Plants that allow **easy transformation** make great candidates for potential edible vaccines. In that case, plants that expression systems have been well understood are preferred as it reduces the risk of low expression levels, which results in weaker immunogenicity.

- The edible portion of the plant should have a **high shelf life**. Plants that can be stored in room temperature for a long time without spoiling are cost effective and would be especially advantageous in developing countries having a high demand for vaccines.
- Plants that can be **eaten raw** is a plus since cooking tends to degrade proteins and lower the bioavailability of the vaccine. If the use of heat is inevitable, the potency of the vaccine would have to be compensated with a higher dosage or added procedures such as bioencapsulation to delay degradation for increased efficacy (Rosales-Mendoza and Salazar-González 2014).

The part of the plant that is utilized to express the desired recombinant antigen as well as the molecular site of expression is important, as the suitability of each plant differs according to the characteristics of their parts. For certain plants, their leaves (e.g. tobacco), fruits (e.g. tomato), tubers (potato), and grains (e.g. rice) are able to demonstrate high levels of expression and thus are widely used in the research of edible vaccines (Streatfield et al. 2015). On a molecular level, the endoplasmic reticulum (ER) and chloroplasts have shown to be great targets of gene delivery into the host plant (Chebolu and Daniell 2009; Lössl and Waheed 2011).

Compatibility of the recombinant antigen with the genetic make-up of the plant is also an important aspect to consider in order to maximize the expression levels of the antigen. Several popular plant candidates are listed in Table 3.1 along with their desirable traits (Mason et al. 1996; Lal et al. 2007; Aryamvally et al. 2017).

Even though it is impossible to find all desirable traits in a single plant, the selection of a few optimum criteria is sufficient to manufacture an efficacious edible plant vaccine. However, it is crucial for the plant to retain its immunomodulatory abilities after transformation, carry an adequate amount of the desired antigen, and not to have effects that counter its immunomodulatory properties (Gunn et al. 2012).

3.4 Production of Edible Vaccine (Plant Transformation)

3.4.1 Chemical-Mediated DNA Transfer Method

Chemical-mediated gene delivery involves the use of natural or synthetic compounds to facilitate the uptake of DNA by target cells. For instance, polyethylene glycol (PEG) is used to induce protoplasts transfection by disrupting the membrane structure of plants and allowing the entry of DNA into the cell. Organic nanoparticles such as liposomes and

Table 3.1 Some popular plant candidates and their desirable traits.

	Potato	Rice	Banana	Tomato	Lettuce	Tobacco	Soybean/ alfalfa	Legumes/ cereals
Fast growing	✓			✓	✓	✓	✓	✓
Affordable	✓		✓					
Easy transformation	✓					✓	✓	✓
Stable shelf life	✓	✓				✓		✓
Direct consumption			✓	✓	✓	✓	✓	

calcium phosphate are introduced as nonviral vectors due to their nonimmunogenic properties and ability to protect plasmid DNA from the enzymatic reaction.

3.4.1.1 Polyethylene Glycol (PEG)-Mediated DNA Transfer Method

Polyethylene glycol (PEG)-mediated DNA transfer method aims to alter the membrane structure of plants, enabling the entry of foreign DNA into the cells. Although this method is easy and efficient, its application is confined to protoplasts (Low et al. 2018). Protoplasts can be produced via enzymatic digestion by disrupting the cell wall, followed by staining procedures such as fluorescein diacetate to illustrate the integrity of cell membrane and cell viability (Gallego 2021a). When the individual plant cell is exposed to PEG, the cell membrane is damaged and is more readily to take up DNA from the medium. It should be noted that PEG plays an important role in promoting the association of compact DNA to the cell membrane by changing DNA conformation (Gallego 2021a). A high concentration of PEG (>25% w/v) may cause the clumping of protoplasts, and therefore inhibit the transfection of oil palm protoplasts (Masani et al. 2014).

The addition of divalent cations (e.g. Mg^{2+} and Ca^{2+} ions) provides a stable environment for protoplasts and promotes its adhesion to each other resulting from high osmotic pressure in polycation solution (Shillito 1999). It is proven that the concentration of Mg^{2+} ions (50 mM $MgCl_2$) had a significant impact on the transfection efficiency of oil palm protoplasts, promoting more efficient uptake of exogenous DNA through the PEG-mediated method (Masani et al. 2014). Once the exogenous DNA is steadily integrated into the plant chromosomal DNA, protoplasts are then cultured under optimal conditions, allowing the growth of cell walls and regeneration of these protoplasts to calluses and whole plants (Narusaka et al. 2012). While this method is easy to perform with relatively low cost and does not require specific equipment, it is only limited to protoplasts, and it takes time for the whole plants to regenerate (Hwang et al. 2017). This method has been widely expanded to protoplasts of several plants, including soybean (*Glycine max*), rice, and banana (Lee et al. 2013; Wu et al. 2020).

PEG-mediated transformation is generally preferred over physical methods such as biolistic bombardment for stable transformation of protoplasts. According to Wu et al., the banana protoplast regeneration efficiency was greatly increased along with the efficiency of delivering Cas9-ribonucleoprotein (RNP) into protoplasts using the PEG-mediated method (Wu et al. 2020). This is further supported by Malnoy et al., using PEG to deliver ribonucleoproteins into protoplasts of grape and apple with a higher mutation efficiency compared to that of using particle bombardment method (Malnoy et al. 2016). The crucial components for the production of transgenic plants mainly rely on stable gene expression, the efficiency of transformation techniques to transport the genes into the plants, as well as the regeneration competence of the transformed protoplasts (Sharma et al. 2005). These factors account for the safe development and categorization of transgenic plants as edible vaccines.

3.4.1.2 Liposome-Mediated DNA Transfer Method

Liposome, on the other hand, encapsulates and transports genetic materials into the protoplast without being affected by nuclease activity. Some of the important characteristics of an effective nonviral vector to transport the desired gene into the cell nucleus include

(i) camouflage the negative charge of the DNA, (ii) promote DNA condensation, and (iii) protection of DNA molecules from nuclease activity present in the plant cell media (Attar 2017). As liposomes are generally made up of more than one phospholipid bilayer, each with hydrophilic heads and hydrophobic tails connected through a backbone linker, they normally present as a spherical vesicle used to transport drugs or genetic material to the target cell.

In addition, the lipophilic feature of liposomes enhances the delivery of encapsulated DNA through interactions of liposomes with cells. This can be done either through simple adsorption or endocytosis (Akbarzadeh et al. 2013). The fusion of liposomes with plant protoplasts is stimulated by chemicals such as PEG. During liposome–cell interaction, the cationic liposome can bind specifically or nonspecifically to the negatively charged cell membrane. Following with the binding to the cell surface, the cationic liposome is engulfed into the cell through endocytosis and accompanied by enzymatic digestion, which in turn promotes intracellular distribution of liposomal components containing DNA materials to cytosol (Daraee et al. 2014).

This technique is of advantageous as it protects the introduced DNA from nuclease digestion as well as the acidic pH surrounding that may degrade the cargo and therefore reduce the transfection efficiency of liposomes (Alghuthaymi et al. 2021). Furthermore, this method is widely applicable to a wide range of plant cell types, and encapsulation of nucleic acids ensures DNA stability during storage (Sternberg 1998). It is worth noting that the surface properties of liposomes, the degree of PEGylation, as well as the location of targeted tissues may have a great impact on the gene delivery efficiency (Xia et al. 2016). A physical method such as biolistic gun and electroporation can improve liposomal gene delivery to target cells (Spiller et al. 1998).

3.4.1.3 Calcium Phosphate Coprecipitation

The principle of calcium phosphate (CaP) coprecipitation transfection incorporates mixing DNA with calcium chloride in an isotonic phosphate solution to generate calcium phosphate DNA coprecipitate. Insoluble Ca-DNA complex is attached to the cell surface of protoplasts and taken up by cells through endocytosis (Kofron and Laurencin 2004). According to Saba et.al., DNA is greatly protected from cellular nucleases when it is encapsulated into the CaP nanoparticles with the size ranging from 20 to 50 nm in diameter, and the transformation efficiency was found to be much higher compared to *Agrobacterium tumafaciens*-mediated genetic transformation technique with 80.7% transformation efficiency using transgenic GUS (β-glucuronidase) Naqvi et al. (2012). Incubation of protoplasts with a plasmid containing insoluble Ca-P complex promotes the genetic transformation of *Nicotiana tabacum*, and the introduced DNA was found to be integrated into cellular DNA through an analysis using Southern blot hybridization (Hain et al. 1985).

Calcium phosphate-based nanoparticles have been long established as a nonviral gene delivery vehicle, but their inconsistency and low transfection efficiencies limit their progress to emerge as an effective gene delivery alternative (Lee et al. 2012). Through the adjustment in calcium: phosphate ratio and mode of mixing, variability in the inconsistency and transfection efficiencies can be optimized (Lee et al. 2012) Although the calcium phosphate coprecipitation method shows low transfection efficiency, they are not prone to microbial attack, can be easily prepared, have low toxicity, and ensure DNA stability during storage (Sokolova and Ukraine 2006).

3.4.1.4 Diethylaminoethyl (DEAE) – Ddextran-mediated DNA Transfer Method

In the early 1980s, the DEAE-dextran method has been widely used due to its efficiency and reproducibility of the procedure (Lalani and Misra 2011). Through electrostatic interaction between DEAE-dextran and DNA, the positively charged complexes bind to the negatively charged plasma membrane and are engulfed into the cell through endocytosis for transient expression (Mahato et al. 1999). Cells are more permeable to the uptake of DNA complexes when being exposed at a high concentration of dimethyl sulfoxide (DMSO). While this cationic polymer has higher transfection efficiency than the calcium phosphate coprecipitate approach, it is forbidden in *in vivo* practice owing to its low transfection efficiency and high cellular cytotoxicity (Mahato et al. 1999).

3.4.2 Direct Gene Delivery Method (Physical)

Nonviral gene delivery system is used as an alternative to replace viral-based systems as it promotes the transfer of exogenous genes into target cells without triggering immune response or causing insertion mutation in the host. In general, physical methods such as gene gun, chemicals, electroporation, ultrasound, and hydrodynamic therapy incorporate an external physical force to allow the desired gene to enter the cell by increasing the cell membrane permeability.

3.4.2.1 Biolistic Transfection

Genetic transformation using a biolistic gun involves the bombardment of plant cells with the high velocity of tungsten or gold particles with a diameter of 0.45–1.5 μm coated with DNA. The coated DNA penetrates different layers of targeted plant cells such as the cytosol and nucleus, depending on the discharge pressure of helium gas. The particles penetrate through the cell wall and enter the plant cells, leaving the DNA particle inside the cells which in turn allows random bombardment of metal particles with the cells. This method was developed to overcome the barrier when the transformation of a particular species was tough to perform (Das et al. 2020). The efficiency of genetic transformation using a biolistic gun depends on various factors, including the bombardment of helium pressure, the density of carrier particles and macromolecules, thickness of targeted tissues, as well as the method of coating gold onto the DNA (Das et al. 2020).

In spite of the fact that this method requires a high cost, the benefits of this biolistic method are vast. This includes its feasibility with different cell types, in particular nondividing cells that are difficult to transfect as well as its ability to accommodate and transport different sizes of RNA and DNA (Belyantseva 2016). Although biolistic gun evidently proved to be a strong candidate in genetic transformation, a few factors need to be taken into consideration. As the particles penetrate into the plant cells, bombarded tissues may require a few days to recover from the mechanical damage resulting from the high speed of bombardment. Besides, it has low transfection efficiency and the cellular target (i.e. nucleus, cytoplasm and plastid) that are uncontrollable (Baltes et al. 2017).

In transgenic plants, there are two types of antigen expression that can be accomplished by biolistic gun, namely nuclear and chloroplast transformation. Among these two types of transformations, chloroplast transformation offers enormous advantages and is widely used in the production of plant-based vaccines. While nuclear transformation allows the integration of the desired gene into the nucleus of plant cells through nonhomologous combination, chloroplast transformation promotes the incorporation of a gene of interest into the chloroplast genome via homologous combination at a precise location of the plastid, which eliminates the possibility of gene silencing and accumulation of foreign proteins resulting from the high level of transgene expression (Adem et al. 2017).

3.4.2.2 Electroporation

Similarly, electroporation uses short high-voltage pulses to temporarily permeabilize the protoplast plasma membrane and allow the exogenous DNA to diffuse through the pores in the membrane. Protoplasts and selected molecules (i.e. DNA) are suspended in a buffer and an electrical circuit is set up around the mixture (Potter and Heller 2003). In order to disrupt the membrane structure, an electrical pulse with optimized voltage is applied through the cell suspension from a few microseconds up to a millisecond, and this allows the DNA to be driven across the membrane through the pores (Potter and Heller 2003). It should be noted that the voltage is optimized based on protoplast density and its size (Bolhassani et al. 2014). The interaction between plasmid DNA and protoplast is further enhanced by the repeated pulses, varying from 100 V/cm to 2 kV/cm to bring the plasmid closer to the cell membrane (Bolhassani et al. 2014). Electroporation has been used for transient and stable expression of protoplast from a wide range of species, such as *Arabidopsis thaliana* and mulberry (Razdan and Thomas 2021; Zeng et al. 2021).

Plasmid DNA concentration, heat shock, as well as the addition of PEG may enhance stable and transient gene expression (Joersbo and Brunstedt 1996). Previous studies by Furugata et al. have suggested that the biophysical features of protein such as the size and surface charges may affect the gene delivery efficiency (Furuhata et al. 2019). The protein, Cre recombinase (35 kDa), can be delivered into *A. thaliana* without removing the cell wall using the electroporation method, and the transfection efficiency was approximately 83% (Furuhata et al. 2019). Besides, electric field strength, the concentration of vector, and carrier DNA, as well as the pH of electroporation buffer, are responsible for the transfection efficiency of protoplasts (Niedz et al. 2003). According to Randall et al., all these variables were optimized as such: (i) 375–450 V/cm of electric field strength and (ii) 100 μg/ml of DNA concentration; electroporation buffer at pH 8, to ensure high transfection efficiency of *Citrus sinensis* (L.) Osbeck (Niedz et al. 2003).

There is no doubt that electroporation is a well-established method that has brought several advantages to produce transgenic plants. Once the optimal parameters of the electroporation are identified, its high efficiency and ability to work with any cell type have encouraged its wide application in gene transformation (DePalma 2018). Nevertheless, this method may lead to potential cell death, owing to permeabilization of protoplast plasma membrane using high voltage pulses (DePalma 2018).

3.4.2.3 Sonication

Similar to electroporation, sonication increases the cell membrane permeability to DNA. It does so by creating transient pores in cell membranes for the transfer of exogenous DNA using ultrasonic frequencies. The entry of DNA is driven by passive diffusion, and its transfer efficiency mainly relies on the ultrasound frequency and intensity (Tomizawa et al. 2013). Through mechanical perturbation, the cell membrane is fractionated, and the resulting cavitation bubbles cause membrane poration which in turn allows the entry of genes into the cells (Miller et al. 2002). It has been proposed that high temperatures allied with inertial cavitation increase the fluidity of phospholipid membrane and promote the fusion of microbubbles with the cell membrane that improves the effects of delivering genes to targeted sites (Moran 2011). To reduce cell damage, optimization of the technique can be done on the pulse intensity, frequency, and treatment duration (Abinaya and Viswanathan 2021).

In spite of its low efficiency, sonication has gained interest as it is a noninvasive and easy procedure (do Minh et al. 2019). This alternative is usually combined with other chemical methods to enhance the transfection efficiency. For instance, the combined effect of sonication and polyethylenimine (PEI) nanoparticles increased the transfection efficiency of DNA into *Crocus sativus* with two times higher than the results obtained by PEI or sonication, respectively (Firoozi et al. 2018).

3.4.2.4 Microinjection

Microinjection refers to the transfer of genetic materials into a living cell using glass micropipettes or metal microinjection needles. It is the most direct alternative used to inject DNA into the cytoplasm or nucleus under microscopical procedures to avoid any cell damage (Neuhaus and Spangenberg 1990). Several plant species have been successfully transformed using this approach including *Nicotina tabacum* and rapeseed plants (Neuhaus et al. 1987; Kost et al. 1995). This technique is slow and requires expensive microinstruments to stop the cells from moving using a holding pipette and gentle suction (Rivera et al. 2012). Still, its high efficiency and precise delivery of genes to target location enabled its application in stable plant transformation (Rivera et al. 2012). In a study by Shen et al., the introduction of a gene into maize ovaries through microinjection overcomes the limitation of *in vivo* seed culture and improves the ratio of transformation (Shen et al. 2001).

3.4.3 Indirect Gene Delivery

Transgenic plants are produced via two methods that are direct and indirect gene transfer. The latter requires plasmid as a vector and the construction of vector is based on Ti and Ri plasmids. The *Agrobacterium*-mediated gene transfer method has been widely promoted to transform both dicot and monocot species in the last few years. It is well known to be the most preferred method to produce transgenic plants.

3.4.3.1 *Agrobacterium*-Mediated Gene Transfer

Agrobacterium is a gram-negative soil bacterium that uses horizontal gene transfer to cause tumors in plants. Its unique properties to introduce part of its DNA into plant cells makes it highly valuable in plant biology and genetic engineering (Mehrotra and Goyal 2012). Among all the other *Agrobacterium* species, *Agrobacterium tumefacians,* and *Agrobacterium rhizognes* are generally used in the integration of genes into the host plant. The main difference between these two *Agrobacterium* sp is the plasmids. *Agrobacterium rhizogens* carries root-inducing (Ri) plasmid, whereas *A. tumefacians* carries tumor-inducing (Ti) plasmid.

Agrobacterium rhizogens is responsible for hairy root disease in dicotyledonous plants. Infection in plants is initiated by the random integration of the transferred (T)-DNA region of the root-inducing (Ri) plasmid of *A. rhizogens* into the host plant (Chen and Otten 2017). This leads to abnormal root growth and production of opines, which are taken up by bacterium and used for its growth. The hairy root expression was suggested to be used in the production of recombinant therapeutic proteins such as reporter proteins (e.g. green fluorescent protein), antibodies, and antigens (e.g. hepatitis B surface antigen) (Sunil Kumar et al. 2006; Aarrouf et al. 2011). Similar to *A.tumefacians*, *A.rhizogens* transforms plants by introducing T-DNA from Ri plasmid to the host plant through conjugation.

On the other hand, *A. tumefacians* is the causal agent for crown gall disease at the wound site of infected dicot plants (Gohlke and Deeken 2014). It infects the cells through the delivery of T-DNA from bacteria into the plant nucleus and integrates into the host plant chromosomal DNA. Two main components are required in the *A. tumefacians*-mediated plant genetic transformation process. The first key component is the transferred (T)-DNA and the second is the virulence region (*vir*). The latter is composed of seven major loci ranging from *virA* to *virG*, and each encodes for the bacterial protein machinery process. *A. tumefacians*-mediated plant transformation begins with the attachment of *A. tumefacians* to the plant cells followed by the transduction of plant signals (Hwang et al. 2017). This allows the T-DNA and virulence proteins to be transported from the bacterial cells into the host plant cells, resulting in the integration and expression of T-DNA in the plant genome (Hwang et al. 2017). After the expression of the foreign gene in the host cells, it results in neoplastic growth of the tumors and promotes opine synthesis which is a main carbon and nitrogen source for *Agrobacterium* growth (Yildiz et al. 2016). T-DNA, containing genes coding for phytohormones synthesis (i.e. auxin and cytokinin), causes infected plants to suffer from uncontrolled cell division. It was then proposed to use Ti plasmid as a vector to introduce foreign genes into the host plant (Gelvin 2003).

As wild-type Ti plasmids are large and they lack specific restriction endonuclease sites, the Ti plasmid is engineered to delete tumor-induced genes and replace them with foreign genes of interest (Gelvin 2003). In the co-integrating vector system, the intermediate plasmid is conjugated with *Agrobacterium* having a disarmed Ti plasmid. In the disarmed Ti plasmid containing the *vir* functions, the oncogenes located in the T-DNA region are replaced by the gene of interest. An intermediate plasmid that contains border sequences, such as left border (LB) and right border (RB), is replicated in *E. coli* (Kiyokawa et al. 2009). As both the intermediate vector and disarmed Ti plasmid consist of common sequences, they are combined through homologous recombination (Kiyokawa et al. 2009). The resulting cointegrated plasmid contains the *vir* genes, LB and RB, the gene of interest, and selectable markers (Kiyokawa et al. 2009). This cointegrative vector will be introduced back into the *Agrobacterium* for transgenic transformation. This method is advantageous as the genetic material would not replicate itself in *Agrobacterium*, reducing the possibility of gene silencing. Although cointegrated vectors are designed to promote site-specific recombination, they are generally less popular due to relatively inefficient gene transfer compared to binary vectors, and the need for sophisticated protocols to introduce the gene of interest into the bacterial plasmid followed by conjugation, which is a complex procedure. Therefore, this method became less preferred ever since the binary vector system was introduced (Low et al. 2018).

In the binary vector system, helper plasmid and binary vector are working together to achieve gene transformation in plants. The T-binary vector contains foreign DNA in place of T-DNA, LB, and RB, as well as markers for selection in both *E. coli* and *A. tumefacians* (Komori et al. 2007). In *A. tumefacians*, the *vir* helper plasmid carries the intact *vir* region, which is crucial in the cleavage of T-DNA. The disarmed strain carrying a Ti plasmid without the wild-type T-DNA and an artificial T-DNA can be combined and replicated in both *E. coli* and *A. tumefaciens* (Komori et al. 2007). The recombinant binary vector is then transferred to *A. tumefacians* carrying helper plasmid through bacterial conjugation or direct transfer (Lee and Gelvin 2008). Although this binary vector system is widely used in gene transformation due to its high transformation efficiency and cloning capacity, it is limited by the instability of plasmids with two different origins of replications in *E. coli* as well as the occurrence of gene silencing (Gallego 2021b; Frandsen 2011).

Several plant-derived vaccines have been developed using various plant expression systems, such as potato, tomato, and maize. As reported by Rao et al., the VP1 protein from the foot-and-mouth disease virus (FMDV) was found in transgenic *Crotalaria juncea* using the *Agrobacterium*-mediated gene transfer method (Rao et al. 2012). Besides, Joung et al. successfully transformed the gene encoding for surface S and preS2 antigen of recombinant hepatitis B virus (HBV) into potato using *Agrobacterium*-mediated approach and the mice fed with potato presented an immune response against HBV S antigen (Joung et al. 2004).

3.4.3.2 Genetically Engineered Plant Virus

The unique properties of plant viruses, including low risk of pre-existing immunity and flexibility in vaccine construction, make them highly suitable and exceptional as a versatile tool in vaccine development (Balke and Zeltins 2020). In general, there are two main strategies to produce vaccines using plant viruses. Plant viruses can be engineered to produce viral coat chimeric gene, allowing the display of heterologous epitope on the chimeric viral particle surface; or the targeted antigen is engineered as the discrete reading frame in the viral genome (Yusibov et al. 2006; Kurup and Thomas 2020). In the former, the viral particle is the end-product, and the chimeric particles can be extracted from the plant and be easily purified (Yusibov et al. 2006). This method is advantageous as it intensifies adaptive immune response through the presentation of various epitopes on the viral particle surface (Masani et al. 2014). In the second strategy, the heterologous protein is the product, and the virus particles with other plant proteins will be removed (Yusibov et al. 2006). Also, plant viruses act as a vector to transport genetic materials to the host plant cells. Some examples of engineered RNA virus vectors being used for these purposes include tobacco mosaic virus (TMV), alfalfa mosaic virus (A1MV), and plum pox virus (PPV) (Yusibov et al. 2006).

TMV is an RNA virus that infects plants, and it has been widely used in vaccine development owing to the suitability of its virions as nanoparticles (McCormick and Palmer 2014). The TMV RNA was first transferred to the *Nicotina benthamiana* plant, and the virus was harvested for 10 days before purification and conjugation (Mansour et al. 2018). As proposed by Mansour et al., a triantigen TMV vaccine presents 50% protection in immunized mice against respiratory tularemia, and 100% protection was spotted with intranasal immunization of tetra-antigen TMV vaccine conjugated with OmpA, DnaK, Tul4, and SucB proteins followed by two boosters (Mansour et al. 2018). The development of this multivalent subunit vaccine lays the foundation for vaccine discovery that can be broadly expanded to many other bacterial pathogens. In addition, different epitopes have been expressed in TMV vectors, such as foot-and-mouth disease virus (FMDV), influenza A, and murine hepatitis virus (Koo et al. 1999; Wu et al. 2003; Petukhova et al. 2013). As such, engineered viruses do not reproduce in mammalian cells, they appear as a sustainable vector for vaccine production, and the vaccine antigen acquired is proven to be protective against the disease (Kurup and Thomas 2020).

There are two groups of plant-infecting DNA viruses referred to as single-stranded DNA viruses (i.e. Geminiviruses) and double-stranded DNA viruses (i.e. Caulimoviruses) (Shafiq et al. 2020). Geminiviruses have been explored as potential expression vectors in vaccine production on account of their ability to replicate in high copy numbers in infected cells and leads to high gene expression levels (Hefferon 2014). *Agrobacterium*-mediated viral vectors were developed after realizing the need to overcome the complex process of producing recombinant geminiviruses *in vitro* in a large scale (Yang et al. 2016). The process is done in such a way that the desired geminivirus genome was put into the tumor-inducing (Ti) plasmid and the two intergenic regions in the dimeric clone recombined to form a circular viral genome (Buragohain et al. 1994). The agrobacteria suspension was injected into the intercellular space or to the stem, leading to infection by geminiviruses replicon containing Ti plasmid and expression of foreign DNA in plants via multiplication (Yang et al. 2016). Vaccine expression vectors using other geminiviruses are still in progress. For instance, Norwalk virus capsid protein and hepatitis B core antigen were expressed using bean yellow dwarf virus (BeYDV)-derived expression vector (Chen et al. 2011). Another example is the subcutaneous injection of mice with HPV-16PsVs DNA vaccine using geminivirus vector was proven to elicit a stronger immune response than using naked DNA vaccination through the gene gun method (Lamprecht et al. 2016).

In contrast to geminiviruses, cauliflower mosaic viruses (CaMV) do not rely on host DNA replication apparatus and are able to replicate via reverse transcription within an infected cell (Turri et al. 2020). This virus predominantly targets radish, canola, cauliflower, broccoli, and cabbage. After entry into the host cell, decapsidation occurs and the viral DNA is transcribed into mRNA through two promoters, namely P19S and P35S (Bak and Emerson 2020). This CaMV 35S leads to low expression in monocots but works well in dicots. Some of the vaccine examples using caulimoviruses as vector include H5N1, hepatitis B, and *S. mutans* surface antigen (Ramírez et al. 2000; Sharp and Doran 2001; Phan et al. 2020).

3.4.3.3 Virus-Like Particles (VLPs)

Plant viruses have also been evolved into virus-like particles (VLPs) and virus nanoparticles (VNPs). VLP has the ability to self-assemble, imitating the form and size of a virus particle without the viral genetic material (Nooraei et al. 2021). This makes them an ideal source in the field of preventive medicine as they are not capable of infecting the host cell and are highly immunogenic. They induce both antibody- and cell-mediated immune responses. VLPs serve two purposes: (i) as a scaffold for molecules conjugation and (ii) as epitope display systems in vaccine production (Venkataraman and Hefferon 2021). Till now, VLP remains the leading candidate for vaccine production and several viruses such as HBV, HIV, foot and mouth viruses, as well as rotaviruses have been well explored for the formation of VLPs (Scotti and Rybicki 2014). This offers a typical instance of plant-derived VLP vaccines carrying influenza hemagglutinin (HA) triggered a strong immune response in both preclinical and clinical trials (Makarkov et al. 2019). Application of VLPs promotes antigen uptake and cross-presentation of peptides (Makarkov et al. 2019). A transmission-blocking vaccine against malaria with Pfs25 surface protein was produced in *N. benthamiana*, and this plant-derived Pfs25-VLP was proven to be safe in the clinical trial with antibody production in patients in a dose-dependent manner (Chichester et al. 2018).

3.5 Plant Species Used as Vaccine Models

3.5.1 Potato

Potato is one of the world's most important food crops, and it is an essential part of the world's food supply today. Through the introduction of potatoes from the New World to the Old World, a significant increase in population and urbanization was observed during the eighteenth and nineteenth centuries (Nunn and Qian 2011). Potatoes has greatly increased food productivity and provided more calories, vitamins, and nutrients relative to pre-existing Old World staple crops (Nunn and Qian 2011). Potato is also widely studied in the recent application of biotechnology as it is an ideal crop to transform with the use of biotechnology (Halterman et al. 2016; Bruce and Shoup Rupp 2019). The potato was one of the first crops to be genetically modified – the first successful potato gene transformation was reported in 1983 by using *Agrobacterium*-mediated transformation (Ooms et al. 1983). The ease of propagation in tissue culture and relative simplicity of genetic manipulation in potatoes make it a suitable model to integrate desired genes and recover plants from transformed tissue (Chakravarty et al. 2007; Halterman et al. 2016). Although various approaches such as particle bombardment, protoplast transformation, and microinjection have been successful, the most prevalent way of stable transformation in potatoes is still the *Agrobacterium*-mediated transformation (Halterman et al. 2016).

The first potato vaccine was reported in 1998, developed by Mason et al. to combat enteritis caused by *E. coli* strain (Mason et al. 1998). The heat-labile enterotoxin B subunit (LT-B) gene was introduced to potato plants by the leaf disc cocultivation method with *Agrobacterium tumefaciens* (Mason et al. 1998). The anti-LT-B fecal IgA and anti-LT-B serum IgG were successfully induced in mice fed with transformed tubers; however, no mouse was completely protected from the toxin challenge (Mason et al. 1998). The same year, a clinical study was conducted and proved humans would also develop mucosal B-cell priming after ingesting the potato vaccine (Tacket et al. 1998). A second-phase clinical study on Norwalk virus capsid protein (NVCP) express transgenic potato was reported by Tacket et al. (2000), showing that 95% (19 of 20) volunteers developed an immune response after ingesting the transgenic potato, however with only a modest increase in serum antibody (Tacket et al. 2000). Furthermore, oral immunogenicity of the hepatitis B surface antigen (HBsAg) expressed in transgenic potato has also been demonstrated in NMRI mice as well as humans (Thanavala and Lugade 2010; Rukavtsova et al. 2015). Edible potato vaccines against *Vibrio cholerae* enterotoxin and rabbit hemorrhagic virus have also been studied, and promising results were obtained (Arakawa et al. 1998; Castañón et al. 1999).

Advantages of producing edible vaccines from potatoes include the ease of transformation, propagation, and storage without refrigeration. However, the consumption of potatoes generally required cooking, which may denature the antigens and decrease immunogenicity.

3.5.2 Rice

Rice is a primary staple food for more than half of the world's population, providing up to half of the dietary caloric supply for millions living especially in Asia (Muthayya et al. 2014). Improved technology of rice fortification is widely focused to address the vitamin and mineral deficiencies for the population that subsist on rice (Muthayya et al. 2014). Thus, studies on transgenic rice have been extensively carried out.

The first transgenic rice to express antigenic protein was reported in 2007. The embryogenic rice callus (*Oryza sativa* L.) was transformed by microprojectile bombardment to express a fusion protein consisting of the LT-B and a synthetic core-neutralizing epitope (COE) of porcine epidemic diarrhea virus (PEDV) (Oszvald et al. 2007). This study indicated that the expression system in rice was capable of generating a significant amount of antigen, demonstrating the suitability of the plant as a promising candidate for edible vaccine development. The same year, oral immunization with transgenic rice expressing the infectious bursal disease virus (IBDV) host-protective immunogen VP2 protein was proven to be an effective, safe, and inexpensive vaccine against IBDV in chicken (Wu et al. 2007). In 2008, the modified HBsAg gene SS1 was expressed by the transgenic rice plant via *Agrobacterium*-mediated transformation to test the potential of the plant-based hepatitis B vaccine (Qian et al. 2008). The rice-derived SS1 protein has successfully induced the immunological responses in BALB/c mice, indicating that this rice-derived SS1 protein could be developed as an alternative oral vaccine against hepatitis B infection (Qian et al. 2008). Besides, according to Oszvald et al. (2008), the synthetic cholera toxin B subunit (CTB) gene was introduced into the rice plant via the biolistic-mediated transformation method; the expression was confirmed by PCR amplification analysis (Oszvald et al. 2008).

The advantages of rice as a vaccine candidate include the low allergenic potential, high expression of the protein, ease of storage, and expression of the heat-stable protein, while the disadvantages are the slow growth rate and that the plant requires glasshouse condition to grow (Kurup and Thomas 2020).

3.5.3 Banana

Banana is one of the ideal plant models for edible vaccine development as it does not require cooking, is inexpensive, and is grown widely in many countries (Kurup and Thomas 2020). The first report on the expression of HBsAg in transgenic bananas was published by Kumar et al. (2005), in which the HBsAg was transformed into banana embryogenic cells using the *Agrobacterium*-mediated transformation method (Kumar et al. 2005). The expression of the antigen was verified by PCR, Southern hybridization, and reverse transcription (RT)-PCR, while the maximum expression was noted in the leaves (Kumar et al. 2005). Besides, a recent paper hypothesized an increase in the expression level of the transgene by using pBIN19, a binary vector for delivering the gene into banana plants, but it required further investigation (Maji et al. 2016). Nevertheless, the long mature time for the tree to develop remains the biggest hassle for the banana to be used in vaccine development.

3.5.4 Tomato

The first tomato vaccine, reported in 2006, was studied in response to the outbreak of severe acute respiratory syndrome (SARS). The SARS-coronavirus (CoV) spike protein (S protein) was expressed in tomato via the *Agrobacterium*-mediated transformation, and its transcription was verified by PCR and RT-PCR analysis (Pogrebnyak et al. 2005). A significant increase of SARS-CoV-specific IgA in mice was obtained after the oral ingestion of the transgenic tomato fruits. This demonstrated the effectiveness of plant-derived antigen to induce the systemic and mucosal immune responses in mice (Pogrebnyak et al. 2005). However, the plant-derived material was not immunogenic by itself and required a booster immunization (Pogrebnyak et al. 2005). Furthermore, the tomato plant has also been extensively transformed to express the HBsAg gene in combat of hepatitis B virus (HBV) (Guan et al. 2010). Most of the tomato transformations were using leaf disk cocultivation with the stable expression of the antigen. However, none of the studies were carried out to evaluate the immunological response of the tomato-derived S protein (Ying et al. 2002; Lou et al. 2007; Srinivas et al. 2008; Guan et al. 2010). According to Zhang et al. in 2006, the tomato plant was used to express the recombinant Norwalk virus (rNV) capsid protein, and the immunogenicity of the dried transgenic tomato fruit was tested in mice (Zhang et al. 2006). The air-dried transgenic tomato fruit was found to be a more potent immunogen than dried potato, suggesting the use of stabilized, dried tomato fruit for oral subunit vaccine development (Zhang et al. 2006). The plant-based subunit vaccine against pneumonic and bubonic plague was also developed from transgenic tomatoes that express the F1-V antigen of *Yersinia pestis* (Alvarez et al. 2006). In recent news, researchers from a Mexican university are working on the development of a transgenic tomato as an edible vaccine against COVID-19, which expresses the SARS-CoV-2 antigen (Uthaya Kumar et al. 2021). The tomato's rapid growth rate, broad cultivation, and abundance of Vitamin A enable it to be a promising plant candidate for edible vaccine development (Goyal et al. 2007; Kurup and Thomas 2020).

3.5.5 Lettuce

Lettuce serves as a great model system against *E. coli*-associated enteric diseases in both animals and humans. The synthetic LT-B (sLTB) was transformed into lettuce cells (*Lactuca sativa*) via *Agrobacterium tumefaciens*-mediated manipulation (Kim et al. 2007). The expression of the sLTB gene was detected in transgenic lettuce extract using Western blot analysis (Kim et al. 2007). On the other hand, the transgenic lettuce expressing E2 glycoprotein of Classical Swine Fever Virus (CSFV) induced sufficient immune responses in mice after oral ingestion of the transgenic lettuce (Legocki et al. 2005). Besides, the transgenic lettuce to produce an antigenic $C_4(V_3)6$ multiepitopic protein using *Agrobacterium*-mediated transformation was reported in 2012 against the human immunodeficiency virus (HIV) (Govea-Alonso et al. 2013). Expression of the transgene was confirmed using PCR, and it was found that approximately 1 g of freeze-dried transgenic lettuce leaves contained 240 μg of the $C_4(V_3)6$ protein (Govea-Alonso et al. 2013).

Lettuce is a fast-growing plant, which is usually consumed directly without cooking. It is highly feasible to be used as an edible plant vaccine to induce mucosal immunity.

3.5.6 Maize

As one of the main grain crops of the world, maize has been employed extensively for genetic manipulation to improve its traits. To date, several maize-based vaccines have been developed. Transgenic maize expressing the ApxllA antigen from *Actinobacillus pleuropneumoniae* was produced via *Agrobacterium* selection (Kim et al. 2010). The Apxll-specific immune responses induced in mice has further confirmed the immunogenicity of the transgenic maize (Shin et al. 2011). The M protein from the porcine reproductive and respiratory syndrome virus (PRRSV) expressed in the transgenic maize using the biolistic transformation method induced significant immune responses in immunized mice through oral administration of the transgenic plant tissue (Hu et al. 2012). The transgenic maize expressing HBsAg was also showing promising results as the long-term memory in the orally immunized mice was observed, and the immune responses after re-boosting at 47 and 50 weeks were significant (Hayden et al. 2015). Furthermore, the use of transgenic maize as the expression system against rabies, swine transmissible gastroenteritis virus, Newcastle disease virus, foot-and-mouth disease virus (FMDV), and influenza virus have also been studied (Rosales-Mendoza et al. 2017). The advantages of using maize as an expression system are that it is easy to transform and propagate, it is broadly cultivated, and it grows fast.

3.5.7 Carrot

Carrot is a highly nutritious root vegetable belonging to the umbellifer family, *Apiaceae*. Several transgenic carrot plant-based vaccines have been developed to combat emerging infectious diseases. By using *Agrobacterium*-mediated transformation, recombinant urease B (Ure B) protein combined with CTB was introduced into the carrot plant expression system to develop an edible plant vaccine against *Helicobacter pylori* infection. The expression of the transgene in the carrot plant was satisfying, which yielded up to 25 μg/g of recombinant protein in the root (Zhang et al. 2010). Mucosal immune response was observed in the mice fed with the transgenic carrot (Zhang et al. 2010). Besides, the carrot genome was transformed with the *Mycobacterium tuberculosis* genes *cfp10*, *esat6*, and human deltaferon (*dIFN*) gene via *Agrobacterium*-mediated transformation to express the fusion protein CFP10-ESAT6-dIFN (Uvarova et al. 2013; Permyakova et al. 2015). The transgenic carrot induced both cell-mediated and humoral immunity in orally immunized mice (Uvarova et al. 2013; Permyakova et al. 2015). Carrot is also a low-cost vaccine candidate against synucleinopathies, *Chlamydia trachomatis*, HIV type 1, rabies virus, and porcine cysticercosis (Rojas-Anaya et al. 2009; Kalbina et al. 2011; Lindh et al. 2014; Monreal-Escalante et al. 2016; Arevalo-Villalobos et al. 2020).

3.5.8 Alfalfa

Alfalfa is one of the major leafy crops widely used for recombinant protein production. Alfalfa is an important forage crop in many countries; thus, it is widely used to develop edible vaccines mainly for veterinary purposes. The production of transgenic alfalfa plants to express the polyprotein P1 and the protease 3C of FMDV was reported by Dus Santos et al. in 2005, while the immunogenicity of the proteins expressed was determined in the mouse model (Dus Santos et al. 2005). The immunized mice showed great immune responses and were completely protected when challenged with the virulent virus (Dus Santos et al. 2005). A transgenic alfalfa vaccine candidate against severe acute rotavirus-induced diarrhea was produced by inserting the human group A rotavirus codon-optimized gene (sVP6) of the VP6 protein into the alfalfa genome via *Agrobacterium*-mediated transformation. Oral immunization with the transgenic alfalfa showed positive results – passive heterotypic protection was observed in the pups (Dong et al. 2005). Moreover, a study on the transgenic alfalfa vaccine containing the Eg95-EgA31 fusion gene of *Echinococcus granulosus* also showed positive results, confirming the potential of the transgenic alfalfa in vaccine development (Ye and Li 2010). The advantages of using the alfalfa plant as an edible vaccine candidate include the ease of transformation, ease of propagation, and large harvests.

Table 3.2 summarizes the plant species that are used as vaccine models.

Table 3.2 Plant species used as vaccine models.

Plant	Applications	Method	Protein	References
Potato	Enteritis by *E. coli*	*Agrobacterium*-mediated transformation	Heat-labile enterotoxin B subunit	(Mason et al. 1998; Tacket et al. 1998)
	Acute gastroenteritis by Norwalk virus	-	Norwalk virus capsid protein	(Tacket et al. 2000)
	Hepatitis B	Agrobacterial transformation	Hepatitis B surface antigen	(Rukavtsova et al. 2015)
	Cholera	*Agrobacterium*-mediated potato-leaf transformation	Cholera toxin B subunit (CTB)	(Arakawa et al. 1998)
	Rabbit hemorrhagic disease	Agrobacterial transformation	VP60	(Castañón et al. 1999)
Rice	Acute enteritis in pigs	Microprojectile bombardment	LTB–COE fusion protein	(Oszvald et al. 2007)
	Infectious bursal disease virus in chickens	Agrobacterial transformation	VP2 protein	(Wu et al. 2007)
	Hepatitis B	*Agrobacterium*-mediated transformation	Hepatitis B virus surface antigen fused with preS1 epitopes	(Qian et al. 2008)
	Cholera	Biolistic-mediated transformation	cholera toxin B subunit (CTB)	(Oszvald et al. 2008)
Banana	Hepatitis B	*Agrobacterium*-mediated transformation	Hepatitis B surface antigen	(Kumar et al. 2005)

Table 3.2 (Continued)

Plant	Applications	Method	Protein	References
Tomato	Severe acute respiratory syndrome (SARS)	*Agrobacterium*-mediated transformation	SARS-coronavirus (CoV) spike protein (S protein)	(Pogrebnyak et al. 2005)
	Hepatitis B	Agroinfiltration	Hepatitis B surface antigen	(Srinivas et al. 2008)
	Acute gastroenteritis by Norwalk virus	Agrobacterial transformation	Recombinant Norwalk virus (rNV) capsid protein	(Zhang et al. 2006)
	Pneumonic and bubonic plague	*Agrobacterium*-mediated Transformation	*Y. pestis* F1-V antigen fusion protein	(Alvarez et al. 2006)
Lettuce	*E. coli*-associated enteric diseases	*Agrobacterium*-mediated transformation	*Escherichia coli* heat-labile enterotoxin B subunit	(Kim et al. 2007)
	Classical Swine Fever Virus (CSFV)	-	E2 glycoprotein of CSFV	(Legocki et al. 2005)
	Human immunodeficiency virus (HIV)	*Agrobacterium*-mediated transformation	Antigenic C4(V3)6 multiepitopic protein	(Govea-Alonso et al. 2013)
Maize	Porcine pleuropneumonia by *Actinobacillus (A.) pleuropneumoniae*	*Agrobacterium*-mediated transformation	ApxllA antigen	(Kim et al. 2010; Shin et al. 2011)
	Porcine reproductive and respiratory syndrome virus (PRRSV)	Biolistic transformation	M protein	(Hu et al. 2012)
	Hepatitis B	*Agrobacterium*-mediated transformation	Hepatitis B surface antigen	(Hayden et al. 2015)
	Influenza virus	*Agrobacterium*-mediated transformation	Nucleoprotein (NP)	(Rosales-Mendoza et al. 2017)
	Foot-and-mouth disease virus	Biolistic transformation	VP1 protein	(Rosales-Mendoza et al. 2017)
	Newcastle disease virus (NDV)	-	Fusion protein of NDV	(Rosales-Mendoza et al. 2017)
	Rabies	Biolistic transformation	G protein from Rabies virus	(Rosales-Mendoza et al. 2017)
	Swine transmissible gastroenteritis virus (TGEV)	*Agrobacterium*-mediated transformation	B-subunit from *E. coli* heat-labile enterotoxin and the S protein from TGEV	(Rosales-Mendoza et al. 2017)

(Continued)

Table 3.2 (Continued)

Plant	Applications	Method	Protein	References
Carrot	*Helicobacter pylori* infection	*Agrobacterium*-mediated transformation	Urease B (Ure B) protein combined with CTB	(Zhang et al. 2010)
	Tuberculosis	*Agrobacterium*-mediated transformation	Fusion protein CFP10-ESAT6-dIFN	(Uvarova et al. 2013; Permyakova et al. 2015)
	Chlamydia trachomatis (Ct) infection	*Agrobacterium*-mediated transformation	Major outer membrane protein (MOMP) of *C. trachomatis*	(Kalbina et al. 2011)
Alfalfa	Foot and mouth disease	*Agrobacterium tumefaciens*-mediated transformation	Polyprotein P1 and the protease 3C of FMDV	(Dus Santos et al. 2005)
	Severe acute rotavirus-induced diarrhea	*Agrobacterium*-mediated transformation	VP6 protein	(Dong et al. 2005)
	Echinococcus granulosus protoscoleces	-	Eg95-EgA31	(Ye and Li 2010)

3.6 Challenges

Plant edible vaccines hold great importance especially in countries where access to essential health care remains restricted to a local community suffering from diseases such as hepatitis and malaria. The emergence of plant edible vaccines provides a cost-effective alternative to overcome this predicament. Till now, the main challenge faced by plant edible vaccines is acceptance by the population. The public holds different views on edible vaccines as a therapeutic alternative, mainly owing to its potential risks to the society as well as the environment. It is suggested that the stability of the vaccine is unknown since the protein present in certain crops such as potatoes might undergo denaturation during cooking, which in turn affects its effectiveness (Khan et al. 2019). For edible vaccine production, banana is a commonly used plant species as it does not require cooking, and it is more economical compared to other crops (Kurup and Thomas 2020). However, banana spoils rapidly after ripening, and the protein content is relatively low (Pantazica et al. 2021). As mentioned, the selection of plant expression system is crucial as not all antigens are compatible with the selected host plants. Optimum selection of the system ensures safety and stability of the plant vaccine.

Another issue is the dosage consistency as it varies from plant to plant and selection of the most suitable plant candidate could be difficult (Pantazica et al. 2021). It is also challenging to assess the required dosage of vaccine to each patient as the immune responses vary in different individuals. The dosage of vaccine needs to be optimized as low concentration may not induce antibody production and too high of a concentration can cause tolerance (Maji et al. 2016). In order to achieve the required immunity, large quantities of a genetically modified product may need to be consumed. It should be noted that not all vaccine proteins are highly immunogenic, and the secondary metabolites may diminish the ability of vaccine protein to induce an immune response (Sahai et al. 2013).

In addition to that the transgene may induce hypersensitive reaction of an individual having pollen or food allergy. Also, cross-contamination between genetically modified plant and nongenetically modified plants during pollination can be harmful to the wildlife. It is crucial to monitor the growing plants for plant vaccine production as the genetically modified plants may become aggressive and the release of DNA or transgenic protein into a water source can cause environmental contamination (Pantazica et al. 2021). Of these, it should be noted that these recombinant plant-based vaccines should undergo a tight regulatory process to monitor the containment of the recombinant material before commercialization (Parvathy and Parvathy 2020). In January 2005, the World Health Organization introduced a report on the execution of good agricultural practices to produce plant-derived vaccines (van der Laan et al. 2006). This report discusses the existing regulatory perspectives including quality control for medicinal plants, clinical evaluation of vaccines, and the need to work closely with other regulatory authorities, such as national regulatory authority of the country in which a trial would be executed, the US FDA, as well as the European Medicinal Evaluation Agency to ensure that all appropriate studies are effectuated (van der Laan et al. 2006). Despite the above concerns that have been addressed, edible vaccines provide a great opportunity to protect an individual from diseases if there is promising evidence of its safety and efficiency in clinical trials.

3.7 Conclusion

Vaccines have always played a major role in disease prevention, and the discovery of edible plant vaccines provide an opportunity to make immunization possible despite the limitations that come with conventional and injectible vaccines. As such, research in this field has gained much attention since its first application in 1990, giving rise to many plants being successfully transformed and developed into edible vaccines. Though several challenges remain to persist in the current society, edible plant vaccines continue to show great promise in making immunization safer and more accessible worldwide in the future.

Ackowledgments

The authors wish to thank Prof. Hoe I. Ling of Columbia Univiersity (New York, USA) for his editorial input.

References

Aarrouf, J., Castro-Quezada, P., Mallard, S. et al. (2011). Agrobacterium rhizogenes-dependent production of transformed roots from foliar explants of pepper (*Capsicum annuum*): a new and efficient tool for functional analysis of genes. *Plant Cell Reports* 31 (2): 391–401.

Abinaya, R.V. and Viswanathan, P. (2021). Biotechnology-based therapeutics. In: *Translational Biotechnology* (ed. Y. Hasija), 27–52. Academic Press.

Adem, M., Beyene, D., and Feyissa, T. (2017). Recent achievements obtained by chloroplast transformation. *Plant Methods* 13 (1): 1–11.

Akbarzadeh, A., Rezaei-Sadabady, R., Davaran, S. et al. (2013). Liposome: classification, preparation, and applications. *Nanoscale Research Letters* 8 (1): 1–9.

Alghuthaymi, M.A., Ahmad, A., Khan, Z. et al. (2021). Exosome/Liposome-like nanoparticles: new carriers for CRISPR genome editing in plants. *International Journal of Molecular Sciences* 22 (14): 7456.

Alvarez, M.L., Pinyerd, H.L., Crisantes, J.D. et al. (2006). Plant-made subunit vaccine against pneumonic and bubonic plague is orally immunogenic in mice. *Vaccine* 24 (14): 2477–2490.

Arakawa, T., Chong, D.K.X., and Langridge, W.H.R. (1998). Efficacy of a food plant-based oral cholera toxin B subunit vaccine. *Nature Biotechnology* 16 (3): 292–297.

Arevalo-Villalobos, J.I., Govea Alonso, D.O., and Rosales-Mendoza, S. (2020). Using carrot cells as biofactories and oral delivery vehicles of LTB-Syn: a low-cost vaccine candidate against synucleinopathies. *Journal of Biotechnology* 309: 75–80.

Aryamvally, A., Gunasekaran, V., Narenthiran, K.R., and Pasupathi, R. (2017). New strategies toward edible vaccines: an overview. *Journal of Dietary Supplements* 14 (1): 101–116.

Attar, A. (2017). Gen terapisiyöntemleri: fizikselvekimyasalmetotlar gene therapy techniques: physical and chemical methods. *Turk Hij Den BiyolDerg* 74 (1): 103–112.

Bak, A. and Emerson, J.B. (2020). Cauliflower mosaic virus (CaMV) biology, management, and relevance to GM plant detection for sustainable organic agriculture. *Frontiers in Sustainable Food Systems* 4: 21.

Balke, I. and Zeltins, A. (2020). Recent advances in the use of plant virus-like particles as vaccines. *Viruses* 12 (3): 270.

Baltes, N.J., Gil-Humanes, J., and Voytas, D.F. (2017). Genome engineering and agriculture: opportunities and challenges. *Progress in Molecular Biology and Translational Science* 149: 1–26.

Baylor, N.W. (2016). The regulatory evaluation of vaccines for human use. In: *Methods in Molecular Biology* (ed. S. Thomas), 773–787. New York: Springer Science+Business Media.

Belyantseva, I.A. (2016). Helios® gene gun-mediated transfection of the inner ear sensory epithelium: recent updates. *Methods in Molecular Biology* 1427: 3–26.

Bhairy, S.R. and Hirlekar, R.S. (2017). Edible vaccines: an advancement in oral immunization. *Asian Journal of Pharmaceutical and Clinical Research* 10 (2): 71.

Bolhassani, A., Khavari, A., and Orafa, Z. (2014). Electroporation – advantages and drawbacks for delivery of drug, gene and vaccine. In: *Application of Nanotechnology in Drug Delivery* (ed. A.D. Sezer). IntechOpen.

Bruce, M.A. and Shoup Rupp, J.L. (2019). Agrobacterium-mediated transformation of *Solanum tuberosum* L., Potato. *Methods in Molecular Biology* 1864: 203–223.

Buragohain, A.K., Sung, Y.K., Coffin, R.S., and Coutts, R.H.A. (1994). The infectivity of dimeric potato yellow mosaic geminivirus clones in different hosts. *Journal of General Virology* 75 (10): 2857–2861.

Castañón, S., Marín, M.S., Martín-Alonso, J.M. et al. (1999). Immunization with potato plants expressing VP60 protein protects against rabbit hemorrhagic disease virus. *Journal of Virology* 73 (5): 4452–4455.

Chakravarty, B., Wang-Pruski, G., Flinn, B. et al. (2007). Genetic transformation in potato: approaches and strategies. *American Journal of Potato Research* 84 (4): 301–311.

Chebolu, S. and Daniell, H. (2009). Chloroplast-derived vaccine antigens and biopharmaceuticals: expression, folding, assembly and functionality. *Current Topics in Microbiology and Immunology* 332: 33–54.

Chen, K. and Otten, L. (2017). Natural agrobacterium transformants: recent results and some theoretical considerations. *Frontiers in Plant Science* 8: 1600.

Chen, Q., He, J., Phoolcharoen, W., and Mason, H.S. (2011). Geminiviral vectors based on bean yellow dwarf virus for production of vaccine antigens and monoclonal antibodies in plants. *Human Vaccines* 7 (3): 331–338.

Chichester, J.A., Green, B.J., Jones, R.M. et al. (2018). Safety and immunogenicity of a plant-produced Pfs25 virus-like particle as a transmission blocking vaccine against malaria: a Phase 1 dose-escalation study in healthy adults. *Vaccine* 36 (39): 5865–5871.

Criscuolo, E., Caputo, V., Diotti, R.A. et al. (2019). Alternative methods of vaccine delivery: an overview of edible and intradermal vaccines. *Journal of Immunology Research* 2019: 1–13.

Daraee, H., Etemadi, A., Kouhi, M. et al. (2014). Application of liposomes in medicine and drug delivery. *Artificial Cells, Nanomedicine, and Biotechnology* 44 (1): 381–391.

Das, A., Thakur, S., Shukla, A. et al. (2020). Genetic transformation. In: *Chickpea: Crop Wild Relatives for Enhancing Genetic Gains* (ed. M. Singh), 205–224. Academic Press.

DePalma, A. (2018). Electroporation and Competing Transfection Methods. https://www.genengnews.com/insights/electroporation-and-competing-transfection-methods/ (accessed 4 December 2021).

Diaz-Dinamarca, D.A., Hernandez, C., Escobar, D.F. et al. (2020). Mucosal vaccination with lactococcus lactis-secreting surface immunological protein induces humoral and cellular immune protection against Group B streptococcus in a murine model. *Vaccine* 8 (2): 146.

Dong, J.-L., Liang, B.-G., Jin, Y.-S. et al. (2005). Oral immunization with pBsVP6-transgenic alfalfa protects mice against rotavirus infection. *Virology* 339 (2): 153–163.

Dus Santos, M.J., Carrillo, C., Ardila, F. et al. (2005). Development of transgenic alfalfa plants containing the foot and mouth disease virus structural polyprotein gene P1 and its utilization as an experimental immunogen. *Vaccine* 23 (15): 1838–1843.

Dutta, A.K. (2020). Vaccine against Covid-19 disease – present status of development. *The Indian Journal of Pediatrics* 87 (10): 810–816.

Firoozi, B., Nasser, Z., Sofalian, O., and Sheikhzade-Mosadegh, P. (2018). Enhancement of the transfection efficiency of DNA into *Crocus sativus* L. cells via PEI nanoparticles. *Journal of Integrative Agriculture* 17 (8): 1768–1778.

Frandsen, R.J.N. (2011). A guide to binary vectors and strategies for targeted genome modification in fungi using *Agrobacterium tumefaciens*-mediated transformation. *Journal of Microbiological Methods* 87 (3): 247–262.

Furuhata, Y., Sakai, A., Murakami, T. et al. (2019). A method using electroporation for the protein delivery of Cre recombinase into cultured Arabidopsis cells with an intact cell wall. *Scientific Reports* 9 (1): 1–10.

Gallego, A. (2021a). Deep dive into plant transformation protocols: protoplast-mediated, biolistic-mediated and agrobacterium-mediated gene transfer. https://www.goldbio.com/articles/article/Deep-Dive-Into-Plant-Transformation-Protocols-Protoplast-Mediated-Biolistic-Mediated-Agrobacterium-Mediated-Gene-Transfer (accessed 6 December 2021).

Gallego, A. (2021b). How to choose the right Agrobacterium vector for plant transformation. https://www.goldbio.com/articles/article/How-to-choose-the-right-Agrobacterium-vector (accessed 6 December 2021).

Gelvin, S.B. (2003). Agrobacterium -mediated plant transformation: the biology behind the "Gene-Jockeying" tool. *Microbiology and Molecular Biology Reviews* 67 (1): 16–37.

Gohlke, J. and Deeken, R. (2014). Plant responses to *Agrobacterium tumefaciens* and crown gall development. *Frontiers in Plant Science* 5: 155.

Govea-Alonso, D.O., Gómez-Cardona, E.E., Rubio-Infante, N. et al. (2013). Production of an antigenic C4(V3)6 multiepitopic HIV protein in bacterial and plant systems. *Plant Cell, Tissue and Organ Culture (PCTOC)* 113 (1): 73–79.

Lal, P., Ramachandran, V., Goyal, R., and Sharma, R. (2007). Edible vaccines: current status and future. *Indian Journal of Medical Microbiology* 25 (2): 93.

Guan, Z., Guo, B., Huo, Y. et al. (2010). Overview of expression of hepatitis B surface antigen in transgenic plants. *Vaccine* 28 (46): 7351–7362.

Gunasekaran, B. and Gothandam, K.M. (2020). A review on edible vaccines and their prospects. *Brazilian Journal of Medical and Biological Research* 53 (2): e8749.

Gunn, K.S., Singh, N., Giambrone, J., and Wu, H. (2012). Using transgenic plants as bioreactors to produce edible vaccines. *Journal of Biotech Research* 4 (1): 92–99.

Hain, R., Stabel, P., Czernilofsky, A.P. et al. (1985). Uptake, integration, expression and genetic transmission of a selectable chimaeric gene by plant protoplasts. *Molecular and General Genetics* 199 (2): 161–168.

Halterman, D., Guenthner, J., Collinge, S. et al. (2016). Biotech potatoes in the 21st century: 20 years since the first biotech potato. *American Journal of Potato Research* 93 (1): 1–20.

Hanley, K.A. (2011). The double-edged sword: how evolution can make or break a live-attenuated virus vaccine. *Evolution: Education and Outreach* 4 (4): 635–643.

Hayden, C.A., Fischer, M.E., Andrews, B.L. et al. (2015). Oral delivery of wafers made from HBsAg-expressing maize germ induces long-term immunological systemic and mucosal responses. *Vaccine* 33 (25): 2881–2886.

Hefferon, K.L. (2014). DNA virus vectors for vaccine production in plants: spotlight on geminiviruses. *Vaccine* 2 (3): 642–653.

Hu, J., Ni, Y., Dryman, B.A. et al. (2012). Immunogenicity study of plant-made oral subunit vaccine against porcine reproductive and respiratory syndrome virus (PRRSV). *Vaccine* 30 (12): 2068–2074.

Hwang, H.-H., Yu, M., and Lai, E.-M. (2017). Agrobacterium-mediated plant transformation: biology and applications. *The Arabidopsis Book* 15: e0186.

Janeway, C.A. Jr., Travers, P., Walport, M., and Shlomchik, M.J. (2001). *Immunobiology: The Immune System in Health and Disease*, 5e. New York: Garland Publishing.

Joersbo, M. and Brunstedt, J. (1996). Electroporation and transgenic plant production. In: *Electrical Manipulation of Cells* (ed. P.T. Lynch and M.R. Davey), 201–222. Chapman & Hall.

Joung, Y.H., Youm, J.W., Jeon, J.H. et al. (2004). Expression of the hepatitis B surface S and preS2 antigens in tubers of *Solanum tuberosum*. *Plant Cell Reports* 22 (12): 925–930.

Kalbina, I., Wallin, A., Lindh, I. et al. (2011). A novel chimeric MOMP antigen expressed in Escherichia coli, *Arabidopsis thaliana*, and *Daucus carota* as a potential *Chlamydia trachomatis* vaccine candidate. *Protein Expression and Purification* 80 (2): 194–202.

Khan, A., Khan, A., Khan, I. et al. (2019). A review on natural way of vaccination: plant derived edible vaccines. *Journal of Vaccines and Immunology* 5 (1): 11–17.

Kim, T.-G., Kim, M.-Y., Kim, B.-G. et al. (2007). Synthesis and assembly of *Escherichia coli* heat-labile enterotoxin B subunit in transgenic lettuce (*Lactuca sativa*). *Protein Expression and Purification* 51 (1): 22–27.

Kim, H.-A., Yoo, H.-S., Yang, M.-S. et al. (2010). The development of transgenic maize expressing *Actinobacillus pleuropneumoniae*ApxIIA gene using Agrobacterium. *Journal of Plant Biotechnology* 37 (3): 313–318.

Kim, S.-H., Lee, K.-Y., and Jang, Y.-S. (2012). Mucosal immune system and M cell-targeting strategies for oral mucosal vaccination. *Immune Network* 12 (5): 165.

Kiyokawa, K., Yamamoto, S., Sakuma, K. et al. (2009). Construction of disarmed Ti plasmids transferable between *Escherichia coli* and *Agrobacterium* species. *Applied and Environmental Microbiology* 75 (7): 1845–1851.

Kiyono, H. and Fukuyama, S. (2004). NALT- versus PEYER'S-patch-mediated mucosal immunity. *Nature Reviews Immunology* 4 (9): 699–710.

Knoop, K.A. and Newberry, R.D. (2018). Goblet cells: multifaceted players in immunity at mucosal surfaces. *Mucosal Immunology* 11 (6): 1551–1557.

Kofron, M.D. and Laurencin, C.T. (2004). Development of a calcium phosphate co-precipitate/poly(lactide-co-glycolide) DNA delivery system: release kinetics and cellular transfection studies. *Biomaterials* 25 (13): 2637–2643.

Komori, T., Imayama, T., Kato, N. et al. (2007). Current status of binary vectors and superbinary vectors. *Plant Physiology* 145 (4): 1155–1160.

Koo, M., Bendahmane, M., Lettieri, G.A. et al. (1999). Protective immunity against murine hepatitis virus (MHV) induced by intranasal or subcutaneous administration of hybrids of tobacco mosaic virus that carries an MHV epitope. *Proceedings of the National Academy of Sciences* 96 (14): 7774–7779.

Kost, B., Galli, A., Potrykus, I., and Neuhaus, G. (1995). High efficiency transient and stable transformation by optimized DNA microinjection into *Nicotiana tabacum* protoplasts. *Journal of Experimental Botany* 46 (9): 1157–1167.

Kumar, G.B.S., Ganapathi, T.R., Revathi, C.J. et al. (2005). Expression of hepatitis B surface antigen in transgenic banana plants. *Planta* 222 (3): 484–493.

Kurup, V.M. and Thomas, J. (2020). Edible vaccines: promises and challenges. *Molecular Biotechnology* 62 (2): 79–90.

van der Laan, J.W., Minor, P., Mahoney, R. et al. (2006). WHO informal consultation on scientific basis for regulatory evaluation of candidate human vaccines from plants. Geneva, Switzerland, 24–25 January 2005.

Lalani, J. and Misra, A. (2011). Gene delivery using chemical methods. In: *Challenges in Delivery of Therapeutic Genomics and Proteomics* (ed. A. Misra), 127–206. Elsevier.

Lamichhane, A., Azegami, T., and Kiyono, H. (2014). The mucosal immune system for vaccine development. *Vaccine* 32 (49): 6711–6723.

Lamprecht, R.L., Kennedy, P., Huddy, S.M. et al. (2016). Production of Human papillomavirus pseudovirions in plants and their use in pseudovirion-based neutralisation assays in mammalian cells. *Scientific Reports* 6 (1): 1–10.

Lee, L.Y. and Gelvin, S.B. (2008). T-DNA binary vectors and systems. *Plant Physiology* 146 (2): 325–332.

Lee, D., Upadhye, K., and Kumta, P.N. (2012). Nano-sized calcium phosphate (CaP) carriers for non-viral gene deilvery. *Materials Science and Engineering B* 177 (3): 289–302.

Lee, H., Park, S.-Y., and Zhang, Z.J. (2013). An overview of genetic transformation of soybean. In: *A Comprehensive Survey of International Soybean Research – Genetics, Physiology, Agronomy and Nitrogen Relationships* (ed. J.E. Board). IntechOpen.

Legocki, A.B., Miedzinska, K., Czaplińska, M. et al. (2005). Immunoprotective properties of transgenic plants expressing E2 glycoprotein from CSFV and cysteine protease from *Fasciola hepatica*. *Vaccine* 23 (15): 1844–1846.

Li, Y., Jin, L., and Chen, T. (2020). The effects of secretory IgA in the mucosal immune system. *BioMed Research International* 2020: 1–6.

Lindh, I., Bråve, A., Hallengärd, D. et al. (2014). Oral delivery of plant-derived HIV-1 p24 antigen in low doses shows a superior priming effect in mice compared to high doses. *Vaccine* 32 (20): 2288–2293.

Lössl, A.G. and Waheed, M.T. (2011). Chloroplast-derived vaccines against human diseases: achievements, challenges and scopes. *Plant Biotechnology Journal* 9 (5): 527–539.

Lou, X.-M., Yao, Q.-H., Zhang, Z. et al. (2007). Expression of the human hepatitis B virus large surface antigen gene in transgenic tomato plants. *Clinical and Vaccine Immunology* 14 (4): 464–469.

Low, L.-Y., Yang, S.-K., Kok, D.-X.A. et al. (2018). Transgenic plants: gene constructs, vector and transformation method. In: *New Visions in Plant Science* (ed. Ö. Çelik). IntechOpen.

Lycke, N.Y. and Bemark, M. (2017). The regulation of gut mucosal IgA B-cell responses: recent developments. *Mucosal Immunology* 10 (6): 1361–1374.

Mahato, R.I., Smith, L.C., and Rolland, A. (1999). Pharmaceutical perspectives of nonviral gene therapy. *Advances in Genetics* 41 (C): 95–156.

Maji, A., Sachin, M., Mohana Srinivasan, V., and Subathra Devi, C. (2016). Banana as edible vaccine asgainst hepatitis-B: a theoretical model. *Immunology, Endocrine & Metabolic Agents in Medicinal Chemistry* 16 (2): 129–133.

Makarkov, A.I., Golizeh, M., Ruiz-Lancheros, E. et al. (2019). Plant-derived virus-like particle vaccines drive cross-presentation of influenza A hemagglutinin peptides by human monocyte-derived macrophages. *npj. Vaccines* 4 (1): 1–12.

Malnoy, M., Viola, R., Jung, M.H. et al. (2016). DNA-free genetically edited grapevine and apple protoplast using CRISPR/Cas9 ribonucleoproteins. *Frontiers in Plant Science* 7: 1904.

Mansour, A.A., Banik, S., Suresh, R.V. et al. (2018). An improved tobacco mosaic virus (TMV)-conjugated multiantigen subunit vaccine against respiratory tularemia. *Frontiers in Microbiology* 9: 1195.

Marasini, N., Skwarczynski, M., and Toth, I. (2014). Oral delivery of nanoparticle-based vaccines. *Expert Review of Vaccines* 13 (11): 1361–1376.

Masani, M.Y.A., Noll, G.A., Parveez, G.K.A. et al. (2014). Efficient transformation of oil palm protoplasts by PEG-mediated transfection and DNA microinjection. *PLoS One* 9 (5): e96831.

Mason, H.S., Ball, J.M., Shi, J.J. et al. (1996). Expression of Norwalk virus capsid protein in transgenic tobacco and potato and its oral immunogenicity in mice. *Proceedings of the National Academy of Sciences of the United States of America* 93 (11): 5335–5340.

Mason, H.S., Haq, T.A., Clements, J.D., and Arntzen, C.J. (1998). Edible vaccine protects mice against *Escherichia coli* heat-labile enterotoxin (LT): potatoes expressing a synthetic LT-B gene. *Vaccine* 16 (13): 1336–1343.

McCormick, A.A. and Palmer, K.E. (2014). Genetically engineered Tobacco mosaic virus as nanoparticle vaccines. *Expert Review of Vaccines* 7 (1): 33–41.

Mehrotra, S. and Goyal, V. (2012). Agrobacterium-mediated gene transfer in plants and biosafety considerations. *Applied Biochemistry and Biotechnology* 168 (7): 1953–1975.

Miller, D.L., Pislaru, S.V., and Greenleaf, J.F. (2002). Sonoporation: mechanical DNA delivery by ultrasonic cavitation. *Somatic Cell and Molecular Genetics* 27 (1): 115–134.

Do Minh, A., Sharon, D., Chahal, P., and Kamen, A.A. (2019). Cell transfection. In: *Comprehensive Biotechnology* (ed. M. Moo-Young), 383–390. Pergamon.

Monreal-Escalante, E., Govea-Alonso, D.O., Hernández, M. et al. (2016). Towards the development of an oral vaccine against porcine cysticercosis: expression of the protective HP6/TSOL18 antigen in transgenic carrots cells. *Planta* 243 (3): 675–685.

Moran, C.M. (2011). Ultrasonic contrast agents. *Clinical Ultrasound* 1: 77–89.

Muthayya, S., Sugimoto, J.D., Montgomery, S., and Maberly, G.F. (2014). An overview of global rice production, supply, trade, and consumption. *Annals of the New York Academy of Sciences* 1324 (1): 7–14.

Naqvi, S., Maitra, A.N., Abdin, M.Z. et al. (2012). Calcium phosphate nanoparticle mediated genetic transformation in plants. *Journal of Materials Chemistry* 22 (8): 3500–3507.

Narusaka, Y., Narusaka, M., Yamasaki, S., and Iwabuchi, M. (2012). Methods to transfer foreign genes to plants. In: *Transgenic Plants - Advances and Limitations* (ed. Y.O. Çiftçi). IntechOpen.

Neuhaus, G. and Spangenberg, G. (1990). Plant transformation by microinjection techniques. *Physiologia Plantarum* 79 (1): 213–217.

Neuhaus, G., Spangenberg, G., Mittelsten Scheid, O., and Schweiger, H.G. (1987). Transgenic rapeseed plants obtained by the microinjection of DNA into microspore-derived embryoids. *Theoretical and Applied Genetics* 75 (1): 30–36.

Niedz, R.P., McKendree, W.L., and Shatters, R.G. (2003). Electroporation of embryogenic protoplasts of sweet orange (*Citrus sinensis* (L.) Osbeck) and regeneration of transformed plants. *in vitro Cellular & Developmental Biology – Plant* 39 (6): 586–594.

Nooraei, S., Bahrulolum, H., Hoseini, Z.S. et al. (2021). Virus-like particles: preparation, immunogenicity and their roles as nanovaccines and drug nanocarriers. *Journal of Nanobiotechnology* 19 (1): 1–27.

Nunn, N. and Qian, N. (2011). The potato's contribution to population and urbanization: evidence from A historical experiment. *The Quarterly Journal of Economics* 126 (2): 593–650.

Ooms, G., Karp, A., and Roberts, J. (1983). From tumour to tuber; tumour cell characteristics and chromosome numbers of crown gall-derived tetraploid potato plants (*Solanum tuberosum* cv. 'Maris Bard'). *Theoretical and Applied Genetics* 66 (2): 169–172.

Oszvald, M., Kang, T.-J., Tomoskozi, S. et al. (2007). Expression of a synthetic neutralizing epitope of porcine epidemic diarrhea virus fused with synthetic B subunit of *Escherichia coli* heat labile enterotoxin in rice endosperm. *Molecular Biotechnology* 35 (3): 215.

Oszvald, M., Kang, T.-J., Tomoskozi, S. et al. (2008). Expression of cholera toxin B subunit in transgenic rice endosperm. *Molecular Biotechnology* 40 (3): 261–268.

Pantazica, A.M.M., Cucos, L.M., Stavaru, C. et al. (2021). Challenges and prospects of plant-derived oral vaccines against hepatitis B and C viruses. *Plants* 10 (10): 2037.

Parvathy, S.T. and Parvathy, S.T. (2020). Engineering plants as platforms for production of vaccines. *American Journal of Plant Sciences* 11 (5): 707–735.

Permyakova, N.V., Zagorskaya, A.A., Belavin, P.A. et al. (2015). Transgenic carrot expressing fusion protein comprising *M. tuberculosis* antigens induces immune response in mice. *BioMed Research International* 2015: 1–11.

Petukhova, N., Gasanova, T., Stepanova, L. et al. (2013). Immunogenicity and protective efficacy of candidate universal influenza A nanovaccines produced in plants by Tobacco mosaic virus-based vectors. *Current Pharmaceutical Design* 19 (31): 5587–5600.

Phan, H.T., Pham, V.T., Ho, T.T. et al. (2020). Immunization with plant-derived multimeric H5 hemagglutinins protect chicken against highly pathogenic avian influenza virus H5N1. *Vaccine* 8 (4): 593.

Pogrebnyak, N., Golovkin, M., Andrianov, V. et al. (2005). Severe acute respiratory syndrome (SARS) S protein production in plants: development of recombinant vaccine. *Proceedings of the National Academy of Sciences* 102 (25): 9062–9067.

Pollard, A.J. and Bijker, E.M. (2021). A guide to vaccinology: from basic principles to new developments. *Nature Reviews Immunology* 21 (2): 83–100.

Potter, H. and Heller, R. (2003). Transfection by electroporation. *Current Protocols in Molecular Biology* 62 (1): 9.3.1–9.3.6.

Powell, N., Canavan, J.B., MacDonald, T.T., and Lord, G.M. (2010). Transcriptional regulation of the mucosal immune system mediated by T-bet. *Mucosal Immunology* 3 (6): 567–577.

Qian, B., Shen, H., Liang, W. et al. (2008). Immunogenicity of recombinant hepatitis B virus surface antigen fused with preS1 epitopes expressed in rice seeds. *Transgenic Research* 17 (4): 621–631.

Ramírez, N., Lorenzo, D., Palenzuela, D. et al. (2000). Single-chain antibody fragments specific to the hepatitis B surface antigen, produced in recombinant tobacco cell cultures. *Biotechnology Letters* 22 (15): 1233–1236.

Rao, J.P., Agrawal, P., Mohammad, R. et al. (2012). Expression of VP1 protein of serotype A and O of foot-and-mouth disease virus in transgenic sunnhemp plants and its immunogenicity for guinea pigs. *Acta Virologica* 56 (2): 91–99.

Rather, R.A., Islam, T., Rehman, I.U., and Pandey, D. (2021). Development of vaccine against coronavirus disease 2019 (COVID-19) in India. *Asian Journal of Advances in Medical Science* 3 (2): 13–21.

Razdan, M.K. and Thomas, D.T. (2021). *Mulberry: Genetic Improvement in Context of Climate Change*. CRC Press.

Rivera, A.L., Gómez-Lim, M., Fernández, F., and Loske, A.M. (2012). Physical methods for genetic plant transformation. *Physics of Life Reviews* 9 (3): 308–345.

Rojas-Anaya, E., Loza-Rubio, E., Olivera-Flores, M.T., and Gomez-Lim, M. (2009). Expression of rabies virus G protein in carrots (*Daucus carota*). *Transgenic Research* 18 (6): 911–919.

Rosales-Mendoza, S. and Salazar-González, J.A. (2014). Immunological aspects of using plant cells as delivery vehicles for oral vaccines. *Expert Review of Vaccines* 13 (6): 737–749.

Rosales-Mendoza, S., Sández-Robledo, C., Bañuelos-Hernández, B., and Angulo, C. (2017). Corn-based vaccines: current status and prospects. *Planta* 245 (5): 875–888.

Rukavtsova, E.B., Rudenko, N.V., Puchko, E.N. et al. (2015). Study of the immunogenicity of hepatitis B surface antigen synthesized in transgenic potato plants with increased biosafety. *Journal of Biotechnology* 203: 84–88.

Sabado, R.L., Balan, S., and Bhardwaj, N. (2017). Dendritic cell-based immunotherapy. *Cell Research* 27 (1): 74–95.

Sahai, A., Shahzad, A., and Shahid, M. (2013). Plant edible vaccines: a revolution in vaccination. In: *Recent Trends in Biotechnology and Therapeutic Applications of Medicinal Plants* (ed. M. Shahid, A. Shahzad, A. Malik and A. Sahai), 225–252. Dordrecht: Springer.

Sahoo, A., Mandal, A.K., Dwivedi, K., and Kumar, V. (2020). A cross talk between the immunization and edible vaccine: current challenges and future prospects. *Life Sciences* 261: 118343.

Scotti, N. and Rybicki, E.P. (2014). Virus-like particles produced in plants as potential vaccines. *Expert Review of Vaccines* 12 (2): 211–224.

Shafiq, M., Qurashi, F., Mushtaq, S. et al. (2020). DNA plant viruses: biochemistry, replication, and molecular genetics. In: *Applied Plant Virology* (ed. L.P. Awasthi), 169–182. Academic Press.

Shah, T.R. and Misra, A. (2011). Proteomics. In: *Challenges in Delivery of Therapeutic Genomics and Proteomics* (ed. A. Misra), 387–427. Elsevier.

Shakoor, S., Rao, A.Q., Shahid, N. et al. (2019). Role of oral vaccines as an edible tool to prevent infectious diseases. *Acta Virologica* 63 (3): 245–252.

Sharma, K.K., Bhatnagar-Mathur, P., and Thorpe, T.A. (2005). Genetic transformation technology: status and problems. *Vitro Cellular & Developmental Biology – Plant* 41 (2): 102–112.

Sharma, O., Sultan, A.A., Ding, H., and Triggle, C.R. (2020). A review of the progress and challenges of developing a vaccine for COVID-19. *Frontiers in Immunology* 11: 585354.

Sharp, J.M. and Doran, P.M. (2001). Strategies for enhancing monoclonal antibody accumulation in plant cell and organ cultures. *Biotechnology Progress* 17 (6): 979–992.

Shen, S.-H., Zhang, X.-J., Guo, Y.-M., and Jing, Y.-X. (2001). Microinjection of genes into in vitro ovaries of maize and identification of transformed plants. *Journal of Integrative Plant Biology* 43 (10): 1055–1057.

Shillito, R. (1999). Methods of genetic transformation: electroporation and polyethylene glycol treatment. In: *Molecular Improvement of Cereal Crops*, Advances in Cellular and Molecular Biology of Plants, vol. 5 (ed. I.K. Vasil). Dordrecht: Springer.

Shin, M.-K., Jung, M.H., Lee, W.-J. et al. (2011). Generation of transgenic corn-derived *Actinobacillus pleuropneumoniae*ApxIIA fused with the cholera toxin B subunit as a vaccine candidate. *Journal of Veterinary Science* 12 (4): 401.

Sijmons, P.C., Dekker, B.M.M., Schrammeijer, B. et al. (1990). Production of correctly processed human serum albumin in transgenic plants. *Bio/technology (Nature Publishing Company)* 8 (3): 217–221.

Sohrab, S.S. (2020). An edible vaccine development for coronavirus disease 2019: the concept. *Clinical and Experimental Vaccine Research* 9 (2): 164.

Sokolova, V., and Ukraine, K. (2006) Synthesis, characterization and application of calcium phosphate nanoparticles for the transfection of cells. Doctoral dissertation, Duisburg, Essen, Univ., Diss.

Sonia, S. and Negi, N.P. (2021). Production and challenges of plant based vaccines. *Annals of the Romanian Society for Cell Biology* 25 (1): 3625–3639.

Spiller, D.G., Giles, R.v., Grzybowski, J. et al. (1998). Improving the intracellular delivery and molecular efficacy of antisense oligonucleotides in chronic myeloid leukemia cells: a comparison of streptolysin-O permeabilization, electroporation, and lipophilic conjugation. *Blood* 91 (12): 4738–4746.

Srinivas, L., Sunil Kumar, G.B., Ganapathi, T.R. et al. (2008). Transient and stable expression of hepatitis B surface antigen in tomato (*Lycopersicon esculentum* L.). *Plant Biotechnology Reports* 2 (1): 1–6.

Sternberg, B. (1998). Ultrastructural morphology of cationic liposome-DNA complexes for gene therapy. *Medical Applications of Liposomes* 5: 395–427.

Streatfield, S.J., Kushnir, N., and Yusibov, V. (2015). Plant-produced candidate countermeasures against emerging and reemerging infections and bioterror agents. *Plant Biotechnology Journal* 13 (8): 1136–1159.

Sunil Kumar, G.B., Ganapathi, T.R., Srinivas, L. et al. (2006). Expression of hepatitis B surface antigen in potato hairy roots. *Plant Science* 170 (5): 918–925.

Tacket, C.O., Mason, H.S., Losonsky, G. et al. (1998). Immunogenicity in humans of a recombinant bacterial antigen delivered in a transgenic potato. *Nature Medicine* 4 (5): 607–609.

Tacket, C.O., Mason, H.S., Losonsky, G. et al. (2000). Human immune responses to a novel Norwalk virus vaccine delivered in transgenic potatoes. *The Journal of Infectious Diseases* 182 (1): 302–305.

Tanne, A. and Bhardwaj, N. (2017). Dendritic cells: general overview and role in autoimmunity. In: *Kelley and Firestein's Textbook of Rheumatology*, 10e (ed. G.S. Firestein, R.C. Budd, S.E. Gabriel, et al.), 126–144.e6. Elsevier.

Thanavala, Y. and Lugade, A.A. (2010). Oral transgenic plant-based vaccine for hepatitis B. *Immunologic Research* 46 (1–3): 4–11.

Tomizawa, M., Shinozaki, F., Motoyoshi, Y. et al. (2013). Sonoporation: gene transfer using ultrasound. *World Journal of Methodology* 3 (4): 39–44.

Turri, V., Latinovic, O.S., Bonafè, M. et al. (2020). Cauliflower mosaic virus TAV, a plant virus protein that functions like ribonuclease H1 and is cytotoxic to glioma cells. *BioMed Research International* 2020: 7465242.

Uthaya Kumar, A., Kadiresen, K., Gan, W.C., and Ling, A.P.K. (2021). Current update and research on plant-based vaccines for coronavirus disease 2019. *Clinical and Experimental Vaccine Research* 10 (1): 13–23.

Uvarova, E.A., Belavin, P.A., Permyakova, N.V. et al. (2013). Oral immunogenicity of plant-made *Mycobacterium tuberculosis* ESAT6 and CFP10. *BioMed Research International* 2013: 1–8.

Venkataraman, S. and Hefferon, K. (2021). Application of plant viruses in biotechnology, medicine, and human health. *Viruses* 13 (9): 1697.

World Health Organisation (WHO) (2021). Vaccines and immunization. https://www.who.int/health-topics/vaccines-and-immunization#tab=tab_1 (accessed 22 October 2021).

Wosen, J.E., Mukhopadhyay, D., Macaubas, C., and Mellins, E.D. (2018). Epithelial MHC Class II expression and its role in antigen presentation in the gastrointestinal and respiratory tracts. *Frontiers in Immunology* 9: 2144.

Wu, L., Jiang, L., Zhou, Z. et al. (2003). Expression of foot-and-mouth disease virus epitopes in tobacco by a tobacco mosaic virus-based vector. *Vaccine* 21 (27–30): 4390–4398.

Wu, J., Yu, L., Li, L. et al. (2007). Oral immunization with transgenic rice seeds expressing VP2 protein of infectious bursal disease virus induces protective immune responses in chickens. *Plant Biotechnology Journal* 5 (5): 570–578.

Wu, S., Zhu, H., Liu, J. et al. (2020). Establishment of A PEG-mediated protoplast transformation system based on DNA and CRISPR/Cas9 ribonucleoprotein complexes for banana. *BMC Plant Biology* 20 (1): 1–10.

Xia, Y., Tian, J., and Chen, X. (2016). Effect of surface properties on liposomal siRNA delivery. *Biomaterials* 79: 56–68.

Yang, Q.Y., Yang, X.L., Ding, B., and Zhou, X.P. (2016). The application of geminiviruses in biotechnology. *SCIENTIA SINICA Vitae* 46 (5): 524–534.

Ye, Y. and Li, W. (2010). The study on immune responses in BALB/c mice by immunization with the transgenic Alfalfa(Medicago sativa) containing Eg95-EgA31 fusion gene of *Echinococcus granulosus*. *Xi bao yu fen zi mianyixue za zhi = Chinese Journal of Cellular and Molecular Immunology* 26 (8): 750–753.

Yildiz, M., Aycan, M., and Park, S. (2016). New approaches to *Agrobacterium tumefaciens*-mediated gene transfer to plants. In: *Genetic Engineering – An Insight into the Strategies and Applications* (ed. F. Jamal). IntechOpen.

Ying, M., SHunquan, L., Yi, G. et al. (2002). Transformation of HBsAg (hepatitis B virus surface antigen) into tomato plants. *Journal of Fujian Agriculture and Forestry University (Natural Science Edition)* 31 (2): 223–227.

Yusibov, V., Rabindran, S., Commandeur, U. et al. (2006). The potential of plant virus vectors for vaccine production. *Drugs in R and D* 7 (4): 203–217.

Zakir, F., Islam, F., Jabeen, A., and Sivagurunathan Moni, S. (2019). Vaccine development: a historical perspective. *Biomedical Research* 30 (3): 452–455.

Zeng, Y., Ji, C., Lin, Y., and Jiang, L. (2021). Transient expression of fluorescent fusion proteins in arabidopsis protoplasts. *Methods in Molecular Biology* 2200. 157–165.

Zhang, X., Buehner, N.A., Hutson, A.M. et al. (2006). Tomato is a highly effective vehicle for expression and oral immunization with Norwalk virus capsid protein. *Plant Biotechnology Journal* 4 (4): 419–432.

Zhang, H., Liu, M., Li, Y. et al. (2010). Oral immunogenicity and protective efficacy in mice of a carrot-derived vaccine candidate expressing UreB subunit against Helicobacter pylori. *Protein Expression and Purification* 69 (2): 127–131.

Zhang, L., Wang, W., and Wang, S. (2015). Effect of vaccine administration modality on immunogenicity and efficacy. *Expert Review of Vaccines* 14 (11): 1509–1523.

4

Plant Cell Culture for Biopharmaceuticals

Zeuko'o Menkem Elisabeth[1,2] *and Rufin Marie Kouipou Toghueo*[2]

[1] *Department of Biomedical Sciences, University of Buea, Buea, Cameroon*
[2] *Antimicrobial & Biocontrol Agents Unit (AmBcAU), Laboratory for Phytobiochemistry and Medicinal Plant Studies, Department of Biochemistry, University of Yaoundé I, Yaoundé, Cameroon*

4.1 Introduction

Plant culture is the culture of plant parts (cells, tissues, organs, or whole plants) using *in vitro* aseptic techniques in controlled nutritional and environmental conditions, mostly to produce genetically modified plants (Ma and Gomez-Cadenas 2012; Espinosa-Leal et al. 2018). The controlled conditions provide a conducive environment for the growth and multiplication of the plants, which varies concerning the objective of the culture. The conducive conditions for the proper growth of plants include nutrient supply, pH, temperature, gases, and water (Hussain et al. 2012; Meskaaoui 2013). Plant tissue culture technology is widely used for the production of a large-scale variety of plants. The pieces of tissues (explants) are used to produce many plants in successive conditioning. The single explant is propagated into many copies of plants relatively quickly and under controlled conditions using different techniques, including micropropagation (Moon et al. 2019). Micropropagation is used to grow and conserve endangered, threatened, and rare species due to its high coefficient of multiplication with small demands on the initial plants and space used (Iliev et al. 2010; Meskaaoui 2013). In addition, micropropagation technology has vast potential to produce plants of high quality with the isolation of useful variants in well-adapted high-yield genotypes with better disease resistance and stress tolerance capacities (Iliev et al. 2010). However, plant tissue culture is considered an efficient crop improvement and production technology. The callus cultures provide clones with inheritable characteristics different from those of the parent plants (Meskaaoui 2013), leading to the development of improved varieties. These culture techniques are becoming a major aspect in the area of plant multiplication, disease eradication, plant amelioration, and the production of secondary metabolites. Biopharmaceuticals are one of the best accomplishments of modern science (Warzecha 2008; Agrawal 2015) (Jozala et al. 2016), with pharmaceuticals produced in biotechnological processes using some biological

Plants as Bioreactors for Industrial Molecules, First Edition.
Edited by Santosh Kumar Upadhyay and Sudhir P. Singh.

methods (Daniell et al. 2009; Dixit et al. 2021a,b; Upadhyay and Singh 2021). These biopharmaceuticals target specific molecules, rarely causing the side effects associated with conventional small-molecule drugs (Agrawal 2015) with high specificity and activity (Jozala et al. 2016), making them the mainstay products of the biotechnology market representing the fastest growing and most exciting sector within the pharmaceutical industry. Moreso, the market for biopharmaceuticals is steadily growing, and currently, nearly 150 biopharmaceuticals have gained approval for general human use (Eibl and Eibl 2009). Over this period, the production capacities for biopharmaceuticals with bioreactors would be a bottleneck, with worldwide fermentation capacities being limited (Goldstein and Thomas 2004). These rapid multiplication processes provide disease-free (transgenic) plants to produce biopharmaceuticals (Stoger et al. 2002), with some medicinal plants being used. *Corydalis yanhusuo* is an example of a medicinal plant that is propagated using somatic embryogenesis to produce virus-free tubers and plants (Nielsen 2013). Moreso, some meristem tip cultures of banana plants without the banana bunchy top virus (BBTV) and brome mosaic virus (BMV) I are also produced (Shargool 1985). Increased yields of virus-free potatoes are obtained by culturing pathogen-free germplasm in controlled experimental conditions (Giddings et al. 2000). Therefore, the objective of this chapter is to describe the tissue culture techniques, various developments, present and future trends, and their application in various fields, especially in the production of biopharmaceuticals.

4.2 Plant Cultures

Plant culture is the culture of any part of a plant (cells, tissues, or organs) in artificial media, aseptic conditions, and under controlled environments (Espinosa-Leal et al. 2018). This set of techniques is an experimental approach to demonstrate the cell theory, which establishes that all living organisms are composed of cells, which are the basic unit structure and reproduction (Pierik 1997a). These different attempts were conducted by several researchers to investigate the conditions to initially achieve the growth of organs (Ogita 2015) or tissues (Poehlman 1987) in an artificial nutrient culture medium (Saad and Elshahed 2012) rather than isolated cells because of the complex nutritional and hormonal requirements. The nutrient solutions alone or supplemented with natural extracts were used as starting culture media, and some important results were reported (Saad and Elshahed 2012); however, the discovery of plant growth regulators was determinant for the successful implementation of the *in vitro* plant tissue culture techniques (Eibl and Eibl 2009). One key advance in plant tissue culture was the control of morphogenesis using different levels and combinations of growth regulators (Pierik 1997a) to allow the regeneration of entire plants. This provides the possibility of using *in vitro* systems to study fundamental aspects of cell differentiation and development for the application of tissue culture for different purposes. Thus, the totipotency concept, which is the genetic potential of a cell to generate an entire multicellular organism, is also applied (Pierik 1997a).

4.2.1 Plant Cell Cultures

Plant cell culture enables the preparation of artificial seeds for seedling production, increasing the potential of producing various kinds of useful plants in field cultivation and agriculture (Tulecke 1963). Some artificial seeds provide genetically homogeneous plants and help enable easy control of planting and agriculture. The use of plant cell culture needs to compete with the production of intact plants cultivated in fields (Ogita 2015), leading to the production of secondary metabolites such as alkaloids and anthocyanins. The biotransformation of some biological compounds, such as terpenoids or steroids, is also possible using plant cells. Moreso, some proteins are produced using plants, particularly genetically engineered plants, with little risk of human infection of proteins by viruses or pathogens compared to other methods using animal cells or microorganisms (Loyola-Vargas and Ochoa-Alejo 2018). The specific cells use different techniques and could be elaborated with specific cells, including the callus and cell suspension culture.

A callus is an undifferentiated mass of cells developed in culture from any kind of explant. The cell suspension cultures are rapidly divided into homogenous suspensions of cells grown in liquid nutrient media (Behbahani et al. 2011). The cell suspension is then placed on a shaker to allow the cell aggregates to disperse to form smaller clumps, and single cells are equally distributed and continuously grow throughout the liquid media (Loyola-Vargas and Ochoa-Alejo 2018). Thus, plant suspension cultures provide a valuable platform for the production of high-value secondary metabolites and other substances of commercial interest (Sogutmaz Ozdemir and Budak 2018). The production of secondary metabolites from callus culture has been reported (Shargool 1985; Fowler 1986; Pant 2014).

Many successful attempts to produce valuable pharmaceuticals in relatively large quantities by callus and cell cultures have been achieved (Bonfill et al. 2015). The plant *Papaver somniferum* was used in morphologically undifferentiated cultures of callus and suspension cultures for the production of compounds such as morphine and codeine (Shargool 1985). The various *Taxus* species cells in cultures are one of the most explored areas of plant cell cultures for the production of Taxol compounds due to their enormous commercial value, scarcity of the parent tree, and high cost of the synthetic process (Veeresham et al. 2003; Sreekanth 2009). Taxol is an anticancer drug approved for the clinical treatment of ovarian and breast cancer by the Food and Drug Administration (Li et al. 2014). It also has significant activity against malignant melanoma, lung cancer, and other solid tumors (Sergio n.d.). However, plant cell culture technology has the potential to meet the continuously growing demand for bioactive natural compounds in the near future (Krasteva et al. 2021).

4.2.2 Plant Tissue Culture

Different organs, such as shoots, roots, and other plant organs, have been developed for the production of desired compounds that require cell differentiation (Ahsan 2014). Although the production of secondary metabolites in organized tissue cultures differs from that in intact plants, stable growth and consistent secondary metabolite production equivalents or higher concentrations are being observed in shoot and root cultures of different species (Vasil and Thorpe 1994; Fu et al. 1999; Meskaaoui 2013). In some cases, compounds not found in the intact plants are detected in cultured tissue. An example is the alkaloid levels of cultured roots of *Stephania cepharantha*, which were much higher than those of the plants, and aromoline was not present in the original plant but was present in the cultured plants (Loyola-Vargas and Ochoa-Alejo 2018). Adventitious root cultures of *Salvia miltiorrhiza* produced a high yield of tanshinone alkaloids under optimized culture conditions. A standardized protocol for complete plant regeneration via somatic embryogenesis from tuber-derived callus of *Corydalis yanhusuo*, a rich source of several pharmacologically important alkaloid-containing plants, is available (Bonfill et al. 2015). They have reported the production of higher concentrations of bioactive compounds such as D, L-tetrahydropalmatine and D-corydaline from the tubers of somatic embryo-derived plants (Shargool 1985). The content of gentiopicroside and *swertiamarin* in the aerial and underground parts of *Gentiana davidii* var. *formosana* was higher than that in the marketed crude drug (underground parts of *G. scabra*) *davidii* var. *formosana* (Bonfill et al. 2015).

4.2.3 Plant Organ Cultures

Plant cell and tissue culture offers an alternative source for rapid plant propagation and controlled production of disease-preventive phytochemicals or food ingredients in medicinal plants (Tulecke 1963; Meskaaoui 2013). Many plant species are regenerated *in vitro* through several approaches, such as from a single cell that is reproduced in a tissue or organ part or a cut-out piece of differentiated tissue (or organ) known as an explant, as shown in Figure 4.1. These technologies are established under sterile conditions from any vegetative parts or organs of plants, such as meristems, nodes, leaves, stems, roots, axillary or adventitious shoot buds, endosperm, and embryos, for unlimited plant multiplication and metabolite production (Ma and Gomez-Cadenas 2012).

The use of the different plant parts for culture requires a well-conditioned and controlled environment for its success. Therefore, specific conditions and factors are necessary for the proper performance of the culture.

4.3 Conditions for Plant Cell, Tissue, and Organ Culture

In plant culture, the cells, tissues, and organs are grown *in vitro* on synthetic media under sterile and controlled conditions. This technique depends mainly on the concept of totipotentiality of plant cells (Ma and Gomez-Cadenas 2012), which is the potential of a single cell to express its full genome through cell division and multiplication. Additionally, in addition to the totipotent potential of the plant, other factors are necessary and primordial to regenerate the entire plant, including the capacity of cells to alter their metabolism, growth, and development (Hughes 1981). These specific conditions include the following:

4.3.1 Culture Medium

Plant tissue culture medium requires nutrients for the normal growth, improvement, and development of plants. The different components are grouped into different categories and include:

- ✓ **Macronutrients**: Nitrogen (N), phosphorus (P), potassium (K), calcium (Ca), magnesium (Mg), and sulfur (S).
- ✓ **Micronutrients**: Iron (Fe), manganese (Mn), zinc (Zn), boron (B), copper (Cu), and molybdenum (Mo).

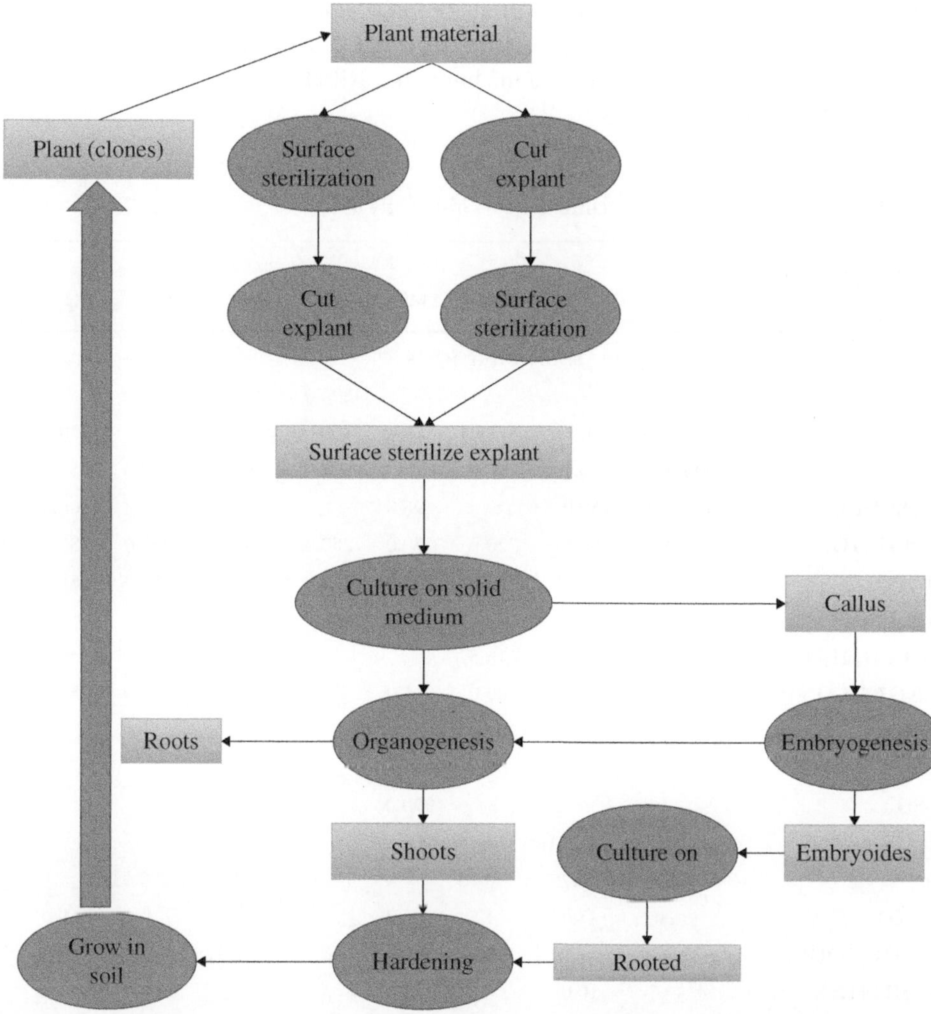

Figure 4.1 Flow chart summarizing the tissue culture experiments.

- ✓ **Vitamins**: thiamine (B1), nicotinic acid, pyridoxine (B6), biotin, folic acid, ascorbic acid, pantothenic acid, tocopherol (vitamin E), riboflavin, and p-amino-benzoic acid.
- ✓ **Amino acids**: Glycine at 2 mg/l, glutamine up to 8 mM, asparagine at 100 mg/l, L-arginine and cysteine at 10 mg/l, and L-tyrosine at 100 mg/l or nitrogen supplements.
- ✓ **Undefined organic components**: Protein hydrolysates, coconut milk, yeast extract, malt extract, ground banana, orange juice, and tomato juice.
- ✓ **Plant growth regulators**: Auxins, cytokinins, gibberellins, and abscisic acid (Yamaguchi and Kamiya 2000; DeMason 2005; Kieber and Schaller 2018).
- ✓ **Carbon sources**: sucrose, glucose, lactose, galactose, maltose, and starch.
- ✓ **Solidifying or gelling agents:** Agar, agarose, and gellan gum in the case of solid medium (Saad and Elshahed 2012).

Some frequently used culture media with their composition for the *in vitro* multiplication of many plant species (Table 4.1) are Murashige and Skoog (MS) medium, Linsmaier and Skoog (LS) medium, Gamborg (B5) medium, and Nitsch and Nitsch (NN) medium (Shahina and Anwar 2014).

Table 4.1 Composition of synthetic culture media frequently used.

Medium Components (mg/l)	MS	G5	W	LM	VW	km	M	NN
				Macronutrients				
$Ca_3(PO_4)_2$					200.0			
NH_4NO_3	1650.0				400.0			720.0
KNO_3	1900.0	2500.0	80.0		525.0	180.0	180.0	950.0
$CaCl2.2H_2O$	440.0	150.0		96.0				166.0
$MgSO4.7H_2O$	370.0	250.0	720.0	370.0	250.0	250.0	250.0	185.0
KH_2PO_4	170.0			170.0	250.0	150.0	150.0	68.0
$(NH_4)_2SO_4$		134.0			500.0	100.0	100.0	
$NaH_2PO_4.H_2O$		150.0	16.5					
$CaNO3.4H_2O$			300.0	556.0		200.0	200.0	
Na_2SO_4			200.0					
KCl			65.0					
K_2SO_4				990.0				
				Micronutrients				
KI	0.83	0.75	0.75			80.0	0.03	
H3BO3	6.20	3.0	1.5	6.2		6.2	0.6	10.0
MnSO4.4H2O	22.30		7.0		0.75	0.075		25.0
MnSO4.H2O		10.0		29.43				
ZnSO4.7H2O	8.6	2.0	2.6	8.6			0.05	10.0
Na2MoO4.2H2O	0.25	0.25		0.25		0.25	0.05	0.25
CuSO4.5H2O	0.025	0.025		0.25		0.025		0.025
CoCl2.6H2O	0.025	0.025				0.025		
Co(NO3)2.6H2O						0.05		
Na2EDTA	37.3	37.3		37.3		74.6	37.3	37.3
FeSO4.7H2O	27.8	27.8		27.8		25.0	27.8	27.8
MnCl2							3.9	0.4
Fe(C4H4O6)3.2H2O					28.0			
				Vitamins and other supplements				
Inositol	100.0	100.0		100.0				100.0
Glycine	2.0	2.0	3.0	2.0				2.0
Thiamine HCl	0.1	10.0	0.1	1.0		0.3	0.3	0.5

Table 4.1 (Continued)

Medium Components (mg/l)	MS	G5	W	LM	VW	km	M	NN
Pyridoxine HCl	0.5		0.1	0.5		0.3	0.3	0.5
Nicotinic acid	0.5		0.5	0.5			1.25	5.0
Ca-panthothenate	0.3		1.0					
Cysteine HCl			1.0					
Riboflavin						0.3	0.05	
Biotin							0.05	0.05
Folic acid							0.3	0.5

Source: Saad and Elshahed (2012).
Legend: MS, Murashige and Skoog, G5, Gamborg et al., W, White, LM, Lloyd and McCown, VW, Vaccin and Went, km, Kudson modified, M, Mitra et al., and NN, Nitsch and Nitsch media.

4.3.2 pH

The pH of the culture medium affects both the growth of plants and the activity of plant growth regulators. It ranges between 5.4 and 5.8 in both the solid and liquid medium used during culture (Hughes 1981; Shahina and Anwar 2014).

Plant hormones (plant growth regulators) and nitrogen sources have important effects on the response of the initial explant in culture medium for the determination of the development pathway in plant cells and tissues (Hughes 1981; Shahina and Anwar 2014).

4.3.2.1 Plant Cell Growth Regulators (auxin, cytokinin, and gibberellin)

Auxins, cytokinins, and gibberellins are the most commonly used plant growth regulators. The species of the plant, the tissue or organ cultured, and the overall goal of the experiment depend on the type and concentration of hormones (Saad and Elshahed 2012).

4.3.2.2 Auxins

Auxins are key regulators of almost all stages of plant growth and development, including the formation of flower organs, vascular tissues, shoot growth, phototropism, and gravitropism (Rozov 2013). The common auxins used are indole-3-acetic acid (IAA), indole-3-butyric acid (IBA), 2,4-dichlorophenoxy-acetic acid (2,4-D), and naphthalene-acetic acid (NAA). IAA is the only natural auxin occurring in plant tissues. There are other synthetic auxins used in culture media, such as 4-chlorophenoxy acetic acid or p-chloro-phenoxy acetic acid (4-CPA, pCPA), 2,4,5-trichloro-phenoxy acetic acid (2,4,5 T), 3,6-dichloro-2-methoxy-benzoic acid (dicamba), and 4-amino-3,5,6-trichloro-picolin ic acid (picloram) (Saad and Elshahed 2012). The concentration of auxins and cytokinins determines the type of culture established or regenerated in the plant tissue culture technique.

4.4.4 Meristem Tip Culture

Meristem tip culture is the excision of the organized apex of the shoot from a selected donor plant for subsequent *in vitro* culture. The conditions of culture are regulated to allow only organized outgrowth of the apex directly into a shoot, without the intervention of any adventitious organs (Grout 1999). This is an advantageous culture type derived from meristem tip explants (meristem tip consists of the meristematic dome and usually a pair of leaf primordia). This technique is used for virus elimination or axillary shoot multiplication purposes (Grout 1999; Mangal et al. 2002; Quiroz et al. 2017). It is a useful technique to maintain superior mother plant varieties that accumulate high amounts of the targeted metabolites. Liquid medium conditions are occasionally used in this technique (Mangal et al. 2002).

4.4.5 Organogenesis

The formation of shoots and roots is initiated from cultured plant cells or tissues. The induction of different adventitious organs of the plant from cultured tissues under *in vitro* conditions is known as organogenesis (Perianez-Rodriguez et al. 2014). The organogenesis process includes two steps, callogenesis (indirect organogenesis) and rhizogenesis (direct organogenesis) (Perianez-Rodriguez et al. 2014; Ogita 2015). Cultured shoots sometimes show heterogeneous differentiation levels, with the culture conditions being carefully controlled to select a suitable culture. After culture condition optimization, the handling procedure for organogenesis is easy and is used for metabolic engineering purposes (Rahamooz-Haghighi et al. 2020).

4.4.6 Callus Culture (Callogenesis)

Callus refers to disorganized tumor-like masses of plant cells; calli proliferate in irregular tissue masses and vary widely in texture, appearance, and growth rate. The process of callus formation is also known as callogenesis (Benjamin et al. 2019). The histological and physiological characteristics of calli vary depending on the explant source and culture medium composition. The callus culture is usually initiated on solidified medium by inoculation of small explants or sections from established organ cultures (Benjamin et al. 2019; Fehér 2019). The handling procedure of callus culture is easy and is used for metabolic engineering purposes when the culture conditions are optimized (Behbahani et al. 2011).

4.4.7 Adventitious Root/Hairy Root Culture (rhizogenesis)

This is a useful technique to produce targeted metabolites. The isolated root tips of apical or lateral origin may produce *in vitro* root systems with indeterminate growth habits (Ogita 2015). The process of *in vitro* root formation is also called rhizogenesis. The root-inducing (Ri) plasmid of *Agrobacterium rhizogenes* is used for hairy root culture induction (Chung et al. 2006). Since cultured roots can sometimes show heterogeneous differentiation levels, culture conditions are carefully controlled to select a suitable culture (Efferth 2019). The dark condition is preferred to maintain rhizogenesis. A liquid culture system is applicable to target metabolite production. The bioreactor system is also used to scale up the targeted metabolite production (Veng 1998; Verma et al. 2007; Alok et al. 2018).

4.4.8 Suspension Culture

This is a type of cell culture in which single cells or small aggregates of cells are allowed to function and multiply in an agitated growth medium, thus forming a suspension. The explants, or calli derived from these, are transferred to liquid medium, and the cultures are then agitated on a mechanical shaker (Morais-Lino et al. 2008). The cells and small cell aggregates that show homogenous growth in liquid medium are useful for metabolic engineering. This is an advantageous technique to maintain cells and small cell aggregates suspended in a liquid medium. A liquid culture system is applicable to target metabolite production (Morais-Lino et al. 2008; Ogita 2015). The bioreactor system is also used to scale up the targeted metabolite production (Fowler 1986).

4.4.9 Protoplast Fusion

Protoplasts are the cells from which the cell walls are removed and the cytoplasmic membrane is the outermost layer in such cells. Protoplasts are obtained by specific lytic enzymes to remove the cell wall (Verma et al. 2008). Somatic hybridization is an important tool of plant breeding and crop improvement by the production of interspecific and intergeneric hybrids. The technique involves the fusion of protoplasts of two different genomes followed by the selection of the desired somatic hybrid cells with hybrid plant regeneration (Ahmed et al. 2021). Protoplast fusion provides an efficient means of gene transfer from one species to another with an increasing impact on crop improvement (Gamborg et al. 1974). The electrofusion treatment methods produced somatic hybrids produced by the fusion of protoplasts from rice and ditch reed. The *in vitro* fusion of protoplasts provides a means for the development of unique hybrid plants by defeating the barriers of sexual incompatibility (Ogita 2015). The technique is applicable in the agricultural industry to create new hybrids with increased fruit yield and better resistance to diseases (Ahmed et al. 2021). The potential of somatic hybridization in crop plants is described by the production of intergeneric hybrid plants among the members of Brassicaceae (Gamborg et al. 1974). A protocol was developed for the production of somatic hybrid plants using two types of wheat protoplasts as recipients and *Haynaldia villosa* protoplasts as fusion donors to remediate the problem of loss of chromosomes and decreased regeneration capacity. Additionally, it is used as an important gene source in wheat improvement (Verma et al. 2008; Gamborg et al. 1974; Ahmed et al. 2021).

4.4.10 Haploid Production

Haploids are plants with a gametophytic chromosome number, and doubled haploids are haploids that have undergone chromosome duplication. The production of haploids and doubled haploids (DHs) through gametic embryogenesis allows a single-step development of complete homozygous lines from heterozygous parents, shortening the time required to produce homozygous plants compared to conventional breeding methods with several generations of self-fertilization (Germana 2011). The production of haploid plants from young pollen cells and anther culture without undergoing fertilization is called androgenesis. The principle involved in the process is to half the development of pollen cells into a gamete and induce it in a suitable environment to develop a haploid plant (Meyers 1988). Tissue culture techniques enable the production of homozygous plants in a short time using protoplast, anther, and microspore cultures (Germana 2011) through spontaneous or induced chromosome doubling. The multiplication of chromosomes restores the fertility of plants, resulting in the production of double haploids with the potential to become pure breeds of new cultivars (Hussain et al. 2012). Haploidy technology is becoming an integral part of plant breeding programs, speeding up the production of inbred lines and controlling the constraints of seed dormancy and embryo nonviability (Germana 2011). This technique has an impact on genetic transformation for the production of haploid plants with reduced resistance to various biotic and abiotic stresses (Hussain et al. 2012). The insertion of genes with new traits at the haploid state with chromosome doubling produced double haploid inbred wheat and drought-tolerant plants (Meyers 1988; Hussain et al. 2012; Pant 2014).

4.4.11 Germplasm Conservation

The conservation of plant genetics is one of the main areas that is refined and revolutionized repeatedly with knowledge. Germplasm conservation is increasingly becoming an essential activity with the high rate of disappearance of plant species, increasing the need for safeguarding plant biodiversity in countries (Quazi 2021). Tissue culture protocols are used for the preservation of vegetative tissues when the targets for conservation are clones instead of seeds to maintain the genetic background of crops, avoiding the loss of conserved patrimony due to natural disasters (Quazi et al. 2021). The plant species not producing seeds (sterile plants) or having "recalcitrant" seeds that cannot be stored for longer periods are preserved via *in vitro* techniques for the maintenance of gene banks, creating an alternative source for the conservation of endangered genotypes (Quazi et al. 2021). The *in vitro* conservation of essential biological material and genetic resources for a long-term period could use cryopreservation (Ogita 2015). It involves the storage of *in vitro* cells or tissues in liquid nitrogen, resulting in cryo-injury upon exposure of tissues to biotic and abiotic stresses. Successful cryopreservation is often ascertained by cell and tissue survival and the ability to regrow or regenerate into complete plants, forming new colonies (Hussain et al. 2012). It is therefore necessary to assess the genetic integrity of the recovered germplasm to determine cell integrity following cryopreservation (Quazi et al. 2021). The fidelity of recovered plants is assessed at phenotypic, histological, cytological, biochemical, and molecular levels (Hussain et al. 2012; Pant 2014; Quazi et al. 2021), although there are advantages and limitations of the various approaches. Cryobionomics is a new approach to study genetic stability in cryopreserved plant materials, with embryonic tissues being cryopreserved for future use (Ogita 2015; Quazi et al. 2021).

4.5 The Techniques Used in Plant Culture

Plant cultures are performed in the laboratory and environment. However, plant cultures involve various steps necessary for concrete results (Meskaaoui 2013).

✓ The selection of an appropriate explant from a healthy and vigorous plant,
✓ The elimination of microbial contamination from the surface of the explant,

✓ The inoculation of the explant in an adequate culture medium, and
✓ The provision of the explant in culture with the proper controlled environmental conditions.

In the case of *in vitro* regenerated plants, they are subjected to an adaptation process (acclimatization) in the greenhouse before being transferred to *ex vitro* conditions (Efferth 2019). Depending on the part of the plant that is cultured, the different appellation is used: cell culture (gametic cells, cell suspension, and protoplast culture), tissue culture (callus and differentiated tissues), and organ culture (any organ such as zygotic embryos, roots, shoots, and anthers). Each of these cultures is used for different basic and biotechnological applications (Moon et al. 2019). Some of these techniques are described as follows:

4.5.1 Micropropagation in Medicinal Plants

Micropropagation or clonal propagation is the vegetative propagation of plants *in vitro,* involving the multiplication of genetically identical copies of cultivars through asexual reproduction (Tabin et al. 2018). These techniques provide the potential to produce thousands, or even billions, of plants per year (Pant 2014). However, it is used for clone selection with the rapid and mass production of disease-free materials of desired species in various organizations or institutes for mass-scale production of superior plants (Fowler 1986). Micropropagation protocols have been developed for many important medicinal plant species, including *Swertia chirayita, Cathranthus roseus, Panax, Stevia rebaudiana, Artemisia annua-aremisin, Elettaria cardamomum, Allium chinense,* and *Camellia sinensis (*Hussain et al. 2012*)*. These medicinal herbs have been successfully propagated *in vitro,* either by organogenesis (Shargool 1985) or somatic embryogenesis. The *in vitro* propagation of these medicinal species was found to be uniform, showing less variation in their secondary metabolite content than their wild/cultivated counterparts (Popper 2011). The micropropagation technique was developed in various plant species for the improvement of phytopharmaceutical preparations through optimized techniques for the production of high-quality plant material (Hussain et al. 2012). The micropropagation techniques involve stages and are as follows.

4.5.1.1 Stage 0: Preparation of the Donor Plant

This stage involves the parent plant being cultivated *ex vitro* under controlled conditions to minimize contamination during the *in vitro* culture, enhancing the probability of a successful experiment (Hussain et al. 2012).

4.5.1.2 Stage I: Initiation Stage

In this stage, an explant is sterilized using chemical solutions to remove contaminants with minimal damage to plant cells and then transferred into a nutrient medium. The bactericide and fungicide products are combined and used depending on the type of explant to be introduced (Iliev et al. 2010). The most commonly used disinfectants are sodium hypochlorite, calcium hypochlorite, ethanol, and mercuric chloride ($HgCl_2$) (Sharma et al. 2015). The cultures are incubated in a growth chamber either under light or dark conditions due to the propagation technique used (Shargool 1985; Iliev et al. 2010; Sharma et al. 2015).

4.5.1.3 Stage II: Multiplication Stage

This phase aims to increase the number of propagules by multiplying them with repeated subcultures until the planned number of plants is attained (Iliev et al. 2010).

4.5.1.4 Stage III: Rooting Stage

The rooting stage could occur simultaneously in the same culture media used for the multiplication of the explants (Saad and Elshahed 2012). However, in some cases, it is necessary to change media with nutritional modification and growth regulator composition to induce rooting and growth development (Hussain et al. 2012; Saad and Elshahed 2012).

4.5.1.5 Stage IV: Acclimatization Stage

Here, the plants are weaned and hardened. The hardening process is performed gradually in the variation of light intensity, i.e. high to low humidity and from low to high light intensity. The plants are then transferred to an appropriate substrate (sand, peat, and compost) and gradually hardened under a greenhouse until stable growth (Shargool 1985; Sharma et al. 2015; Sogutmaz Ozdemir and Budak 2018).

4.5.2 Elicitation

Elicitation is one of the most effective techniques currently used for improving the biotechnological production of secondary metabolites. Elicitors are compounds that stimulate any type of plant defence, promoting secondary metabolism to protect the cell and the whole plant (Ramirez-Estrada et al. 2016). According to their origin, elicitors are divided into different types: (i) biotic and (ii) abiotic. Abiotic elicitors are substances of nonbiological origin and are predominantly inorganic compounds such as salts or physical factors, as shown in Figure 4.2 (Singh 1999). Biotic elicitors are living organisms that can be bacteria or fungi (Naik and Al-Khayri 2016).

Elicitors are a chemically diverse group of biotic or abiotic signals that, when applied in low amounts to a living cell system, induce or enhance the biosynthesis of specific secondary metabolites by inducing defense or stress-induced responses (Halder et al. 2019). Additionally, they stimulate the synthesis of phytoalexins and various secondary metabolites in plants. The cell cultures are exposed to biotic elicitors, such as *Aspergillus niger* and crude chitin, and abiotic elicitors, such as mannitol and methyl jasmonate, to induce metabolite production. Elicitors such as fungal materials, plant and microbial polysaccharides, and some chemicals increase secondary metabolite production in various plant cell and tissue culture systems (Halder et al. 2019).

Methyl jasmonate is a proven signal compound and the most effective elicitor in the production of Taxol and goinsenoside in *Taxus chinensis* and *P. ginseng*, respectively (Singh 1999). Additionally, biotic polysaccharide elicitor, chitosan, increased the production of anthraquinone in *Rubia akane*. Several studies have shown the increased effect of elicitation on the production yield of many food ingredients and secondary metabolites in various types of plant cell and tissue cultures (Shargool 1985). The enhanced rate of secondary metabolite production due to elicitation treatment activates novel genes that encode proteins with no similarities to those of known genes. It is now known that elicitation

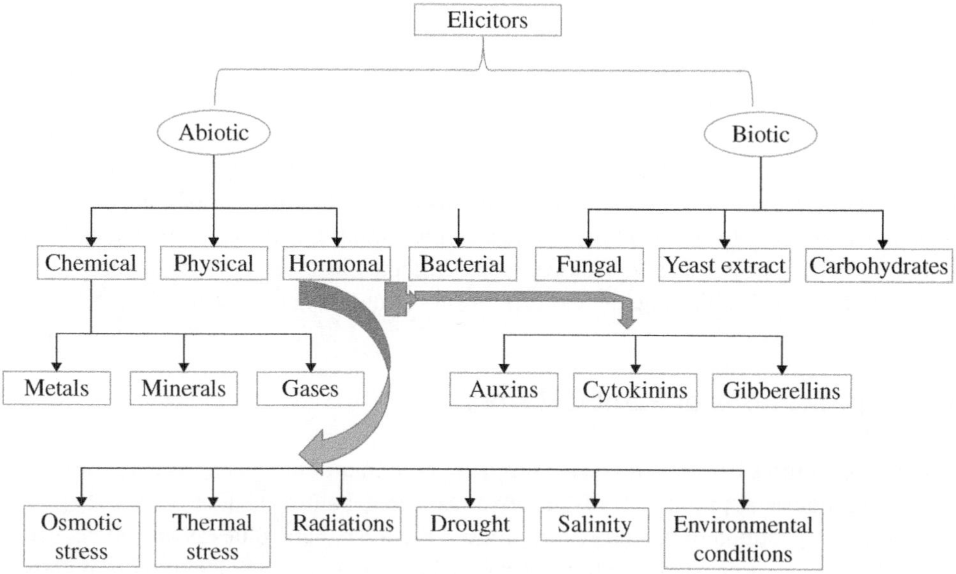

Figure 4.2 Schematic representation of elicitors in plant cells.

ultimately affects the transcription rates of genes encoding enzymes involved in plant defence pathways (Ramirez-Estrada et al. 2016).

4.5.3 Transformed Tissue Cultures

The achievement of recombinant DNA technology has contributed greatly to the welfare of mankind, introducing the technology beyond the imagination. *Agrobacterium rhizogenes*-mediated transformation is used to produce transgenic hairy root cultures for secondary metabolite production (Fowler 1986). The transgenic hairy root cultures improved secondary metabolite production in plant culture. Their genetic and biosynthetic stability, faster growth rate, and easy maintenance make them unique. The infection of wounded plants by *A. rhizogenes* bacteria results in hairy root disease, which is characterized by rapidly growing and highly branched roots at the host wound sites (Chung et al. 2006). The T-DNA of the bacterial Ri (root-inducing) plasmid, which contains genes encoding endogenous hormones, is integrated into the plant genome and allows the proliferation of these adventitious roots. Similarly, the transformation of plant tissue with T-DNA from *A. tumefaciens* leads to the formation of adventitious shoots (Shargool 1985). A wide range of plant metabolites and food ingredients were synthesized using this methodology (Shanmugaraj et al. 2020). The hairy root cultures grow faster than the untransformed roots or shoots, and in many cases, their growth rates approach those of cell suspension cultures with stable genetic and biochemical characteristics (Moon et al. 2019). The transformed roots produce secondary metabolites equivalent to or higher than those in the parent plants. Additionally, in the clone-to-clone variation in some transformation processes, various strategies are investigated for stability(Fowler 1986).

4.5.4 Metabolic Engineering

Metabolic engineering is applied to improve the production of secondary metabolites in different culture systems (Shargool 1985). Metabolic engineering using recombinant DNA technology is the targeted and sequential alteration of metabolic pathways in an organism to achieve new cellular pathways for chemical transformation and regulatory functions of the cells (Popper 2011). It directs cellular metabolism to produce new metabolites by inserting genes that encode specific biosynthetic enzymes into the plant genome (Ogita 2015). More efforts have been made to produce specific metabolites, such as scopolamine, nicotine, and berberine, by altering metabolic pathways (Angelova et al. 2006).

4.6 Applications of Plant Cultures

Plant tissue cultures have potential as a supplement to traditional means of plant propagation in the industrial production of bioactive plant metabolites (Halder et al. 2019). The plant cells being totipotent in culture can produce the same metabolites as the whole plant. Many efforts have been made for the commercial production of medicinal plant metabolites from plant-cell culture (Shargool 1985). This is more convenient, as useful compounds from the cells of plants are produced under controlled conditions independent of environmental conditions, with these metabolites being more easily extractable from the culture cells than from the intact plants. The conventional biotechnological methods for influencing secondary metabolite production include changing nutrient composition, environmental conditions, and growth regulator composition (Shahina and Anwar 2014; Sharma et al. 2015). More recently, the use of specialized techniques such as metabolic engineering, cell immobilization, elicitation, and *in situ* product removal has been investigated and applied to increase the yields of secondary metabolites (Ramirez-Estrada et al. 2016). Medicinal plants in plant cell cultures are being reported for the *in vitro* production of secondary metabolites. Many food ingredients and phytochemicals, including flavours, nutraceuticals, colourants, essential oils, sweeteners, and antioxidants, are also produced in cell culture, (Popper 2011) some of which are commercialized. Advances in plant cell cultures provide alternative methods for the cost-effective, commercial production of rare or exotic plants (Sogutmaz Ozdemir and Budak 2018). Continuation and intensification efforts in this field have led to the manageable biotechnological production of specific new plant chemicals with the following results (Pant 2014; Efferth 2019):

1) Production of plant metabolites
2) Production of genetically modified plants
3) Genetic engineering

4.7 Biopharmaceuticals

The production of pharmaceuticals from plants through genetic engineering is among the most important techniques for the production of a variety of new therapeutic products such as pharmacologically active proteins, including mammalian antibodies, vaccines, hormones, and cytokines (Warzecha 2008). Most genetically engineered plants are accurately used to create recombinant proteins more abundantly than the mammalian cell systems (Agrawal 2015). Due to their freedom in human diseases and mammalian viral vectors, the plant-derived proteins are mostly chosen compared to others (Goldstein and Thomas 2004). Therefore, biopharmaceutical production in plants needs the consideration of three major areas (Warzecha 2008):

i The gene expression system required,
ii The location of gene to be expressed within the plant,
iii The type of plant used.

Many gene expression strategies used for the production of specific proteins during plant culture include transient expression (TE). This process involves the insertion of genes into plant cells using plant viruses or gene-gun without the addition of a new genetic material into the plant chromosome (Pierik 1997b). TE systems produce large amounts of proteins through mitosis or meiosis (Daniell et al. 2009). Additionally, TE systems are very important for research and development in drug production. Also, the plant chromosome could be modified allowing the expression of a particular protein. Some microorganisms such as *Agrobacterium tumefaciens* are used to transfer the genetic material (Niazian et al. 2017) through the insertion of the desired genes into the plant chromosome of different plant species in nature especially dicots (Chung et al. 2006).

This genetic material is covered with small metallic pellets then inserted into the cells using a "gene-gun." It is advantageous due to its stability, and the ongoing protein production with repeated planting is less costly (Shargool 1985) compared to the permanent modification of the plant genome, which is more costly and time consuming. Finally, systems for the modification of chloroplast DNA exist in plants that could lead to hereditary changes during protein expression. These tiny energy-producing organelles are advantageous over nuclear transformation, resulting in the capacity to maintain very high numbers of functional gene copies. Moreso, transgenic tobacco chloroplasts produce human somatotropin and insecticidal protein (about 7 and 45% of the total plant protein production) at a higher rate than their nuclear transgenic counterparts (Pierik 1997b).

The selection of a plant expression system is determined by the site in the plant where a protein is synthesized, the cost, safety, and production factors (Goldstein and Thomas 2004). The current technology allows the use of selective promoter systems for gene expression and protein production in either whole-plant expression (green matter) or selectively in the seed or other tissues. Thus, the stability of protein or peptide controls the green matter production (Loyola-Vargas and Ochoa-Alejo 2018).

In tuber or root production, unlike green matter, seeds generally contain fewer phenolic compounds with a less complex mixture of proteins ensuring successful germination and providing a stable and long-term storage of proteins. The seeds are therefore extremely attractive to the production medium, providing the flexibility to store products for delayed processing (Krasteva et al. 2021).

4.7.1 Biopharmaceuticals from Plants

The production of biopharmaceuticals has progressively grown interest and drawn attention as an alternative source of drugs for the treatment of many ailments. Many techniques

(Figure 4.3) are used to improve, modify, and increase the yields of these compounds (Kesik-Brodacka 2018). Plants have the potential as a virtually unlimited source of compounds categorized as vaccines, secondary metabolites, recombinant proteins, and plantibodies (Meskaaoui 2013). The plant-derived antibodies have many applications including its attachment to pathogenic organisms, serum or body fluid of effector proteins (interleukins), tumor antigens and cellular receptor site either activating or inactivation the receptors thereby leading to the sorting out of a particular complication (Tilahun et al. 2019).

Genetically engineered plants/transgenic plants used as hosts for plant body production represent a huge prospect for pharmaceuticals. Among all pharmaceutical compounds, the maximum contribution is accompanied by recombinant proteins (Teli and Timko 2004). Presently, clinical trials are going on many plantibodies for their therapeutic role (Stoger et al. 2002). Plants expressing clinical proteins and polypeptides of pharmaceutical

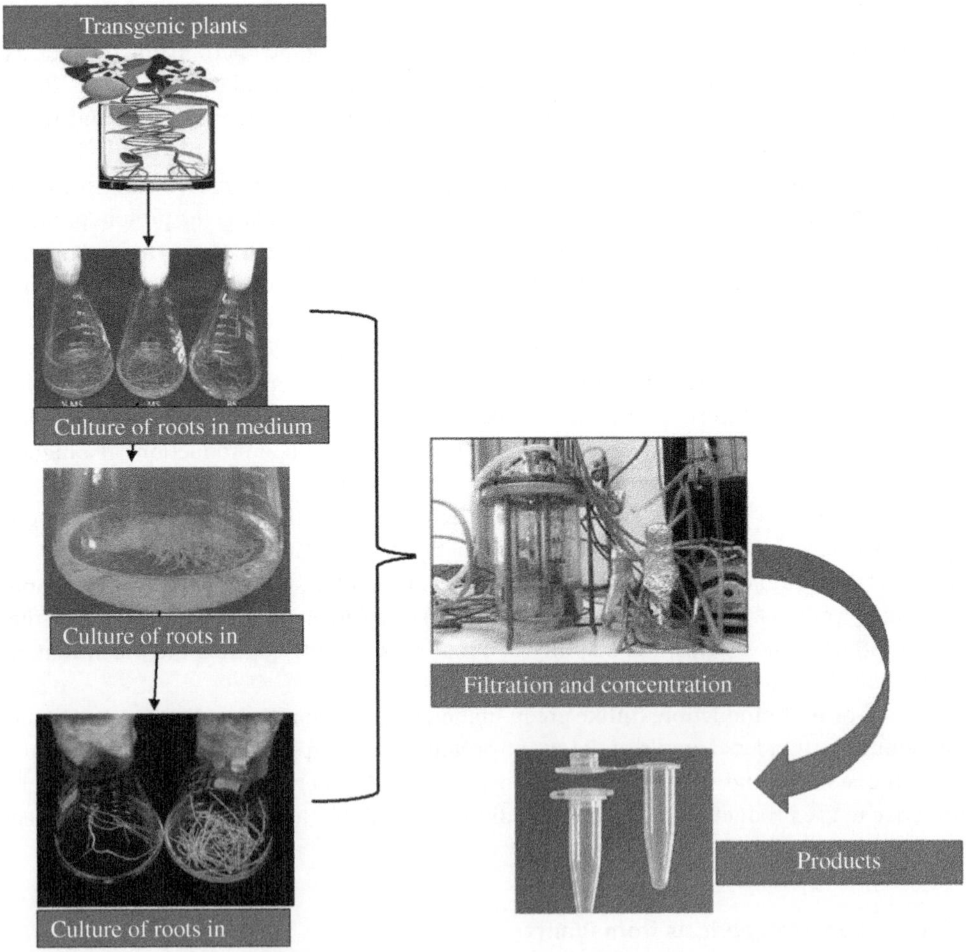

Figure 4.3 Overview of plant culture for biopharmaceuticals. Source: Meskaaoui (2013), Longdom Publishing, CC BY 3.0.

importance, such as human C protein, interferons, hormones, and cytokines, are currently used to provide immunization (Malik 2018). Moreso, vaccines expressed in the edible tissue of plants have proven to be an excellent source for the expression of desirable antigens and their fragments (Nair 2017). This is a cost-effective method to provide immunization (Malik 2018). Over the past two decades, the treatment of many human diseases, such as neurological and genetic illness, diabetes mellitus, blood dyscrasias, growth disorders, and inflammatory diseases, is done by approximately 95 biopharmaceutical products approved by regulatory agencies (Fowler 1986; Goldstein and Thomas 2004).

These biopharmaceuticals need to be produced on large scales to meet the demands of phytochemical products independent of plant availability. The environment in which plant cell tissues and organs grow usually changes when cultures are scaled up from shake flasks to bioreactors, which may result in reduced productivity (Meskaaoui 2013).

4.7.1.1 Scale-up of Secondary Metabolites by Using Different Systems

Advances in tissue culture, combined with genetic engineering and other techniques, have opened new avenues for the high production of pharmaceuticals, nutraceuticals, and other beneficial substances (Shargool 1985). The production of large-scale bioactive compounds uses specific bioreactors systems, as shown in Figure 4.4. These plant cells are cultured in liquid suspension and must be accommodated in large-scale bioreactors for the constant and continuous production of metabolites (Meskaaoui 2013). Some types of bioreactors are designed and used for the commercial production of metabolites (Goldstein and Thomas 2004). Some of the culture systems for the optimization of secondary metabolites, such as cell immobilization techniques, fix plant cells in a suitable matrix. The immobilization of plant cells protects the cells from mechanical stress and facilitates product recovery. In recent years, the preparation and use of immobilized cells have drawn great interest in biotechnology, and their range of applications is still expanding. This is used to synthesize extracellular metabolites and for the biotransformation of many metabolites, observing its positive effects on secondary metabolite production (Table 4.2). Some of the natural compounds localized in morphologically specialized tissues or organs of plants are produced in culture systems by inducing specific organized and undifferentiated cell cultures. The possible use of plant cell cultures for the production of natural compounds has been reported (Moon et al. 2019). However, capsaicin production was increased 1,000-fold after *Capsicum frutescens* cells were immobilized in the pores of a polyurethane foam matrix and cultured in Murashige and Skoog's basal medium containing a specific concentration of calcium chloride ($CaCl_2$) (Lindsey 1985). The production of plumbagin in cell cultures of *Plumbago rosea* in calcium alginate enhanced its production (Lindsey 1985).

Plant tissue culture technology has resulted in the production of many pharmaceutical substances for new therapeutics (Table 4.2). Advances in the area of cell culture have made possible the production of a wide variety of natural compounds, such as alkaloids, terpenoids, steroids, saponins, phenolics, flavanoids, and amino acids (Warzecha 2008). Some widely produced compounds are described as follows.

4.7.1.1.1 Taxol Taxol (paclitaxel) is a complex diterpene alkaloid and one of the most promising anticancer agents known due to its unique mode of action on the microtubular cell system (Li et al. 2014). The production of Taxol in culture is one of the most extensively

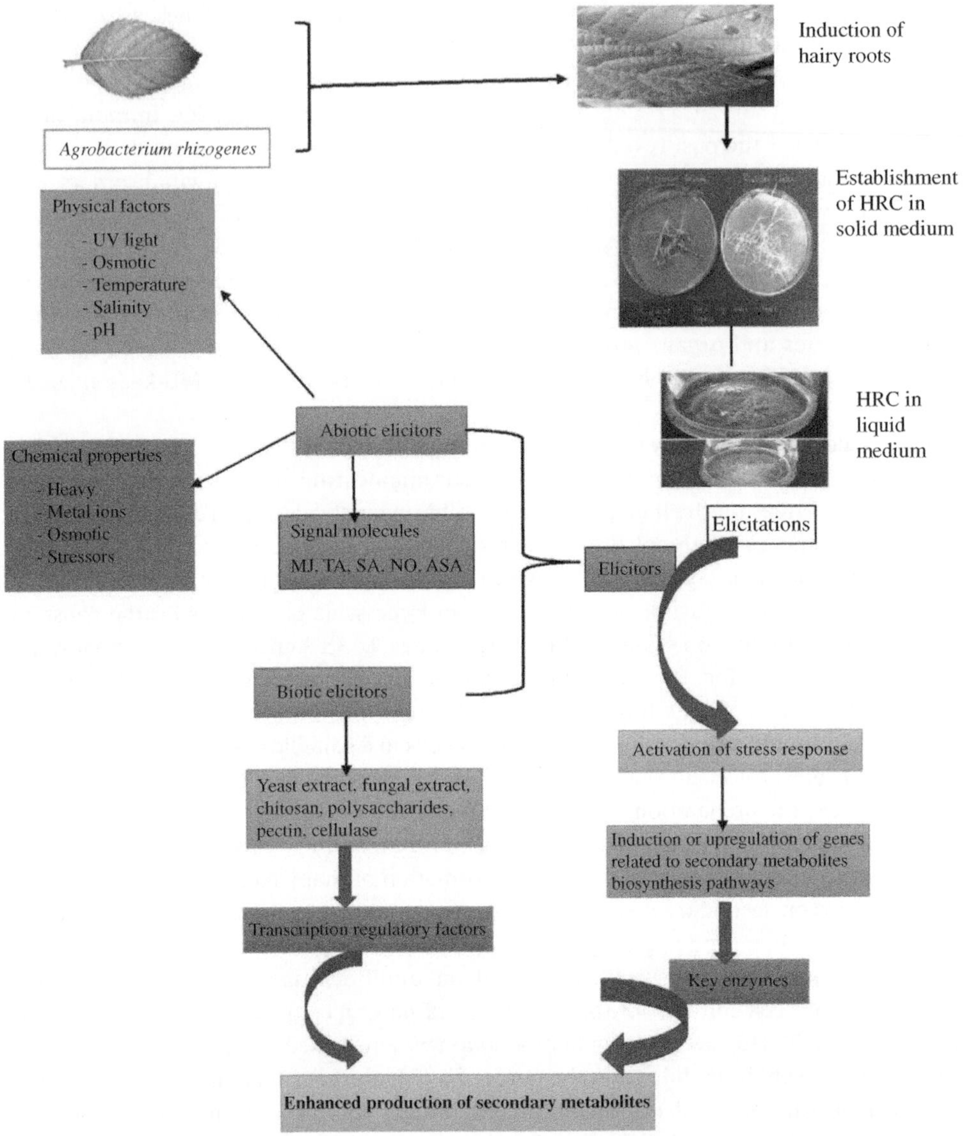

Figure 4.4 Schematic representation of hairy root induction from different explants and elicitation treatment in liquid medium.

explored areas owing to the enormous commercial value of Taxol and the scarcity of the *Taxus* tree with the costly synthetic process. The supplementation of the culture medium with different biotic and abiotic elicitors (amino acids) was studied to improve Taxol production with phenylalanine found to maximize the production in *Taxus cuspidata* cultures (Agrawal 2015).

Table 4.2 Some important plant-derived secondary metabolites of pharmaceutical value.

Active compound	Plant source	Proposed therapeutic use
Alkaloids atropine, hyoscyamine, scopolamine	*Solanaceous* spp.	Anticholinergic
Vinblastine, vincristine	*Catharanthus roseus L.*	Antineoplastic
Nicotine	*Nicotiana* spp.	Smoking cessation
Codeine, morphine	*Papaver somniferum L.*	Analgesic, antitussive
Quinine	*Cinchona* spp.	Antimalarial
Quinidine	*Cinchona* spp.	Cardiac depressant
Terpenes and steroids Artemisinin	*Artemisia annua L.*	Antimalarial
Diosgenin, hecogenin	*Dioscorea* spp.	Oral contraceptives,
Stigmasterol		hormonal drugs
Taxol and other taxoids	*Taxus brevifolia Nutt.*	Antineoplastic
Glycosides digoxin, digitoxin	*Digitalis* spp.	Cardiotonic
Sennosides A and B	*Cassia angustifolia Vahl.*	Laxative
Others Ipecac	*Cephaelis ipecacuanha (Brot.)*	Emetic A. Rich
Podophyllotoxin	*Podophyllum peltatum L.*	Antineoplastic

4.7.1.1.2 Morphine and Codeine Latex is a commercial source of analgesics. The callus and suspension cultures of *P. somniferum* are investigated as an alternative means for the production of morphine and codeine (Benjamin et al. 2019). The maximum codeine and morphine concentrations were 3.0 mg/g dry weight and 2.5 mg/g dry weight, respectively, up to three times higher than those in cultures supplied with hormones. The biotransformation of codeinone to codeine with immobilized cells of *P. somniferum* was reported (Furuya 1972), providing a yield of 70.4% with approximately 88% of the codeine being excreted into the medium.

4.7.1.1.3 L-DOPA L-3,4-dihydroxyphenylalanine is an intermediate of secondary metabolism in higher plants and is known as a precursor of alkaloids, betalain, and melanine, isolated from *Vinca faba*, *Mucuna*, *Baptisia*, and *Lupinus* (Soares et al. 2014). It is also a precursor of catecholamines in animals and is used as a potent drug for Parkinson's disease. The widespread application of this therapy created a demand for large quantities of L-DOPA at a low-cost level, which led to the introduction of cell cultures. The callus tissue of *Mucuna pruriense* accumulated DOPA in the medium containing relatively high concentrations of 2,4-D. Additionally, the induced callus tissues of *Mucuna hassjoo*, *M. Pruriense*, and *Mucuna deeringiana* optimized the culture conditions to produce high concentrations of DOPA (Bapat et al. 2000).

4.7.1.1.4 Diosgenin Diosgenin is a precursor for the chemical synthesis of steroidal drugs. The use of cell cultures of *Dioscorea deltoidea* for the production of diosgenin identified carbon and nitrogen levels (Zhang et al. 2009). Dioscorea cell cultures were established, with reticulated polyurethane foam shown to stimulate diosgenin production, increasing the cellular concentration (Zhang et al. 2009; Li et al. 2011; Mou et al. 2015).

4.7.1.1.5 Capsaicin Capsaicin, an alkaloid, is a food additive obtained from fruits of green pepper (*Capsicum* spp.). Capsaicin is also used as a digestive stimulant and for rheumatic disorders. Suspension cultures of *Capsicum frutescens* produce low levels of capsaicin, but immobilizing the cells in reticulated polyurethane foam increased its production approximately 100-fold (Ochoa-Alejo 2005). Biotechnological processes were developed for the production of capsaicin from *C. frutescens* cells with the use of elicitors (Kashyap et al. 2016). The effects of nutritional stress on capsaicin production in immobilized cell cultures of *Capsicum annum* have been studied thoroughly (Hou et al. 2018).

4.7.1.1.6 Camptothecin Campothecin, a potent antitumor alkaloid, was isolated from *Camptotheca acuminata*. *C. acuminata* callus was used to induce the production of camptothecin using MS medium supplemented with gibberellin and l-tryptophan (Asano et al. 2009).

4.7.1.1.7 Berberine Berberine is an isoquinoline alkaloid found in the roots of *Coptis japonica* and the cortex of *Phellondendron amurense* (Chintalwar et al. 2003; Narasimhan and Nair 2004). The productivity of berberine was increased in cell cultures by optimizing the nutrients in the growth medium with phytohormones (Narasimhan and Nair 2004). The use of elicitors (yeast polysaccharides) in cultures improves the yield of the compounds (Chintalwar et al. 2003).

4.7.1.2 Vaccines

Plant-made vaccines were developed for two decades, with none being processed to pass the phase I tests. However, some biopharmaceuticals are now progressing through phase II and phase III human clinical trials (Daniell et al. 2009). Vaccines are antigenic substance(s) capable of controlling an immune response before an infection or disease manifestation. The vaccination can be through different routes (Table 4.3). Oral immunization requires large amount of soluble antigen to initiate a response compared to parenteral immunization which is more efficient. Presently, considerable interest is in the exploitation of the capacity of transgenic plants to express antigenic proteins in human disease-causing agents (Daniell et al. 2009).

There is much interest in developing low-cost, edible vaccines such as polio, which use the entire or inactivated organisms to cause both local membrane (Ig-A-mediated) and systemic (Ig-G-mediated) immunity. Plant vaccines can exhibit specific proteins with the main immunogenic proteins or antigenic sequences being synthesized in plant tissues. The mucosal immune system induces a protective immune response against pathogens or toxins. The mucosal regions are the site for the production of secretory Ig-A (sIg-A) and specific immune lymphocytes. In addition to the low intrinsic cost of production, some

Table 4.3 Some recombinant vaccines expressed in plants.

Vaccine antigen	Species
Malaria parasite antigen	Virus particle[a]
Rabies virus glycoprotein	Tomato
Human rhinovirus 14 (HRV-14) and human immunodeficiency virus type (HIV-1) epitopes	Virus particle[a]
Norwalk virus capsid protein	Tobacco, potato
Diabetes-associated autoantigen	Tobacco, potato, carrot
Hepatitis B surface proteins	Potato
Mink enteritis virus epitope	Virus particle[a]
Rabies and HIV epitopes	Virus particle[a]
Escherichia coli heat-labile enterotoxin	Tobacco, potato
Rabies virus	Virus particle[a]
Cholera toxin B subunit	Potato
Human insulin-cholera toxin B subunit fusion protein	Potato
Foot and mouth disease virus VP1 structural protein	Alfalfa, Arabidopsis
Hepatitis B virus surface antigen	Yellow lupin, lettuce, tobacco
Human cytomegalovirus glycoprotein B	Tobacco
Dental caries (*S. mutans*)	Tobacco
Respiratory syncytial virus	Tomato

[a] Plant virus can be expressed in multiple plant species.

advantages of plant-based vaccines include increased solidity, durability, effectiveness, and innocuous. Plants producing vaccines can be grown locally where needed, circumventing the costs of storage and transportation. The oral vaccines are can be injected directly in the food product eliminating the costs purification. Plants transformed to express antigenic portions of a pathogen may reduce immunotoxicity and other harmful effects. Finally, the development of multicomponent vaccines is possible through the cross-breeding of transgenic lines and the insertion of multiple genetic elements from various pathogenic organisms. However, some limitations associated with the use of transgenic plants for vaccine production include the expression of recombinant antigens in transgenic plants conferring total immunity. The regulation of yields of expression is important to reduce the variability assuring consistent effective immunization (Goldstein and Thomas 2004; Loza-Rubio and Gómez-Lim 2009; Nielsen 2013).

During the last decade, about a dozen vaccine antigens have been exhibited in plants (Table 4.4). The production of enterotoxigenic *E. coli* heat-labile enterotoxin B subunit antigens from transgenic potatoes protects viruses and bacteria causing diarrhea. Other, edible vaccines are still being processed for foot and mouth disease (veterinary), cholera, rabies, measles virus, diabetes, and respiratory diseases. The hepatitis B surface antigen is expressed in transgenic lupin and lettuce plants.

Table 4.4 Antigens expressed in plants.

Antigen	Pathogen or disease	Plant species
Hepatitis B surface antigen (Hbs Ag)	Hepatitis B	Tobacco
LT-B	*E. coli* heat labile	Potato, tobacco
Enteroxin B	subunit SIgA-G *Streptococcus mutans*	Tobacco
Glycoprotein G	Rabies virus	Tomato
Capsid protein	Norwalk virus	Tobacco, potato
CT-B	*Vibrio cholerae*	Potato
Insulin	Diabetes (autoimmune)	
Structural protein	VP 1 Foot and mouth disease	Arabidopsis
Spike (S) glycoprotein	Swine transmissible gastroenteritis coronavirus (TGEV)	
s-LT-B	Heat-labile enterotoxin B	Potato
Structural protein	VP60 Rabbit hemorrhagic disease virus (RHDV)	
HBsAg	Hepatitis B	Lupin, lettuce
Glycoprotein B	Human cytomegalovirus (HCMV)	Tobacco
HBsAg	Hepatitis B	Potato
F protein	Respiratory syncytial virus	Tomato
(RSV) S-glycoprotein	TGEV	Tobacco
Capsid protein	2L21 Canine parvovirus	Arabidopsis
Hemagglutinin	protein measles virus (MV)	Tobacco
Human acetyl-choline esterase	Organophosphate poisoning	Tomato
(AchE) LT-B	Heat-labile enterotoxin	Corn
B subunit S	TGEV	
CT-A2: CFA/I-CT-B: NSP4 Cholera,	enterotoxigenic *E. coli*	Potato
Protective antigen	*Bacillus anthracis*	Tobacco

The development of plant-based oral subunit vaccine for the respiratory syncytial virus (RSV) using either apple or tomato is still under process. The plant species (alfalfa, apple, asparagus, banana, barley, cabbage, canola, cantaloupe, carrots, cauliflower, cranberry, cucumber, eggplant, flax, grape, kiwi, lettuce, lupins, maize, melon, papaya, pea, peanut, pepper, plum, potato, raspberry, rice, serviceberry, and soybean) are selected for the production and delivery of an oral vaccines to achieve the desired goals (Tables 4.5 and 4.6). Each plant has advantages, of which corn is a good candidate for vaccine production since it is a major component in the diet of the domestic animals. Banana is a

Table 4.5 Human pathogen antigens expressed in transgenic plants.

Expressed protector antigen	Causal agent	Plant	Immunogenic capacity	References
Subunit B of the heat sensible toxin	*E. coli* enterotoxigenic	Tobacco	The intact protein forms multimerics and is immunogenic administered orally	(Krasteva et al. 2021)
Subunit B of the heat sensible toxin	*E. coli* enterotoxigenic	Potato	Immunogenic, protector, and of binding activity administered orally	(Shahina and Anwar 2014; Kieber and Schaller 2018; Krasteva et al. 2021)
Subunit B of the heat sensible toxin	*E. coli* enterotoxigenic	Corn	Immunogenic and protector administered orally	(Sah et al. 2016)
Subunit B of the heat sensible toxin	*Vibrio cholerae*	Potato	The intact protein forms multimerics, has the activity of binding to the receptor, and it is immunogenic and protector administered orally	(Loyola-Vargas and Ochoa-Alejo 2018; Fehér 2019)
Main antigen	Anthrax	Tobacco	Biological activity	(Tabin et al. 2018)
Superficial protein of the cover	Hepatitis B virus	Potato	Forms virus-type particles and the extracted protein is immunogenic administered parenterally	(Yamaguchi and Kamiya 2000; Ahsan 2014)
Superficial protein of the cover	Hepatitis B virus	Lupine (*Lupinus* spp)	Immunogenic administered orally	(Efferth 2019)
Superficial protein of the cover	Hepatitis B virus	Lupine (Lupinus spp)	Immunogenic administered orally	(Rozov 2013)
Protein of the capsid	Norwalk virus	Tobacco	Forms virus-type particles, immunogenic administered orally	(Vasil and Thorpe 1994)
Protein of the capsid	Norwalk virus	Potato	Forms virus-type particles, immunogenic administered orally	(Vasil and Thorpe 1994; Morais-Lino et al. 2008)
Glycoprotein	Rabies virus	Tomato	Intact protein	(Fu et al. 1999)
Glycoprotein B	Human cytomegalovirus	Tobacco	Immunologically related protein	(Veng 1998)
Protein F	Respiratory syncytial virus	Tomato	Immunogenic administered orally	(Popper 2011)
HPV 11	Papillomavirus	Potato	Immunogenic administered orally	(Sharma et al. 2015)

(*Continued*)

Table 4.5 (Continued)

Expressed protector antigen	Causal agent	Plant	Immunogenic capacity	References
Principle protein of the capsid L 1	Papillomavirus	Norwalk virus	Immunogenic administered orally	(Ramirez-Estrada et al. 2016)
HPV-type 16 – L 1	Papillomavirus	*Nicotiana tabacum*	Immunogenic administered orally	(Singh 1999)
VP 7	Rotavirus	Potato	Immunogenic administered orally	(Ramirez-Estrada et al. 2016)

Table 4.6 Antigens of animal pathogens expressed in transgenic plants.

Protector antigen expressed in plants	Causal agent	Plant	Immunogenic capacity	References
VP60	Hemorrhagic virus of rabbits	Potato	Immunogenic and protector administered orally	(Schmülling 2004; Halder et al. 2019)
VP1	Foot and mouth disease	Arabidopsis	Immunogenic and protector administered orally	
VP1	Norwalk virus Foot and mouth disease	Alfalfa	Immunogenic and protector administered orally	(Daniell et al. 2009)
VP135 – 160	Foot and mouth disease epitope highly immunogenic		Immunogenic and protector administered orally	(Chung et al. 2006)
VP1	Foot and mouth disease	Tobacco	Immunogenic and protector administered orally	(Halder et al. 2019)
Glycoprotein S	Transmissible gastroenteritis in pigs (TGE)	Arabidopsis	Immunogenic and protector administered parenterally	(DeMason 2005)
Glycoprotein S	TGE	Corn	Protector per oral	(Sah et al. 2016)
Hemagglutinin	Rinderpest virus	Peanut	Seroconversion when administered parenterally with plant derivatives	(Niazian et al. 2017)

common constituent of many infant diets thath is consumed ripe, eliminating the possibility of protein denaturation in high temperatures. Unfortunately, it is difficult to create transgenic bananas making the production period longer than that for other food crops. Cereals and edible plants are advantageous for vaccine production due to the lower levels of toxic metabolites present. Many strategies to identify and develop low-cost plant-derived vaccine materials exist (Goldstein and Thomas 2004; Daniell et al. 2009).

4.7.1.3 Plantibodies

Antibodies or immunoglobulins are vital components of the adaptive immune system in mammals. These are mainly present in the body fluid and constitute an assembly of glycoproteins released by B cells. B cells constitute the humoral type of adaptive immune system. They are highly specific to their mode of action. They recognize their target specifically, bind to the antigen/toxin of the pathogen, and produce and elicit the immune response. These features permit immunoglobulins to be employed in a diverse range of applications, such as diagnosis, prevention, and treatment (Stoger et al. 2002). The production of a defective antibody response results in increased vulnerability to pyogenic infections due to impairment in B-cell function, e.g. in hyper-IgM syndrome as a result of improper signaling between B and T cells and in X-linked agammaglobulinemia. To overcome these defects, the constant region of a human immunoglobulin is modified by a transgenic approach to create recombinant antibodies (Nair 2017). With this advancement, a new expectation of a reliable, inexpensive cure for diseases such as diabetes and cancer were awakening, but the major drawback was its cost-ineffectiveness.

Genetically engineered plants/transgenic plants used as hosts for plantibody production represent a huge prospect for pharmaceuticals. Among all the pharmaceutical compounds, the maximum contribution is accompanied by recombinant proteins. Presently, clinical trials are going on several plantibodies for their therapeutic role (Malik 2018). Among therapeutics, CaroRx®, which was the first plantibody produced from tobacco, was used. It is an anti-Streptococcus mutants secretory antibody and protects against dental caries. Another plantibody was developed in soybean against the herpes simplex virus (Stoger et al. 2002). Plants expressing clinical proteins and polypeptides of pharmaceutical importance, such as human C protein, interferons, hormones, and cytokines, are currently used to provide immunization, for example, oral immunization. Edible vaccines expressed in the edible tissue of plants have proven to be an excellent source for the expression of desirable antigens and their fragments (Table 4.7). This is a cost-effective method to provide immunization (Malik 2018). Genetically engineered plants proved to be a superior system for vaccination in humans (Sogutmaz Ozdemir and Budak 2018).

4.7.1.4 Proteins

Tobacco suspension cells, particularly those from the closely related cultivars BY-2 and NT-1, are frequently chosen as host cell lines because transformation and propagation are simple and well established with favorable growth characteristics (Yarra et al. 2019). Although intact plants are a useful source of suspension cells, freshly initiated plant suspension cultures may take a long time to acquire the favorable growth characteristics of BY-2 and NT-1. Other plant suspension cultures used for the production of recombinant proteins include rice, soybean, and tomato. Some cell lines were studied because of the possibility that they could be more favorable than tobacco in terms of by-product levels and since they are derived from food crops' regulatory compliance (Hellwig et al. n.d.). Other anticipated benefits of exploring diverse cell lines include faster growth, higher expression levels, more efficient secretion, and other advantages concerning process compatibility.

Table 4.7 Plantibodies for therapeutic and diagnostic use.

Application and specificity	Antibody type	Plant	References
Dental cavities; Antigen I or II of *S. mutants*	IgA	*Nicotiana tabacum*	(Gamborg et al. 1974; Germana 2011)
Diagnosis; antihuman	IgG IgG	Alfalfa	(Goldstein and Thomas 2004)
Cancer treatment; carcinoembryonic antigen	Simple chain from antibody FvScFv	Wheat	(Loyola-Vargas and Ochoa-Alejo 2018)
Cancer treatment; carcinoembryonic antigen	ScFv	Rice	(Loyola-Vargas and Ochoa-Alejo 2018)
Cancer treatment; carcinoembryonic antigen	ScFv	Rice	(Loyola-Vargas and Ochoa-Alejo 2018)
Cancer treatment: carcinoembryonic antigen	IgG	*Nicotiana tabaccum*	(Veeresham et al. 2003)
B cells lymphoma treatment; ideotype vaccine	scFv	*Nicotiana benthamiana*	(Quazi et al. 2021)
Herpes simples virus 2	IgG	Soy	(Fowler 1986)
B-hepatitis surface	Antiantigen of Hepatitis B	*Nicotiana tabacum*	(Tulecke 1963)
Anti RH factor	Anti RH-D	Arabidopsis	(Quazi 2021)
Antirabies treatment	IgG	*Nicotiana tabacum*	(Quazi et al. 2021)
Polyepitope of cells B against measles	scFv	Carrot	(Moon et al. 2019)

Some researchers focus on plants with higher protein content, for example, soybean and lupin, assuming that these might more readily facilitate higher expression levels (Moon 2020).

The design of the construct used to express the recombinant protein is a key factor in determining yield. The promoter choice affects yield by determining the rate of transcription. The most commonly used promoter is the cauliflower mosaic virus (CaMV) 35S promoter or its enhanced version (Table 4.8), but several alternative constitutive promoters are used, including the hybrid (ocs)3mas promoter (constructed from octopine synthase (ocs) and mannopine synthase (mas) promoter sequences) and the ubiquitin promoters from maize and *A. thaliana* (Yarra et al. 2019). In contrast to these constitutive promoters, the rice α-amylase RAmy3D promoter is induced by sugar deprivation. Another important aspect of the designed construct is the presence or absence of a leader peptide, which directs the recombinant protein to the secretory pathway. The leader peptides from plant and nonplant proteins appear to function equivalently, and many human secreted proteins are expressed using their endogenous leaders (Table 4.8). The proteins directed to the secretory pathway in cultured cells eventually reach the apoplast, from where they diffuse through the cell wall and into the culture medium unless they become trapped in the cell wall matrix (Fowler 1986). This depends on the size of the protein as

Table 4.8 Biopharmaceutical proteins produced in transgenic plants.

Protein(s)	Plant host
a-Trichosanthin	Tobacco
Hirudin Brassica	Seeds
a-Interferon	Rice
Erythropoietin	Tobacco
Glucocererbosidase	Tobacco
Protein-C	Tobacco
Avidin	Maize
Ribosome-inactivating protein	Tobacco
Granulocyte-macrophage	Tobacco
Colony-stimulating factor b-glucuronidase	Corn
Lactoferrin and b-casein	Potato
Human collagen I [Pro1a (I)]	Tobacco
Phytase	Canola
Thermostable endo-1,4 b-D-glucanase	Arabidopsis
Human a-antitrypsin	Rice
GM-CS factor	Tobacco
Human somatotropin	Tobacco
Nuclear expression	Tobacco
Human erythropoietin	Tobacco
Human encephalin	Arabidopsis
Interferon-β	Tobacco
Human serum albumin	Tobacco
Human homotrimetric collagen	Tobacco
Human α-1-antitrypsin	Rice
Human aprotinine	Corn
Human lactoferrin	Potato
Angiotensin converter enzyme	Tobacco, tomato
α-Tricosanthin ofTMV-U1 protein of the subgenomic capsid	*Nicotiana bethamiana*
IL-12	Tobacco

well as other physicochemical properties. The proteins of less than 30 kDa tend to be secreted into the medium, whereas larger proteins are quantitatively retained, but large proteins, including full-sized antibodies, are secreted efficiently, whereas some small proteins remain trapped, suggesting that charge and/or hydrophobicity may also be important determinants.

Finally, a transiently expressed tumor-specific vaccine was produced using tobacco plant for the eradication of lymphoma. Currently, seven plant-derived antibodies have reached advanced stages of clinical product processing, most for the treatment and diagnosis of cancer, dental caries, herpes simplex virus, and RSV. Currently, no plantibodies have reached the commercialized stage. The elevated levels of formation, refinement, efficiency, and small immunogenicity of recombinant human antibodies derived from plants render them important for the future production of monoclonal antibodies (Nair 2017).

4.7.2 The Effects of Production, Safety, and Efficacy

All biopharmaceuticals produced need to be safe and efficient in the diverse applications used with a unique interdisciplinary process leading to the progress of recent remedial agents for various unexamined conditions (Goldstein and Thomas 2004). The search for recent products with the perceived feasibility of technological successes is conditioned by the curative efficiency and stability parameters triggered by a health condition (Warzecha 2008).

However, both Good Laboratory Practice and Good Manufacturing Practice are essential for the production of biopharmaceuticals. These new pharmaceutical agents are submitted to the same quality control and safety standards as those derived from microorganisms and mammalian cell systems (Goldstein and Thomas 2004).

Plant biotechnology research has faced several challenges in some areas, among which the destruction of experimental sites raises more reliability considerations. These concerns are tackled through grassland monitoring and safety with the use of encircled surroundings for low-scale performances (Goldstein and Thomas 2004). Additionally, plant cells are able of assembling complex proteins, avoiding the appearance of endotoxins. Therefore, there is safety and value in using plants as the precursor of transformed proteins (Yarra et al. 2019).

However, suitable actions must be taken to eradicate unpleasant plant-derived proteins or other biomolecules conditioning the presence of fungal toxins or pesticides used in plant extension. Plant-produced pharmaceuticals have possible repercussions on untargeted classes such as butterflies, honeybees, and other wildlife around or besides the culture sites. Many biopharmaceutical proteins (antibodies) are very precise in their actions, creating the probability of their entry into the human food chain in the use of food crops (Goldstein and Thomas 2004).

4.8 Conclusion

The use of plants as reservoirs for the production of biopharmaceuticals (novel vaccines, antibodies, and other therapeutic proteins) is increasing. However, molecular techniques are used for the production of a wide variety of new biopharmaceuticals. More efforts are needed to focus on increasing the yields and scaling up the production, distribution, and handling of transgenic plant material with techniques unique to biopharmaceuticals. Therefore, there is a need for continued surveillance to guard the surroundings, ensuring that products have no secondary effects on untargeted organisms. Gene containment methodologies must pursue to be established with precautions on the overexpression of inherently unfavorable proteins, necessitating more work with the use of transgenic food plants.

References

Agrawal, P. (2015). Biopharmaceuticals: an emerging trend in drug development. *SOJ Pharm. Pharm. Sci. [Internet]* 2 (1): 2–1. http://www.symbiosisonlinepublishing.com/pharmacy-pharmaceuticalsciences/pharmacy-pharmaceuticalsciences20.php (accessed 26 September 2021).

Ahmed, M.A.A., Miao, M., Pratsinakis, E.D. et al. (2021). Protoplast isolation, fusion, culture and transformation in the woody plant *Jasminum* spp. *Agriculture* 11 (8): 699.

Ahsan, A. (2014). Tissue culture and some of the factors affecting them and the micropropagation of strawberry. *Life Sci. J.* 11 (8): 484–493.

Alok, A., Kumar, J., and Upadhyay, S.K. (2018). Engineering in hairy roots using CRISPR/Cas9-mediated editing. In: *Hairy Roots* (ed. V. Srivastava, S. Mehrotra, and S. Mishra). Singapore: Springer. https://doi.org/10.1007/978-981-13-2562-5_15.

Angelova, Z. et al. (2006). Elicitation of plants. *Biotechnol. Biotechnol. Equip.* 20 (2): 72–83. https://doi.org/10.1080/13102818.2006.10817345.

Asano, T., Sudo, H., Yamazaki, M., and Saito, K. (2009). Camptothecin production by in vitro cultures and plant regeneration in *Ophiorrhiza* species. In: *Protocols for in vitro Cultures and Secondary Metabolite Analysis of Aromatic and Medicinal Plants [Internet]*, Methods in Molecular Biology, vol. 547 (ed. S.M. Jain and P.K. Saxena), 337–345. Totowa, NJ: Humana Press http://link.springer.com/10.1007/978-1-60327-287-2_27 (accessed 6 February 2022).

Bapat, V.A., Suprasanna, P., Ganapathi, T.R., and Rao, P.S. (2000). In vitro production of L-dopa in tissue cultures of banana. *Pharm. Biol. Former Int. J. Pharmacogn.* 38 (4): 271–273.

Behbahani, M., Shanehsazzadeh, M., and Hessami, M.J. (2011). Optimization of callus and cell suspension cultures of *Barringtonia racemosa* (Lecythidaceae family) for lycopene production. *Sci. Agric.* 68 (1): 69–76.

Benjamin, E.D., Ishaku, G.A., Peingurta, F.A., and Afolabi, A.S. (2019). Callus culture for the production of therapeutic compounds. *Am. J. Plant Biol.* 4 (4): 76.

Binenbaum, J., Weinstain, R., and Shani, E. (2018). Gibberellin localization and transport in plants. *Trends Plant Sci.* 23: 410–421.

Bonfill, M., Cusidó, R.M., Lalaleo, L., and Palazón, J. (2015). Plant cell and organ cultures as a source of phytochemicals. 17. In: *Recent Advances in Pharmaceutical Sciences V* (ed. D. Muñoz-Torrero, M.P. Vinardell and J. Palazón), 33–49.

Breyer, D., Kopertekh, L., and Reheul, D. (2014). Alternatives to antibiotic resistance marker genes for *in vitro* selection of genetically modified plants – scientific developments, current use, operational access and biosafety considerations. *Crit. Rev. Plant Sci.* 33 (4): 286–330.

Bukovinszki, Á., Divéki, Z., Csányi, M. et al. (2007). Engineering resistance to PVY in different potato cultivars in a marker-free transformation system using a 'shooter mutant' *A. tumefaciens*. *Plant Cell Rep.* 26 (4): 459–465.

Chintalwar, G.J., Gupta, S., Roja, G., and Bapat, V.A. (2003). Protoberberine alkaloids from callus and cell suspension cultures of *Tinospora cordifolia*. *Pharm. Biol.* 41 (2): 81–86.

Chung, S.-M., Vaidya, M., and Tzfira, T. (2006). Agrobacterium is not alone: gene transfer to plants by viruses and other bacteria. *Trends Plant Sci.* 11 (1): 1–4.

Daniell, H., Singh, N.D., Mason, H., and Streatfield, S.J. (2009). Plant-made vaccine antigens and biopharmaceuticals. *Trends Plant Sci. [Internet]* 14 (12): 669–679. https://linkinghub.elsevier.com/retrieve/pii/S1360138509002520 (accessed 6 November 2021).

DeMason, D.A. (2005). Auxin–cytokinin and auxin–gibberellin interactions during morphogenesis of the compound leaves of pea (*Pisum sativum*). *Planta* 222 (1): 151–166.

Dixit, S., Shukla, A., Singh, V., and Upadhyay, S.K. (2021a). Bioprospecting of natural compounds for industrial and medical applications. In: *Bioprospecting of Plant Biodiversity for Industrial Molecules* (ed. S.K. Upadhyay and S.P. Singh). John Wiley & Sons Ltd. https://doi.org/10.1002/9781119718017.ch3.

Dixit, S., Shukla, A., Singh, V., and Upadhyay, S.K. (2021b). Engineering of plant metabolic pathway for nutritional improvement. In: *Genome Engineering for Crop Improvement* (ed. S.K. Upadhyay). John Wiley & Sons Ltd. https://doi.org/10.1002/9781119672425.ch20.

Efferth, T. (2019). Biotechnology applications of plant callus cultures. *Engineering* 5 (1): 50–59.

Eibl, R. and Eibl, D. (2009). (Principles and Practice)Plant cell-based bioprocessing. In: *Cell and Tissue Reaction Engineering [Internet]* (ed. Institute of Biotechnology, Zurich University of Applied Sciences and Department of Life Sciences and Facility Management). Berlin, Heidelberg: Springer Berlin Heidelberg, 315–56 http://link.springer.com/10.1007/978-3-540-68182-3_8 (accessed 23 November 2021).

Espinosa-Leal, C.A., Puente-Garza, C.A., and García-Lara, S. (2018). In vitro plant tissue culture: means for production of biological active compounds. *Planta* 248 (1): 1–18.

Fehér, A. (2019). Callus, dedifferentiation, totipotency, somatic embryogenesis: what these terms mean in the era of molecular plant biology? *Front. Plant Sci.* 10: 536.

Fowler, M.W. (1986). Biotechnological applications of plant tissue cultures. In: *Perspectives in Biotechnology and Applied Microbiology [Internet]* (ed. D.I. Alani and M. Moo-Young), 295–315. Dordrecht: Springer Netherlands http://link.springer.com/10.1007/978-94-009-4321-6_23 (accessed 03 December 2021).

Fu, T.-J., Singh, G., and Curtis, W.R. (ed.) (1999). *Plant Cell and Tissue Culture for the Production of Food Ingredients [Internet]*. Boston, MA: Springer US http://link.springer.com/10.1007/978-1-4615-4753-2 (accessed 11 January 2022).

Furuya, T. (1972). Alkaloids from callus tissue of *Papaver somniferum*. *Phytochemistry* 11: 3041–3044.

Gamborg, O.L., Constabel, F., Fowke, L. et al. (1974). Protoplast and cell culture methods in somatic hybridization in higher plants. *Can. J. Genet. Cytol.* 16 (4): 737–750.

Germana, A. (2011). Anther culture for haploid and doubled haploid production. *Plant Cell Tissue Organ. Cult.* 104: 283–300.

Giddings, G., Allison, G., Brooks, D., and Carter, A. (2000). Transgenic plants as factories for biopharmaceuticals. *Nat. Biotechnol.* 18 (11): 1151–1155.

Goldstein, D.A. and Thomas, J.A. (2004). Biopharmaceuticals derived from genetically modified plants. *QJM [Internet]* 97 (11): 705–716. https://academic.oup.com/qjmed/article-lookup/doi/10.1093/qjmed/hch121 (accessed 19 November 2021).

Grout, B.W.W. (1999). Meristem-tip culture for propagation and virus elimination. In: *Plant Cell Culture Protocols [Internet]*, 115–126. New Jersey: Humana Press http://link.springer.com/10.1385/1-59259-583-9:115 (accessed 02 February 2022).

Halder, M., Sarkar, S., and Jha, S. (2019). Elicitation: a biotechnological tool for enhanced production of secondary metabolites in hairy root cultures. *Eng. Life Sci.* 19 (12): 880–895. https://onlinelibrary.wiley.com/doi/10.1002/elsc.201900058 (accessed 11 January 2022).

Hellwig, S., Drossard, J., Twyman, R.M., and Fischer, R. Plant cell cultures for the production of recombinant proteins. *Nat. Biotechnol.* 22: 1415–1422.

Hou, B.-Z., Li, C.-L., Han, Y.-Y., and Shen, Y.-Y. (2018). Characterization of the hot pepper (*Capsicum frutescens*) fruit ripening regulated by ethylene and ABA. *BMC Plant Biol.* 18 (1): 162.

Hughes, K.W. (1981). In vitro ecology: exogenous factors affecting growth and morphogenesis in plant culture systems. *Environ. Exp. Bot.* 21 (3–4): 281–288.

Hussain, A., Ahmed, I., Nazir, H., and Ullah, I. (2012). Plant tissue culture: current status and opportunities. In: *Recent Advances in Plant in vitro Culture [Internet]* (ed. A. Leva). IntechOpen http://www.intechopen.com/books/recent-advances-in-plant-in-vitro-culture/plant-tissue-culture-current-status-and-opportunities (accessed 23 January 2022).

Iliev, I., Gajdoov, A., Libiakov, G., and Jain, S.M. (2010). Plant micropropagation. In: *Plant Cell Culture [Internet]* (ed. M.R. Davey and P. Anthony), 1–23. Chichester, UK: John Wiley & Sons, Ltd https://onlinelibrary.wiley.com/doi/10.1002/9780470686522.ch1 (accessed 25 January 2022).

Jozala, A.F., Geraldes, D.C., Tundisi, L.L. et al. (2016). Biopharmaceuticals from microorganisms: from production to purification. *Braz. J. Microbiol. [Internet]* 47: 51–63. https://linkinghub.elsevier.com/retrieve/pii/S1517838216310413 (accessed 07 December 2021).

Kashyap, S., Kapoor, N., and Kale, R.D. (2016). *Coscinium fenestratum*: Callus and suspension cell culture of the endangered medicinal plant using vermicompost extract and coelomic fluid as plant tissue culture media. *Am. J. Plant Sci.* 7 (6): 899–906.

Kesik-Brodacka, M. (2018). Progress in biopharmaceutical development. *Biotechnol. Appl. Biochem.* 65 (3): 306–322.

Kieber, J.J. and Schaller, G.E. (2018). Cytokinin signaling in plant development. *Development* 145 (4): dev149344.

Krasteva, G., Georgiev, V., and Pavlov, A. (2021). Recent applications of plant cell culture technology in cosmetics and foods. *Eng. Life Sci. [Internet]* 21 (3–4): 68–76. https://onlinelibrary.wiley.com/doi/10.1002/elsc.202000078 (accessed 11 January 2022).

Li, P., Mao, Z., Lou, J. et al. (2011). Enhancement of diosgenin production in *Dioscorea zingiberensis* cell cultures by oligosaccharides from its endophytic fungus *Fusarium oxysporum* Dzf17. *Molecules* 16 (12): 10631–10644.

Li, Y., Zhang, G., and Pfeifer, B.A. (2014). Current and emerging options for taxol production. In: *Biotechnology of Isoprenoids [Internet]*, Advances in Biochemical Engineering/Biotechnology, vol. 148 (ed. J. Schrader and J. Bohlmann), 405–425. Cham: Springer International Publishing http://link.springer.com/10.1007/10_2014_292 (accessed 31 January 2022).

Lievens, L., Pollier, J., Goossens, A. et al. (2017). Abscisic acid as pathogen effector and immune regulator. *Front. Plant Sci. [Internet]* 8: http://journal.frontiersin.org/article/10.3389/fpls.2017.00587/full (accessed 2 February 2022).

Lindsey, K. (1985). Manipulation, by nutrient limitation, of the biosynthetic activity of immobilized cells of *Capsicum frutescens* Mill. cv. annuum. *Planta* 165: 126–133.

Loyola-Vargas, V.M. and Ochoa-Alejo, N. (2018). An introduction to plant tissue culture: advances and perspectives. In: *Plant Cell Culture Protocols [Internet]*, Methods in Molecular Biology, vol. 1815 (ed. V.M. Loyola-Vargas and N. Ochoa-Alejo). New York: Springer New York, 3–13 http://link.springer.com/10.1007/978-1-4939-8594-4_1 (accessed 19 November 2021).

Loza-Rubio, E., Gómez-Lim, M.Á., Miguel Ángel (2009). Producción de vacunas y otros compuestos biológicos en plantas transgénicas. Veterinaria México, vol. 37, núm. 4, octubre-diciembre, 2006, pp. 441–455 Universidad Nacional Autónoma de México Distrito Federal, México.

Ma, R. and Gomez-Cadenas, A. (2012). In vitro tissue culture, a tool for the study and breeding of plants subjected to abiotic stress conditions. In: *Recent Advances in Plant in vitro Culture [Internet]* (ed. A. Leva). InTechOpen http://www.intechopen.com/books/recent-advances-in-plant-in-vitro-culture/in-vitro-tissue-culture-a-tool-for-the-study-and-breeding-of-plants-subjected-to-abiotic-stress-cond (accessed 24 January 2022).

Malik, A. (2018). Plantibodies: a new approach for immunomodulation in human health. *Biomed J Sci Tech Res [Internet]* 11 (2): https://doi.org/10.26717/BJSTR.2018.11.002061.

Mangal, M., Bhardwaj, S.V., Sharma, D.R. et al. (2002, 2002). Use of meristem tip culture to eliminate carnation latent virus from carnation plants. *Indian J. Exp. Biol.* 40 (1): 119–122.

Meskaaoui, A.E. (2013). Plant cell tissue and organ culture biotechnology and its application in medicinal and aromatic plants. *Med. Aromat. Plants [Internet]* 2 (3): e147. https://doi.org/10.4172/2167-0412.1000e147.

Meyers, R.A. (1988). *Haploid Piants from Tissue Culture: New Plant Variries in a Shofflned Time Frame*. Nature Publishing Group.

Moon, K.-B. (2020). Development of systems for the production of plant-derived biopharmaceuticals. *Plan. Theory* 9: 30.

Moon, K.-B., Park, J.-S., Park, Y.-I. et al. (2019). Development of systems for the production of plant-derived biopharmaceuticals. *Plan. Theory* 9 (1): 30.

Morais-Lino, L.S., Santos-Serejo, J.A.d., Silva, S.d.O.e. et al. (2008). Cell suspension culture and plant regeneration of a Brazilian plantain, cultivar Terra. *Pesqui Agropecuária Bras.* 43 (10): 1325–1330.

Mou, Y., Zhou, K., Xu, D. et al. (2015). Enhancement of diosgenin production in plantlet and cell cultures of *Dioscorea zingiberensis* by palmarumycin C_{13} from the endophytic fungus, *Berkleasmium* sp. Dzf12. *Trop. J. Pharm. Res.* 14 (2): 241.

Naik, P.M. and Al-Khayri, J.M. (2016). Abiotic and biotic elicitors–role in secondary metabolites production through in vitro culture of medicinal plants. In: .

Nair, B.J. (2017). Plantibodies: paving novel avenues for immunotherapy. *MOJ Surg. [Internet]* 4 (4): 75–77. https://doi.org/10.15406/mojs.2017.04.00078.

Narasimhan, S. and Nair, G.M. (2004, 2004). Effect of auxins on berberine synthesis in cell suspension culture of *Coscinium fenestratum* (Gaertn.) Colebr-A critically endangered medicinalliana of Western Ghats. *Indian J. Exp. Biol.* 42 (6): 616–619.

Niazian, M., Sadat Noori, S.A., Galuszka, P., and Mortazavian, S.M.M. (2017). Tissue culture-based Agrobacterium-mediated and in planta transformation methods. *Czech J. Genet. Plant Breed.* 53 (4): 133–143.

Nielsen, J. (2013). Production of biopharmaceutical proteins by yeast: advances through metabolic engineering. *Bioengineered* 4 (4): 207–211.

Ochoa-Alejo, N. (2006). Capsaicin accumulation in *Capsicum* spp. suspension cultures. *Methods Mol. Biol.* 318: 327–334. https://doi.org/10.1385/1-59259-959-1:327.

Ogita, S. (2015). Plant cell, tissue and organ culture: the most flexible foundations for plant metabolic engineering applications. *Nat. Prod. Commun. [Internet]* 10 (5): 1934578X1501000. http://journals.sagepub.com/doi/10.1177/1934578X1501000527 (accessed 23 November 2021).

Okere, A.U. and Adegeye, A. (2011). In vitro propagation of an endangered medicinal timber species *Khaya grandifoliola* C. Dc. *Afr. J. Biotechnol.* 10 (17): 3335–3339.

Pant, B. (2014). Application of plant cell and tissue culture for the production of phytochemicals in medicinal plants. In: *Infectious Diseases and Nanomedicine II [Internet]*, Advances in Experimental Medicine and Biology, vol. 808 (ed. R. Adhikari and S. Thapa), 25–39. New Delhi: Springer India http://link.springer.com/10.1007/978-81-322-1774-9_3 (accessed 31 January 2022).

Perianez-Rodriguez, J., Manzano, C., and Moreno-Risueno, M.A. (2014). Post-embryonic organogenesis and plant regeneration from tissues: two sides of the same coin? *Front. Plant Sci. [Internet]* 5: 219. https://doi.org/10.3389/fpls.2014.00219.

Pierik, R.L.M. (1997a). *In vitro Culture of Higher Plants*, 4e, 348. Dordrecht: Kluwer CrossRefGoogle Scholar - Google Search [Internet]. https://www.google.com/search?q=Pierik%2C+R.L.M.+(1997)+In+vitro+culture+of+higher+plants.+4th+ed.+Kluwer%2C+Dordrecht.+348+p.CrossRefGoogle+Scholar&oq=Pierik%2C+R.L.M.+(1997)+In+vitro+culture+of+higher+plants.+4th+ed.+Kluwer%2C+Dordrecht.+348+p.CrossRefGoogle+Scholar&aqs=chrome..69i57.1599j0j7&sourceid=chrome&ie=UTF-8 (accessed 12 January 2022).

Pierik, R.L.M. (1997b). *In vitro Culture of Higher Plants [Internet]*. Dordrecht: Springer Netherlands http://link.springer.com/10.1007/978-94-011-5750-6 (accessed 12 January 2022).

Poehlman, J.M. (1987). Plant cell and tissue culture applications in plant breeding. In: *Breeding Field Crops [Internet]*, 148–170. Dordrecht: Springer Netherlands http://link.springer.com/10.1007/978-94-015-7271-2_8 (accessed 8 December 2021).

Popper, Z.A. (ed.) (2011). *The Plant Cell Wall Methods and Protocols*. Humana Press.

Purkayastha, J., Sugla, T., Paul, A. et al. (2010). Efficient *In vitro* plant regeneration from shoot apices and gene transfer by particle bombardment in *Jatropha curcas*. *Biol. Plant.* 54 (1): 13–20.

Quazi, S., Golani, T., and Martino Capuzzo, A. (2021). Germplasm conservation. In: *Endangered Plants [Internet]* (ed. S. Kumar). IntechOpen https://www.intechopen.com/books/endangered-plants/germplasm-conservation (accessed 2 February 2022).

Quiroz, K.A., Berríos, M., Carrasco, B. et al. (2017). Meristem culture and subsequent micropropagation of Chilean strawberry (*Fragaria chiloensis* (L.) Duch.). *Biol. Res.* 50 (1): 20.

Rahamooz-Haghighi, S., Bagheri, K., Danafar, H., and Sharafi, A. (2020). Tissue culture, in vitro organogenesis and regeneration of *Plantago lanceolata*. *J. Appl. Biotechnol. Rep. [Internet]* 7 (4): 258–265. https://doi.org/10.30491/jabr.2021.122254.

Ramirez-Estrada, K., Vidal-Limon, H., Hidalgo, D. et al. (2016). Elicitation, an effective strategy for the biotechnological production of bioactive high-added value compounds in plant cell factories. *Molecules* 21 (2): 182.

Rozov, S.M. (2013). Auxins: biosynthesis, metabolism, and transport. *Biol. Bull. Rev.* 133 (1): 50–62.

Saad, A.I.M. and Elshahed, A.M. (2012). Plant tissue culture media. In: *Recent Advances in Plant in vitro Culture [Internet]* (ed. A. Leva). IntechOpen http://www.intechopen.com/books/recent-advances-in-plant-in-vitro-culture/plant-tissue-culture-media (accessed 31 January 2022).

Sah, S.K., Reddy, K.R., and Li, J. (2016). Abscisic acid and abiotic stress tolerance in crop plants. *Front. Plant Sci. [Internet]* 7: 571. https://doi.org/10.3389/fpls.2016.00571.

Schmülling, T. (2004). Cytokinin. In: *Encyclopedia of Biological Chemistry* (ed. W. Lennarz and M.D. Lane). Academic Press/Elsevier Science.

Sergio, G.r. (n.d.). Plant Tissue Culture.pdf [Internet]. https://www.apsnet.org/edcenter/disimpactmngmnt/labexercises/PlantBiotechnology/Documents/PlantTissueCulture.pdf.

Shahina, P. and Anwar, S. (2014). Factors affecting in vitro plant regeneration from cotyledonary node explant of *Senna sophera* (L.) Roxb. A highly medicinal legume. *Afr. J. Biotechnol.* 13 (3): 413–422.

Shanmugaraj, B.I., Bulaon, C.J., and Phoolcharoen, W. (2020). Plant molecular farming: a viable platform for recombinant biopharmaceutical production. *Plants [Internet]* 9 (7): 842. https://www.mdpi.com/2223-7747/9/7/842 (accessed 02 February 2022).

Shargool, P.D. (1985). Future uses of plant cell culture techniques. *Biochem. Educ.* 13 (2): 50–53. http://linkinghub.elsevier.com/retrieve/pii/0307441285900020 (accessed 03 December 2021).

Sharma, G.K., Jagetiya, S., and Dashora, R. (2015). *General Techniques of Plant Tissue Culture*, 28pp. Raleigh, NC: Lulu Press Inc.

Singh, G. (1999). Elicitation—manipulating and enhancing secondary metabolite production. In: *Plan Cell and Tissue Culture for the Production of Food Ingredients* (ed. T.J. Fu, G. Sing and W. Curtis), 121–128. New York: Kluwer Academic.

Soares, A.R., Marchiosi, R., Siqueira-Soares, R.d.C. et al. (2014). The role of L-DOPA in plants. *Plant Signal. Behav.* 9 (4): e28275.

Sogutmaz Ozdemir, B. and Budak, H. (2018). Application of tissue culture and transformation techniques in model species *Brachypodium distachyon*. In: *Brachypodium Genomics [Internet]*, Methods in Molecular Biology, vol. 1667 (ed. G. Sablok, H. Budak and P.J. Ralph), 289–310. New York: Springer New York http://link.springer.com/10.1007/978-1-4939-7278-4_18.

Sreekanth, D. (2009). Production, purification, and characterization of taxol and 10-DABIII from a new endophytic fungus *Gliocladium* sp. isolated from the Indian yew tree, *Taxus baccata*. *J. Microbiol. Biotechnol. [Internet]* 19 (11): 1342–1347. https://doi.org/10.4014/jmb.0904.4041.

Stoger, E., Sack, M., Fischer, R., and Christou, P. (2002). Plantibodies: applications, advantages and bottlenecks. *Curr. Opin. Biotechnol.* 13 (2): 161–166. https://linkinghub.elsevier.com/retrieve/pii/S0958166902003038 (accessed 12 January 2022).

Tabin, S., Kamili, A.N., and Gupta, R.C. (2018). Micropropagation and conservation of *Rheum webbianum* collected from Zanaskar valley via tissue culture. *Ann. Plant Sci.* 7 (4): 2187–2203.

Teli, N.P. and Timko, M.P. (2004). Recent developments in the use of transgenic plants for the production of human therapeutics and biopharmaceuticals. *Plant Cell Tissue Organ. Cult.* 79 (2): 125–145.

Tilahun, H., Negash, G., and Fesseha, H. (2019). An insight review on application of plantibodies. *Indian J. Pure Appl. Biosci. [Internet]* 7 (6): 29–41. http://www.ijpab.com/vol7-iss6a5.php (accessed 12 January 2022).

Tulecke, W. (1963). Plant tissue and organ culture. *Am. Biol. Teach.* 25 (2): 90–91.

Upadhyay, S.K. (ed.) (2021). *Genome Engineering for Crop Improvement*. John Wiley & Sons Ltd. https://doi.org/10.1002/9781119672425.

Upadhyay, S.K. and Singh, S.P. (eds.) (2021). *Bioprospecting of Plant Biodiversity for Industrial Molecules*. John Wiley & Sons Ltd. https://doi.org/10.1002/9781119718017.

Vasil, I.K. and Thorpe, T.A. (ed.) (1994). *Plant Cell and Tissue Culture [Internet]*. Dordrecht: Springer Netherlands http://link.springer.com/10.1007/978-94-017-2681-8 (accessed 22 January 2022).

Veeresham, C., Mamatha, R., Babu, C.P. et al. (2003). Production of taxol and its analogues from cell cultures of *Taxus wallichiana*. *Pharm. Biol.* 41 (6): 426–430.

Veng, C. (1998). Bioreactor technology for production of valuable algal products. *Chin. J. Oceanol. Limnol.* 16: 7.

Verma, N., Bansal, M.C., and Kumar, V. (2008). Protoplast fusion technology and its biotechnological applications. *Chem. Eng. Trans.* 8: 113–118.

Verma, P.C., Trivedi, I., Singh, H. et al. (2009). Efficient production of gossypol from hairy rootculture of cotton (*Gossypium hirsutum*). *Curr. Pharm. Biotechnol.* 10: 691–700. https://doi.org/10.2174/138920109789542048.

Warzecha, H. (2008). Biopharmaceuticals from plants: a multitude of options for posttranslational modifications. *Biotechnol. Genet. Eng. Rev.* 25 (1): 315–330. http://www.tandfonline.com/doi/abs/10.5661/bger-25-315 (accessed 19 November 2021).

Yamaguchi, S. and Kamiya, Y. (2000). Gibberellin biosynthesis: its regulation by endogenous and environmental signals. *Plant Cell Physiol.* 41: 251–257.

Yarra, R., Jin, L., Zhao, Z., and Cao, H. (2019). Progress in tissue culture and genetic transformation of oil palm: an overview. *Int. J. Mol. Sci.* 20 (21): 5353.

Zhang, R., Li, P., Xu, L. et al. (2009). Enhancement of diosgenin production in *Dioscorea zingiberensis* cell culture by oligosaccharide elicitor from its dndophytic fungus *Fusarium oxysporum* Dzf17. *Nat. Prod. Commun.* 4 (11): 1934578X0900401.

5

Microalgal Bioreactors for Pharmaceuticals Production

Rufin Marie Kouipou Toghueo

Antimicrobial & Biocontrol Agents Unit (AmBcAU), Laboratory for Phytobiochemistry and Medicinal Plants Studies, Department of Biochemistry, University of Yaoundé I, Yaoundé, Cameroon

5.1 Introduction

The production of biomolecules with a potential pharmaceutical application is a tedious and complex task where a slight variation in production conditions can significantly influence the quality, efficacy, safety, and yield of the target product (Kesik-Brodacka 2018). Plant-based biopharmaceuticals have attracted great interest in the past decades as an alternative to mammalian cell systems, susceptible to pathogens contamination (Lomonossoff and D'Aoust 2016; Dixit et al. 2021a). The significant increase in demand for new and more efficient biomolecules led pharmaceutical industries to look at marine organisms such as algae as potential bioreactors for producing biologically active molecules (Kumar et al. 2020; Malla et al. 2021; Upadhyay and Singh 2021). Indeed, algae offer an exceptional advantage over terrestrial plants as a rich source of various biomolecules that have a significant role in diseases prevention and treatment (Gomez-Zavaglia et al. 2019). They are photosynthetic organisms, considered the oldest members of the plant kingdom, dating back three billion years (Delwiche and Cooper 2015). Algae are distributed worldwide from the polar region to tropical areas and from nutrient-rich coastal seas to oligotrophic open oceans and are responsible for approximately 40–50% of the photosynthesis that occurs on Earth each year (Falkowski 1994). They grow in a variety of aquatic habitats (lakes, oceans, pounds, rivers, and wastewater) and can also withstand a wide range of conditions in either reservoirs or deserts including temperatures, salinities, pH values, and different light intensities. They can grow alone or in association (symbiosis) with other organisms (Barsanti et al. 2008). Algae are broadly classified by size as macroalgae or microalgae. Macroalgae are large-size multicellular organisms, while microalgae comprising a diverse group of prokaryotic and eukaryotic natures are single microscopic cells (Khan et al. 2018).

Microalgae can produce a variety of metabolites under various growth environments including heterotrophic, mixotrophic, and photoautotrophic conditions. They are rapidly becoming promising natural biofactories due to several key advantages, including their simple growth requirements, faster growth rate, higher biomass productivity, and ability to

Plants as Bioreactors for Industrial Molecules, First Edition.
Edited by Santosh Kumar Upadhyay and Sudhir P. Singh.

synthesize complex molecules (Kumar et al. 2020; Mutanda et al. 2020). Microalgae can utilize light energy and inorganic nutrients to synthesize numerous high-value natural biomolecules (Nur and Buma 2019). They also can adjust their biochemical composition depending on cultivation conditions. Microalgae can alter their biomass composition under stress conditions and accumulate not only lipids or carbohydrates, which could be used for biofuel production (Brennan and Owende 2010; Bhatia et al. 2021), but also specific high-value metabolites (such as pigments, proteins, vitamins, etc.) with significant importance in food, cosmetic, and pharmaceutical industries (Yan et al. 2016; Ahmed 2018). Biopharmaceuticals synthesized by microalgae include growth and blood-clotting factors, immune regulators, hormones, enzymes, monoclonal antibodies, viral vaccines, and other natural products, including anticancer, antimicrobial, antiviral, antidiabetic, antioxidant agents, to name but a few (Mutanda et al. 2020). Due to their outstanding biosynthetic ability, microalgae have been continuously exploited to produce biomolecules, and the microalgal biotechnology industry has significantly expanded in recent years. Microalgae are now well established as economical and practical bioreactors for obtaining high-value pharmaceutical compounds to improve human health (Walker et al. 2005; Xu et al. 2009; Yan et al. 2016; Khan et al. 2018). The present chapter aims to provide a brief overview of the potential of microalgae as promising bioreactors and the pharmaceutical potential of their bioproducts.

5.2 Microalgae Strains Selection

Microalgae are numerous. They are estimated between 200 000 and several million species (Norton et al. 1996). Their genetic and phenotypic diversity is evident in their growth, nutrient requirement, biomass production, metabolites production, and ubiquitous distribution in the biosphere (Singh and Saxena 2015). Like any other investigation aiming to search for bioactive natural products, the success of the production of the target biopharmaceutical depends mainly upon the characteristics of the microalgal strains. Therefore, the successful selection of the microalgae strains is of paramount importance (Figure 5.1). Thus, since the number of microalgae species in the world is so important, creative and carefully designed screening strategies must be used to quickly narrow the search for microalgae displaying desired characteristics needed for the effective mass production of the target bioactive compounds (Mutanda et al. 2011; Gifuni et al. 2018). The selection of suitable microalgae strain or strains can focus on several key parameters such as biomass and product productivity and the harvestability and recovery of the target products. Other intrinsic characteristics include their ability to resist and adapt to various conditions including a wide range of light intensity, nutrient loading, pH, salinity, and temperature (Kim et al. 2018; Mutanda et al. 2020).

The starting point for microalgae bioprospecting is the sourcing of microalgae strains. This can be achieved via direct purchasing from available repositories (Table 5.1) or by isolation from the natural environments. Microalgae strains from culture collection centers have the advantage of being fully characterized in term of their identity, growth, and nutritional requirements to facilitate their mass production. However, a serious drawback of using these strains is the scrupulous respect of their strict storage conditions because any deviation can lead to the inactivation or death of the cultures. They also have strict nutritional and environmental condition requirements and may not adapt quickly to new

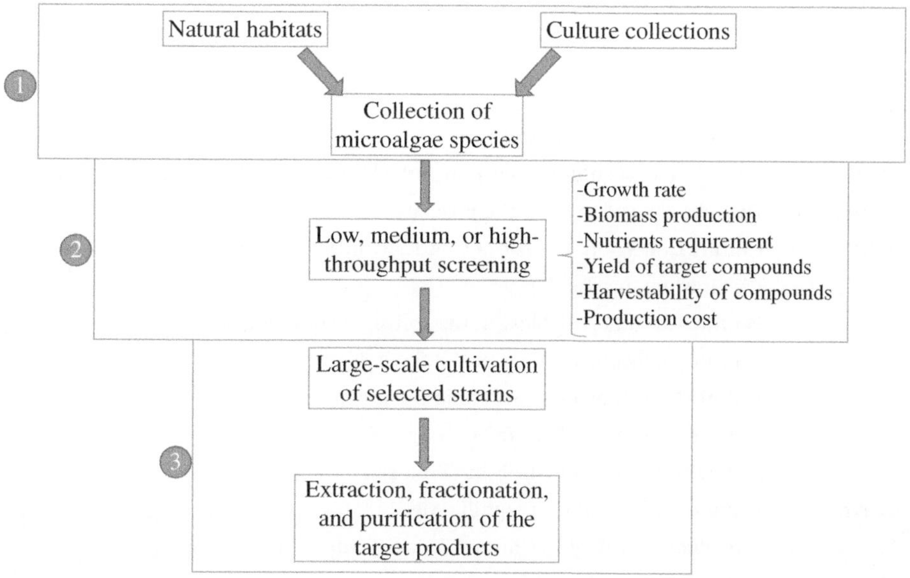

Figure 5.1 Successful selection of microalgae strains for biopharmaceuticals production. A collection of microalgae can be obtained from cultural collections or directly isolated from their natural environment (1). To select the best-performing microalgae strains (be it from the culture collection or the freshly isolated) needed for the mass production of the desired compound(s), a rigorous and carefully conducted screening against a set of parameters is necessary (2). Selected strain(s) are after that culture on a large scale, and the target product(s) is obtained after a series of extraction, fractionation, and purification (3).

environmental conditions (Mutanda et al. 2020). Conversely, indigenous microalgae isolates are highly adaptable to local environmental and climatic conditions and can constitute good candidates for bioprospecting studies. Indeed, the adaptability of strains to the local environment should be considered one of the critical factors when selecting appropriate strains for biomass production (Lee et al. 2014; Duong et al. 2015; Thao et al. 2017). However, the investigation of microalgae strains from natural habitats is also associated with challenges. For instance, defining the most appropriate growth conditions and the feasibility for the mass production of the desire metabolites by experimental techniques is cumbersome, labor intensive, and costly (Rawat et al. 2013). Regardless of the origin of microalgae strains, it is imperative to submit all strains to screening against several wells and carefully defined parameters to select the best-suited strain or strains with the required abilities to mass produce the target pharmaceutical metabolites.

5.3 Microalgae Cultivation

Over the past decades, the industrial cultivation of microalgae to produce high-value products has dramatically increased. Microalgae cultivation depends on the species. Several essential factors and the type of bioreactor systems used can affect the growth and productivity of microalgae species (Plaza et al. 2009).

Table 5.1 A selected list of cultures collections available worldwide.

N°	Acronyms	Full names
1	ACOI	Coimbra Collection of Algae
2	ALGOBANK	The microalgal culture collection at the University of Caen Basse-Normandie
3	BCCM/DCG	Diatoms Collection at Gent University
4	CAUP	Culture Collection of Algae at Charles University Prague
5	CC	Chlamydomonas Center
6	CCAC	Culture Collection of Algae at the University of Cologne
7	SAG	Culture Collection of Algae at Göttingen University
8	CCALA	Culture Collection of Autotrophic Organisms
9	CCAM	Culture Collection of Algae Marburg
10	CCAP	Culture Collection of Algae and Protozoa
11	CCCryo	Culture Collection of Cryophilic Algae
12	CPCC	Canadian Phycological Culture Centre (former UTCC)
13	CSIRO	CSIRO Collection of Living Microalgae
14	KU-MACC	Kobe University Macro-Algal Culture Collection
15	MZCH-SVCK	Microalgae and Zygnematophyceae Collection Hamburg
16	NCMA	Provasoli – Guillard National Center for Marine Algae and Microbiota (former CCMP)
17	NIES	Microbial Culture Collection at the National Institute for Environmental Studies
18	PCC	Pasteur Culture Collection of Cyanobacterial Strains
19	PLY	The Plymouth Culture Collection of Marine Microalgae
20	RCC	The Roscoff Culture Collection, Brittany, France
21	UTEX	The Culture Collection of Algae at The University of Texas

5.3.1 Factors Affecting the Growth and Productivity of Microalgae

Parameters that greatly influence the overall yield of microalgae biomass and their bioproducts include nutrient availability, temperature, pH, salinity, carbon, oxygen, light intensity, CO_2, and mixing (Richmond 2007; Khan et al. 2018).

5.3.1.1 Nutrients

Although different microalgae species may vary in their nutritional needs, all species have the exact basic nutrient requirements, including essential nutrients, carbon source (organic or inorganic), iron, nitrogen, and phosphorus (Juneja et al. 2013). The appropriate concentration of macronutrients such as nitrogen and phosphorus needed for the growth and production of the target metabolites may vary for different species of microalgae. Daliry et al. (2017) reported that *Chlorella vulgaris* exhibited optimal growth with nitrogen concentration at 0.5 g/l. The growth, morphological characteristics, and metabolites profiles of *P. ellipsoidea* were found to be significantly affected by the absence of nitrogen in the growth media (Ito et al. 2013). The production of Domoic acid by *Nitzschia pungens* forma *multiseries* was found to depend on the nitrogen content, preferably ammonium and light (Bates et al. 1991; Bates et al. 1993).

However, several studies also show that nitrogen limitation can boost the production of carbohydrates, lipids, and other active metabolites. For instance, Moore et al. (1986) reported that nutrient limitation, primarily nitrogen and phosphorus limitation, was necessary for producing high levels of acutiphycin in *Oscillatoria acutissima*. The same observation was later made by Rapala et al. (1993) to produce biotoxin by planktonic cyanobacteria. Converti et al. (2009) also found that a 75% decrease of the nitrogen concentration in the medium increased the lipid fractions of *N. oculata* (from 7.90 to 15.31%) and *C. vulgaris* (from 5.90 to 16.41%). *Spirulina maxima* increased the *Spirulina* lipids content by 3 times with the decrease of nitrogen content (de Macedo and Alegre 2001). Illman et al. (2000) also found that lipid content in all five *Chlorella* strains significantly increased with the decrease in nitrogen concentration in the growth medium.

In addition to macronutrients, several micronutrients such as B, Co, Fe, K, Mg, Mn, Mo, and Zn only required in minute quantity have a substantial impact on microalgae growth via their influence on many enzymatic activities in cells (Song et al. 2012; Aslam et al. 2021). Carbon is also crucial for growth and biomass production and can be added to the culture in organic (glycerol) or inorganic (CO_2) forms. Glucose is the most commonly used carbon source for heterotrophic cultures of microalgae. However, several other sugar molecules have also proven their effectiveness in improving microalgae's growth. For instance, from the screening of the effect of various mono and disaccharides on the growth of *C. zofingiensis*, glucose, fructose, mannose, and sucrose were identified as efficient for the rapid growth, and the effect was found to be concentration dependent. The higher concentration resulted in the higher cell density while lowering the specific growth rate due to the substrate inhibition (Venkata Mohan et al. 2015). In general, the use of environmental CO_2 as a carbon source for large-scale microalgae cultivation is cheaper and environmentally friendly as it adds to the benefit of CO_2 mitigation (Prathima Devi and Venkata Mohan 2012). Overall, the supply of the nutrients to the microalgae growth medium must strongly depend on the microalgae species and the target objective to be achieved.

5.3.1.2 Temperature

The growth temperature of microalgae species is also another critical parameter that has direct influences not only on the growth but also on several biochemical processes. An elevation or diminution of the temperature far beyond the optimal point can significantly influence the growth and metabolism of microalgae species. While cold temperatures affect photosynthesis by reducing carbon assimilation activity, high temperatures reduce cell size and respiration and limit photosynthesis via the inactivation of the photosynthetic proteins that lead to the decrease in growth rate (Salvucci and Crafts-Brandner 2004). Microalgae grow exponentially and produce high biomass when cultured at their optimal growth temperature, which usually varies between 20 and 24 °C. However, most microalgae can tolerate temperatures ranging from 16 to 35 °C (Bitog et al. 2011). The optimum growth temperature for *Nannochloropsis*, *Tetraselmis,* and *Isochrysis* was reported to be between 19–21, and 24–26 °C, respectively (Kunjapur and Eldridge 2010). In another investigation, *Euglena gracilis* was reported to grow optimally in the range of temperature between 27 and 31 °C (Kitaya et al. 2005). On the other hand, the temperature can also be manipulated to induce the production of valuable metabolites. According to Gromov et al. (1991), the antibiotic cyanobacterin LU1 was induced by growing *Nostoc linckia* at low temperatures. Converti et al.(2009) found that increasing the temperature from 20 to 25 °C practically magnified the lipid content (from 7.90 to 14.92%) of *N. oculata*, while changing the temperature from 25 to 30 °C significantly reduced the lipid content of *C. vulgaris* (from 14.71 to 5.90%). This influence of the temperature on the quality and quantity of metabolites produced by microalgae suggest that for each species, carefully conducted screenings for growth and productivity conditions are necessary to harness their biosynthetic capabilities.

5.3.1.3 pH, Salinity, and Pressure

Like other factors such as nutrients and temperature, each algal strain also has an optimal pH range for growth and biomass accumulation. Maintaining the pH of the growth medium in the optimal range is crucial because significant variation in this parameter has an impact on both the chemistry of the growth medium and the availability of many nutrients (CO_2, iron, and organic acids). Moreover, pH variation influences the electrical charge of the cell wall surface, the transport systems at the plasmalemma, and the membrane potentials (Suh and Lee 2003). Most microalgae are pH sensitive and grow well in the pH range of 6.0–8.76 (Daliry et al. 2017); nevertheless, it is possible to culture them in the pH range between 7 and 9 (Wang et al. 2012; Huang et al. 2017). Khalil et al.(2010) reported that the dry weight gain and the biochemical components of *Dunaliella bardawil* and *Chlorella ellipsoidea* were greatly enhanced at pH 7.5 and 9, respectively. Interestingly, the protein content of *C. ellipsoidea* recorded its highest value at pH 4, while the pigment content of the same alga was highest at pH 4, 6, and 7.5 and decreased at alkaline pH. Overall, the increasing pH often leads to the increase of salinity of the culture media, which is very harmful to algae cells (Juneja et al. 2013). To solve this issue, the standard practice in conventional bioreactors consists of supplementing the growth medium with chemical substances such as sodium bicarbonate to control the pH and keep it from rising too fast (Huang et al. 2017). High salinity affects the growth and the cell composition of microalgae. Like pH, every microalga also have their preferred salinity range. Therefore, to avoid substantial variation of salt concentrations in ponds, the easiest way for salinity control is by adding freshwater or salt as required (Mata et al. 2010). Along with pH and salinity, pressure is also another parameter that significantly impacts the solubility of gases essential for microalgae, and consequently, could indirectly effect the growth and productivity (Yegneswaran et al. 1990).

5.3.1.4 Light

Light intensity is one of the most important factors in microalgae cultivation to such a degree that the growth rate and biomass production can be directly predicted by the degree of culture exposure to light (Huesemann et al. 2013). It is well established that light intensities vary and are reduced in culture depth. Therefore, assuring the perfect availability of light to each algae cell in culture ponds is the fundamental principle in designing a photobioreactor (Borowitzka 1999; Kunjapur and Eldridge 2010). The light requirements for maximum growth and biomass accumulation vary from one microalgae species to another. The duration and intensity level of light is crucial because they have a direct effect on photosynthesis and influence the biomass yield and the biochemical composition of microalgae (Krzemińska et al. 2014). For instance, Amini Khoeyi et al. (2012) previously reported the significant influence of the intensity of light and photoperiod on the biomass and the fatty acid composition of *C. vulgaris*. A similar experiment also showed that various light cycles (night/day) had an essential influence on CO_2 fixation and biomass production of *Aphanothece microscopica nägeli* (Jacob-Lopes et al. 2009). Previous studies have shown that 16 hours of light and 8 hours of dark is optimal for microalgae growth. Overall, when the light intensity is below the optimal level required, photolimitation occurs and culture die. On the other hand, too much light can increase the photosynthetic rate to some maximum point, but the photoinhibition will occur to limit microalgae growth (Wang et al. 2012; Huang et al. 2017). Given the importance of light in microalgae cultivation, an appropriate cell density should be maintained to create a well-aerated culture, avoiding mutual shading between cells (Park and Lee 2001; Huang et al. 2017; Grubišić et al. 2019). The biofilm formation on the surface of photobioreactors is also a factor drastically reducing the light transmission in the culture ponds. Therefore, appropriate materials (glass, plexiglas, poly(vinyl chloride) (PVC), acrylic-PVC, and polyethylene) that can satisfy the transparency requirement and prevent the formation of biofilms are more often used for the construction of photobioreactors (Grubišić et al. 2019).

In addition to the duration and intensity, the light spectral quality is also vital for successful microalgae cultivation. The photosynthetically active radiation useful for microalgae growth range from 400 to 700 nm from light radiation, corresponding to 50% of solar radiation and intensity from 800 to 1000 W/m^2 (Zhang et al. 2015; Huang et al. 2017). The growth and carbohydrates contents of *M. aeruginosa* were found to be enhanced at 25 °C under a red LED light of 90 $\mu mol/m^2/s$ intensity (Khan et al. 2016). In contrast, *Dunaliella salina* was found to show the highest biomass and carotenoid productivity when 75% of red light (wavelength about 700 nm) and 25% of blue light (wavelength about 400 nm) are used together, compared with the only red light (Fu et al. 2013). At the same time, the photosynthetic photon flux of about 100 $\mu mol/m^2/s$ was reported to be the most efficient for biomass production by *E. gracilis* (Kitaya et al. 2005). For *C. vulgaris*, the optimum light intensity for the maximum growth rate and lipid production was reported to be in the range of 5000–7000 lx (Daliry et al. 2017). Another *Chlorella* sp. cultured in LED-lighted 800 L raceway ponds was found to grow optimally in the presence of constant light intensity of 1650 $\mu mol/m^2/s$ (Huesemann et al. 2013).

5.3.1.5 Mixing

The culture medium's effective stirring (mechanical mixing or air bubbles mixing) helps maintain cells in suspension and homogenizes their distribution, eliminates thermal separation, distributes nutrients, and increases gas exchange efficiency (Khan et al. 2018; Dolganyuk et al. 2020). Stirring the culture is also responsible for the possibility of CO_2 capture from the atmosphere and facilitates the transfer of biosynthesized O_2 from the liquid phase to the gaseous one, which stimulates photosynthesis in microalgae culture (Wang and Lan 2018). Mixing the culture appropriately also enables the penetration and uniform distribution of light inside the pond and prevent the biomass from settling and causing aggregation (Show et al. 2017). Mixing also prevents the accumulation of cells in low-oxygen area, which can cause anaerobic decomposition and downgrade the quality of the product (Suh and Lee 2003; Huang et al. 2017). Besides preventing cell sedimentation, stirring also limits the biofilm formation by preventing cells from attaching to the walls of the photobioreactor (Huang et al. 2017; Grubišić et al. 2019). The optimum level of stirring without causing cell death is strain-dependent and should be investigated to avoid the decline in productivity (de Barbosa 2003). For instance, the microalgae culture *Isochrysis galbana* mixed by bubbling carbon dioxide into the pond at a flow rate of 10 ml/min was found to produce a biomass two times higher than a system without stirring (Sánchez et al. 2013). *C. vulgaris* cultivated in a bubble column photobioreactor was also found to produce the optimal biomass at the aeration flow rate of 200 ml/min (Daliry et al. 2017).

5.3.2 Methods and Systems for Microalgae Cultivation

5.3.2.1 Methods

The culture of microalgae has over 50 years of history, and their multiple advantages have led to an increase in commercial and biotechnological interest. Microalgae can adapt to strong fluctuating conditions during biosynthesis and are capable of a metabolic shift to respond to the changes in their environment. Therefore, can be cultured by different methods (Table 5.2) including photoautotrophic, heterotrophic, mixotrophic, and photoheterotrophic (Mata et al. 2010; Chen et al. 2011; Zhan et al. 2017; Khan et al. 2018; Grubišić et al. 2019; Dolganyuk et al. 2020).

In photoautotrophic cultivation, microalgae produce chemical energy through photosynthetic reactions using light and inorganic carbon (e.g. carbon dioxide) as energy source and carbon source, respectively. This is the most commonly used method for the cultivation microalgae (Huang et al. 2010; Dolganyuk et al. 2020; Osama et al. 2021). With heterotrophic cultivation, microalgae can grow by utilizing only organic compounds such as glucose, acetate, fructose, sucrose, lactose, galactose, glycerol, and mannose as carbon and energy source. This method can increase biomass concentration and productivity, but the requirement for light is eliminated (Hu et al. 2018). However, this system relying on sugars as carbon source and often suffers from problems of contamination (Chen et al. 2011; Venkata Mohan et al. 2015). In mixotrophic cultivation, microalgae are grown under the most suitable autotrophic and heterotrophic conditions, combining the advantages of both cultivation methods. Both inorganic carbon to be fixed via photosynthesis and some organic carbon sources, such as glucose, glycerol, and acetate, are supplied to the culture. Microalgae in mixotrophic cultures grow faster and produce high biomass concentration and synthesize compounds using both autotrophic and heterotrophic pathways. This method offers the advantages of limited photo-inhibition and photo-oxidative damage. Moreover, the cost of light energy (in comparison with autotrophic culture) and organic compounds (compared with heterotrophic culture) is reduced (Chojnacka and Kinetic 2003; Cerón Garc et al. 2005; Zhan et al. 2017; Rincon et al. 2019). In photoheterotrophic cultivation, also known as photoganitrophy, photoassimilation, and photometabolism, microalgae require sugar and light simultaneously. Here, microalgae need light while using organic compounds as carbon source (Chojnacka and Kinetic 2003; Mata et al. 2010; Chen et al. 2011).

The photoheterotrophic and mixotrophic metabolisms are not well distinguished but can be defined according to a difference in the energy source required to perform growth and specific metabolite production (Mata et al. 2010). Overall, many microalgae can be grown

Table 5.2 The cultivation methods and their differences.

Conditions	Carbon source	Energy source	Light requirement	Metabolism	Bioreactor system	Advantage	Disadvantage
Heterotrophic	Organic carbon	Organic carbon	Not require	switch	Stirred tank	1) The growth rate high 2) The biomass and lipid accumulation is high 3) Bioreactors design with little constraints	1) Higher cost 2) Need for sterile media 3) Easy contamination 4) Production of CO_2 5) Diminish pigmentation and phytochemicals production
Mixotrophic	CO_2 and organic carbon	Light and organic carbon	Not obligatory	Simultaneous utilization	Closed photobioreactor	1) Higher growth rate 2) Higher biomass and lipid accumulation 3) Sustain of pigmentation and phytochemicals production 4) Less CO_2 production	1) Higher cost 2) Easy contamination 3) Low energy conversion efficiency
Photoautotrophic	CO_2	Light	Obligatory	No switch between source	Open and closed photobioreactor	1) Low cost 2) High production of pigmentation and phytochemicals	1) Low growth rate 2) Low biomass and lipid accumulation 3) Need special bioreactors 4) High dependence of weather conditions
Photoheterotrophic	Organic	Light	Obligatory	Switch between source	Closed photobioreactor	1) Higher biomass production	1) Higher cost 2) Easy contamination

under different conditions, while others prefer certain conditions over others. For instance, *Arthrospira (Spirulina) platensis* Gomont, *C. vulgaris*, and *Haematococcus pluvialis* Flotow can grow under photoautotrophic, heterotrophic, and mixotrophic conditions. At the same time, *Scenedesmus acutus* Meyen and *Selenastrum capricornutum* Printz are a few examples of microalgae strains that grow preferably in photoautotrophic, heterotrophic, or photoheterotrophic conditions (Chojnacka and Kinetic 2003; Mata et al. 2010; Grubišić et al. 2019).

5.3.2.2 Microalgae Cultivation Systems

One of the most critical parameters in microalgae cultivation is the type of bioreactor. Intrinsic properties of microalgae and the purpose of the culture are the main guiding parameters in the choice of the culture system. Moreover, the cost, availability of water, and the light can also be considered. Overall, microalgae can be grown in open-culture systems such as lakes or ponds and in highly controlled closed-culture systems called photobioreactors (Table 5.3).

Open pond systems are cost effective for large-scale microalgae cultivation, and the great majority of commercial cultivation of photosynthetic cells (around 90%) is done in open ponds (Płaczek et al. 2017). Artificial open ponds include circular controlled ponds, controlled raceway ponds, and unstirred open ponds, (Płaczek et al. 2017; Lafarga 2020). Sunlight is used as the only energy source (Kannan and Venkat 2019; Mutanda et al. 2020). Because optical absorption and self-shading by microalgal cells limit light penetration in deep, the ponds are usually kept shallow to increase the exposure of all cells to the light and

Table 5.3 Overview of some characteristic differences between open and closed culture systems.

	Open ponds system	Closed photobioreactors
Productivity	Low	High (3–5 times more productive)
Investment cost	1) Investment and operative cost is low 2) Require important space	1) Investment and operative cost is high 2) Require less space
Cultivation parameters	1) Sterility of the growth medium very difficult to achieve 2) High risk of contamination 3) Temperature control not achievable 4) Increase evaporation 5) Very poor stirring of the medium 6) Light utilization efficiency very low 7) Microalgae density very low	1) Sterility of the medium achievable 2) Reduce risk of contamination 3) Temperature control achievable 4) Evaporation of the medium reduce 5) Effective mixing of the medium 6) High Light utilization efficiency 7) Culture density very high
Cultivation process	The overall cultivation process is very difficult to control	The cultivation process is very easy to control
Biopharmaceuticals	Less suitable for the production of some high-value pharmaceutical compounds	Very suitable for the production of all high-value biopharmaceuticals

therefore increase their productivity (Chisti 2016). For instance, raceway ponds are usually designed with flat bottom and vertical walls; operate at water depths between 10 and 50 cm. In the case of open spherical circulating ponds, a rotating arm performs the agitation. Compared with closed systems, open-culture systems are less expensive to build and operate, more durable with a large production capacity, and require simple cleaning up procedures. Moreover, the direct exposure of the growth medium to the atmosphere facilitates liquid evaporation, therefore helping to regulate the bioprocess temperature (Jorquera et al. 2010; Grubišić et al. 2019). These systems have several disadvantages. There is little control of process parameters since the cultures are subjected to weather conditions. Only a limited number of microalgae strains can be cultivated this way. Other problems include the diffusion of CO_2 into the atmosphere and the tremendous evaporative losses. Due to the low productivity, these systems require a large land area, and the direct contact of cultures with the atmosphere makes them susceptible to contamination (by other microalgae or bacteria) and predators (Deruyck et al. 2019; Grubišić et al. 2019; Mutanda et al. 2020). Therefore, open-culture systems are not suitable for producing food ingredients and biopharmaceuticals (Vuppaladadiyam et al. 2018; Grubišić et al. 2019).

To overcome challenges associated with open ponds systems, much attention was invested in designing and developing closed photobioreactor systems (PRB). PBRs systems are flexible, and growth parameters (pH, temperature, mixing, CO_2 and O_2) can be configured for each microalgae species being cultivated according to their biological and physiological characteristics (Hu et al. 2018). These systems allow many species of interest that do not grow optimally in highly selective environments and cannot be grown in open ponds. PRBs systems allow continuous operation, offer higher biomass productivity and quality, higher photosynthesis efficiency, and reduce CO_2 losses. Liquid evaporation is minimized, and a contamination-free environment is assured for producing high-value biopharmaceuticals products (Acién et al. 2017; Dolganyuk et al. 2020). PBRs main limitations include overheating, bio-fouling, oxygen accumulation, difficulty in scaling up, the high cost of building, operating and algal biomass cultivation, cell damage by shear stress, and deterioration of material used for the photostage (Huang et al. 2017; Mutanda et al. 2020). Several intensive reviews are available and can be consulted for in-depth information on different microalgae cultivation systems (Xu et al. 2009; Grubišić et al. 2019; Osama et al. 2021).

5.4 Acquiring Biopharmaceuticals from Microalgae's

To access all high-value target compounds accumulated by microalgae after successful cultivation in photobioreactors, microalgae biomass processing involving a series of sequential steps is required.

5.4.1 Microalgae Harvesting

One of the foremost steps in processing microalgae is the biomass recovering process. This step is essential and costly since it accounts for about 20–30% of the total production cost due to high energy demand and capital cost (Barros et al. 2015). Given that several harvesting techniques are available, it is vital to select the adequate harvesting method taking into account the characteristics of the microalgae and type and the value of the target products (Barros et al. 2015; Singh and Patidar 2018). Numerous harvesting methods have been used to collect biomass, including biomass aggregation (flocculation and ultrasound), centrifugation, filtration, and flotation. In some cases, combinations of two or more techniques are used to increase the harvesting efficiency. Overall, all these harvesting methods aim to remove as much culture media as possible from microalgae biomass to facilitate the extraction and purification of high-value molecules (Barros et al. 2015; Singh and Patidar 2018; Tan et al. 2020). All these methods, their advantages and inconvenience were intensively discussed by Laamanen et al. (2016), Barros et al. (2015) and Singh and Patidar (2018).

5.4.1.1 Flocculation and Ultrasound

For the biomass aggregation using the flocculation technique, free-floating microalgae cells are aggregated together as single particle known as floc by adding multivalent cations and cationic polymers to the medium. These agents act by neutralizing the surface charge of cells to facilitate their assemblage (Muradov et al. 2015). Flocculating agents can be chemical and biological.

Chemical flocculants such as iron and aluminum salts are widely used in industry because of their availability and low cost. However, these chemicals are highly toxic and must be removed using different treatment processes, which add to the production cost (Bracharz et al. 2018; Tan et al. 2020). On the contrary, bioflocculants are much safer and ecofriendly when compared to their chemical counterparts. Most of the bioflocculants used are acrylic acid and chitosan, which are biopolymers existing naturally or produced artificially. Not only they are cheaper, but they also do not require further treatment before the following downstream processing of microalgae biomass (Nguyen et al. 2019; Pugazhendhi et al. 2019).

Another biomass aggregation method is the use of ultrasound. Here, the aggregation followed the sedimentation augmentation that facilitated microalgae biomass harvesting. This method is beneficial in maintaining the integrity of valuable products since it does not produce shear stress on the biomass (Bosma et al. 2003).

5.4.1.2 Centrifugation

The centrifugation process separates microalgae cells from the culture media using centrifugal force, separating each component based on their density and size. The harvesting of microalgae by centrifugation is rapid, easy, and efficient. However, due to the high energy input and maintenance required, this method can be costly (Najjar and Abu-Shamleh 2020). Another disadvantage of this technique is that high gravitational force might cause the internal damage of cells, causing a loss of sensitive nutrients, making it unfavorable for specific applications (Heasman et al. 2000). In industry, centrifugal systems such as decanters, disk stack centrifuges, imperforated basket centrifuges, hydro-cyclones, and perforated basket centrifuges have been used for microalgae harvesting (Singh and Patidar 2018).

5.4.1.3 Filtration

In the filtration process, a semipermeable membrane with various particle size ranges is used to retain microalgae on the membrane while allowing the liquid media to pass through. The varying pore size of the filter membrane enables the system to be adjusted to different species of microalgae (Najjar and Abu-Shamleh 2020). For instance, the conventional filtration method is used to harvest large size microalgae (>70 μm) such as *Coelastrum* and *Spirulina*, while microfiltration and ultrafiltration methods are used to harvest smaller size microalgae, equal to the size of the bacteria (Gultom and Hu 2013). This technique is also suitable for handling the more delicate microalgae species prone to damage due to shearing. Although efficient, this method is prone to fouling and clogging, require constant monitoring and replacement of filter membrane, which significantly increase the processing cost. Fortunately, an alternative filter membrane efficient (harvesting efficiency of 66–93%) and cheaper was developed to solve this particular problem (Tan et al. 2020).

5.4.1.4 Flotation

Flotation is often defined as "inverted" sedimentation, where tiny bubbles are used to promote the floatation of cells at the surface of the culture medium for easy harvesting without any addition of chemicals. This technique has a relatively high harvesting efficiency, easy operating procedure, high processing throughput, and low operational costs (Laamanen et al. 2016). Flotation is more effective and beneficial in microalgal harvesting than in sedimentation (Hanotu et al. 2012). Flotation is traditionally done through three main mechanisms. The air can be added through a diffuser (dispersed air flotation), through pressurization (dissolved air flotation), or by the use of electrodes (electrolytic flotation). While dissolved air flotation system generates air bubbles through saturating the culture with compressed air and then discharging the culture at atmospheric conditions, dispersed air flotation uses a sparger to generate air bubbles. On the other hand, the electroflotation method applies electrolysis operation to generate microbubbles from its electrode to trap free-floating microalgae. Overall, flotation is advantageous due to the easy operating procedure, biomass harvesting, and low operational efficiency (Alkarawi et al. 2018). One of the main downsides of flotation is the frequent replacement of electrodes due to fouling and the extremely high-energy consumption.

5.4.2 Biomass Dehydratation

Microalgae biomass is immediately processed following the complete separation from the growth medium to extend its shelf life and prevent spoilage. In industry, drying microalgae biomass often poses a significant economic constraint in that it may constitute between 70 and 75% of the processing cost (Show et al. 2013). The different drying or dehydration process types, including drum drying, freeze-drying, spray drying, and sun drying, differed in the extent of capital investment and energy requirements. The drying method's selection depends not only on the scale of operation but also on the desired final products and the intended usage (Show et al. 2013; Balasubramaniam et al. 2021).

Freeze-drying is a natural dehydration process of frozen products using the mechanism of sublimation. This method is too expensive for large-scale commercial recovery of microalgal products; therefore, it has been widely used only for drying microalgae in research laboratories (Molina Grima et al. 2003). The drying of biomass using solar energy (sun drying) is the cheapest method available compared to the other techniques. However, this technique has several limitations regarding weather conditions, long drying period, and the large drying area needed. Moreover, the process is difficult to control, and problems such as overheating, change of texture, color, and taste of the microalgae may occur (Show et al. 2013). Spray drying is the method of choice for high-value products (>\$1000 ton^{-1}). With this method, the operation is continued and produced a good quality product. However, some algal components such as pigments can be significantly deteriorated and is it also very expensive (Molina Grima et al. 2003; Show et al. 2013). Of note, although the extraction of many bioactive metabolites from microalgae involves drying, the extraction of some proteins requires the biomass that has not been previously dried (Bermejo Román et al. 2002).

5.4.3 Cell Disruption for Bioproducts Extraction

Microalgae are organisms composed of many very important biomolecules, including carbohydrates, lipids, proteins, minerals, and many other compounds with various applications. Therefore, to recover these targeted metabolites, microalgae cell walls need to be pretreated to enable all the desired components to be extracted. Therefore, many cell disruption techniques can be employed to release all metabolites of interest from microalgae cells (Lafarga 2020). Mechanical disruption (e.g. cell homogenizers, bead mills, ultrasounds, autoclave, and spray drying) or nonmechanical disruption (e.g. freezing, organic solvents and osmotic shock and acid, base, and enzyme reactions) techniques can be used depending on the microalgae characteristics and the nature of the desired final products (Mata et al. 2010; Lafarga 2020).

In general, because their effectiveness is independent of the microalgae species, mechanical techniques are often employed to disrupt microalgae cells (Lee et al. 2012). However, less aggressive than mechanical methods, nonmechanical disruption methods are often employed to extract compounds susceptible to degradation under the harsh conditions of mechanical disruption (Günerken et al. 2015). In the case of proteins, for instance, although the protein recovery strategies employed highly depend on the quality of the final product and the starting material (wet or dry biomass of microalgae), the process always involves several steps going from cell disruption to concentration (Grossmann et al. 2020). Nowadays, countless proteins have been purified using various extraction and purification strategies (Safi et al. 2014; Safi et al. 2017; Grossmann et al. 2018; Buchmann et al. 2019; Grossmann et al. 2020). However, the first report on protein extraction from microalgae *Scenedesmus obliguus* was achieved using a simple isoelectric point precipitation method (Hedenskog et al. 1970). By contrast, the manothermosonication method, combining pressure, temperature and ultrasound, was recently proposed as an alternative to the simple ultrasonic treatment. According to the authors, combining these three elements increases the effectiveness of cells disruption and intensifies the material transfer process, leading to an increase of protein extraction yield by 229% (Vernès et al. 2019).

Similarly, numerous strategies based on various cell disruption mechanisms (Mechanical and nonmechanical) have been devised to extract other bioactive compounds from microalgae biomass. Taking the example of the astaxanthin recovery, different strategies were investigated, and the most efficient extraction method was by combining autoclaving microalgae biomass with mechanical disruption (Richmond 2004; Richmond 2007). Another investigation obtained the efficient recovery of pigments from *Nannochloropsis oceanica* by combining pulsed electric fields (PEF) pretreatment with the supercritical CO_2 extraction method (Gianpiero et al. 2019). Similarly, another investigation led to identifying the best extraction condition for lutein and β-carotene from the microalga *Desmodesmus* sp. It was found that the best solvent extractor was hexane: ethanol mixture in a 1 : 1 (v/v) proportion under ultrasound treatment (Soares et al. 2016).

5.5 Microalgal Compounds and their Pharmaceutical Applications

Microalgae have grown in importance as rich sources of structurally diverse and novel bioactive compounds. These bioproducts belong to several structural classes including polyketide synthase (PKS), nonribosomal polypeptide synthetase (NRPS), and the hybrid PKS-NRPS. These high-value compounds (Figure 5.2) can be carotenoids, polysaccharides, proteins, polyunsaturated fatty acids, vitamins, etc. (Morales-Sánchez et al. 2017; Dolganyuk et al. 2020; Balasubramaniam et al. 2021; Khavari et al. 2021).

5.5.1 Carotenoids

Microalgae produce an abundance of high-value carotenoids (Table 5.4). Depending on the microalgae species, they can be stored in oil droplets, in the stroma of the chloroplast, or

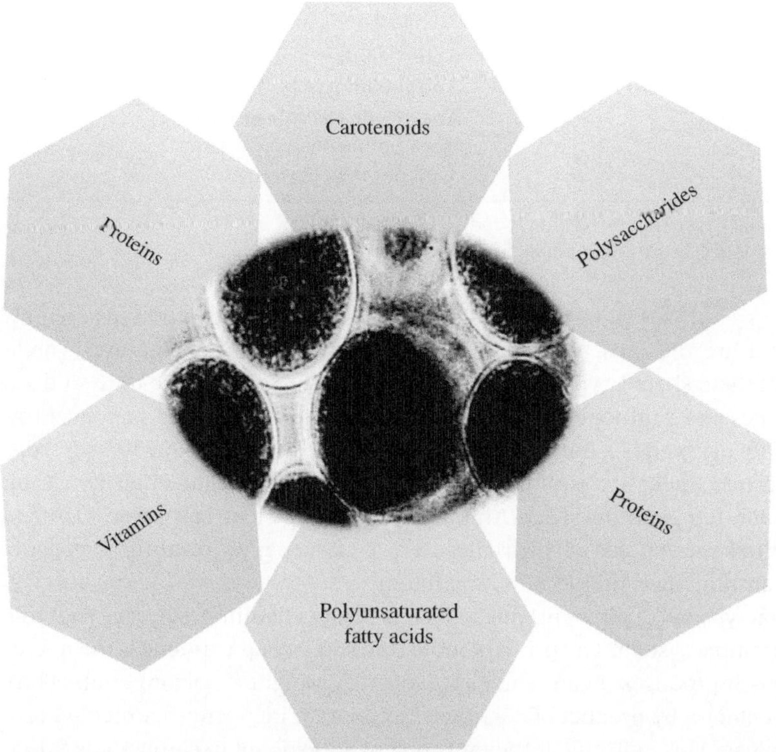

Figure 5.2 Some high-value compounds produced by microalgae.

Table 5.4 Examples of some high-value carotenoids and their biological properties.

Compounds	Market price (US$)[a]	Producing organism	Biological activity	References
Astaxanthin	2500/kg	*H. pluvialis*, *C. zofigiensis*, *C. vulgaris*	Antitumoralantioxidantanti-inflammatory	Lorenz and Cysewski (2000)
β-Carotene	300–3000/kg	*D. salina*	Antioxidant Anti-inflammatory	Lorenz and Cysewski (2000)
Fucoxanthin	70–150/kg	*P. tricornutum*	Antioxidant Anti-inflammatory Anticancer	Mikami and Hosokawa (2013) and Talero et al. (2015)
Lutein	30–$220/kg	*D. salina*, *C. pyrenoidosa*, *C. protothecoids* *Muriellopsis* sp.	AMD, Atheroscelerosis, Retinal neural damage	Talero et al. (2015) and Manayi et al. (2016)
Violaxanthin	NF	*D. tertiolecta*, *C. ellipsoidea*	Anti-inflammatory, Anticancer	Talero et al. (2015)
Zeaxanthin	NF	*D. salina*, *P. cruentum*, *C. protothecoids*	Anitoxidant, Cataracts, Anti-inflammatory	Talero et al. (2015)
Phycobiliproteins	3.0–25.0/mg	*Spirulina* spp	Antioxidant, Anti-inflammatory, Anticancer, Neuroprotective Hepatoprotective Antiviral, Antibacterial,	Wu et al. (2016) and Hu (2019)

AMD, age-related macular degeneration; NF, not found.
[a] The market price mentioned in the present table fluctuate.

the cytosol (Mulders et al. 2013). Carotenoids are lipophilic colored compounds (yellow, orange, or red) and are the most widespread and diverse pigments in nature (Varela et al. 2015). Most of them share a typical C40 backbone structure of isoprene units and are divided into carotenes and xanthophylls. These molecules are often present in concentrations (below 0.5% g/g dry weight), but some *Chlorophyta* can produce about 10% μg^{-1} dry weight when cultivated under stressful conditions, including high light intensity (Kang et al. 2007), extreme temperatures (Ben-Amotz 1996), or nutrient starvation (Lamers et al. 2012). Essential carotenoids of microalgae are β-carotene, astaxanthin, lycopene, zeaxanthin, violaxanthin, phycobiliproteins, and lutein.

β-carotene (orange-yellowish pigment) due to its relation to vitamin A play a crucial role in vision and the immune system. The primary industrial source of β-carotene is the microalgae *D. salina*, which produces β-carotene above 14% dry weight (Spolaore et al. 2006). Astaxanthin (rosy color), a by-product of β-carotene, is another high-value carotenoid produced from microalgae. This molecule is the most potent antioxidant in carotenoids exhibiting several fold more robust antioxidant activity than vitamin E and β-carotene. Among microalgae strains, the higher producer is *Haematococcus pluvialis*, accumulating up to 1.5–3% per dry weight of astaxanthin. This compound has achieved commercial success and sells for about US$2500 kg^{-1} with a worldwide market estimated at US$200 million each year (Lorenz and Cysewski 2000).

Lutein and zeaxanthin are also becoming increasingly important. Lutein is clinically proven to prevent cataracts and macular degeneration (Manayi et al. 2016), the progression of early atherosclerosis (Dwyer et al. 2001). Among microalgae species, *Muriellopsis* sp. and other *chlorophycean* species can accumulate a high amount of lutein (Del Campo 2000). Phycobiliproteins, classified as phycocyanobilin (blue), phycoerythrobilin (red), phycouroblin (yellow), and phycobiviolin (purple), are used as natural colorants, fluorescent agents. They are used as antioxidant, anti-inflammatory, neuroprotective, and hepatoprotective agents (Sekar and Chandramohan 2008). Due to their potent antioxidant properties in mediating free radicals' harmful effects, carotenoids can potentially protect humans from premature ageing, cardiovascular diseases, arthritis, immune disorders, diabetes, cataract, macular degeneration, and neurodegeneration various cancers. Their multiple health benefits and pharmaceuticals application are summarized in many reviews for more details (Del Campo et al. 2007; Varela et al. 2015; D'Alessandro and Antoniosi Filho 2016; Gong and Bassi 2016).

Several investigations using genetic engineering to direct microalgae species toward desired quantity and quality of specific metabolites have also been reported. For instance, introducing an additional copy of the gene encoding gateway enzyme of the terpenoid pathway, 1-deoxy-D-xylulose 5-phosphate synthase (dxs) in *P. tricornutum* was found to significantly enhance the accumulation of fucoxanthin (24.2 mg/g dry weight)(Eilers et al. 2016). In another investigation, knocking out the zeaxanthin epoxidase by preassembling DNA-free CRISPR-Cas9 ribonucleoproteins in the *C. reinhardtii* CC-4349 strain significantly increase the zeaxanthin content (56-fold) and productivity (47-fold) (Baek et al. 2018). Previously, the overexpression of *Haematococus pluvialis* gene encoding β-carotene ketolase (4,4′β-oxygenase) in *D. salina* was found to induce the production of astaxanthin (3.5 μg/g DW) and canthaxanthin (1.9 μg/g DW) (Anila et al. 2016). The endogenous phytoene desaturase coding sequence (*pds*) optimized and overexpressed in the chloroplast of *H. pluvialis* under the control of the *psbA* promoter/UTR led to the accumulation of astaxanthin over 67% higher than reported in the wild-type strain(Galarza et al. 2018). Overall, this genetic engineering of potent strains could be one of the most viable options to improve the production of targeted compounds by microalgae while improving the production cost in the process.

5.5.2 Polyunsaturated Fatty Acids

Polyunsaturated fatty acids (PUFAs) are essential in tissue integrity and have beneficial health effects. Although the traditional source of PUFAs is mainly from fish oil, microalgae are the primary producers in the oceanic environment (Wang et al. 2015). PUFAs with the most important being eicosapentaenoic acid (EPA; 20 : 5 $n-3$) and docosahexaenoic acid (DHA; 22 : 6 $n-3$) have recently gained significant interest in the pharmaceutical industry as their therapeutic benefits include prevention of inflammatory diseases, heart problems, arthritis, asthma, headache, benefits in brain development, and the preservation of mental health (Ruxton et al. 2004).

Microalgae are now recognized as the superior and sustainable route for producing PUFAs as an alternative to extraction from fish and herbs (Doughman et al. 2007). Many microalgae species have been reported to have these valuable fatty acids. The main microalgal EPA-producers strains have been reported to be *Nannochloropsis* sp. (Chini Zittelli et al. 1999) and *Phaeodactylum tricornutum* (Belarbi et al. 2000). For docosahexaenoic acid (DHA), the leading producer strains were reported to be *C. cohnii*, *Schizochytrium*, or *Pavlova lutheri* (Winwood 2013).

Growth conditions greatly influence the production of these omega-3 polyunsaturated fatty acids. Conditions such as nitrogen (Mujtaba et al. 2012) and phosphate starvation (Khozingoldberg and Cohen 2006), limited illumination (Forján et al. 2011), and low temperatures (Jiang and Gao 2004) are known to induce lipid accumulation – for instance, the screening of optimum conditions for the production of DHA by the strain *Schizochytrium* sp. S31 reveals a maximum DHA production (0.516 g/l) in conditions such as 27.98 g glucose/l, 4.52 g yeast extraction/l, 24.82 g sodium chloride/l, pH 6.96 incubation for four days at 30 °C (Wu and Lin 2003). In another investigation, the aeration of microalgae culture was shown to influence the production of fatty acids. Ren et al. found that varying aeration rates in the culture medium of *Schizochytrium* sp. led to high cell density (71 g/l), high lipid content (35.75 g/l), and high DHA percentage (48.95%). Moreover, the DHA productivity reached 119 mg/l h, 11.21% higher than that produced in culture with a constant aeration rate (Ren et al. 2010). The combination of the appropriate inoculum size and the carbon: nitrogen ratio (C:N 55 : 1) was also reported as a suitable strategy to obtain high DHA yields from the culture of *Aurantiochytrium limacinum* SR21 (Rosa et al. 2010).

The addition of chemical elicitors to the growth medium has also yielded interesting results. The addition of kinetin and gibberellic acid (GA3) to the culture of the microalgae *Nannochloropsis oceanica* CASA CC201 was found to increase the amount of omega 3 fatty acid fourfold and twofold in comparison with the control. Moreover, the combination of kinetin and GA3 exhibited a synergistic effect on the total lipid yield of 246.25 mg/l compared with the control (121.5 mg/l) (Udayan et al. 2018). Following their investigation, the exogenous application of methyl jasmonate (MeJA) and salicylic acid (SA) on the growth and production of fatty acid by *Nannochloropsis oceanica CASA CC201* was investigated. The authors found that higher doses of MeJA promoted microalgal growth and monounsaturated fatty acid production, mainly oleic acid (C18 : 1) at the early stationary growth phase, while the treatment with SA (40 ppm) induces essential omega 3 fatty acid production (EPA, C20 : 5)(Udayan et al. 2020).

The metabolic engineering of microalgae PUFA-producing strains has been considered as the most promising strategy for the sustainable production of PUFAs. For instance, using UV-light as mutagenic agent to induce the mutation of *Pavlova lutheri*, Meireles et al. increase EPA and DHA contents by 32.8 and 32.9% (of DW) higher than those of the native strain (Meireles et al. 2003). The twofold overexpression of an endogenous thioesterase in *P. tricornutum* was found to increase the total fatty acid contents by 72% (Gong et al. 2011). The metabolic engineering of *P. tricornutum* by expressing the heterologous elongase resulted in an eightfold increase in DHA content in the transgenic *P. tricornutum strain* (Hamilton et al. 2014). For more information on PUFA produced by microalgae, please consult reviews by Chauton et al. (2015) and Koller et al. (2014).

5.5.3 Polysaccharides, Vitamins, and Minerals

Complex polysaccharides compose the significant structural features of the microalgal cell wall. They play significant roles in various functions in the cells of different organisms and serve mainly as storage and structural molecules in microalgae (Arad and Levy-Ontman 2010). These macromolecules are produced in various forms according to the group of microalgae. For instance, diatoms (*Bacillariophycae, Heterokontophyta*) produce chrysolaminarin, a linear polymer of β(1,3) and β(1,6) linked glucose units (Gügi et al. 2015), and *Chlorophyta* synthesizes starch in the form of two glucose polymers, amylopectin and amylose (Busi et al. 2014). Polysaccharides have been reported to exhibit pharmacological properties such as anticancer (Kurd and Samavati 2015), antioxidant (Yu et al. 2019), antiviral (Huleihel et al. 2001), and immunomodulatory (Parages et al. 2012) activities.

In addition to their polysaccharides content, microalgae also constitute important source vitamins (A, B1, B2, B6, B12, biotin, C, E, folic acid, pantothenic acid, etc.) and various essential minerals (potassium, sodium, magnesium, iron, and calcium, etc.). The type and amount of each vitamin depend on the microalgae species and the growth conditions. *Spirulina* was reported as an excellent source of vitamin A (Annapurna et al. 1991) and was recently reported as a potential vegetarian source of bioavailable vitamin B12 (Madhubalaji et al. 2019). A Moroccan *Spirulina* (*Arthrospira platensis*) was reported to contain a large amount essential minerals (20.91 ± 0.88%) in addition to protein (76.65 ± 0.15%), carbohydrates (6.46 ± 0.32%), crude fiber (4.07 ± 1.42%), lipids (2.45 ± 0.82%), and ash (14.56 ± 0.74)(Seghiri et al. 2019). Methylcobalamin, a vitamin B12, was identified in *C. vulgaris* (Kumudha et al. 2015). *Spirulina platensis* (120.5 μg/g), *I. galbana* (115.5 μg/g), *Porphyridium cruentum* (184.7 μg/g), and *Tetraselmis suecica* (159.8 μg/g) were reported to produce a significant amount of vitamin E with antioxidant activity (López-Hernández et al. 2020). The diatom *Skeletonema marinoi* cultivation was reported to produce Vitamin C. Moreover, the light manipulation regimes significantly improved the yield (Smerilli et al. 2017). *Spirulina, Isocrysis, Nannochloropsis, Schizochytrium, Haematococcus, Chlorella,* and *Dunaliella* have also been reported to produce high amounts (4.0–471.61% dwt) of vitamins B1, B3, B6, B12, and E (Dineshbabu et al. 2019). Microalgae are also rich in minerals like calcium, phosphorous, magnesium, potassium, sodium, zinc, iron, copper, and sulfur, and their content has been found to be between 2.2 and 4.8% of total dry weight (Dineshbabu et al. 2019).

5.5.4 Proteins

In microalgae, proteins are present in different cell parts like cytoplasm, organelles, plastids, cell walls, and nucleus (Safi et al. 2014; Safi et al. 2017). The protein content significantly varies from one species to another. For example, *Haematococcus pluvialis* (Ba et al. 2016) and *Porphyridium cruentum* (Becker 2007) show a protein content of 26 and 28–39% on a dry weight basis, while microalgae like *Arthrospira* (El-Kassas et al. 2015), *Chlorella* (Liu and Hu 2013), and *Synechococcus* sp.(Becker 2007) can show high contents of total proteins (60–70% on a dry weight basis). Several growth parameters significantly influence the protein yield in microalgae farms, including the carbon supply and source, climate condition, light quality and intensity, nutrients, temperature, and halostress, with the most critical parameter for high-protein production being nitrogen (Soares et al. 2018).

Proteins from microalgae are valuable in the food industry for their high nutritional quality and techno-functional properties. Indeed, microalgal protein content has been suggested to be comparable to other sources such as soy, milk, and animal meat. For example, *Arthorspira maxima*, *C. vulgaris*, *D. salina*, *H. pluvialis*, *Scenedesmus obliquus*, and *Synechococcus* sp. produce a very high content of proteins and similar energetic values in comparison to soybean, corn, and wheat (Lum et al. 2013; Amorim et al. 2021). The microalga *Spirulina* (*A. Platensis*) and *C. vulgaris*, with their protein content between 51 and 63% dry matter (Becker 2007), are promising food ingredients that have been cultivated and marketed in the form of tablets and pills as protein supplements for human consumption (Grahl et al. 2018). Nevertheless, the commercial expansion of microalgae as protein supplements for human consumption has been obstructed because many microalgae cellulosic cell walls are not digested by human enzymes (Grahl et al. 2018; Fasolin et al. 2019).

On the other hand, bioactive peptides obtained from microalgae protein have been reported as functional foods for promoting cardiovascular health (Ejike et al. 2017). Their consumption was associated with many beneficial effects such as antiatherosclerosis and antihypertensive, antioxidant, anticancer, hepatoprotective, and immunosuppressive activities (Bleakley and Hayes 2017). Two cyclic depsipeptides molecules isolated from *Microcystic ichthyoblabe* were also reported for their antiviral activity against influenza A virus (Zainuddin et al. 2007).

Microalgae have also been proposed as a potential next-generation platform as the expression system for various industrial, therapeutic, and diagnostic recombinant proteins. They combine all benefits of eukaryotic expression systems in addition to their high growth rates. Since the first demonstration of the efficient expression of a unique large single-chain (lsc) antibody in the green alga, *Chlamydomonas reinhardtii* (Mayfield et al. 2003), several investigations have reported successful production of several monoclonal antibodies and vaccines in microalgae. Barrera et al. reported the production of camelid VHH antitoxins, capable of neutralizing botulinum neurotoxin in *C. reinhardtii* (Barrera et al. 2015). The antibody 83K7C, a full-length IgG1 human monoclonal antibody (mAb), derived from a human IgG1 directed against anthrax protective antigen 83 (PA83) (Tran et al. 2009), a small monomeric and larger dimeric immunotoxins (Tran et al. 2013), a vascular endothelial growth factor (VEGF), and high mobility group protein B1 (HMGB1) (Rasala et al. 2010), and E7GGG protein, a mutated, attenuated form of the E7 oncoprotein of the human papillomavirus (HPV) (Demurtas et al. 2013) have also been produced in *C. reinhardtii*.

Phaeodactylum tricornutum has also emerged as a suitable host for producing recombinant proteins. Using this microalga, Hempel et al. demonstrate the expression of a monoclonal human IgG antibody against the hepatitis B surface protein (Hempel et al. 2011). A human antihepatitis B monoclonal antibody was also produced in *P. tricornutum* (Vanier et al. 2015). Hempel et al. also engineered *P. tricornutum* to produce a monoclonal IgG antibody against the nucleoprotein of Marburg virus, the principal causative agent of hemorrhagic fever in western Africa (Hempel et al. 2017).

Besides antibodies and vaccines, many other recombinant proteins have been successfully expressed in microalgae, including luciferase (Mayfield and Schultz 2004; Matsuo et al. 2006), green fluorescent protein (GFP) (Franklin et al. 2002), and β-glucuronidase(Ishikura et al. 1999). We also have important therapeutic molecules such as Kunitz trypsin inhibitor of soybean (SKTI) used against inflammation (Chai et al. 2013), the mammary-associated serum amyloid (M-SAA) (Manuell et al. 2007), and the tumour necrosis factor-related apoptosis-inducing ligand (TRAIL) a therapeutic protein that induces selective apoptosis in various tumor cells and virus-infected cells (Yang et al. 2006).

5.6 Conclusions

Microalgae are gifted natural bioreactors that can be successfully harnessed to produce many valuable compounds to improve the health and well-being of humanity. The successful exploitation of microalgae for biopharmaceutical production requires several key strategies, including a robust and careful selection of the best suitable microalgae candidates, and the implementation of effective cultivation, harvesting, drying, and cell disruption methods. Owing to their multiple advantages, microalgae have gained tremendous attention over the past decades and because of their abundant biodiversity, their potential for new products, derived products and applications is important. Indeed, microalgae are potent producers of a great and complex diversity of molecules accessible from their primary and secondary metabolism, including carotenoids, proteins, polysaccharides, PUFAs, etc. with potential applications in biomedical and pharmaceutic industries. Their biologically active substances exhibit various activities, including antioxidant, anticancer, anti-inflammatory, antibacterial, antiviral, antitumor, regenerative, antihypertensive, neuroprotective, immunostimulating, etc. However, several challenges, including high production costs, limit the large-scale production of biomolecules from microalgae. Another challenge in downstream processing of microalgal biomass is the lack of low-cost, more efficient, and highly sensitive methods for extracting compounds. Therefore, further investigation into microalgal composition and disruption methods is still needed to develop large-scale and energy-efficient production of bioactive compounds from microalgae.

References

Acién, F.G., Molina, E., Fernández-Sevilla, J.M. et al. (2017). Economics of microalgae production. In: *Microalgae-Based Biofuels and Bioproducts*, 485–503. Elsevier ISBN 978-0-08-101023-5.

Ahmed, I. (2018). Microalgae as a source of high-value bioactive compounds. *Frontiers in Bioscience* 10: 197–216. https://doi.org/10.2741/s509.

Alkarawi, M.A.S., Caldwell, G.S., and Lee, J.G.M. (2018). Continuous harvesting of microalgae biomass using foam flotation. *Algal Research* 36: 125–138. https://doi.org/10.1016/j.algal.2018.10.018.

Amini Khoeyi, Z., Seyfabadi, J., and Ramezanpour, Z. (2012). Effect of light intensity and photoperiod on biomass and fatty acid composition of the microalgae, *Chlorella vulgaris*. *Aquaculture International* 20: 41–49. https://doi.org/10.1007/s10499-011-9440-1.

Amorim, M.L., Soares, J., Coimbra, J.S.d.R. et al. (2021). Microalgae proteins: production, separation, isolation, quantification, and application in food and feed. *Critical Reviews in Food Science and Nutrition* 61: 1976–2002. https://doi.org/10.1080/10408398.2020.1768046.

Anila, N., Simon, D.P., Chandrashekar, A. et al. (2016). Metabolic engineering of *Dunaliella salina* for production of ketocarotenoids. *Photosynthesis Research* 127: 321–333. https://doi.org/10.1007/s11120-015-0188-8.

Annapurna, V.V., Deosthale, Y.G., and Bamji, M.S. (1991). Spirulina as a source of vitamin A. *Plant Food for Human Nutrition* 41: 125–134. https://doi.org/10.1007/BF02194081.

Arad, S.(.M.). and Levy-Ontman, O. (2010). Red microalgal cell-wall polysaccharides: biotechnological aspects. *Current Opinion in Biotechnology* 21: 358–364. https://doi.org/10.1016/j.copbio.2010.02.008.

Aslam, A., Rasul, S., Bahadar, A. et al. (2021). Effect of micronutrient and hormone on microalgae growth assessment for biofuel feedstock. *Sustainability* 13: 5035. https://doi.org/10.3390/su13095035.

Ba, F., Ursu, A.V., Laroche, C., and Djelveh, G. (2016). *Haematococcus Pluvialis* soluble proteins: extraction, characterization, concentration/fractionation and emulsifying properties. *Bioresource Technology* 200: 147–152. https://doi.org/10.1016/j.biortech.2015.10.012.

Baek, K., Yu, J., Jeong, J. et al. (2018). Photoautotrophic production of macular pigment in a *Chlamydomonas reinhardtii* strain generated by using DNA-free CRISPR-Cas9 RNP-mediated mutagenesis. *Biotechnology and Bioengineering* 115: 719–728. https://doi.org/10.1002/bit.26499.

Balasubramaniam, V., Gunasegavan, R.D.-N., Mustar, S. et al. (2021). Isolation of industrial important bioactive compounds from microalgae. *Molecules* 26: 943. https://doi.org/10.3390/molecules26040943.

de Barbosa, M.J.G., V. (2003). *Microalgal Photobioreactors: Scale-up and Optimisation.* Wageningen Univ ISBN 978-90-5808-898-7.

Barrera, D.J., Rosenberg, J.N., Chiu, J.G. et al. (2015). Algal chloroplast produced camelid V_H H antitoxins are capable of neutralizing botulinum neurotoxin. *Plant Biotechnology Journal* 13: 117–124. https://doi.org/10.1111/pbi.12244.

Barros, A.I., Gonçalves, A.L., Simões, M., and Pires, J.C.M. (2015). Harvesting techniques applied to microalgae: a review. *Renewable and Sustainable Energy Reviews* 41: 1489–1500. https://doi.org/10.1016/j.rser.2014.09.037.

Barsanti, L., Coltelli, P., Evangelista, V. et al. (2008). Oddities and curiosities in the Algal World. In: *Algal Toxins: Nature, Occurrence, Effect and Detection*, NATO Science for Peace and Security Series A: Chemistry and Biology (ed. V. Evangelista, L. Barsanti, A.M. Frassanito, et al.), 353–391. Dordrecht: Springer Netherlands ISBN 978-1-4020-8479-9.

Bates, S.S., de Freitas, A.S.W., Milley, J.E. et al. (1991). Controls on domoic acid production by the diatom *Nitzschia pungens* f. *Multiseries* in culture: nutrients and irradïance. *Canadian Journal of Fisheries and Aquatic Sciences* 48: 1136–1144. https://doi.org/10.1139/f91-137.

Bates, S.S., Worms, J., and Smith, J.C. (1993). Effects of ammonium and nitrate on growth and domoic acid production by *Nitzschia pungens* in batch culture. *Canadian Journal of Fisheries and Aquatic Sciences* 50: 1248–1254. https://doi.org/10.1139/f93-141.

Becker, E.W. (2007). Micro-algae as a source of protein. *Biotechnology Advances* 25: 207–210. https://doi.org/10.1016/j.biotechadv.2006.11.002.

Belarbi, E.H., Molina, E., and Chisti, Y. (2000). A process for high yield and scaleable recovery of high purity eicosapentaenoic acid esters from microalgae and fish oil. *Enzyme and Microbial Technology* 26: 516–529. https://doi.org/10.1016/S0141-0229(99)00191-X.

Ben-Amotz, A. (1996). effect of low temperature on the stereoisomer composition of Beta-carotene in the halotolerant alga *Dunaliella bardawil* (Chlorophyta). *Journal of Phycology* 32: 272–275. https://doi.org/10.1111/j.0022-3646.1996.00272.x.

Bermejo Román, R., Alvárez-Pez, J.M., Acién Fernández, F.G., and Molina Grima, E. (2002). Recovery of pure B-phycoerythrin from the microalga *Porphyridium cruentum*. *Journal of Biotechnology* 93: 73–85. https://doi.org/10.1016/S0168-1656(01)00385-6.

Bhatia, S.K., Mehariya, S., Bhatia, R.K. et al. (2021). Wastewater based microalgal biorefinery for bioenergy production: progress and challenges. *Science of the Total Environment* 751: 141599. https://doi.org/10.1016/j.scitotenv.2020.141599.

Bitog, J.P., Lee, I.-B., Lee, C.-G. et al. (2011). Application of computational fluid dynamics for modeling and designing photobioreactors for microalgae production: a review. *Computers and Electronics in Agriculture* 76: 131–147. https://doi.org/10.1016/j.compag.2011.01.015.

Bleakley, S. and Hayes, M. (2017). Algal proteins: extraction, application, and challenges concerning production. *Food* 6: 33. https://doi.org/10.3390/foods6050033.

Borowitzka, M.A. (1999). Commercial production of microalgae: ponds, tanks, tubes and fermenters. *Journal of Biotechnology* 70: 313–321. https://doi.org/10.1016/S0168-1656(99)00083-8.

Bosma, R., van Spronsen, W.A., Tramper, J., and Wijffels, R.H. (2003). Ultrasound, a new separation technique to harvest microalgae. *Journal of Applied Phycology* 15: 143–153. https://doi.org/10.1023/A:1023807011027.

Bracharz, F., Helmdach, D., Aschenbrenner, I. et al. (2018). Harvest of the oleaginous microalgae *Scenedesmus obtusiusculus* by flocculation from culture based on natural water sources. *Frontiers in Bioengineering and Biotechnology* 6: 200. https://doi.org/10.3389/fbioe.2018.00200.

Brennan, L. and Owende, P. (2010). Biofuels from microalgae – a review of technologies for production, processing, and extractions of biofuels and co-products. *Renewable and Sustainable Energy Reviews* 14: 557–577. https://doi.org/10.1016/j.rser.2009.10.009.

Buchmann, L., Brändle, I., Haberkorn, I. et al. (2019). Pulsed electric field based cyclic protein extraction of microalgae towards closed-loop biorefinery concepts. *Bioresource Technology* 291: 121870. https://doi.org/10.1016/j.biortech.2019.121870.

Busi, M.V., Barchiesi, J., Martín, M., and Gomez-Casati, D.F. (2014). Starch metabolism in green algae. *Starch – Stärke* 66: 28–40. https://doi.org/10.1002/star.201200211.

Cerón García, M.C., Sánchez Mirón, A., Fernández Sevilla, J.M. et al. (2005). Mixotrophic growth of the microalga *Phaeodactylum tricornutum*. *Process Biochemistry* 40: 297–305. https://doi.org/10.1016/j.procbio.2004.01.016.

Chai, X.-J., Chen, H.-X., Xu, W.-Q., and Xu, Y.-W. (2013). Expression of soybean *Kunitz trypsin* inhibitor gene SKTI in *Dunaliella salina*. *Journal of Applied Phycology* 25: 139–144. https://doi.org/10.1007/s10811-012-9847-8.

Chauton, M.S., Reitan, K.I., Norsker, N.H. et al. (2015). A techno-economic analysis of industrial production of marine microalgae as a source of EPA and DHA-rich raw material for aquafeed: research challenges and possibilities. *Aquaculture* 436: 95–103. https://doi.org/10.1016/j.aquaculture.2014.10.038.

Chen, C.-Y., Yeh, K.-L., Aisyah, R. et al. (2011). Cultivation, photobioreactor design and harvesting of microalgae for biodiesel production: a critical review. *Bioresource Technology* 102: 71–81. https://doi.org/10.1016/j.biortech.2010.06.159.

Chini Zittelli, G., Lavista, F., Bastianini, A. et al. (1999). Production of eicosapentaenoic acid by *Nannochloropsis* sp. cultures in outdoor tubular photobioreactors. In: *Progress in Industrial Microbiology*, vol. 35 (ed. R. Osinga), 299–312. Elsevier ISBN 978-0-444-50387-9.

Chisti, Y. (2016). Large-Scale Production of Algal Biomass: Raceway Ponds. In: *Algae Biotechnology*, Green Energy and Technology (ed. F. Bux and Y. Chisti), 21–40. Cham: Springer International Publishing ISBN 978-3-319-12333-2.

Chojnacka, K. and Marquez-Rocha, F.-J. (2003). Kinetic and Stoichiometric relationships of the energy and carbon metabolism in the culture of microalgae. *Biotechnology* 3: 21–34. https://doi.org/10.3923/biotech.2004.21.34.

Converti, A., Casazza, A.A., Ortiz, E.Y. et al. (2009). Effect of temperature and nitrogen concentration on the growth and lipid content of *Nannochloropsis oculata* and *Chlorella vulgaris* for biodiesel production. *Chemical Engineering and Processing: Process Intensification* 48: 1146–1151. https://doi.org/10.1016/j.cep.2009.03.006.

D'Alessandro, E.B. and Antoniosi Filho, N.R. (2016). Concepts and studies on lipid and pigments of microalgae: a review. *Renewable and Sustainable Energy Reviews* 58: 832–841. https://doi.org/10.1016/j.rser.2015.12.162.

Daliry, S., Hallajsani, A., Mohammadi Roshandeh, J. et al. (2017). Investigation of optimal condition for *Chlorella vulgaris* microalgae growth. *GJESM* 3: https://doi.org/10.22034/gjesm.2017.03.02.010.

Del Campo, J. (2000). Carotenoid content of *Chlorophycean* microalgae: factors determining lutein accumulation in *Muriellopsis* Sp. (Chlorophyta). *Journal of Biotechnology* 76: 51–59. https://doi.org/10.1016/S0168-1656(99)00178-9.

Del Campo, J.A., García-González, M., and Guerrero, M.G. (2007). Outdoor cultivation of microalgae for carotenoid production: current state and perspectives. *Applied Microbiology and Biotechnology* 74: 1163–1174. https://doi.org/10.1007/s00253-007-0844-9.

Delwiche, C.F. and Cooper, E.D. (2015). The evolutionary origin of a terrestrial flora. *Current Biology* 25: R899–R910. https://doi.org/10.1016/j.cub.2015.08.029.

Demurtas, O.C., Massa, S., Ferrante, P. et al. (2013). A chlamydomonas-derived human papillomavirus 16 E7 vaccine induces specific tumor protection. *PLoS One* 8: e61473. https://doi.org/10.1371/journal.pone.0061473.

Deruyck, B., Thi Nguyen, K.H., Decaestecker, E., and Muylaert, K. (2019). Modeling the impact of rotifer contamination on microalgal production in open pond, photobioreactor and thin layer cultivation systems. *Algal Research* 38: 101398. https://doi.org/10.1016/j.algal.2018.101398.

Dineshbabu, G., Goswami, G., Kumar, R. et al. (2019). Microalgae–nutritious, sustainable aqua- and animal feed source. *Journal of Functional Foods* 62: 103545. https://doi.org/10.1016/j.jff.2019.103545.

Dixit, S., Shukla, A., Singh, V., and Upadhyay, S.K. (2021a). Bioprospecting of natural compounds for industrial and medical applications. In: *Bioprospecting of Plant Biodiversity for Industrial Molecules* (ed. S.K. Upadhyay and S.P. Singh). John Wiley & Sons Ltd. https://doi.org/10.1002/9781119718017.ch3.

Dolganyuk, V., Belova, D., Babich, O. et al. (2020). Microalgae: a promising source of valuable bioproducts. *Biomolecules* 10: 1153. https://doi.org/10.3390/biom10081153.

Doughman, S.D., Krupanidhi, S., and Sanjeevi, C.B. (2007). Omega-3 fatty acids for nutrition and medicine: considering microalgae oil as a vegetarian source of EPA and DHA. *CDR* 3: 198–203. https://doi.org/10.2174/157339907781368968.

Duong, V.T., Ahmed, F., Thomas-Hall, S.R. et al. (2015). High protein- and high lipid-producing microalgae from Northern Australia as potential feedstock for animal feed and biodiesel. *Frontiers in Bioengineering and Biotechnology* 3: https://doi.org/10.3389/fbioe.2015.00053.

Dwyer, J.H., Navab, M., Dwyer, K.M. et al. (2001). Oxygenated carotenoid lutein and progression of early atherosclerosis: the Los Angeles atherosclerosis study. *Circulation* 103: 2922–2927. https://doi.org/10.1161/01.CIR.103.24.2922.

Eilers, U., Bikoulis, A., Breitenbach, J. et al. (2016). Limitations in the biosynthesis of Fucoxanthin as targets for genetic engineering in *Phaeodactylum tricornutum*. *Journal of Applied Phycology* 28: 123–129. https://doi.org/10.1007/s10811-015-0583-8.

Ejike, C.E.C.C., Collins, S.A., Balasuriya, N. et al. (2017). Prospects of microalgae proteins in producing peptide-based functional foods for promoting cardiovascular health. *Trends in Food Science & Technology* 59: 30–36. https://doi.org/10.1016/j.tifs.2016.10.026.

El-Kassas, H.Y., Heneash, A.M.M., and Hussein, N.R. (2015). Cultivation of *Arthrospira* (Spirulina) platensis using confectionary wastes for aquaculture feeding. *Journal, Genetic Engineering & Biotechnology* 13: 145–155. https://doi.org/10.1016/j.jgeb.2015.08.003.

Falkowski, P.G. (1994). The role of phytoplankton photosynthesis in global biogeochemical cycles. *Photosynthesis Research* 39: 235–258. https://doi.org/10.1007/BF00014586.

Fasolin, L.H., Pereira, R.N., Pinheiro, A.C. et al. (2019). Emergent food proteins – towards sustainability, health and innovation. *Food Research International* 125: 108586. https://doi.org/10.1016/j.foodres.2019.108586.

Forján, E., Garbayo, I., Henriques, M. et al. (2011). UV-A mediated modulation of photosynthetic efficiency, xanthophyll cycle and fatty acid production of *Nannochloropsis*. *Marine Biotechnology* 13: 366–375. https://doi.org/10.1007/s10126-010-9306-y.

Franklin, S., Ngo, B., Efuet, E., and Mayfield, S.P. (2002). Development of a GFP reporter gene for *Chlamydomonas reinhardtii* chloroplast. *The Plant Journal* 30: 733–744. https://doi.org/10.1046/j.1365-313X.2002.01319.x.

Fu, W., Guðmundsson, Ó., Paglia, G. et al. (2013). Enhancement of carotenoid biosynthesis in the green microalga *Dunaliella salina* with light-emitting diodes and adaptive laboratory evolution. *Applied Microbiology and Biotechnology* 97: 2395–2403. https://doi.org/10.1007/s00253-012-4502-5.

Galarza, J.I., Gimpel, J.A., Rojas, V. et al. (2018). Over accumulation of astaxanthin in *Haematococcus pluvialis* through chloroplast genetic engineering. *Algal Research* 31: 291–297. https://doi.org/10.1016/j.algal.2018.02.024.

Gianpiero, P., Daniele, C., and Giovanna, F. (2019). PEF-assisted supercritical CO_2 extraction of pigments from microalgae *Nannochloropsis oceanica* in a continuous flow system. *Chemical Engineering Transactions* 74: 97–102. https://doi.org/10.3303/CET1974017.

Gifuni, I., Olivieri, G., Pollio, A., and Marzocchella, A. (2018). Identification of an industrial microalgal strain for starch production in biorefinery context: the effect of nitrogen and carbon concentration on starch accumulation. *New Biotechnology* 41: 46–54. https://doi.org/10.1016/j.nbt.2017.12.003.

Gomez-Zavaglia, A., Prieto Lage, M.A., Jimenez-Lopez, C. et al. (2019). The potential of seaweeds as a source of functional ingredients of prebiotic and antioxidant value. *Antioxidants* 8: 406. https://doi.org/10.3390/antiox8090406.

Gong, M. and Bassi, A. (2016). Carotenoids from microalgae: a review of recent developments. *Biotechnology Advances* 34: 1396–1412. https://doi.org/10.1016/j.biotechadv.2016.10.005.

Gong, Y., Guo, X., Wan, X. et al. (2011). Characterization of a novel thioesterase (PtTE) from *Phaeodactylum tricornutum*. *Journal of Basic Microbiology* 51: 666–672. https://doi.org/10.1002/jobm.201000520.

Grahl, S., Strack, M., Weinrich, R., and Mörlein, D. (2018). Consumer-oriented product development: the conceptualization of novel food products based on *Spirulina* (*Arthrospira Platensis*) and resulting consumer expectations. *Journal of Food Quality* 2018: 1–11. https://doi.org/10.1155/2018/1919482.

Gromov, B.V., Vepritskiy, A.A., Titova, N.N. et al. (1991). Production of the antibiotic cyanobacterin LU-1 by *Nostoc linckia* CALU 892 (Cyanobacterium). *Journal of Applied Phycology* 3: 55–59. https://doi.org/10.1007/BF00003919.

Grossmann, L., Ebert, S., Hinrichs, J., and Weiss, J. (2018). Effect of precipitation, lyophilization, and organic solvent extraction on preparation of protein-rich powders from the microalgae chlorella protothecoides. *Algal Research* 29: 266–276. https://doi.org/10.1016/j.algal.2017.11.019.

Grossmann, L., Hinrichs, J., and Weiss, J. (2020). Cultivation and downstream processing of microalgae and cyanobacteria to generate protein-based technofunctional food ingredients. *Critical Reviews in Food Science and Nutrition* 60: 2961–2989. https://doi.org/10.1080/10408398.2019.1672137.

Grubišić, M., Šantek, M.I., and Šantek, B. (2019). Potential of microalgae for the production of different biotechnological products. *Chemical and Biochemical Engineering Quarterly (Online)* 33: 161–181. https://doi.org/10.15255/CABEQ.2019.1657.

Gügi, B., Le Costaouec, T., Burel, C. et al. (2015). Diatom-specific oligosaccharide and polysaccharide structures help to unravel biosynthetic capabilities in diatoms. *Marine Drugs* 13: 5993–6018. https://doi.org/10.3390/md13095993.

Gultom, S. and Hu, B. (2013). Review of microalgae harvesting via co-pelletization with filamentous fungus. *Energies* 6: 5921–5939. https://doi.org/10.3390/en6115921.

Günerken, E., D'Hondt, E., Eppink, M.H.M. et al. (2015). Cell disruption for microalgae biorefineries. *Biotechnology Advances* 33: 243–260. https://doi.org/10.1016/j.biotechadv.2015.01.008.

Hamilton, M.L., Haslam, R.P., Napier, J.A., and Sayanova, O. (2014). Metabolic engineering of *Phaeodactylum tricornutum* for the enhanced accumulation of omega-3 long chain polyunsaturated fatty acids. *Metabolic Engineering* 22: 3–9. https://doi.org/10.1016/j.ymben.2013.12.003.

Hanotu, J., Bandulasena, H.C.H., and Zimmerman, W.B. (2012). Microflotation performance for algal separation. *Biotechnology and Bioengineering* 109: 1663–1673. https://doi.org/10.1002/bit.24449.

Heasman, M., Diemar, J., O'connor, W. et al. (2000). Development of extended shelf-life microalgae concentrate diets harvested by centrifugation for *Bivalve molluscs* – a summary: extended shelf-life microalgae concentrate diets *M. Heasman* et al. *Aquaculture Research* 31: 637–659. https://doi.org/10.1046/j.1365-2109.2000.318492.x.

Hedenskog, G., Mogren, H.Å.R., and Enebo, L. (1970). A method for obtaining protein concentrates from microorganisms. *Biotechnology and Bioengineering* 12: 947–959. https://doi.org/10.1002/bit.260120607.

Hempel, F., Lau, J., Klingl, A., and Maier, U.G. (2011). Algae as protein factories: expression of a human antibody and the respective antigen in the diatom *Phaeodactylum tricornutum*. *PLoS One* 6: e28424. https://doi.org/10.1371/journal.pone.0028424.

Hempel, F., Maurer, M., Brockmann, B. et al. (2017). From hybridomas to a robust microalgal-based production platform: molecular design of a diatom secreting monoclonal antibodies directed against the marburg virus nucleoprotein. *Microbial Cell Factories* 16: 131. https://doi.org/10.1186/s12934-017-0745-2.

Hu, I.-C. (2019). Production of potential coproducts from microalgae. In: *Biofuels from Algae* (ed. A. Pandey, J.-S. Chang, C.R. Soccol, et al.), 345–358. Elsevier ISBN 978-0-444-64192-2.

Hu, J., Nagarajan, D., Zhang, Q. et al. (2018). Heterotrophic cultivation of microalgae for pigment production: a review. *Biotechnology Advances* 36: 54–67. https://doi.org/10.1016/j.biotechadv.2017.09.009.

Huang, G., Chen, F., Wei, D. et al. (2010). Biodiesel production by microalgal biotechnology. *Applied Energy* 87: 38–46. https://doi.org/10.1016/j.apenergy.2009.06.016.

Huang, Q., Jiang, F., Wang, L., and Yang, C. (2017). Design of photobioreactors for mass cultivation of photosynthetic organisms. *Engineering* 3: 318–329. https://doi.org/10.1016/J.ENG.2017.03.020.

Huesemann, M.H., Van Wagenen, J., Miller, T. et al. (2013). A screening model to predict microalgae biomass growth in photobioreactors and raceway ponds. *Biotechnology and Bioengineering* 110: 1583–1594. https://doi.org/10.1002/bit.24814.

Huleihel, M., Ishanu, V., Tal, J., and Arad, S.(.M.). (2001). Antiviral effect of red microalgal polysaccharides on *Herpes simplex* and *Varicella zoster* viruses. *Journal of Applied Phycology* 13: 127–134. https://doi.org/10.1023/A:1011178225912.

Illman, A.M., Scragg, A.H., and Shales, S.W. (2000). Increase in Chlorella strains calorific values when grown in low nitrogen medium. *Enzyme and Microbial Technology* 27: 631–635. https://doi.org/10.1016/S0141-0229(00)00266-0.

Ishikura, K., Takaoka, Y., Kato, K. et al. (1999). Expression of a foreign gene in *Chlamydomonas Reinhardtii* chloroplast. *Journal of Bioscience and Bioengineering* 87: 307–314. https://doi.org/10.1016/S1389-1723(99)80037-1.

Ito, T., Tanaka, M., Shinkawa, H. et al. (2013). Metabolic and morphological changes of an oil accumulating *Trebouxiophycean alga* in nitrogen-deficient conditions. *Metabolomics* 9: 178–187. https://doi.org/10.1007/s11306-012-0463-z.

Jacob-Lopes, E., Scoparo, C.H.G., Lacerda, L.M.C.F., and Franco, T.T. (2009). Effect of light cycles (Night/Day) on CO_2 fixation and biomass production by microalgae in photobioreactors. *Chemical Engineering and Processing: Process Intensification* 48: 306–310. https://doi.org/10.1016/j.cep.2008.04.007.

Jiang, H. and Gao, K. (2004). Effects of lowering temperature during culture on the production of polyunsaturated fatty acids in the marine diatom *Phaeodactylum tricornutum* (Bacillariophyceae)1: PUFAs in *Phaeodactylum tricornutum*. *Journal of Phycology* 40: 651–654. https://doi.org/10.1111/j.1529-8817.2004.03112.x.

Jorquera, O., Kiperstok, A., Sales, E.A. et al. (2010). Comparative energy life-cycle analyses of microalgal biomass production in open ponds and photobioreactors. *Bioresource Technology* 101: 1406–1413. https://doi.org/10.1016/j.biortech.2009.09.038.

Juneja, A., Ceballos, R., and Murthy, G. (2013). Effects of environmental factors and nutrient availability on the biochemical composition of algae for biofuels production: a review. *Energies* 6: 4607–4638. https://doi.org/10.3390/en6094607.

Kang, C.D., Lee, J.S., Park, T.H., and Sim, S.J. (2007). Complementary limiting factors of Astaxanthin synthesis during photoautotrophic induction of *Haematococcus pluvialis*: C/N ratio and light intensity. *Applied Microbiology and Biotechnology* 74: 987–994. https://doi.org/10.1007/s00253-006-0759-x.

Kannan, D.C. and Venkat, D. (2019). An open outdoor algal growth system of improved productivity for biofuel production: an improved productivity algal growth system. *Journal of Chemical Technology and Biotechnology* 94: 222–235. https://doi.org/10.1002/jctb.5768.

Kesik-Brodacka, M. (2018). Progress in biopharmaceutical development. *Biotechnology and Applied Biochemistry* 65: 306–322. https://doi.org/10.1002/bab.1617.

Khalil, Z.I., Asker, M.M.S., El-Sayed, S., and Kobbia, I.A. (2010). Effect of PH on growth and biochemical responses of *Dunaliella bardawil* and *Chlorella ellipsoidea*. *World Journal of Microbiology and Biotechnology* 26: 1225–1231. https://doi.org/10.1007/s11274-009-0292-z.

Khan, M.I., Lee, M.G., Seo, H.J. et al. (2016). Enhancing the feasibility of *Microcystis aeruginosa* as a feedstock for bioethanol production under the influence of various factors. *BioMed Research International* 2016: 1–9. https://doi.org/10.1155/2016/4540826.

Khan, M.I., Shin, J.H., and Kim, J.D. (2018). The promising future of microalgae: current status, challenges, and optimization of a sustainable and renewable industry for biofuels, feed, and other products. *Microbial Cell Factories* 17: 36. https://doi.org/10.1186/s12934-018-0879-x.

Khavari, F., Saidijam, M., Taheri, M., and Nouri, F. (2021). Microalgae: therapeutic potentials and applications. *Molecular Biology Reports* 48: 4757–4765. https://doi.org/10.1007/s11033-021-06422-w.

Khozingoldberg, I. and Cohen, Z. (2006). The effect of phosphate starvation on the lipid and fatty acid composition of the fresh water eustigmatophyte *Monodus subterraneus*. *Phytochemistry* 67: 696–701. https://doi.org/10.1016/j.phytochem.2006.01.010.

Kim, H.S., Devarenne, T.P., and Han, A. (2018). Microfluidic systems for microalgal biotechnology: a review. *Algal Research* 30: 149–161. https://doi.org/10.1016/j.algal.2017.11.020.

Kitaya, Y., Azuma, H., and Kiyota, M. (2005). Effects of temperature, CO_2/O_2 concentrations and light intensity on cellular multiplication of microalgae, *Euglena gracilis*. *Advances in Space Research* 35: 1584–1588. https://doi.org/10.1016/j.asr.2005.03.039.

Koller, M., Muhr, A., and Braunegg, G. (2014). Microalgae as versatile cellular factories for valued products. *Algal Research* 6: 52–63. https://doi.org/10.1016/j.algal.2014.09.002.

Krzemińska, I., Pawlik-Skowrońska, B., Trzcińska, M., and Tys, J. (2014). Influence of photoperiods on the growth rate and biomass productivity of green microalgae. *Bioprocess and Biosystems Engineering* 37: 735–741. https://doi.org/10.1007/s00449-013-1044-x.

Kumar, G., Shekh, A., Jakhu, S. et al. (2020). Bioengineering of microalgae: recent advances, perspectives, and regulatory challenges for industrial application. *Frontiers in Bioengineering and Biotechnology* 8: 914. https://doi.org/10.3389/fbioe.2020.00914.

Kumudha, A., Selvakumar, S., Dilshad, P. et al. (2015). Methylcobalamin – a form of vitamin B12 identified and characterised in *Chlorella vulgaris*. *Food Chemistry* 170: 316–320. https://doi.org/10.1016/j.foodchem.2014.08.035.

Kunjapur, A.M. and Eldridge, R.B. (2010). Photobioreactor design for commercial biofuel production from microalgae. *Industrial and Engineering Chemistry Research* 49: 3516–3526. https://doi.org/10.1021/ie901459u.

Kurd, F. and Samavati, V. (2015). Water soluble polysaccharides from *Spirulina platensis*: extraction and in vitro anti-cancer activity. *International Journal of Biological Macromolecules* 74: 498–506. https://doi.org/10.1016/j.ijbiomac.2015.01.005.

Laamanen, C.A., Ross, G.M., and Scott, J.A. (2016). Flotation harvesting of microalgae. *Renewable and Sustainable Energy Reviews* 58: 75–86. https://doi.org/10.1016/j.rser.2015.12.293.

Lafarga, T. (2020). Cultured microalgae and compounds derived thereof for food applications: strain selection and cultivation, drying, and processing strategies. *Food Reviews International* 36: 559–583. https://doi.org/10.1080/87559129.2019.1655572.

Lamers, P.P., Janssen, M., De Vos, R.C.H. et al. (2012). Carotenoid and fatty acid metabolism in nitrogen-starved *Dunaliella salina*, a unicellular green microalga. *Journal of Biotechnology* 162: 21–27. https://doi.org/10.1016/j.jbiotec.2012.04.018.

Lee, A.K., Lewis, D.M., and Ashman, P.J. (2012). Disruption of microalgal cells for the extraction of lipids for biofuels: processes and specific energy requirements. *Biomass and Bioenergy* 46: 89–101. https://doi.org/10.1016/j.biombioe.2012.06.034.

Lee, K., Eisterhold, M., Rindi, F. et al. (2014). Isolation and screening of microalgae from natural habitats in the Midwestern United States of America for biomass and biodiesel sources. *Journal of Natural Science, Biology and Medicine* 5: 333. https://doi.org/10.4103/0976-9668.136178.

Liu, J. and Hu, Q. (2013). *Chlorella*: industrial production of cell mass and chemicals. In: *Handbook of Microalgal Culture* (ed. A. Richmond and Q. Hu), 327–338. Oxford, UK: John Wiley & Sons, Ltd ISBN 978-1-118-56716-6.

Lomonossoff, G.P. and D'Aoust, M.-A. (2016). Plant-produced biopharmaceuticals: a case of technical developments driving clinical deployment. *Science* 353: 1237–1240. https://doi.org/10.1126/science.aaf6638.

López-Hernández, J.F., García-Alamilla, P., Palma-Ramírez, D. et al. (2020). Continuous microalgal cultivation for antioxidants production. *Molecules* 25: 4171. https://doi.org/10.3390/molecules25184171.

Lorenz, R.T. and Cysewski, G.R. (2000). Commercial potential for haematococcus microalgae as a natural source of Astaxanthin. *Trends in Biotechnology* 18: 160–167. https://doi.org/10.1016/S0167-7799(00)01433-5.

Lum, K.K., Kim, J., and Lei, X.G. (2013). Dual potential of microalgae as a sustainable biofuel feedstock and animal feed. *Journal of Animal Science and Biotechnology* 4: 53. https://doi.org/10.1186/2049-1891-4-53.

de Macedo, R.V.T. and Alegre, R.M. (2001). INFLUÊNCIA DO TEOR DE NITROGÊNIO NO CULTIVO DE Spirulina Maxima EM DOIS NÍVEIS DE TEMPERATURA: Parte II: PRODUÇÃO DE LIPÍDIOS. *Ciencia y Tecnologia de Alimentos* 21: 183–186. https://doi.org/10.1590/S0101-20612001000200011.

Madhubalaji, C.K., Rashmi, V., Chauhan, V.S. et al. (2019). Improvement of vitamin B12 status with *Spirulina* supplementation in wistar rats validated through functional and circulatory markers. *Journal of Food Biochemistry* 43: https://doi.org/10.1111/jfbc.13038.

Malla, A., Rosales-Mendoza, S., Phoolcharoen, W., and Vimolmangkang, S. (2021). Efficient transient expression of recombinant proteins using DNA viral vectors in freshwater microalgal species. *Frontiers in Plant Science* 12: 650820. https://doi.org/10.3389/fpls.2021.650820.

Manayi, A., Abdollahi, M., Raman, T. et al. (2016). Lutein and cataract: from bench to bedside. *Critical Reviews in Biotechnology* 36: 829–839. https://doi.org/10.3109/07388551.2015.1049510.

Manuell, A.L., Beligni, M.V., Elder, J.H. et al. (2007). Robust expression of a bioactive mammalian protein in *Chlamydomonas* chloroplast. *Plant Biotechnology Journal* 5: 402–412. https://doi.org/10.1111/j.1467-7652.2007.00249.x.

Mata, T.M., Martins, A.A., and Caetano, N.S. (2010). Microalgae for biodiesel production and other applications: a review. *Renewable and Sustainable Energy Reviews* 14: 217–232. https://doi.org/10.1016/j.rser.2009.07.020.

Matsuo, T., Onai, K., Okamoto, K. et al. (2006). Real-time monitoring of chloroplast gene expression by a luciferase reporter: evidence for nuclear regulation of chloroplast circadian period. *Molecular and Cellular Biology* 26: 863–870. https://doi.org/10.1128/MCB.26.3.863-870.2006.

Mayfield, S.P. and Schultz, J. (2004). Development of a luciferase reporter gene, *LuxCt*, for *Chlamydomonas reinhardtii* chloroplast. *The Plant Journal* 37: 449–458. https://doi.org/10.1046/j.1365-313X.2003.01965.x.

Mayfield, S.P., Franklin, S.E., and Lerner, R.A. (2003). Expression and assembly of a fully active antibody in algae. *Proceedings of the National Academy of Sciences* 100: 438–442. https://doi.org/10.1073/pnas.0237108100.

Meireles, L.A., Guedes, A.C., and Malcata, F.X. (2003). Increase of the yields of eicosapentaenoic and docosahexaenoic acids by the microalga *Pavlova lutheri* following random mutagenesis. *Biotechnology and Bioengineering* 81: 50–55. https://doi.org/10.1002/bit.10451.

Mikami, K. and Hosokawa, M. (2013). Biosynthetic pathway and health benefits of fucoxanthin, an algae-specific xanthophyll in brown seaweeds. *IJMS* 14: 13763–13781. https://doi.org/10.3390/ijms140713763.

Molina Grima, E., Belarbi, E.-H., Acién Fernández, F.G. et al. (2003). Recovery of microalgal biomass and metabolites: process options and economics. *Biotechnology Advances* 20: 491–515. https://doi.org/10.1016/S0734-9750(02)00050-2.

Moore, R.E., Patterson, G.M.L., Mynderse, J.S. et al. (1986). Toxins from cyanophytes belonging to the scytonemataceae. *Pure and Applied Chemistry* 58: 263–271. https://doi.org/10.1351/pac198658020263.

Morales-Sánchez, D., Martinez-Rodriguez, O.A., and Martinez, A. (2017). Heterotrophic cultivation of microalgae: production of metabolites of commercial interest: heterotrophic cultivation of microalgae: products. *Journal of Chemical Technology and Biotechnology* 92: 925–936. https://doi.org/10.1002/jctb.5115.

Mujtaba, G., Choi, W., Lee, C.-G., and Lee, K. (2012). Lipid production by *Chlorella vulgaris* after a shift from nutrient-rich to nitrogen starvation conditions. *Bioresource Technology* 123: 279–283. https://doi.org/10.1016/j.biortech.2012.07.057.

Mulders, K.J.M., Weesepoel, Y., Lamers, P.P. et al. (2013). Growth and pigment accumulation in nutrient-depleted *Isochrysis aff.* Galbana T-ISO. *Journal of Applied Phycology* 25: 1421–1430. https://doi.org/10.1007/s10811-012-9954-6.

Muradov, N., Taha, M., Miranda, A.F. et al. (2015). Fungal-assisted algal flocculation: application in wastewater treatment and biofuel production. *Biotechnology for Biofuels* 8: 24. https://doi.org/10.1186/s13068-015-0210-6.

Mutanda, T., Ramesh, D., Karthikeyan, S. et al. (2011). Bioprospecting for hyper-lipid producing microalgal strains for sustainable biofuel production. *Bioresource Technology* 102: 57–70. https://doi.org/10.1016/j.biortech.2010.06.077.

Mutanda, T., Naidoo, D., Bwapwa, J.K., and Anandraj, A. (2020). Biotechnological applications of microalgal oleaginous compounds: current trends on microalgal bioprocessing of products. *Frontiers in Energy Research* 8: 598803. https://doi.org/10.3389/fenrg.2020.598803.

Najjar, Y.S.H. and Abu-Shamleh, A. (2020). Harvesting of microalgae by centrifugation for biodiesel production: a review. *Algal Research* 51: 102046. https://doi.org/10.1016/j.algal.2020.102046.

Nguyen, T.D.P., Le, T.V.A., Show, P.L. et al. (2019). Bioflocculation formation of microalgae-bacteria in enhancing microalgae harvesting and nutrient removal from wastewater effluent. *Bioresource Technology* 272: 34–39. https://doi.org/10.1016/j.biortech.2018.09.146.

Norton, T.A., Melkonian, M., and Andersen, R.A. (1996). Algal biodiversity. *Phycologia* 35: 308–326. https://doi.org/10.2216/i0031-8884-35-4-308.1.

Nur, M.M.A. and Buma, A.G.J. (2019). Opportunities and challenges of microalgal cultivation on wastewater, with special focus on palm oil mill effluent and the production of high value compounds. *Waste and Biomass Valorization* 10: 2079–2097. https://doi.org/10.1007/s12649-018-0256-3.

Osama, A., Hosney, H., and Moussa, M.S. (2021). Potential of household photobioreactor for algae cultivation. *Journal of Water and Climate Change* 12: 2147–2180. https://doi.org/10.2166/wcc.2021.261.

Parages, M.L., Rico, R.M., Abdala-Díaz, R.T. et al. (2012). Acidic polysaccharides of *Arthrospira* (Spirulina) platensis induce the synthesis of TNF-α in RAW macrophages. *Journal of Applied Phycology* 24: 1537–1546. https://doi.org/10.1007/s10811-012-9814-4.

Park, K.-H. and Lee, C.-G. (2001). Effectiveness of flashing light for increasing photosynthetic efficiency of microalgal cultures over a critical cell density. *Biotechnology and Bioprocess Engineering* 6: 189–193. https://doi.org/10.1007/BF02932549.

Płaczek, M., Patyna, A., and Witczak, S. (2017). Technical evaluation of photobioreactors for microalgae cultivation. *E3S Web of Conferences* 19: 02032. https://doi.org/10.1051/e3sconf/20171902032.

Plaza, M., Herrero, M., Cifuentes, A., and Ibáñez, E. (2009). Innovative natural functional ingredients from microalgae. *Journal of Agricultural and Food Chemistry* 57: 7159–7170. https://doi.org/10.1021/jf901070g.

Prathima Devi, M. and Venkata Mohan, S. (2012). CO_2 Supplementation to domestic wastewater enhances microalgae lipid accumulation under mixotrophic microenvironment: effect of sparging period and interval. *Bioresource Technology* 112: 116–123. https://doi.org/10.1016/j.biortech.2012.02.095.

Pugazhendhi, A., Shobana, S., Bakonyi, P. et al. (2019). A review on chemical mechanism of microalgae flocculation via polymers. *Biotechnology Reports* 21: e00302. https://doi.org/10.1016/j.btre.2018.e00302.

Rapala, J., Sivonen, K., Luukkainen, R., and Niemelä, S.I. (1993). Anatoxin-a concentration inAnabaena andAphanizomenon under different environmental conditions and comparison of growth by toxic and non-toxicAnabaena-strains – a laboratory study. *Journal of Applied Phycology* 5: 581–591. https://doi.org/10.1007/BF02184637.

Rasala, B.A., Muto, M., Lee, P.A. et al. (2010). Production of therapeutic proteins in algae, analysis of expression of seven human proteins in the chloroplast of *Chlamydomonas reinhardtii*: production of therapeutic proteins in algae. *Plant Biotechnology Journal* 8: 719–733. https://doi.org/10.1111/j.1467-7652.2010.00503.x.

Rawat, I., Ranjith Kumar, R., Mutanda, T., and Bux, F. (2013). Biodiesel from microalgae: a critical evaluation from laboratory to large scale production. *Applied Energy* 103: 444–467. https://doi.org/10.1016/j.apenergy.2012.10.004.

Ren, L.-J., Ji, X.-J., Huang, H. et al. (2010). Development of a stepwise aeration control strategy for efficient docosahexaenoic acid production by *Schizochytrium* Sp. *Applied Microbiology and Biotechnology* 87: 1649–1656. https://doi.org/10.1007/s00253-010-2639-7.

Richmond, A. (2004). *Handbook of Microalgal Culture: Biotechnology and Applied Phycology*. Oxford Ames: Blackwell Science ISBN 978-0-632-05953-9.

Rincon, S.M., Urrego, N.F., Avila, K.J. et al. (2019). Photosynthetic activity assessment in mixotrophically cultured *Chlorella vulgaris* biofilms at various developmental stages. *Algal Research* 38: 101408. https://doi.org/10.1016/j.algal.2019.101408.

Rosa, S.M., Soria, M.A., Vélez, C.G., and Galvagno, M.A. (2010). Improvement of a two-stage fermentation process for docosahexaenoic acid production by *Aurantiochytrium limacinum* SR21 applying statistical experimental designs and data analysis. *Bioresource Technology* 101: 2367–2374. https://doi.org/10.1016/j.biortech.2009.11.056.

Ruxton, C.H.S., Reed, S.C., Simpson, M.J.A., and Millington, K.J. (2004). The health benefits of omega-3 polyunsaturated fatty acids: a review of the evidence. *Journal of Human Nutrition and Dietetics* 17: 449–459. https://doi.org/10.1111/j.1365-277X.2004.00552.x.

Safi, C., Ursu, A.V., Laroche, C. et al. (2014). Aqueous extraction of proteins from microalgae: effect of different cell disruption methods. *Algal Research* 3: 61–65. https://doi.org/10.1016/j.algal.2013.12.004.

Safi, C., Olivieri, G., Campos, R.P. et al. (2017). Biorefinery of microalgal soluble proteins by sequential processing and membrane filtration. *Bioresource Technology* 225: 151–158. https://doi.org/10.1016/j.biortech.2016.11.068.

Salvucci, M.E. and Crafts-Brandner, S.J. (2004). Relationship between the heat tolerance of photosynthesis and the thermal stability of rubisco activase in plants from contrasting thermal environments. *Plant Physiology* 134: 1460–1470. https://doi.org/10.1104/pp.103.038323.

Sánchez, Á., Maceiras, R., Cancela, Á., and Pérez, A. (2013). Culture aspects of *Isochrysis galbana* for biodiesel production. *Applied Energy* 101: 192–197. https://doi.org/10.1016/j.apenergy.2012.03.027.

Seghiri, R., Kharbach, M., and Essamri, A. (2019). Functional composition, nutritional properties, and biological activities of moroccan *Spirulina* microalga. *Journal of Food Quality* 2019: 1–11. https://doi.org/10.1155/2019/3707219.

Sekar, S. and Chandramohan, M. (2008). Phycobiliproteins as a commodity: trends in applied research, patents and commercialization. *Journal of Applied Phycology* 20: 113–136. https://doi.org/10.1007/s10811-007-9188-1.

Show, K.-Y., Lee, D.-J., and Chang, J.-S. (2013). Algal Biomass Dehydration. *Bioresource Technology* 135: 720–729. https://doi.org/10.1016/j.biortech.2012.08.021.

Show, P., Tang, M., Nagarajan, D. et al. (2017). A holistic approach to managing microalgae for biofuel applications. *IJMS* 18: 215. https://doi.org/10.3390/ijms18010215.

Singh, G. and Patidar, S.K. (2018). Microalgae harvesting techniques: a review. *Journal of Environmental Management* 217: 499–508. https://doi.org/10.1016/j.jenvman.2018.04.010.

Singh, J. and Saxena, R.C. (2015). An introduction to microalgae. In: *Handbook of Marine Microalgae* (ed. S.-K. Kim), 11–24. Elsevier ISBN 978-0-12-800776-1.

Smerilli, A., Orefice, I., Corato, F. et al. (2017). Photoprotective and antioxidant responses to light spectrum and intensity variations in the coastal diatom *S Keletonema Marinoi*: photoprotective network in *Skeletonema marinoi*. *Environmental Microbiology* 19: 611–627. https://doi.org/10.1111/1462-2920.13545.

Soares, A.T., Marques Júnior, J.G., Lopes, R.G. et al. (2016). Improvement of the extraction process for high commercial value pigments from *Desmodesmus* Sp. microalgae. *Journal of the Brazilian Chemical Society* https://doi.org/10.5935/0103-5053.20160004.

Soares, J., Kriiger Loterio, R., Rosa, R.M. et al. (2018). *Scenedesmus* Sp. cultivation using commercial-grade ammonium sources. *Annales de Microbiologie* 68: 35–45. https://doi.org/10.1007/s13213-017-1315-x.

Song, L., Qin, J.G., Su, S. et al. (2012). Micronutrient requirements for growth and hydrocarbon production in the oil producing green Alga *Botryococcus braunii* (Chlorophyta). *PLoS One* 7: e41459. https://doi.org/10.1371/journal.pone.0041459.

Spolaore, P., Joannis-Cassan, C., Duran, E., and Isambert, A. (2006). Commercial applications of microalgae. *Journal of Bioscience and Bioengineering* 101: 87–96. https://doi.org/10.1263/jbb.101.87.

Suh, I.S. and Lee, C.-G. (2003). Photobioreactor engineering: design and performance. *Biotechnology and Bioprocess Engineering* 8: 313–321. https://doi.org/10.1007/BF02949274.

Talero, E., García-Mauriño, S., Ávila-Román, J. et al. (2015). Bioactive compounds isolated from microalgae in chronic inflammation and cancer. *Marine Drugs* 13: 6152–6209. https://doi.org/10.3390/md13106152.

Tan, J.S., Lee, S.Y., Chew, K.W. et al. (2020). A review on microalgae cultivation and harvesting, and their biomass extraction processing using ionic liquids. *Bioengineered* 11: 116–129. https://doi.org/10.1080/21655979.2020.1711626.

Thao, T., Linh, D., Si, V. et al. (2017). Isolation and selection of microalgal strains from natural water sources in Viet Nam with potential for edible oil production. *Marine Drugs* 15: 194. https://doi.org/10.3390/md15070194.

Tran, M., Zhou, B., Pettersson, P.L. et al. (2009). Synthesis and assembly of a full-length human monoclonal antibody in algal chloroplasts. *Biotechnology and Bioengineering* 104: 663–673. https://doi.org/10.1002/bit.22446.

Tran, M., Van, C., Barrera, D.J. et al. (2013). Production of unique immunotoxin cancer therapeutics in algal chloroplasts. *Proceedings of the National Academy of Sciences* 110: E15–E22. https://doi.org/10.1073/pnas.1214638110.

Udayan, A., Kathiresan, S., and Arumugam, M. (2018). Kinetin and gibberellic acid (GA3) act synergistically to produce high value polyunsaturated fatty acids in *Nannochloropsis oceanica* CASA CC201. *Algal Research* 32: 182–192. https://doi.org/10.1016/j.algal.2018.03.007.

Udayan, A., Sabapathy, H., and Arumugam, M. (2020). Stress hormones mediated lipid accumulation and modulation of specific fatty acids in *Nannochloropsis oceanica* CASA CC201. *Bioresource Technology* 310: 123437. https://doi.org/10.1016/j.biortech.2020.123437.

Upadhyay, S.K. and Singh, S.P. (eds.) (2021). *Bioprospecting of Plant Biodiversity for Industrial Molecules*. John Wiley & Sons Ltd. https://doi.org/10.1002/9781119718017.

Vanier, G., Hempel, F., Chan, P. et al. (2015). Biochemical characterization of human anti-hepatitis B monoclonal antibody produced in the microalgae *Phaeodactylum tricornutum*. *PLoS One* 10: e0139282. https://doi.org/10.1371/journal.pone.0139282.

Varela, J.C., Pereira, H., Vila, M., and León, R. (2015). Production of carotenoids by microalgae: achievements and challenges. *Photosynthesis Research* 125: 423–436. https://doi.org/10.1007/s11120-015-0149-2.

Venkata Mohan, S., Rohit, M.V., Chiranjeevi, P. et al. (2015). Heterotrophic microalgae cultivation to synergize biodiesel production with waste remediation: progress and perspectives. *Bioresource Technology* 184: 169–178. https://doi.org/10.1016/j.biortech.2014.10.056.

Vernès, L., Abert-Vian, M., El Maâtaoui, M. et al. (2019). Application of ultrasound for green extraction of proteins from Spirulina. Mechanism, optimization, modeling, and industrial prospects. *Ultrasonics Sonochemistry* 54: 48–60. https://doi.org/10.1016/j.ultsonch.2019.02.016.

Vuppaladadiyam, A.K., Prinsen, P., Raheem, A. et al. (2018). Microalgae cultivation and metabolites production: a comprehensive review. *Biofuels, Bioproducts and Biorefining* 12: 304–324. https://doi.org/10.1002/bbb.1864.

Walker, T.L., Purton, S., Becker, D.K., and Collet, C. (2005). Microalgae as bioreactors. *Plant Cell Reports* 24: 629–641. https://doi.org/10.1007/s00299-005-0004-6.

Wang, C. and Lan, C.Q. (2018). Effects of shear stress on microalgae – a review. *Biotechnology Advances* 36: 986–1002. https://doi.org/10.1016/j.biotechadv.2018.03.001.

Wang, B., Lan, C.Q., and Horsman, M. (2012). Closed photobioreactors for production of microalgal biomasses. *Biotechnology Advances* 30: 904–912. https://doi.org/10.1016/j.biotechadv.2012.01.019.

Wang, J., Wang, X.-D., Zhao, X.-Y. et al. (2015). From microalgae oil to produce novel structured triacylglycerols enriched with unsaturated fatty acids. *Bioresource Technology* 184: 405–414. https://doi.org/10.1016/j.biortech.2014.09.133.

Winwood, R.J. (2013). Recent developments in the commercial production of DHA and EPA rich oils from micro-algae. *OCL* 20: D604. https://doi.org/10.1051/ocl/2013030.

Wu, S.-T. and Lin, L.-P. (2003). Application of response surface methodology to optimize docosahexaenoic acid production by *Schizochytrium* Sp. S31. *Journal of Food Biochemistry* 27: 127–139. https://doi.org/10.1111/j.1745-4514.2003.tb00271.x.

Wu, Q., Liu, L., Miron, A. et al. (2016). The antioxidant, immunomodulatory, and anti-inflammatory activities of *Spirulina*: an overview. *Archives of Toxicology* 90: 1817–1840. https://doi.org/10.1007/s00204-016-1744-5.

Xu, L., Weathers, P.J., Xiong, X.-R., and Liu, C.-Z. (2009). Microalgal bioreactors: challenges and opportunities. *Engineering in Life Sciences* 9: 178–189. https://doi.org/10.1002/elsc.200800111.

Yan, N., Fan, C., Chen, Y., and Hu, Z. (2016). The potential for microalgae as bioreactors to produce pharmaceuticals. *IJMS* 17: 962. https://doi.org/10.3390/ijms17060962.

Yang, Z., Chen, F., Li, D. et al. (2006). Expression of human soluble trail in *Chlamydomonas reinhardtii* chloroplast. *Chinese Science Bulletin* 51: 1703–1709. https://doi.org/10.1007/s11434-006-2041-0.

Yegneswaran, P.K., Gray, M.R., and Thompson, B.G. (1990). Kinetics of CO_2 hydration in fermentors: PH and pressure effects. *Biotechnology and Bioengineering* 36: 92–96. https://doi.org/10.1002/bit.260360112.

Yu, M., Chen, M., Gui, J. et al. (2019). Preparation of *Chlorella vulgaris* polysaccharides and their antioxidant activity *in vitro* and *in vivo*. *International Journal of Biological Macromolecules* 137: 139–150. https://doi.org/10.1016/j.ijbiomac.2019.06.222.

Zainuddin, E.N., Mentel, R., Wray, V. et al. (2007). Cyclic depsipeptides, ichthyopeptins A and B, from *Microcystis ichthyoblabe*. *Journal of Natural Products* 70: 1084–1088. https://doi.org/10.1021/np060303s.

Zhan, J., Rong, J., and Wang, Q. (2017). Mixotrophic cultivation, a preferable microalgae cultivation mode for biomass/bioenergy production, and bioremediation, advances and prospect. *International Journal of Hydrogen Energy* 42: 8505–8517. https://doi.org/10.1016/j.ijhydene.2016.12.021.

Zhang, X., Tang, X., Zhou, B. et al. (2015). Effect of enhanced UV-B radiation on photosynthetic characteristics of marine microalgae *Dunaliella salina* (Chlorophyta, Chlorophyceae). *Journal of Experimental Marine Biology and Ecology* 469: 27–35. https://doi.org/10.1016/j.jembe.2015.04.002.

6

Micropropagation for the Improved Production of Secondary Metabolites

Rupasree Mukhopadhyay

Department of Genetics and Biotechnology, University College for Women, Hyderabad, Telangana, India

6.1 Introduction

From time immemorial, plants have played a prominent role in traditional as well as modern medicine to cure diseases and ailments naturally. This is based on the fact that higher plants produce a diverse range of low molecular-weight chemical compounds with miraculous properties, which are termed as secondary metabolites (Narayani and Srivastava 2017; Upadhyay and Singh 2021; Dixit et al. 2021a). These phytochemicals do not contribute to primary metabolic processes (e.g. respiration, photosynthesis, protein and lipid biosynthesis, etc.), which are directly required for the basic growth and development of the plant. The secondary metabolites are usually produced by specific secondary biosynthetic pathways and play a major role in plant–environment interactions like protection to UV light, defense against herbivores and pathogens, attracting pollinators, repellants to predators that are vital for their survival and existence (Rao and Ravishankar 2002; Pagare et al. 2015; Dixit et al. 2019; Singh et al. 2019). Apart from their role as plant-defense compounds, they also help to induce flowering, fruit setting, and abscission and to maintain perennial growth of plants. While the primary metabolites are regarded as low-value and high-bulk compounds, the secondary metabolites are regarded as high-value and low-bulk compounds due to their complex structures and wide commercial use. These compounds are usually produced and accumulated in specific tissues, structures, and organs (morphological differentiation) like trichomes, specialized glands, and vacuoles. For example, essential oils accumulate in glandular scales and secretory ducts (*Mentha piperita*), tobacco alkaloids are synthesized in roots, and quinine is found in the bark of *Cinchona* trees. The production of secondary metabolites in plants occurs only during certain developmental stages and is often affected by several factors, like their genotype, plant physiology, climatic, geographical and environmental conditions, and endemic pathogens (Shitan 2016; Isah et al. 2018).

Secondary metabolites have gained great attention because of their human-friendly biological activities (e.g. antibiotic, antimicrobial, insecticidal, molluscicidal, hormonal properties, valuable pharmacological and pharamaceutical activities, cosmetic and use as flavors,

Plants as Bioreactors for Industrial Molecules, First Edition.
Edited by Santosh Kumar Upadhyay and Sudhir P. Singh.

fragrances, etc.). Some of these compounds are also highly toxic in nature. A large variety of compounds like alkaloids, terpenoids, bibenzyls, phenyl propanoids, and stilbenoids are solely derived from plants (Table 6.1). These metabolites are biosynthesized from acetyl coenzyme A, mevalonic acid, shikimic acid, deoxyxylulose 5-phosphate, or combined pathways by diverting energy-generating routes in metabolic pathways like photosynthesis, glycolysis, and Krebs cycle into various biosynthetic intermediates. Based on their biosynthetic origins, plant natural secondary metabolites are categorized into three major groups: the terpenoids, the phenolic compounds, and the alkaloids [*N*- and *S*-containing compounds] (Figure 6.1). Each group of secondary metabolites is contributed by a complex network of

Table 6.1 Some examples of secondary metabolites obtained from plants.

Compounds	Examples
Pigments	Carotenoids, anthocyanins, etc.
Terpenoids	Menthol, camphor, lemon grass oil, rubber (polyterpene), etc.
Alkaloids	Codeine, morphine, nicotine, berberine, cocaine, quinine, etc.
Phenylpropanoids	Flavonoids, isoflavonoids, coumarins, stilbenes, tannins, etc.
Quinones	Anthraquinones, benzoquinones, naphthoquinones, etc.
Steroids	Sterols, diosgenin, ferruginol, etc.
Toxins	Abrin, ricin, etc.
Drugs	Vinblastine, curcumin, ginseng, etc.

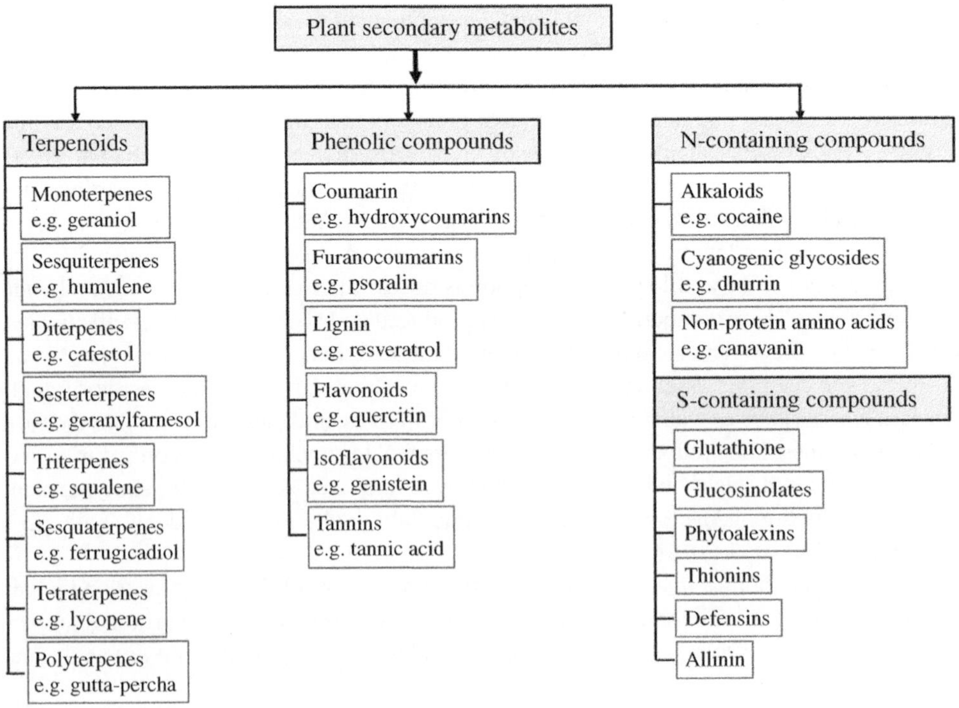

Figure 6.1 Plant secondary metabolites classification.

precursors, enzyme, and cofactors. The terpenoids or isoprenoids constitute a major group of natural compounds produced as a result of defense mechanism to insects/predators that attack plants. Examples of terpenoids include rubber, essential oils, camphor, menthol, and carotenoid pigments. An interesting example of terpenoids is phorbol, a diterpene ester produced by *Euphorbiaceae* species that acts as a skin irritant and toxin to grazing animals (Croteau et al. 2000). Certain phenolic compounds that include a phenolic group in their structure are produced by plants as a defense mechanism to pests or nematodes. This group includes flavonoids, isoflavonoids, lignins, and tannins. Flavonoids like vitamin p or citrin produced by plants possess anti-inflammatory properties (Thirumurugan et al. 2018). The nitrogen and sulfur-containing compounds (glycosylates, alkaloids, phytoalexins, defensins, etc.) are produced by plants against microbial pathogens. Despite being produced as a defense mechanism, these metabolites find use as dietary supplements, in pharmaceutical industries and also to treat cancer (Yang et al. 2018).

Most of the conventional secondary metabolites are produced naturally by the plants, but their commercial production is hindered due to environmental and regional constraints (Yue et al. 2016). Hence, efforts are being made toward extraction, structural elucidation, bioactivity evaluation, and mass-scale production of important high-value compounds. But, this process is greatly limited by the rapid destruction of natural habitats of a large number of plants thereby leading to extinction of many valuable and endemic plant species. Moreover, the conventional methods are time consuming as plant takes several years to grow and reach the point for desired metabolite production. The complexity in structures of many of these compounds makes their chemical synthesis difficult and economically less preferable (Alamgir 2018). In this process, many of the potentially useful compounds remain undiscovered and unidentified. The emergence and advancements in plant tissue culture techniques for active production of secondary metabolites offer a promising alternative to overcome this problem. Production of plant metabolites by callus and cell suspension cultures can be traced back to a period as early as the 1950s (West and Mika 1957). Currently, many advanced *in vitro* plant tissue culture techniques like micropropagation, elicitation, hairy root culture, and biotransformation are being used for large-scale production of secondary metabolites (Verma et al. 2007; Wilson and Roberts 2012; Alok et al. 2018). Additionally, the advancements in biotechnology research tend to provide a new dimension to *in vitro* plant culture by not only improving the yields of the valuable metabolites but also producing multiple and novel products using genetically engineered plants.

6.2 Micropropagation for Production of Secondary Metabolites

Plant tissue culture *in vitro* involves the clonal propagation of new cells, tissues, and organs derived from meristematic parts of the plant (explant) using an artificial nutrient medium under aseptic environment. This method of artificially producing plants vegetatively through tissue culture or cell culture techniques is referred to as micropropagation. The main types of micropropagation methods include callus culture (dedifferentiated callus developed from explants), embryo culture (culture of isolated and rescued embryos), cell suspension culture isolated cells or small callus aggregates dispersed uniformly in liquid media), organ culture (culture of plant organs), meristem culture (culturing shoot buds/meristems to produce disease-free plants), and protoplast culture (isolation and culture of plant protoplasts). The general procedure of micropropagation is a step-wise process involving five stages (Figure 6.2). The composition of nutrient medium along with all culture conditions requires to be optimized and thus plays a major role in the total process.

A wide range of valuable phytochemicals have been produced through callus and suspension cultures till date. The isolated cells of the whole plants or explant are surface sterilized and cultivated under appropriate physiological conditions aseptically, and the desired secondary product is extracted from the cultured cells. The yields of plant secondary metabolites are usually affected by the genetic background, the geographic location, and climatic conditions at the cultivation site, along with the potential effect of harvest and transport methods. By employing micropropagation techniques, it becomes possible to control the quality and quantity of the compound of interest by optimizing the factors affecting its synthesis and accumulation. The use of *in vitro* techniques for secondary metabolite production as an alternative to their extraction from field-grown plants offers many advantages as follows:

1) Metabolites can be produced independently from all environmental, climatic, edaphic, seasonal, and geographical constraints (Karuppusamy 2009).
2) It reduces overexploitation and threat of extinction of natural populations of plant species.

Stages of micropropagation

Stage	Methodology
Stage 0	Selection of mother plant and its maintenance ↓
Stage I	Initiation and establishment of culture ↓
Stage II	Multiplication of shoots/somatic embryo formation ↓
Stage III	Rooting of shoots/*in vitro* germination of somatic embryos ↓
Stage IV	Transfer of plantlets to sterile soil for hardening under greenhouse environment

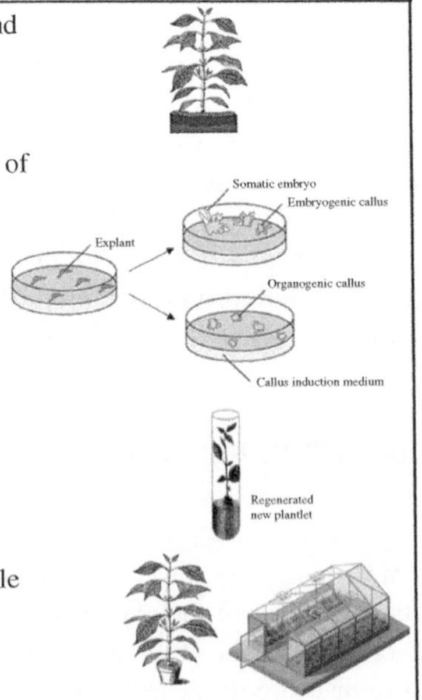

Figure 6.2 Stages of micropropagation.

3) Cultured cells are free of microbe or pathogen contaminations (Murthy et al. 2014; Ochoa-Villarreal et al. 2016).
4) Important compounds with big market demands can be produced under controlled conditions by developing step-by-step protocols.
5) Any cell or plant material can be used as an explant to propagate and produce specific metabolites (Verpoorte et al. 1999).
6) Production is more reliable, simpler, and the resultant metabolites are more predictable.
7) A continuous supply and a consistent product quality and quantity can be ascertained by using improved cell lines.
8) The metabolic pathways can be regulated and controlled to improve the productivity thus reducing cost and labor.
9) In some cases, cultured cells produce higher levels of the metabolite as compared to the differentiated plant tissues.
10) Novel compounds and new routes of synthesis not previously found in whole plants can be recovered from genetically modified cell lines.
11) Cell cultures can be directed to carry out biotransformations to convert specific substrates into more valuable products by single/multistep chemical reactions.
12) The metabolites can be extracted and isolated more rapidly and efficiently as compared to extraction from organized, complex tissues of whole plants, thus reducing the time for product recovery (Verpoorte et al. 2002).

Secondary metabolites can be produced by using micropropagation techniques using callus cultures, cell suspension cultures, and/or organ cultures. Nevertheless, such techniques also have certain limitations like low concentration of the metabolite produced, slow growth, instability of cell lines, lack of basic knowledge of the biosynthetic routes, and high production costs (Ravishankar and Venkataraman 1993). To enhance the yield of metabolites for commercial purposes, several strategies have been adapted (Figure 6.3). Efforts are being made to achieve optimal culture conditions, selection of high-producing strains, and improvizing the existing micropropagation method by introducing organ cultures, precursor feeding, elicitation, immobilization, hairy root cultures, biotransformations, metabolic engineering, and scaling-up using bioreactors (DiCosmo and Misawa 1995).

6.3 Strategies to Improve Secondary Metabolite Production

6.3.1 Optimizing Culture Conditions

The composition of artificial nutrient medium is detrimental for secondary metabolite production. The culture system is usually a one-stage or two-stage process. In one-stage culture system, both growth and production steps occur in the same medium (e.g. berberine production from *Thalictrum minus*) (Nakagawa et al. 1986), while in two-stage culture system, the cell biomass is first grown on a standard growth medium and then transferred into a separate production medium for synthesis of the desired secondary metabolite (Chawla 2009). The synthesis of secondary products varies with the growth phases in a batch or continuous culture system. The product formation is found to be the lowest during

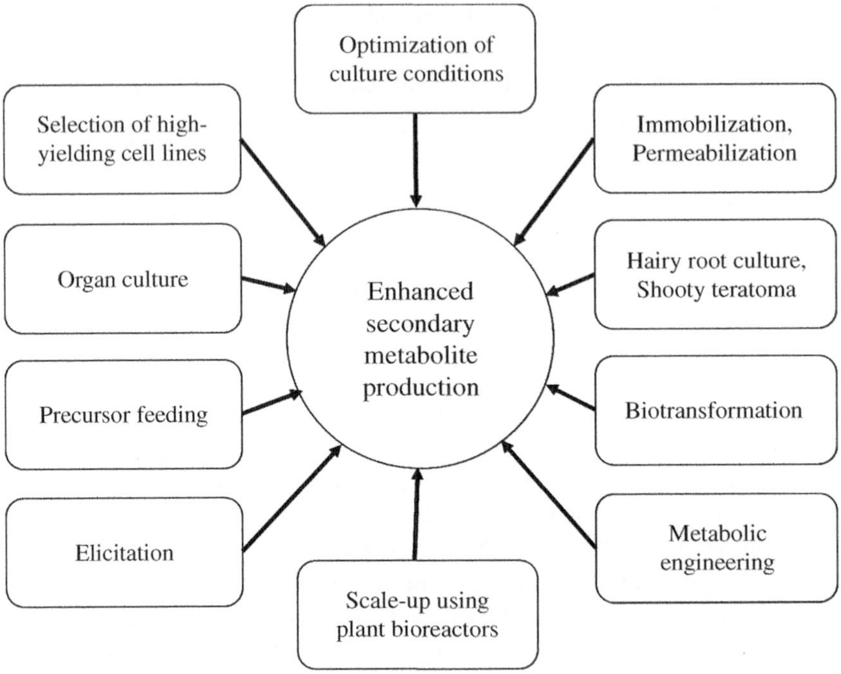

Figure 6.3 Strategies to enhance the production of secondary metabolites.

lag and early exponential phases, while the product conversion and accumulation begins in late exponential phase and reaches its peak during stationary phase. Therefore, it is preferable to produce biomass on a large-scale and then transfer it into an optimized production medium to promote secondary metabolite production.

For achieving a higher yield of the secondary product, several physical and chemical factors like composition of nutrient medium, pH, plant growth regulators, light, and temperature need to be studied and optimized (Wang et al. 1999; Lee and Shuler 2000). The most commonly used nutrient medium for production of secondary metabolites is Murashige and Skoog (MS) medium with an appropriate carbon source like sucrose and glucose and optimized levels of growth regulators. The modification of nutrients, sugar, and phytohormones in the production medium brings about dramatic changes in secondary product synthesis as reported in *Lithospermum erythrorhizon* cell cultures for shikonin production (Tabata et al. 1974). Manipulation of culture conditions and nutrients in media also showed higher accumulation of metabolites in cultured cells than in intact plants as exemplified by rosmarinic acid production in *Coleus bluemei* cultures (Ulbrich et al. 1985), berberine by *Coptis japonica* cultures (Matsubara et al. 1989), and ginsenoside accumulation in *Panax ginseng* cultures (Choi et al. 1994; Franklin and Dixon 1994). Apart from the macro- and micronutrient constitution, the amount of nitrogen also affects secondary metabolite production. A lower ammonia: nitrate ratio in the media promotes shikonin and betacyanin production, while a high ratio of ammonia: nitrate increases the production of berberine and ubiquinone (Ikeda et al. 1977; Fujita et al. 1981; Nakagawa et al. 1984; Bohm and Rink 1988). Phosphates also play an important role in secondary metabolite production.

For example, in *Digitalis purpurea* cultures, increasing phosphate concentration led to an increase in production of digitoxin (Hagimori et al. 1982a). Besides media constituents, physical factors like variation in light and temperature contributes to the ability of plant cell cultures to accumulate secondary metabolites. Light stimulates the production of secondary compounds like carotenoids, flavonoids, polyphenols, and plastoquinones. The accumulation of α- and β-pinene was reported to be highly influenced by light exposure in *Pinus radiata* callus cultures (Banthorpe and Njar 1984). Temperature ranging between 17 and 25 °C is normally used for induction of callus and growth of cultured cells. However, a low temperature treatment to growing calli sometimes show a significant increase in secondary product accumulation as seen in calycosin production by *Astragalus membraceous* seedlings (Haiyun et al. 2007) and bioflavonoid in *Cyclopia subternata* (Kokotkiewicz et al. 2014). Contrastingly, tobacco cell cultures produced higher yield of ubiquinone at 32 °C rather than either 24 or 28 °C (Ikeda et al. 1977). Fluctuations in alkaloid accumulation in *Pappaver* sp. were observed in response to temperature changes (Bernáth and Tétényi 1981).

6.3.2 Selecting High-Producing Cell Lines

An important prerequisite to secondary metabolite production includes selection of high-yielding plant cell lines, which can accumulate the product in increased amounts as compared to the intact plants. High-yielding lines are usually isolated using clonal selection by adapting approaches like single-cell cloning or cell-aggregate cloning. Single-cell cloning involves isolation of single cells from suspension cultures and growing them individually on suitable nutrient media to produce cell populations with altered phenotypes, which can be screened for increased yields. In cell-aggregate cloning, calli obtained from suspension cultures are grown as cell aggregates on solid media, which are divided into two halves – one half for continuing callus growth and the other half for assessing and analyzing the level of secondary product formation. This enables identification of cells showing increased biosynthesis of specific secondary products. Many high-yielding cell lines were established using cell cloning methods. For example, selection of a *Euphorbia milli* strain showing a sevenfold increase in anthocyanin accumulation was reported (Yamamoto et al. 1982). Cell-aggregate cloning was employed to obtain a *Coptis japonica* strain producing higher amounts of berberine (Yamada and Sato 1981).

6.3.3 Organ Cultures

The technique of *in vitro* organ culture is widely used for increasing metabolite yield. In some cases, the metabolite production requires morphological differentiation of tissues (organogenesis) with accumulation of the products in specialized organs. Callus cultures often lack this secondary metabolic activity from physiological differentiation due to its dedifferentiated state. Organ cultures thus provide an advantage of having sites of synthesis and storage of the secondary metabolites in separate compartments. Also, the differentiated cells or organs in culture show greater stability without undergoing rapid genetic variations as the unorganized cells. In most of the cases, differentiated shoots or root cultures are able to produce higher level of metabolites in comparison to native plants

(Karuppusamy 2009). For example, ginseng roots are cultured on large scale *in vitro* as saponins and other valuable metabolites accumulate only in the roots (Kim et al. 2004). *In vitro* organ culture of *Fritillaria unibracteata* rapidly propagates from small cuttings of the bulb and produces higher content of alkaloids than the wild bulb (Gao et al. 2004). Other examples include *in vitro* organ cultures of *Papaver somniferum* (Yoshikawa and Furuya 1985) and *Digitalis purpurea* (Hagimori et al. 1982b), which successfully produced increased amounts of the desired secondary products.

6.3.4 Precursor Feeding

The supplementation of nutrient media with certain precursors or related intermediate compounds from a secondary metabolic biosynthetic pathway, noticeably stimulates secondary metabolite production, as the presence of such compounds increases the yield of the final product (Mulabagal and Tsay 2004). An effective increase in production of plant secondary metabolites by supplying precursor or intermediate compounds has been reported in many plant species. Addition of ferulic acid to *Vanilla planifolia* cultures showed increase in vanillin accumulation (Romagnoli and Knorr 1988). Addition of 500 mM tropic acid to culture medium of *Scopolia japonica* led to 14 folds higher accumulation of alkaloids (Fontanel and Tabata 1987). Precursor feeding can be more useful, economical, and advantageous if the precursors are inexpensive. Amino acids have been widely used as inexpensive precursors in many cell cultures. The incorporation of amino acids valine and isoleucine in media enhanced the biosynthesis of hyperforin and adhyperforin in *Hypericum perforatum* shoot cultures (Karppinen et al. 2007). Similarly, an exogenous feeding of biosynthetic precursor phenylalanine to the culture medium was reported to increase shikonin production in *L. erythrorhizon*, rosmarinic acid yield in *Salvia officinalis,* and taxol production in *Taxus* suspension cultures (Ellis and Towers 1970; Mizukami et al. 1977; Fett-Neto et al. 1993).

6.3.5 Elicitation

Elicitation has been proven to be one of the most effective strategies to improve the production of bioactive secondary compounds using plant tissue culture. Elicitors are defined as signals or compounds that trigger the formation of secondary metabolites. They are grouped into abiotic and biotic elicitors based on their nature (Namdeo 2007). Abiotic elicitors, usually nonbiological in nature, include various abiotic factors like UV radiation, heat, ethylene, fungicides, antibiotics and metal ions like aluminum, lithium, magnesium, calcium, chromium, manganese, and iron. Addition of selenium and $NiSO_4$ for increased ginseng production and increased vincristine and vinblastine production in *Catharanthus roseus* by using NaCl are examples of abiotic elicitation (Jeong and Park 2006; Samar et al. 2015). A rapid accumulation of sesquiterpenoid defensive compounds on treating root cultures of *Datura stramonium* with copper and cadmium salts (Furze et al. 1991). Biotic elicitors composed of substances of biological nature like fungal, yeast, or bacterial cultures, fungal mycellial extracts, culture filtrates, fraction or compounds obtained microbial cell walls (chitin or glucans), and polysaccharides derived from plant cell walls (pectin or cellulose). Based on origin, they are further categorized into exogenous (originate outside the cell) such as fatty acids, polysaccharides, polyamines and endogenous (originate inside the cell) like galacturonide and hepta-β-glucosides. Improved levels of shikonin were reported by using endogenous polysaccharides in cell cultures of *L. erythrorhizon* and using fungal elicitors in *Arnebia euchroma* (Fukui et al. 1990; Fu and Lu 1999). Chitosan, a biotic elicitor brought about an increase in anthraquinone production in *Rubia* akane cell culture (Jin et al. 1999). Treating *P. somniferum* cell suspensions with *Botrytis* mycelium homogenate resulted in high accumulation of sanguinarine (Constabel 1990). A few more examples of elicitor-mediated secondary metabolite production are listed in Table 6.2. Sometimes, metal ions prove to act as a better elicitor than their biotic counterparts because of their ready availability, ease of use, cheaper, and chemically defined nature. For a successful elicitation process, several factors are to be considered like sensitivity of cell cultures strains, elicitor specificity, concentration of the elicitor, and time and duration of elicitor exposure.

Table 6.2 Few examples of elicitors used for secondary metabolite production.

Abiotic elicitor	**Plant species**	**Product(s)**	**References**
Arachidonic acid	*Capsicum annuum*	Capsidiol, Rishitin	Hoshino et al. (1994)
Amino acid starvation	*Arabidopsis*	Camalexin	Zhao et al. (1998)
Copper chloride	*Matricaria chamomilla*	Hemiarin, Umbelliferone	Eliasova et al. (2004)
Cu^{2+}, Cd^{2+}	*Atropa belladona*	Tropane alkaloids	Lee et al. (1998)
Curdlan, Xanthan	*Capsicum frutescens*	Capsaicin	Johnson et al. (1991)
Electromagnetic treatment	*Ammi majus* L.	Umbelliferone	Krolicka et al. (2006)
Salicylic Acid	*Daucus carota*	Chitinase	Muller et al. (1994)
Vanadium sulphate	*Catharanthus roseus*	Catharanthine	Smith et al. (1987)
Biotic elicitor	**Plant species**	**Product(s)**	**References**
Exogenous elicitors			
Cellulase	*C. annuum*	Capsidol	Patrica et al. (1996)
Chitosan	*Ocimum basilicum*	Rosmarinic acid, Eugenol	Kim et al. (2005)
Hemicellulase	*Brugmansia candida*	Hyosujamine	Sandra et al. (1988)
Endogenous elicitors			
Methyl jasmonate	*Hyoscyamus albus*	Phytoalexins	Kuroyanagi et al. (1998)
Alginate oligomer	*Catharanthus roseus*	Ajmalicine	Akimoto et al. (1999)
Whole organism			
Fungal elicitor	*Cupressus lucitanica*	Beta-thujaplicin	Zhao et al. (2001)
Trichoderma viridae	*C. roseus*	Ajmalicine	Namdeo et al. (2002)
Yeast elicitor	*Medicago truncatula*	Beta-amvrin	Broeckling et al. (2005)

6.3.6 Immobilization

There is a tremendous impact of immobilization of plant cells on the enhanced accumulation of secondary metabolites. The plant cells are usually entrapped/immobilized by various techniques onto a fixed support and allowed to grow immersed in liquid nutrient media. The cells may be entrapped in polymer gels like calcium alginate, agarose, agar, carrageenan, gelatin or polyacrylamide, etc. by ionotropic gelation due to ionic cross-linking of polyionic chains with multivalent counter-ions. Alternatively, the cultured cells may be fixed onto matrices/supports like polyurethane foam, glass, sponge, lignocellulosic materials and/or hollow fibre membranes. This technique exploits the ability of suspended cells to become entrapped within interspaces of polymers or meshes which do not interfere with culture system. The immobilized cells show longer viability and improved biosynthetic ability, and can be grown in continuous culture systems to yield higher levels of secondary products. Immobilization of plant cells using alginate offers a simple and non-toxic technique and has been used on large scale. Immobilization of *L. erythrirhizon* cells using calcium alginate showed 2.5 times increase in shikonin production (Kim and Chang 1990). Besides individual cells, it is also possible to immobilize aggregate cells or even calli. Immobilization of homogenous suspensions of cells offers a better cell-to-cell contact, and the cells are protected from high liquid shear stresses. All this helps in the maximal production the secondary metabolite. Moreover, if the product is released into the medium instead of being stored intracellularly, it offers an added advantage to continuous cultures and reduces the steps of downstream processing. It was observed that 50 times more of capsaicin was produced in a continuous culture system using immobilized cell cultures of *Capsicum frutescens* on polyurethane foam under *in vitro* controlled conditions, and leaching out of capsaicin into the medium facilitated product recovery and purification with ease (Johnson and Ravishankar 1996). Similarly, *C. roseus* cells immobilized into polysaccharide beads (Brodelius and Nilsson 1980) and on polyacrylamide sheets (Lambe and Rosevear 1982) produced higher amount of alkaloids with no loss of cell viability. Some more examples of immobilized cell cultures for increased production of metabolites is mentioned in Table 6.3. For continuous culture system, the immobilized of plant cells require to at least partially secrete the products into the medium. An effective large-scale process is possible only if immobilization can be combined with other techniques like elicitation or biotransformation to increase the product yield to greater manifolds.

Table 6.3 Immobilized cell cultures used for secondary metabolite production.

Plant species	Type of immobilization	Product	References
Plumbago rosea	Calcium alginate	Plumbagin	Vanisree et al. (2004)
Glycine max	Hollow fibers	Phenolics	Prenosil and Pedersen (1983)
Dioscoria deltoidea	Polyurethane foam	Diosgenin	Ishida (1988)
Tag. patula	Natural glass	Thiophenes	Hulst and Tramper (1989)
Coffea arabica	Gel	Methylxanthin	Haldimann and Brodelius (1987)
Catharanthus roseus	Membrane	Total alkaloids	Payne et al. (1988)

6.3.7 Permeabilization

In many cases, the desired secondary product is accumulated intracellularly, and efforts are made to increase the permeability of the cells to release out these metabolites, which maybe inhibitory to their own production. Various mechanical and nonmechanical methods have been employed to initiate the release of products from plant cells. The mechanical methods like ultrasonication, electroporation, heat/cold treatments, and high pressure may often cause cell damage. The recent technique of freeze-shattering method retains the integrity of cells and tissues. Electroporation is also a frequently used tool for permeabilization of biological membranes. Among nonmechanical methods, the commonly used techniques include chemical permeabilization using organic solvents (toluene, DMSO, chloroform, benzene, etc.), antibiotics, surfactants (Triton), chaotropic agents and enzymatic permeabilization using enzymes like β-(1-6) and β-(1-3) glucanases, proteases, and mannases. However, the permeabilization of cell membranes is associated with the loss of viability of cells due to formation of pores in one or more of the membrane systems. In *Chenopodium rubrum* cell cultures, high electric field pulses less than 0.75 kV/cm resulted in high levels of permeabilization and product release, but an electric pulse beyond this value led to loss of cell viability (Knorr et al. 1993). Ultrasonication of *Beta vulgaris* cells for 20–60 seconds released the betanin with no loss of cell viability (Kilby and Hunter 1990). High hydrostatic pressure (50 MPa) was applied to increase accumulation of amarantin and antraquinones in cell cultures of *C. rubrum* and *Morinda citrifolia*, respectively (Knorr et al. 1993). A treatment with chitosan-induced pore formation in plasma membrane of cell cultures of *C. rubrum*, which led to increased release of amarantin (Dornenburg and Knorr 1997). *B. vulgaris* cell cultures treated with 0.7 mM Triton-X-100 for 15 minutes induced the release of 30% betacyanines without losing cell viability (Trejo-Tapia et al. 2007).

6.3.8 Genetic Transformation: Hairy Root Cultures and Shooty Teratomas

Genetic transformation of plant cell cultures using *Agrobacterium tumefaciens* and *Agrobacterium rhizogenes* not only offers a promising approach to production of transgenic plants but also aids to increase the manufacture of various important secondary metabolites (Saito et al. 1992). The insertion of *Ri* plasmid from *A. rhizogenes* into a wounded tissue induces the growth of adventitious hairy roots, which can be cultured *in vitro* in a hormone-free medium to yield increased amounts of secondary products (Guillon et al. 2006). This technique was accepted widely because of its advantages like genetic stability, rapid hormone-independent growth of roots, easy scale-up, and the direct introduction of foreign genes to induce the biosynthetic capacity of the cultures (Shanks and Morgan 1999). It was observed that the metabolites produced by hairy root cultures are chemically and functionally identical to those synthesized by the parent plants (Sevon and Oksman-Caldentey 2002). The transformed roots also produced higher yields of secondary metabolites than their parent plants (Giri and Narasu 2000). Secondary compounds like azadirachtin, betalain, camptothecin, and whitanolide A were produced in higher quantities using hairy root cultures (Pavlov et al. 2005; Shajahan et al. 2017; Thakore and Srivastava 2017; Wetterauer et al. 2018). Some more examples include production of tropane alkaloids such as scopolamine in *Hyoscamus muticus* (Jouhikainen et al. 1999) and hyoscyamine in *Hyoscyamus reticulatus* (Khezerluo et al. 2018) hairy root cultures, ginsenosides from ginseng hairy root lines (Woo et al. 2004), and solanoside by *Physalis* sp. (Putalun et al. 2004). Coupling the hairy root culture systems with elicitation may drastically influence the accumulation of secondary products as seen in the increased production of isoflavonoid in *Pueraria candollei* hairy roots after elicitation by the yeast extract (Udomsuk et al. 2011). Introduction of foreign genes to enhance production of metabolites using transgenic hairy root culture technique has been widely applied to many plant species. For example, introduction of 6-Hydroxylase gene from *Hysocyamus muticus* into *Atropa belladonna* through *A. rhizogenes*-mediated transformation led to an increased production of scopolamine (Jeong and Park 2006). Apart from hairy root cultures, transgenic shoot cultures termed as "shooty teratomas" are obtained after infection with nopaline strains of *Agrobacterium tumefaciens* containing modified Ti plasmid lacking auxin genes (Alvarez et al. 1994; Subroto et al. 1996). This promotes uncontrolled shoot proliferation after infection since the cytokinin gene (*ipt*, isopentenyl transferase) is retained in the T-DNA (Rhodes et al. 1992). Genetic manipulation of T-DNA also shows that *onc* genes involved in teratoma development may modify the resulting shoot phenotype. The growth of shooty teratomas does not require exogenous growth regulators unlike untransformed shoots, and they serve as a good source for producing metabolites usually synthesized in leaves and aerial parts of plants. Due to the difficulty in establishing rapid growth of shooty teratomas in liquid medium, there are very limited reports of their application for secondary metabolite production. In a study, oil glands containing significant quantities of mint oil were reported in shooty teratomas induced in *Mentha citrata* (Spencer et al. 1990; Spencer et al. 1993). *Agrobacterium tumefaciens* C58-induced shooty teratomas of *C. roseus* successfully exhibited 10 folds increase in vincristine production as compared to the untransformed control (Begum et al. 2009). Similar studies were reported in *Atropa belladonna*, *Nicotiana tabacum*, and *Solanum tuberosum* teratomas for manufacture of tropane, nicotine, and steroidal alkaloids, respectively (Saito et al. 1991).

6.3.9 Biotransformation

Biotransformation or bioconversion is defined as the ability of plant cell cultures to catalyze the conversion of readily available inexpensive precursors into a more valuable final product through chemical reactions catalyzed by cells, organs, or enzymes. This process explores the unique properties of plant cells (biocatalysts), especially their stereo- and regiospecificity (Hamada and Furuya 1996) and their ability to carry out several reactions such as, oxidation, hydroxylation, reduction, methylation, aminoacylation, glucosylation-acylation at nonextreme pH values, and temperatures (Rao and Ravishankar 2002). An added advantage of this technique is that it may be used to produce novel chemicals or in combination with organic synthesis utilizing the potential of plant cells to transform cheap and plentiful substances, such as industrial by-products, into rare and expensive products (Meyer et al. 1997). The production of digoxin, a cardiac glycoside, by conversion of β-methyl-digitoxin to β-methyl-digoxin using *Digitalis lanata* cell cultures is a classic example of biotransformation on a large scale (Hamada et al. 1996). Based on the reactions involved, biotransformations are grouped into pathway and nonpathway type. The pathway/specific biotransformations exploit a characteristic biosynthetic pathway of the plant

Table 6.4 Production of secondary metabolites in plant cell cultures using biotransformations.

Plant species	Substrate	Product	References
Catharanthus roseus	Vindoline, Vinblastine	Vincristine	DiCosmo and Misawa (1995)
Eucalyptus perriniana	Taxol	Taxol derivatives	Hamada et al. (1996)
Coffea arabica	Vanillin	Vanillin glucosides	Johnson and Ravishankar (1996)
Mentha spp	Menthol	Neomenthol	Dornenburg and Knorr (1997)
Mucuna pruriens	Tyrosine	DOPA	Pras et al. (1993)
Papaver somniferuem	Thebaine	Codeine	Wilhelm and Zenk (1997)
Vanilla planifolia	Ferulic acid	Vanillin	Romagnoli and Knorr (1988)

or use a natural intermediate of the normal biosynthetic pathway such as the bioconversion of digitoxin and digitoxigenin in cultures of *Digitalis purpurea*. The nonpathway/nonspecific biotransformation may not utilize an intermediary pathway of the cell culture and may involve biochemical transformation of exogenously added substrates. The reaction type and stereochemistry depend upon functional groups in the substrate and the structural moieties in the vicinity of functional group. A list of secondary metabolites produced by biotransformation is given in Table 6.4. Biotransformations using immobilized plant cells using single and multistep conversions of precursors into high-value products have proven beneficial for secondary metabolite production. The use of both free suspension cultures and immobilized cultures of *C. frutescens* used for bioconversion of protocatechuic aldehyde and caffeic acids to vanillin and capsaicin showed that the yield was more in immobilized system rather than free cells. The conversion of isoeugenol to vanillin flavor metabolites and capsaicin was further enhanced by β-cyclodextrin and fungal elicitor (Rao and Ravishankar 2002). Consequently, biotransformation studies also provide the basic information needed to elucidate the biosynthetic pathway, and catalysis can be carried out under mild conditions, thus reducing undesired by-products, energy, safety, and costs.

6.3.10 Metabolic Engineering

Metabolic engineering of the biosynthetic pathways is a promising strategy to produce a secondary metabolite constitutively in larger quantities. To enhance the secondary metabolite production, there is a need to study the biosynthetic pathways and its regulatory steps, which are detrimental for overexpressing the regulatory genes (Verpoorte et al. 1999; O'Connor 2015; Dixit et al. 2021b). As most of the secondary metabolic pathways are poorly defined and understood, efforts are made to elucidate these pathway genes along with their control elements and also identify the rate-limiting steps (Kolewe et al. 2008). This is followed by overexpression/suppression of the genes using recombinant DNA technology and more recent technologies like omics and integrated network analysis. Various strategies are

Table 6.5 Use of metabolic engineering in plant secondary metabolite production.

Plant species	Technique	Gene(s)/ enzyme(s)	Target compounds	References
Catharanthus roseus	Transcription factor	ORC A	Increase of terpenoid indole alkaloid	Memelink and Gantet (2007)
Papaver somniferum	Single transgene	Codeinone reductase (COR)	Increase of morphine	Philip et al. (2007)
Nicotiana tabacum	Overexpression of target gene	Avian farnesyl diphosphate synthase, Terpene synthase	Increase of sesquiterpenes patchoulol and amorpha-4,11-diene, monoterpene limonene	Wu et al. (2006)
Medicago sativa	Transcription factor	LAP1 MYB	High accumulation of anthocyanins	Peel et al. (2009)
Papaver somniferum	RNAi	Salutaridinol 7-*O*-acetyl transferase (SalAT)	Accumulation of salutaridinol	Allen et al. (2008)
Eschscholzia californica	RNAi	Berberine bridge enzyme (BBE)	Accumulation of (s)-reticuline	Fujii et al. (2007)

employed to overexpress the gene product that limits the specific pathway, overproduce precursors of secondary metabolites, downregulate the existing reaction using antisense methods, create a new branch for a preexisting pathway, manipulate regulatory genes viz. transcriptional activators or repressors, and select the regulatory mutants with increased tissue-independent expression for *in vitro* production of metabolites. Sometimes, lack of knowledge of the enzymes and genes involved limits the success of the process, which can be overcome by utilizing current day technologies like genomics, proteomics, and metabolomics. These provide knowledge about the genetic makeup of the organism along with the role of factors, enzymes, and biosynthetic pathways for enhanced metabolite production, which in turn helps in understanding various secondary metabolic pathways. Further, microarray analysis aids to obtain a profile of transcriptome and identify the key genes involved in the production of secondary metabolites. Thus, enhanced levels of secondary metabolites using metabolic engineering can be achieved by adopting simple steps like increasing precursors flux, introducing new routes of metabolism, blocking competitive pathways, overexpressing regulatory genes or transcription factors that induce the pathway, increasing the concentration of specialized cells producing the metabolite, inhibiting, or limiting catabolism, and overcoming rate-limiting steps (Koffas et al. 1999; Gómez-Galera et al. 2007; Upadhyay 2021). Many plant species have been targeted to produce important metabolites in improved quantities by employing these approaches (Table 6.5).

6.3.11 Plant Bioreactors and Scale-up

Large-scale production of secondary metabolites essentially requires stable and genetically identical plant cell cultures that can be grown uninterruptedly in continuous system. The application of immobilized cells and plant bioreactors combined with improvized and efficient micropropagation methods leads toward commercial production of plant secondary products on industrial scale. Plant cell suspensions grown in liquid media provide perfect inoculum for many different bioreactor types designed specifically for scale-up manufacture. Further, the drawbacks of solid medium growth and labor-intensive conventional methods can be overcome by mechanization and automation of the micropropagation process using diverse bioreactor systems, not only to scale-up the process but also to improve the productivity and economic benefits (Steingroewer et al. 2013). Additionally, plant bioreactors also aid in faster growth of cell cultures, uniform transfer of nutrients and gaseous components, and easy scaling-up procedures. The use of specially designed plant bioreactors for micropropagation was first reported in *Begonia* cultures (Takayama and Misawa 1981). Since then, various bioreactors have been designed keeping in view the parameters like agitation and aeration without damaging the cells in culture. Especially in hairy root cultures, mechanical agitation results in shear stress that causes wounding and callus formation, thus lowering the productivity. Hence to attain best growth characteristics, special bioreactors of varying sizes and features were designed avoiding mechanical stirring. These include the airlift bioreactors, bubble-column bioreactors, mist reactors, wave reactors, packed-bed, and fluidized-bed bioreactors (Figure 6.4) (Eibl and Eibl 2007). With the evolution in technology, bioreactors are now commonly categorized as mechanically agitated, pneumatically agitated, nonagitated, and temporary immersion systems (TIS). Usually, a two-stage cultivation process involving optimization of growth parameters and induction of product formation with subsequent harvest and extraction is preferred to produce large amounts of biomass and increased metabolite biosynthesis (Murthy et al. 2014). Various important metabolites were produced at commercial scale using scale-up procedures like shikonin, berberine, scopolamine, ginsenosides, and rosmarinic acid (Kolewe et al. 2008; Isah et al. 2018). One of the most successful examples of commercial production of secondary metabolites using scale-up process is the production of taxol by Phyton Biotech Company, Germany in 2002 (Wink et al. 2005).

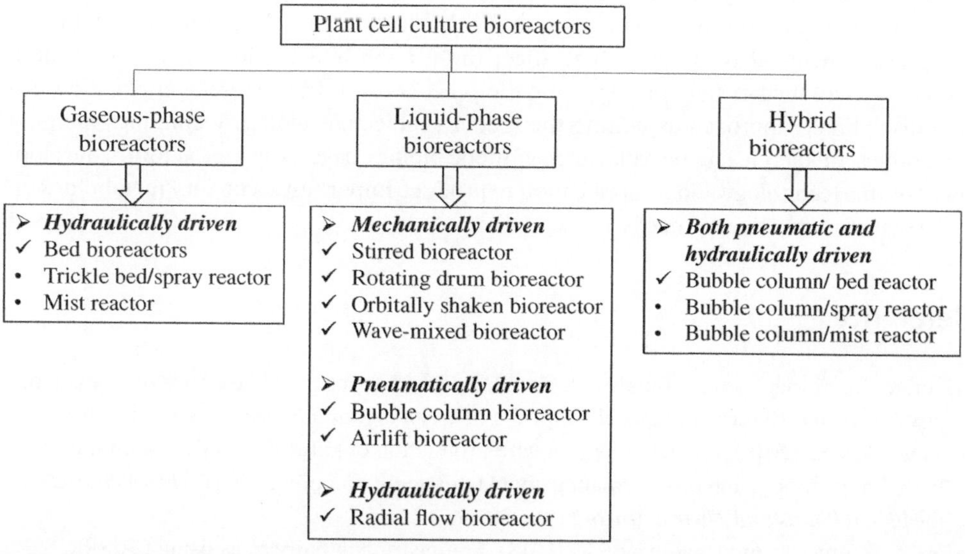

Figure 6.4 Classification of plant cell bioreactors.

6.4 Conclusions

Micropropagation techniques have proven to be a promising alternative to the conventional methods for the production of important plant-based phytochemicals. The *in vitro* propagation of plant cells and tissues in aseptic culture systems in large provide a continuous source of sustainable and economical production of secondary metabolites, independent of geographical and climatic constraints, and eliminates contamination of any form. Further, it permits the rapid clonal multiplication of plants on mass scale, within a short duration of time, which is highly necessary for conservation and extraction of important chemicals from endangered plant species. The increasing demands of these cost-effective natural phytochemicals have directed the researchers to focus on devising strategies to improve the yields by adapting various methods like optimizing the growth conditions by manipulation of nutrients as well as culture conditions and employing different biotechnological methods like precursor feeding, elicitation, immobilization, genetic manipulation, and metabolic engineering. It has been observed that the efficient selection of high-yielding species, the types of explants chosen, and the culture conditions are the decisive factors influencing the final yield of the product and play a major role in scaling-up process. Also, the innate potential of plant cells to act as biocatalysts and transform available inexpensive substrates into high-value compounds can lead to the discovery of many novel commercially important secondary metabolites in addition to an enhanced production of the already existing ones. Despite all the successful research over the years, the large-scale production of important natural metabolite still requires further optimizations, specially with regard to the low yield, incomplete knowledge of the pathways, and product recovery methods. To overcome the hurdles of low productivity and difficulties in scaling up, a clear understanding of the secondary metabolism and associated biosynthetic pathways within the cells/organelles is absolutely essential. Biotechnological approaches such as increase in number of cloned biosynthetic genes, improvized transformation and expression systems, and engineering biosynthetic pathways, in combination with a deeper understanding of the physicochemical parameters that affect the growth and production levels inside a bioreactor, will further help in upgrading the process toward commercial application. An interdisciplinary approach exploiting the modern molecular biology techniques merging with omics, metabolic engineering, and synthetic biology paves way for the future development of this technology and its application to produce important secondary metabolites on large scale to serve human needs.

References

Akimoto, C., Aoyagi, H., and Tanaka, H. (1999). Endogenous elicitor-like effects of alginate on physiological activities of plant cells. *Applied Microbiology and Biotechnology* 52: 429–436.

Alamgir, A.N.M. (2018). Biotechnology, *in vitro* production of natural bioactive compounds, herbal preparation, and disease management (treatment and prevention). *Therapeutic Use of Medicinal Plants and Their Extracts* 2 (74): 585–664.

Alok, A., Kumar, J., and Upadhyay, S.K. (2018). Engineering in hairy roots using CRISPR/Cas9-mediated editing. In: *Hairy Roots* (ed. V. Srivastava, S. Mehrotra, and S. Mishra). Singapore: Springer. https://doi.org/10.1007/978-981-13-2562-5_15.

Allen, R.S., Miller, J.A.C., Chitty, J.A. et al. (2008). Metabolic engineering of morphinan alkaloids by over-expression and RNAi suppression of salutaridinol 7-*O*-acetyltransferase in opium poppy. *Plant Biotechnology Journal* 6: 22–30.

Alvarez, M.A., Talou, J.R., Paniego, N.B., and Giulietti, A.M. (1994). Solasodine production in transformed organ cultures (roots and shoots) of *Solanum eleagnifolium* Cav. *Biotechnology Letters* 16: 393–396.

Banthorpe, D.V. and Njar, V.C.O. (1984). Light-dependent monoterpene synthesis in *Pinus radiata* cultures. *Phytochemistry* 23: 295–299.

Begum, F., Nageswara Rao, S.S., Rao, K. et al. (2009). Increased vincristine production from *Agrobacterium tumefaciens* C58 induced shooty teratomas of *Catharanthus roseus* G. Don. *Natural Product Research* 23 (11): 973–981.

Bernáth, J. and Tétényi, P. (1981). The effect of environmental factors on growth, development and alkaloid production of poppy (*Papaver somniferum* L.): II. Interaction of light and temperature. *Biochemie und Physiologie der Pflanzen* 176 (7): 599–605.

Bohm, H. and Rink, E. (1988). Betalaines. In: *Cell Culture and Somatic Cell Genetics of Plants* (ed. F. Constabel and I. Vasil). New York: Academic Press.

Brodelius, P. and Nilsson, K. (1980). Entrapment of plant cells in different matrices. *FEBS Letters* 122: 312–316.

Broeckling, C.D., Huhman, D.V., Farag, M.A. et al. (2005). Metabolic profiling of *Medicago truncatula* cell cultures reveals the effects of biotic and abiotic elicitors on metabolism. *Journal of Experimental Botany* 56: 323–336.

Chawla, H.S. (2009). *Introduction to Plant Biotechnology*, 3e. New Delhi: Oxford & IBH Publishing Co. Pvt. Ltd.

Choi, K.T., Ahn, I.O., and Park, J.C. (1994). Production of ginseng saponin in tissue culture of ginseng (*Panax ginseng* C.A. Mayer). *Russian Journal of Plant Physiology* 41: 784–788.

Constabel, F. (1990). Medicinal plant biotechnology. *Planta Medica* 56: 421–425.

Croteau, R., Kutchan, T.M., and Lewis, N.G. (2000). Natural products (secondary metabolites). *Biochemistry and Molecular Biology of Plants* 24: 1250–1319.

DiCosmo, F. and Misawa, M. (1995). Plant cell and tissue culture: alternatives for metabolite production. *Biotechnology Advances* 13: 425–435.

Dixit, S., Shukla, A., Singh, V., and Upadhyay, S.K. (2021a). Bioprospecting of natural compounds for industrial and medical applications. In: *Bioprospecting of Plant Biodiversity for Industrial Molecules* (ed. S.K. Upadhyay and S.P. Singh), 351–379. John Wiley & Sons Ltd. https://doi.org/10.1002/9781119718017.ch3.

Dixit, S., Shukla, A., Singh, V., and Upadhyay, S.K. (2021b). Engineering of plant metabolic pathway for nutritional improvement. In: *Genome Engineering for Crop Improvement* (ed. S.K. Upadhyay), 351–379. John Wiley & Sons Ltd. https://doi.org/10.1002/9781119672425.ch20.

Dixit, S., Shukla, A., Upadhyay, S.K., and Verma, P.C. (2019). Mode of communication between plants during environmental stress. In: *Molecular Approaches in Plant Biology and Environmental Challenges*. (ed. S. Singh, S.K. Upadhyay, A. Pandey and S. Kumar). Energy, Environment, and Sustainability. Singapore: Springer. https://doi.org/10.1007/978-981-15-0690-1_7.

Dornenburg, H. and Knorr, D. (1997). Challenges and opportunities for metabolite production from plant cell and tissue cultures. *Food Technology* 51: 47–54.

Eibl, R. and Eibl, D. (2007). Design of bioreactors suitable for plant cell and tissue cultures. *Phytochemistry Reviews* 7: 593–598.

Eliasova, A., Repcak, M., and Pastirova, A. (2004). Quantitative changes of secondary metabolites of *Matricaria chamomilla* by abiotic stress. *Zeitschrift für Naturforschung* 59: 543–548.

Ellis, B.E. and Towers, G.H.N. (1970). Biogenesis of rosamrinic acid in *Mentha*. *Journal of Biochemistry* 118: 291–297.

Fett-Neto, A.G., Melanson, S.J., Sakata, K., and Di-Cosmo, F. (1993). Improved growth and taxol yield in developing calli of *Taxus cuspidata* by medium composition modification. *Nature Biotechnology* 11: 731–734.

Fontanel, A. and Tabata, M. (1987). Production of secondary metabolites by plant tissue and cell cultures – present aspects and prospects. In: *Nestlé Reserach News*, 93–103. Vevey, Switzerland: Nestec Ltd.

Franklin, C.I. and Dixon, R.A. (1994). Initiation and maintenance of callus and cell suspension cultures. In: *Plant Cell Culture-A Practical Approach*, 2e (ed. R.A. Dixon and R.A. Gonzales), 1–25. Oxford: IRL Press.

Fu, X.-Q. and Lu, D.-W. (1999). Stimulation of shikonin production by combined fungal elicitation and in situ extraction in suspension cultures of *Arnebia euchroma*. *Enzyme and Microbial Technology* 24: 243–246.

Fujii, N., Inui, T., Iwasa, K. et al. (2007). Knockdown of berberine bridge enzyme by RNAi accumulates (*S*)-reticuline and activates a silent pathway in cultured California poppy cells. *Transgenic Research* 16 (3): 363–375.

Fujita, Y., Hara, Y., Ogino, T., and Suga, C. (1981). Production of shikonin derivatives by cell suspension cultures of *Lithospermum erythrorhizon*. *Plant Cell Reports* 1: 59–60.

Fukui, H., Tani, M., and Tabata, M. (1990). Induction of shikonin biosynthesis by endogenous polysaccharides in *Lithospermum erythrorhizon* cell suspension cultures. *Plant Cell Reports* 9: 73–76.

Furze, J.M., Rhodes, M.J.C., Parr, A.J. et al. (1991). Abiotic factors elicit sesquiterpenoid phytoalexin production but not alkaloid production in transformed root cultures of *Datura stramonium*. *Plant Cell Reports* 10: 111–114.

Gao, S.L., Zhu, D.N., Cai, Z.H. et al. (2004). Organ culture of a precious Chinese medicinal plant – *Fritillaria unibracteata*. *Plant Cell, Tissue and Organ Culture* 59: 197–201.

Giri, A. and Narasu, M.L. (2000). Transgenic hairy roots: recent trends and applications. *Biotechnology Advances* 18: 1–22.

Gómez-Galera, S., Pelacho, A., Gené, A. et al. (2007). The genetic manipulation of medicinal and aromatic plants. *Plant Cell Reports* 26 (10): 1689–1715.

Guillon, S., Tremouillaux-Guiller, J., Pati, P.K. et al. (2006). Hairy root research: recent scenario and exciting prospects. *Current Opinion in Plant Biology* 93: 341–346.

Hagimori, M., Matsumoto, T., and Obi, Y. (1982a). Studies on the production of *Digitalis purpurea* cardenolides by plant tissue culture. *Plant Physiology* 69: 653–656.

Hagimori, M., Matsumoto, T., and Obi, Y. (1982b). Studies on the production of *Digitalis* cardenolides by plant tissue cultures. *Plant & Cell Physiology* 23: 1205–1211.

Haiyun, P., Changming, F., Tong-Shui, Z. et al. (2007). Accumulation of calycosin and its 7-*O*-beta-*D*-glucoside and related gene expression in seedlings of *Astragalus membranaceus* Bge. var. mongholicus (Bge.) Hsiao induced by low temperature stress. *Plant Cell Reports* 26: 1111–1120.

Haldimann, D. and Brodelius, P. (1987). Redirecting cellular metabolism of cultured plant cells: a model study with *Coffea arabica*. *Phytochemicals* 26: 1431–1434.

Hamada, H. and Furuya, T. (1996). Recent advances in plant biotransformations. *Plant Tissue Culture and Biotechnology* 2: 52–59.

Hamada, H., Sanada, K., Furuya, T. et al. (1996). Biotransformation of paclitaxel by cell suspension cultures of *Eucalyptus perriniana*. *Natural Product Letters* 9: 47–52.

Hoshino, T., Chida, M., Yamaura, T. et al. (1994). Phytoalexin induction in green pepper cultures treated with arachidonic acid. *Phytochemistry* 36: 1417–1419.

Hulst, A.C. and Tramper, J. (1989). Immobilized plant cells: a literature survey. *Enzyme and Microbial Technology* 11: 546–558.

Ikeda, T., Matsumoto, T., and Noguchi, M. (1977). Effects of inorganic nitrogen source and physical factors on the formation of ubiquinone by tobacco plant cells in suspension culture. *Agricultural and Biological Chemistry* 41: 1197–1201.

Isah, T., Umar, S., Mujib, A. et al. (2018). Secondary metabolism of pharmaceuticals in the plant in vitro cultures: strategies, approaches, and limitations to achieving higher yield. *Plant Cell Tissue and Organ Culture* 132: 239–265.

Ishida, B. (1988). Improved diosgenin production in *Dioscorea deltoidea* cell cultures by immobilized in polyurethane foam. *Plant Cell Reports* 7: 270–273.

Jeong, G.T. and Park, D.H. (2006). Enhanced secondary metabolite biosynthesis by elicitation in transformed plant root system: effect of abiotic elicitors. *Applied Biochemistry and Biotechnology* 129–132: 436–446.

Jin, J.H., Shin, J.H., Kim, J.H. et al. (1999). Effect of chitosan elicitation and media components on the production of anthraquinone colorants in madder (*Rubia akane* Nakai) cell culture. *Biotechnology and Bioprocess Engineering* 4: 300–304.

Johnson, T.S. and Ravishankar, G.A. (1996). Precursor biotransformation in immobilized placental tissues of *Capsicum frutescens* Mill: I. Influence of feeding intermediate metabolites of the capsaicinoid pathway on capsaicin and dihydrocapsaicin accumulation. *Journal of Plant Physiology* 147: 481–485.

Johnson, T.S., Ravishankar, G.A., and Venkataraman, L.V. (1991). Elicitation of capsaicin production in freely suspended cells and immobilized cell cultures of *Capsicum frutescens*. *Food Biotechnology* 5: 197–205.

Jouhikainen, K., Lindgren, L., Jokelainen, T. et al. (1999). Enhancement of scopolamine production in *Hyoscyamus muticus* L. hairy root cultures by genetic engineering. *Planta* 208: 545–551.

Karppinen, K., Hokkanen, J., Tolonen, A. et al. (2007). Biosynthesis of hyperforin and adhyperforin from amino acid precursors in shoot cultures of *Hypericum perforatum*. *Phytochemistry* 68 (7): 1038–1045.

Karuppusamy, S. (2009). A review on trends in production of secondary metabolites from higher plants by *in vitro* tissue, organ and cell cultures. *Journal of Medicinal Plant Research* 3 (13): 1222–1239.

Khezerluo, M., Hosseini, B., and Amiri, J. (2018). Sodium nitroprusside stimulated production of tropane alkaloids and antioxidant enzymes activity in hairy root culture of *Hyoscyamus reticulatus* L. *Acta Biologica Hungarica* 69: 437–448.

Kilby, N.J. and Hunter, C.S. (1990). Towards optimisation of the use of 1.02-MHz ultrasound to harvest vacuole-located secondary product from in vitro grown plant cells. *Applied Microbiology and Biotechnology* 34: 478–480.

Kim, D.J. and Chang, H.N. (1990). Enhanced shikonin production from *Lithospermum erythrorhizon* by *in situ* extraction and calcium alginate immobilization. *Biotechnology and Bioengineering* 36 (5): 460–466.

Kim, Y.S., Hahn, E.J., Murthy, H.N., and Paek, K.Y. (2004). Adventitious root growth and ginsenoside accumulation in *Panax ginseng* cultures as affected by methyl jasmonate. *Biotechnology Letters* 26 (21): 1619–1622.

Kim, H.J., Chen, F., Wang, X., and Rajapakse, N.C. (2005). Effect of chitosan on the biological properties of sweet basil (*Ocimum basilicum* L.). *Journal of Agricultural and Food Chemistry* 53: 3696–3701.

Knorr, D., Caster, C., Dörnenburg, H. et al. (1993). Biosynthesis and yield improvement of food ingredients from plant cell and tissue culture. *Food Technology* 47 (12): 57–63.

Koffas, M., Roberge, C., Lee, K., and Stephanopoulos, G. (1999). Metabolic engineering. *Annual Review of Biomedical Engineering* 1 (1): 535–557.

Kokotkiewicz, A., Bucinski, A., and Luczkiewicz, M. (2014). Light and temperature conditions affect bioflavonoid accumulation in callus cultures of *Cyclopia subternata* Vogel (honeybush). *Plant Cell, Tissue and Organ Culture* 118 (3): 589–593.

Kolewe, M.E., Gaurav, V., and Roberts, S.C. (2008). Pharmaceutically active natural product synthesis and supply via plant cell culture technology. *Molecular Pharmaceutics* 5: 243–256.

Krolicka, A., Kartanowicz, R., Wosinski, S.A. et al. (2006). Induction of secondary metabolite production in transformed callus of *Ammi majus* L. grown after electromagnetic treatment of the culture medium. *Enzyme and Microbial Technology* 39: 1386–1391.

Kuroyanagi, M., Arakava, T., Mikami, Y. et al. (1998). Phytoalexins from hairy root culture of *Hyoscyamus albus* treated with methyl jasmonate. *Journal of Natural Products* 61: 1516–1519.

Lambe, C.A. and Rosevear, A. (1982). Production of chemical compounds from viable cells. U.K. Patent No. GB2, 096, 169 A.

Lee, C.W.T. and Shuler, M.L. (2000). The effect of inoculum density and conditioned medium on the production of ajmalcine and catharanthine from immobilized *Catharanthus roseus* cells. *Biotechnology and Bioengineering* 67: 61–71.

Lee, T.K., Yamakawa, T., Kodama, T., and Shimomura, K. (1998). Effects of chemicals on alkaloid production by transformed roots of *Atropa belladona*. *Phytochemistry* 49: 2343–2347.

Matsubara, K., Shigekazu, K., Yoshioka, T. et al. (1989). High density culture of *Coptis japonica* cells increases berberine production. *Journal of Chemical Technology and Biotechnology* 46: 61–69.

Memelink, J. and Gantet, P. (2007). Transcription factors involved in terpenoid indole alkaloid biosynthesis in *Catharanthus roseus*. *Phytochemistry* 6 (2): 353–362.

Meyer, H.P., Kiener, A., Imwinkelried, R., and Shaw, N. (1997). Biotransformations for fine chemical production. *Chimia International Journal for Chemistry* 51: 287–289.

Mizukami, H., Konoshima, M., and Tabata, M. (1977). Effect of nutritional factors on shikonin derivative formation in *Lithospermum* callus cultures. *Phytochemistry* 16: 1183–1186.

Mulabagal, V. and Tsay, H. (2004). Plant cell cultures – an alternative and efficient source for the production of biologically important secondary metabolites. *International Journal of Applied Science and Engineering* 2: 29–48.

Muller, S.S., Kurosaki, F., and Nishi, A. (1994). Role of salicylic acid and intrercelluar Ca^{2+} in the induction of chitinase activity in carrot suspension culture. *Physiological and Molecular Plant Pathology* 45: 101–109.

Murthy, H.N., Lee, E.J., and Paek, K.Y. (2014). Production of secondary metabolites from cell and organ cultures: strategies and approaches for biomass improvement and metabolite accumulation. *Plant Cell Tissue and Organ Culture* 118: 1–16.

Nakagawa, K., Konagai, A., Fukui, H., and Tabata, M. (1984). Release and crystallization of berberine in the liquid medium of *Thalictrum minus* cell suspension cultures. *Plant Cell Reports* 3: 254–257.

Nakagawa, K., Fukui, H., and Tabata, M. (1986). Hormonal regulation of berberine production in cell suspension cultures of *Thalictrum minus*. *Plant Cell Reports* 5: 69–71.

Namdeo, A.G. (2007). Plant cell elicitation for production of secondary metabolites: a review. *Pharmacognosy Reviews* 1: 69–79.

Namdeo, A., Patil, S., and Fulzele, D.P. (2002). Influence of fungal elicitors on production of ajmalicine by cell cultures of *Catharanthus roseus*. *Biotechnology Progress* 18: 159–162.

Narayani, M. and Srivastava, S. (2017). Elicitation: a stimulation of stress in in vitro plant cell/tissue cultures for enhancement of secondary metabolite production. *Phytochemistry Reviews* 16: 1227–1252.

O'Connor, S.E. (2015). Engineering of secondary metabolism. *Annual Reviews in Genetics* 49: 71–94.

Ochoa-Villarreal, M., Howat, S., Hong, S. et al. (2016). Plant cell culture strategies for the production of natural products. *BMB Reports* 49: 149–158.

Pagare, S., Bhatia, M., Tripathi, N. et al. (2015). Secondary metabolites of plants and their role: overview. *Current Trends in Biotechnology and Pharmacy* 9 (3): 293–304.

Patrica, M., Moctezuma, L., and Gloria, L.E. (1996). Biosynthesis of sesqutiterpenic phytoalaxin captidiol in elicited root cultures of chilli peppers (*C. annum*). *Plant Cell Reports* 15: 360–366.

Pavlov, A., Georgiev, V., and Ilieva, M. (2005). Betalain biosynthesis by red beet (*Beta vulgaris* L.) hairy root culture. *Process Biochemistry* 40: 1531–1533.

Payne, G.F., Payne, N.N., Shuler, M.L., and Asada, M. (1988). In situ adsorption for enhanced alkaloid production by *Catharanthus roseus*. *Biotechnology Letters* 10: 187–192.

Peel, G.J., Pang, Y., Modolo, L.V., and Dixon, A. (2009). The LAP1 MYB transcription factor orchestrates anthocyanidin biosynthesis and glycosylation in *Medicago*. *The Plant Journal* 59: 136–149.

Philip, J.L., James, A.C.M., Robert, S.A. et al. (2007). Increasing morphinan alkaloid production by over – expressing codeinone reductase in transgenic *Papaver somniferum*. *Plant Biotechnology Journal* 5 (1): 26–37.

Pras, N., Woerdenbag, H.J., Batterman, S. et al. (1993). *Mucuna pruriens*: improvement of the biotechnological production of the anti-Parkinson drug L-DOPA by plant cell selection. *Pharmacy World & Science* 15: 263–268.

Prenosil, J.E. and Pedersen, H. (1983). Immobilized plant cell reactors. *Enzyme and Microbial Technology* 5: 323–331.

Putalun, W., Prasarnsiwamai, P., Tanaka, H., and Shoyama, Y. (2004). Solasodine glycoside production by hairy root cultures of *Physalis minima* Linn. *Biotechnology Letters* 26: 545–548.

Rao, S.R. and Ravishankar, G.A. (2002). Plant cell cultures: chemical factories of secondary metabolites. *Biotechnology Advances* 20 (2): 101–153.

Ravishankar, G.A. and Venkataraman, L.V. (1993). Role of plant cell culture in food biotechnology: current trends, limitations and future prospects. In: *Plant Biotechnology: Commercial Prospects and Problems* (ed. J. Prakash and R.L.M. Pierik), 255–274. New Delhi: Oxford IBH Press.

Rhodes, M.J.C., Spencer, A., Hamill, J.D., and Robins, R.J. (1992). Flavour improvement through plant cell culture. In: *Bioformation of Flavours* (ed. R.L.S. Patterson, B.V. Charlwood, G. MacLeod and A.A. Williams), 42–64. Cambridge: Royal Society of Chemistry.

Romagnoli, L.G. and Knorr, D. (1988). Effects of ferulic acid treatment on growth and flavor development of cultured *Vanilla planifolia* cells. *Food Biotechnology* 2 (1): 93–104.

Saito, K., Yamazaki, M., Kawaguchi, A., and Murakoshi, I. (1991). Metabolism of solanaceous alkaloids in transgenic plant teratomas integrated with genetically engineered genes. *Tetrahedron* 47: 5955–5968.

Saito, K., Yamazaki, M., and Murakoshi, I. (1992). Transgenic medicinal plants: agrobacterium-mediated foreign gene transfer and production of secondary metabolites. *Journal of Natural Products* 55 (2): 149–162.

Samar, F., Abdul, M., and Dipti, T. (2015). NaCl amendment improves vinblastine and vincristine synthesis in *Catharanthus roseus*: a case of stress signalling as evidenced by antioxidant enzymes activities. *Plant Cell, Tissue and Organ Culture* 121 (2): 1–15.

Sandra, I., Alvarez, P., and Giulietti, A.M. (1988). Novel biotechnological approaches to obtain scopolamine and tryosumine: the influence of biotic elicitors and stress agents on cultures of transformed rats of *Bygumansia candida*. *Phytotherapy Research* 12: 518–520.

Sevon, N. and Oksman-Caldentey, K.M. (2002). *Agrobacterium rhizogenes* mediates transformation: root cultures as a source of alkaloids. *Planta Medica* 68: 859–868.

Shajahan, A., Thilip, C., Faizal, K. et al. (2017). An efficient hairy root system for withanolide production in *Withania somnifera* (L.) Dunal. In: *Production of Plant Derived Natural Compounds through Hairy Root Culture*, 133–143. Cham: Springer International Publishing.

Singh, S.P., Upadhyay, S.K., Pandey, A., and Kumar, S. (2019). Molecular approaches in plant biology and environmental challenges. In: *Molecular Approaches in Plant Biology and Environmental Challenges* (ed. S. Singh, S.K. Upadhyay, A. Pandey, and S. Kumar). Energy, Environment, and Sustainability. Singapore: Springer. https://doi.org/10.1007/978-981-15-0690-1_1.

Shanks, J.V. and Morgan, J. (1999). Plant hairy root culture. *Current Opinion in Biotechnology* 10: 151–155.

Shitan, N. (2016). Secondary metabolites in plants: transport and self-tolerance mechanisms. *Bioscience, Biotechnology, and Biochemistry* 80: 1283–1293.

Smith, J.I., Smart, N.J., Misawa, M. et al. (1987). Increased accumulation of indole alkaloids by some cell lines of *Catharanthus roseus* in response to addition of vanadyl sulphate. *Plant Cell Reports* 6: 142–145.

Spencer, A., Hamill, J.D., and Rhodes, M.J.C. (1990). Production of terpenes by differentiated shoot cultures of *Mentha citrata* transformed with *Agrobacterium tumefaciens* T37. *Plant Cell Reports* 8: 601–604.

Spencer, A., Hamill, J.D., and Rhodes, M.J.C. (1993). *In vitro* biosynthesis of monoterpenes by *Agrobacterium* transformed shoot cultures of two *Mentha* species. *Phytochemistry* 32: 911–919.

Steingroewer, J., Bley, T., Georgiev, V. et al. (2013). Bioprocessing of differentiated plant in vitro systems. *Engineering in Life Sciences* 13: 26–38.

Subroto, M.A., Hamill, J.D., and Doran, P.M. (1996). Development of shooty teratomas from several solanaceous plants: growth kinetics, stoichiometry and alkaloid production. *Journal of Biotechnology* 45: 45–57.

Tabata, M., Mizukami, H., and Hiraoka, N. (1974). Pigment formation in callus culture of *Lithospermum erythrorhizon*. *Phytochemistry* 13: 927–932.

Takayama, S. and Misawa, M. (1981). Mass propagation of *Begonia* × *hiemalis* plantlets by shake culture. *Plant and Cell Physiology* 22: 461–467.

Thakore, D. and Srivastava, A.K. (2017). Production of biopesticide azadirachtin using plant cell and hairy root cultures. *Engineering in Life Sciences* 17: 997–1005.

Thirumurugan, D., Cholarajan, A., Raja, S.S., and Vijayakumar, R. (2018). An introductory chapter: secondary metabolites. In: *Secondary Metabolites: Sources and Applications* (ed. R. Vijayakumar and S.S.S. Raja), 1–21. IntechOpen.

Trejo-Tapia, G., Cuevas-Celis, J., Salcedo-Morales, G. et al. (2007). *Beta vulgaris* suspension cultures permeabilized with Triton x-100 retain cell viability and betacyanines production ability: a digital image analysis study. *Biotechnology Progress* 23 (2): 359–363.

Udomsuk, L., Jarukamjorn, K., Tanaka, H., and Putalun, W. (2011). Improved isoflavonoid production in *Pueraria candollei* hairy root cultures using elicitation. *Biotechnology Letters* 33: 369–374.

Ulbrich, B., Weisner, W., and Arens, H. (1985). Large-scale production of rosmarinic acid from plant cell cultures of *Coleus blumei* benth. In: *Primary and Secondary Metabolism of Plant Cell Cultures* (ed. K.H. Neumann, W. Barz and E. Reinhard), 293–303. Berlin, Heidelberg: Springer.

Upadhyay, S.K. (ed.) (2021). *Genome Engineering for Crop Improvement*. John Wiley & Sons, Ltd. https://doi.org/10.1002/9781119672425.

Upadhyay, S.K. and Singh, S.P. (eds.) (2021). *Bioprospecting of Plant Biodiversity for Industrial Molecules*, 1–431. John Wiley & Sons Ltd. https://doi.org/10.1002/9781119718017.

Vanisree, M., Chen, Y.L., Shu-Fung, L. et al. (2004). Studies on the production of some important secondary metabolites from medicinal plants by plant tissue cultures. *Botanical Bulletin of Academia Sinica* 45: 1–22.

Verma, P.C., Trivedi, I., Singh, H. et al. (2009). Efficient production of gossypol from hairy root culture of cotton (*Gossypium hirsutum*). *Current Pharmaceutical Biotechnology* 10: 691–700. https://doi.org/10.2174/138920109789542048.

Verpoorte, R., Heijden, R.V.D., Ten-Hoopen, H.J.G., and Memelink, J. (1999). Metabolic engineering of plant secondary metabolite pathways for the production of fine chemicals. *Biotechnology Letters* 21: 467–479.

Verpoorte, R., Contin, A., and Memelink, J. (2002). Biotechnology for the production of plant secondary metabolites. *Phytochemistry Reviews* 1: 13–25.

Wang, H.Q., Yu, J.T., and Zhong, J.J. (1999). Significant improvement of taxane production in suspension cultures of *Taxus chinensis* by sucrose feeding strategy. *Process Biochemistry* 35: 479–483.

West, F.R. Jr. and Mika, E.S. (1957). Synthesis of atropine by isolated roots and root-callus cultures of *Belladonna*. *Botanical Gazette* 119: 50–54.

Wetterauer, B., Wildi, E., and Wink, M. (2018). Production of the anticancer compound camptothecin in root and hairy root cultures of *Ophiorrhiza mungos* L. In: *Biotechnological Approaches for Medicinal and Aromatic Plants* (ed. N. Kumar), 303–341. Singapore: Springer Singapore.

Wilhelm, R. and Zenk, M.H. (1997). Biotransformation of thebaine by cell cultures of *Papaver somniferum* and *Mohonia nervosa*. *Phytochemicals* 46: 701–708.

Wilson, S.A. and Roberts, S.C. (2012). Recent advances towards development and commercialization of plant cell culture processes for the synthesis of biomolecules. *Plant Biotechnology Journal* 10 (3): 249–268.

Wink, M., Alfermann, A.W., Franke, R. et al. (2005). Sustainable bioproduction of phytochemicals by plant in vitro cultures: anticancer agents. *Plant Genetic Resources: Characterisation and Utilisation* 3: 90–100.

Woo, S.S., Song, J.S., Lee, J.Y. et al. (2004). Selection of high ginsenoside producing ginseng hairy root lines using targeted metabolic analysis. *Phytochemistry* 65: 2751–2761.

Wu, S., Schalk, M., Clark, A. et al. (2006). Redirection of cytosolic or plastidic isoprenoid precursors elevates terpene production in plants. *Nature Biotechnology* 24 (11): 1441–1447.

Yamada, Y. and Sato, F. (1981). Production of berberine in cultured cells of *Coptis japonica*. *Phytochemistry* 20: 545–547.

Yamamoto, Y., Mizuguchi, R., and Yamada, Y. (1982). Selection of a high and stable pigment-producing strain in cultured *Euforbia millii* cells. *Theoretical and Applied Genetics* 61: 113–116.

Yang, L., Wen, K.S., Ruan, X. et al. (2018). Response of plant secondary metabolites to environmental factors. *Molecules* 23 (4): 762.

Yoshikawa, T. and Furuya, T. (1985). Morphinan alkaloid production by tissues differentiated from cultured cells of *Papaver somniferum*. *Planta Medica* 2: 110–113.

Yue, W., Ming, Q.-l., Lin, B. et al. (2016). Medicinal plant cell suspension cultures: pharmaceutical applications and high-yielding strategies for the desired secondary metabolites. *Critical Reviews in Biotechnology* 36 (2): 215–232.

Zhao, J., Williams, C.C., and Last, R.L. (1998). Induction of *Arabidopsis* tryptophan pathway enzymes and camalexin by amino acid starvation, oxidative stress and an abiotic elicitor. *Plant Cell* 10: 359–370.

Zhao, J., Fujita, K., Yamada, J., and Sakai, K. (2001). Improved beta-thujaplicin production in *Cupressus lusitanica* suspension cultures by fungal elicitor and methyl jasmonate. *Applied Microbiology and Biotechnology* 55: 301–305.

7

Metabolic Engineering for Carotenoids Enrichment of Plants

Monica Butnariu

Chemistry & Biochemistry Discipline, University of Life Sciences "King Mihai I" from Timisoara, Timisoara, Timis, Romania

7.1 Background

Metabolomics is one of the "OMIC" sciences that has as its key components the following two major parts: chemical analysis of all metabolites in a plant extract, followed by a second, calculation and interpretation of huge data sets obtained from analyses.

In order to meet these requirements, the rapidly evolving field of metabolomics combines sophisticated analytical technologies with the most advanced statistical methods such as chemometry, for information collection and data interpretation.

In the past 10 years, sensitive, high-resolution methods suitable for determining as large a set of metabolites as possible have been developed in accordance with the two major approaches to metabolomics: target and nontarget analyses. This complex approach makes metabolomics an important tool used to explain, understand, and receive information in many areas of research, such as human disease and nutrition, the discovery of new drugs, plant physiology, the food industry, and more (Kusmita et al. 2022).

Carotenoid pigments represent all plant CARs. These are the most common pigments in the plant kingdom. In the animal kingdom, these pigments come from plant foods. CARs are pigments colored yellow, orange, red, purple, etc. In the plant body, it is free or in combination with haloproteins and carbohydrates (carotenoproteins, carotenoid glycosides). CAR pigments are spread in all plant organs with or without chlorophyll (leaves, fruits, stem, bulb, seeds, etc.). The content of CAR pigments depends on the nature of the species and the influence of environmental conditions.

CARs are substances of a terpenoid nature, with a number of 40-carbon in the molecule, yellow, orange, or red, which have strong lipophilic properties. They are found in many species, ranging from colored pigments in bacteria to most higher plants.

The structure of CARs is characterized by the fact that they have 40-carbon in the molecule and a considerable number of conjugated double bonds.

Plants as Bioreactors for Industrial Molecules, First Edition.
Edited by Santosh Kumar Upadhyay and Sudhir P. Singh.

There are several types of classifications, the simplest being those that group these compounds into colorless and colored, respectively into CARs (for CAR hydrocarbons) and xanthophylls (for oxygenated compounds).

The typical acyclic structure is the structure of lycopene (lycopine), from tomato fruits, tetraterpenoid hydrocarbon with 13 double bonds. The numbering is done according to a symmetrical system that assumes the formula as a bis-diterpene.

By closing the end of the polyene chain, due to the pseudo-cyclic structure and as a result of a cyclization process, the β-ionic structure is formed, typical for the CAR group.

For the name of the CARs (which are around 600), the nonsystematic nomenclature is usually used, which most often uses the source of obtaining (CAR comes from carrot – carrot), but a systematic nomenclature has been developed that takes into account the type of the terminal group at the end of the polyene chain (there are seven such terminal groups).

CARs are a class of yellow-orange natural pigments found mainly in fruits and vegetables. Due to its structure, lycopene is insoluble in water, but poorly soluble in oil at room temperature, which makes it difficult to incorporate into food matrices. In addition, CARs are sensitive to light, heat, and oxygen, which limits their use in the food industry (Sandmann 2021).

For example, α-CAR is, according to this nomenclature, β, ε-CAR.

CARs can be separated by extraction with nonpolar organic solvents (benzene, petroleum ether, and ethyl ether) and distillation to obtain crude CARs, which also contain other lipoids. Purification is done by basic hydrolysis of lipoids and CAR esters and crystallization of CARs from petroleum ether-methanol. A more advanced purification is done by fractionation on a chromatographic column. In its pure state, CARs are solids with a melting point close to 200 °C (β-CAR has a melting range of 181–184 °C). Their color varies in solution, depending on the concentration, from yellow to red. Pure hydrocarbon CARs are red, and xanthophylls have various shades of yellow. In general, CARs are strongly lipophilic, very labile, easily oxidizable, especially in the presence of air, heat, and light. Various color reactions (blue staining with antimony trichloride in chloroform solution), chromatography in thin layer or on paper impregnated with aluminum oxide, UV, IR, ^{1}H-NMR, and MS spectroscopies are used to identify CARs (Sreekumar et al. 2022).

There is no general method for dosing, there is a volumetric method for the determination of β-CAR in plant products. An efficient separation of β-CAR is obtained by passing the extractive solution on a column of aluminum oxide and eluting the area corresponding to the CAR, then photometry (Keawmanee et al. 2022).

The methods used today are the spectrophotometric ones, where the problem of high purity standards is raised.

7.2 Classification of Carotenoid Pigments

A more natural system, of structural consideration but also of classification, is the one that distinguishes between acyclic, monocyclic, and bicyclic CARs, regardless of whether they are hydrocarbons or oxygenated compounds because the biosynthesis line being the same, oxygenation is a collateral stage for any of the basic hydrocarbon structures depending on the chemical structure:

- CAR or CAR hydrocarbons;
- oxygenated hydrocarbon derivatives or xanthophylls;
- esters, ethers, glycosides of xanthophylls, and epoxides of hydrocarbons;
- keto- and aldo-CARs;
- CAR acids;
- CARs with aromatic rings;
- acetylene CARs;
- allenic CARs;
- pentacycle CARs.

Depending on the number of carbon atoms in the molecule:

- with 40–50 carbon atoms: homo-CARs or higher CARs;
- with 40-carbon: CARs;
- with 20–40-carbon: degraded apo-CARs or CARs;
- with 30 carbon atoms: diapo-CARs.

Depending on the nature of the polyene system:

- with normal polyene system;
- with retro-shaped polyene system.

CAR pigments represent all plant CARs. These are the most common pigments in the whole plant kingdom. In the animal kingdom, these pigments come from plant foods. CARs are pigments colored in yellow, orange, red, purple, etc. In the plant body, it is free or in combination with holoproteins and carbohydrates (CAR proteins and CAR glycosides) (Kupka et al. 2022). The classification of carotenoids can be seen in Figure 7.1.

CAR pigments are spread in all plant organs with or without chlorophyll (leaves, fruits, stem, bulb, seeds, etc.). Due to the hydrocarbon structure, CAR pigments are hydrophobic substances, soluble only in organic solvents, oils, and fats.

From a chemical point of view, CAR pigments are characterized by a conjugated double bond structure, which determines the unsaturated character and therefore the possibility of oxidation and self-oxidation reactions (in the presence of air), the ability to absorb light radiation, etc.

The biochemical role of CAR pigments is determined by their chemical structure and the properties mentioned above.

CARs, like terpenes, sterols, phytol, vitamins K, vitamins E, etc., have as a repetitive structural unit the activated isoprene, which can form CAR, steroid, etc. hydrocarbon chains. CARs are divided into: CAR hydrocarbons and oxygenated derivatives of CAR hydrocarbons. CAR hydrocarbons are CARs with 40-carbon of the crude formula $C_{40}H_{56}$.

The most important of these are lycopene, α-CAR, β-CAR, γ-CAR [$C_{40}H_{56}$].

Lycopene is a crystalline substance, red-purple in color, insoluble in water, and soluble in organic solvents. It is the coloring of fruits and tomatoes.

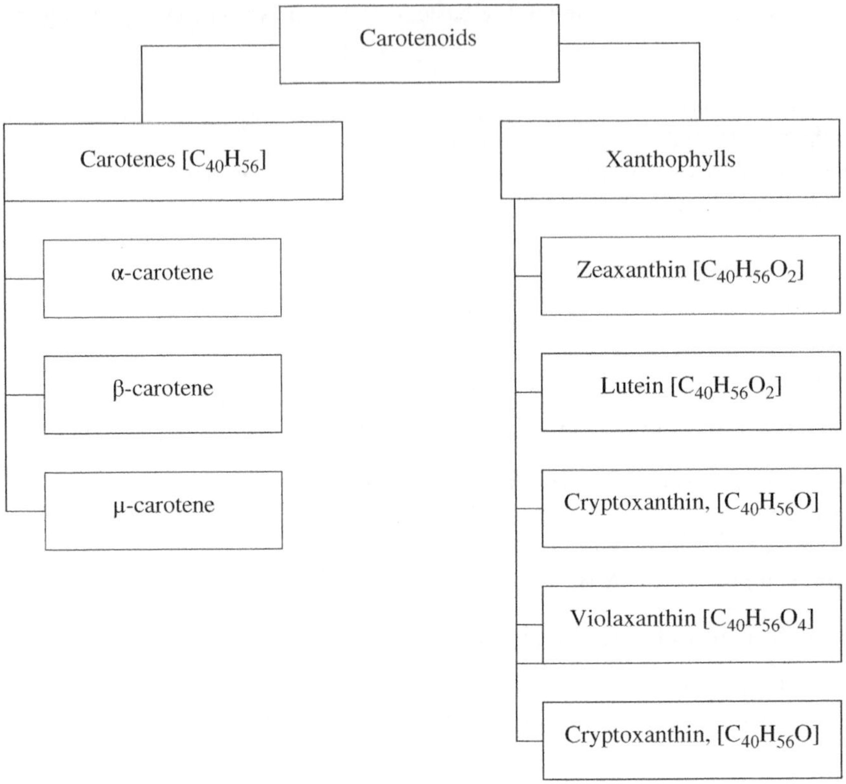

Figure 7.1 Schematic illustration of carotenoids classifications.

α-CAR is a copper-colored crystalline substance, soluble in organic solvents. By heating, α-CAR can be converted to β-CAR. In plants, it is found in smaller amounts than β-CAR.

β-CAR is in the form of violet crystals, soluble in organic solvents. It is very widespread throughout the plant kingdom, constantly accompanying chlorophyll. By oxidative enzymatic hydrolysis, β-CAR is transformed into two vitamin A1 molecules. β-CAR is the main provitamin A.

γ-CAR is in the form of red crystals with blue reflections, soluble in organic solvents. γ-CAR is not widespread in the plant kingdom. It is mostly found in carrots (*Daucus carota* L.) (Gupta and Hirschberg 2022).

Oxygenated derivatives of CARs are alcohols, ketones, acids, etc., derived from CAR hydrocarbons. Many representatives of this category of compounds are known, of which representatives would be:

Of the CAR acids, with less than 40-carbon, the most important are bixin and crocetin, yellow pigments. CAR pigments intervene in the process of photosynthesis, having a role both in absorbing light energy and in protecting against self-destruction of chlorophyll molecules or other active substances (cytochromes, peroxidases, catalases, flavonoid pigments, vitamin B12, vitamins E, vitamins K, etc.).

Because they can fix oxygen by forming unstable oxygenated compounds, CAR pigments are involved in redox processes.

They can form intermediate metabolites that stimulate or inhibit plant growth. In the animal body, these pigments play the role of provitamin A. They are also important for their action in Phototropism and Phototaxis. CARs have many uses in the pharmaceutical, cosmetic, food industry (as antioxidants and natural dyes), and animal husbandry (animal feed), etc.

CARs are the most important compounds in the food industry, being naturally fat-soluble pigments, most of which are C_{40} terpenoids (Olmedilla-Alonso et al. 2021).

The most important CARs obtained by microbial fermentation are α-CAR, β-CAR, lycopene [$C_{40}H_{56}$], torulene [$C_{40}H_{54}$], and torularhodin [$C_{40}H_{52}O_2$].

They act as antioxidants that protect the membrane, effectively removing $^1O_2$ and peroxyl radicals; their antioxidant effectiveness is apparently related to their structure.

CARs are a class of yellow-orange natural pigments found mainly in fruits and vegetables. Due to its structure, lycopene is insoluble in water, but poorly soluble in oil at room temperature, which makes it difficult to incorporate into food matrices.

In addition, CARs are sensitive to light, heat, and oxygen, which limits their use in the food industry. CARs are an important group of natural and fat-soluble pigments in plants and microorganisms, which are yellow, orange, or red.

They are divided into two groups: CAR (carbon and hydrogen molecules), such as beta-CAR and lycopene, and xanthophylls (carbon, hydrogen and oxygen molecules), such as lutein, zeaxanthin [$C_{40}H_{56}O_2$], and violaxanthin [$C_{40}H_{56}O_4$].

Among the known plant pigments, CARs are the most abundant and differ in terms of structural diversity and a wide range of biological action.

In higher plants, CARs are biosynthesized and localized in cellular plastids, where they are bound in light-sensitive complexes, participating in the process of photosynthesis and protecting plants from OSs caused by excessive light.

Of the 700 known CARs, 40 are constantly present in human food; only β-CAR, alpha-CAR, and cryptoxanthin have provitamin (A) activity in mammals.

CARs are considered to be one of the most powerful single oxygen eliminators. The antioxidant properties of these compounds largely determine their biological activity (Kabir et al. 2022).

Although CARs are present in many traditional foods, the richest sources for humans are brightly colored vegetables, fruits, and juices, the yellow-orange vegetables and fruits providing most of the intake of β- and α-CAR.

Oranges are sources of α-cryptoxanthin, dark green vegetables are sources of lutein, and chili peppers and plants from the genus Capsicum are sources of capsanthin, which is the mixture of esters of capsanthin, capsorubin, carotenoids, cryptocapsin, cryptoxanthin, violaxanthin, and zeaxanthin (Hien et al. 2022).

According to the level of CAR accumulation between vegetable crops, spinach, rich in lutein and zeaxanthin, are leaders, as well as representatives of the fat pepper genus containing capsanthin and capsorubin in fruits.

Among the exogenous factors, the growth temperature, the light intensity, the duration of the light period, and the use of fertilizers have a significant effect on the accumulation of CARs. So, it is known that in the shade, the content of lutein and β-CAR in plants is lower than in light, and in summer grown cabbage has higher concentrations of these CARs than when grown in winter.

As the plant grows, the CAR content of the leaves increases and decreases during the aging stage, i.e. the amount of CARs in the plant depends on the time of harvest.

Experimental studies confirm that organic farming offers the highest accumulation of red and yellow pigments in sweet pepper fruits.

Due to their antioxidant properties, CARs are especially attractive in the fight against chronic diseases such as cancer, cardiovascular disease, diabetes, and osteoporosis.

The most important biological function of CARs in the human body is the activity of provitamin (A) (Mavrommatis et al. 2022).

CARs with this activity:

- support the differentiation of healthy epithelial cells;
- normalizes the reproduction functions; and
- sight.

Vitamin A is part of the visual pigment rhodopsin, which explains the important role in maintaining vision of β-CAR, α-CAR, and β-Cryptoxanthin.

In particular, vitamin A deficiency in food can lead to the development of so-called "chicken blindness," characterized by a significant decrease in retinal sensitivity to twilight and, in severe cases, to the development – called "tubular" vision, when cells light-sensitive peripheral retina ceases to function.

Lutein and zeaxanthin are two of the 7 CARs found in blood plasma and are the only CARs in the retina and lens. In the retina, lutein and zeaxanthin are responsible for yellow pigmentation and are called macular pigments.

This area occupies only 2% of the entire surface of the retina and consists exclusively of conical cells responsible for color vision.

It has been suggested that macular pigments are involved in photoprotection and that low levels of lutein and zeaxanthin may be associated with retinal damage.

Increasing the amount of these pigments can be achieved by increasing the intake of antioxidants, vegetables and fruits, food CARs, normalizing body mass index, and quitting smoking. Many of these factors are also associated with a reduced risk of senile macular degeneration, suggesting a causal relationship (Takayanagi et al. 2021).

Research shows that increasing the proportion of lutein and zeaxanthin, as well as lycopene, reduces the risk of macular degeneration.

It should be noted in particular that the high levels of consumption of various vegetables by ensuring the intake of a variety of CARs in the body reduces the risk of eye disease more strongly than the consumption of individual CARs.

CARs are yellow, orange, or red pigments that are universally spread in both the plant and animal kingdoms. The human body does not have the ability to synthesize CARs; all CARs found in animals are of plant origin and are introduced into the body with nutrients.

CARs that are not metabolized in the human body in vitamin A are stored in various tissues in the case of ruminants – CAR accumulates in the storage fats and milk fat in these fats are also present CAR, cryptoxanthin, and lutein.

Birds preferentially store oxy-CARs in the liver, egg, body fat, skin, feathers in the natural materials in which they are found, CARs appear in low concentrations (less than 0.1% in carrots). They self-oxidize easily.

Naturally, natural materials contain more CARs with much more stunning properties. Their separation is relatively difficult in most cases by the chromatographic method (Dong et al. 2022).

For the detection of CARs, the color reactions (dark blue or purple blue) that they give with strong acids, such as concentrated sulfuric acid, hydrochloric acid, and perichloric acid, as well as arsenic and antimony trichlorides are useful. These reactions are nonspecific, being found in other pollen. Absorption spectra are very useful for CAR characterization. About 80 CARs have been isolated so far and the structure of about 50 of them has been determined. Most have molecules made up of 40-carbon and only a few have smaller molecules. These are probably products of oxidative degradation of the CARs themselves. CARs are classified as CAR hydrocarbons, $C_{40}H_{56}$, and their oxygen derivatives (alcohols, epoxides, furanoid oxides, and ketones); a third group also comprises carboxylic acids, containing in the molecule less than 40-carbon. CARs or xanthophylls (originally phylloxanthins), isolated from hen's eggs Willstätter are considered pure substances, are, as later work has shown, mixtures of several similar isomeric substances. The name xanthophylls (introduced by Berzelius) has been retained to denote the group of alcohols with CAR skeleton (Bae et al. 2021).

Xanthophylls are usually found in plants, esterified with higher fatty acids.

Therefore, after extraction of the dry natural material, in an inert gas atmosphere with solvents (benzene, alcohol ether, methanol, ethanol, etc.) and removal of the solvent, the residue is subjected to hydrolysis with methanolic KOH.

7.2.1 Carotenoid Hydrocarbons

Lycopene, $C_{40}H_{56}$, dark red-violet crystals (by dark red-brown spray), with melting point (m.p.) 175 °C, soluble in benzene and carbon sulfide, almost insoluble in ethanol, is the dye in red aubergines but has also been identified in over 70 of plant species, especially in fruit, as well as in some animal materials (butter, blood serum, and human liver).

For example, in the case of oranges, more than 50% of oxy-CARs are in the form of esters, 25% in the unesterified form, and 10% in the form of CARs (Tarshish et al. 2022).

β-CAR, $C_{40}H_{56}$, can crystallize from benzene-methanol in dark purple hexagonal prisms and from petroleum ether in dark red rhombic plates of 183 °C. β-CAR is extremely widespread in nature, being contained in all green plants, as a permanent companion of chlorophyll (along with lutein, lutein epoxide, and sometimes α-CAR). Apricots are an excellent source of CAR, in pears, β-CAR represents only 10% of total CARs. Spinach (leaves) contains up to 40 mg/kg β-CAR, and lutein, zeaxanthin, violaxanthin, neoxanthin are also present.

α-CAR, $C_{40}H_{56}$, forms violet crystals with m.p. 187 °C. It is almost as common in animal and plant materials as β-carotine, but is always found to be less than it (traces, up to 25%, compared to β-carotine).

μ-CAR, $C_{40}H_{56}$, red crystals with blue surface luster, m.p. 178 °C, was discovered by chromatography of crude CAR (in carrots). It is relatively rare in nature.

The types of CARs differ from plant to plant. Thus, CARs are also found more in fruits than in leaves. β-CAR represents up to 60% of all CARs in fruits. μ-Carotine comes in the form of red crystals with blue reflections, soluble in organic solvents.

μ-CAR is not widespread in the plant kingdom. It is mostly found in *D. carota* L. Oxygen derivatives of CARs are alcohols, ketones, acids, etc., derived from CAR hydrocarbons. Many representatives of these derivatives are known: xanthophylls, CAR ketones, CAR acids, etc. Corn contains zeaxanthin, zeaxanthin isomers, lutein, cryptoxanthin, neo-cryptoxanthin, α, β, μ-CAR, and α-CAR hydroxy (Kapoor et al. 2022).

Of the vegetable oils, crude palm oil has the highest concentration of CARs (500 mg/kg), with α and β-CARs being found in the highest amount and in a 3 : 2 ratio. CARs also predominate in butter (6 mg/kg), while oxy-CARs (xanthophylls) do not exceed 1 mg/kg. Egg yolk contains up to 90 mg/kg CARs, the most important being lutein, zeaxanthin, cryptoxanthin, and in smaller amounts β-CAR (Al-Hassan et al. 2021).

In higher plants, CARs are present in chloroplasts. Fruit CARs are associated with proteins.

7.2.2 Xanthophylls

Representative are xanthophylls, CAR ketones, CAR acids, etc.

They are hydroxyl derivatives of carotenoid hydrocarbons. Several xanthophylls are known, the most important of which are lutein and zeaxanthin.

Lutein is a yellow crystalline substance with a purple sheen. It accompanies β-CAR and chlorophyll in all green plants. It is the yellow dye of flowers. It is also found in egg yolks, animal tissues, and especially in the ovary. Zeaxanthin is in the form of orange crystals. Structurally, it differs from lutein by the position of a double bond (2 β-ion cycles).

Lutein (xanthophyll), $C_{40}H_{56}O_2$, forms transparent yellow crystals, with metallic violet surface luster, m.p. 193 °C, strong dextrose. After β-CAR, lutein is the most widespread CAR in nature. It insolence CAR and chlorophyll in all green plants. Lutein also appears in many yellow and red flowers. It is also found in many animal tissues, egg yolk, ovaries, and canary feathers. In the ripening fruits, there is a decrease in the chlorophyll content and an increase in the CAR content, as well as the CAR / xanthophyll ratio (Cao et al. 2021).

In some roots (*D. carota* L.), the CARs are also located in plastids in lipophilic or globule paintings, such as filaments or crystals. Not all CARs in the plant and animal kingdoms can be considered as precursors of vitamin A. In order to have vitamin A activity in the animal body, a CAR compound must have at least one ionic ring and a polyene chain attached to it. The other end of the polyene chain may have a cyclic or acyclic structure. The polyene chain must have at least 11 carbon atoms.

Zeaxanthin, $C_{40}H_{56}O_2$, yellow plates with m.p. 215 °C, was discovered not only in corn but also in many other vegetables.

Cryptoxanthin, $C_{40}H_{56}O$, has been found in peppers, egg yolks, butter, and various plants (Nishida et al. 2021).

7.2.3 Carotenoid Ketones

Among the CAR ketones, we mention rhodoxanthine and astacine.

Rhodoxanthine is a bluish-red substance found in aquatic plants and conifers. It is a CAR diketone. Astacine is a purple substance. Chemically, it is a CAR tetracetone.

Rhodoxanthine, $C_{40}H_{50}O_2$, found in the red-brown leaves of the aquatic slope (*Potamogeton Natans*, as well as in various conifers, m.p. 219 °C. Astacina, $C_{40}H_{58}O_4$ acicular crystals with purple surface luster, m.p. 243 °C, is the red dye of lobster shell and other crustaceans. Astaxanthin, $C_{40}H_{52}O_4$, crystals with m.p. 216 °C, is contained in the outer skin and in the eggs of crustaceans under the form of a combination with a protein (a chromoprotein), ovoverdin, green-brown or blue-black. The change in color from dark green to red, which occurs when crayfish boil in water, is due to protein denaturation and the release of the CAR prosthetic group (Liu et al. 2022).

Apart from crustaceans, astaxanthin is found in many other classes of animals, from protozoa to higher animals, fish, birds, and mammals.

7.2.4 Carotenoid Acids

Bixin (apo-CAR), $C_{25}H_{30}O_4$, or norbixin $C_{24}H_{28}O_4$, is a yellow dye isolated from a tropical plant, i.e. seeds of the annatto tree (*Bixa orellana* L.) and confers and widely used before the advent of synthetic dyes.

Crocetine, $C_{20}H_{24}O_4$, is found in the form of an ester or with two molecules of gentiobiosis, crocin, in saffron, a product isolated from the stigmas of *Crocus sativus* flowers in southern and eastern Europe. Factors that influence the content of CARs in plant products are genetic and environmental factors (oxygen, light, and temperature), the post-germination period of seeds, the ripening period of fruits and vegetables, and post-harvest storage conidia.

Some technological processes (grinding cereals and bleaching flours) lead to a decrease in the amount of CARs, either by passing them into by-products or by oxidizing them (the case of flour bleaching) (Flegler and Lipski 2021).

CARs, such as provitamin A / vitamin A, are slightly yellowish in color, but are not identical to CAR, which is strongly colored. Due to the hydrocarbon structure, CAR pigments are hydrophobic substances, soluble only in organic solvents, oils, and fats.

From a chemical point of view, CAR pigments are characterized by a structure with conjugated double readings, which determines the unsaturated character and therefore the possibility of oxidation and autooxidation reactions (in the presence of air), the ability to absorb light radiation, etc.

The biochemical role of CAR pigments is determined by their chemical structure and their representative properties. CARs, like terpenes, sterols, phytol, vitamins K, vitamins E, etc., have as a repetitive structural unit the activated isoprene, which can form CAR, steroid, etc. hydrocarbon chains.

Capsantine $C_{40}H_{56}O_3$ is a viscous red liquid, in its pure state it is presented in the form of carmine-red crystals; m.p.: 175–176 °C. Capsantine is a product of red pepper (*Capsicum annuum* L.). *Capsicum anuumm* is a species of the genus *Capsicum* in southern North America and northern South America.

This species is the most common and most cultivated of 5 species of domestic *Capsicum*. The medicinal properties of peppers are generally limited to their vitaminizing properties (especially as regards vitamins A and C) (Shilpa et al. 2021).

Pepper is very good in food, but in addition to its intense taste, it also offers many benefits to the body, such as the destruction of cancer cells. Capsanthine is an oil that can be combined with an essential oil that contains a relatively large amount of CARs, commonly used in the food industry as a colorant for sauces, soups, or meat products (Chen et al. 2021).

Capsantine's ability to color is due to its CAR content, which is also the specific pigment of red pepper. It includes 7 CARs, and capsanthin, epoxy-capsanthin, and capsorubin are exclusively biosynthesized in red pepper.

The source of a pepper's spice (the Scoville count-measurement of pepper pungency) is a compound 8-methyl-N-vanillyl-6-nonenamide called capsaicin, an natural active compounds of chili peppers (Novikov et al. 2022).

CAR is a pigment present in orange-red plant products. It is a fat-soluble phytonutrient, originally used as a dye. The color is due to the large number of conjugated double bonds. They have about 40-carbon. There are about 600 CARs (natural pigments).

7.3 Aspects of the Mechanism of Carotenoid Biosynthesis

In plants, CARs have a role in light capture, photo protective, and as a precursor for the synthesis of abscisic acid. Plants attract insects and other animals to act as pollinators and vehicles for transporting the seed using stimuli generated by the CAR pigment, including red, orange, and yellow, which are present in fruits and flowers.

Carotenoid pigments are biosynthesized into chloroplasts and chromoplasts, through the mevalonate cycle, of acetyl CoA resulting from the biodegradation of carbohydrates and lipids. The biosynthesis process comprises numerous intermediate compounds, the most important of which are mevalonate, isopentyl pyrophosphate, geranyl pyrophosphate, phytoene, lycopene, and carotene.

Many of the genes that are included in CAR biosynthesis have now been cloned from various plants such as *Arabidopsis*, tomato, pepper, daffodil, and marigold.

These CAR genes are in regular development and also directly connected with the color phenotype of these plants, as shown in the tomato fruit and marigold petals.

In the metabolic cycle, β-CAR is converted by cyclization of lycopene. Enzyme inhibitors or stressors can stop lycopene cyclization and therefore β-CAR formation.

In the past four decades, there has been a growing interest in sexually selected traits as honest indicators of individual quality and health; however, the mechanisms that link the expression of the ornament to the individual condition remain evasive (Schmidt et al. 2021).

Increasingly, redox processes (and especially OSs) have been hypothesized to provide this link and thus maintain honesty in many sexual signals.

Oxygen and reactive nitrogen species (ROS), the source of OSs and damage, are inevitable and ever-present by-products of an oxidizing metabolism. ROS can also occur by activating immunity or by polluting the environment.

The need to prevent OSs and maintain redox homeostasis may therefore provide a universal mechanism linking the expression of the ornament to immune activation and other metabolically demanding processes (e.g. stress response, sperm molding, or production).

To produce high-quality lycopene, lycopene cyclase inhibitors should have the following characteristics: minimum required, good inhibitory effect, water solubility, low toxicity, and rapid biodegradability. 2-Isopropylimidazole has shown numerous advantages over piperidine, nicotine, imidazole, pyridine, and 2-methylimidazole as a lycopene cyclase inhibitor.

First, 2-isopropylimidazole has a higher selectivity (lycopene accounts for 96% of total CARs) than other lycopene cyclase inhibitors.

Second, only a small amount of 2-isopropylimidazole is required (300 mg/l) for effective use (Kiziler et al. 2021).

Third, 2-isopropylimidazole is soluble in water, allowing it to be easily eliminated from lycopene and reducing the risk of food toxicity.

Fourth, 2-isopropylimidazole improves the metabolic flow of lycopene and the proportion of cis-lycopene, accounting for approximately 35% of total lycopene.

Finally, 2-isopropylimidazole can be completely biodegraded by microorganisms, thus eliminating the risk of environmental pollution.

Chemically, lycopene is a polyenic hydrocarbon, an open chain unsaturated acyclic CAR having 13 double bonds, of which 11 are conjugated double bonds arranged in a linear series, having the molecular formula $C_{40}H_{56}$. Two central methyl groups are in position 1.6, while the rest of the methyl groups are in position 1.5 relative to each other. A series of conjugated double bonds is a chromatograph of variable length. The coloring and antioxidant activities of lycopene are a consequence of its unique structure, of an extensive system of conjugated double bonds. Lycopene owes its reddish color to its conjugated polyethylene structure. In nature, lycopene exists in trans-trans form and seven of these bonds can be isomerized from trans to mono or poly-cis under the influence of heat, light or certain chemical reactions. Lycopene does not act as provitamin A due to the lack of a β-ionic ring structure. The stereoisomeric forms of lycopene are due to its light-absorbing properties in relation to its molecular structures. Lycopene is also very sensitive to light, heat, oxygen, and degrading acids and several metal ions, such as Cu^{2+}, Fe^{3+}, catalyze its oxidation (Narang et al. 2021).

CARs are traditionally considered to be important antioxidants due to their ability to quench ROS *in vitro*. Therefore, a compromise has been proposed in the allocation of CARs between ornamentation and antioxidant defense as a possible mechanism for maintaining signal honesty (the "allocation allocation hypothesis").

The importance of CARs as antioxidants *in vivo* has recently been questioned, however, resulting in several alternative hypotheses. The "protective hypothesis" does not imply any CAR antioxidant function, instead proposing that redox homeostasis be signaled by oxidation and cleavage of ROS-induced CARs, leading to discoloration that can be prevented by an effective antioxidant system. An extension of this hypothesis is the "handicap" hypothesis, which proposes that high levels of CARs directly aggravate OSs through the pro-oxidant activity of CAR cleavage products, thus presenting a direct physiological handicap.

More recently, structural and functional similarities between CARs and ubiquinones have led to the hypothesis that red CAR synthesis may occur within the inner mitochondrial membrane (MIM) of hepatocytes, with CARs, redox-coupled with ubiquinone, exhibiting antioxidant and modulator functions in the chain (Gea-Botella et al. 2021).

The hypothesis proposes that the synthesis of red CARs be related to the maintenance of MIM potential, thus linking the expression of the ornament not only to redox homeostasis but also to the efficacy of mitochondrial respiration (MIM CAR oxidation hypothesis).

Clearly, knowledge of the true role of CARs in redox homeostasis is crucial to test hypotheses about the mechanisms of CAR-based signaling honesty.

In vivo studies of antioxidant CAR function have yielded mixed results, however, leading to a low overall association of CAR levels with AC in a recent meta-analysis.

We suggest that the current controversy over CAR antioxidant function could result from the use of inadequate methods for assessing AC. Due to their lipophilic nature, CARs are located inside the bilateral layers of an organism and other lipid compartments (Meléndez-Martínez et al. 2021).

Figure 7.2 shows that lycopene is the branch point for the delta and gamma-carotene pathways, which usually end in lutein and zeaxanthin, respectively.

Their grouping and path are shown in Figure 7.2.

CARs are an important group of natural, fat-soluble pigments in plants and microorganisms that are yellow, orange, or red. They are divided into two groups: carotene (molecules of carbon and hydrogen), such as beta-carotene and lycopene, and xanthophylls (molecules of carbon, hydrogen, and oxygen), such as lutein, zeaxanthin, and violaxanthin.

In the metabolic pathway, beta-carotene is converted by cyclization of lycopene.

Enzyme inhibitors or stressors can stop lycopene cyclization and therefore beta-carotene formation. Thus, laboratory tests are very important at this stage to identify the stressors that prevent lycopene cyclization and the production of large amounts of beta-carotene.

Lycopene accumulation can be promoted by the addition of inhibitors that prevent chain cyclization during beta-carotene biosynthesis.

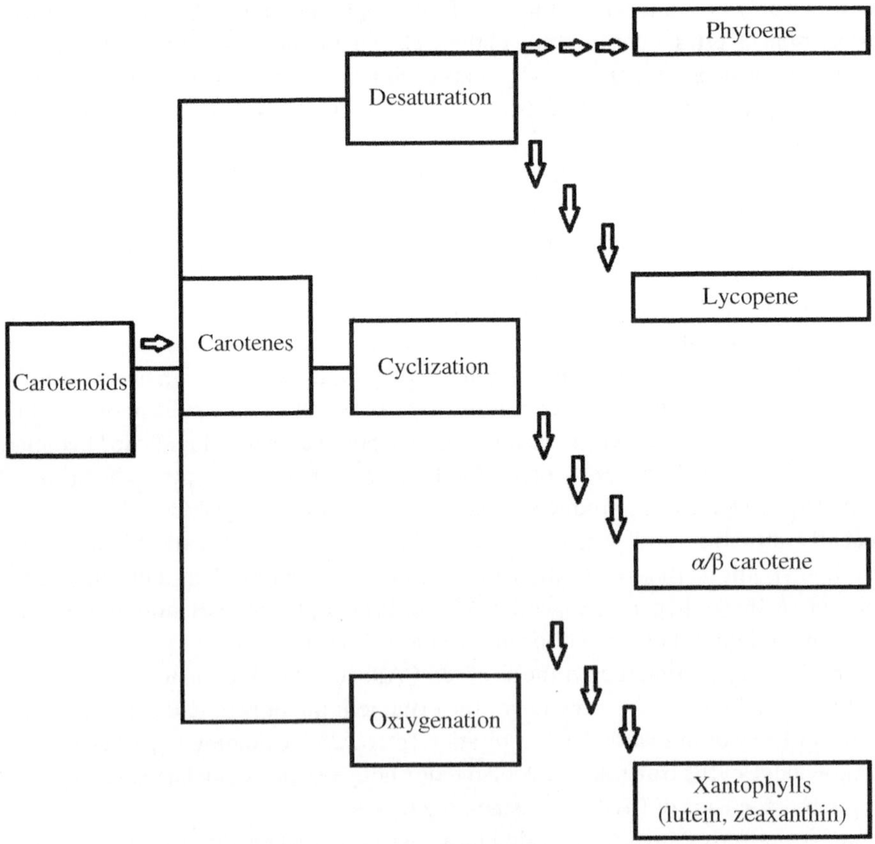

Figure 7.2 Schematic illustration of carotenoid grouping and metabolic pathway.

Because lipid bonds (e.g. mitochondrial membranes) are considered a major source of very harmful ROS, CARs have a high potential to play a role in redox homeostasis. The most commonly used antioxidant tests (e.g. Ferric Reducing Antioxidant Power (FRAP), Oxygen Radical Absorbance Capacity (ORAC), oxy-adsorbent test (OXY), Trolox Equivalent Antioxidant Capacity (TEAC), or Total radical-trapping antioxidant parameter (TRAP)) measure AC only in aqueous media, making them unsuitable for the evaluation of lipophilic antioxidants, such as CARs. In view of this, we decided to use a new marker of lipophilic AC, i.e. the ratio between the stereoisomeric Z (German, zusammen for together) or E (German, entgegen for opposite) Z and E forms of hydroxyacecadienic acid (HODE). Z- and E-OHE are formed as oxidizing products of linoleic acid, the proportion of Z-OHE to total HODE (Z / tHODE) increasing with higher concentrations of hydrogen-donating lipophilic antioxidants, such as vitamin E or coenzyme Q, an integrated measure of their *in vivo* activity (Liu et al. 2022).

Because ROS-CAR uptake activity is most likely mediated by radical addition rather than hydrogen donation, the CAR antioxidant action is probably not directly measured by Z / tHODE. Despite this, the lipophilic nature of this measure of AC makes it a much better candidate for detecting the possible effects of CARs on the body's antioxidant system than tests based on traditional hydrophilicity.

CAR-based redox assumptions were evaluated based on signaling honesty by testing the importance of CARs for redox homeostasis and the role of redox homeostasis in constraining CAR signal expression. Using a 2×2 factorial design, the effects of the oxidative challenge of diquat dibromide and CAR supplements on beak coloration, circulating CAR levels, and redox homeostasis in adult zebra males (*Taeniopygia guttata*) were analyzed (Roca and Pérez-Gálvez 2021).

Diquat dibromide [$C_{12}H_{12}Br_2N_2$], a bipyridyl compound renowned to initiate superoxide anions *in vivo* by redox cycles, has recently been recognized as a suitable OSs inducer in environmental studies as well as in laboratory models of Parkinson's disease.

If homeostasis redox really constrains signal expression, beak coloration should be reduced after diquat treatment (De Cosmi et al. 2021).

Because a change in the intensity of OSs can occur either as a change in oxidative damage or as antioxidant activity (or, indeed, both together), we anticipate that CAR supplementation will result in

- a reduction in oxidative damage and / or antioxidant activity other than CARs due to their reduced need (reduced Z / thode ratio) if CARs act as antioxidants;
- an increase in each or both of the parameters if the CARs act as antioxidants; or
- no effect on any parameter if the CARs have no influence on redox homeostasis.

Given the prevalence of strong genetic manipulation platforms, recent advances in genomic analysis, the rapid development of synthetic biology, and the variety of tools and strategies currently available for metabolic engineering, synthetic optimization of target biosynthetic metabolic cycles could be streamlined.

Lycopene is always biosynthesized by Geranylgeranyl diphosphate (GGPP) or Farnesyl pyrophosphate (FPP), which in turn are biosynthesized by GGPP synthase (GGPPS / CrtE) {(CRT gene cluster consists of twenty-five genes; responsible for the biosynthesis of carotenoids)} or FPP synthesis (IspA) {(Protein names: Farnesyl diphosphate synthase)}. Certain proteins perform both IspA and CrtE functions, such as the gps (protein pathway suppressor) gene for *Archaeoglobus fulgidus*. Condensation of two GGPP molecules by 15-cis-phytoene synthase (crtB) leads to the formation of colorless phytonene compounds (phytoene is 40-carbon intermediate in the biosynthesis of CARs) (Igreja et al. 2021).

Lycopene is the first colored compound produced during CAR biosynthesis, it is then biosynthesized by the catalytic activity of the enzyme EC 1.3.99.31 – phytoene desaturase (lycopene-forming) (CrtI).

Recent studies have shown that the synthesis of GGPP is a step that limits the production of lycopene, and the considerable diversity of sequences among the crtE genes indicates that these enzymes have different levels of activity.

Therefore, it is essential to choose an effective crtE gene to remove limitations in GGPP synthesis. Even so, various IspA proteins have been evaluated for enhanced substrate improvement. In particular, the elimination of genes involved in the synthesis of lycopene has led to an improved synthesis of lycopene by increasing the supply of precursors (Canonico et al. 2021).

Some researchers performed a balanced analysis of the stoichiometric balance of the genome by calculating metabolic flows and then scanned the genome for single and multiple genes that produce improved product yield.

A deletion encoding the strain in gdhA (Protein names: NADP-specific glutamate dehydrogenase), aceE (Protein names: Pyruvate dehydrogenase E1 component), and fdhF {(Protein names: Formate dehydrogenase H)} (DgdhA, DaceE, DfdhF) showed a lycopene production per dry matter amount of 6.6 mg/g, an increase of approximately 40% over the modified parental strain. In previous studies, promoters (agents) with varying resistance have been used to quantitatively assess the effects of distinct levels of gene expression on product growth and formation.

Regarding the number of children of plasmids constructed, while most studies found that a higher number of children leads to higher levels of production, no linear relationship was found between these factors. Because gene order can also affect lycopene production levels, assessing gene order permutations in an operon provides an additional strategy for manipulating transcription levels.

To date, engineering strategies have most often been implemented in *E. coli*.

Host selection depends on a number of different factors, including genetic tractability, native ability to produce the target compound, potential toxicity of the final product, competition with inherent essential requirements, and potential / ability to perform biotransformations / modifications (desired or harmful) (Moopantakath et al. 2021).

Lycopene production capacities vary widely between strains. The first attempt to produce lycopene using yeast was performed by introducing the CAR genes of *Pantoea ananatis* (formerly *Erwinia uredovora*) into *Saccharomyces cerevisiae*.

Meanwhile, in a subsequent study, genetically engineered *Candida* yeast produced 0.7–6 mg/g lycopene per amount of wet matter, indicating that this organism is another promising source for this valuable compound.

Given that lycopene production is limited by enzymatic and regulatory factors, many of which may still be unknown, different types of combinations may be used in terms of enzymatic activity, gene expression, genetic modification, metabolic flow, and improved lycopene production.

Oxidative challenge and CAR intake had opposite effects on beak ornamentation, plasma CAR levels, and lipophilic AC in RBC.

Despite the considerable negative effect of the oxidative challenge on both signal strength and circulating CAR levels, only a negligible marginal increase in oxidative damage to the blood resulted.

Although the condition of the blood redox is usually used as an estimate of the general redox condition of the body, we cannot completely rule out increased oxidative damage in certain organs, such as the liver or kidneys.

However, the oxidative challenge resulted in a significant increase in both hydrophilic and lipophilic AC, suggesting the mobilization of antioxidants to prevent oxidation damage.

Because increased AC accompanied by stable (or increased) oxidative damage should be interpreted as increased OSs, this result supports the pro-oxidant effect of diquat treatment (Schmeisser et al. 2021).

Importantly, CARs counteracted the effect of oxidative challenge on hydrogen-donating lipophilic antioxidants.

Chemically, lycopene is a polyene hydrocarbon, an open-chain unsaturated acyclic CAR having 13 double bonds, of which 11 are conjugated double bonds arranged in a linear series. Two central methyl groups are in position 1.6, while the rest of the methyl groups are in position 1.5 relative to each other. A series of conjugated double bonds is a chromatograph of variable length. The coloring and antioxidant activities of lycopene are a consequence of its unique structure, of an extensive system of conjugated double bonds. Lycopene owes its reddish color to its conjugated polyethylene structure. In nature, lycopene exists in trans-trans form and seven of these bonds can be isomerized from trans to mono or poly-cis under the influence of heat, light, or certain chemical reactions. Lycopene does not act as provitamin A due to the lack of a β-ionic ring structure. The stereoisomeric forms of lycopene are due to its light-absorbing properties in relation to its molecular structures.

Lycopene is also very sensitive to light, heat, oxygen, and degrading acids, and several metal ions, such as Cu^{2+}, Fe^{3+}, catalyze its oxidation. Like all CARs, lycopene is a polyunsaturated hydrocarbon (an alkene substituent). Structurally, this is a tetraterpen consisting of eight isoprene units, composed entirely of carbon and hydrogen, and is insoluble in water. Due to its strong color and nontoxicity, lycopene is a useful food coloring (registered as E160d) and is approved for use in the United States, Australia, and New Zealand (registered as 160d) and the EU. The eleven conjugated double bonds, made of lycopene, give an intense red color and are responsible for the antioxidant activity. Lycopene is a systemic tetraterpene assembled from 8 isoprene units. It is a member of the CAR family, and because it is made up entirely of carbon and hydrogen, it is also a carotene. Lycopene isolation procedures were reported in 1910, and the structure of the molecule was determined in 1931.

In its natural, all-trans form, the molecule is long and straight, due to its eleven conjugate double bonds. Each extension in this conjugate system reduces the energy required for electrons to transition to high energy states, allowing the molecule to absorb visible light from progressively longer wavelengths. Lycopene absorbs all wavelengths, but the longest wavelengths of visible light, so it looks red.

Photosynthetic plants and bacteria naturally produce all-trans lycopene, a total of 72 geometric isomers of the molecule are possible from a steric point of view.

When exposed to light or heat, lycopene may undergo isomerization to any of these cis isomers, which have an inclination rather than a linear shape.

Different isomers have been shown to have different stabilities due to their molecular energy (best stability: 5-cis ≥ all-trans ≥ 9-cis-cis 13> 15-cis> 7-cis> 11-cis: most little).

The dark color of lycopene was due to the large number of conjugated double bonds. It consists of identical, symmetrical halves. Each half is composed of four isoprene residues, tied to the tail; in the middle of the molecule, the way of binding the isoprene residues is reversed. This constructive principle is found in all CARs and suggests the hypothesis that plants biosynthesize CAR molecules by combining two molecules of a diterpenoid.

Lycopene is found in tomatoes, melons, and some apricots. Fruit oxy-CARs are in esterified form. Lycopene is insoluble in water and can only be dissolved in organic solvents and oils. Due to its nonpolarity, lycopene in food will stain any sufficiently porous material, including most plastics. While a red stain can be quite easily removed from fabrics (provided the stain is fresh), lycopene diffuses into the plastic, making it impossible to remove with warm water or detergent. Lycopene is an acyclic unsaturated hydrocarbon containing 13 double bonds, 11 of which are conjugated. It does not have in the chemical structure the beta-ion rings that are found in the structure of retinol and therefore does not have the specific activity of provitamin A. Natural lycopene contains the isomer of all-trans-lycopene in a proportion of 94–96%. Lycopene isomers are much more polar in nature, more soluble in oils, and other hydrocarbon-type solvents; solubilization is effective in lipophilic solutions because it is less prone to crystallization. Degradation of lycopene due to isomerization and oxidation affects the bioavailability and decreases the functionality of the molecule for the benefit of health.

The physico-chemical parameters that contribute to the degradation of lycopene are time, temperature, reaction medium, environmental conditions, etc. However, the most important factors that contribute to the degradation of lycopene are heat, light, and oxygen. Lycopene self-oxidation is irreversible. Synthetic lycopene is a mixture of geometric isomers of lycopene and is produced by Wittig condensation of synthetic intermediates commonly used in the production of other CARs used in food. Synthetic lycopene consists predominantly of all-trans lycopene mixed with 5-cis-lycopene and minor amounts of other isomers.

Commercial lycopene preparations for use in food are formulated as suspensions in edible oils or as water-dispersible or water-soluble powder. *In vitro* studies have shown that the antioxidant effect of lycopene is twice that of beta-carotene and ten times that of alpha-tocopherol (expressed in neutralizing the oxygen radical). The cis isomer of lycopene has been shown to be much more stable with a strong antioxidant effect compared to all-trans-lycopene. However, the biological significance of lycopene isomerization is not fully understood.

Studies have shown the anticancer properties of lycopene against many cancer cells, including cancer cells in the prostate, stomach, lungs, colon, and skin. There are many studies on the effect of lycopene on cancer. Most of the *in vitro* experiments using prostate cancer culture cells demonstrate a protective effect. Lycopene also has antimutagenic action against chemically induced DNA damage (Takemura et al. 2021).

Such reallocation of resources can be applied not only to antioxidants (possibly including CARs) but also to other micronutrients needed for CAR metabolism (e.g. cofactors) or energy, as the absorption and conversion of CARs are considered to require energy.

The findings supporting the CAR antioxidant function are also consistent with the hypothesis of oxidation of the CAR MIM. In this case, the expression of the reduced signal after diquat exposure is interpreted as a result of reduced red keto-CAR production due to mitochondrial dysfunction, rather than reallocation of resources to antioxidant defense.

However, the synthesis of the red keto-CAR is considered to take place in the beak tissue at the zebra finishes and not in the liver mitochondria, as is assumed in the hypothesis of oxidation of the CAR MIM.

In addition, the synthesis of the red keto-CAR has recently been proposed to be catalyzed by cytochrome P450 2J2 in zebra fungi. Inhibited from human studies, this cytochrome is probably located in the endoplasmic reticulum and not in the mitochondria. These two pieces of evidence suggest that the "MIM CAR oxidation hypothesis" may be a less likely explanation for the honesty of CAR signaling, at least in this model species (Li et al. 2021).

The reduced ornamental expression after diquat exposure observed in our experiment is the first demonstration of an adverse effect of a redox cycle agent on a CAR signal in adult animals. Because the induction of OSs is a primary mechanism of diquat toxicity, this result supports the role of redox homeostasis as a major constraint that ensures sincerity in CAR-based sexual signaling. No loss of body weight was observed in our birds exposed to diquat, which indicates that the reduced ornamentation was not a consequence of reduced food intake or total intestinal absorption affected by a possible diquat toxic effect on the wall intestinal.

Therefore, our results support the idea that conditions of stress and energy stress, immune activation, or environmental pollution (e.g. urban environment, heavy metals or pesticides) could reduce CAR-based ornamentation through high OSs.

Previous studies have directly tested the effect of oxidative state on CAR-based ornament expression, either by specifically increasing the oxidative load by cyclooxygenating agents or by reducing antioxidant protection, and these have produced contrasting results. In these studies, while weaker pigmentation and skin pigmentation developed due to redox disturbance in red-footed *Alectoris rufa*, no effect was observed on feather coloration in adult *Parus major*, chlorine, or *Haemorhous mexicanus*.

One possible explanation for this discrepancy could be a difference in compromise solutions between developing and adult birds, with the latter investing more in breeding than in self-maintenance.

Demonstration of such an effect in sexually active adults in finch zebra, however, does not support this explanation. To date, the adverse effect of the oxidative challenge has been observed only on the bare parts only, probably suggesting a higher sensitivity to OSs in the more dynamic lower parts, which may change color over a period of only a few weeks compared to ornamentation of semi-static feather (Sun et al. 2022).

Therefore, future studies should test the possible differences between honesty mechanisms and signaling content between semi-static and dynamic CAR signals and explore possible differences in these mechanisms between different taxa and life histories.

Isoprenoises and their derivatives, such as sterols, terpenes, dolichols, and quinines, are the most abundant class of secondary metabolites, with various functions in microbial, plant, and animal metabolism.

The building block of isoprenoid-derived compounds is isopentenyl diphosphate (PPI), which is biosynthesized by two nonhomologous metabolic cycles: the mevalonate metabolic cycle and 1-deoxy-d-xylulose 5-phosphate / 2-*C*-methyl-d-teritol 4-phosphate (DOXP). / MEP) path.

CARs are synthesized only by plants. All CARs, including lycopene, are synthesized from the isopentenyl pyrophosphate precursor (PPI) which takes the pathway of mevalonic acid (MVA) under the action of acetylcoenzyme A. Lycopene is the precursor of cyclic CARs and similar to cholesterol and triglycerides. Lycopene from chylomicrons in the cellular mucosa is secreted into the lymphatic system and then into the blood. It is found in human plasma between 0.2 and 1.0 μmol/l. More studies are needed to understand the intracellular metabolism, the bioavailability of lycopene, and whether oxidation products are absorbed and have a biological effect.

Terpenoids and CARs are the most important isoprenoid derivatives from an industrial point of view, which are used as cosmetics, flavors, dyes and pharmaceutical compounds.

There has been a great deal of research in genetic engineering on the increase in the amount of bioactive substances in the microbial mass, even of compounds in the isoprenoids (biosyntetized from isopentenyl diphosphate [IPP]) and polyketides (biosyntetized by polyketide synthases) classes. Most of these studies focused on providing an improved set of precursors, such as PPIs to heterologous hosts, including *E. coli* bacteria and yeast.

Other research has aimed to improve metabolic flow for the biosynthesis of enhanced bioactive substances, which play a role in plant protection, avoiding the development of harmful intermediates. Biotechnological research of this type has been optimized to keep under control the enzymes of metabolic cycles, in transcriptional relationship, but also in relation to translation, with parts of genetic bioengineering, even target precursors, ribosome-binding sites, and genetic cycles.

CARs are important compounds in the food industry, being naturally fat-soluble pigments, most of which are C_{40} terpenoids. The most important corotenoids obtained by microbial fermentation are lycopene, beta-carotene, torulene, and astaxanthin. They act as antioxidants that protect the membrane, effectively removing $^1O_2$ and peroxyl radicals; their antioxidant effectiveness is apparently related to their structure. There are several types of yeasts (mainly of the genus *Rhodotorula*) and fungi (*Blakeslea trispora*) that can be used to convert crude glycerol into lycopene, beta-carotene, and other imported CARs for the food and pharmaceutical industries.

These microorganisms can assimilate different carbon sources such as glucose, cellulose, crude glycerol, and sorbitol, and therefore, various wastes can be used as cheap substrates for their cultivation. Among the CARs, lycopene is an important compound with a growing industrial value. The microbial production of lycopene, an important commercial and medical compound, is an alternative process to the exploitation of tomatoes given the placement of this product derived from the *B. trispora* fungus on the European market as a new and recently accepted food ingredient called "E 160d."

Fungal cells can produce this acyl precursor from cyclic CARs in the presence of specific chemical compounds that inhibit the activity of the lycopene cyclase enzyme. Compared to other microorganisms, *B. trispora* fungus has the advantage of large-scale aerobic cultivation in bioreactors with a fast growth rate. To date, the development of bioprocesses for the production of lycopene in bioreactors is in its infancy, as research is quite limited, especially compared to that for the production of β-carotene by *B. trispora*. In this context, it can be concluded that crude glycerol can be used successfully in fermentation processes in order to obtain value-added products such as lycopene (Zhou et al. 2021).

CARs are derived from isoprenoids that have various roles in nature. To date, more than 1000 CARs have been insulated and characterized. CARs have gained significant consideration in research in the past due to their use as natural pigments; however, their high antioxidant and anticarcinogenic activities and potential use as cosmetic or pharmaceutical compounds have only recently been revealed. The main difference between CAR and xanthophyll is that CAR gives it an orange color, while xanthophyll gives it a yellow color. Moreover, CAR is a hydrocarbon that does not contain an oxygen atom in its structure, while xanthophyll is a hydrocarbon that contains an oxygen atom in its structure. CAR and xanthophyll are the two classes of CARs, which are tetraterpene plant pigments that serve as accessory pigments in photosynthesis. They are responsible for giving red-orange yellow color, especially to fruits and vegetables. CAR is one of the two types of CARs present in plants responsible for the orange color of the plant. In general, CARs are organic pigments produced only by photosynthetic organisms, including plants, algae, and bacteria. The main function of CARs is to serve as accessory pigments in photosynthesis. Although animals cannot synthesize CARs from their bodies, these compounds play a key role as antioxidants and anti-inflammatory molecules. Animals get CARs through their diet, by eating fruits and vegetables. After oxidative cleavage of dietary CARs (including xanthine's), bioactive cleavage products, such as apo-CARs, are used in retinal formation and activation of the transcription system. In addition, apo-CARs are known to act as anticancer agents and cellular modulators of retinoic acid receptors, retinoid X receptors, active peroxisome proliferator receptors, and estrogen receptors. A recent study suggested an anticancer mechanism, consisting of two major CARs, crocin and crocetine, found in the dark red stigmas of *Crocus sativus*, which has traditionally been used in ancient Asia. Southwest for the treatment of diseases and a flavoring and coloring agent. These two CARs are C_{20} structures generated by the cleavage of C_{40} CARs, the most abundant CAR structures in nature, including lycopene, β-CAR, and astaxanthin. Some bacteria produce C_{50} or C_{30} CARs; however, they are structurally less diverse than C_{40} CARs. To date, only a few natural sources of C_{30} CARs are known, including *Staphylococcus aureus*, *Methylomonas* sp., and some strains of the genus *Bacillus* (Pang et al. 2021).

Several studies have successfully discovered new CARs by applying a combinatorial biosynthesis approach with *in vitro* evolved carotenogenic enzymes.

The *in vitro* evolution of key enzymes, which determine the structure of the CAR in the first stage, has led to the generation of new series of CARs in *E. coli* bacteria. Recently, some authors created non-natural C_{50} CARs by assembling moderately selective enzymes, designed by directed evolution in *E. coli* bacteria.

Enzymes typically have substrate specificity; however, carotenogenic enzymes have relatively promiscuous substrate specificity, giving rise to new CAR structures in a heterologous host.

Phytoene synthase – EC 2.5.1.32 (CrtB), phytoene desaturase (ζ-CAR-forming) – EC 1.3.99.29 (CrtI), lycopene β-cyclase – EC 5.5.1.19 (CrtY), 4,4′-diapophytoene synthase or dehydrosqualene synthase, C_{30} CAR synthase, DAP synthase – EC 2.5.1.96, (CrtM), and 4,4′-diapophytoene desaturase – EC 1.3.8.2 (4,4′-diapolycopene-forming) (CrtN) were evolved *in vitro*, allowing the biosynthesis of various CARs in *E. coli* bacteria.

Although extensive work has been done on the diversification and identification of CAR structures, the existence of cyclic C_{30} CARs in nature has not been reported; 4,4-diapotorene was first reported to be monosyclic with the metabolic C_{30} CAR structure of *E. coli* bacteria biosynthesized.

In this study, we created 13 structurally new C_{30} CARs, including acyclic, monocyclic, and bicyclic structures, and one new C_{35} CAR, by combinatorial biosynthesis with natural CAR biosynthetic enzymes from various microorganisms and evolved enzymes, CrtN and CrtY, in *E. coli* bacteria (Zhang et al. 2022).

Furthermore, we tested the neuronal differentiation and differentiation activities of C_{30} CARs on 2,2-diphenyl-1-picrylylhydrazyl (DPPH) to explore their potential as bioactive compounds.

Moreover, the main structural feature of the CARs used to distinguish them from xanthophyll is the absence of any oxygen atom in the molecule. There are also three main types of CAR such as β-CAR, α-CAR, and lycopene. Mainly β-CAR and, to some extent, α-CAR are responsible for the synthesis of vitamin A in the animal's body. Analyzing the sources, β-CAR occurs in cantaloupe, mango, papaya, carrots, sweet potatoes, spinach, kale, and pumpkin, while α-CAR occurs in pumpkin, carrots, tomatoes, coleslaw, tangerines, winter squash, and peas. Lycopene appears in watermelon, tomatoes, guavas, and grapefruit (Tkáčová et al. 2022).

Xanthophyll is the second type of CAR found in plants, giving the plant a yellow color. However, the structure of xanthophyll contains a single oxygen atom in contrast to CAR. However, like CAR, xanthophyll appears in the leaves of plants in large quantities.

In addition, animal bodies contain xanthophyll, which is herbal. As examples, egg yolk, adipose tissue, and skin contain plant-derived xanthophyll. First of all, lutein is the form of xanthophylls found in chicken egg yolk (Li et al. 2022).

Also, two types of xanthophylls known as lutein and zeaxanthin occur in spotted clay or in the yellow spot in the retina of the human eye. They are responsible for the central vision of the eye. Moreover, they protect the eye from blue light.

Therefore, kale, spinach, greens, summer squash, pumpkin, paprika, yellow fruit, avocado, and egg yolk are good sources of lutein and zeaxanthin.

In general, these two xanthophylls are effective in age-related macular degeneration, leading to blindness. Furthermore, lutein is known to prevent the formation of atherosclerosis due to its antioxidant effect on cholesterol, which in turn prevents the accumulation of cholesterol in the arteries (Lem et al. 2021).

CAR and xanthophyll are the two classes of CARs. Hydrocarbons are tetraterpenes-related unsaturated substances having the same formula C_{40} H x. Both also give a red-orange-yellow color to the plant parts. And, both appear in chloroplasts. Animals are not capable of producing them. Moreover, they serve as accessory pigments in photosynthesis, capturing sunlight and passing on chlorophyll a. However, the wavelengths absorbed by them are different from the wavelengths absorbed by chlorophylls. I also absorb light in the UV range.

Moreover, these pigments serve as antioxidants, deactivating free radicals.

They have anti-inflammatory effects; therefore, they perform an immune function in the body. In addition, both types of CARs are known to reduce the risk of cardiovascular disease and cancer (Nwoba et al. 2021).

CAR refers to an orange or red vegetable pigment, including β-CAR found in carrots and many other plant structures, while xanthophyll refers to a yellow or brown vegetable pigment that causes the autumn colors of the leaves. Thus, this is the main difference between CAR and xanthophyll. Eta-CAR, which is a CAR, absorbs 450 nm wavelengths, while lutein and vioxanthan, which are xanthophylls, absorb 435 nm. Therefore, this is another difference between CAR and xanthophyll (Ouyang et al. 2022).

Also, the color produced by each is another difference between CAR and xanthophyll. While CAR gives off an orange color, xanthophyll gives it a yellow color.

Moreover, another difference between CAR and xanthophyll is the presence of oxygen atoms in their structure. CAR does not contain an oxygen atom in its structure, while xanthophyll contains an oxygen atom in its structure.

CAR appears mainly in cantaloupe, mango, papaya, carrots, sweet potatoes, spinach, kale, and pumpkin, while xanthophyll appears in kale, spinach, turnip greens, summer squash, pumpkin, paprika, yellow fruit, avocado, and egg yolk. This is also a difference between CAR and xanthophyll.

CAR is one of the two types of CARs that occur mainly in plant parts, including fruits and vegetables. It is responsible for giving the plant an orange color. In addition, it does not contain oxygen atoms in its structure. Instead, xanthophyll is the other type of CAR responsible for giving the plant a yellow color. Significantly, it contains a single oxygen atom in its structure. Both CAR and xanthophyll serve as accessory pigments that capture and pass sunlight to chlorophyll a. However, the main difference between CAR and xanthophyll is the color they give to the plant (Manochkumar et al. 2021).

CARs are molecules with a precise three-dimensional structure, which is crucial for their physico-chemical and nutritional properties. In the spatial representation, it is observed that the isoprenoid chain is linear, but the two cycles in terminal positions are not in the same plan. Different geometric cis isomers (at different levels of the molecule) have different solubility and stability properties compared to the trans isomer.

Most known CARs have at least one asymmetric carbon, which can induce various optical isomers, including enantiomers. Most CARs carry α- or β-rings in terminal positions. These structural changes determine the nature of the various CARs: the position of the double bond, the opening rings, the presence or absence of hydroxyl (−OH) or oxo (>C = O) groups. The class of CARs with oxygen atoms are xanthophylls, while those composed only of carbon and hydrogen atoms they are CARs. Color is the hallmark of CARs ranging from yellow to red. These molecules owe their color to their conjugated double bond system creating a chromophore and thus allowing the absorption of visible light between 400 and 500 nm. The degree of conjugation of the chromophore determines the absorption characteristics of the CARs (Liu et al. 2021).

CARs are insoluble in water, but are soluble in nonpolar solvents and food fats. The spectrum of a CAR has three maxima, whose wavelengths depend on the number of conjugated double bonds. The absorption wavelengths of colored CARs are between 400 and 700 nm, in the visible light spectrum. Among the physico-chemical factors responsible for the change of the initial color, an important role is played by the heavy metals that form with the pigments very stable chelating complex combinations. Light also gives rise to color distortions. Thus, under the action of light, the color is reduced due to the intensification of the oxidation processes of CARs. At the same time, isomerization processes take place, the transform passes into the cis. These reactions are found in canned green beans, peas, and spinach. In the case of tomato paste, it is found browning, due to the degradation of lycopene. When vegetable products are frozen, CARs discolor as a result of their degradation under the action of lipooxidase, according to a coupled oxidation mechanism in which linoleic acid also participates. And in the case of dehydrated products, there is a change in color. Thus, carotenes, in the presence of oxygen, are transformed into ß-ionone, as a result of the oxidation process that takes place, resulting in a discoloration of the product to white. Carotenes are oxidized to monohydroxy-β carotene, dihydroxy-β carotene, and 5.6 or 5.8 corresponding epoxides. CARs can also suffer from epoxy isomerization in some plant products, which leads to a decrease in color. The CAR content is even lower, when at the same time it has the oxidation of unsaturated fatty acids (Kang et al. 2021).

Different groups of CARs have different abilities to absorb sunlight. But there are some properties that unite them:

- As part of protein complexes, CARs become more stable;
- CARs do not dissolve in water;
- In pure form, CARs have a great lability: they are very sensitive to sunlight, sensitive to oxygen, not resistant to strong heat, acids, and alkalis;
- They are selectively absorbed on mineral absorbents, this property being used for their separation by chromatography;
- They have a good solubility in organic solvents: benzene, hexane, chloroform;
- Under the influence of these negative factors, the CAR dye is destroyed.

Despite the fact that all substances are in the same group and have a similar structure, they are classified according to color pigmentation into two groups:

- CARs. These are orange hydrocarbons. There are no oxygen atoms in the structure;
- Xanthophylls – painted in various colors, ranging from yellow to red.

CARs are:

- Alpha-CAR. In large quantities found in orange vegetables. Entering the body can become vitamin A. Lack of alpha-CAR leads to the development of cardiovascular diseases.
- β-CAR. Contained in yellow fruits and vegetables. Protects the body from the harmful effects of free radicals. It is a powerful antioxidant, which can be called a defender of the immune system.
- Lutein. It protects the health of the eye's retina, protecting it from the harmful effects of ultraviolet radiation. Regular use reduces the risk of cataracts by 25%. Lots of lutein is found in spinach, cabbage, zucchini, and carrots.
- β-cryptoxanthin. Reduces the risk of inflammatory diseases, especially rheumatoid arthritis and other joint diseases. In large quantities found in citrus, pumpkin, and sweet pepper.
- Lycopene. It has a direct role in normalizing cholesterol metabolism, prevents the development of atherosclerosis, helps fight excess weight, and suppression of the development of pathogenic intestinal microflora. The sources of lycopene are tomatoes, tomato paste, melons.

In photosynthetic organisms, CARs play a central role in the process of photosynthesis: on the one hand, they participate in the energy transport chain, and on the other hand, they protect the oxidation reaction center. However, in nonphotosynthetic organisms, these molecules appear to play an important role in antioxidative mechanisms.

CARs have many physiological properties and have important effects in both plants and other organisms. Due to their particular molecular structure, they are able to bind and eliminate free radicals and, in this sense, play an important role in the immune system of vertebrates. Animals are not able to synthesize CARs, except for aphids, and must necessarily take them through their diet. In addition to the physiological effects described earlier, they are also important in some superior animals for the ornamental aspects related to the yard.

The pink flamingo livery and some salmon species, for example, and the red color of the lobsters are due to CARs. It is likely that the importance of CARs as an ornamental factor is related to their physiological properties and is an indicator of an individual's health, thus providing a useful reference during the yard for choosing a partner.

Some CAR decomposition products are important perfumes and are therefore actively used in the perfume industry. Many CARs are actually compounds that are usually responsible for the scent of flowers, and the typical smells of tobacco, tea, and many fruits depend on the molecules derived from the breakdown of CARs.

CAR assimilation in raw carrots is 4–5%. Being fat soluble, the bioavailability increases up to 5 times in the presence of prolonged cooking (as in the case of steam) and more if you add modest amounts of fatty acids (3–5 g per table of oil or fat or avocado). In contrast, the consumption of plant phytosterols reduces the level of beta-carotene in the blood (Yu et al. 2021).

CARs are also important for photosynthesis. They are located in light harvesting complexes to help the pants get solar energy for photosynthesis. CARs like lycopene are important in preventing cancer and heart disease. They are also precursors to many compounds, which give flavor and aroma. CAR pigments are synthesized by plants, bacteria, fungi, and lower algae, while some animals obtain them through diet. All CAR pigments have two carbon rings at the ends, which are linked by a chain of carbon and hydrogen atoms. These are relatively nonpolar. As mentioned above, CAR is nonpolar compared to xanthophylls. Xanthophylls contain oxygen atoms, which give them a polarity.

The human body does not have the ability to synthesize CARs; all CARs found in animals are of plant origin and are introduced into the body with food. CARs that are not metabolized in the human body into vitamin A are stored in various tissues. In the case of ruminants, β-CAR accumulates in storage of fats and milk fat.

β-CAR, cryptoxanthin, and lutein are also present in these fats. Birds preferentially store oxycartinoids in the liver, eggs, body fat, skin, and feathers. The human body needs continuous ingestion of beta-CAR (β-CAR) due to the products because it does not know how to produce it.

CAR acts as a natural precursor to vitamin A – one of the most important substances for the proper functioning and health of our body. In the human body, β-CAR is not assimilated. The pigment is dissolved by the fat medium so that it is absorbed with sufficient fat.

According to biochemical availability, 6 g of pure β-CAR corresponds to 1 g of vitamin A in the form of retinol. Assimilation of the substance strongly affects the presence of fat in the conversion of β-CAR to retinol:

- p-CAR puree, dissolved in a fatty medium, will be taken with 50%;
- natural β-CAR, extracted from the body by the product, will be required by 8.3%;
- other CARs from a natural source, including alpha-CAR and gamma CAR, are digested with 4.16%.

The physiological norm of β-CAR is estimated at 5–7 mg/d for adults. The body of children needs an intake of 1.8–3 mg of substances.

The upper limit is not set – organic CAR even when used in high doses does not give negative symptoms.

For the detection of CARs, the color reactions (dark blue or purple blue) that they give with strong acids, such as concentrated sulfuric acid, hydrochloric acid, and perichloric acid, as well as arsenic and antimony trichlorides are useful. These reactions are nonspecific, being found in other pollen. Absorption spectra are very useful for CAR characterization.

7.3.1 Premises of Metabolic Engineering

Metabolic engineering allows the creation of unnatural metabolic cycles for new structures of biotechnological importance, especially by exploiting the promiscuity of biosynthetic metabolic cycle enzymes and directed evolution.

Recent research interest in phytochemicals has consistently led efforts in the field of metabolic engineering to the microbial production of various isoprenoids.

Despite systematic studies on the microbial production of CARs, similar to other isoprenoids, such as terpenoids, the possibility of using C_{30} CARs has not been explored so far, which have not been reported in plant sources because they are biologically functional compounds. Studies shed light on the generation of a new series of short C_{30} CAR structures, as well as on their potential biological properties, such as radical scavenging activity and neuronal differentiation activity.

CARs including apo-CARs are known as antioxidant compounds and are also interested in other beneficial biological activities, including anticancer activity. Although some studies have been done to understand the relationship between the structure of CARs and their anticancer effect, there is still a lack of knowledge about how CARs are transferred and modified into apo-CARs in the cell.

However, complex CAR structures with longer CBD and / or more functional groups are known to show higher antioxidant activity, which was consistent with our radical escape test (Huang et al. 2021).

The structural new CAR extension fund will open up possibilities for the discovery of bioactive compounds, which are potential candidates for the development of drugs and cosmetic applications. However, as this is the first report on the generation of new C_{30} and C_{35} CAR structures, a more detailed and comprehensive investigation of the biological and physiological activity of C_{30} CARs is needed.

To address this, more quantitative and systemic studies are needed to selectively direct the metabolic flow to the formation of new structures and to increase the titer of these new structures in *E. coli* bacteria. Finally, we believe that this paper may serve as a basis for investigators treating various CARs as candidates for biologically active compounds in the cosmetics, pharmaceutical, feed, and fine chemicals industries.

Using a new measure of lipophilic AC, we demonstrate that CARs can greatly inhibit the effect of oxidative challenge on redox status, thus supporting the recent antioxidant function of CARs *in vivo*. This result is not consistent with both the "protection" and the "disability" hypotheses, but provides circumstantial support for the "allocation" hypothesis, as CAR antioxidant function is a key hypothesis of the latter (Rischer et al. 2022).

However, other mechanisms, such as the regulation of CAR absorption and / or metabolism, cannot be ruled out, and studies using isotopic labeling of ingested CARs are likely to be needed to determine the relative involvement of the allocation compromise and other mechanisms in signaling. CAR-based honesty.

Moreover, the antioxidant effect of CARs is missed when a hydrophilic antioxidant test is used. The data also demonstrate the body's impressive ability to adjust its antioxidant mechanisms and maintain a stable level of oxidative damage to the blood, despite large fluctuations in oxidative load or CAR intake. As a combination of oxidative destruction markers and hydrophilic antioxidant assays is commonly used to assess redox status, these two observations provide a possible explanation for the weak overall support for *in vivo* CAR antioxidant function reported in a recent meta-analysis (Coe et al. 2021).

We suggest that the assessment of lipophilic AC will allow a deeper understanding of redox processes in lipophilic compartments, such as lipid bills and their ecological and evolutionary importance.

7.4 Concluding Remarks and Future Perspectives

The group of CARs has been and is being studied by many specialists who, through their research, have made truly valuable contributions. However, these organic molecules continue to be a little-known group, which greatly contributes to the balance and maintenance of body functions. CARs belong to the group of terpenoids, which are a series of compounds derived from mevalonic acid (derived from acetyl CoA).

Terpenes are derivatives of isoprene, a hydrocarbon made up of five carbon atoms. Specifically, CARs are tetraterpenes and are made up of forty carbon atoms. These atoms form conjugated chains that can end in carbon rings, substituted and unsaturated at each end. They have an isoprenoid structure, which means that they have a variable number of conjugated double bonds. This is important because it determines the wavelength of light that the molecule will absorb. Depending on the type of light it absorbs, it will give a specific color to the vegetable or plant in which it is located.

Molecules with few double bonds absorb light with a shorter wavelength.

For example, there is a molecule that contains only three conjugate bonds, so it can only capture ultraviolet light, it is colorless. There is another type of CAR that contains in its structure a total of eleven conjugated double bonds and absorbs even red. CARs are fat-soluble pigments, which means they are very soluble in oils and fats. In the same way, they are not synthetic, but are naturally produced by plants, some photosynthetic bacteria and algae.

They are also soluble in organic solvents such as ketones, diethyl ether, methanol, and chloroform, among many others. When in contact with an acid, CARs are extremely unstable.

This causes cyclization or isomerization reactions. Because they are hydrophobic, CARs will be found in lipid-related environments, such as inside cell membranes. Due to the presence of double bonds in their chemical structure, these compounds are very sensitive to certain elements in the environment, such as oxygen, peroxides, metals, acids, and light and heat, among others. Also, given their chemical structure, many of the CARs that exist in nature are precursors of vitamin A. For a CAR to be a precursor of vitamin A, there must be two conditions: the presence of ß-ionone and the capacity in the animal's body to convert it to retinol. Among the CARs that can act as precursors of vitamin A, we can mention: α-CAR, ß-zeaCAR, and ß-cryptoxanthin, among many others (about 50). CARs are chemical compounds that perform certain functions, including

- They are involved in the process of photosynthesis. This is because they are pigments present in plants capable of absorbing light of different wavelengths.
- CARs have a function of provitamin A. This means that some CARs, such as CARs, are precursor forms of retinol (Vitamin A). Once in the body, through various biochemical mechanisms inside the cells, they are converted into retinol, which has many benefits for humans. Especially at the level of the sense of sight.
- They are extremely beneficial for the human being, as they help to maintain good health, helping to prevent various diseases such as cancer and eye diseases, among others.

To understand the antioxidant effect of CARs, it is important to consider some knowledge about the molecular biology of the body. It has several mechanisms to purify the so-called free radicals, which cause significant damage. CARs have been shown to be extremely effective agents in inactivating O^{-2}, largely preventing photoxidative damage of this molecule to tissues.

This damage is caused by the action of light, which acts on certain molecules, causing the formation of compounds that can be harmful to cells. Considering that some of the CARs are precursors of retinol (vitamin A), they are an excellent source for the body to get the required amount. Retinol is a chemical compound that acts on the retina to optimize the functioning of the eye receptors and to greatly improve visual acuity, especially when it comes to night vision.

The exact mechanism by which CARs contribute to the health of the cardiovascular system is still unclear. However, doctors agree that a balanced diet should include CAR-containing foods, whether CAR or xanthophyll. Some CARs, such as lycopene, reduce the incidence of cancers, such as cancer of the prostate, lung, and digestive tract.

In the same vein, CARs have compounds known as acetylenics, which are known to help prevent the development of tumors. However, this is an area where much remains to be studied. The World Health Organization says that the claim that CARs protect against cancer is "possible but insufficient," so we must still wait for the results of many studies that are still ongoing. Despite this, everything seems to indicate that the results will be favorable and that CARs play an important role in preventing this terrible disease.

References

Al-Hassan, A., Vyas, R., Zhang, Y. et al. (2021). Assessment of maternal macular pigment optical density (MPOD) as a potential marker for dietary carotenoid intake during lactation in humans. *Nutrients* 14 (1): 182. https://doi.org/10.3390/nu14010182.

Bae, J.Y., Park, W.S., Kim, H.J. et al. (2021). Protective effect of carotenoid extract from orange-fleshed sweet potato on gastric ulcer in mice by inhibition of NO, IL-6 and PGE_2

production. *Pharmaceuticals (Basel, Switzerland)* 14 (12): 1320. https://doi.org/10.3390/ph14121320.

Canonico, M., Konert, G., Crepin, A. et al. (2021). Gradual response of cyanobacterial thylakoids to acute high-light stress-importance of carotenoid accumulation. *Cell* 10 (8): 1916. https://doi.org/10.3390/cells10081916.

Cao, Y., Yang, L., Qiao, X. et al. (2021). Dietary astaxanthin: an excellent carotenoid with multiple health benefits. *Critical Reviews in Food Science and Nutrition* 1–27. https://doi.org/10.1080/10408398.2021.1983766.

Chen, X., Deng, Z., Zheng, L. et al. (2021). Interaction between flavonoids and carotenoids on ameliorating oxidative stress and cellular uptake in different cells. *Foods (Basel, Switzerland)* 10 (12): 3096. https://doi.org/10.3390/foods10123096.

Coe, K.M., Ellison, S., Senalik, D. et al. (2021). The influence of the Or and Carotene Hydroxylase genes on carotenoid accumulation in orange carrots [*Daucus carota* (L.)]. *TAG. Theoretical and Applied Genetics. Theoretische und angewandte Genetik* 134 (10): 3351–3362. https://doi.org/10.1007/s00122-021-03901-3.

De Cosmi, V., Parazzini, F., Agostoni, C. et al. (2021). Antioxidant vitamins and carotenoids intake and the association with poor semen quality: a cross-sectional analysis of men referring to an Italian fertility clinic. *Frontiers in Nutrition* 8: 737077. https://doi.org/10.3389/fnut.2021.737077.

Dong, C., Wang, J., Hu, Y. et al. (2022). Analyses of key gene networks controlling carotenoid metabolism in Xiangfen 1 banana. *BMC Plant Biology* 22 (1): 34. https://doi.org/10.1186/s12870-021-03415-6.

Flegler, A. and Lipski, A. (2021). The C_{50} carotenoid bacterioruberin regulates membrane fluidity in pink-pigmented Arthrobacter species. *Archives of Microbiology* 204 (1): 70. https://doi.org/10.1007/s00203-021-02719-3.

Gea-Botella, S., Moreno-Chamba, B., de la Casa, L. et al. (2021). Carotenoids from persimmon (*Diospyros kaki* Thunb.) byproducts exert photoprotective, antioxidative and microbial anti-adhesive effects on HaCaT. *Pharmaceutics* 13 (11): 1898. https://doi.org/10.3390/pharmaceutics13111898.

Gupta, P. and Hirschberg, J. (2022). The genetic components of a natural color palette: a comprehensive list of carotenoid pathway mutations in plants. *Frontiers in Plant Science* 12: 806184. https://doi.org/10.3389/fpls.2021.806184.

Hien, H., Thom, L.T., Pham, Q.H., and Hong, D.D. (2022). Genome analysis of the heterotrophic microalga *Thraustochytrium* sp. TN22 to identify genes involved in exopolysaccharide and carotenoid biosynthesis pathways. *Marine Genomics* 61: 100918. https://doi.org/10.1016/j.margen.2021.100918.

Huang, Z., Guo, Y., Xiao, Q. et al. (2021). *Aegicerativicinus sediminis* gen. nov., sp. nov., a novel carotenoid-producing marine bacterium in the family Flavobacteriaceae. *Antonie Van Leeuwenhoek* 114 (10): 1551–1563. https://doi.org/10.1007/s10482-021-01621-8.

Igreja, W.S., Maia, F.A., Lopes, A.S., and Chisté, R.C. (2021). Biotechnological production of carotenoids using low cost-substrates is influenced by cultivation parameters: a review. *International Journal of Molecular Sciences* 22 (16): 8819. https://doi.org/10.3390/ijms22168819.

Kabir, M.T., Rahman, M.H., Shah, M. et al. (2022). Therapeutic promise of carotenoids as antioxidants and anti-inflammatory agents in neurodegenerative disorders. *Biomedicine &*

Pharmacotherapy=Biomedecine & Pharmacotherapie 146: 112610. https://doi.org/10.1016/j.biopha.2021.112610.

Kang, M., Chhetri, G., Kim, J. et al. (2021). *Sphingomonas sabuli* sp. nov., a carotenoid-producing bacterium isolated from beach sand. *International Journal of Systematic and Evolutionary Microbiology* 71 (7): https://doi.org/10.1099/ijsem.0.004896.

Kapoor, L., Simkin, A.J., George Priya Doss, C., and Siva, R. (2022). Fruit ripening: dynamics and integrated analysis of carotenoids and anthocyanins. *BMC Plant Biology* 22 (1): 27. https://doi.org/10.1186/s12870-021-03411-w.

Keawmanee, N., Ma, G., Zhang, L. et al. (2022). Exogenous gibberellin induced regreening through the regulation of chlorophyll and carotenoid metabolism in Valencia oranges. *Plant Physiology and Biochemistry: PPB* 173: 14–24. https://doi.org/10.1016/j.plaphy.2022.01.021.

Kiziler, M.E., Orak, T., Doymus, M. et al. (2021). Farnesol and tyrosol: novel inducers for microbial production of carotenoids and prodigiosin. *Archives of Microbiology* 204 (1): 107. https://doi.org/10.1007/s00203-021-02742-4.

Kupka, T., Gajda, T., Ochędzan-Siodłak, W. et al. (2022). On the impact of side methyl groups on the structure and vibrational properties of β-carotenoids. The case of butadiene and isoprene. *Food Chemistry* 369: 130880. https://doi.org/10.1016/j.foodchem.2021.130880.

Kusmita, L., Franyoto, Y.D., Mutmainah, M. et al. (2022). *Bixa Orellana* L. carotenoids: antiproliferative activity on human lung cancer, breast cancer, and cervical cancer cells *in vitro*. *Natural Product Research* 1–7. https://doi.org/10.1080/14786419.2022.2036144.

Lem, D.W., Gierhart, D.L., and Davey, P.G. (2021). A systematic review of carotenoids in the management of diabetic retinopathy. *Nutrients* 13 (7): 2441. https://doi.org/10.3390/nu13072441.

Li, T., Deng, Y.J., Liu, J.X. et al. (2021). DcCCD4 catalyzes the degradation of α-carotene and β-carotene to affect carotenoid accumulation and taproot color in carrot. *The Plant Journal: For Cell and Molecular Biology* 108 (4): 1116–1130. https://doi.org/10.1111/tpj.15498.

Li, X., Li, N., Zhao, L. et al. (2022). Tissue distribution and seasonal accumulation of carotenoids in Yesso scallop (*Mizuhopecten yessoensis*) with orange adductor muscle. *Food Chemistry* 367: 130701. https://doi.org/10.1016/j.foodchem.2021.130701.

Liu, Z., Natalizio, F., Dragone, G., and Mussatto, S.I. (2021). Maximizing the simultaneous production of lipids and carotenoids by *Rhodosporidium toruloides* from wheat straw hydrolysate and perspectives for large-scale implementation. *Bioresource Technology* 340: 125598. https://doi.org/10.1016/j.biortech.2021.125598.

Liu, J., Liu, D., Bi, J. et al. (2022). Micelle separation conditions based on particle size strongly affect carotenoid bioaccessibility assessment from juices after in vitro digestion. *Food Research International (Ottawa, Ont.)* 151: 110891. https://doi.org/10.1016/j.foodres.2021.110891.

Liu, B., Zhang, X., You, X. et al. (2022). Hydrogen sulfide improves tall fescue photosynthesis response to low-light stress by regulating chlorophyll and carotenoid metabolisms. *Plant Physiology and Biochemistry: PPB* 170: 133–145. https://doi.org/10.1016/j.plaphy.2021.12.002.

Manochkumar, J., Doss, C., El-Seedi, H.R. et al. (2021). The neuroprotective potential of carotenoids *in vitro* and *in vivo*. *Phytomedicine: International Journal of Phytotherapy and Phytopharmacology* 91: 153676. https://doi.org/10.1016/j.phymed.2021.153676.

Mavrommatis, A., Zografaki, M.E., Marka, S. et al. (2022). Effect of a carotenoid extract from *Citrus reticulata* by-products on the immune-oxidative status of broilers. *Antioxidants (Basel, Switzerland)* 11 (1): 144. https://doi.org/10.3390/antiox11010144.

Meléndez-Martínez, A.J., Benítez, A., Corell, M. et al. (2021). Screening for innovative sources of carotenoids and phenolic antioxidants among flowers. *Foods (Basel, Switzerland)* 10 (11): 2625. https://doi.org/10.3390/foods10112625.

Moopantakath, J., Imchen, M., Kumavath, R., and Martínez-Espinosa, R.M. (2021). Ubiquitousness of *Haloferax* and carotenoid producing genes in Arabian sea coastal biosystems of India. *Marine Drugs* 19 (8): 442. https://doi.org/10.3390/md19080442.

Narang, P.K., Dey, J., Mahapatra, S.R. et al. (2021). Genome-based identification and comparative analysis of enzymes for carotenoid biosynthesis in microalgae. *World Journal of Microbiology and Biotechnology* 38 (1): 8. https://doi.org/10.1007/s11274-021-03188-y.

Nishida, Y., Nawaz, A., Hecht, K., and Tobe, K. (2021). Astaxanthin as a novel mitochondrial regulator: a new aspect of carotenoids, beyond antioxidants. *Nutrients* 14 (1): 107. https://doi.org/10.3390/nu14010107.

Novikov, V.S., Kuzmin, V.V., Darvin, M.E. et al. (2022). Relations between the Raman spectra and molecular structure of selected carotenoids: DFT study of α-carotene, β-carotene, γ-carotene and lycopene. *Spectrochimica Acta. Part A, Molecular and Biomolecular Spectroscopy* 270: 120755. https://doi.org/10.1016/j.saa.2021.120755.

Nwoba, E.G., Rohani, T., Raeisossadati, M. et al. (2021). Monochromatic light filters to enhance biomass and carotenoid productivities of *Dunaliella salina* in raceway ponds. *Bioresource Technology* 340: 125689. https://doi.org/10.1016/j.biortech.2021.125689.

Olmedilla-Alonso, B., Benítez-González, A.M., Estévez-Santiago, R. et al. (2021). Assessment of food sources and the intake of the colourless carotenoids phytoene and phytofluene in Spain. *Nutrients* 13 (12): 4436. https://doi.org/10.3390/nu13124436.

Ouyang, M., Huang, Y., Wang, Y. et al. (2022). Stability of carotenoids and carotenoid esters in pumpkin (*Cucurbita maxima*) slices during hot air drying. *Food Chemistry* 367: 130710. https://doi.org/10.1016/j.foodchem.2021.130710.

Pang, Y., Chen, M., Lu, W. et al. (2021). Comparative genome characterization of *Echinicola* marina sp. nov., isolated from deep-sea sediment provide insight into carotenoid biosynthetic gene cluster evolution. *Scientific Reports* 11 (1): 24188. https://doi.org/10.1038/s41598-021-03683-0.

Rischer, H., Nohynek, L., Puupponen-Pimiä, R. et al. (2022). Plant cell cultures of Nordic berry species: phenolic and carotenoid profiling and biological assessments. *Food Chemistry* 366: 130571. https://doi.org/10.1016/j.foodchem.2021.130571.

Roca, M. and Pérez-Gálvez, A. (2021). Metabolomics of chlorophylls and carotenoids: analytical methods and metabolome-based studies. *Antioxidants (Basel, Switzerland)* 10 (10): 1622. https://doi.org/10.3390/antiox10101622.

Sandmann, G. (2021). Diversity and origin of carotenoid biosynthesis: its history of coevolution towards plant photosynthesis. *The New Phytologist* 232 (2): 479–493. https://doi.org/10.1111/nph.17655.

Schmeisser, J., Verlhac-Trichet, V., Madaro, A. et al. (2021). Molecular mechanism involved in carotenoid metabolism in Post–Smolt Atlantic Salmon: astaxanthin metabolism during flesh pigmentation and its antioxidant properties. *Marine Biotechnology (New York, N.Y.)* 23 (4): 653–670. https://doi.org/10.1007/s10126-021-10055-2.

Schmidt, M., Hopfhauer, S., Schwarzenbolz, U., and Böhm, V. (2021). High-pressure processing of kale: effects on the extractability, in vitro bioaccessibility of carotenoids & vitamin E and the lipophilic antioxidant capacity. *Antioxidants (Basel, Switzerland)* 10 (11): 1688. https://doi.org/10.3390/antiox10111688.

Shilpa, S., Shwetha, H.J., Perumal, M.K. et al. (2021). Turmeric, red pepper, and black pepper affect carotenoids solubilized micelles properties and bioaccessibility: capsaicin/piperine improves and curcumin inhibits carotenoids uptake and transport in Caco-2 cells. *Journal of Food Science* 86 (11): 4877–4891. https://doi.org/10.1111/1750-3841.15926.

Sreekumar, J., Muhammed Sadiq, P.A., Raju, S., and Mukherjee, A. (2022). *In silico* analysis of carotenoid biosynthesis pathway in cassava (*Manihot esculenta* Crantz). *Journal of Genetics* 101: 2.

Sun, X., Ma, L., Lux, P.E. et al. (2022). The distribution of phosphorus, carotenoids and tocochromanols in grains of four Chinese maize (*Zea mays* L.) varieties. *Food Chemistry* 367: 130725. https://doi.org/10.1016/j.foodchem.2021.130725.

Takayanagi, Y., Obana, A., Muto, S. et al. (2021). Relationships between skin carotenoid levels and metabolic syndrome. *Antioxidants (Basel, Switzerland)* 11 (1): 14. https://doi.org/10.3390/antiox11010014.

Takemura, M., Takagi, C., Aikawa, M. et al. (2021). Heterologous production of novel and rare C_{30}-carotenoids using *Planococcus* carotenoid biosynthesis genes. *Microbial Cell Factories* 20 (1): 194. https://doi.org/10.1186/s12934-021-01683-3.

Tarshish, E., Hermoni, K., Sharoni, Y., and Muizzuddin, N. (2022). Effect of Lumenato oral supplementation on plasma carotenoid levels and improvement of visual and experiential skin attributes. *Journal of Cosmetic Dermatology* https://doi.org/10.1111/jocd.14724. Advance online publication. https://doi.org/10.1111/jocd.14724.

Tkáčová, J., Zara, G., Ianiri, G. et al. (2022). Impairment of carotenoid biosynthesis through CAR1 gene mutation results in CoQ_{10}, sterols, and phytoene accumulation in *Rhodotorula mucilaginosa*. *Applied Microbiology and Biotechnology* 106 (1): 317–327. https://doi.org/10.1007/s00253-021-11673-5.

Yu, Y., Yu, J., Wang, Q. et al. (2021). Overexpression of the rice ORANGE gene OsOR negatively regulates carotenoid accumulation, leads to higher tiller numbers and decreases stress tolerance in Nipponbare rice. *Plant Science: An International Journal of Experimental Plant Biology* 310: 110962. https://doi.org/10.1016/j.plantsci.2021.110962.

Zhang, J., Guan, X., Lu, Y. et al. (2022). Titanium dioxide-mediated fatty acids promoted carotenoid synthesis in Phaffia rhodozyma PR106 analyzed whole genome. *Bioresource Technology* 347: 126699. https://doi.org/10.1016/j.biortech.2022.126699.

Zhou, W., Zhao, S., Xu, M. et al. (2021). Identification of key genes controlling carotenoid metabolism during apricot fruit development by integrating metabolic phenotypes and gene expression profiles. *Journal of Agricultural and Food Chemistry* 69 (32): 9472–9483. https://doi.org/10.1021/acs.jafc.1c00496.

8

Plant Genome Engineering for Improved Flavonoids Production

Monica Butnariu

Chemistry & Biochemistry Discipline, University of Life Sciences "King Mihai I" from Timisoara, Timisoara, Timis, Romania

8.1 Background

Flavonoids are a class of compounds with a polyphenolic structure of plant origin, considered to be secondary metabolites of plants. The name flavonoid comes from the Latin word flavus, which means yellow, one of the roles of these substances being the pigmentation of flower petals. Flavonoids are universal in the plant kingdom and occur in seed plants as well as in mosses and ferns. The formation of flavonoids is known from only a few microorganisms, such as watering the mold of *Aspergillus candidus*. Animals cannot produce flavonoids.

The appearance in some animal species, for example, in the wings of some butterflies, is due to the ingestion of plant flavonoids with food and their storage in the body. According to other information, flavonoids are limited to plants.

More than 9000 flavonoids are currently known, but this number is growing year by year. The special attention that these compounds receive is mainly due to the relatively high spectrum of biological activities. Various studies have shown that flavonoids have properties like antioxidant, antitumor, antimicrobial, antiviral, anthelmintic, and so on (Shan et al. 2020).

Flavonoids are a group of phytochemicals to which a large part of flower pigments belong. They are chemically derived from the basic structure of Flavan (2-phenylchroman) and consist of two aromatic rings attached to a tetrahydropyran ring according for Figure 8.1.

There are about 8000 compounds in nature, the diversity of which is created by different oxidation states in the ring containing oxygen, various substitutions on the aromatic rings, and the addition of sugars (formation of glycosides).

Biosynthesis takes place via shikimic acid.

Flavonoids are universally present in plants, so in the human diet, having antioxidant properties. Many plants that contain flavonoids are used medicinally.

Plants as Bioreactors for Industrial Molecules, First Edition.
Edited by Santosh Kumar Upadhyay and Sudhir P. Singh.

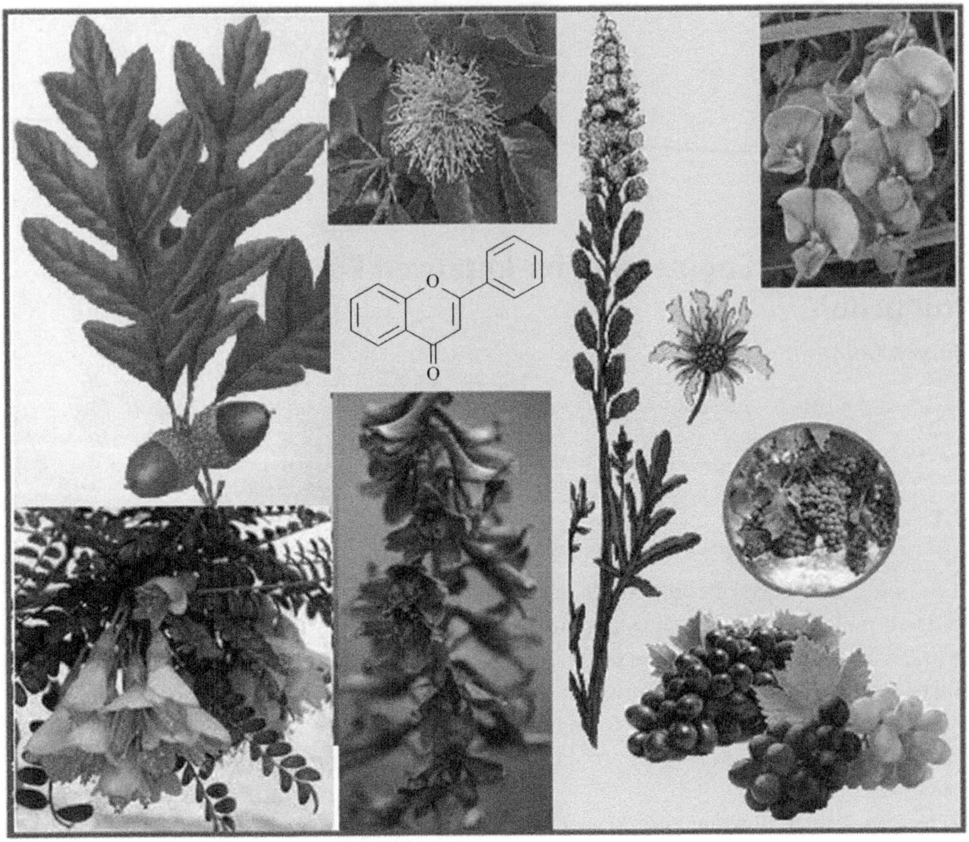

Figure 8.1 Schematic of the structure of the flavone backbone (2-phenyl-1,4-benzopyrone).

Flavonoids were discovered in the 1930s by Nobel laureate Albert von Szent-Györgyi Nagyrapolt and were originally mentioned as vitamin P. "P" in vitamin P means "permeability factor."

Some plants such as the painter's oak (*Quercus tinctoria*), the painter's wall (*Reseda lutea*), or the painter's mulberry (*Maclura tinctoria* content Macluracochinones) have been used for yellowing in the past.

After identifying their ingredients, this group of dyes was named flavone, after the Latin word flavus for yellow.

When it was recognized that a lot of ingredients had the same structure, many of which are differently colored or colorless, the group of substances was called flavonoids.

Flavonoids are a group of natural substances of plant origin, derived from benzopyran, have a substituted phenyl radical in position 2, and are part of the colored pigments in flowers and fruits. These compounds are included in the class of plant polyphenols (Bamdad et al. 2022).

Numerous flavonic substances have been identified; thus, only in the species of the genus Prunus, 32 types are described, and in the species of the genus *Citrus*, they contain 98 types. They are abundant substances in higher plants, which may contain one or more flavonoids.

In the cells, they are dissolved in vacuolar juice, by drying they are absorbed on the membrane walls.

The richest families in flavonoids are *Amentaceae*, *Urticaceae*, *Euphorbiaceae*, *Ericaceae*, *Compositae*, *Polygonaceae*, *Cariophylaceae*, *Cruciferae*, *Apocynaceae*, *Malvaceae*, *Sterculiaceae*, *Rubiaceae*, etc.

Pteridophytes and phyllines are known to produce flavonoids, such as fern cirthominetin.

Citrus flavanones play an important role in generating taste. Flavones play a role in the formation of the color of plant tissue and participate in the formation of taste (Zaragozá et al. 2022).

Flavonols, represented by chempferol, are widespread in fruits, leafy vegetables, spices, etc.

Anthocyanins are a mixture of compounds with different structures, which are found in plants such as shock, blackberries, and cherries, often used in the food industry as a food coloring.

Flavans are the most complex compounds, identified as mono-, bi-, tri-, and polyflavans.

Isoflavonoids are flavonoid compounds found in legumes, beans, and sunflower seeds, primarily known for their estrogenic effects.

8.2 Structure, Diversity, and Subgroups

The basic structure of flavonoids consists of two aromatic rings that are connected by a C3 bridge. Ring A mainly shows the floroglucin substitution pattern, which indicates the origin of acetogenin.

The B ring and the C3 bridge come from shikimic acid, the B ring is mostly hydroxylated at 4′, often at 3′ or 3′ and 5′. In the vast majority of flavonoids (except for templates), the C3 bridge is closed to form an O-heterocyclic ring (C ring).

The basic structure of flavonoids is therefore flavan (2-phenylchroman).

Ring B is rarely moved to position 3 (isoflavan, derived from isoflavones) or 4 (neoflavan) (Al-Maharik et al. 2022).

Based on groups and based on place of B-ring location, compounds of the flavonoid class can be classified according to Figure 8.2.

Flavonoids are phenolic substances, type C 6-C3-C6, derivatives of 2-phenyl-benzopyran (flavan) or 3-phenylbenzopyran (isoflavan).

Some of them represent yellow pigments (flavonols, flavones, chalcones), reds, violets, blues (anthocyanins), brown-browns (tannins), flowers, fruits, and other plant organs, others (flavanones and colorless flavanonols) are co-pigments.

From the IUPAC point of view, they are classified as follows:

- Flavonoids derived from 2 phenyl chromene 4 -one (structure 2 phenyl 1,4 benzopyronic);
- Isoflavonoids, derivatives of 3 phenyl chromen4-one (structure 3 phenyl 1,4 benzopyronic);
- Neoflavonoids derived from 4 phenyl coumarin (4 phenyl 1,2 benzopyronic).

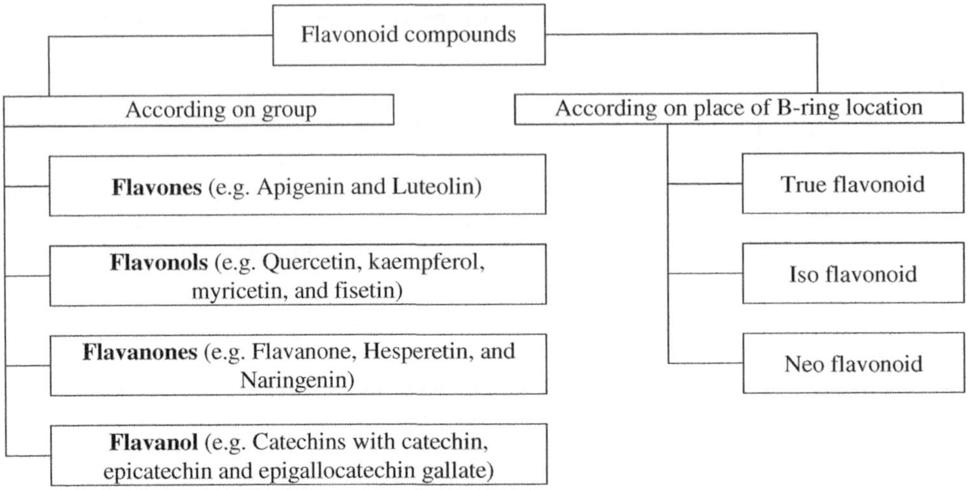

Figure 8.2 Schematic illustration of classes of flavonoid compounds.

A total of over 8000 different flavonoids have already been described (as of 2006). The design, especially the degree of oxidation at the C3 bridge, serves to subdivide the flavonoids into different subgroups (Mattia et al. 2022; Yang et al. 2022).

Flavonoids are built from the same basic backbone, C6-C3-C6, and two aromatic nuclei joined together by a chain of three carbon atoms. The vast majority of flavonoids are derived from phenylbenzopyran.

Depending on the position in which the rest of the phenyl binds, three broad classes are defined, namely flavonoids, derived from 2-phenylbenzopyran, isoflavonoids, derived from 3-phenylbenzopyran, and neoflavonoids, derived from 4-phenylbenzopyran. In addition to these three major classes, flavonoids are found in nature, and the structure of which does not derive from phenyl benzopyran.

These are called minor flavonoids and fall into the category of flavonoids due to the basic skeleton C6-C3-C6. Examples of this are calcones and aurons.

Flavonoids are divided into different groups depending on the degree of oxidation of the C3 bridge. Six major subgroups are found in most higher plants: calcones, flavones, flavonols, flavandiols, anthocyanins, and condensed tannins.

Aurons are very common, but not ubiquitous.

Isoflavones (especially in *Fabaceae*) and 3-deoxy-anthocyanidins, which are used as precursors of flobafen, for example, are limited to a few groups *Vitis vinifera*, *Arachis hypogaea*, and *Pinus sylvestris* are formed.

The structural diversity of flavonoids is due to the large number of substitution patterns on rings A and B, as well as the fact that flavonoids are not usually free, but rather glycosides.

More than 80 different sugars that are components of the flavonic C-glycoside subclass have been identified. Totally, 179 different glycosides were described for quercetin alone (Ble-González et al. 2022).

The structural feature of flavonoids is the presence in their molecule of the substituted benzopyran nucleus. Benzopyran substituted (chroman) with phenyl in position 2 (flavan) is found in the structure of the vast majority of flavonoids.

If the substitution is made in 3, it will generate the isoflavone series.

Flavonoids are phenolic pigments that contain in their molecule a pyranic or furanic heterocycle condensed with a benzene ring.

Another benzene ring is coupled to the heterocycle. The rings have hydroxyl groups, which determines the phenolic character of these pigments.

Flavonoids are plant pigments that predominate in higher plants. They are found in flowers, fruits, leaves, stems, roots, bark, etc. They are found in small amounts in some algae, microorganisms, and some insects (flavones).

Examples of basic structure are presented in Table 8.1.

Most flavonoids are colored and largely contribute to the formation of the color of flowers and fruits. They are found in nature in the free state, but especially in the form of glycosides. Chemically flavonoid pigments are water-soluble phenolic glycosides.

They are found in vacuolar juice and chromoplasts. Six types of flavonoids are known: flavans, anthocyanidins, flavones, flavanones, calcones, and aurons.

These groups differ in the type of heterocycle and in the number and position of hydroxyl and methoxyl groups related to benzene rings (Zeng et al. 2022).

Flavans (chromania) are pigments derived from flavan (2-phenyl-benzopyran, chroman). They have a benzopyran ring in the molecule. They tend to polymerize and form catechins that form catechin tannins.

Anthocyanins and anthocyanins. Anthocyanidins are pigments derived from 2-phenyl-benzopyrene (2-phenyl-chromene). They are the main pigments that give flowers

Table 8.1 Examples of flavonoids.

No	Class	Examples
	Chalcones	Butein ($C_{15}H_{12}O_5$), Cardamonin ($C_{16}H_{14}O_4$), Isobavachalcone ($C_{20}H_{20}O_4$), Isoliquiritigenin ($C_{15}H_{12}O_4$), Xanthohumol ($C_{21}H_{22}O_5$)
	Flavones	Luteolin ($C_{15}H_{10}O_6$), Apigenin ($C_{15}H_{10}O_5$), Baicalein ($C_{15}H_{10}O_5$), Crysoeriol ($C_{16}H_{12}O_6$), Scutellarin ($C_{21}H_{18}O_{12}$), Wogonin ($C_{15}H_{10}O_5$)
	Flavonols	Morin ($C_{15}H_{10}O_7$), Quercetin ($C_{15}H_{10}O_7$), Rutin ($C_{27}H_{30}O_{16}$), Kaempferol ($C_{15}H_{10}O_6$), Myricetin ($C_{15}H_{10}O_8$), Isorhamnetin ($C_{16}H_{12}O_7$), Fisetin ($C_{15}H_{10}O_6$)
	Flavanols	Catechin ($C_{15}H_{14}O_6$), Galocatechin ($C_{15}H_{14}O_7$), Epicatechin ($C_{15}H_{14}O_6$), Epigallocatechin gallate or Epigallocatehin-3-gallate ($C_{22}H_{18}O_{11}$)
	Flavanones	Hesperetin ($C_{16}H_{14}O_6$), Naringenin ($C_{15}H_{12}O_5$), Eriodictyol ($C_{15}H_{12}O_6$)
	Flavanols	Taxifolin ($C_{15}H_{12}O_7$)
	Isoflavone	Genistein ($C_{15}H_{10}O_5$), Daidzein ($C_{15}H_{10}O_4$), Licoricidin ($C_{26}H_{32}O_5$)
	Anthocyanins (Anthocyanins)	Cyanidin ($C_{15}H_{11}O_6$), Dolphinidin ($C_{15}H_{11}O_7$), Malvidin ($C_{17}H_{15}O_7$), Pelargonidine ($C_{15}H_{11}O_5$), Peonidine ($C_{16}H_{13}O_6$), Petunidine ($C_{16}H_{13}O_7$)
	Aurones	Aureusidin ($C_{15}H_{10}O_6$), 4′-chloro-2-hydroxyaurone ($C_{15}H_{11}O_3Cl$), and 4′-chloroaurone ($C_{15}H_9O_2Cl$)

and fruits their red and blue color. They are usually found in nature as glycosides, which are called anthocyanins. The most important anthocyanins are pelargonidine, cyaniding, and dolphinidine, which differ in the number and position of the hydroxyl groups on the benzene ring C (Peng et al. 2022).

Pelargonidine is found in geranium flowers, cyanidin in chicory and centaur flowers, and dolphinidine in flowers *Delphinium consolida*.

In anthocyanins, carbohydrates (monoglycerides, diglucides) bind to anthocyanidins (aglycone), usually to the C-3 hydroxyl on the heterocycle.

Monoglycoside anthocyanins predominate in nature, but diglycoside anthocyanins are also known.

By methylation of the mentioned anthocyanidins, new pigments with different colors are obtained such as peonidine, pentunidine, and malvidin.

Anthocyanins are found in flowers alone and especially when mixed with other pigments, forming a wide variety of colors. The pink, red, bright red flowers contain predominantly pelargonidine, the purple and cherry flowers contain cyanidin.

Anthocyanins are soluble in water and alcohol, hardly soluble in ether, benzene, and chloroform. Extract with water or alcohol in hydrochloric acid medium.

From the extract obtained, if ether is added, anthocyanin chloride precipitates. With mineral acids, anthocyanins form red salts that are stable when diluted (unlike flavones).

Anthocyanidins with adjacent hydroxyl groups form blue complexes with metals (Al, Fe). Anthocyanins change color depending on pH, which is why they are used as acid–base indicators (Liu et al. 2022).

Flavones and isoflavones. They are yellow pigments derived from 2-phenyl-benzopyron (2-phenyl-chromone) and 3-phenyl-benzopyron (3-phenyl-chromene), respectively. In nature, flavones and isoflavones are usually found in the form of glycosides and are widespread.

More than a hundred flavones are known to be found in flowers, fruits, leaves, wood, and tree bark. Flavones and isoflavones contain hydroxyl and methoxyl groups on the A and C rings.

If there are hydroxyl groups on the heterocyclic ring, the pigments are called flavanols and isoflavanols. The most common flavones include apigenin, luteolin, and quercitrin. Flavones are crystalline, yellow substances, soluble in water and alcohol. In an alkaline environment, the pyran ring opens and diketones are formed.

Flavones have absorption maxima between 335 and 350 nm and flavanols between 360 and 380 nm. With metals, flavones form complexes. It dissolves in concentrated sulfuric acid giving yellow solutions, due to the formation of flavili salts.

Flavones have two or three characteristic ultraviolet absorption bands. They protect the body from oxidizing vitamin C and adrenaline. It absorbs ultraviolet radiation and protects the cytoplasm and chlorophyll from this radiation. They are found in greater numbers in tropical plants and in those of mountainous and alpine regions. Flavanols that have a large number of hydroxyl and methoxyl groups on ring A are intensely yellow.

Such is the case with gosipetin, the quercetagetin that imprints the yellow color on cotton flowers.

The white flowers have a low content of flavones and flavanols (Hua et al. 2022).

Flavanone,calcone,andaurone.Flavanonesarederivedfrom2-phenyl-dihydrobenzo-piron, benzylidene-acetone-phenone calcones, and 2-benzylidene-couaranone aurones.

Flavones are colorless pigments (due to disruption of the chromoform system) that accompany flavones. In the basic environment, they are transformed into chalcones (by the noise of the heterocycle), and these in an acidic environment are cyclized and the flavones are restored. By oxidation they turn into aurons, which have an intense golden-yellow color and are found predominantly in the flowers of some composites.

Flavanones, calcones, and aurons also have hydroxyl groups on the A, C, and even heterocycle rings. They are usually found in nature as glycosides.

Auronas and calcones usually appear in flowers and flavanones in flowers, leaves, wood, etc.

The presence of flavanones in plant organisms increases their resistance to attack by insects and microorganisms. They influence the taste, aroma, and stability of some preparations extracted from plants. Some flavonoids are used as antioxidants in fat preservation, others are used as pharmaceuticals.

Flavonoids can form redox systems in cells, which play a significant role in the body. Some flavonoids are used as vitamin P in therapy. Through different oxidation, reduction, isomerization reactions, etc., different groups of flavonoids can transform into each other, which is a great advantage for plants.

Depending on the needs of the plants, some flavonoids can be transformed into others, making good use of these pigments.

Calcons that are not actually considered flavones, or even flavonoids, because from a biogenetic point of view they represent the direct precursors of flavones (Wang et al. 2021; Zhang et al. 2022).

Flavonoids are a class of plant secondary metabolites known for their antioxidant properties. The name flavonoids was first given in 1895 by the Swiss chemist S. Kostanetsky and comes from the Latin flavum – yellow because the first such substances were yellow.

Flavonoids distributed in plants, being responsible for the yellow color, also have a protective role against microbes and insects.

Due to their wide distribution of their variety and low toxicity compared to other secondary metabolites, they cause their ingestion by humans and animals in large quantities.

Due to their obvious ability to modify certain reactions of the body to allergens, viruses, and carcinogens, they are also called natural modifiers of the biological response.

Starting from the rich literature on this topic, we tried a synthetic theoretical approach to flavonoids, presenting general information about flavones, plant products with flavones, pharmacological properties and mechanisms, drugs and food supplements containing flavonoids (Wei et al. 2022).

To date, about 3000 such compounds are known, of which about 500 are free in higher plants (vascular cryptogams, gymnosperms, angiosperms); very rarely found in algae (e.g. peonidol in *Chlamidomonas* species).

They have been reported in insects and butterflies where they come from the pollen or nectar of the flowers they feed on.

The most common are flavonosides, anthocyanins, and catechols.

Heterosides are found dissolved in cell juice, and aglycones predominate in lignified tissues (wooden vessels, fibers, ideoblasts, cuticle, etc.).

The flavonoids are based on 2-phenylbenzopyran or flavan, rarely 3-phenylbenzopyran or isoflavan, in various stages of oxidation. Most derivatives have on the nucleus A, at C5 and C7 hydroxyl alcoholic groups derived from the carbonyl function of malonylCoA and phenolic oxidryl groups in 4′or 3′-4′, 3′-4′-5′, derived from phenyl propane compounds.

These hydroxyl groups may be methoxylated, glycosidated, or prenylated.

Depending on the degree of oxidation of the nucleus, the saturation of the C3 segment, the C3 substituent and the place of insertion of the phenyl, several groups of flavonoid compounds are known.

These are: biflavonosides, flavonoid dimers; chalcone, flavonoids having a split pyronic cycle; aurons, benzylidene couaranone derivatives; isoflavones (isoflavones proper), derivatives of 3-phenyl-y-benzopyrone; pterocaptans, rotenones, isoflavone derivatives; anthocyanosides (anthocyanin pigments), derivatives of 2-phenylchromene (2-phenylbenzopyrrile); proanthocyanidols or flavan-3,4-diols; catechols or flavan-3-oils.

Flavanones are fairly common flavonoids in the plant kingdom. They are found in nature both as free aglycones, but especially as glycosides.

Flavanonols are 3-hydroxylated derivatives of flavanones, more widespread compounds than flavanones and whose biosynthesis is preceded by the compound: hydroxy-phenyl-propionyl-CoA.

Terpenyl flavonoids, among the flavonoid substances, lipophilic combinations have been identified in plants that have been shown to be substituted with chains of 5 carbon atoms, derived from hemiterpenic biogenetic units.

This structure gives these substances strong lipophilic properties, which makes them localized in the lipid deposits of plants. Because of this they are found more like aglycones and very rarely glycosides (Li et al. 2022).

Isoflavones are derivatives of 3-phenylbenzopyran. Due to the position of the substituted phenyl ring, they can lead to the closure of new rings, either pyranic in position 2–3 or in position 7–8.

Pterocarpans are cyclization products of isoflavones, in which a new furan cycle closes between the C6 of the substituted phenyl ring and the C4 ketone group of the pyran ring. They are known in nature as phytoalexins.

Rotenoids are a special type of isoflavone, with much more complicated structures. They are isoflavones with a furanic or pyranic ring in position 7-8, of hemiterpenic origin and which close a second pyranic cycle between the benzopyranic nucleus and the substituted phenyl, from position 3.

Biflavonoids are a large number of biflavonoids that are produced by the condensation of leucoanthocyanins with catechins, but also dimers composed of flavonic molecules or flavones and flavanones, which are isolated from plants.

Flavonic alkaloids represented by ficin, one of these alkaloids, is a condensation product of erysol with n-methylpyrrole (Rauf et al. 2022).

The biosynthesis of flavonoids in plants is preceded by the amino acid phenylalanine and involves a sequence of complex reactions. Flavones are solids, crystallized as aglycones.

They are yellow, tasteless, and odorless, have characteristic fluorescence in UV light and absorption spectra in UV and IR.

Chemical characteristics represented by the identification reactions such as cyanidine reaction; citro-boric reaction (Wilson); oxalo-boric reaction (Taubck); reaction with $SbCl_5$ (Marini–Bettolo); coloration in an alkaline environment; reaction with the salts of some plurivalent metals; fluorescence in the light of the Wood lamp; paper and thin layer chromatography (Kumar et al. 2022).

Biochemical characteristics: antioxidant effect and / or fixation of free radicals, anti-inflammatory effect and influence of the functioning of the immune system, effect against asthma and allergies; modification, generally inhibition of enzyme function; effect against viruses and bacteria; estrogen / antiestrogen effect characteristic of isoflavonoids only; effect that influences the mutation in genetic matter (DNA); modifying effect, mainly inhibiting the processes related to carcinogenic malformations; liver protection features / characteristics; effect that influences the functioning, the condition of the vascular system, primarily of the capillaries. The above characteristics are interdependent in several cases: the hepatic protective effect with the characteristic of fixing free radicals, in many cases, the antioxidant effect and the antiasthma effect with the characteristic that influences the functioning of different enzymes.

8.3 Flavonoid Biosynthesis

Flavonoid biosynthesis is continuously considered genuine to only plants. Flavonoids are synthesized by the phenylpropanoid metabolic pathway in which the amino acid phenylalanine is used to produce 4-cumaroyl-CoA.

This can be combined with malonyl-CoA to produce the true backbone of flavonoids, a group of compounds called calcones, which contain two phenyl rings.

The conjugated closure of the ring with chalcone results in the familiar shape of flavonoids, the three-ring structure of a flavone.

The metabolic pathway (Shikimate pathway) continues through a series of enzymatic changes to produce flavanones → dihydroflavonols → anthocyanins (Rauf et al. 2022).

The shikimic acid cycle has as its substrate erythroso-4-phosphate, produced in the pentosophosphate cycle. The result is phenols, lignins, anthocyanins, and phenolic amino acids such as tryptophan from which auxin is biosynthesized. It is estimated that under normal conditions about 20% of the fixed carbon of plants is used in the shikimat cycle.

During this cycle, the phosphoenolpyruvate and erythroso-4-phosphate condense into a composed of 7 carbon atoms: 3 deoxy-D-arabino-heptulosonate-7-phosphate (DHAP), which results in chorismate, as follows:

DAHP → 3 dihydroquinate → 3 dihydroshikimate →shikimat → corismat

Chorismate is the initial compound for three cycles:

- the first results in phenylalanine, lignins, and flavonoids;
- from the second tryptophan, auxins, glucosinolates, phytoalexins, and alkaloids;
- from the third tyrosine and melanin.

The key enzyme in the shikimic acid cycle is phenylalanammonium lyase (PAL), which binds to the membrane of the endoplasmic reticulum, chloroplasts, mitochondria, and plasma membranes. The enzymes that catalyze the reactions in this cycle are synthesized in the ribosomes of the cytoplasm. The activity of these enzymes has been shown in chloroplasts and their identification in the cytoplasm is uncertain.

Some of the amino acids that form in this cycle are in turn precursors of other substances. Tryptophan is the precursor of the growth hormone auxin, phenylalanine is the precursor of flavonoid pigments and lignins, and tyrosine is the precursor of ubiquinone, a substance that carries electrons in the process of respiration. In the shikimic acid cycle, which takes place in the endoplasmic reticulum and in the cytoplasm, phenolic acids are also biosynthesized: p-coumaric, cinnamic, caffeic, ferulic, chlorogenic, as well as galotanins.

Phenolic substances can stimulate or inhibit the action of hormones, inhibit the synthesis of ATP in mitochondria, as well as the activity of enzymes, or cytoplasmic currents in the cells of absorbent hairs. Some phenolic substances (ferulic acid, lunular acid, chlorogenic acid, and catechins) have an inhibitory effect on seed germination (Kumar et al. 2022; Zhang et al. 2022).

Cinnamic acid derivatives have been identified in vacuoles and chloroplasts. Enzymes involved in the biosynthesis of flavonoids: chalcone-flavon isomerase and flavonoid hydrolase have been identified in the endoplasmic reticulum and chloroplasts, and flavonoids have been identified in vacuoles, extraplasmic space, and chloroplasts.

The biosynthesis of flavonoids is genetically coordinated.

In plants, the shikimate pathway first leads to the formation of chorismate, which is the precursor of phenylalanine, tyrosine, and tryptophan.

These aromatic amino acids are derivatives of many secondary metabolites, all essential for the biological functions of the plant, such as the hormones salicylate and auxin. This pathway contains enzymes that can be regulated by inhibitors, which can stop the production of chorismate and, finally, the biological functions of the body.

Many products can be formed along this path, including flavonols, flavan-3-ols, proanthocyanidins (tannins), and a number of other polyphenols (Fang et al. 2022).

So, the precursor for flavonoid biosynthesis is the aromatic amino acid phenylalanine, which is formed via shikimic acid. Phenylalanine is converted to phenylalanine ammonia binder (PAL) transformed into cinnamic acid. In turn, it is hydroxylated into p-coumaric acid by cinnamic acid 4-hydroxylase. This pathway is common to all phenylpropanoids. P-coumaric acid is activated in cinnamoyl coenzyme A.

In the next step, the second aromatic ring is formed: the enzyme calcone synthase (CHS) forms the calcone from cinnamoyl-CoA and three molecules of malonyl-coenzyme A, which come from the synthesis of fatty acids. Calcone is in equilibrium with flavanone by the action of calcone isomerase (CHI). This closes the third ring.

The three key enzymes (PAL, CHS, and CHI) as well as partially the enzymes of the subsequent synthesis steps are present as enzymatic complexes. The complex is probably located on the cytosolic side of the endoplasmic reticulum (Lu et al. 2022).

The different pathways range from flavanone to flavones, flavonols, isoflavones, and anthocyanidins.

Flavonoid biosynthesis is light-induced; storage takes place mainly in vacuole.

Most of the enzymes involved in flavonoid biosynthesis come from three classes of enzymes found in all organisms: oxoglutarate-dependent dioxygenases, NADPH-dependent reductases, and cytochrome P450 hydroxylases.

The two key enzymes CHS and CHI belong to different families. CHI is likely to be unique to plants in terms of both sequence and three-dimensional structure. CHS, in turn, belongs to the superfamily of plant polyhydric synthetases.

The C6-C3 segment (nucleus A) has as biosynthetic precursor three molecules of acetyl-CoA or malonyl-CoA, while the C6 segment (nucleus B) is formed by shikimic acid, p-cumaroyl-CoA, or another derivative hydroxylated phenylpropane (cinnamic acid, etc.). By condensing them, a chalcone results, through which cyclization the flavones are formed.

Under normal physiological conditions, chalcones tend to isomerize spontaneously into racemic flavones. Stereospecific cyclization of chalcones occurs in the presence of chalcom-isomerase which induces a cycle closure, with the formation of a (2-S)-flavanone.

The introduction of the hydroxyl group to C3 is done under the catalytic action of a dioxygenase. Thus, (2S)-naringethol is converted to (2R, 3R)-dihydrocaemferrol, (2S)-heriodictiol to (2R, 3R)-dihydroquercetol. Flavonols appear to be derived by enzymatic dehydration of a 2-hydroxylated intermediate (Sun et al. 2022).

The mechanism that leads to the conversion of flavanones to flavones has not yet been elucidated. Considering chalcones as biogenetic precursors of flavonoids.

The physicochemical properties, action, and uses of these compounds will be presented in each class.

Plant metabolism (e.g. sugar metabolism, glycolysis, TCA cycle, and shikimate-phenylpropanoid metabolism) is responsible for the production of accessible energy, endurance metabolites, and carbon skeletons for plant life activities during growth and development. C metabolism is closely related to plant growth and senescence and can be divided into C assimilation, C transformation, and C accumulation between different physiological processes of plants (e.g. photosynthesis and respiration). Glucose and fructose, two important reducing monosaccharides, are the degradation products of sucrose and starch. A sharp decrease in glucose and fructose content was observed after the square stage in plants grown in Dali and in the flowering stage in cultivated plants, indicating that the transition from C uptake to C accumulation occurred later in cultivated plants. Interestingly, sucrose, a cytoplasmic product produced during photosynthesis in plant cells by sucrose synthase, can be used as transport sugar for glycolysis entry and the TCA cycle for the production of ATP and NADH (Sharifi-Rad et al. 2022).

The shikimate-phenylpropanoid pathway generates numerous antioxidants (e.g. flavonoids, phenols, and lignins) and their precursors (e.g. aromatic amino acids and shikimic acid) to escape or inhibit specific reactive oxygen synthesis (ROS) in plant cells under biotic or abiotic stress, thus protecting cellular proteins, membrane lipids, DNA, and other cellular components from serious damage. The increased content of defense-related metabolites (e.g. dopamine, chlorogenic acid, scopoline, and tocopherol) and their precursors (e.g. phenylalanine, shikimic acid, and tryptamine) in plant shikimate-phenylpropanoid metabolism could be closely associated with high levels of temperature and less abundant rainfall experienced during the mature period (Marques et al. 2022).

The increased content of defense-related metabolites (e.g. dopamine, chlorogenic acid, scopoline, and tocopherol) and their precursors (e.g. phenylalanine, shikimic acid and tryptamine) in plant shikimate-phenylpropanoid metabolism could be closely associated with elevated temperature and less abundant rainfall experienced during the mature period.

Flavonoids that are plant pigments responsible for the yellow coloration of flowers, fruits, sometimes even leaves. The presence of flavonoids in algae has not yet been demonstrated. They are common in *Bryophyrae*, *Pteridophyrae*, *Gymnospermae*, and *Angiospermae*, where the structural diversity of these compounds is maximum.

Abundant in *Apiaceae, Asteraceae, Fabaceae, Laminaceae, Liliaceae, Malvaceae, Polygonaceae, Rosaceae, Rutaceae, Scholphulariaceae*, etc.

They are present in all plant organs in glycosidic or aglycone form. Their existence in the leaf cuticle and in the epidermal cells of the leaves ensures the protection of the cuticle against the harmful effects of UV radiation.

Heteroside (water-soluble) forms accumulate in vacuoles, concentrate in the epidermis of the leaves, or are distributed between the epidermis and the mesophile.

In the case of flowers, they are stored in epidermal cells. As aglycones, they are distributed in the cuticle of the leaves and in the wood.

Flavonoid aglycones are based on the 2-phenylbenzoyl-pyronic (2-phenylchromatic) ring on which hydroxyl, methoxyl, dimethylallyl, etc. groups are grafted. Most compounds are hydroxylated in C5, C7 on the A nucleus (Guo et al. 2019).

1–3 phenolic groups at C4′, 3′, and 5′ can be found grafted on the B core. Depending on the degree of oxidation of the C3 segment and the C3 substituents, we distinguish: flavones, flavonols, flavanones, and flavanonols. Flavones always have a double bond between C2 and C3.

Flavonols are 3-hydroxylated flavones (also known as acylated flavonols-tilyroside from *Tiliae flores*). Flavanones are 2,3-dihydrogenated flavones. Flavanonols are 2,3-dihydrogenated flavonols.

From *Sophora subprostrate* Chun. Several terpenyl flavanones (sophoranone, sophoradine, etc.) have been isolated with antiulcer action. Flavanonols often form dimers with lignans (flavanonol-lignans: silibin, silidianin, and silicristin).

Flavononol – lignans are combinations of lignans with flavonols.

Lignans are compounds resulting from the condensation of two to five molecules of phenylpropane derivatives (C6-C3) *n*, linked by side chains, acyclic or cyclic, often lactonized, β-glucosidated, or nonglucosidated.

Flavanonols (flavononol-lignans) may have a benzodioxane (silibin), 2,3-benzofuranic (silicristin) or lactonic (silidianine) structure.

They are solid, crystallized, colorless, optically active, water-soluble, dilute alcohol, dilute acetone (heteroside forms) or dilute alcohol, acetone, benzene, ether, chloroform, etc., slightly soluble in petroleum ether (nonglycosidized forms) (Seitz et al. 2007).

Forms with alkaline hydroxides, water-soluble phenols, hardly soluble in alcohol. Under the same conditions, flavononol lactone lignans generate hydroxy acids. In acidic or basic medium, it is easily isomerized.

The extraction is performed with alkaline hydroxides. Purification can be performed by various chromatographic processes (HPLC, CSS, column chromatography, etc.). The identification of flavononols is done with iron (III) chloride when it turns green.

Flavononol lignans are colored green or blue with iodine green, yellow with sulfuric aniline, and red hydrochloride with hydrochloric acid.

Flavononol lignans – Americanin A ($C_{18}H_{16}O_6$), B, D from *Phytolacca americana* L., Silibinin ($C_{25}H_{22}O_{10}$), Silidianin ($C_{25}H_{22}O_{10}$), Silychristin also known as silichristin ($C_{25}H_{22}O_{10}$) from *Silybum marianum*, Wuweizisu C ($C_{22}H_{24}O_6$) from *Schizandra chinensis*, and *Korean red ginseng* have hepatoprotective action.

The carbohydrate part, consisting of one or more oases, rarely uronic acids (baicaloside), can be attached to any of the hydroxyl groups on the molecule.

One ozone molecule (quercetol-3-galactoside = hyperoside, quervetol-4′-glucoside = spireoside) or several (quercetol-3-glucoramnoside = rutoside) can be attached to the same hydroxyl group, and flavones are also known as di- and triglycosides, wherein the ozone binds to several hydroxyl groups on the nucleus (luteol-7-rutinoside-4′-glucoside = cinnarotrioside from *Cynarae folium*) (Austin and Noel 2003).

Most are O-heterosides, but there are also about 300 C-heterosides.

These include apigenol-8-C-glucoside (vitexin), apinogel-5-C-glucoside (isovitexin), luteol-8-Cglucoside (orientin) from *Crataegi fructus* and flores; 7-O-glucosyl-isovitexin (saponarin) from *Saponaria officinalis* L.

The bond between aglycone and oasis is determined by the anomeric carbon of the oasis and the 6 or 8 carbon of the aglycone (flavone, flavonols, and chalcone).

There are mono-C-glucosyl-flavone 8 (vitexin), di-C-glucosyl-flavone, C-glucosyl-O-glucosyl-flavone (saponarin), and acyl-C-flavone.

In the case of C-glucosyl-flavones-5-hydroxylated, the heterocycle opens easily, in acidic medium, facilitating the isomerization 6↔8, 8↔6 (Wessely-Moser isomerism).

The reaction has applications in determining the structure of these compounds.

Biflavonoids. These are flavonoid dimers. The monomers can bind to each other through carbon atoms in positions 6 or 8, which are highly reactive. The established links may be C-C (amentoflavone, agathisflavone, bilobetol, and ginkgetol) or C-O-C (hinokiflavone) (Ko et al. 2020).

The two constituent units of biflavonoids can be the same type (biflavone, biflavanone) or different types (flavone-flavanone). Hydroxyl groups may be free or, more often, methylated.

Biflavonoids are characteristic of Gymnospermae. In Angoispermae, their distribution is sporadic.

Chalcones result from the opening of the central pyranic ring, characterized by the presence of an unsaturated a↔b ketone segment, consisting of three carbon atoms. Nucleus A carries the same substituents as other flavonoids. Core B may or may not be substituted.

Neoflavonoids that are compounds of type C6-C3-C6, with the phenyl radical arranged in position 4 and the ketone at C2. They are actually phenylcoumarins.

They were isolated from *Fabaceae* and *Clusiaceae*; recently they have been highlighted in some *Rubiaceae* (dalbergina of species belonging to the genus *Dalbergia*; *D. melaxylon*, *D. nigra*, *D. latifolia*). They are responsible for the dermatitis caused to the workers who handle the wood of these species (Oyeleke and Owoyele 2022).

Aurone. They are yellow pigments, like flavones, derivatives of 2-benzylidene couaranone. Examples: hispidol ($C_{15}H_{10}O_4$) from *Glycine maxima*, aureusidine ($C_{15}H_{10}O_6$) from *Antirrhinum*, *Oxalis* and *Linaria* species.

Heterosides are extracted with water or dilute alcohol. Purification of the solution is carried out by precipitation with lead acetate (in the case of hydroalcoholic solutions, before purification, the alcohol is removed by distillation or evaporation).

Under these conditions, however, flavonic heterosides precipitate which form chelates with Pb^{2+}; their regeneration is done by treating the precipitate with sulfuric acid, removing excess lead with sodium hydrogen carbonate, neutralizing the solution with sodium carbonate. The solution thus purified is concentrated and subjected to repeated crystallization or extraction with solvents of different polarities.

The extraction of aglycones is performed by solubilizing them in an apolar solvent; the extract solution is concentrated to remove the solvent; the residue is then extracted with solvents of different coatings for purification.

Characterization is done by elemental analysis, NMR spectroscopy, alkaline melting, and chromatography (thin layer, gas, and HPLC) (Lu et al. 2021).

Flavonosides intervene (*physiological role*) in the cellular respiration of plants, especially those with peroxidases, their recovery taking place based on vitamin C.

Many of them have the behavior of coenzymes, chelating microelements from plants that they mobilize in various enzyme systems. Protects plant tissues from the harmful effects of ultraviolet radiation.

They are solid, crystallized substances, colored in yellow of different intensities.

Heterosides are soluble in water, concentrated and diluted alcohol, specifically soluble in ethyl acetate, acetone, and insoluble in apolar solvents.

Aglycones are soluble in alcohol, ethyl acetate, apolar solvents, and insoluble in water. Due to phenolic hydroxyls, they are also soluble in alkaline solutions.

Heterosides hydrolyze only in the presence of mineral acids (do not hydrolyze enzymatically), under different conditions, depending on the bonding atom between the aglycone and the carbohydrate side: a longer time, at high pressure and in the presence of catalysts.

Other chemical properties are imprinted by the C4 ketone group and the phenolic OH groups. Due to the C4 ketone group, flavones can be reduced with hydrogen at birth in a hydroalcoholic environment when anthocyanidols are formed (Shibata reaction).

Their color can be red (flavonols), orange (flavones).

Flavanones, chalcones, and aurons do not stain.

Due to the presence of phenolic oxidril groups, flavones can be transformed into phenoxides, in an alkaline environment (yellow more intense than the initial flavone); with concentrated sulfuric acid or concentrated hydrochloric acid is converted to yellow oxonium salts (Reyes Jara et al. 2022).

By coupling with diazonium salts, yellow-orange compounds are obtained, the phenolic oxidril groups give the flavors a reducing character (they reduce the Fehling, Tollens, and Folin-Ciocâlteu rectifiers).

The presence of hydroxyl-alcohol groups in the vicinity of the keto group allows the formation of chelates with bi- and trivalent metals Al (III), Sb (III), Zn (II), Mg (II), colored yellow, more intense than the initial flavone.

Flavones with C3 or C5-free OH groups together with the C4 keto group chelate in the ortho (I) or peri (II) position.

These compounds can also be obtained when there are two neighboring OH groups on the B nucleus, and the reaction medium is alkaline because under these conditions the OH group of 4′ is transformed into a ketone group by tautomerism-para (III) position.

The chelates obtained are yellow, intense, and have a yellow-green or blue fluorescence in the UV. It also forms complex combinations, oxalo-boric or citroboric with yellow-green fluorescence in UV.

Flavanones by dissolving in dilute alkalis give colorless or pale-yellow solutions which when heated turn yellow to red.

The color change is due to the isomerization of flavanone in chalcone. By dissolving in sulfuric acid, the flavanones form salts of chalcones of orange to red color.

The color is more intense when conjugation with a C4′-hydroxyl group (ion II) is possible. The 5 hydroxylated derivatives, through the hydrogen bonds they establish with the ketone group of 4, stabilize the pyranic cycle (Moore et al. 2002).

If this oxidril is glycosidated, the equilibrium is shifted to the chalconic form.

The transformation of flavanones into chalcone and the reverse reaction can take place even during the extraction of plant products, often obtaining artifacts. Identification: is performed using the cyanidol reaction.

Reactions can be made with alkaline hydroxides, bi- and trivalent metal salts. With the Fe (III) ion, the flavones form green chelates. To establish the structure of heterosides, acid hydrolysis is done first, then the extraction of aglycones with apolar solvent and their characterization by physico-chemical methods.

In acidic aqueous media, ozone is highlighted. Dosage: gravimetric, colorimetric (most often) based methods can be used either to obtain chelates with aluminum chloride according to the Roman Pharmacopoeia (FR X) (*Farmacopeea Romana X / Romanian Pharmacopoeia* 1993) or with antimony trichloride (yellow color) or on coupling with aluminum chloride with p-sulfobenzene diazonium chloride or p-nitrobenzene diazonium. The most modern method is fast HPLC.

In vitro studies indicate a wide range of biological activities for different flavonoids.

These studies were performed primarily with flavonic aglycones or heterosides. Until recently, flavonoid metabolites were rarely used because their identity data were limited, and metabolite standards were less commercially available.

8.4 The Mechanism of Action of Flavonoids

These mainly refer to quercetol, which is the most studied flavonoid.

Quarcetol has antioxidant, anticarcinogenic, anti-inflammatory, antiplatelet, and vasodilator actions. Behind these effects are mechanisms that are largely unknown. For example, the antioxidant effect could result from the chelation of metals, the capture of free radicals, the inhibition of enzymes, or the induction of enzyme expression.

Regarding the anticarcinogenic action, it should be noted that in the 1970s, quercetol was actually considered carcinogenic because the compound was shown to be mutagenic in the Ames test. However, a number of subsequent long-term animal studies have shown that quercetol is not carcinogenic.

In contrast, quercetol has been shown to inhibit carcinogenesis in laboratory animals. In recent years, some reports on the biological activities of quercetol metabolites have been published (Choi et al. 2021).

In these studies, the type of conjugation has been shown to be very important in antioxidant activity. Flavanones have gained little interest over flavonoids. Great attention has been paid to their anticarcinogenic properties.

Hesperidin (and orange juice) has been shown to inhibit carcinogenesis in laboratory animals. Hesperidin also has some antioxidant activity, although this activity is poorer compared to many other polyphenols.

Other effects of hesperedin and naringenol are on lipid metabolism. They act on apolipoprotein B, secreted by HepG2 cells, possibly by inhibiting cholesterol synthesis and inhibiting 3-hydroxy-3-methyl-tilglutaryl-coenzyme A reductase and acyl-coenzyme A.

In addition, lower LDL and cholesterol levels have been reported in rabbits fed high cholesterol. According to some studies, these results have not been confirmed or denied in clinical trials (Li et al. 2021).

Pharmacodynamic actions vitamin P or factor P. The main action attributed to flavonosides is that of vitamin P or factor P (permeability factor). It works by binding intracellular proteins, lowering the permeability of blood capillaries, and increasing their resistance (they are potentially venoactive).

The notion of factor P is related to observations on the treatment of some forms of scurvy, with ascorbic acid as such (vitamin C) or lemon juice.

In the first case, the scurvy manifestations could not be alleviated, while by treatment with lemon juice they disappeared. It was therefore noted that ascorbic acid can only act in combination with another factor (later called factor P).

Thus, the antioxidant role of flavonic derivatives in the protection of vitamin C and in cellular oxidation processes was discovered, with Huszak having the merit of determining which of the flavones have this action (dehydroxylated derivatives in ortho on the B nucleus). The name P factor is now given to flavonoids, anthocyanosides, and proanthocyanins.

Some flavonoids (apigenol, chrysol, taxifolol, and gosipine) have anti-inflammatory action *in vitro* due to their influence on the metabolism of arachidonic acid, by blocking cyclooxygenase and / or lipoxygenase, enzymes involved in the biosynthesis of pro-inflammatory prostaglandins and blood clotting; others may be antiallergic (isobutyrin, hispidulin), hepatoprotective (flavanonol lignanii-silibin, silidianin, and silicristin), antispasmodic (liquiritigenol), and hypocholesterolemic (Haida et al. 2022).

Some compounds have shown uricosuric and antinephritic action (camphorol diramnoside from Lespedesa capitata); antiulcer (kaempferol) due to inhibition of factor PAF (involved in the production of ulcers of the lining of the digestive tract) and leukotrienes (LTC4), while increasing the synthesis of antiulcer prostaglandins (PGE).

Flavonoids have diuretic, antibacterial, antiviral, and antifungal action; decrease the time of bleeding and blood clotting. A small number of them have *in vitro* antifungal properties.

Others have antioxidant properties or capture-free radicals formed in various pathological conditions: anoxia (superoxide radical generation occurs), inflammation (superoxide anions are produced by NADPH-membrane oxidation of leukocytes, hydroxyl radicals, and other reactive radicals that form during phagocytosis), and lipid self-oxidation due to hydroperoxide and alkoxyl lipophilic radicals.

Flavonoids and especially tocopherols react with free radicals, thus delaying or stopping degradation related to their intense activity in membrane phospholipids (Abe and Morita 2010).

In vitro flavonoids are enzyme inhibitors:

- inhibits bacterial succinoxidase and cholinacetylase by quinone formation (they are antibacterial and antiviral); potentiates the action of vitamin C (they are carriers of it), and stabilizes it by blocking metal ions (Cu^{2+}) that catalyze self-oxidation;

- inhibits elastase, collagenase, histidine decarboxylase, and hyaluronidase, which preserves the integrity of the fundamental substance of the cell wall;
- inhibits lipoxygenase and / or cyclooxygenase, enzymes involved in the inflammatory process (flavonol monomers and biflavonoids);
- nonspecifically inhibits catechol-O-methyltransferase (COMT), which leads to an increase in the amount of catecholamines (adrenaline, noradrenaline) available, which cause an increase in vascular resistance; inhibits cAMP phosphodiesterase, which could explain, among other things, the antiplatelet activity (baicaloside from *Scutellaria baicalensis*).

Rarely, flavonoids stimulate enzymatic activity, which is observed in proline hydroxylase (Jan et al. 2021).

This promotes the establishment of bridges between the collagen fibers, solidifying them. Proanthocyanidols act in this sense. Anthocyanosides are also involved in the protection of collagen fibers, inhibiting the degradative process by capturing the superoxide anionic radical, which appears to be involved in nonenzymatic proteolysis of collagen *in vitro*.

Flavonoids protect animal organisms from the harmful effects of X-rays or ultraviolet radiation.

The bioavailability of flavones is generally low, and the stated actions not always being found *in vivo*. Certain results were obtained with hydrolyzed heterosides in the digestive tract.

Although there are numerous studies on the pharmacological potential of these compounds, it has not been possible to arrive at a precise relationship between structure and activity.

Flavonoids are constantly present in human diet (about 1 g/day) because they are not toxic, mutagenic, or carcinogenic.

The routes of administration of flavonoid-containing medicines are oral route (tablets, capsules, and dragees); cutaneous route (gels) (Thapa et al. 2012).

Flavonoid absorption has been a hotly debated issue. It was first thought that the absorption of flavonoids takes place only in the large intestine where under the action of bacterial flora the heterosides are cleaved into aglycones.

It was not until 1990 that it was shown that both aglycones and heteroids appear in the plasma after the consumption of favonoid heteroside foods, which means that absorption also takes place in the small intestine.

Regarding quercetol, various studies have been done with different pure compounds that have given accurate information on its absorption and kinetics.

These indicate that quercetol and quercetol glycosides are absorbed from the upper gastrointestinal tract, probably the duodenum, while quercetol-3-rutinoside is absorbed from the distal parts, probably the colon.

Several studies have been performed on the urinary excretion of quercetol. In these urinary excretion studies, the percentage ranged from 0.07 to 0.3% of the ingested dose. From urinary excretion data, it cannot be concluded that 3% of quercetol is available.

Biliary excretion has been shown to be the most important route of elimination of quercetol in rats. In rats fed a diet containing 0.25% quercetol, the concentrations of quercetol metabolites were approximately threefold higher in bile compared to urine.

The molecular weight of quercetol and sulfuronoconjugate glucuronide conjugates and their binding to extended proteins could promote their biliary excretion. Regarding hesperidin and naringin, urinary excretion was based on various animal and human studies, and the results were varied (Kim et al. 2013).

For hesperetin, urinary retrieval was 3% in a subject in which 500 mg naringol and 500 mg hesperidin were ingested once and 24% in subjects in whom 1250 ml of grapefruit juice and 1250 ml of grapefruit juice were ingested, and oranges per day for four weeks.

For naringenol, individual urinary findings of 5–59% (6 subjects), 5% (1 subject), 14–15% (2 subjects), and 1–6% (6 subjects) were reported after ingestion of 214–700 mg of naringenol as a supplement or in juice. The half-life for naringeninol conjugates in urine was estimated to be 2–6 hours.

Gastrointestinal side effects (abdominal pain, heartburn, nausea, vomiting, and diarrhea); skin or systemic hypersensitivity reactions; sleepiness; altralgii; coordination and balance disorders; megaloblastic anemia due to folic acid deficiency; muscle contractions; and muscular cramps.

They are used alone or in combination, in hypertension, in phlebology as vasoprotective and venotonic (the most active are biflavonoids); treatment of symptoms accompanying venous lymphatic insufficiency (heavy legs, paresthesias, cramps, pain and other functional signs, intermittent claudication, edema); treatment of capillary fragility disorders in the skin (bruises, petechiae) and mucous membranes (gingivoragias, epistaxis); treatment of functional signs related to the hemorrhoidal crisis, metrorrhagia due to contraception by intrauterine devices, retinal and / or choroidal circulation disorders (Song et al. 2005).

For the diuretic effect they can be administered as adjuvants in the treatment of hypertension and urinary retention.

Flavanonol-lignans are used in degenerative liver disease or as hepatoprotectants, when given drugs with hepatotoxic potential.

Some of the plant products are used for the industrial extraction of flavones (*Sophorae flores* and *Fagopyri herba* rutoside; *Citri pericarpium* hesperidosis; *Citrus diosmetoside, Hyssopi herba, Bursae pastoris* herba, *Scrophulariae herba*; *Citrus* citro flavonoids).

Others are used in the form of standardized extracts (*Ginkgo* and *Crataegus*). They are also used as natural dyes and antioxidants or as chelating reagents in analytical chemistry.

Properties and mode of action of flavonoids at the cellular level

In the plant kingdom, one of the most widespread molecular groups is polyphenolic compounds, of which the main classes are flavonoids and phenolic acids (phenols).

Both flavonoids and phenolic acids are introduced into the body through the daily diet (Badshah et al. 2021).

Several subclasses of flavonoids have been described based on their chemical structure, including flavones, flavonols, flavanones, flavanols (catechins), anthocyanidins, and isoflavones.

Fruits, vegetables, and plant-based beverages such as wine and tea are important sources of flavonoids. Most commonly, flavonoids are found in the leaves, seeds, bark, and flowers of plants. To date, more than 4000 flavonoids have been identified.

The chemical structures of flavonoids are polycyclic, with two aromatic rings (A and B) linked to heterocyclic rings (C) and are most commonly classified into six subclasses, based on the binding position of rings B and C, and the degree of C-ring saturation in the form of flavones, flavonols, flavanones, flavanols (catechins), anthocyanidins, and isoflavones.

Because flavonoid biosynthesis is stimulated by light, these molecules accumulate in the outer tissues, in the leaf shell (Chiou et al. 2021).

Notable differences can be seen in the amount of flavonoids even on different parts of the same fruit, depending on sunlight exposure. In nature, most flavonoids exist as glycosides.

Flavonols are less common than fruit and vegetable flavonols. The main flavones in food are luteolin and apigenin.

Flavones have been identified in sweet red pepper (luteolin) and celery (apigenin). Flavanones are found in tomatoes and certain herbs, such as mint, but are present in high concentrations only in citrus fruits.

The most consumed flavanone is orange hesperidin.

Anticoagulants (anthocyanidin glycosides) are pigments from red fruits such as cherries, plums, strawberries, raspberries, grapes, red currants, and blackcurrants.

The most common anthocyanidins in edible parts of plants are cyanidin, pelargonidine, peonidine, dolphinidine, petunidine, and malvidin.

Isoflavones have a limited distribution in nature. For example, soybeans, which belong to the legume family, are essentially the only natural source of these compounds of nutritional importance.

The primary isoflavones in soy are the glycosides genistin and daidzine and their aglycones. Catechins are the main components of tea.

They make up about 30% of the dry mass of green tea and 90% of the dry mass of black tea. They usually appear as aglycones.

In green tea, catechins, namely catechin, epicatechin (EC), epigallocatechin (EGC), epicatechin gallate (ECG), and epigallocatechin galactate (EGCG), account for 80–90% of all flavonoids. Flavonoids have recently attracted a lot of interest from the scientific world due to their health benefits (Murugan et al. 2022).

A series of researches have been carried out on their antioxidant, antiestrogenic, antiproliferative activities, and a relationship has been highlighted between the intake of flavonoids and cardiovascular and cancerous diseases.

8.5 The Role of Flavonoids in Food and Medicine

The various flavonoids perform a variety of functions in plants. Flavonoids form the most important group of flower pigments and serve to attract pollinators. Anthocyanidins come in a variety of colors, from orange to red to blue.

All anthocyanidin-containing flowers also contain flavones and / or flavonols, which serve to stabilize anthocyanidins but, in higher concentrations, cause the flowers to change color in the blue zone.

The color of the yellow flower is less caused by flavonoids.

Flavonols such as gossypetin and quercetagetin are for the color of the yellow flower *Gossypium hirsutum*, *Primula vulgaris*, and some composites like *Chrysanthemum segetum* responsible.

Chalcone and Aurone cause yellow flowers in other asterites, such as *Coreopsis* and *Dalie*, and nine other plant families.

Yellow flavonoids are often found in the daisy family along with yellow carotenoids. Totally, 95% of the color of the white flower is caused by flavonoids: flavones such as luteolin and apigenin and flavonols such as kaempferol and quercetin, the flavonols being absorbed somewhat further in the long wave range.

Condensed tannins interact with the glycoproteins in herbivore saliva and have an astringent effect. They reduce the digestibility of plants and thus discourage many potential herbivores.

Other flavonoids act as protection against herbivores (repellents). For specialized insects, such flavonoids are in turn stimulants for feeding (Calland et al. 2012).

In particular, flavone and flavonol glycosides, for example based on rutin, quercitrin, and isoquercitrin, are toxic to insects, while they are not toxic to higher animals.

The growth of various caterpillars is dramatically reduced in the presence of, for example, dietary isoquercitrine in 10% of control groups.

These flavonoids are found mainly in herbaceous plants and should replace the condensed tannins of woody plants.

Flavones and flavonols act mainly as protection against UV radiation and shortwave light. They are deposited on the surface of the leaves in the free form of plants in extreme locations, such as arid or alpine areas, often in the form of flour-like coatings.

In this way, they prevent the photooxidative destruction of membranes and photosynthetic pigments. Due to their lipophilicity, they also reduce the colonization of the leaf surface by microorganisms.

Flavonoids also have direct antiviral, antibacterial, and antifungal effects.

Certain plant flavonoids play a role in regulating gene expression in *Rhizobium* nodule bacteria.

Highly methoxylated flavonoids are often found in bud exudates and other lipophilic secretions. They have a fungicidal effect, just like nobiletin from *Citrice*-Sul (Messaoudi et al. 2021).

Flavonoids serve as a structural leitmotif for the development of selective GABAA receptor ligands.

About two-thirds of the approximately one gram of phenolic substances that humans consume are flavonoids. It is believed that due to their antioxidant effects, which *in vitro*, e.g. stronger than that of known antioxidants, such as vitamin E, have a significant influence on human health.

Epidemiological studies have shown a lower risk of developing various diseases with a higher intake of flavonoids, including mortality from cardiovascular disease.

Flavonoids act on the metabolism of arachidonic acid and thus on blood clotting. Epidemiological studies have shown no link to cancer, except for lung cancer, whose risk is reduced mainly by the intake of flavonoids from apples.

A mutagenic or genotoxic effect has been demonstrated for some compounds in *in vitro* tests. However, there is no evidence of human toxicity, and animal experiments did not show any carcinogenic effects of flavonoids.

Some flavonoids lead to a strong inhibition of cytochrome P450-dependent monooxygenases (phase I enzymes), while others lead to activation.

Dose-dependent activation of phase II enzymes may also occur. All of these can lead to drug interactions, such as grapefruit.

Several drugs that contain flavonoids are used therapeutically, as well as some pure substances. They are used as venous agents due to their vascular protective effect, edema-protective, as cardiovascular agents due to their positive inotropic effect, antihypertensive, as diuretics, as spasmolytics for gastrointestinal disorders, and as liver therapies. Its effect is mainly attributed to its antioxidant properties and inhibition of enzymes (Bouyahya et al. 2017).

Epidemiologically, most *in vivo* studies indicate that flavonoids have a positive influence on various cardiovascular diseases.

Traditionally, these effects have only been attributed to their antioxidant activities. However, in addition to the direct binding of reactive oxygen species (ROS), there are a number of other effects that, in pharmacologically achievable concentrations, also have a positive effect on the cardiovascular influence of flavonoids, such as, e.g. taxifoline may be responsible.

These include, in particular, inhibition of ROS-forming enzymes, inhibition of platelet function, inhibition of leukocyte activation, and vasodilating properties.

Among the many effects of flavonoids that have been demonstrated in *in vitro* and *in vivo* tests, the most important are antiallergic and anti-inflammatory effect; antiviral and antimicrobial effects; antioxidant effect; and antiproliferative and anticancer effects

Flavonoids work through several mechanisms of action. The focus is on the interaction with DNA and enzymes, the activation of cells, their free radical scavenging property, and the influence of different signal transduction pathways in cells (NF-κB, MAPK) (De Araújo et al. 2021).

Flavonoids inhibit over thirty enzymes in the human body. They activate different types of cells in the immune system. The last two properties are responsible for the anti-inflammatory effects of flavonoids.

The following flavonoids are used as pure substances as venous agents: citrus bioflavonoids, hesperidin, diosmin, rutin, and hydroxymethyl rutinoside. Among the predominant drugs are those containing flavonol glycosides and flavone glycosyl.

Flavonoids are of particular interest due to their various interactions with the human body. They can act on cancer cells through various mechanisms, such as

triggering the processes leading to apoptosis (programmed cell death);

preventing cell division by depolymerizing microtubules; and / or

inhibition of the angiogenesis process, respectively, the formation of new blood vessels, necessary for the development and proliferation of tumors.

Flavonoids have the ability to regulate the inflammatory response through various mechanisms, such as neutralizing free radicals or regulating factors or cellular processes involved in inflammation. The literature has a plethora of studies on the antioxidant capacity of flavonoids, with an emphasis on extracting them from plants – fruits, tea plants, and plants used for other medicinal purposes – and testing the ability to neutralize various root species.

A number of studies have shown the antiviral properties of some flavonoids, by testing their activity against viruses such as hepatitis C virus, dengue fever virus, or A / H5N1 virus. Some flavonoids have the potential to be used to treat intestinal worm infections (Di Petrillo et al. 2022).

Flavonoids also have antimicrobial potential, acting against both Gram-positive or Gram-negative bacteria and fungi.

8.6 Concluding Remarks and Future Perspectives

Flavonoids are a class of plant secondary metabolites known for their antioxidant properties. Flavonoids are synthesized via phenyl propanol, where the amino acid phenylalanine is used in the production of 4-cumaroyl-CoA. It is condensed with manoyl-CoA to give a group of compounds called calcones that contain two phenyl rings. The conjugation continues until the characteristic three-cycle form is obtained. The metabolic pathway continues with a series of enzymatic changes until flavanones are obtained, hence dihydroflavones and then anthocyanidins. During these processes, the formation of intermediate compounds takes place: flavonols, flavan-3-oils, proanthocyanidins, and several polyphenolic compounds. Flavonoids are distributed in plants, being responsible for the yellow and red / blue color of the flowers, having a protective role against microbes and insects. Due to their wide distribution of their variety and low toxicity compared to other secondary metabolites (alkaloids), they cause their ingestion by humans and animals in large quantities. Due to their obvious ability to modify certain reactions of the body to allergens, viruses, and carcinogens, they are also called natural modifiers of the biological response.

Flavonoids are easy to spot in the leaves, flowers, fruits, and roots of various plant species, including vegetables and grains.

The concentration of flavonoids differs from one species to another, but also varies depending on the area in which they are found on the plant (leaves, fruits, and flowers).

Polyphenolic substances, including flavonoids, are one of the most important classes of natural compounds that have a remarkable biological activity.

Their use is limited by their low stability and low solubility, both in organic solvents and in aqueous solutions. To eliminate these drawbacks, flavonoids can be converted to glycosylated or acylated derivatives by chemical, enzymatic, or chemoenzymatic methods. The enzymatic methods are more efficient because they offer the possibility to exploit the high regioselectivity of these biocatalysts, achieving a selective functionalization of the flavonoid. The transformation of flavonoids into bioconjugates by fatty acid acylation offers the possibility to introduce into their molecule another biologically active function and thus to positively modify not only physical properties such as solubility but also biological activity. The protection and especially the controlled release of various organic molecules is done almost exclusively by means of encapsulated compounds.

References

Abe, I. and Morita, H. (2010). Structure and function of the chalcone synthase superfamily of plant type III polyketide synthases. *Natural Product Reports* 27 (6): 809–838. https://doi.org/10.1039/b909988n.

Al-Maharik, N., Jaradat, N., Bassalat, N. et al. (2022). Isolation, identification and pharmacological effects of *Mandragora autumnalis* fruit flavonoids fraction. *Molecules (Basel, Switzerland)* 27 (3): 1046. https://doi.org/10.3390/molecules27031046.

Austin, M.B. and Noel, J.P. (2003). The chalcone synthase superfamily of type III polyketide synthases. *Natural Product Reports* 20 (1): 79–110. https://doi.org/10.1039/b100917f.

Badshah, S.L., Faisal, S., Muhammad, A. et al. (2021). Antiviral activities of flavonoids. *Biomedicine & Pharmacotherapy=Biomedecine & Pharmacotherapie* 140: 111596. https://doi.org/10.1016/j.biopha.2021.111596.

Bamdad, F., Farrokhpour, H., Ashrafizaadeh, M., and Najafi, B. (2022). Decomposition of the interaction energy of several flavonoids with *Escherichia coli* DNA Gyr using the SAPT (DFT) method: the relation between the interaction energy components, ligand structure, and biological activity. *Biochimica et Biophysica Acta, General Subjects* 1866: 130111. https://doi.org/10.1016/j.bbagen.2022.130111.

Ble-González, E.A., Gómez-Rivera, A., Zamilpa, A. et al. (2022). Ellagitannin, phenols, and flavonoids as antibacterials from *Acalypha arvensis* (Euphorbiaceae). *Plants (Basel, Switzerland)* 11 (3): 300. https://doi.org/10.3390/plants11030300.

Bouyahya, A., Et-Touys, A., Bakri, Y. et al. (2017). Chemical composition of *Mentha pulegium* and *Rosmarinus officinalis* essential oils and their antileishmanial, antibacterial and antioxidant activities. *Microbial Pathogenesis* 111: 41–49. https://doi.org/10.1016/j.micpath.2017.08.015.

Calland, N., Albecka, A., Belouzard, S. et al. (2012). (−)-Epigallocatechin-3-gallate is a new inhibitor of hepatitis C virus entry. *Hepatology (Baltimore, Md.)* 55 (3): 720–729. https://doi.org/10.1002/hep.24803.

Chiou, W.C., Lu, H.F., Hsu, N.Y. et al. (2021). Ugonin J Acts as a SARS-CoV-2 3C-like protease inhibitor and exhibits anti-inflammatory properties. *Frontiers in Pharmacology* 12: 720018. https://doi.org/10.3389/fphar.2021.720018.

Choi, M., Sathasivam, R., Nguyen, B.V. et al. (2021). Expression analysis of phenylpropanoid pathway genes and metabolomic analysis of phenylpropanoid compounds in adventitious, hairy, and seedling roots of tartary buckwheat. *Plants (Basel, Switzerland)* 11 (1): 90. https://doi.org/10.3390/plants11010090.

De Araújo, E., Guerra, G., Andrade, A. et al. (2021). Gastric ulcer healing property of *Bryophyllum pinnatum* leaf extract in chronic model *in vivo* and gastroprotective activity of its major flavonoid. *Frontiers in Pharmacology* 12: 744192. https://doi.org/10.3389/fphar.2021.744192.

Di Petrillo, A., Orrù, G., Fais, A., and Fantini, M.C. (2022). Quercetin and its derivates as antiviral potentials: a comprehensive review. *Phytotherapy Research: PTR* 36 (1): 266–278. https://doi.org/10.1002/ptr.7309.

Fang, Y., Cao, W., Xia, M. et al. (2022). Transport and interactions of co-incubated bi-functional flavonoids through inhibiting the function of P-glycoprotein (P-gp) using KB/multidrug-resistant (MDR) cells and rat everted gut sacs. *Journal of Agricultural and Food Chemistry* 70 (6): 1923–1933. https://doi.org/10.1021/acs.jafc.1c07694.

(1993). *Farmacopeea Romana X / Romanian Pharmacopoeia*, 10e.

Guo, L., Gao, L., Ma, X. et al. (2019). Functional analysis of flavonoid 3′-hydroxylase and flavonoid 3′,5′-hydroxylases from tea plant (*Camellia sinensis*), involved in the B-ring hydroxylation of flavonoids. *Gene* 717: 144046. https://doi.org/10.1016/j.gene.2019.144046.

Haida, Z., Ab Ghani, S., Juju Nakasha, J., and Hakiman, M. (2022). Determination of experimental domain factors of polyphenols, phenolic acids and flavonoids of lemon *(Citrus limon)* peel using two-level factorial design. *Saudi Journal of Biological Sciences* 29 (1): 574–582. https://doi.org/10.1016/j.sjbs.2021.09.022.

Hua, F., Zhou, P., Bao, G.H., and Ling, T.J. (2022). Flavonoids in Lu'an GuaPian tea as potential inhibitors of TMA-lyase in acute myocardial infarction. *Journal of Food Biochemistry* 46 (7): e14110. https://doi.org/10.1111/jfbc.14110.

Jan, R., Kim, N., Lee, S.H. et al. (2021). Enhanced flavonoid accumulation reduces combined salt and heat stress through regulation of transcriptional and hormonal mechanisms. *Frontiers in Plant Science* 12: 796956. https://doi.org/10.3389/fpls.2021.796956.

Kim, M., Kim, S.Y., Lee, H.W. et al. (2013). Inhibition of influenza virus internalization by (−)-epigallocatechin-3-gallate. *Antiviral Research* 100 (2): 460–472. https://doi.org/10.1016/j.antiviral.2013.08.002.

Ko, Y.H., Kim, S.K., Lee, S.Y., and Jang, C.G. (2020). Flavonoids as therapeutic candidates for emotional disorders such as anxiety and depression. *Archives of Pharmacal Research* 43 (11): 1128–1143. https://doi.org/10.1007/s12272-020-01292-5.

Kumar, D., Sachin, K.M., Kumari, N., and Bhattarai, A. (2022). Physico-chemical and spectroscopic investigation of flavonoid dispersed C *n* TAB micelles. *Royal Society Open Science* 9 (1): 210758. https://doi.org/10.1098/rsos.210758.

Li, Z., Zhang, M., Haenen, G. et al. (2021). Flavonoids seen through the energy perspective. *International Journal of Molecular Sciences* 23 (1): 187. https://doi.org/10.3390/ijms23010187.

Li, T., Xiao, Y., Wang, Z. et al. (2022). The mechanism study of common flavonoids on antiglioma based on network pharmacology and molecular docking. *Evidence-based Complementary and Alternative Medicine: Ecam* 2022: 2198722. https://doi.org/10.1155/2022/2198722.

Liu, W., Zheng, W., Cheng, L. et al. (2022). *Citrus* fruits are rich in flavonoids for immunoregulation and potential targeting ACE2. *Natural Products and Bioprospecting* 12 (1): 4. https://doi.org/10.1007/s13659-022-00325-4.

Lu, H.C., Chen, W.K., Wang, Y. et al. (2021). Effect of the seasonal climatic variations on the flavonoid accumulation in *Vitis vinifera* cvs. 'Muscat Hamburg' and 'Victoria' grapes under the double cropping system. *Foods (Basel, Switzerland)* 11 (1): 48. https://doi.org/10.3390/foods11010048.

Lu, S., Zhuge, Y., Hao, T. et al. (2022). Systematic analysis reveals *O*-methyltransferase gene family members involved in flavonoid biosynthesis in grape. *Plant Physiology and Biochemistry: PPB* 173: 33–45. https://doi.org/10.1016/j.plaphy.2022.01.007.

Marques, S., Fernandes, D.A., Teles, Y. et al. (2022). *Sidastrum paniculatum* (L.) Fryxell (Malvaceae): a promising source of bioactive sulfated flavonoids against *Aedes aegypti* L. *Frontiers in Pharmacology* 12: 760156. https://doi.org/10.3389/fphar.2021.760156.

Mattia, M.R., Du, D., Yu, Q. et al. (2022). Genome-wide association study of healthful flavonoids among diverse mandarin accessions. *Plants (Basel, Switzerland)* 11 (3): 317. https://doi.org/10.3390/plants11030317.

Messaoudi, M., Rebiai, A., Sawicka, B. et al. (2021). Effect of extraction methods on polyphenols, flavonoids, mineral elements, and biological activities of essential oil and extracts of *Mentha pulegium* L. *Molecules (Basel, Switzerland)* 27 (1): 11. https://doi.org/10.3390/molecules27010011.

Moore, B.S., Hertweck, C., Hopke, J.N. et al. (2002). Plant-like biosynthetic pathways in bacteria: from benzoic acid to chalcone. *Journal of Natural Products* 65 (12): 1956–1962. https://doi.org/10.1021/np020230m.

Murugan, S.K., Bethapudi, B., Raghunandhakumar, S. et al. (2022). A flavonoid rich standardized extract of *Glycyrrhiza glabra* protects intestinal epithelial barrier function and regulates the tight-junction proteins expression. *BMC Complementary Medicine and Therapies* 22 (1): 38. https://doi.org/10.1186/s12906-021-03500-1.

Oyeleke, M.B. and Owoyele, B.V. (2022). Saponins and flavonoids from *Bacopa floribunda* plant extract exhibit antioxidant and anti-inflammatory effects on amyloid beta 1-42-induced Alzheimer's disease in BALB/c mice. *Journal of Ethnopharmacology* 288: 114997. https://doi.org/10.1016/j.jep.2022.114997.

Peng, F., Yin, H., Du, B. et al. (2022). Anti-inflammatory effect of flavonoids from chestnut flowers in lipopolysaccharide-stimulated RAW 264.7 macrophages and acute lung injury in mice. *Journal of Ethnopharmacology* 115086. https://doi.org/10.1016/j.jep.2022.115086.

Rauf, A., Raza, M., Humayun Khan, M. et al. (2022). *In vitro* and *in silico* studies on clinically important enzymes inhibitory activities of flavonoids isolated from *Euphorbia pulcherrima*. *Annals of Medicine* 54 (1): 495–506. https://doi.org/10.1080/07853890.2022.2033826.

Reyes Jara, A.M., Gómez Lobato, M.E., Civello, P.M., and Martínez, G.A. (2022). Phenylalanine ammonia lyase is more relevant than Chalcone synthase and Chalcone isomerase in the biosynthesis of flavonoids during postharvest senescence of broccoli. *Journal of Food Biochemistry* 46 (2): e14054. https://doi.org/10.1111/jfbc.14054.

Seitz, C., Ameres, S., and Forkmann, G. (2007). Identification of the molecular basis for the functional difference between flavonoid 3′-hydroxylase and flavonoid 3′,5′-hydroxylase. *FEBS Letters* 581 (18): 3429–3434. https://doi.org/10.1016/j.febslet.2007.06.045.

Shan, T., Wang, Y., Wang, S. et al. (2020). A new *p*-terphenyl derivative from the insect-derived fungus *Aspergillus candidus* Bdf-2 and the synergistic effects of terphenyllin. *PeerJ* 8: e8221. https://doi.org/10.7717/peerj.8221.

Sharifi-Rad, J., Quispe, C., Kumar, M. et al. (2022). *Hyssopus* essential oil: an update of its phytochemistry, biological activities, and safety profile. *Oxidative Medicine and Cellular Longevity* 2022: 8442734. https://doi.org/10.1155/2022/8442734.

Song, J.M., Lee, K.H., and Seong, B.L. (2005). Antiviral effect of catechins in green tea on influenza virus. *Antiviral Research* 68 (2): 66–74. https://doi.org/10.1016/j.antiviral.2005.06.010.

Sun, J., Sun, W., Zhang, G. et al. (2022). High efficient production of plant flavonoids by microbial cell factories: challenges and opportunities. *Metabolic Engineering* 70: 143–154. https://doi.org/10.1016/j.ymben.2022.01.011.

Thapa, M., Kim, Y., Desper, J. et al. (2012). Synthesis and antiviral activity of substituted quercetins. *Bioorganic & Medicinal Chemistry Letters* 22 (1): 353–356. https://doi.org/10.1016/j.bmcl.2011.10.119.

Wang, M.Y., Ma, W.Y., Wang, Q.L. et al. (2021). Flavonoid-enriched extract from *Millettia speciosa* Champ prevents obesity by regulating thermogenesis and lipid metabolism in high-fat diet-induced obese C57BL/6 mice. *Food Science & Nutrition* 10 (2): 445–459. https://doi.org/10.1002/fsn3.2664.

Wei, H.T., Hou, D., Ashraf, M.F. et al. (2022). Metabolic profiling and transcriptome analysis reveal the key role of flavonoids in internode coloration of *Phyllostachys violascens* cv. Viridisulcata. *Frontiers in Plant Science* 12: 788895. https://doi.org/10.3389/fpls.2021.788895.

Yang, X., Zhang, H., Li, L. et al. (2022). Proteomic analysis of protective effects of *Epimedium* flavonoids against ethanol-induced toxicity in retinoic acid-treated SH-SY5Y cells. *Molecules (Basel, Switzerland)* 27 (3): 1026. https://doi.org/10.3390/molecules27031026.

Zaragozá, C., Álvarez-Mon, M.Á., Zaragozá, F., and Villaescusa, L. (2022). Flavonoids: antiplatelet effect as inhibitors of COX-1. *Molecules (Basel, Switzerland)* 27 (3): 1146. https://doi.org/10.3390/molecules27031146.

Zeng, X., Zheng, Y., He, Y. et al. (2022). A rapid LC-MS/MS method for simultaneous determination of ten flavonoid metabolites of naringin in rat urine and its application to an excretion study. *Foods (Basel, Switzerland)* 11 (3): 316. https://doi.org/10.3390/foods11030316.

Zhang, Y., Lei, X., Xu, H. et al. (2022). Tissue distribution of total flavonoids extracts of *Drynariae rhizoma* in young and old rats by UPLC-MS/MS determination. *Journal of Analytical Methods in Chemistry* 2022: 2447945. https://doi.org/10.1155/2022/2447945.

9

Antibody Production in Plants

Vipin Kumar Singh[1], *Prashant Kumar Singh*[2], *and Amit Kumar Mishra*[3]

[1] *Department of Botany, K.S. Saket P.G. College, Ayodhya, Uttar Pradesh, India*
[2] *Department of Biotechnology, Pachhunga University College Campus, Mizoram University, Aizawl, Mizoram, India*
[3] *Department of Botany, School of Life Sciences, Mizoram University, Aizawl, Mizoram, India*

9.1 Introduction

It has been a tradition to use microbial systems as a source for commercially vital biomolecules. Nevertheless, with the advent of plant biotechnological techniques, scientists were able to develop transgenic plants that act as alternate systems to produce carbohydrates, lipids, and other nonplant-based proteins. These biomolecules could change the plant phenotype itself. On the other hand, the inimitable properties of plants make them an important source for the large-scale manufacturing of the anticipated biomolecules. During the 1990s, plants were first regarded as an important host for the production of antibodies, and the term "plantibody" was introduced (Wycoff 2005). "Plantibodies" refers to the genetically engineered products of plants that have the ability to express antibodies (ABs). Plants are often used as bioreactors (factories of antibodies) because of this unique ability. The advancement of biotechnological techniques has led to the utilization of the plant's endomembrane and secretory systems for the production of clinically viable proteins (CVP) and their purification from the tissue at later stages. In plants, the expression of the antibodies is achieved either as full-length molecules or as small fragments. In nutshell, "plantibodies" are the antibodies that are manufactured by plants that are genetically modified. The plants do not have the natural ability to manufacture antibodies, but plantibodies work in a similar way as the normal antibodies (Oluwayelu and Adebiyi 2016). For the past two decades, plants are being utilized as a heterologous expression system to produce recombinant antibodies.

As stated earlier, plantibodies function like normal antibodies; however, the utilization of plants to produce antibodies has unique advantages when compared with the conventional methods of antibodies production using mammalian cells. First, plants grow quickly and are widespread and abundant as compared to the animals. They possess the ability to manufacture the product in a short span of time due to their seasonal growth period, which

Plants as Bioreactors for Industrial Molecules, First Edition.
Edited by Santosh Kumar Upadhyay and Sudhir P. Singh.

reduces the cost of antibody production (Sharma and Sharma 2009). Second, utilizing plants to produce plantibodies reduces the screening costs for bacterial toxins, prions, viruses, and other organisms as they hardly introduce harmful pathogens as compared to mammalian cells. When plants are exposed to foreign agents, they do not prompt immune responses; however, animal antibodies are prone. Plants manufacture more antibodies in a shorter time when compared to the animal cells (Fischer et al. 2003). Moreover, plants also possess the ability to produce any type of antibody molecule which ranges from smaller fragments to full-length as well as multimeric ABs (Whitelam et al. 1994). Genetic modification can be done in plants for the efficient manufacturing of the proteins with remarkable lower production costs as compared to the mammalian cell cultures (Chen 2008). The endomembrane system and secretory pathway of plants and human cells are quite similar when compared to bacteria and other prokaryotic systems. Additionally, plants act as an appropriate system for comprehensive post-translational modifications of proteins, which are required for complete biological activity, accompanying the N-glycosylation, in a way similar to mammalian systems (Fischer et al. 2004). During 2005, it was predicted that vaccine use at global level might require low maintenance, reduced cost, and effectual distribution. Koprowski (2005) originated the advancement of plant expression system an alternative approach for the production of vaccines and ABs for cancer and other diseases. Successful production of monoclonal antibodies have been done that neutralizes viruses and controls the growth of cancerous cells (Brodzik et al. 2006). The therapeutic potential of biological drugs is never realized if they are costly. So, the development of production systems having low cost and increased effectiveness is required in the present era of biological therapeutics. One such method is the use of transgenic plants that contains a foreign gene (gene of interest) to produce proteins, vaccines, antibodies, and other biological molecules. This chapter summarizes that how antibody production is done using plants as a source material.

9.2 How Are Antigens Expressed in Plants?

9.2.1 Transient Expression of Antigens

A foreign gene is introduced into a cell's genome using *Agrobacterium*-mediated gene transfer or biolistics methods such as protoplast fusion, bombardment, and microprojection for a stable transformation. Although these gene transfer methods are not only popular in the biotechnology industry but also possess a few limitations. The expression of a particular protein is low (0.01–0.3%) when compared to a plant's total soluble protein (TSP) using the *Agrobacterium*-mediated gene transfer method (Laere et al. 2016). Laere et al. (2016) also suggested that the expression of a foreign gene can be done without integration in the cell's genome using transient expression systems. Usually, the production of stable transgenic lines consume time and clumsy; however, expression using transient systems is simple and rapid (Dubey et al. 2018).

Several attempts have been discovered for the advancement in the vaccines production process using transient gene expression technology (Li et al. 2011; Phoolcharoen et al. 2011; Jones et al. 2013; Hensel et al. 2015). Table 9.1 shows the list of antigenic epitope(s) along

Table 9.1 List of antigenic epitope(s) along with the plant expression system used for their production.

Serial no.	Antigenic determinants (Epitopes)	Plant(s) used as an expression system	Reference
1.	GP1 (Ebola glycoprotein)	*Nicotiana benthamiana*	Phoolcharoen et al. (2011)
2.	Pfs25-CP VLP	*Nicotiana benthamiana*	Jones et al. (2013)
3.	HA (Hemagglutinin)	*Nicotiana benthamiana*	Redkiewicz et al. (2014)
4.	3C protease, and polyprotein P1-2A	Tomato	Pan et al. (2008)
5.	HBsAg (hepatitis B surface antigen)	Tomato	Li et al. (2011)
6.	CT- B (Cholera toxin beta subunit) and heat-labile enterotoxin (LT-B)	Maize	Chikwamba et al. (2002)
7.	2G12 (Anti-HIV)	Maize	Rademacher et al. (2008)
8.	VP1 (virus structural protein)	Tobacco	Li et al. (2006)
9.	CT- B (Cholera toxin beta subunit)	Tobacco	Jani et al. (2002)
10.	Rota virus VP7	Potato	Wu et al. (2003)
11.	HBsAg, and preS2 antigens	Potato	Joung et al. (2004)
12.	E559 (monoclonal antibody, MAb)	Tobacco and maize	Zungu et al. (2008)
13.	*E. coli* heat-labile toxin B subunit, and Tuberculosis antigen	*Arabidopsis thaliana*	Rigano et al. (2004)
14.	2G12 (Anti-HIV-1 MAb)	Barley	Hensel et al. (2015)
15.	LT-B (heat-labile enterotoxin)	Carrot	Marquet-Blouin et al. (2003)

with the plant expression system used for their production. Plants such as tobacco, tomato, potato, carrot, maize, barley, *Nicotiana*, and *Arabidopsis* have been used as expression systems for the development of "plantibodies." Previous studies described the manufacturing of a surface protein (0.02% of TSP) and a heat-liable protein (0.05% of TSP) from *Streptococcus mutans* and *Escherichia coli*, respectively, in tobacco (Curtiss and Cardineau 1997; Rosales-Mendoza et al. 2011). For the treatment of diarrhea, expression of a chlolera toxin (CT) and an enterotoxin subunit from *Vibrio cholerae* and *E. coli* were done using maize, rice, and tobacco plants (Daniell et al. 2001; Chikwamba et al. 2002; Soh et al. 2015). Zungu et al. (2008) used tobacco and maize plants as an expression systems for the development of E559 rabies antibody.

9.2.2 Plant Virus Fusion Proteins

Antigenic epitopes can be expressed using plant viral vectors (PVV) as they possess viral peptides with self-assembling ability (Cañizares et al. 2005). These PVV can induce easy and fast infections that allow large production of recombinant proteins in suitable hosts, which are prone to diseases. Several techniques are utilized for transient transformation such as cowpea mosaic virus (CPMV), cucumber mosaic virus (CMV), tobacco mosaic virus (TMV), and potato virus X (PVX) (Dubey et al. 2018). These techniques represent recombinant plant viruses that act as protein expression vectors that are widely used for transient transformation. Successful transformation of CT-B (it is one of the component of CT of *V. cholerae* related to serum and mucosal immunity) genes was done in tobacco plant, which was cloned using the plant expression vectors (Mikschofsky et al. 2009).

Expression of a foreign protein possessing a viral coat is another approach that consist of purification of modified virus particles from the infected tissues (Lomonossoff and Johnson 1996). One advantage of this approach includes presence of several copies of an antigenic peptide over the macromolecule carrier's surface (Dubey et al. 2018). Zhang et al. (2006) conducted a successful expression of a capsid protein in tomato from Norwalk virus. A viral coat protein gene was utilized to determine the fusion of recombinant antigens using CPMV (Liu et al. 2005). In cowpea leaves, the expression of a recombinant CPMV bonded with HRV (human rhinovirus) has been reported by Porta et al. (2003). Production of NVCP (Norwalk virus capsid protein) protein has been reported by transformation in tobacco and potato using 35S CaMV (Cauliflower mosaic virus) and potato-specific patatin promoters, respectively, together with *Agrobacterium tumefaciens* (Herbst-Kralovetz et al. 2010). Successful expression of HbaAg (hepatitis B virus surface antigen) has been reported in tomato, potato, and tobacco (Richter et al. 2000; Lou et al. 2007). Ortega-Berlanga et al. (2016) demonstrated successful transformation of RhoA peptide in *Nicotiana benthamiana* that led to decreased syncytial virus load in respiratory tract.

9.3 Plant-Derived Antibodies: Are There any Alternative Approaches?

The exponentially rising human population has accelerated the demand for antibodies to cure human health disorders. However, the increased cost of antibodies recovered from diverse sources is one of the important challenges, suggesting the search for newer sources. In this context, plant-derived biopharmaceuticals including antibodies may be more attractive because of ease in large-scale production, storage, and safety in comparison to animal-derived ones (Daniell et al. 2001). Many of the plant-originated recombinant antibodies, i.e. plantibodies, although under clinical assessments (Fischer et al. 2003), may have multifaceted applications in human disease management.

Long developmental phase and expensiveness is the major limitation of transgenic plants applied for antibody production. To date, various alternative strategies have been utilized to minimize the plant development to days and weeks. Transient expression is one of the many approaches considered for antibody production (Pogue et al. 2010; Castillo-Esparza and Gómez-Lim 2021) before proceeding with the transgenic approach. The transient expression generally assesses the activity of the designed construct or evaluates the functional properties of recombinant protein. However, the application of the method for antibody production substantially depends on the amount of protein produced (Fischer et al. 2003). Transient expression based on the bacterium *Agrobacterium tumefaciens* known as "agroinfiltration" is a commonly used technique.

The expression is fast ranging from a few days to weeks and enough desired protein can be recovered after each harvest. However, the scale-up process is expensive, limiting industrial application (Vaquero et al. 1999). Another alternative approach is concerned with the application of plant viruses as suitable vectors (Soleimanizadeh et al. 2022). Like agroinfiltration, the viral infection does not lead to the stable transformation of plant, but the commencement of recombinant antibody expression is comparatively quicker. The yield of recombinant products is characteristically high in the case of viral vectors because of increased expression and systemic dissemination of viruses and manual scale-up could be performed easily by simple inoculation procedures. Researchers have demonstrated the expression of fragments and complete antibodies in tobacco using the TMV and potato virus X (PVX) vectors (Verch et al. 1998; Franconi et al. 1999; Hendy et al. 1999; McCormick et al. 1999). The expression of a complete immunoglobulin molecule, however, relied on two different TMV vectors expressing heavy and light chains. Interestingly, simultaneous inoculation of tobacco with viral vectors resulting in the expression of functional antibody chain has also been demonstrated (Verch et al. 1998).

Apart from using the whole plant as a platform, the potential of plant cell cultures can also be employed for producing antibodies (Sharp and Doran 2001; Efferth 2019; Van Giap et al. 2019). The technique is simple and not time consuming. The plant cells are transformed with the desired gene directing the synthesis of antibodies. The expression of protein commences rapidly and the quantity of recombinant antibodies reaching up to 25 mg/l could be expected. The cost of tissue culture operation may rise when there is a need for a reactor system and skilled person compared to using the whole plant as the platform. However, there would be additional benefits like the ease in isolation and purification of recombinant proteins and maintenance of sterile environments, essentially required for bioactive products important for human health. The production of recombinant antibodies has been registered in suspension cultures of rice as well as tobacco (Fischer et al. 1999; Torres et al. 1999).

Genome editing has important opportunities for gene manipulation at specified points or rearrangement of available nucleotide sequences for various purposes (Gaj et al. 2016; Gupta and Shukla 2017; Basu et al. 2018; Upadhyay 2021). The common strategy for the purpose lies in the application of nucleases like zinc finger, clustered regularly interspaced short palindromic repeats (CRISPR)-associated protein 9, and effector nucleases (Bortesi and Fischer 2015; Gaj et al. 2016; Sushmita et al. 2021). Genome editing technology has been practiced in manipulating cell lines and organisms via knockout, sequence deletion, repairing, inversion, and translocation (Hou et al. 2013; Mali et al. 2013; Ran et al. 2013; Torres et al. 2014; Alok et al. 2021). In addition, nucleases have been employed for genetic manipulation in *Triticum* (Upadhyay et al. 2013; Wang et al. 2014), *Glycine max* (Haun et al. 2014), *Brassica oleracea* (Lawrenson et al. 2015), *Solanum lycopersicum* (Čermák et al. 2015), and *Arabidopsis* (Wang et al. 2015). Although editing in the genome may also occur through mutagenesis during cell culture, however, such events in transgenic plants for antibody production are scarce.

Apart from different plant sources, antibodies can be produced and recovered from bacteria (Lee and Jeong 2015), yeast (Lee and Jeong 2015; Wang et al. 2021), green algae (Yusibov et al. 2016; Vanier et al. 2018), and animal cell cultures (Chen et al. 2001; Dhara et al. 2018; Carrara et al. 2021; Kamle et al. 2022). However, the search for newer host sources and alternative approaches with efficient accumulation and stability of the desired antibodies are still being conducted by researchers worldwide.

9.4 Antibody Production in Plants: Advantages and Concerns

Plants have been regarded to possess numerous characteristic features desired for the generation of pharmaceutical products including antibodies (Fischer and Emans 2000; Daniel et al. 2001; Giddings 2001; Stoger et al. 2002; Dixit et al. 2021a, 2021b; Upadhyay and Singh 2021; Chen 2022; De Greve 2022). So far, antibody production in plants has been reported in species namely wheat (Stöger et al. 2000), rice (Torres et al. 1999; Stöger et al. 2000), alfalfa (Khoudi et al. 1999), soybean (Zeitlin et al. 1998), and different species of *Nicotiana* (Ma et al. 1995; Ma et al. 1998; Verch et al. 1998; McCormick et al. 1999; Vaquero et al. 1999). The important advantage of exploiting plants lies in the reduced production expenses, indicating conventional agronomic practices and unskilled farmers are appropriate to maintain and harvest the cultivated transgenic crops. Moreover, industrial-scale processes are already established for multiple plant species. In many cases, scaling is fast, effective, and requires additional cultivation areas. Strikingly, the establishment of the found herd as required for the transgenic animal is not necessary as transgenic plant species can be stored satisfactorily in form of seeds. Furthermore, the risks of contamination with pathogenic microorganisms are considerably minimized (Daniel et al. 2001). In addition, the desired molecule can be directed to express in specific plant organelles.

Because of conserved protein synthesis patterns in animals and plants, successful folding and assembling of complete serum immunoglobulin (Hiatt et al. 1989) as well as secretory IgAs (Ma et al. 1995) can be achieved in plant cells. In the second case, four separate moieties would be needed to bring together in a plant cell, whereas the animal system will need only two different subunits to generate the functional product. Because of alterations in post-translational processing of proteins carried out in plant and animal systems, certain minor structural changes in plant-specific moieties like α 1,3-fucose and β 1,2-xylose may be observed (Cabanes-Macheteau et al. 1999). Experimental investigations have suggested few changes in plant isolated recombinant IgG; however, both antibodies and glycans were nonimmunogenic (Chargelegue et al. 2000). Additionally, full-length antibodies as well as different biologically active antibody-derived products including fusion protein have been satisfactorily synthesized in plants. Thus, in forthcoming years, pharmaceutically important recombinant antibodies could be successfully and safely produced in transgenic plant systems (Fischer and Emans 2000).

Despite successful antibody production in the selected host plant, there may be certain limitations as listed below.

a) The patients receiving plant-derived antibodies may have a negative impact upon exposure to plant species allergens like protein glycans, other antigens (Ferrante and Simpson 2001), and chemical contaminants like herbicides employed during plant cultivation.

b) Considerable differences in posttranslational modification of expressed proteins may be observed.
c) The production of antibodies may be substantially affected by the host species selected, prevailing environmental conditions, and plant antigens. Moreover, there may be allergenic responses to mycotoxins and plant metabolites (Hashemzade et al. 2014)
d) There may be gene silencing and species-to-species variations in patterns of glycosylation (Stoger et al. 2002).
e) Regulatory constraints are a major factor when considering the application of therapeutic plantibodies for human health management (Rani and Usha 2013).
f) The amount of plantibodies produced may be insufficient in some cases, requiring multiple optimizations for yield enhancement, putting additional expenses.
g) Concerns have been raised about the contamination of food crop strains with those exploited for expressing desired antibodies (Oluwayelu and Adebiyi 2016) through cross-pollination (Sinaga et al. 2021), thereby suggesting the use of nonfood crops for antibody synthesis. The contamination may also be prevented by cultivating the antibody generating plant in specified locations such as greenhouses. Furthermore, the possibilities in the transfer of toxic substances from pesticides and chemical fertilizers applied to plants cannot be neglected absolutely (Daniell et al. 2001).
h) Recombinant antibodies can be directed to seeds and tubers (Artsaenko et al. 1998; Conrad et al. 1998; Smith and Glick 2000; Stöger et al. 2000; Hernández-Velázquez et al. 2015) facilitating the storage, transport, and administration in desired plant tissues.

9.5 Conclusion and Prospects

The antibody is a proteinaceous biomolecule with wide applications starting from disease diagnosis to treatment. Most of the platforms currently used for antibody production are animal-based, indicating the need for alternative approaches. The application of plants to produce antibodies (plantibodies) is an attractive approach to meet the demand of the exponentially rising human population. So far, successful production of numbers of antibodies as whole or fragments thereof using diverse plants or cultures have been reported by different researchers across the globe. Despite multiple advantages of using plants to produce antibodies, limitations also exist. The major limitations include the extraction and purification cost, the long developmental phases of plants, the amount of antibodies produced, and the possibility of contamination with microbial toxins. The major issue is concerned with biosafety, contamination of food crop strains by cross-pollination, and the strict regulatory laws hindering market entry and application for intended purposes. The influence of environmental factors on antibody production in plants is another important challenge in the technology. However, fully controlled conditions under greenhouses may be used as an alternative strategy.

A detailed understanding of the mechanism of expression of varied antigens and fusion proteins in transgenic lines could resolve the associated challenges. The selection of a suitable host plant to produce antibodies in desired amounts needs to be investigated comprehensively. The selection of common food crops for the production of antibodies has the risk of contamination of natural crops. In-depth, studies are needed to reduce the developmental phases of plants and direct the accumulation of desired plantibodies in specific organelles. Commercialization of plant-derived antibodies is an important aspect (Dubey et al. 2018) and needs special attention to overcome the existing market hurdles. The performance and safety of the antibody derived from plants should be like those derived from conventional approaches. Therefore, extensive investigations both at the laboratory as well as clinical levels are inevitable to achieve the success of technology developed. Appropriate purification strategies need to be developed to ensure the presence of the desired antibody only and eliminate the presence of any unwanted plant metabolites of antigenic nature and toxic substances if any (Giddings et al. 2000) that could interfere with the antibody functionality. In general, the nature and extent of glycosylation in plant-derived pharmaceutical products are somewhat different from animal-originated ones (Whitelam 1995; Ganz et al. 1996; Ma and Hein 1996; Cabanes-Macheteau et al. 1999; Smith and Glick 2000). As some of the plant-specific carbohydrate moieties may provoke antigenic responses by the human immune system, proper elimination of such residues using state-of-the-art technologies becomes crucial for biopharmaceutical products. The development of transgenic plant lines expressing the desired antibody with the capability to grow efficiently under natural environmental conditions should be tried to harness the potentiality at a larger scale. Such advancements would help commercialize the generated products.

References

Alok, A., Chauhan, H., Upadhyay, S.K. et al. (2021). Compendium of plant-specific CRISPR vectors and their technical advantages. *Lifestyles* 11: 1021. https://doi.org/10.3390/life11101021.

Artsaenko, O., Kettig, B., Fiedler, U. et al. (1998). Potato tubers as a biofactory for recombinant antibodies. *Molecular Breeding* 4: 313–319.

Basu, S., Rabara, R.C., Negi, S., and Shukla, P. (2018). Engineering PGPMOs through gene editing and systems biology: a solution for phytoremediation? *Trends in Biotechnology* 36 (5): 499–510.

Bortesi, L. and Fischer, R. (2015). The CRISPR/Cas9 system for plant genome editing and beyond. *Biotechnology Advances* 33 (1): 41–52.

Brodzik, R., Glogowska, M., Bandurska, K. et al. (2006). Plant-derived anti-Lewis Y mAb exhibits biological activities for efficient immunotherapy against human cancer cells. *Proceedings of the National Academy of Sciences of the United States of America* 103: 8804–8809.

Cabanes-Macheteau, M., Fitchette-Lainé, A.C., Loutelier-Bourhis, C. et al. (1999). N-Glycosylation of a mouse IgG expressed in transgenic tobacco plants. *Glycobiology* 9 (4): 365–372.

Cañizares, M.C., Nicholson, L., and Lomonossoff, G.P. (2005). Use of viral vectors for vaccine production in plants. *Immunology and Cell Biology* 83: 263–270.

Carrara, S.C., Ulitzka, M., Grzeschik, J. et al. (2021). From cell line development to the formulated drug product: the art of manufacturing therapeutic monoclonal antibodies. *International Journal of Pharmaceutics* 594: 120164.

Castillo-Esparza, J. and Gómez-Lim, M.A. (2021). Transient expression in cytoplasm and apoplast of rotavirus VP6 protein fused to anti-DEC205 antibody in *Nicotiana benthamiana* and *Nicotiana sylvestris*. *Molecular Biotechnology* 63 (10): 973–982.

Čermák, T., Baltes, N.J., Čegan, R. et al. (2015). High-frequency, precise modification of the tomato genome. *Genome Biology* 16 (1): 1–15.

Chargelegue, D., Vine, N.D., van Dolleweerd, C.J. et al. (2000). A murine monoclonal antibody produced in transgenic plants with plant-specific glycans is not immunogenic in mice. *Transgenic Research* 9 (3): 187–194.

Chen, Q. (2008). Expression and purification of pharmaceutical proteins in plants. *Biological Engineering Transactions* 1: 291–321.

Chen, Q. (2022). Development of plant-made monoclonal antibodies against viral infections. *Current Opinion in Virology* 52: 148–160.

Chen, K., Liu, Q., Xie, L. et al. (2001). Engineering of a mammalian cell line for reduction of lactate formation and high monoclonal antibody production. *Biotechnology and Bioengineering* 72 (1): 55–61.

Chikwamba, R., Cunnick, J., Hathaway, D. et al. (2002). A functional antigen in a practical crop: LT-B producing maize protects mice against *Escherichia coli* heat labile enterotoxin LT and cholera toxin CT. *Transgenic Research* 11: 479–493.

Conrad, U., Fiedler, U., Artsaenko, O., and Phillips, J. (1998). High-level and stable accumulation of single-chain Fv antibodies in plant storage organs. *Journal of Plant Physiology* 152: 708–711.

Curtiss, R. and Cardineau, G.A. (1997). Oral immunization by transgenic plants. Patent No. 5, 654, 184 A. Washington, DC: U.S. Patent and Trademark Office.

Daniel, H., Streatfield, S.J., and Wycoff, K. (2001). Medical molecular farming: production of antibodies biopharmaceutical and edible vaccines in plants. *Trends in Plant Science* 6 (5): 219–226.

Daniell, H., Lee, S.B., Panchal, T., and Wiebe, P.O. (2001). Expression of the native cholera toxin B subunit gene and assembly as functional oligomers in transgenic tobacco chloroplasts. *Journal of Molecular Biology* 311: 1001–1009.

De Greve, H. (2022). Production of designer VHH-based antibodies in plants. In: *Single-Domain Antibodies* (ed. G. Hussack and K.A. Henry), 205–230. New York: Humana.

Dhara, V.G., Naik, H.M., Majewska, N.I., and Betenbaugh, M.J. (2018). Recombinant antibody production in CHO and NS0 cells: differences and similarities. *BioDrugs* 32 (6): 571–584.

Dixit, S., Shukla, A., Singh, V., and Upadhyay, S.K. (2021a). Bioprospecting of natural compounds for industrial and medical applications; current scenario and bottleneck. In: *Bioprospecting of Plant Biodiversity for Industrial Molecules* (ed. S.K. Upadhyay and S.P. Singh), 53–57. John Wiley & Sons Ltd. https://doi.org/10.1002/9781119718017.ch3.

Dixit, S., Shukla, A., Singh, V., and Upadhyay, S.K. (2021b). Engineering of plant metabolic pathway for nutritional improvement; recent advances and challenges. In: *Genome Engineering for Crop Improvement* (ed. S.K. Upadhyay), 351–379. John Wiley & Sons Ltd. https://doi.org/10.1002/9781119672425.ch20.

Dubey, K.K., Luke, G.A., Knox, C. et al. (2018). Vaccine and antibody production in plants: developments and computational tools. *Briefings in Functional Genomics* 17: 295–307.

Efferth, T. (2019). Biotechnology applications of plant callus cultures. *Engineering* 5 (1): 50–59.

Ferrante, E. and Simpson, D. (2001). A review of the progression of transgenic plants used to produce plantibodies for human usage. *J Young Investigators* 4 (1): 11.

Fischer, R. and Emans, N. (2000). Molecular farming of pharmaceutical proteins. *Transgenic Research* 9 (4): 279–299.

Fischer, R., Liao, Y.C., and Drossard, J. (1999). Affinity-purification of a TMV-specific recombinant full-size antibody from a transgenic tobacco suspension culture. *Journal of Immunological Methods* 226 (1–2): 1–10.

Fischer, R., Twyman, R.M., and Schillberg, S. (2003). Production of antibodies in plants and their use for global health. *Vaccine* 21: 820–825.

Fischer, R., Stoger, E., Schillberg, S. et al. (2004). Plant-based production of biopharmaceuticals. *Current Opinion in Plant Biology* 7: 152–158.

Franconi, R., Roggero, P., Pirazzi, P. et al. (1999). Functional expression in bacteria and plants of an scFv antibody fragment against tospoviruses. *Immunotechnology* 4 (3–4): 189–201.

Gaj, T., Sirk, S.J., Shui, S.L., and Liu, J. (2016). Genome-editing technologies: principles and applications. *Cold Spring Harbor Perspectives in Biology* 8 (12): a023754.

Ganz, P.R., Dudani, A.K., Tackaberry, E.S. et al. (1996). Expression of human blood proteins in transgenic plants: the cytokine GM-CSF as a model protein. In: *Transgenic Plants: A Production System for Industrial and Pharmaceutical Proteins* (ed. M.R.L. Owen and J. Pen), 281–297. London, UK: John Wiley & Sons Ltd.

Giddings, G. (2001). Transgenic plants as protein factories. *Current Opinion in Biotechnology* 12 (5): 450–454.

Giddings, G., Allison, G., Brooks, D., and Carter, A. (2000). Transgenic plants as factories for biopharmaceuticals. *Nature Biotechnology* 18: 1151–1155.

Gupta, S.K. and Shukla, P. (2017). Gene editing for cell engineering: trends and applications. *Critical Reviews in Biotechnology* 37 (5): 672–684.

Hashemzade, H., Zebarjadi, A., Akhshi, N., and Aghaee, K. (2014). Application of transgenic plants as factories for producing biopharmaceuticals. *Journal of Biodiversity and Environmental Sciences* 4 (4): 58–74.

Haun, W., Coffman, A., Clasen, B.M. et al. (2014). Improved soybean oil quality by targeted mutagenesis of the fatty acid desaturase 2 gene family. *Plant Biotechnology Journal* 12 (7): 934–940.

Hendy, S., Chen, Z.C., Barker, H. et al. (1999). Rapid production of single-chain Fv fragments in plants using a potato virus X episomal vector. *Journal of Immunological Methods* 231 (1–2): 137–146.

Hensel, G., Floss, D.M., Arcalis, E. et al. (2015). Transgenic production of an anti HIV antibody in the barley endosperm. *PLoS One* 10: e0140476.

Herbst-Kralovetz, M., Mason, H.S., and Chen, Q. (2010). Norwalk virus-like particles as vaccines. *Expert Review of Vaccines* 9: 299–307.

Hernández-Velázquez, A., López-Quesada, A., Ceballo-Cámara, Y. et al. (2015). Tobacco seeds as efficient production platform for a biologically active anti-HBsAg monoclonal antibody. *Transgenic Research* 24 (5): 897–909.

Hiatt, A., Cafferkey, R., and Bowdish, K. (1989). Production of antibodies in transgenic plants. *Nature* 342 (6245): 76–78.

Hou, Z., Zhang, Y., Propson, N.E. et al. (2013). Efficient genome engineering in human pluripotent stem cells using Cas9 from *Neisseria meningitidis*. *Proceedings of the National Academy of Sciences* 110 (39): 15644–15649.

Jani, D., Meena, L.S., Rizwan-ul-Haq, Q.M. et al. (2002). Expression of cholera toxin B subunit in transgenic tomato plants. *Transgenic Research* 11: 447–454.

Jones, R.M., Chichester, J.A., Mett, V. et al. (2013). A plant-produced Pfs25 VLP malaria vaccine candidate induces persistent transmission blocking antibodies against Plasmodium falciparum in immunized mice. *PLoS One* 8: e79538.

Joung, Y.H., Youm, J.W., Jeon, J.H. et al. (2004). Expression of the hepatitis B surface S and preS2 antigens in tubers of *Solanumtu berosum*. *Plant Cell Reports* 22: 925–930.

Kamle, S., Li, D., Lee, C.G., and Elias, J.A. (2022). Methods for transient expression and purification of monoclonal antibodies in mammalian cells. In: *Advances in Protein Molecular and Structural Biology Methods* (ed. R. Donev), 31–39. Academic Press.

Khoudi, H., Laberge, S., Ferullo, J.M. et al. (1999). Production of a diagnostic monoclonal antibody in perennial alfalfa plants. *Biotechnology and Bioengineering* 64 (2): 135–143.

Koprowski, H. (2005). Vaccines and sera through plant biotechnology. *Vaccine* 23: 1757.

Laere, E., Ling, A.P.K., Wong, Y.P. et al. (2016). Plant-based vaccines: production and challenges. *Journal of Botany* 2016: 1–11.

Lawrenson, T., Shorinola, O., Stacey, N. et al. (2015). Induction of targeted, heritable mutations in barley and *Brassica oleracea* using RNA-guided Cas9 nuclease. *Genome Biology* 16 (1): 1–13.

Lee, Y.J. and Jeong, K.J. (2015). Challenges to production of antibodies in bacteria and yeast. *Journal of Bioscience and Bioengineering* 120 (5): 483–490.

Li, Y., Sun, M., Liu, J. et al. (2006). High expression of foot-and-mouth disease virus structural protein VP1 in tobacco chloroplasts. *Plant Cell Reports* 25: 329–333.

Li, T., Kuan Sun, J., Hua Lu, Z., and Liu, Q. (2011). Transformation of HBsAg (Hepatitis B surface antigen) gene into tomato mediated by *Agrobacterium tumefaciens*. *Czech Journal of Genetics and Plant Breeding* 47: 69–77.

Liu, L., Cañizares, M.C., Monger, W. et al. (2005). Cowpea mosaic virus-based systems for the production of antigens and antibodies in plants. *Vaccine* 23: 1788–1792.

Lomonossoff, G.P. and Johnson, J.E. (1996). Use of macromolecular assemblies as expression systems for peptides and synthetic vaccines. *Current Opinion in Structural Biology* 6: 176–182.

Lou, X.M., Yao, Q.H., Zhang, Z. et al. (2007). Expression of the human hepatitis B virus large surface antigen gene in transgenic tomato plants. *Clinical and Vaccine Immunology* 14: 464.

Ma, J.K.C. and Hein, M.B. (1996). Antibody production and engineering in plants. In: *Engineering Plants for Commercial Products and Applications* (ed. G.B. Collins and R.J. Sheperd), 72–81. New York: Academy of Sciences.

Ma, J.K., Hiatt, A., Hein, M. et al. (1995). Generation and assembly of secretory antibodies in plants. *Science* 268 (5211): 716–719.

Ma, J.K.C., Hikmat, B.Y., Wycoff, K. et al. (1998). Characterization of a recombinant plant monoclonal secretory antibody and preventive immunotherapy in humans. *Nature Medicine* 4 (5): 601–606.

Mali, P., Aach, J., Stranges, P.B. et al. (2013). CAS9 transcriptional activators for target specificity screening and paired nickases for cooperative genome engineering. *Nature Biotechnology* 31 (9): 833–838.

Marquet-Blouin, E., Bouche, F.B., Steinmetz, A. et al. (2003). Neutralizing immunogenicity of transgenic carrot *Daucus carota* L.-derived measles virus hemagglutinin. *Plant Molecular Biology* 51: 459–469.

McCormick, A.A., Kumagai, M.H., Hanley, K. et al. (1999). Rapid production of specific vaccines for lymphoma by expression of the tumor-derived single-chain Fv epitopes in tobacco plants. *Proceedings of the National Academy of Sciences* 96 (2): 703–708.

Mikschofsky, H., König, P., Keil, G.M. et al. (2009). Choleratoxin B CTB is functional as an adjuvant for cytoplasmatic proteins if directed to the endoplasmatic reticulum ER, but not to the cytoplasm of plants. *Plant Science* 177: 35–42.

Oluwayelu, D.O. and Adebiyi, A.I. (2016). Plantibodies in human and animal health: a review. *African Health Sciences* 16 (2): 640.

Ortega-Berlanga, B., Musiychuk, K., Shoji, Y. et al. (2016). Engineering and expression of a RhoA peptide against respiratory syncytial virus infection in plants. *Planta* 243: 451–458.

Pan, L., Zhang, Y., Wang, Y. et al. (2008). Foliar extracts from transgenic tomato plants expressing the structural polyprotein, P1-2A, and protease, 3C, from foot-and-mouth disease virus elicit a protective response in guinea pigs. *Veterinary Immunology and Immunopathology* 121: 83–90.

Phoolcharoen, W., Bhoo, S.H., Lai, H. et al. (2011). Expression of an immunogenic Ebola immune complex in *Nicotiana benthamiana*. *Plant Biotechnology Journal* 9: 807–816.

Pogue, G.P., Vojdani, F., Palmer, K.E. et al. (2010). Production of pharmaceutical-grade recombinant aprotinin and a monoclonal antibody product using plant-based transient expression systems. *Plant Biotechnology Journal* 8 (5): 638–654.

Porta, C., Spall, V.E., Findlay, K.C. et al. (2003). Cowpea mosaic virus-based chimaeras. Effects of inserted peptides on the phenotype, host range, and transmissibility of the modified viruses. *Virology* 310: 50–63.

Rademacher, T., Sack, M., Arcalis, E. et al. (2008). Recombinant antibody 2G12 produced in maize endosperm efficiently neutralizes HIV-1 and contains predominantly single-GlcNAc *N*-glycans. *Plant Biotechnology Journal* 6: 189–201.

Ran, F.A., Hsu, P.D., Lin, C.Y. et al. (2013). Double nicking by RNA-guided CRISPR Cas9 for enhanced genome editing specificity. *Cell* 154 (6): 1380–1389.

Rani, S.J. and Usha, R. (2013). Transgenic plants: types, benefits, public concerns and future. *Journal of Pharmacy Research* 6 (8): 879–883.

Redkiewicz, P., Sirko, A., Kanem, K.A., and Góra-Sochacka, A. (2014). Plant expression systems for production of hemagglutinin as a vaccine against influenza virus. *Acta Biochimica Polonica* 61: 551–560.

Richter, L.J., Thanavala, Y., Arntzen, C.J., and Mason, H.S. (2000). Production of hepatitis B surface antigen in transgenic plants for oral immunization. *Nature Biotechnology* 18: 1167–1171.

Rigano, M.M., Alvarez, M.L., Pinkhasov, J. et al. (2004). Production of a fusion protein consisting of the enterotoxigenic *Escherichia coli* heat-labile toxin B subunit and a tuberculosis antigen in *Arabidopsis thaliana*. *Plant Cell Reports* 22: 502–508.

Rosales-Mendoza, S., Soria-Guerra, R.E., Moreno-Fierros, L. et al. (2011). Immunogenicity of nuclear-encoded LTB:ST fusion protein from *Escherichia coli* expressed in tobacco plants. *Plant Cell Reports* 30: 1145–1152.

Sharma, A.K. and Sharma, M.K. (2009). Plants as bioreactors: recent developments and emerging opportunities. *Biotechnology Advances* 27: 811.

Sharp, J.M. and Doran, P.M. (2001). Strategies for enhancing monoclonal antibody accumulation in plant cell and organ cultures. *Biotechnology Progress* 17 (6): 979–992.

Sinaga, D.S., Ho, S.L., Lu, C.A. et al. (2021). Knockdown expression of a MYB-related transcription factor gene, OsMYBS2, enhances production of recombinant proteins in rice suspension cells. *Plant Methods* 17 (1): 1–10.

Smith, M.D. and Glick, B.R. (2000). The production of antibodies in plants. *Biotechnology Advances* 18: 85–89.

Soh, H.S., Chung, H.Y., Lee, H.H. et al. (2015). Expression and functional validation of heat-labile enterotoxin B (LTB) and cholera toxin B (CTB) subunits in transgenic rice (*Oryza sativa*). *Springerplus* 4: 148.

Soleimanizadeh, M., Jalali Javaran, M., Bagheri, A., and Behdani, M. (2022). Apoplastic production of recombinant AntiVEGF protein using plant-virus transient expression vector. *Molecular Biotechnology* 64: 1–9.

Stöger, E., Vaquero, C., Torres, E. et al. (2000). Cereal crops as viable production and storage systems for pharmaceutical scFv antibodies. *Plant Molecular Biology* 42 (4): 583–590.

Stoger, E., Sack, M., Fischer, R., and Christou, P. (2002). Plantibodies: applications, advantages and bottlenecks. *Current Opinion in Biotechnology* 13 (2): 161–166.

Sushmita, Kaur, G., Upadhyay, S.K., and Verma, P.C. (2021). An overview of genome-engineering methods. In: *Genome Engineering for Crop Improvement* (ed. S.K. Upadhyay), 1–21. John Wiley & Sons Ltd https://doi.org/10.1002/9781119672425.ch1.

Torres, E., Vaquero, C., Nicholson, L. et al. (1999). Rice cell culture as an alternative production system for functional diagnostic and therapeutic antibodies. *Transgenic Research* 8 (6): 441–449.

Torres, R.A.U.L., Martin, M.C., Garcia, A. et al. (2014). Engineering human tumour-associated chromosomal translocations with the RNA-guided CRISPR–Cas9 system. *Nature Communications* 5 (1): 1–8.

Upadhyay, S.K. (ed.) (2021). *Genome Engineering for Crop Improvement*. John Wiley & Sons Ltd. https://doi.org/10.1002/9781119672425.

Upadhyay, S.K. and Singh, S.P. (eds.) (2021). *Bioprospecting of Plant Biodiversity for Industrial Molecules*. John Wiley & Sons Ltd. https://doi.org/10.1002/9781119718017.

Upadhyay, S.K., Kumar, J., Alok, A., and Tuli, R. (2013). RNA guided genome editing for multiple target gene mutations in wheat. *G3: Genes, Genomes, Genetics* 3 (12): 2233–2238.

Van Giap, D., Jung, J.W., and Kim, N.S. (2019). Production of functional recombinant cyclic citrullinated peptide monoclonal antibody in transgenic rice cell suspension culture. *Transgenic Research* 28 (2): 177–188.

Vanier, G., Stelter, S., Vanier, J. et al. (2018). Alga-made anti-hepatitis B antibody binds to human Fcγ receptors. *Biotechnology Journal* 13 (4): 1700496.

Vaquero, C., Sack, M., Chandler, J. et al. (1999). Transient expression of a tumor-specific single-chain fragment and a chimeric antibody in tobacco leaves. *Proceedings of the National Academy of Sciences* 96 (20): 11128–11133.

Verch, T., Yusibov, V., and Koprowski, H. (1998). Expression and assembly of a full-length monoclonal antibody in plants using a plant virus vector. *Journal of Immunological Methods* 220 (1–2): 69–75.

Wang, Y., Cheng, X., Shan, Q. et al. (2014). Simultaneous editing of three homoeoalleles in hexaploid bread wheat confers heritable resistance to powdery mildew. *Nature Biotechnology* 32 (9): 947–951.

Wang, Z.P., Xing, H.L., Dong, L. et al. (2015). Egg cell-specific promoter-controlled CRISPR/Cas9 efficiently generates homozygous mutants for multiple target genes in Arabidopsis in a single generation. *Genome Biology* 16 (1): 1–12.

Wang, Y., Li, X., Chen, X. et al. (2021). Expression of antibody fragments in *Saccharomyces cerevisiae* strains evolved for enhanced protein secretion. *Microbial Cell Factories* 20 (1): 1–17.

Whitelam, G.C. (1995). The production of recombinant proteins in plants. *Journal of the Science of Food and Agriculture* 68: 1–9.

Whitelam, G.C., Cockburn, W., and Owen, M.R.L. (1994). Antibody production in transgenic plants. *Biochemical Society Transactions* 22: 940–944.

Wu, Y.Z., Li, J.T., Mou, Z.R. et al. (2003). Oral immunization with rotavirus VP7 expressed in transgenic potatoes induced high titers of mucosal neutralizing IgA. *Virology* 313: 337–342.

Wycoff, K.L. (2005). Secretory IgA antibodies from plants. *Current Pharmaceutical Design* 11: 2429–2437.

Yusibov, V., Kushnir, N., and Streatfield, S.J. (2016). Antibody production in plants and green algae. *Annual Review of Plant Biology* 67: 669–701.

Zeitlin, L., Olmsted, S.S., Moench, T.R. et al. (1998). A humanized monoclonal antibody produced in transgenic plants for immunoprotection of the vagina against genital herpes. *Nature Biotechnology* 16 (13): 1361–1364.

Zhang, X., Buehner, N.A., Hutson, A.M. et al. (2006). Tomato is a highly effective vehicle for expression and oral immunization with Norwalk virus capsid protein. *Plant Biotechnology Journal* 4: 419–432.

Zungu, N., Lotter, T., Dube, N., et al. (2008). Expression of Rabies antibodies in tobacco and maize. In: *SASBMB Conference*, 1 (23–25 January 2008). Grahamstown, Eastern Cape province, South Africa. South African Society of Biochemistry and Molecular Biology.

10

Metabolic Engineering of Essential Micronutrients in Plants to Ensure Food Security

Swarnavo Chakraborty and Aryadeep Roychoudhury

Post Graduate Department of Biotechnology, St. Xavier's College (Autonomous), Kolkata, West Bengal, India

10.1 Introduction

The deficiency of micronutrients happens to be a rather widespread phenomenon, affecting a huge number of people worldwide. Micronutrient deficiency in the human diet can be linked to an array of disorders, most prevalently occurring in developing nations. The human body demands certain essential micronutrients for the sustenance of normal physiological mechanisms required for their proper mental and physical health. However, humans are unable to produce these nutrients within their system, hence have to fully depend on their dietary sources in order to procure optimal levels of these essential molecules. Probably due to the malabsorption of these micronutrients and feeding of monotonous dietary items, deficiency of these nutrients has become a rather widespread event among the human population globally (Tulchinsky 2010). This has often been referred to as "hidden hunger" and seems to be prevalent in rural and poor populations in developing sectors of the globe. Techniques associated with the processing, preparation, and storage of various food crops and grains generally results in substantial lowering of micronutrient levels, for instance, milling of rice in order to increase shelf life, leads to loss of the majority of micronutrients associated with the aleurone and outer kernel layers, thereby depreciating the nutritional value (Blancquaert et al. 2017). Deficiency of certain key vitamins like folate (vitamin B_9), cobalamin (vitamin B_{11}), vitamin A and some important trace elements like iron, zinc, and iodine in regular diet have evoked serious health issues worldwide.

The most widespread nutritional disorder in this globe is the deficiency of the key micronutrient, viz., iron (Blancquaert et al. 2017). Hence, anemia is found to be the most commonly occurring heath issue, affecting prevalently women and children even in developed nations. Moreover, nutritional deficiency of iron hinders optimal cognitive and physical development, reduces work capacity, hampers the normal outcome of pregnancy, and even increases premature morbidity risks. The deficiency of vitamin A primarily affects pregnant women and preschool children, particularly those residing in developing countries. The common deficiency disorders are associated with the eye, including night blindness,

Plants as Bioreactors for Industrial Molecules, First Edition.
Edited by Santosh Kumar Upadhyay and Sudhir P. Singh.

keratomalacia, xerophthalmia, and even complete blindness (McLaren et al. 1966). The deficiency of vitamin B_9 has been commonly used as a parameter to understand the folate status across a particular population, since insufficiency of this vitamin in mother's diet leads to neural tube defects in babies. Rice serves as the primary diet of the Asian population, where approximately 3,00,000 neural tube defects occur per year. Moreover, the occurrence of vitamin B_9 deficiency affects the world population in epidemic proportions, hitting almost billions across the globe (Blancquaert et al. 2014).

In order to address the problem of deficiency of micronutrients, dietary diversification will serve as the obvious solution. The most beneficial strategy for the generation of diversity in regular diet is by the biofortification of essential micronutrients in important crops. This strategy will serve as a great solution for addressing the hidden hunger, particularly among the poor developing populations across the world. The biofortification strategy can be carried out in two ways, i.e. metabolic engineering and conventional breeding. However, the process of conventional breeding is rather incapable to generate the desired dietary diversity, as it is associated with the generation of natural variation within sexually compatible crop plants, in order to bring about variation in the levels of target micronutrients via a time-consuming and lengthy process (Blancquaert et al. 2010; De Steur et al. 2015). Basically, the conventional breeding technique is a rather tiresome procedure as it involves a number of rounds of crossing, followed by screening and the rate of success is quite limited. Hence, metabolic engineering has proved to be a better alternative, due to its higher success rates in fortifying target crops, cost-effectiveness, and sustainability (Farré et al. 2014; De Steur et al. 2015; Upadhyay 2021). Metabolic engineering can be employed to increase the nutritional parameters of important crops, through the use of genetic modification in order to engineer important metabolic pathways in plants. This strategy particularly deals with the alteration of endogenous metabolic cascades or the improvization of one or a series of heterologous mediators, in order to increase the rate of generation of certain target compounds, or decrease the concentration of undesirable products, or alter the flux for enhanced accumulation of products with greater activity, stability, and bioavailability (Roychoudhury and Bhowmik 2020; Dixit et al. 2021). The process of metabolic engineering demands a great deal of knowledge involving the modulation of endogenous metabolic cascades, such that a successful engineering process can be developed, whereby certain key enzymes can be up or down regulated for the increased bioaccumulation of essential micronutrients, without having any adverse effects on the crop growth and yield parameters as a whole (Blancquaert et al. 2017).

10.2 Metabolic Engineering of Crops for Increased Nutritional Value

10.2.1 Iron

Iron forms the essential component of a number of enzymes and proteins, including hemoglobin. The most commonly associated disorder due to iron deficiency in the human diet is anemia, leading to millions of deaths per year (WHO 2002). In addition, the mechanism of iron uptake and acquisition in plants have been meticulously studied, due to widespread deficiency symptoms and broadly described physiological aspects associated with iron metabolism in plants. Rhizospheric iron is absorbed via the epidermis of the roots and then undergoes transportation via the xylem, followed by its relocation to different plant tissues and stored there until future demand. The most common cause of iron deficiency in plants is the lower solubility of iron, particularly Fe (III), mainly in neutral and alkaline soil. Hence, plants have come up with two strategies to promote iron solubility. Both of these systems are found to be induced by lower availability of iron in soil and involve the implementation of high-affinity uptake mechanisms. However, when iron levels are optimum, the uptake mechanisms switch to low-affinity systems. The strategy I plants deals with the process of soil acidification via H^+-ATPases, which function via excretion of citrate and malate, serving as small chelators and also by pumping of protons into the rhizosphere and regions of the apoplast, thereby leading to effective solubilization and chelation of Fe(III) (Blancquaert et al. 2017; Upadhyay 2022; Shumayla and Upadhyay 2022). This strategy is particularly observed in nongraminaceous monocots and eudicots. The bound form of iron undergoes reduction by plasma membrane-associated ferric chelate reductases at the root surfaces. The ferric reductase oxidase (FRO) family in *Arabidopsis* comprises of eight members, among which *AtFRO2* products display the highest reductase potential (Wu et al. 2005). This gene has been found to be associated with the roots, thus playing a major role in rhizospheric iron reduction. Overexpression of this gene in soybean plants led to almost fivefold rise in the levels of iron in the pods and leaf tissues. However, seeds demonstrated only a 10% rise in iron levels, hinting at the involvement of other crucial factors mediating iron translocation in seeds (Vasconcelos et al. 2014). Followed by reduction, ferrous moieties are transported by iron-regulated transporter 1 (IRT1) into the roots. The IRT1 belongs to the ZIP metal transporter family (zinc-regulated transporter, IRT like protein). These transporters can serve as potential targets of metabolic engineering to promote the biofortification of iron.

Members of the grass family tend to mostly follow strategy II, involving the release of phytosiderophores for the chelation of ferric moieties from the rhizosphere before its uptake. Nicotinamine, produced via condensation of 3 SAM (*S*-adenosyl methionine), serves as the precursor of phytosiderophores, and the key enzyme responsible for production is nicotinamine synthase (NAS). Another enzyme named nicotinamine aminotransferase catalyzes the next step of nicotinamine conversion to deoxymugineic acid, which is further converted to mugineic acid and other associated derivatives, functioning as phytosiderophores. Nicotinamine serves as a general metal chelator in higher plants (Stephan and Scholz 1993). The iron-phytosiderophore complex enters the roots via the yellow stripe 1 (YS1). Interestingly, strategy I plant, *Arabidopsis* contains eight homologs of the gene. Although the generation of phytosiderophores is not an intrinsic property of the strategy I plants, yet several strategy I plants synthesize these proteins in order to transport nicotinamine-metal complexes within the plant system. Rice cultivation demands anoxic waterlogged soil for optimal growth and yield. In such anaerobic conditions, the solubilization of Fe (III) is not required, since Fe (II) is readily available. Therefore, rice plants have been found to adhere to a combination of both the strategies occasionally. Rice plants maximally express genes like *OsIRT1* and *2* (strategy I) in order to compensate the lesser levels of phytosiderophores (strategy II), unlike maize and barley plants (Morrissey and Guerinot 2009). Iron is transported from the cortex of the roots to the pericycle via symplastic transport, whereby this metal undergoes xylem loading to ensure subsequent transport throughout the plant system. In the vasculature, iron moieties are most likely chelated by citrate, to prevent their precipitation (Curie et al. 2009; Morrissey and Guerinot 2009). In rice plants, Fe (II) loading into the phloem takes place via OsIRT1, followed by the formation of iron-nicotinamine complexes. Next, the seed-specific delivery of iron is carried out by the Yellow Stripe Like (YSL) proteins. Hence, overexpression of *NAS* genes and Fe transporter genes, i.e. *OsYSL* and *OsIRT* genes has been successful in iron biofortification in important food crops. Transgenic rice plants overexpressing *HvNAS1* under the control of *OsActin1* promoter display almost 5–10 folds greater accumulation of nicotinamine in shoots and seeds, eventually resulting in higher Fe concentration within the system (Masuda et al. 2009). Similarly, overexpression of *OsIRT1* and *OsYSL15* in transgenic rice plants led to much higher folds of Fe accumulation, as compared to the nontransgenic ones. Several members of the Graminae family tend to produce phytosiderophores belonging to the mugineic acid (MA) family, acting as chelators of ferric moieties form the rhizosphere. These iron-MA complexes have also been found to be absorbed through the roots via the same group of YSL proteins. In addition, the introduction of this *phytosiderophore synthase* (*MA synthase*) gene (*IDS3*) resulted in about 1.4 folds higher iron content in transgenic plants, as compared to the nontransgenic ones.

Interestingly, iron is found to be bound to myo-inositol hexakisphosphate (phytate) in the developing seeds of *Arabidopsis* and is likely to be stored in the vacuoles (Morrissey and Guerinot 2009). In addition, limited portions of the iron in seeds have been observed to be bound by ferritin in the plastids (Ravet et al. 2009a). The principal seed storage form of phosphorus is phytate, which accounts for about 65–80% of the total phosphorus, maximally present in the outer layers of seeds (Raboy et al. 2000). Phytate can effectively chelate a range of divalent cations, i.e. magnesium, iron, zinc, etc. leading to the formation of phytin (mixed salt). Therefore, phytate has been designated as an antinutrient, causing depreciation of the nutritional value of foods. The enzyme phytase is responsible for the breakdown of the antinutrient, phytate, thereby resulting in the release of adequate amounts of phosphorus and bound metals (Cromwell et al. 1993). However, this enzyme is absent in monogastric animals including humans, poultry animals, fish, etc. Thus, it is a necessity to reduce the levels of phytate in important crop plants for ensuring enhancement in the nutritional value. One of the most widely practised strategies to lower plant phytate levels without depreciation of crop yield is via seed-specific knock down of biosynthetic genes involved in phytate production. The phytate biosynthetic pathway initiates with the generation of a myo-inositol backbone due to the action of the enzyme, myo-inositol 3-phosphate synthase, later this backbone is phosphorylated to produce phytate. Hence, manipulation of the seed-specific expression of this gene has been performed to lower phytate levels in important crop plants, including rice (Kuwano et al. 2006). However, this strategy has an added disadvantage: lowering of myo-inositol levels can hamper important plant biosynthetic cascades, thereby generating adverse side effects. Therefore, it is rather judicious to target the downstream enzymes associated with the biosynthetic pathways involved in phytate production. Seed-specific silencing of the enzyme (inositol phosphate kinase), involved in the last step of phytate biosynthesis, has resulted in almost 69% lowering of the phytate pool and about 80% rise in the iron content of rice seeds (Ali et al. 2013; Roychoudhury 2021a).

In addition, the enhancement of ferritin levels in the seeds of crop plants holds great promise as an iron biofortification strategy in plants. Generally, ferritin comprises an apoprotein coat and a core containing about 4500 atoms of iron (Zielinska-Dawidziak 2015). They serve as iron donors and acceptors and can also render protection to plants against free iron toxicity, by functioning as an iron storage form (Ravet et al. 2009a). Iron bound to ferritin displays a wide range of bioavailability during the digestion process. However, ferritin overexpression in plants led to only a moderate rise in the iron content of seeds, due to the fact that ferritin accumulation particularly takes place in the seed vacuoles, thereby resulting in maximum iron storage in plastids, instead of vacuoles (Ravet et al. 2009b). Thus, for successful biofortification, this approach of overexpressing ferritin in seeds must be tagged with an effective iron translocation strategy from the vacuoles to the plastids. Certain transporters like AtNRAMP3 and 4, and AtVIT1 serve as important members mediating iron transport to seed vacuoles, so impairment of the import of iron into vacuoles and its subsequent accumulation in plastids can make the ferritin-based strategy successful in all regards.

Hence, it is evident that biofortification of crops with iron can be achieved via a number of means (represented in Figure 10.1), encompassing root-mediated iron uptake till its storage in edible plant parts. However, for the successful implementation of the biofortification, all of these strategies have to be properly orchestrated via multigene approaches using tissue-specific promoters. In rice, one of the most important food crops, the best results were achieved via manipulation of the translocation of iron (via NAS), along with enhancement of endosperm-associated iron accumulation (YSL2), and also the promotion of ferritin-mediated sequestration of iron, thereby resulting in an overall sixfold enhancement in iron content in seeds (Masuda et al. 2012).

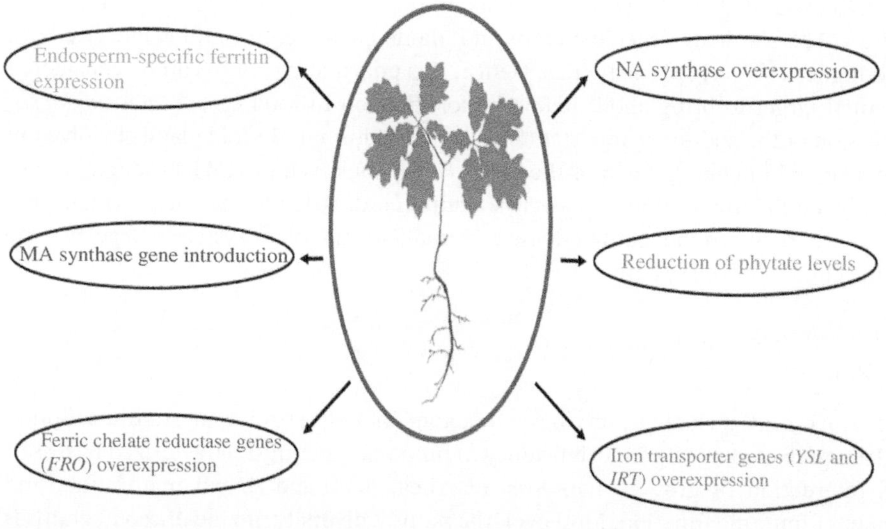

Figure 10.1 Approaches to promote iron biofortification in plants.

10.2.2 Iodine

Iodine serves as a beneficial micronutrient, regulating the functioning of the thyroid gland. Insufficiency of this nutrient in a regular diet can trigger a series of health issues, associated with alteration of the entire endogenous metabolism, affecting all age groups worldwide. Although iodine has not been classified as an essential nutrient for the plant system, iodine uptake can occur via plant roots from soil or via leaves from the atmosphere. Iodine can promote plant growth when present at very low levels, but tends to be phytotoxic at higher doses. Iodine-fortified food crops can act as an efficient vehicle to treat iodine deficiency across the global population. Administration of iodine in soil or spraying on leaves (agronomic approach) has been feasible in leafy vegetables (spinach, lettuce, etc.) and also in fruits and tubers (tomato, potato, etc.); however, this strategy is almost impracticable for plants whose grains serve as the edible product, as the extent of accumulation of iodine in grains is much below the optimal dietary needs (Gonzali et al. 2017). Hence, identification and subsequent modulation of genetically controlled traits associated with iodine mobilization and uptake by the plant system and the development of strategies to ensure reduction of the volatilization of iodine via leaves have become the key focus of biotechnologists to promote successful iodine biofortification.

The mechanisms involving the uptake of iodine from the atmosphere or soil are rather partially deciphered. Several hypotheses have been laid on the basis of the mode of absorption of other halogens, like chlorine via roots and their subsequent loading in the xylem (White and Broadley 2009). In addition, there has been no report till date indicating the presence of iodine-specific transporters or the storage and mobilization of iodine species through the plant system. However, the mechanism behind the mobilization of iodine in the form of methyl iodide from plant roots and leaves has been characterized in a better way. In this regard, the existence of HMT (S-adenosyl-L-methionine-dependent halide methyltransferase) / HTMT (S-adenosyl-L-methionine-dependent halide/thiol methyltransferase) enzymatic activity has been observed in several plant species (Redeker et al. 2000; Rhew et al. 2003; Nagatoshi and Nakamura 2007; Itoh et al. 2009; Takekawa and Nakamura 2012). Nonleafy vegetable crops and plants having edible grains like rice have displayed higher HMT/HTMT activities. Hence, this enzymatic system can be considered as a potential target to bring about iodine biofortification in food crops. In *Arabidopsis*, overexpression of the sodium symporter (NIS) from the human thyroid glands involved in iodine uptake and knocking down of the gene (*HOL1*) involved in HMT production have considerably led to enhanced iodine levels (Landini et al. 2012). Hence, it is evident that the fine balance between the iodine release and intake by the plant system will govern the final iodine content of the system.

10.2.3 Zinc

Zinc is also an essential element, serving as cofactors for a range of important transcription factors (Palmgren et al. 2008). Zinc deficiency in humans is mainly characterized by loss of appetite, retardation of growth, hair loss, diarrhea, disrupted sexual maturation, and impairment of immune function. Moreover, the natural diversity of zinc in cereal grains is quite limited. Hence, increment of the zinc content in these important cereal crops will serve as a successful approach to improve human metabolism and nutrition. As compared to iron, manipulating zinc content in crops is not so straight forward, and there has been reports indicating a strong correlation between zinc content and the total protein content, similar to iron (Ozturk et al. 2006). A wheat quantitative trait locus, involved in increased protein content of grains [Grain protein content B1 (Gpc-B1)], has also been found to be associated with the rise in the levels of zinc and iron within the system. Introduction of this locus from *Triticum turgidum* ssp. dicoccoides, the wild tetraploid wheat into cultivated wheat varieties resulted in almost 10–34% rise in the levels of these essential micronutrients, probably due to the remobilization of these micronutrients along with proteins by Gpc-B1 (Distelfeld et al. 2007). Upon overexpression of zinc translocation and mobilization genes, without imposition of any yield penalty upon the plant system, considerable enhancement of the zinc levels associated with cereal grains occurs (Borrill et al. 2014). Although a series of rice cation transporters have been identified, only a limited number of those transporters have been characterized on the basis of their cellular localization, substrate specificity, and patterns of expression. The well-characterized members involved in zinc uptake and transport are the members of the CDF (cation diffusion facilitator) and ZIP (ZRT, IRT-related protein) families. IRT1 protein, a member of the ZIP family, potentially mediates the uptake of zinc via *Arabidopsis thaliana* roots (Korshunova et al. 1999; Sharma et al. 2022). In addition, two- to threefold increase in zinc levels has been reported in rice upon overexpression of *NAS* under the control of *CaMV35S* promoter (Lee et al. 2009). A similar two- to threefold increase in zinc accumulation has also been observed upon overexpressing *HvNAS1* via *actin1* promoter in rice (Masuda et al. 2009). Transgenic rice plants overexpressing soybean/rice ferritin and *OsNAS2* have successfully demonstrated higher zinc and iron accumulation, as compared to nontransgenic ones. Hence, nicotinamine has been found to be an extraordinary target to aid in zinc biofortification in plants.

10.2.4 Vitamin A

Vitamin A displays a series of aspects controlling proper human growth and development, particularly regulating vision, immune functions, differentiation of epithelial cells, reproduction, etc. Vitamin A (retinoids) belongs to a group of molecules harboring a retinyl moiety as a common skeleton. The retinyl group is made of an isoprenoid side chain and a β-ionone ring. Plant carotenoids serve as one of the major sources of vitamin A, among which β-carotene displays the greatest pro-vitamin A activity (Blancquaert et al. 2017). Followed by ingestion, vitamin A precursors like retinol and provitamin A get converted to their functional forms, i.e. retinoic acid and retinal. Plants generally synthesize four provitamin A carotenoids (PACs) including α, β, and ϒ-carotene and also β-cryptoxanthine. All of these molecules contain a minimum of one retinyl group. Carotenoids belong to the tetraterpenoid group, containing a 40 carbon-containing linear backbone. The biosynthetic cascade depicted in Figure 10.2 involving provitamin A initiates with condensation of two molecules of GGPP (geranylgeranyl pyrophosphate) to generate 15-cis-phytoene mediated by the enzyme phytoene synthase (PSY), which is encoded by the gene *crtB* in bacteria. Three molecules of isopentenyl diphosphate and one molecule of dimethylallyl diphosphate undergo condensation to form one molecule of GGPP, via the action of the enzyme, GGPP synthase, in the plastids (Chappell 1995). The carotenoid chromophore is

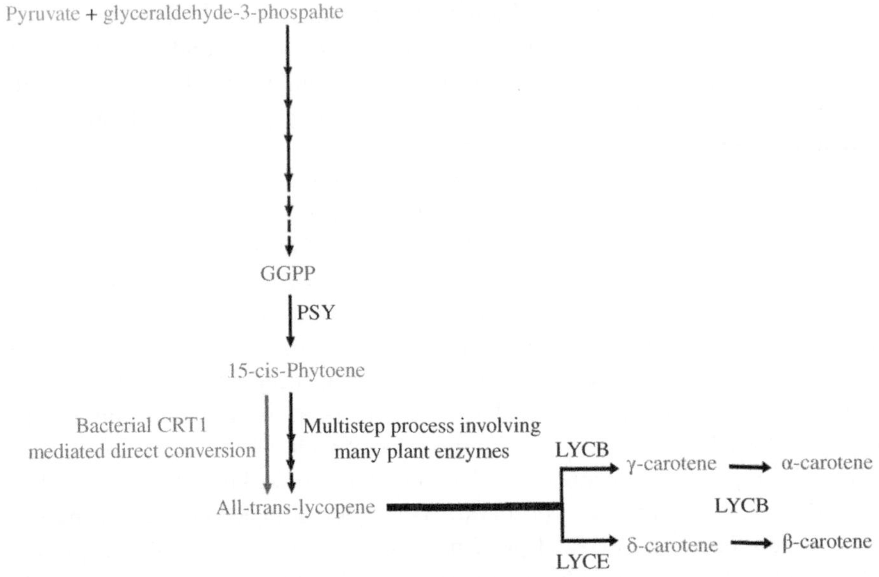

Figure 10.2 Flow chart depicting the carotenoid biosynthetic pathway in plants.

synthesized via four desaturation reactions, and these reactions have been found to be mediated by phytoene desaturase and ζ-carotene desaturase in plants. Subsequently, carotenoid isomerase results in the generation of all-trans-lycopene (Isaacson et al. 2004; Breitenbach and Sandmann 2005; Li et al. 2010). However, in the bacterial system, direct conversion of 15-cis-phytoene to all-trans-lycopene takes place by CRT1. All-trans-lycopene serves as the substrate of both lycopene β-cyclase (LYCB) and lycopene ε-cyclase (LYCE; bacterial CRTY). Lycopene β-cyclase catalyzes the formation of ϒ-carotene via the addition of a β-ring to lycopene. The addition of another β-ring leads to the formation of β-carotene. Generally, the accumulation of ϒ-carotene never occurs due to rapid conversion of all ϒ-carotene to β-carotene (Farré et al. 2010; Zhu et al. 2010). However, lycopene ε-cyclase mediates the formation of δ- carotene, having no provitamin A activity, but lycopene β-cyclase can lead to the addition of a β-ring to δ-carotene, generating α-carotene, displaying provitamin A activity. α- and β-carotene can also be converted to zeaxanthin and lutein under the action of the enzyme β-carotene hydroxylase (bacterial CRTZ). These pigments have immense antioxidative power, but fail to demonstrate any sort of provitamin A activity.

Rice was the first crop to be biofortified with vitamin A. The endosperm of rice grains does not possess vitamin A naturally. Hence, metabolic engineering served as the primary way for the enhancement of vitamin A content in rice kernels. This transgenic crop has been named "Golden rice," by virtue of the yellow color of the kernels. For the development of the first generation Golden rice, a transgene from bacteria *Erwinia uredovora*, i.e. *crtI* and two other genes from daffodil, i.e. *psy* and *lycb* have been used (Ye et al. 2000), resulting in a higher accumulation of total carotenoids, ranging around 1.6 μg/g dry weight. However, replacement of the daffodil *psy* gene by the corn *psy1* gene, in conjunction with the *crtI* gene, led to even higher folds of β-carotene levels, reaching up to 30 μg/g dry weight (Paine et al. 2005). The development of "Golden rice" simplified the path associated with the generation of other "Golden" crops, including corn, potato, wheat, sorghum, cassava, etc. (Roychoudhury 2020). The similar genes and strategy involving the production of "Golden rice" have been administered in the case of wheat and corn (Cong et al. 2009; Naqvi et al. 2009). Overexpression of a particular protein associated with the flux emanating toward GGPP, named 1-deoxyxylulose-5-phosphate synthase (AtDXS) from *Arabidopsis*, along with *crtB* resulted in the rise in β-carotene levels in cassava. The *psy1* and *crtI* genes from corn also increased β-carotene content in sorghum (Sayre et al. 2011; Lipkie et al. 2013). In addition, successful results were generated upon the introduction of the *Erwinia uredovora*-associated genes, *crtI*, *crtB*, and *crtY*, in potato plants (Diretto et al. 2007).

10.2.5 Vitamin B_6

The major functions of vitamin B_6 involve the reduction in the risks associated with a number of deadly human disorders like cardiovascular disease, diabetes, hypertension, pellagra, kidney disorders, neurological diseases, and epilepsy (Hellmann and Mooney 2010). Six vitamers have been found to constitute this vitamin, including pyridoxal, pyridoxamine, pyridoxine, and their phosphorylated forms. Among all these vitamers, pyridoxal-5-phosphate is found to be the most important one, by virtue of its role as a cofactor in a series of important chemical reactions in the cell (Hellmann and Mooney 2010). Thus, the primary target of the biofortification strategy deals with the increased accumulation of this important vitamer. An elaborate cross-talk exists between the biosynthetic pathways involving vitamin B_6 and other salvage pathways, enabling the efficient interconversion between these vitamers (Tanaka et al. 2005; Roychoudhury 2021b).

In plants, the de novo vitamin B_6 biosynthesis is regulated by two key enzymes, pyridoxal-5-phosphate synthase 1 (PDX1) and pyridoxal 5-phosphate synthase 2 (PDX2), completely independent of the bacterial DXP (deoxyxylulose-5-phosphate) pathway (Ehrenshaft et al. 1999; Mittenhuber 2001). Plants like cassava, rice, and *Arabidopsis* have one homolog of the *PDX2* gene and three homologs of the *PDX1* gene (Ehrenshaft et al. 1999; Tambasco-Studart et al. 2005; Prochnik et al. 2012). Overexpression of both *PDX1* and *2* from a fungal pathogen *Cercospora nicotianae* led to almost 20% rise in the vitamin B_6 content in tobacco plants. However, this overexpression had an adverse effect on the plant growth, yield, and germination parameters (Herrero and Daub 2007). Interestingly, the homologs, *PDX1.1* and *1.3*, displayed the maximum importance in the vitamin B_6 biosynthetic cascades. Seed-specific overexpression of *AtPDX1.1* and *AtPDX2* in *Arabidopsis* improved vitamin B_6 levels in a better way as compared to constitutive overexpression, without having any ill effects on the plant performance. Seed-specific overexpression of these two genes increased the vitamin B_6 pool by threefolds. In addition, constitutive overexpression of *AtPDX2*, along with another homolog, *AtPDX1.3* in *Arabidopsis* enhanced the production of pyridoxal-5-phosphate by fivefolds, thereby pointing to the importance of choosing correct homologs for designing metabolic engineering strategies (Raschke et al. 2011). Moreover, this biofortification resulted in an increment in the size of aerial organs and seeds. Overexpression of *AtPDX1.1* and *AtPDX2* in cassava led to a rise in vitamin B_6 levels by several folds, particularly the unphosphorylated vitamers in leaf and root tissues (Li et al. 2015).

10.2.6 Vitamin B_9

Vitamin B_9 comprises a collection of water-soluble B vitamins, commonly named as folates, having a crucial function as cofactors associated with one-carbon metabolic pathways, functioning inside living cells (Blancquaert et al. 2010). A range of metabolic cascades including DNA synthesis, DNA methylation, etc. involves donation or accepting of single carbon (C1) units present in methyl, methylene, formyl, or methenyl groups. The basic backbone consists of a para-aminobenzoate group, a pterin moiety, and a glutamate tail. The folates essentially differ from each other in the length of the glutamate tail and the associated C1 units. The tail groups might comprise of the most reduced form, i.e. tetrahydrofolate (THF) or its other derivatives like 5-formyl THF, 5-methyl THF, 5,10-methylene THF, and 10-formyl THF, thereby generating monoglutamates or polyglutamates. Apart from the differences in the lengths of these glutamate tails, each of these molecules has a specific role in the C1 metabolic pathways (Ravanel et al. 2011). Deficiency of folate in the mother's diet can lead to neural tube defects (NTDs), i.e. cleft spine in the new born (Geisel 2003). Moreover, insufficiency of folate in the diet increases the risks of serious disorders like megaloblastic anemia, coronary diseases, cardiovascular ailments, stroke, cancers, and Alzheimer's disease (Scott and Weir 1996; Seshadri et al. 2002; Li et al. 2003; Stanger 2004; Endres et al. 2005). Nowadays, the fully oxidized form, i.e. folic acid has been used in supplementation and fortification strategies. However, excess of folic acid in the human diet can lead to detrimental outcomes on human health, and also the folic acid needs to be reduced to its active forms, i.e. THF and dihydrofolate (DHF) for optimal functioning (Blancquaert et al. 2014).

The biosynthetic pathway involved in folate synthesis is associated with three steps occurring in three subcellular compartments. The first step involves the synthesis of para-aminobenzoate in the plastids and the synthesis of the pterin group in the cytosol, followed by the condensation, reduction, and glutamylation of these two precursors inside the mitochondrial apparatus. The first metabolic engineering strategy involved two key enzymes associated with the folate biosynthesis pathway, i.e. the first step in the pterin branch, GTPCHI (GTP cyclohydrolase I; G-line) and ADCS (aminodeoxychorismate synthase; A-line) governing the first step in para-aminobenzoate branch. Overexpression of either ADCS or GTPCHI dramatically enhanced para-aminobenzoate or pterin content. However, this strategy resulted in depleted levels of other nontargeted precursors, along with only a moderate rise in folate content, as observed in *Arabidopsis*, tomato, lettuce, rice, potato, and corn G line plants (Díaz de la Garza et al. 2004; Hossain et al. 2004; Díaz de la Garza et al. 2007; Storozhenko et al. 2007; Blancquaert et al. 2013; Dong et al. 2014). Implementation of single-locus T-DNA insertion (developing GA lines) or crossing A and G lines in rice and tomato plants, respectively, led to the high dose of folate accumulation (Storozhenko et al. 2007; Nunes et al. 2009). Huge folate accumulation, ranging up to 1723 μg/g of fresh weight, has been observed in GA lines (Storozhenko et al. 2007). However, this approach failed in *Arabidopsis* and potato plants, indicating at some probable bottleneck in the pathway involved in folate production. Also pterin and para-aminobenzoate levels have been reported to be much higher in these plants, thereby resulting in such bottleneck effects. Recently, combined ADCS and GTPCHI overexpression, in conjunction with the overexpression of a fusion protein generated from *Arabidopsis* carbonic anhydrase 2 and a synthetic protein depicting folate-binding potential in rice, resulted in as much as 50% higher folate levels, as compared to nonengineered rice plants (Blancquaert et al. 2015). This phenomenon focuses on the efficient sequestration power of this chimeric protein, leading to enhanced folate levels in rice seeds.

10.2.7 Vitamin E

Lipophilic vitamin E functions as a potent antioxidant and in the prevention of a series of human diseases resulting from oxidative degeneration (Ricciarelli et al. 2001). Tocochromanols can be defined as the collective terminology associated with eight categories of tocotrienols and tocopherols, mainly having difference in the number and the position of methyl units and displaying varying levels of vitamin E activity. The maximum vitamin E activity has been reported to be displayed by α-tocopherol. The vitamin E biosynthesis starts with the conversion of hydroxyphenylpyruvate to homogentisate (HGA), via the enzyme hydroxyphenylpyruvate dioxygenase (HPPD). Following the synthesis of HGA, the production of tocochromanol branching occurs toward the formation of tocotrienols and tocopherols. In the tocopherol branch, phosphorylation of phytol to phytyl monophosphate is mediated by the enzyme phytol kinase (Blancquaert et al. 2017). This product is again phosphorylated to generate phytyl diphosphate (PDP). In the next step, both PDP and HGA undergo condensation to form 2-methyl-6-phytyl-1,4-benzoquinol (MPBQ), via the action of the enzyme, homogentisate phytyltransferase (HPT). The enzyme MPBQ methyltransferase (MPBQ-MT) attaches another methyl unit to MPBQ to produce 2,3-dimethyl-6-phytyl-1,4-benzoquinol (DMPBQ). The enzyme tocopherol cyclase mediates a cyclization reaction, generating ϒ-tocopherol from MPBQ and δ-tocopherol from DMPBQ. Another enzyme, ϒ-tocopherol methyltransferase (ϒ-TMT) catalyzes the second and third methylation step, thereby converting δ-tocopherol to β-tocopherol and ϒ-tocopherol to α-tocopherol. The same set of enzymes functions in the tocotrienol branch, with an exception where geranylgeranyl pyrophosphate is utilized along with HGA to generate the tocotrienol equivalent of MPBQ, i.e. 2-methyl-6-geranylg eranyl-plastoquinol, in the presence of the enzyme, homogentisate geranylgeranyl transferase (Blancquaert et al. 2017).

Several metabolic engineering strategies have been adopted to enhance the tocochromanol content in important food crops or to shift the tocochromanol accumulation in favor of α-tocochromanols. In the seeds of plants like soybean, *Arabidopsis*, and canola, upon overexpression of prephenate dehydrogenase from bacteria (mediating the generation of hydroxyphenylpyruvate from prephenate) in conjunction with *AtHPPD*, an increase in total tocochromanol content to about fourfolds has been observed (Karunanandaa et al. 2005). However, overexpression of *AtHPPD* in seeds of rice plants displayed a rather marginal increment of tocochromanol concentration, with almost no rise in the level of total tocopherols, and also leading to a shift toward α-tocopherol from ϒ-tocopherol (Farré et al. 2012). Moreover, *AtTMT* overexpression in either endosperm-specific or constitutive manner in rice greatly increased the levels of α-tocotrienol, without any massive changes in the total tocopherol and total tocotrienol pool, probably at the expense of δ/ϒ-tocotrienols (Zhang et al. 2013). Overexpression of barley *HGGT* in corn led to the increment in total tocochromanol content by sixfolds. In addition, *AtMPBQ-MT* and *AtHPPD* overexpression in a constitutive manner resulted in tripling of the ϒ-tocopherol content in corn seeds. Hence, it is evident that biofortifying crops with vitamin E demand a multitarget approach, involving enhancement in the rate of generation of tocopherol precursors and also via diversion of the entire flux toward α-tocopherol formation.

10.3 Conclusion and Future Perspectives

Metabolic engineering of crucial plant pathways forms a central theme of a number of plant biotechnological platforms since the past decades, particularly aiming to promote better human health worldwide. Although biofortification of important micronutrients via metabolic engineering can never be the ultimate and only solution to eradicate global malnutrition, this biotechnological approach can surely serve as a successful complementary strategy. In this regard, both metabolic engineering and conventional breeding rather go hand in hand for the development of engineered crops with health benefits. Metabolic engineering forms a better alternative, due to the lesser success rate and time-consuming nature of the conventional breeding approach. Owing to the progressive rise in the accumulation of information and knowledge associated with the regulation of micronutrient metabolic circuitry, plant biotechnologists have been capable of improving the existing biofortification platforms and have successfully developed more robust biofortification strategies. Some of the biofortification strategies have been summarized in Table 10.1. A number of factors, including the bioavailability and stability of micronutrients, have to be taken into account, in order to meet the required levels of these essential nutrients. In order to avoid shooting up of the upper limits of micronutrient intake, the expression of transgenes under the control of semi-strong promoters and also judicious choosing of engineered lines, displaying an optimal expression of the target micronutrient, can prevent the chances of occurrence of severe health issues arising out of overconsumption of micronutrients. In addition, with the advent of newer and improved technologies like genome editing, i.e. CRISPR/Cas and TALENs, this rising field of metabolic engineering will experience a revolutionary boost, which might promote the eradication of "hidden hunger" worldwide in the near future.

Acknowledgments

Financial assistance from Council of Scientific and Industrial Research (CSIR), Government of India, through the research grant [38(1387)/14/EMR-II], Science and Engineering Research Board, Government of India through the grant [EMR/2016/004799] and Department of Higher Education, Science and Technology and Biotechnology, Government of West Bengal, through the grant [264(Sanc.)/ST/P/S&T/1G-80/2017] to Dr. Aryadeep Roychoudhury is gratefully acknowledged.

Table 10.1 Transgenic plants biofortified with different key micronutrients.

Name of plant engineered	Micronutrient	Associated gene and protein	Outcome	Reference
Soybean	Iron	*AtFRO2* (Ferric reductase oxidase 2) overexpression	Fivefold rise in the levels of Fe in pods and leaves	Vasconcelos et al. (2014)
Rice	Iron	*HvNAS1* (Nicotinamine synthase) overexpression	5–10 folds greater accumulation of nicotinamine in shoots and seeds, eventually resulting in higher Fe levels	Masuda et al. (2009)
Rice	Iron	Inositol phosphate kinase silencing	69% lowering of the phytate pool and about 80% rise in Fe content of rice seeds	Ali et al. (2013)
Rice	Iron	*YSL2* (Yellow Stripe Like 2), ferritin, and *NAS* overexpression	Sixfold enhancement in Fe content of seeds	Masuda et al. (2012)
Arabidopsis	Iodine	Human sodium symporter (*NIS*) overexpression; *HOL1* knock-down	Appreciable enhancement in iodine levels	Landini et al. (2012)
Wheat	Zinc, Iron	Grain protein content B1 (Gpc-B1) introgression	10–34% rise in Zn and Fe levels	Distelfeld et al. (2007)
Rice	Zinc	*NAS* overexpression	Two- to threefold increase in Zn levels	Lee et al. (2009)
Rice	Vitamin A	*crtI* (from *Erwinia uredovora*); *psy* and *lycb* (from daffodil) overexpression	Higher accumulation of total carotenoids (1.6 μg/g dry weight)	Ye et al. (2000)
Rice	Vitamin A	*crtI* (from *Erwinia uredovora*);*lycb* (from daffodil); *psy1* (from corn) overexpression	Higher folds of β-carotene levels (30 μg/g dry weight)	Paine et al. (2005)
Tobacco	Vitamin B_6	Overexpression of pyridoxal-5-phosphate synthase genes (*PDX1* and *2* from *Cercospora nicotianae*)	20% rise in the vitamin B_6 content	Herrero and Daub (2007)
Arabidopsis	Vitamin B_6	Constitutive overexpression of *AtPDX2* and *AtPDX1.3*	Increase in production of pyridoxal-5-phosphate by fivefolds, leading to higher vitamin B_6 pool	Raschke et al. (2011)

(*Continued*)

Table 10.1 (Continued)

Name of plant engineered	Micronutrient	Associated gene and protein	Outcome	Reference
Rice	Vitamin B_9	Combined ADCS (aminodeoxychorismate synthase) and GTPCHI (GTP cyclohydrolase I) overexpression; overexpression of a fusion protein generated from *Arabidopsis* carbonic anhydrase 2 and a synthetic protein depicting folate-binding potential in rice	50% higher folate levels	Blancquaert et al. (2015)
Soybean, *Arabidopsis*, canola	Vitamin E	Bacterial prephenate dehydrogenase; *AtHPPD* (hydroxyphenylpyruvate dioxygenase) overexpression	Increase in total tocochromanol content to about fourfolds	Karunanandaa et al. (2005)
Rice	Vitamin E	*AtTMT* (ϒ-tocopherol methyltransferase) overexpression	Increase in the levels of α-tocotrienol, without any massive changes in the total tocopherol and total tocotrienol pool	Zhang et al. (2013)

References

Ali, N., Paul, S., Gayen, D. et al. (2013). Development of low phytate rice by RNAi mediated seed-specific silencing of inositol 1,3,4,5,6-pentakisphosphate 2-kinase gene (IPK1). *PLoS One* 8: e68161.

Blancquaert, D., Storozhenko, S., Loizeau, K. et al. (2010). Folates and folic acid: from fundamental research toward sustainable health. *Critical Reviews in Plant Sciences* 29: 14–35.

Blancquaert, D., Storozhenko, S., Van Daele, J. et al. (2013). Enhancing pterin and *para*-aminobenzoate content is not sufficient to successfully biofortify potato tubers and *Arabidopsis thaliana* plants with folate. *Journal of Experimental Botany* 64: 3899–3909.

Blancquaert, D., De Steur, H., Gellynck, X., and Van Der Straeten, D. (2014). Present and future of folate biofortification of crop plants. *Journal of Experimental Botany* 65: 895–906.

Blancquaert, D., Van Daele, J., Strobbe, S. et al. (2015). Improving folate (vitamin B_9) stability in biofortified rice through metabolic engineering. *Nature Biotechnology* 33: 1076–1078.

Blancquaert, D., De Steur, H., Gellynck, X., and Van Der Straeten, D. (2017). Metabolic engineering of micronutrients in crop plants. *Annals of the New York Academy of Sciences* 1390: 5973.

Borrill, P., Connorton, J.M., Balk, J. et al. (2014). Biofortification of wheat grain with iron and zinc: integrating novel genomic resources and knowledge from model crops. *Frontiers in Plant Science* 21: 53–60.

Breitenbach, J. and Sandmann, G. (2005). ζ-Carotene *cis* isomers as products and substrates in the plant poly-*cis*carotenoid biosynthetic pathway to lycopene. *Planta* 220: 785–793.

Chappell, J. (1995). Biochemistry and molecular biology of the isoprenoid biosynthetic pathway in plants. *Annual Review of Plant Physiology and Plant Molecular Biology* 46: 521–547.

Cong, L., Wang, C., Chen, L. et al. (2009). Expression of phytoene synthase1 and carotene desaturase *crtl* genes result in an increase in the total carotenoids content in transgenic elite wheat (*Triticum aestivum* L.). *Journal of Agricultural and Food Chemistry* 57: 8652–8660.

Cromwell, G.L., Stahly, T.S., Coffey, R.D. et al. (1993). Efficacy of phytase in improving the bioavailability of phosphorus in soybean-meal and corn–soybean meal diets for pigs. *Journal of Animal Science* 71: 1831–1840.

Curie, C., Cassin, G., Couch, D. et al. (2009). Metal movement within the plant: contribution of nicotinamine and yellow stripe 1-like transporters. *Annals of Botany* 103: 1–11.

De Steur, H., Blancquaert, D., Strobbe, S. et al. (2015). Status and market potential of transgenic biofortified crops. *Nature Biotechnology* 33: 25–29.

Díaz de la Garza, R., Quinlivan, E.P., Klaus, S.M. et al. (2004). Folate biofortification in tomatoes by engineering the pteridines branch of folate synthesis. *PNAS USA* 101: 13720–13725.

Díaz de la Garza, R.I., Gregory, J.F. III, and Hanson, A.D. (2007). Folate biofortification of tomato fruit. *PNAS USA* 104: 4218–4222.

Diretto, G., Al-Babili, S., Tavazza, R. et al. (2007). Metabolic engineering of potato carotenoid content through tuber specific overexpression of a bacterial mini-pathway. *PLoS One* 2: e350.

Distelfeld, A., Cakmak, I., Peleg, Z. et al. (2007). Multiple QTL-effects of wheat Gpc-B1 locus on grain protein and micronutrient concentrations. *Plant Physiology* 129: 635–643.

Dixit, S., Shukla, A., Singh, V., and Upadhyay, S.K. (2021). Engineering of plant metabolic pathway for nutritional improvement; recent advances and challenges. In: *Genome Engineering for Crop Improvement* (ed. S.K. Upadhyay), 351–379. John Wiley & Sons Ltd https://doi.org/10.1002/9781119672425.ch20.

Dong, W., Cheng, Z.J., Lei, C.L. et al. (2014). Overexpression of folate biosynthesis genes in rice (*Oryza sativa* L.) and evaluation of their impact on seed folate content. *Plant Foods for Human Nutrition* 69: 379–385.

Ehrenshaft, M., Bilski, P., Li, M.Y. et al. (1999). A highly conserved sequence is a novel gene involved in *de novo* vitamin B6 biosynthesis. *PNAS USA* 96: 9374–9378.

Endres, M., Ahmadi, M., Kruman, I. et al. (2005). Folate deficiency increases post ischemic brain injury. *Stroke* 36: 321–325.

Farré, G., Sanahuja, G., Naqvi, S. et al. (2010). Travel advice on the road to carotenoids in plants. *Plant Science* 179: 28–48.

Farré, G., Sudhakar, D., Naqvi, S. et al. (2012). Transgenic rice grains expressing a heterologous ρ-hydroxyphenylpyruvate dioxygenase shift tocopherol synthesis from the ϒ to the α-isoform without increasing absolute tocopherol levels. *Transgenic Research* 21: 1093–1097.

Farré, G., Blancquaert, D., Capell, T. et al. (2014). Engineering complex metabolic pathways in plants. *Annual Review of Plant Biology* 65: 187–223.

Geisel, J. (2003). Folic acid and neural tube defects in pregnancy: a review. *The Journal of Perinatal & Neonatal Nursing* 17: 268–279.

Gonzali, S., Kiferle, C., and Perata, P. (2017). Iodine biofortification of crops: agronomic biofortification, metabolic engineering and iodine bioavailability. *Current Opinion in Biotechnology* 44: 16–26.

Hellmann, H. and Mooney, S. (2010). Vitamin B_6: a molecule for human health? *Molecules* 15: 442–459.

Herrero, S. and Daub, M.E. (2007). Genetic manipulation of vitamin B_6 biosynthesis in tobacco and fungi uncovers limitationsto up-regulation of the pathway. *Plant Science* 172: 609–620.

Hossain, T., Rosenberg, I., Selhub, J. et al. (2004). Enhancement of folate in plants through metabolic engineering. *PNAS USA* 101: 5158–5163.

Isaacson, T., Ohad, I., Beyer, P., and Hirschberg, J. (2004). Analysis *in vitro* of the enzyme CRTISO establishes a poly-cis carotenoid biosynthesis pathway in plants. *Plant Physiology* 136: 4246–4255.

Itoh, N., Toda, H., Matsuda, M. et al. (2009). Involvement of *S*-adenosylmethionine-dependent halide/thiol methyltransferase (HTMT) in methyl halide emissions from agricultural plants: isolation and characterization of an HTMT coding gene from *Raphanus sativus* (daikon radish). *BMC Plant Biology* 9: 116.

Karunanandaa, B., Qi, Q., Hao, M. et al. (2005). Metabolically engineered oilseed crops with enhanced seed tocopherol. *Metabolic Engineering* 7: 384–400.

Korshunova, Y.O., Eide, D., Clark, W.G. et al. (1999). The IRT1 protein from *Arabidopsis thaliana* is a metal transporter with a broad substrate range. *Plant Molecular Biology* 40: 37–44.

Kuwano, M., Ohyama, A., Tanaka, Y. et al. (2006). Molecularbreeding for transgenic rice with low-phytic-acid phenotype through manipulating myo-inositol 3-phosphate synthase gene. *Molecular Breeding* 18: 263–272.

Landini, M., Gonzali, S., Kiferle, C. et al. (2012). Metabolic engineering of the iodine content in *Arabidopsis*. *Scientific Reports* 2: 338.

Lee, S., Jeon, U.S., Lee, S.J. et al. (2009). Iron fortification of rice seeds through activation of the nicotianamine synthase gene. *PNAS USA* 106: 22014–22019.

Li, G.M., Presnell, S.R., and Gu, L.Y. (2003). Folate deficiency, mismatch repair-dependent apoptosis, and human disease. *The Journal of Nutritional Biochemistry* 14: 568–575.

Li, Q., Farre, G., Naqvi, S. et al. (2010). Cloning and functional characterization of the maize carotenoid isomerase and β-carotene hydroxylase genes and their regulation during endosperm maturation. *Transgenic Research* 19: 1053–1068.

Li, K.T., Moulin, M., Mangel, N. et al. (2015). Increased bioavailable vitamin B_6 in field-grown transgenic cassava for dietary sufficiency. *Nature Biotechnology* 33: 1029–1032.

Lipkie, T.E., De Moura, F.F., Zhao, Z.Y. et al. (2013). Bioaccessibility of carotenoids from transgenic provitamin A biofortified sorghum. *Journal of Agricultural and Food Chemistry* 61: 5764–5771.

Masuda, H., Usuda, K., Kobayashi, T. et al. (2009). Overexpression of the barley nicotinamine synthase gene (*HvNAS1*) increase iron and zinc concentrations in rice grains. *Rice* 2: 155–166.

Masuda, H., Ishimaru, Y., Aung, M.S. et al. (2012). Iron biofortification in rice by the introduction of multiple genes involved in iron nutrition. *Scientific Reports* 2: 543.

McLaren, D.S., Oomen, H.A.P.C., and Escapini, H. (1966). Ocular manifestations of vitamin-A deficiency in man. *Bulletin of the World Health Organisation* 34: 357–361.

Mittenhuber, G. (2001). Phylogenetic analyses and comparative genomics of vitamin B_6 (pyridoxine) and pyridoxal phosphate biosynthesis path ways. *Journal of Molecular Microbiology and Biotechnology* 3: 1–20.

Morrissey, J. and Guerinot, M.L. (2009). Iron uptake and transport in plants: the good, the bad, and the ionome. *Chemical Reviews* 109: 4553–4567.

Nagatoshi, Y. and Nakamura, T. (2007). Characterization of three halide methyltransferases in *Arabidopsis thaliana*. *Plant Biotechnology* 24: 503–506.

Naqvi, S., Zhu, C., Farrè, G. et al. (2009). Transgenic multivitamin corn through biofortification of endosperm with three vitamins representing three distinct metabolic pathways. *PNAS USA* 106: 7762–7767.

Nunes, A.C., Kalkmann, D.C., and Aragão, F.J. (2009). Folate biofortification of lettuce by expression of a codon optimized chicken *GTP cyclohydrolase I* gene. *Transgenic Research* 18: 661–667.

Ozturk, L., Yazici, M.A., Yuce, C. et al. (2006). Concentration and localization of zinc during seed development and germination in wheat. *Physiologia Plantarum* 128: 144–152.

Paine, J.A., Shipton, C.A., Chaggar, S. et al. (2005). Improving the nutritional value of golden rice through increased provitamin A content. *Nature Biotechnology* 23: 482–487.

Palmgren, M.G., Clemens, S., Williams, L.E. et al. (2008). Zinc biofortification of cereals: problems and solutions. *Trends in Plant Science* 13: 464–473.

Prochnik, S., Marri, P.R., Desany, B. et al. (2012). The cassava genome: current progress, future directions. *Tropical Plant Biology* 5: 88–94.

Raboy, V., Gerbasi, P.F., Young, K.A. et al. (2000). Origin and seed phenotype of maize low phytic acid 1-1 and low phytic acid 2-1. *Plant Physiology* 124: 355–368.

Raschke, M., Boycheva, S., Crèvecoeur, M. et al. (2011). Enhanced levels of vitamin B_6 increase aerial organ size and positively affects tress tolerance in *Arabidopsis*. *The Plant Journal* 66: 414–432.

Ravanel, S., Douce, R., and Rèbeillè, F. (2011). Metabolism of folates in plants. *Advances in Botanical Research* 59: 67–106.

Ravet, K., Touraine, B., Boucherez, J. et al. (2009a). Ferritins control interaction between iron homeostasis and oxidative stress in *Arabidopsis*. *The Plant Journal* 57: 400–412.

Ravet, K., Touraine, B., Kim, S.A. et al. (2009b). Post translational regulation of AtFER2 ferritin in response to intracellular iron trafficking during fruit development in *Arabidopsis*. *Molecular Plant* 2: 1095–1106.

Redeker, K.R., Wang, N., Low, J.C. et al. (2000). Emissions of methyl halides and methane from rice paddies. *Science* 290: 966–969.

Rhew, R.C., Østergaard, L., Saltzman, E.S., and Yanofsky, M.F. (2003). Genetic control of methyl halide production in *Arabidopsis*. *Current Biology* 13: 1809–1813.

Ricciarelli, R., Zingg, J.M., and Azzi, A. (2001). Vitamin E: protective role of a Janus molecule. *FASEB Journal* 15: 2314–2325.

Roychoudhury, A. (2020). Agronomic and genetic biofortification of rice grains with microelements to assure human nutritional security. *SF Journal of Agricultural and Crop Management* 1 (1): 1005.

Roychoudhury, A. (2021a). Potentiality of RNA interference technology in enhancing the nutritional status and food value of plant species. *Journal of Agriculture & Forestry Research* 1 (1): 13–26.

Roychoudhury, A. (2021b). Metabolic engineering of water-soluble vitamins in plants for enhanced nutrition. *International Journal on Recent Advancement in Biotechnology & Nanotechnology* 4 (1): 1–10.

Roychoudhury, A. and Bhowmik, R. (2020). Genetic engineering of rice to fortify micronutrients. In: *Rice Research for Quality Improvement: Genomics and Genetic Engineering*, vol. 2 (ed. A. Roychoudhury), 563–579. Singapore: Springer Publishers.

Sayre, R., Beeching, J.R., Cahoon, E.B. et al. (2011). The bio-cassava-plus program: biofortification of cassava for sub-Saharan Africa. *Annual Review of Plant Biology* 62: 251–272.

Scott, J.M. and Weir, D.G. (1996). Homocysteine and cardiovascular disease. *Quarterly Journal of Medicine* 89: 561–563.

Seshadri, S., Beiser, A., Selhub, J. et al. (2002). Plasma homocysteine as a risk factor for dementia and Alzheimer's disease. *The New England Journal of Medicine* 346: 476–483.

Sharma, A., Sharma, H., and Upadhyay, S.K. (2022). Cation transporters in plants: an overview. In: *Cation Transporters in Plants* (ed. S.K. Upadhyay), 1–28. Elsevier. https://doi.org/10.1016/B978-0-323-85790-1.00011-7.

Shumayla and Upadhyay, S.K. (2022). Mechanism of iron uptake and homeostasis in plants. In: *Cation Transporters in Plants* (ed. S.K. Upadhyay), 147–165. Academic Press https://doi.org/10.1016/B978-0-323-85790-1.00015-4.

Stanger, O. (2004). The potential role of homocysteine in percutaneous coronary interventions (PCI): review of current evidence and plausibility of action. *Cellular and Molecular Biology* 50: 953–988.

Stephan, U.W. and Scholz, G. (1993). Nicotinamine: a mediator of transport of iron and heavy metals in the phloem? *Physiologia Plantarum* 88: 522–529.

Storozhenko, S., De Brouwer, V., Volckaert, M. et al. (2007). Folate fortification of rice by metabolic engineering. *Nature Biotechnology* 25: 1277–1279.

Takekawa, Y. and Nakamura, T. (2012). Rice OsHOL1 and OsHOL2 proteins have S-adenosyl-L-methionine-dependent methyltransferase activities toward iodide ions. *Plant Biotechnology* 29: 103–108.

Tambasco-Studart, M., Titiz, O., Raschle, T. et al. (2005). Vitamin B_6 biosynthesis in higher plants. *PNAS USA* 102: 13687–13692.

Tanaka, T., Tateno, Y., and Gojobori, T. (2005). Evolution of vitamin B_6 (pyridoxine) metabolism by gain and loss of genes. *Molecular Biology and Evolution* 22: 243–250.

Tulchinsky, T.H. (2010). Micronutrient deficiency conditions: global health issues. *Public Health Reviews* 32: 243–255.

Upadhyay, S.K. (2021). *Genome Engineering for Crop Improvement*. John Wiley & Sons Ltd https://doi.org/10.1002/9781119672425.

Upadhyay, S.K. (2022). *Cation Transporters in Plants*. Elsevier Science https://doi.org/10.1016/C2020-0-02921-X.

Vasconcelos, M.W., Clemente, T.E., and Grusak, M.A. (2014). Evaluation of constitutive iron reductase (AtFRO2) expression on mineral accumulation and distribution in soybean (*Glycine max* L.). *Frontiers in Plant Science* 5: 112.

White, P.J. and Broadley, M.R. (2009). Biofortification of crops with seven mineral elements often lacking in human diets – iron, zinc, copper, calcium, magnesium, selenium and iodine. *New Phytologist* 182: 49–84.

WHO (2002). *The World Health Report*. Geneva: World Health Organization.

Wu, H., Li, L., Du, J. et al. (2005). Molecular and biochemical characterization of the Fe(III) chelate reductase gene family in *Arabidopsis thaliana*. *Plant and Cell Physiology* 46: 1505–1514.

Ye, X., Al-Babili, S., Klöti, A. et al. (2000). Engineering the provitamin A (β-carotene) biosynthetic pathway into (carotenoid-free) rice endosperm. *Science* 287: 303–305.

Zhang, G.Y., Liu, R.R., Xu, G. et al. (2013). Increased α-tocotrienol content in seeds of transgenic rice overexpressing *Arabidopsis* ϒ-tocopherol methyltransferase. *Transgenic Research* 22: 89–99.

Zhu, C., Bai, C., Sanahuja, G. et al. (2010). The regulation of carotenoid pigmentation in flowers. *Archives of Biochemistry and Biophysics* 504: 132–141.

Zielinska-Dawidziak, M. (2015). Plant ferritin – a source of iron to prevent its deficiency. *Nutrients* 7: 1184–1201.

11

Plant Hairy Roots as Biofactory for the Production of Industrial Metabolites

Nidhi Sonkar[1], *Pradeep Kumar Shukla*[2], *and Pragati Misra*[3]

[1] *Department of Molecular and Cellular Engineering, Jacob Institute of Biotechnology and Bioengineering, Sam Higginbottom University of Agriculture, Technology and Sciences, Prayagraj, Uttar Pradesh, India*
[2] *Department of Biological Sciences, Sam Higginbottom University of Agriculture, Technology and Sciences, Prayagraj, Uttar Pradesh, India*
[3] *Centre for Tissue Culture Technology, Jacob Institute of Biotechnology and Bioengineering, Sam Higginbottom University of Agriculture, Technology and Sciences, Prayagraj, Uttar Pradesh, India*

11.1 Introduction

Plants have always played an outstanding role in traditional as well as modern medicines (Chandran et al. 2020). Plants preoccupy chemical classes with defined functional roles such as maintaining metabolism, physiological role, immunity, and defence of plants system in abundance (Mohammed et al. 2021). Plants have been a vital source of certain chemical substances such as secondary metabolites and phytochemicals having various importance in the pharmaceutical, food, and flavor industries (Khatodia and Biswas 2014; Korde et al. 2016; Dixit et al. 2021; Upadhyay and Singh 2021). There are several plants that produce therapeutic metabolites having pharmacological effects and are also rich in phytochemicals. Most of the compounds obtained from plants act as therapeutic agents, nutritional additives, or supplements and toxic agents on several other species including human. Those plants are known as medicinal plants (Chandran et al. 2020; Mohammed et al. 2021). Medicinal plants are the jewels of nature comprising of almost unlimited resource, diverse, complex, and most valuable naturally occurring compounds comprising of pharmacological properties, which have been used globally by all human beings for maintaining and restoring good health since ancient times (Halder et al. 2021). Medicinal plants have the ability to treat certain human ailments and are also used in numerous industrial sectors such as pharmaceuticals, cosmetics, and nutraceuticals (Chandran et al. 2020, Rawat et al. 2019). The production of secondary metabolites from medicinal plants, also called plant-derived medicinal compounds (PDMC), has been gaining importance since last decade (Cardoso et al. 2019).

Plants as Bioreactors for Industrial Molecules, First Edition.
Edited by Santosh Kumar Upadhyay and Sudhir P. Singh.

11.2 Types of Metabolites and Industrial Metabolites

Plants are a reservoir of natural compounds that they synthesize, and are used as therapeutics for treating a variety of human illnesses. It has been previously estimated that approximately two-thirds of the active ingredients used as anticancer drugs and drugs for infectious diseases are derived by plants (Kolewe et al. 2008; Chandra and Chandra 2011; Ochoa-Villarreal et al. 2016; Wang et al. 2019). Plants have chemical classes with defined functional roles, e.g. maintaining physiological role, metabolism, immunity, and defense. Most of the compounds produced by plants act as therapeutic agents, nutritional additives or supplements and toxic agents on several other species including human (Mohammed et al. 2021; Fazili et al. 2022).

The sum of all the biochemical reactions carried out by an organism is called as metabolism (Hussain et al. 2012; Thirumurugan et al. 2018). The products of a metabolic reaction and intermediate products usually restricted to small molecules are known as metabolites. The term "secondary" was introduced by Kossel in 1891 (Thirumurugan et al. 2018; Badyal et al. 2020). The secondary plant metabolites are produced by plants via different primary metabolic pathways (Badyal et al. 2020). Organic compounds produced by plants that do not directly participate in growth and development of the plants are known as secondary metabolites. They do not have direct role to play in the photosynthesis, growth, and development but indirectly plays a role in the plant survival. These compounds may be secreted by plants as a signaling molecules, have role in preventing herbivores, may act as an antimicrobial, etc., and their use as plant extracts for medicinal purpose in the treatment of various human ailments. Compounds, such as phytosterols, acyl lipids, nucleotides, amino acids, and organic acids, in contrast, are known as primary metabolites (Kolewe et al. 2008; Hussain et al. 2012; Ochoa-Villarreal et al. 2016; Cardoso et al. 2019; Rawat et al. 2019; Badyal et al. 2020). Plant secondary metabolites are a fascinating class of phytochemicals exhibiting immense chemical diversity (Gandhi et al. 2015). The secondary metabolites produced by plants are known to have antifungal, antibiotic, and antiviral ability, due to which it holds a significant position in maintaining the plant's defense system. They produce toxicity that repels other microbes and herbivores, thereby protecting them from any kind of pathogens. Also, they have been an important source of pharmaceuticals that makes them industrially important too (Gandhi et al. 2015; Badyal et al. 2020). Although secondary metabolites are derived from primary metabolism, they do not make up basic molecular skeleton of the organism due to which its absence does not affects the life of an organism. The products observed in primary metabolism overlap with the intermediates of secondary metabolites, which serve as the main difference between primary and secondary metabolites. Amino acids though considered a product of primary metabolite are definitely secondary metabolite as well (Thirumurugan et al. 2018). The end products of primary metabolic pathways are very few, while secondary metabolic pathways result in the production of too many products that are also industrially important. There are three potential pathways for primary metabolism, i.e. the Embden Meyerhof–Parnas Pathway (EMP), the Entner–Dourdorof pathway, and the hexose monophosphate (HMP) pathway, mostly occurred in animal, plant, fungal, yeast, and bacterial cells (Hussain et al. 2012).

Plant secondary metabolites can be classified into four major classes: terpenoids, phenolic compounds, alkaloids, and sulfur-containing compounds. Plant secondary metabolites have numerous applications in both the scientific and industrial field such as pharmaceutics, cosmetics, agrochemical biopesticides, flavorings or food additives, fragrances, and as natural pigments (Zhou and Wu 2006; Verma et al. 2007; Yue et al. 2016; Guerriero et al. 2018; Rawat et al. 2019).

There are several secondary metabolites produced from different plants, belong to a diversified groups of chemicals like alkaloids, steroids, saponins, terpenes, tannins, lignins, and flavanoids. The industrial application of secondary metabolites is highly effective, seeking attention of the manufacturers for its production at commercial levels through hairy root cultures, callus cultures, and cell suspension cultures. Maintaining the similar genetic structure of the cells leads to comparatively low secondary metabolites production through cell cultures. *A. rhizogenes* has been used as a natural vector system that serves as a new method for increasing secondary metabolite production as these hairy roots are able to produce unlimited biomass in the culture media free from the plant growth regulators (Rao and Ravishankar 2002; Abraham and Thomas 2017; Rawat et al. 2019). Therefore, plant cell cultures with the production of high-value secondary metabolites are promising potential alternative sources for the production of pharmaceutical agents of industrial importance (Yue et al. 2016). Table 11.1 depicts the information of the Classes of secondary metabolites produced by plant tissue culture.

Table 11.1 Classes of secondary metabolites produced by plant tissue culture.

S. No.	Alkaloids	Terpenoids	Steroids	Quinones	Phenylpropanoids
1.	Acridines	Artemisinin	Brassinolide	Aloe-emodin	Anthocyanins
2.	Betalaines	Cucurbitacins	Bufadienolides	Anthraquinones	Caffeic acid
3.	Furoquinolines	Diterpenes	Catasterone	Benzoquinones	Coumarins
4.	Galanthamine	Ginsenosides	Digoxin	Chrysophanol	Eugeno
5.	Harringtonines Isoquinolines	Meroterpenes	Digitoxin	Emodin	Ferulic acid
6.	Lobeline	Monoterpenes	Digitoxigenin	β-Lapachol	Flavonoids
7.	Quinolizidines	Paclitaxel	Ecdysteroids	Naphthoquinones	Hydroxycinnamoyl derivatives
8.	Indole alkaloids	Sesquiterpenes	Helleborin	Phenanthrenequinone	Isoflavonoids
9.	Isoquinoline alkaloids	Sesterpenes	Physodine	Plumbagin	Lignan
10.	Piperidine	Thapsigargin	Ouabain	Rhein	Phenalinones
11.	Thebaine	Triterpenes	Scillaridine	Shikonin	Proanthocyanidins
12.	Trigonelline	Ursane triterpenoid	Steroidal glycosides	Thymoquinone	Stilbenes
13.	Tropane alkaloids	Withanolides	Steroidal lactones	Uglone	Tannins

Different plants have different medicinal value with an array of secondary metabolites present. Alkaloids, tannins, flavonoids, phenolics, and other aromatic amino acid derivatives constitute the major bulk of these phytobiochemicals

11.3 Secondary Metabolites

There are certain organic compounds that are produced by plants that are not involved in the primary metabolic process but are quintessential for the proper functioning of the plants. Those compounds are known as secondary metabolites. These compounds synthesized by plants provide protection to the plant from many adverse environmental conditions or attack from antagonists (Thirumurugan et al. 2018; Supriya et al. 2020). Medicinal and aromatic plants, both are known to produce secondary metabolites, that find uses as flavoring agents, fragrances, insecticides, dyes, and drugs (Gandhi et al. 2015; Chandran et al. 2020; Supriya et al. 2020). There is a diverse and broad groups of chemicals like alkaloids, steroids, saponins, terpenes, tannins, lignins, and flavanoids that constitute secondary metabolites from different plants. Various kinds of modern medicines are made by the extracts of many plants, and the medicinal properties of the plant thus attribute to the production of secondary metabolites (Rawat et al. 2019). The output products of secondary metabolite, i.e. antibiotics, enzyme inhibitors, immunomodulators, antitumor agents, and plants and animals' growth. It has the potential application of enhancing agricultural productivity as a (pesticides, insecticides, effectors of ecological competition, and symbiosis and pheromones (pigments and nutraceuticals) (Thirumurugan et al. 2018).

Biotechnology has been an interdisciplinary area providing various possibilities through which secondary metabolism in medicinal plants can be modified in innovative ways, for the production of phytochemicals of interest. It may reduce the content of toxic compounds and can even produce novel chemicals (Gandhi et al. 2015). The commercial production of secondary metabolites is very tedious and time taking. It produces low yield in wild plants. The cell and tissue culture technology combined with recent developments in molecular biology offers a viable approach to overcome these difficulties (Supriya et al. 2020). Hairy root (HR) culture technology is a novel biotechnological approach that aims toward the genetic and biochemical improvement of various important medicinal plants (Gantait and Mukherjee 2020). Hairy roots are heterotropic and extensively branched roots produced as a result of infection caused by a gram-negative soil bacterium called *A. rhizogenes* (Supriya et al. 2020). The development in the field of transformation of genes in 1980s with the use of *Agrobacterium* strains has increased the possibility of using plants for the expression of heterologous recombinant proteins. *Agrobacterium* strains can be used to produce hairy roots or to target the expression of heterologous proteins in root tissues. The production of secondary metabolites or newly produced proteins in hairy root cultures has proved to be a new biotechnological tool (Angulo-Bejarano et al. 2019). Hairy roots have been studied as an upcoming platform for the production of a variety of compounds in different plant systems for almost 35 years. It has been considered as an efficient and convenient biotransformation system for the production of new agents with desired physico-chemical properties, sufficient solubility, and low toxicity (Habibi et al. 2018).

Plant tissue culture-based biotechnological processes have been utilized effectively for the conservation of those medicinal plants that are disease-prone, recalcitrant, and at risk of extinction. The enhancement of secondary metabolite production can be done via *in vitro* micropropagation, hairy root culture system, ploidy engineering, metabolic pathway engineering, transgenic plant development, large-scale production in bioreactors, application of zinc-finger nucleases, TALENs, and CRIPR/Cas9 (Alok et al. 2018; Halder et al. 2021; Shushmita et al. 2021; Upadhyay 2021). For scaling up the production of secondary metabolites, bioprocessing in the form of suspension culture, organ culture, or transformed hairy roots has been a huge success in many cases. For getting the enhanced yield of the metabolite of interest, pathway elucidation and metabolic engineering have been found to be most useful approaches. The heterologous expression of claimed plant secondary metabolite biosynthesis genes in a microbe is useful to validate their functions, and in some cases, also, to produce plant metabolites in microbes (Gandhi et al. 2015). The scale-up of the industrial production of plant secondary metabolites can be done using specially designed bioreactors with distinct sizes and features for the purpose, which meets the unstable productivity of plant cells, sensitivity, minimum oxygen requirements, and slow growth rate. Bioreactors are self-contained units intended for optimization and monitoring as well as to provide homogeneous culture conditions like pH, air circulation, dissolved gasses, and temperature for mass propagation of cells, tissues, organogenic propagules, or somatic embryos in a sterile environment. The major attraction behind the use of hairy root technology for the production of important secondary metabolites is their genetic stability, higher biomass production, and enhanced product accumulation (Supriya et al. 2020).

11.4 Importance of Secondary Metabolites

Secondary metabolites are the organic compounds synthesized by plants that do not participate in the growth and development of the plant in a direct way rather shows significance in the interaction between plant–plant, plant–environment, or else plays a defensive role and thus are produced in low quantity. Secondary metabolites have been exploited for the production of industrially important compounds and as signaling molecules in ecological interactions, symbiosis, metal transport, and competition. There are various applications of plant secondary metabolites, e.g. they are used as drugs in pharmaceutical industries, can be used as flavoring agents, pesticides, and chemicals utilized in agricultural industries, has been utilized in producing fragrance, and can be used as food additives. There has been researches on plant hairy roots that show their effective response in fighting with COVID-19. Plant secondary metabolites play an important role in protection of plants from grazing animals, i.e. herbivores and microorganisms present in the vicinity. It produces certain chemicals that have the ability to attract thus influencing the competition among plant species, and protecting the plants by showing themselves as allelopathic agents for pollinators and animals that can disperse the seeds. Primary metabolites are different from secondary metabolites. Due to extensive properties of medicinal plants, they are falling in the endangered category. Industrial usage of plant secondary metabolites has been increased and is increasing day by day thus increasing the demand. The production of certain drugs is dependent on medicinal plants and their extracts (Rao and Ravishankar 2002; Gurunani et al. 2015; Guerriero et al. 2018; Thirumurugan et al. 2018; Fazili et al. 2022). Secondary metabolites are a source of antibiotics, and the metabolite secretions are harvested by human kind to improve their health. Various applications such as antibiotics, enzyme inhibitors, immunomodulators, antitumor agents, and growth promoters of animals and plants have widen the scope of healthy nutrition (pigments and nutraceuticals), enhancing agricultural productivity, e.g. pesticides, insecticides, effectors of ecological competition and symbiosis and pheromones, fragrances, and dyes (Gandhi et al. 2015; Thirumurugan et al. 2018).

Plants possess plenty of chemical classes with defined functional roles such as maintaining physiological role, metabolism, immunity, and defense. Among these, most of the compounds act as therapeutic agents, nutritional additives or supplements, and toxic agents on several other species including human (Mohammed et al. 2021). Secondary metabolites play defensive role between the species and against different kinds of stresses (Fazili et al. 2022).

Plants secondary metabolites are used by pharmaceutical industries as these bioactive compounds trigger a pharmacological or toxicological effect in humans and animals. They have numerous applications in cosmetics, nutrition, for the manufacture of drugs, dyes, fragrances, flavors, and dietary supplements. Plants are renewable resources providing raw material (like lignocellulosic biomass; and phytochemicals (importantly secondary metabolites) for different industrial applications, e.g. in the textile, construction, pharmaceutical, nutraceutical, cosmetic, perfume, dyeing, and flavor industries. The extracts of many plants are constituents of various modern medicines; the medicinal properties being ascribed to the production of secondary metabolites (Gandhi et al. 2015; Guerriero et al. 2018; Rawat et al. 2019).

11.5 Enhancement of Secondary Metabolites

There are various difficulties when doing the field cultivation of plants for the production of secondary metabolites. Due to variability in the geographical regions, variations in seasonal and environmental factors instabilities in the concentrations are observed, and lower yields as output are observed. These are a few limiting factors for the cultivation of plants for secondary metabolite production. As a result, to overcome these difficulties, several technologies such as plant cell culture, tissue culture, and organ cultures came into scene and proved to be an appropriate alternative for the production of secondary metabolites (Rao and Ravishankar 2002). Due to high-cost value and associated side effects, synthetic drugs cannot replace the natural drugs. Thus, the increase in world population has increased the demand of natural drugs globally. The addition of elicitors enhances as well as is responsible for fast growth of the culture systems, the adoption of hairy roots in scientific studies has given attention largely due to properties such as genotype and phenotype stability, biochemical stability, autotrophy in plant hormones, and high levels of secondary metabolites that are produced (Gurunani et al. 2015; Srivastava et al. 2016). Plants have been found to elicit the same response as the pathogen itself when challenged by elicitors, i.e. they produce signals triggering the biosynthesis of the secondary metabolites (Yue et al. 2016). Strain improvement, methods for the selection of high-producing cell lines, and medium optimizations can also lead to an enhancement in secondary metabolite production (Hussain et al. 2012).

For *in vitro* cultures biotic and abiotic elicitors have been used widely for enhancing the yield of secondary metabolites as they show a great impact on indole alkaloid biosynthesis. Carbohydrate and protein are being used as biotic elicitors. In a study, cellulase from *Aspergillus niger* (protein) has been utilised and mannan from *Saccharomyces cerevisiae* (carbohydrate). For enhancing the plant secondary metabolites production, plant cell and tissue cultures methods are the best suitable methods used industrially. It does not depend upon the geographical or seasonal variations by doing a few modifications of different growth parameters. These culture techniques provide excellent way for the improvement of the secondary metabolites production (Srivastava et al. 2016). The treatment of plant cells with biotic and abiotic elicitors has been one of the most effective means for the enhancement of secondary metabolite production in plant tissue cultures (Zhou and Wu 2006). *In vitro* strategies mainly hairy root culture, elicitation, suspension culture system, etc. are used for the improvement of the production of secondary metabolites of plants. The two best methods known are the suspension culture and elicitation and hairy root culture, and other organ cultures have been a satisfied and proven method to meet the demand of secondary metabolites. Manipulation and control of the pathways that produce the plant secondary metabolites is now easier through these techniques (Fazili et al. 2022). Elicitation of hairy roots increases the production of secondary metabolites and also helps in designing of metabolic traps for the adsorption of product, preventing feedback inhibition and protecting the metabolites from degrading in the culture media (Chandra and Chandra 2011). The biosynthesis of a specific compound gets improved and initiated by introducing elicitor in a small concentration to a cell system. The major advantage of adding elicitor and the elicitation process is to increase and enhance the production of plant secondary metabolites by the plants that leads to their survival, persistence, and competitiveness. The major regulatory factors for proper enhancement of production of secondary metabolites includes, at first elicitation is used, the concentration of elicitors that is added and the type of medium used in the enhancement process (Chandra and Chandra 2011). Elicitors when added to the plants systems for the enhancement of secondary metabolites exhibit the same response as the pathogen itself, i.e. signals triggering the biosynthesis of the secondary metabolites (Yue et al. 2016; Ochoa-Villarreal et al. 2016). Secondary metabolites important in pharmaceutical industry is found in hairy root cultures. Growth rate of secondary metabolites is highest in transformed cultures, so hairy roots alone are capable of producing valuable metabolites at commercial level. The combination of elicitation with hairy roots enhances the yield of targeted metabolites (Srivastava et al. 2016).

In the study conducted by Hussain et al. (2012), several abiotic elicitors were applied to enhance the growth and ginseng saponin biosynthesis in the hairy roots of the plant *P. ginseng*. Elicitor treatments were found to inhibit the growth of the hairy roots, although simultaneously enhancing secondary metabolite ginseng "saponin" biosynthesis. Tannic acid profoundly inhibited the hairy root growth during growth period. Elicitors, benzo (1,2,3)-thiadiazole-7-carbothionic acid, S-methyl ester, and autoclaved lysate of cell suspension of bacteria *Enterobacter sakazaki* produced secondary metabolites in callus, cell suspension, and hairy roots of *Ammi majus* L (commonly called "ajwain"). Gas chromatography and gas chromatography-mass spectrometry analysis of chloroform and methanol

Table 11.2 Enhancement of secondary metabolites production by using elicitors.

Plant species	Products	Elicitors
Ammi majus	BION®	Coumarine/Furocoumarine
Arachis hypogaea	Resveratrol, Piceatannol, Arachidin-1 Arachidin-3	Cyclodextrin
Astragalus membranaceus	Isoflavonoid	Methyl jasmonate
Brugmansia candida	Hyoscyamine	CdCl2
Chicorium intybus	Esculin/Esculetin	Phytopthora parasitica filtrate
Echinacea pupurea	Caffeic	Gibberellic acid
Fagopyrum tataricum	Rutin, quercetin	UV-B
Hyoscyamus muticus	Tropane alkaloids	Chitosan
Panax ginseng	Ginsenoside	Methyl jasmonate
Papavar orientale	Morphine	Methyl jasmonate
Plumbago indica	Plumbagin	Jasmonic acid
Psoralea corylifolia	Daidzin	Jasmonate
Salvia castanea	Tanshinone	Methyl jasmonate
Salvia miltiorrhiza,	Tanshinone	Salicylic acid
Scopolia parviflora	Scopolamine	*Bacteria* sp.
Solanum khasianum	Alkaloids	*Aspergillus niger*, cellulase

extracts indicated a higher accumulation of umbelliferone in the elicited tissues than in the control ones. Biotic elicitor, polysaccharide produces chitosan that was eliciting the manifold increase of anthraquinone production in Rubia akane cell culture.

Some of the examples of elicitor, i.e. salicylic acid, methyl jasmonate, chitosan, and heavy metals, when treated with callus cultures, secondary metabolites production is enhanced in many plant species. Tumor-like cultures called teratoma cultures, i.e. shooty teratoma cultures and hairy root cultures in some plant species produces secondary metabolites. Alkaloids are produced at large by hairy root culture and shooty teratomas produce monoterpenes (Fazili et al. 2022). Table 11.2 gives the information for the enhancement of secondary metabolites production using different elicitors.

11.6 Hairy Roots

11.6.1 Hairy Roots

Hairy roots emerge when a plant is infected by a symbiotic bacteria called *R. rhizogenes* (previously referred to as *A. rhizogenes*) bacteria. Hairy roots are produced by a wide variety of plants, producing highly diverse molecules. *A. rhizogenes*-mediated hairy roots formation has become an important method for the production of secondary metabolites that are synthesized in plant roots and is the rapid method with high transformation rate (Hussain et al. 2012; Gutierrez-Valdes et al. 2020; Cheng et al. 2021). *A. rhizogenes*-mediated hairy root transformation is *A. rhizogenes* (or *R. rhizogenes*) is able to transform plant genomes and induce the production of hairy roots (Ron et al. 2014). Hairy root cultures are also known to produce a spectrum of secondary metabolites that are not present in the parent plant (Rawat et al. 2019). The physical characteristics of hairy root formed are identified by fast hormone-independent growth, lack of geotropism, lateral branching, and their stability at genetic level. The genes present in this region are expressed as the endogenous genes of the plant cells. The hairy roots are normally induced on aseptic, wounded parts of plants by inoculating them with *A. rhizogenes*. (Karuppusamy 2009; Hussain et al. 2012; Gutierrez-Valdes et al. 2020)

11.6.2 Hairy Roots in Plants and *In vitro* Production of Secondary Metabolites

Hairy root culture has been a pioneering path for massive production of secondary metabolites *in vitro* and the production of phytochemicals in a very short span of time for continuous supply of improved value products (Korde et al. 2016; Singh et al. 2018; Gutierrez-Valdes et al. 2020; Kowalczyk et al. 2022). Hairy root cultures have the capability to grow in hormone-free media due to which their use in plant biotechnology is being endorsed. They also have certain properties including fast growth, genetic, and biochemical stability (Shanks and Morgan 1999). Hairy root cultures have various applications in tissue culture for high production of valuable secondary metabolites such as pharmaceuticals, coloring, and flavoring agents. Its stability at genetic level and potential of imitating the parent plants for metabolites biosynthesis over plant cell or callus and suspension cultures make it an extraordinary system. These root cultures are cost efficient and are very attractive for mass-producing desired plant metabolites (Habibi et al. 2018; Singh et al. 2018). The synthesis of novel molecules that are required in the pharmaceuticals industry requires the genetic or biochemical stability, the large-scale production of desired metabolites, low-cost cultural requirements, and hormone-independent growth as main characteristics making hairy root as an efficient system. Also, this system contains inherent enzymes, required for biotransformation reactions, including methylation, oxidation, hydroxylation, glycosylation, reduction, isomerization, and esterification. Thus, hairy root platform acts as an effective and convenient biotransformation system for the production of new agents with desired physico-chemical properties, sufficient solubility, and low toxicity (Bhatia and Bera 2015; Habibi et al. 2018). For plant genetic transformation procedure, the most important factor is the type of explant, i.e. different sources of plant cells or tissues (e.g. seeds, roots, shoots, leaves, and shoot or root apical meristems) resulting in different degrees of callus induction, and thus different levels of plant regeneration and genetic transformation efficiency. *A. rhizogenes*-mediated hairy root production allows us to obtain biologically valuable compounds without destroying the entire plant and its habitat, which is extremely important in the era of environmental protection. Plant regeneration from callus cultures has opened the door to the possibilities of their laboratory and industrial application, viz. a large-scale *in vitro* culture of plant cells for the accumulation of secondary metabolites. The development of a complete medium for the culture of plant tissues by Murashige and Skoog by the techniques of culturing plant cells and organs (Kowalczyk et al. 2022).

11.7 Initiation of Hairy Root Cultures

11.7.1 Formation of Highly Proliferative Hairy Roots

Hairy root culture is a novel source for the production of secondary metabolites because of its fast growth without hormones, biochemical, and genetic stability (Korde et al. 2016). Hairy roots are formed when plantlets get wounded after they are infected by a gram-negative soil bacterium, *A. rhizogenes*, is taxonomically renamed as *R. rhizogenes*, which is a bacterium showing symbiotic relationship with plant roots. Hairy root syndrome is a disorder caused by *R. rhizogenes* that produces adventitious roots at the wounded site of plants that are infected (Makhzoum et al. 2013; Gutierrez-Valdes et al. 2020). A bacterial DNA segment, i.e. T-DNA, that consists of a region having 200 kb large and extrachromosomal Ri also called root inducing plasmids is inserted into the nuclear genome of higher plants including dicotyledons, gymnosperms, and some monocotyledon species leading to the transfer of the *A. rhizogenes* T-DNA to the nuclear plant genome (Makhzoum et al. 2013; Supriya et al. 2020). After the bacterial infection, the plant responds to it by initiating and forming various defense-related proteins that can suppress the pathogenicity caused by the bacterium (Gutierrez-Valdes et al. 2020). The production of hairy root culture involves three major steps, i.e. induction of roots, decontamination of roots, and establishing liquid medium (Gurunani et al. 2015). Figure 11.1 depicts the steps involved in the production of hairy root culture. A large number of hairy roots are produced

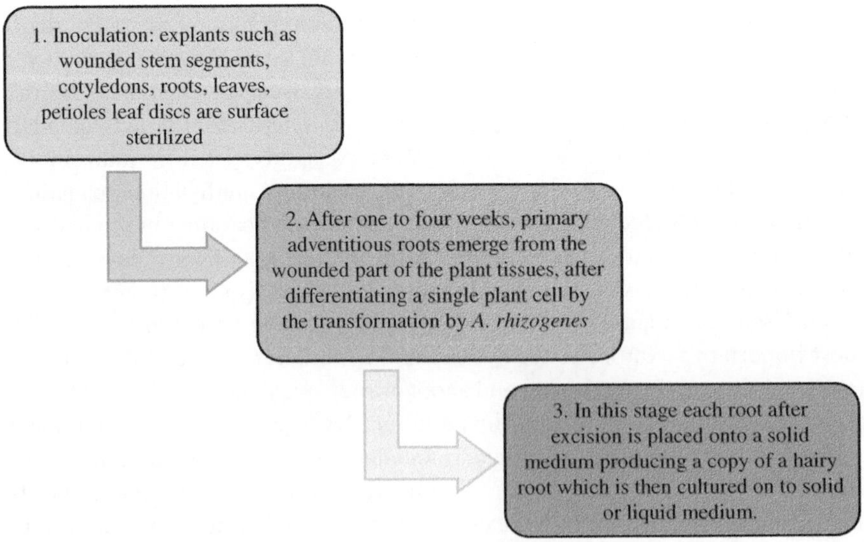

Figure 11.1 Steps involved in the production of hairy root culture.

through *in vitro* condition. The different strains are responsible for formation of hairy roots. Thus, hairy roots can be used as the easiest and most convenient option that can yield important metabolites from the plants. Hairy roots have been used as a supply for various plant products as well as facility for modification of metabolic pathways by the integration of related genes to the plants using *A. rhizogenes* mediated genetic transformation (Gurunani et al. 2015; Korde et al. 2016). Hairy roots can be excised and excess bacteria can be cleared off by using antibiotics and are grown indefinitely *in vitro* by subculture of root tips in liquid medium. Certain techniques of hairy roots culture have been explored for initiation, culture, genetic manipulation, and molecular analysis (Gurunani et al. 2015).

11.7.2 *Agrobacterium rhizogenes* for Hairy Root Production and as a Biotechnology Tools

Agrobacterium rhizogenes is a soil-borne, Gram-negative bacterium, which is responsible for the production of hairy root wounding and infection of monocot and eudicot plants. This biotechnology tool has been explored to discover novel biological insights (Makhzoum et al. 2013). Metabolic enzyme function can be determined by gene overexpression or RNA interference approaches using hairy root transformation (Ron et al. 2014). During infection, when *A. rhizogenes* transfers T-DNA from (Ri) plasmid, it gets integrates with the plant genome. The oncogene present in the T-DNA stimulates the formation of highly proliferative hairy roots. Due to the fast growth and genetic and biochemical stability of hairy roots, this technology is very promising for plant secondary metabolite production (Supriya et al. 2020). A mass of roots is formed, which is a symptom of hairy root disease. A large number of small roots protrude as fine hairs directly from the infection site. Plant roots are the best plant organ for large-scale cultivation, since the roots are the site of synthesis and/or the storage organ of phytochemicals. Figure 11.2 provides the information for induction of hairy roots and its applications in Biotechnology. Many products of interest are also synthesized in organized tissues (roots), but not formed in suspension or callus culture (shoots and leaves) (Mishra and Ranjan 2008). The root-inducing plasmid, containing t-DNA encodes root locus (rol) gene loci (rolA, rolB, and rolC), that are responsible for introducing genetic material into host cells and forms hairy roots. Certain biotechnological approaches like recombinant protein production and metabolic engineering, analyses of rhizosphere physiology and biochemistry have been studied during past 30 years through this technology (Ron et al. 2014). The production of secondary metabolites using hairy root culture guarantees high biomass production in short span of time, for the extraction of industrially important metabolites. Plants systems by the application of hairy root induction for the production of specialized metabolites have some advantages than other biotechnological process (Acharya et al. 2021).

Different metabolic pathways in plants are catalyzed with the involvement of different kinds of enzymes and, hence, inevitable destruction of their habitat, ecosystem, and biodiversity. Such approach has pushed these plants toward endangered category. To meet with the problem of withdrawing of these useful biochemicals having pharmacological

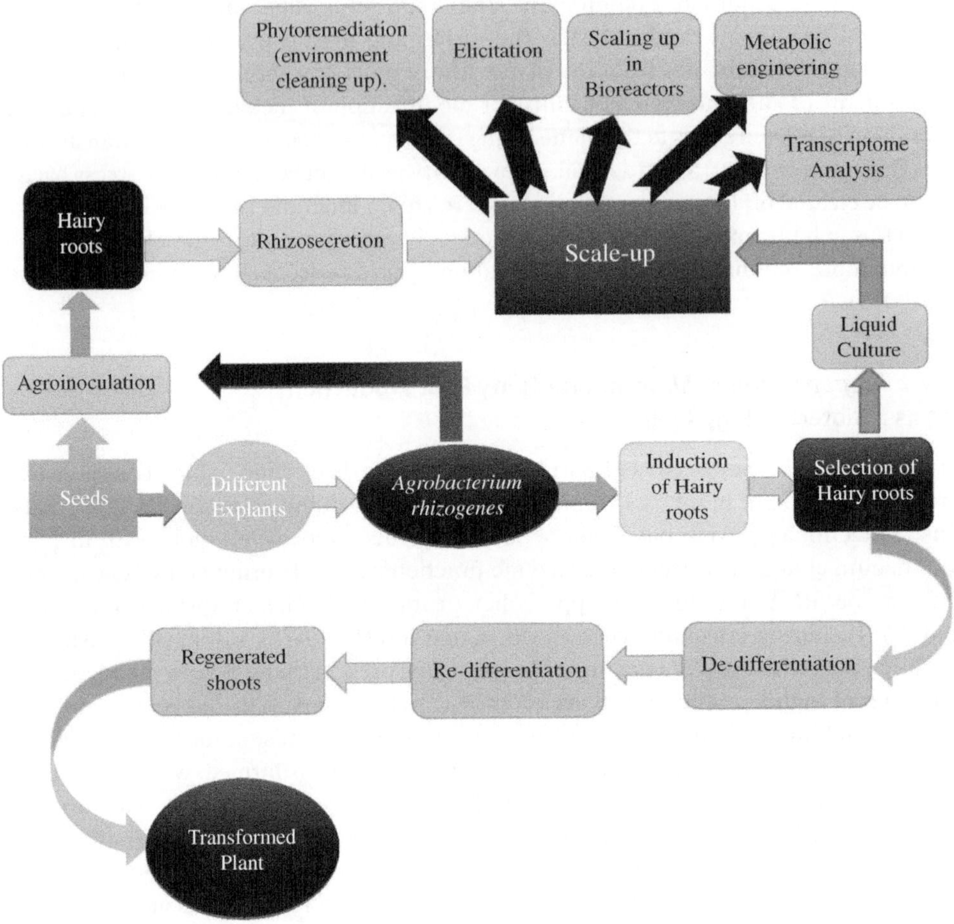

Figure 11.2 Induction of hairy roots and its applications in Biotechnology.

and medicinal uses through conventional field-grown methods involves large-scale uprooting of the plants. Therefore, *in vitro* culture technology could effectively be applied for the large-scale propagation, conservation as well as production of secondary metabolites from these plants (Gantait and Mukherjee 2020). Biotechnological approaches like organ culture, callus culture, and *Rhizobium*-mediated hairy root culture provide an important way out for the procurement of authentic plant material. *R. rhizogenes*-mediated hairy roots represent an important "*in vitro*" method for specialized metabolite biosynthesis (Gutierrez-Valdes et al. 2020). Also, the fast growth and genetic and biochemical stability of hairy roots over a long period made this a promising technology for plant secondary metabolite production. Figure 11.3 shows interaction of *Agrobacterium rhizogenes* with plant roots. Bioreactors can be used for the large-scale production of hairy roots that can be optimized in these bioreactors that are specially designed and the production of the desired metabolite, which can be further enhanced by the addition of suitable elicitors (Supriya et al. 2020).

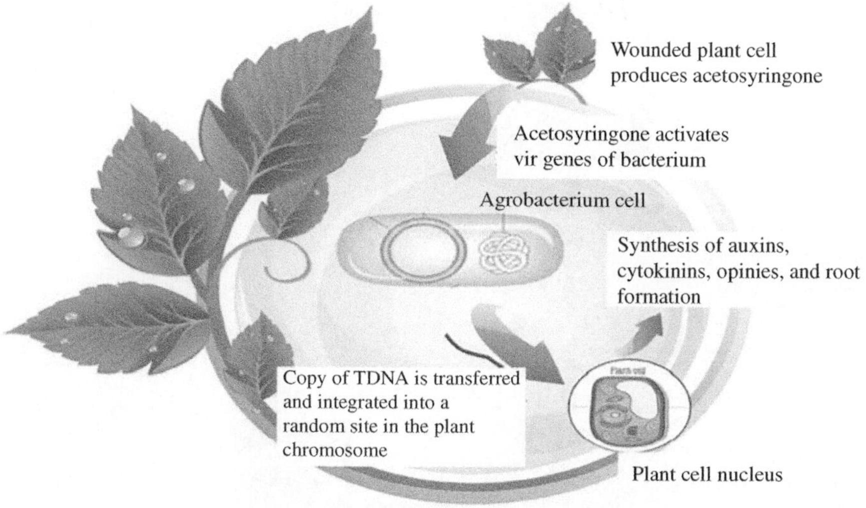

Figure 11.3 Interaction of *Agrobacterium rhizogenes* with plant roots.

11.8 Large-Scale Production of Secondary Metabolites

For the optimum production of specialized metabolites, certain plant natural biosynthesis pathways are required, which are expressed through novel genetic methods (Gutierrez-Valdes et al. 2020). Bioprocessing in the form of suspension culture, organ culture, or transformed hairy root culture has been successful in scaling up secondary metabolite production. For the enhancement in yield of the production of metabolite of interest, or, for producing novel metabolites, metabolic engineering has been successfully found to be useful (Gandhi et al. 2015). Hairy roots culture produces and are found to secrete complex active glycoproteins from a huge spectrum of organisms, which has positioned hairy root platforms as major biotechnological tools. Due to the advancement in the field of production of secondary metabolites by using large-scale bioreactors, these culture systems are used at a large scale in different industries, e.g. pharmaceutical, cosmetics, and food sectors, etc. (Gutierrez-Valdes et al. 2020).

The major motto of the hairy root culture system is the production of secondary metabolites at an industrial extent. Small amounts of hairy roots in a medium without any plant growth regulator grows fast. Due to mechanical disturbance, there could be changes in the morphology and growth of a root thus forming a callus. For continuous and large-scale production, well-maintained equipment and an appropriate batch fermentation process are essential. For unimportant products, the inoculation of a packed bed of roots could be beneficial to start the reactor in an endless stable process (Rawat et al. 2019). Hairy root cultures are the potential source for the production of secondary metabolites as they are able to produce these metabolites for a longer period of time more than cell or callus cultures as they have the important features like genetic stability, and production of high biomass and have an efficient biosynthetic capacity with high growth rate in hormone-free media. Thus, due to these roles, hairy root culture provides promising approach to the biotechnological exploitation of plant cell culture for fine chemical production commercially (Guillon et al. 2008; Gurunani et al. 2015; Häkkinen and Oksman-Caldentey 2018; Rawat et al. 2019). Figure 11.4

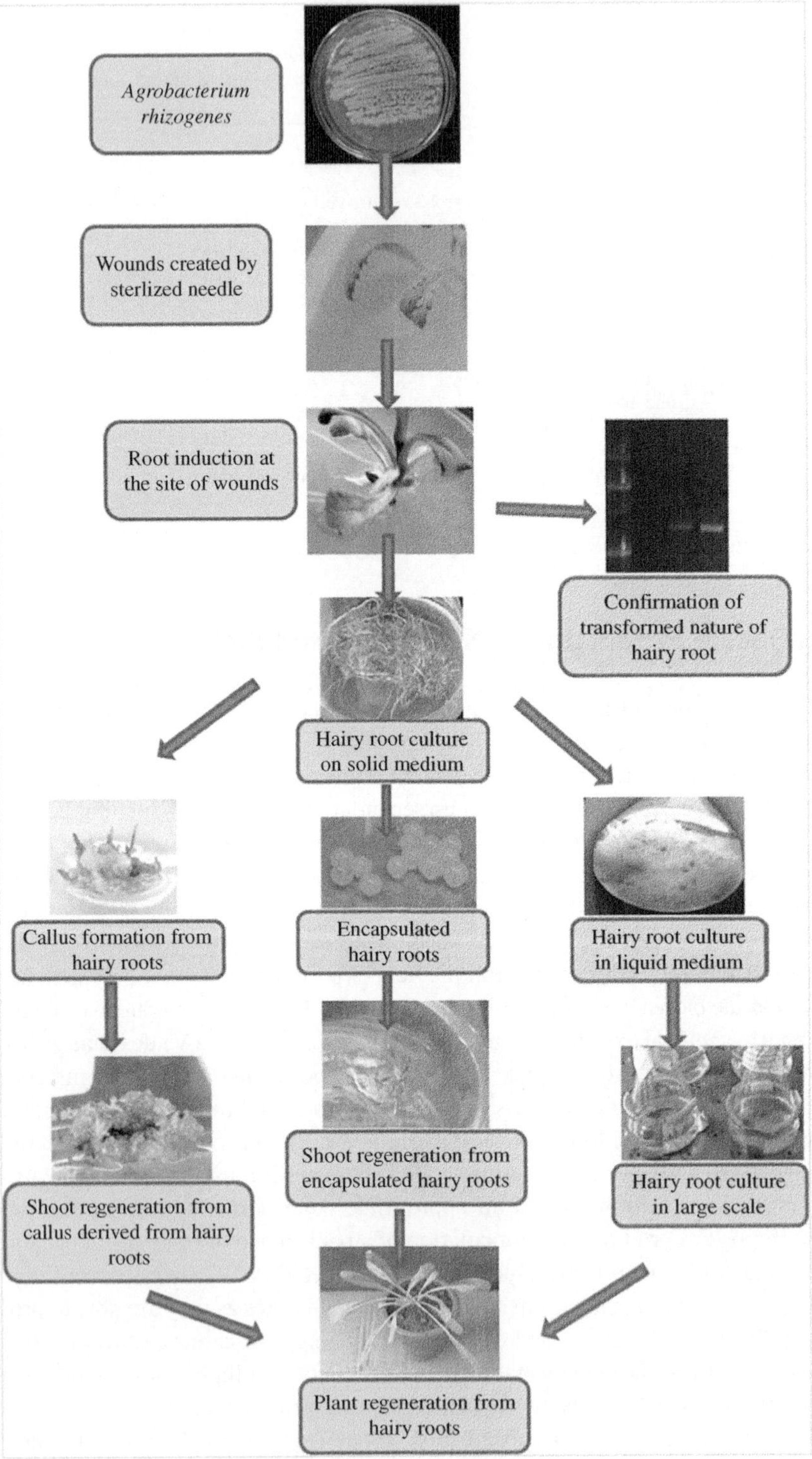

Figure 11.4 Complete protocol of hairy root induction and its biotechnological application courtesy. Source: Rawat et al. (2019), ELSEVIER.

depicts the complete protocol of hairy root induction and its biotechnological application. By using nontransgenic hairy root cultures, specialized metabolites are produced as the plants selected are capable to produce the output naturally. For increasing the amount of a particular secondary metabolites, a transgene can also be used in "wild-type" hairy roots (Gutierrez-Valdes et al. 2020). The root cultures require supply of phytohormones and grow at a very slow rate resulting in a poor or negligible amount of secondary metabolite synthesis. However, when the hairy root cultures are introduced, the plant tissue culture technique is the best accepted technique as a very good alternative approach for secondary metabolite production. Hairy roots that are genetically transformed are characterized by their very fast growth. The commercial production of hairy root cultures is economically feasible as they can produce more than one type of secondary metabolite. Many plant species are there that are responsible for the large-scale production of secondary metabolites (Rawat et al. 2019). Progresses made in the scaling-up of the hairy root cultures has opened the way for industrial exploitation of this system. Biotechnological approaches including hairy root cultures have been reported for the production of anticancerous compounds like resveratrol, podophyllotoxin, and zerumbone. To make the production of secondary metabolites feasible, development of large-scale culture methods using bioreactors has played an important role. Hairy root culture system in several plants for *in vitro* production of secondary metabolites is one of the methods of choice and a promising way to produce secondary metabolites at the commercial level so as to make it an economically viable proposition for *in vitro* production (Singh et al. 2018). Table 11.3 provides examples of plant hairy root cultures and secondary metabolites produced by them.

Certain properties such as genotypic and phenotypic stability, biochemical stability, autotrophy in plant hormones and fast growth by addition of elicitors, high levels of secondary metabolites production have led hairy roots to increasing adoption in scientific studies. Hairy root cultures offer high production and productivity of valuable secondary metabolites. For commercial exploitation of hairy root cultivations system, the development and scaling up of appropriate reactor vessels (bioreactors) that permit the growth of interconnected tissues normally unevenly distributed throughout the vessel is focussed primarily. High stability and productivity of the hairy root cultures have been a valuable biotechnological tool for the production of plant secondary metabolites. Additionally, elicitation methods can also be used to further enhance their accumulation in both small- and large-scale production (Gurunani et al. 2015).

11.9 Strategies Used *In vitro*

The inception of *in vitro* technology has been used to produce secondary metabolites due to the recognition of capacity for plant cell, tissue, and organ cultures. As in today's modern days, there is a huge and growing demand for natural, renewable products that has strongly gained attention on *in vitro* production of plant secondary phytochemical products making a way out for new research for *in vitro* secondary product expression. The advantage of *in vitro* production of secondary metabolites in plant cell culture, rather than *in vivo* as by using *in vitro* method for production can be more reliable, simpler, and more predictable. Also, the isolation of the phytochemical can be rapid and efficient, when compared with

Table 11.3 Examples of plant hairy root cultures and secondary metabolites produced.

S. No.	Plant name	Plants secondary metabolites produced
1.	*Atropa belladonna*	Scopolamine
2.	*Ammi majus*	Furanocoumarins (psoralen, xanthotoxine, bergapten and imperatorin)
3.	*Brugmansia candida*	Tropane alkaloids
4.	*Glycyrrhiza glabra*	Licoagrodin, glycyrrhizin, and isoliquiritigenin
5.	*Ophiorrhiza pumila*	Camptothecin
6.	*Panax ginseng*	Ginsenosid
7.	*Withania somnifera* L.	Withaferin-A
8.	*Beta vulgaris*	Betalain
9.	*Echinacea purpurea*	Cichoric acid
10.	*Fagopyrum esculentum M*	Rutin
11.	*Silybum marianum L*	Silymarin
12.	*Rauvolfa serpentine*	Vomilenine, reserpine
13.	*Angelica gigas Nakai*	Pyranocoumarins
14.	*Arnebia hispidissima*	Shikonin
15.	*Ophiorrhiza alata Craib*	Camptothecin
16.	*Centella asiatica*	Madecassoside, asiaticoside, madecassic acid, and asiatic acid
17.	*Tripterygium wilfordii Hook. F*	Triptolide and Wilforine
19.	*Linum usitatissimum*	Lignan
20.	*Arachis hypogaea*	Resveratrol, Piceatannol, Arachidin-1, Arachidin-3

extraction from complex whole plants. The intermediate compounds interfering in the growth of the field-grown plant can be avoided in cell cultures. Phytochemicals are produced in large volumes through tissue and cell cultures. The cultures, tissue, and cell cultures are a potential model to test elicitation (Hussain et al. 2012).

According to a study conducted by Ghorpade et al. (2011), plant tissue culture and suspension culture are used at a large scale for the production of secondary metabolites. Several plant cells cultures, e.g. that of *Berberis willsoniae, Coptis japonica*, and *Lithospermum erythrorhizon*, etc. produce secondary metabolites in a huge amount while various plant species have been found to be not able to produce a significant amount of secondary metabolites in suspension culture like that of *Atropa belladonna, Duboisia leichhardtii, Cinchona ledgeriana, Digitalis lanata,* etc. (Fazili et al. 2022).

For *in vitro* production of secondary metabolites, hairy root cultures system is one of the methods of choice and commercially important. *In vitro* techniques, when compared to the conventional cultivation of medicinal plants in greenhouse or in the field, have the advantages of producing plant-derived medicinal compounds (PDMC) in sterile and controlled environmental conditions, allowing better control of the developmental processes, such as organogenesis, and producing tissues with high PDMC contents, due to the efficient use of different biotic and abiotic elicitors. *In vitro* culture involves the production of new cells, tissues, and organs derived exclusively from the mitotic cell division, thus generating cloned cells, tissues, and individuals. The use of *in vitro* techniques to produce secondary metabolites has the advantages such as less environmental interference due to the controlled conditions in *in vitro* growing rooms; possibility of more control over the PDMC production by developing step-by-step methods to accelerate and increase PDMC concentration in the tissues. *In vitro* production consists of growing explants from the target medicinal plant. Explants first are surface sterilized and inoculated on *in vitro* conditions, using a previously formulated culture medium, and cultivated for different phases under controlled environmental conditions. The media used contains organo-mineral formulated solutions to nurture the tissues; sugars, such as sucrose, to act as the energy source to a system (*in vitro* conditions) in which light conditions and low CO_2 concentrations severely limit photosynthesis. *In vitro* phase called organogenesis, in which specialized cells resume their capacity of cell division and develop callus (indirect organogenesis) or buds and shoots directly from the original tissue (direct organogenesis) by explants, such as leaf segments, *in vitro* micropropagation, followed by plantlet acclimatization and growth in the field or greenhouse, is a strategy for production of plant secondary metabolites (Cardoso et al. 2019).

11.9.1 Why Hairy Root Culture?

Due to the low yields of the desirable compounds, secondary metabolites production gets affected. Also, availability of the limited knowledge of biosynthetic pathways of the secondary metabolites production and the processes involved in the extraction and collection of these industrially important compounds affects the production yield. Chandra and Chandra 2011; *in their study have talked about* Lupin (legume) cell cultures that failed to produce alkaloids (secondary metabolites) in desirable amounts as the chloroplasts of leaf is the site of biosynthesis of alkaloids and are collected or accumulated in epidermal cells. It has been studied that transformed roots which is hairy roots, have been focussed by researchers from past two decades, using differentiated cultures instead of cell suspension cultures (Chandra and Chandra 2011).

11.10 Plants as Bioreactors

For a long period of time, thousands of plant species are used for medicines and most of the world population uses them to cure acute and chronic health problems. The use of transgenic plants as "Bioreactors" is relatively a new bioscience involving the genetic modification of the host plant through the insertion and expression of new genes. For the production of high-value secondary metabolites of industrial importance, plant cell and transgenic hairy root cultures are being used as a promising potential alternative source (Rao and Ravishankar 2002; Sharma and Sharma 2009). Plant tissue culture is the field in which the growth and propagation of plant cells, tissue or organs is done *in vitro*, i.e. an indispensable tool of plant science research and plant biotechnology, having its potential applications as the mass production of plant-derived pharmaceuticals in bioreactors in a way similar to antibiotic production by microbial fermentation. For the mass production of secondary metabolites, plant tissue culture is the feasible method for rare and slow-growing plant species. For applying the plant tissue and cell cultures technique at large scale an efficient bioreactor design and operation are needed. Plant cells in cultures have different morphological and growth characteristics from those of microorganisms, such as their relatively large size, formation of large aggregates, sensitivity to shear stress, and slow growth (Zhou and Wu 2006; Yue et al. 2016).

In recent years, plant bioreactors have flourished into an exciting field of synthetic biology because of their extremely important features such as product safety, inexpensive production cost, and easy scale-up (Yang et al. 2021). Development of large-scale culture methods using bioreactors has made production of secondary metabolites easy at the industrial scale (Gutierrez-Valdes et al. 2020). Plant bioreactors are the chemical factories in which genetically modified crops are cultivated to produce biological agents with potential commercial values, such as vaccine antigens, antibodies, nutritional supplements, industrial enzymes, medical diagnostics proteins, industrial and pharmaceutical proteins, nutritional supplements like minerals, vitamins, carbohydrates, and biopolymers. Naturalized bioreactors are used for the production of protein products by having in depth knowledge of plant molecular biology and gene engineering (Sharma and Sharma 2009; Yue et al. 2016; Yang et al. 2021). Strategies such as reactor design, cell immobilization, and enzyme elicitation as well as characterizing cellular heterogeneity and variability through flow cytometric techniques, metabolic engineering, and system-wide analysis in which recent researches are directed are applied (Kolewe et al. 2008).

Biotechnology is a diverse field that offers an opportunity to exploit the cell, tissue, organ, or entire organism by growing them *in vitro* and to genetically manipulate them for getting desired compounds (Rao and Ravishankar 2002). In recent years, the use of plants as bioreactors has emerged as an exciting area of research and significant advances have created new opportunities. Low production cost, product safety, and easy scale up are serving as the major forces behind the rapid growth of plant bioreactors. Augmenting tissue-specific transcription, elevating transcript stability, tissue-specific targeting, translation optimization, and subcellular accumulation are some of the strategies employed (Sharma and Sharma 2009). Development of large-scale culture methods using bioreactors has made production of secondary metabolites easy at the industrial scale. For hairy root cultures, gas phase bioreactors and liquid phase bioreactors are used for the culture of plant tissues. These bioreactors are adapted from classical bioreactors. Roots are submerged in the liquid medium in liquid-phase reactors, e.g. stirred tank and airlift and bubble column reactors. Due to its simple design and construct and low risk of contamination with low maintenance are the advantages of this type of bioreactors. The classical downstream processing for the retrieval of biosynthetic compounds produced takes place. The compounds produced are secreted into the media that facilitate the downstream processing. Hairy root cultures system serves as a tool for the study of genes function, studying phytoremediation and molecular breeding as well (Gutierrez-Valdes et al. 2020). New advances in bioreactor systems offer the opportunities for the scaling up of the cultivation of transformed roots from small scale to large scale in an industrial process (Guillon et al. 2008). Since the production of secondary metabolites at a large scale is limited due to several environmental factors such as low propagation rates and restricted cultivation, the use of bioreactors is the last step in the expansion of techniques for secondary metabolite production from *in vitro* plant systems. Chemical and physical parameter should be taken into consideration for a proper development in a bioreactor system. The major chemical factor for scale-up is the nutrient availability which can be maintained with rejuvenated medium. For an effective scale-up in bioreactors, the major issues to be considered are the morphology, physiology, and elevated stress sensitivity of hairy roots. Several difficulties such as the excessive branching of hairy roots can display a resistance to flow since they form an interlocked matrix. Also, one difficulty of scaling up is the delivery of nutrients in gas and liquid phases simultaneously. In liquid medium, the meristem growth of root cultures causes a root ball with young roots growing on the outside and inside older tissues' core. The supply of oxygen creates a major difficulty of hairy roots, that when delivered towards the central mass of tissue, it arises in a pocket of dead tissue; that creates a gap, and a continuous supply of oxygen is needed in the bioreactor. Then, mixing process is applied for the supply of dissolved oxygen and taking out the carbon dioxide. The types of bioreactors applied in plant cell fermentation are stirred-tank bioreactor, bubble column bioreactor and airlift bioreactor. Stirred-tank bioreactors (STB) are commonly used because of advantages: ease of large-scale production, use for high viscously cell culture, high oxygen mass transfer ability and good fluid mixing and alternative impellers (Yue et al. 2016; Angulo-Bejarano et al. 2019).

Plant cells in liquid suspension offer a unique combination of physical and chemical environments that must be accommodated in large-scale bioreactor process. Due to the advances in scale-up approaches and immobilization techniques, a considerable increase in the number of applications of plant cell cultures for the production of compounds with a high added value is present (Hussain et al. 2012). Large-scale culturing of plant cells has been conducted in a variety of bioreactors including standard STR, bubble column bioreactors (BCB), and air-life bioreactors (ALB). The most widely exploited bioreactor is STR, that provides relatively easy scale-up, good fluid mixing and oxygen transfer ability. Bioreactors are distinct in their modes of operation with batch culture systems; typically used in large-scale production and often not cost-effective as repeated sterilization, filling and cleaning of the system vessel is required. These factors drove the development of more advanced culture systems such as: fed-back, two-stage, perfusion, semi-continuous, and continuous cultures. As plant cells can be reused, it maximizes the turnaround time between batches significantly thus increasing productivity. Various types of disposable bioreactors are also developed i.e. wave-mixed, slug bubble, plastic-lined, stirred bag orbital shaken or stirred bioreactors. Wave bioreactors according to studies represents the most successful type of disposable bioreactor. These bioreactors consist of an inflated bag placed on a rocking platform that induces wave action within the liquid medium. Different plant cell species, e.g. *Vitis vinifera*, *Malus domestica*, *Nicotiana tabacum* (BY-2) and hairy root cultures i.e. *Harpagophytum procumbens*, *Hyoscyamus muticus* and *P. ginseng* are used (Ochoa-Villarreal et al. 2016).

11.11 A Case Study

According to the study conducted by Ghorpade et al. (2011); it was studied that a plant cell, tissue, and organ culture have an inherent capacity to manufacture valuable chemical compounds as the parent plant does in nature. *In vitro* plant materials were found to be one of the good sources for the production of secondary metabolites, and elicitation was considered to be used as one of the important tools in order for the improvement and enhancement of the synthesis of these compounds. In a variety of plant cell cultures, it has been observed that elicitors have increased production of terpenoid indole alkaloids, isoflavonoid phytoalexins, serquiterpenoid phytoalexin, coumarins, etc. Although elicitation has been carried out in large number of medicinal plants, Ghorpade et al. (2011); studied it in *Catharanthus roseus*, because of its anticancerous properties and production of anticancer compound called vinblastine (VLB) and vincristine (VCR). The use of elicitor increases the production of phytochemicals (secondary metabolites) to meet the market demands, for reducing production costs and for in-depth investigation of biochemical and metabolic pathways. This information has helped in manipulation of biosynthetic pathways, which can be used as a powerful tool to make natural product-like compounds. For the commercial production of secondary metabolites, *in vitro* studies like plant tissues cultures and suspension cultures are used (Ghorpade et al. 2011). Different media have been employed for different species, and the use of biotic and abiotic elicitors has also been engaged because of their strong and rapid improving effects on indole alkaloid synthesis. In *C. roseus*, enhanced alkaloid biosynthesis in suspension culture was reported with the basic prerequisite in all this, a good yield of the compound, and cheap cost compared to the natural synthesis by the plants (Zhao et al. 2001). Elicitation was observed to be used as one of the important strategies in order to get better productivity of the bioactive secondary and reducing production costs. There are reports where elicitors have increased production of isoflavonoid phytoalexins, serquiterpenoid phytoalexin, podophyllotoxin, azadirachtin, terpenoids, hypericins (hypericin and pseudohypericin), and hyperforin and resistance to pathogens. Besides that, elicitor has been found to be used for the study of the chemical nature and signaling pathways of natural substances discharged by microbes, herbivores, and plants during pathogenic contamination, herbivory, symbiosis, and allelopathic interactions (Siddiqui et al. 2013). Figure 11.5 shows a study conducted by (Sujathaa et al. 2013) for hairy root culture of *Artemisia vulgaris*.

11.12 Conclusion

Plants have always been exploited for many years for synthesizing a large quantity of complex chemicals having various industrial applications such as in the field of medicine, dyes, fragrances, and flavors. Many large industries are dependent on compounds produced by plants, and many regulatory agencies have stopped giving permission to the industries for the addition of synthetic colors or flavors in the food industry. Plant produced metabolites when consumed show certain health benefits. The demand for natural molecule for the consumption has been increased thus creating the need for the development of technological aspects for plant secondary metabolites production, i.e. through plant cell cultue, plant organ culture, and plant hairy roots for secondary metabolite production.

Plant hairy roots are the major source of secondary metabolite production having the properties of high genetic stability and unlimited propagation *in vitro*. Hairy roots are formed by the *A. rhizogenes* transformed roots infected with this bacterium which is T-DNA mediated gene transfer and hairy roots formation. Therefore, hairy roots produce a vast

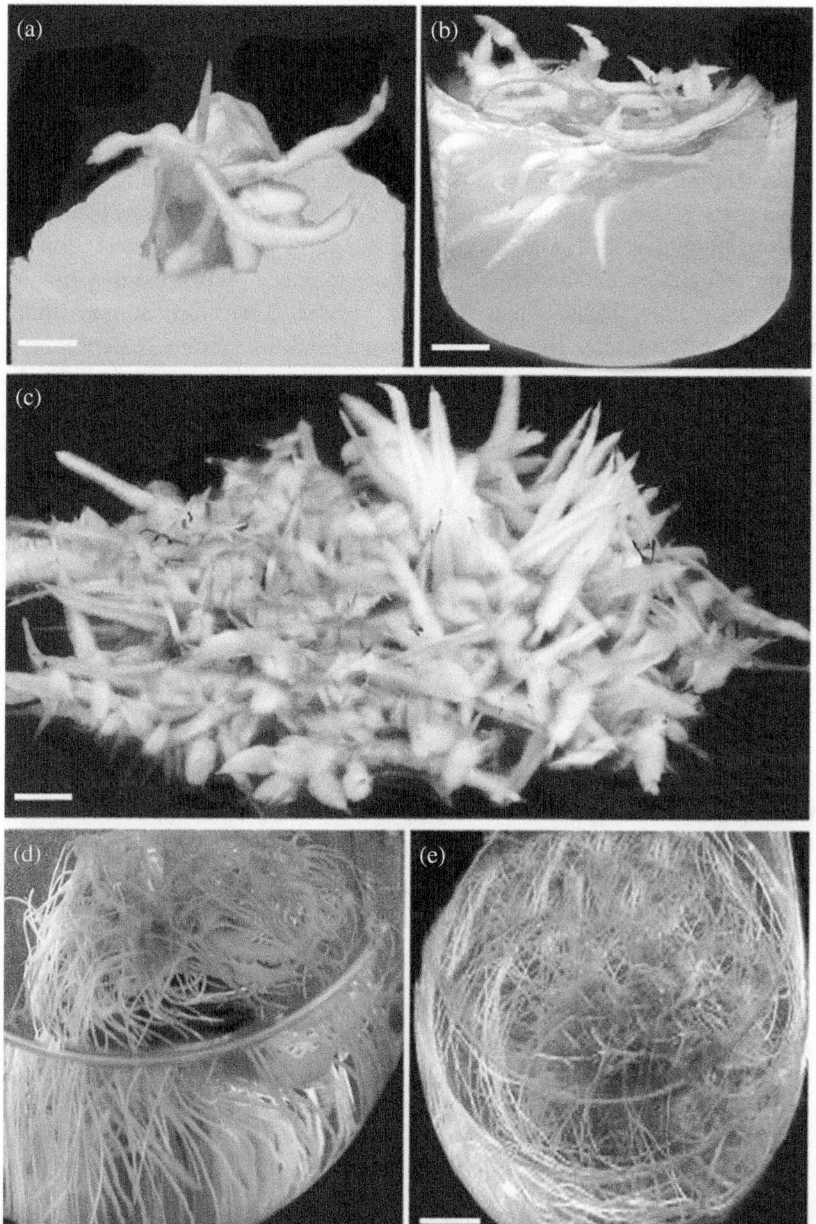

Figure 11.5 Hairy root culture of Artemisia vulgaris. (a) Hairy root initiation from leaf explant cultured on MS + 500 mg/l cefotaxime medium (after two weeks of co-culture) (bar = 2 mm). (b) Hairy root proliferation (after six weeks of culture) (bar = 2 mm). (c) Biomass accumulation of hairy roots cultured on MS basal medium supplemented with 3% sucrose + 0.7% agar and without antibiotic (after 12 weeks) (bar = 10 mm). (d) Growth of hairy roots in ½ MS liquid medium (after seven days of culture) (bar = 5 mm). (e) Proliferative biomass accumulation of hairy roots cultured on ½ MS liquid medium (after 30 days) (bar = 5 mm). Source: Sujathaa et al. (2013), ELSEVIER.

scope for the experimental systems and gene functions and as a research tool for the production of metabolites at a large scale in a controlled environment providing *in vitro* conditions. There has been need for the enhancement of secondary metabolites production, due to limited availability of products obtained thorugh certain culture methods. For this, Elicitation process is being used that increases and enhances the secondary metabolites production. Biotic and abiotic elicitors are used. The enzymes present in the culture system protects the product from getting degraded. This process of elicitation also helps to adsorb the product from culture media. T-DNA activation tagging allows increased production of secondary metabolite in hairy roots. There has been a need to develop a process to generate viable whole plants from hairy roots, a feat that has proven to be more difficult than expected which will serve as the most useful genetic transformation system. Different biotechnological tools such as screening and selection, precursor feeding, optimization of culture conditions and chemical constituents of culture medium, application of elicitor molecules, and metabolic engineering can improve production and yield of secondary metabolites in hairy root cultures are found to be potential for biosynthesizing valuable metabolites of transformed hairy root cultures.

Bioreactors increases the production of plant secondary metabolites at a large scale. They have been found to be involved in the genetic modification of the host plant through the insertion and expression of new genes. Various products, e.g. bioactive peptides, vaccine antigens, antibodies, diagnostic proteins, nutritional supplements, enzymes, and biodegradable plastics have been produced in plants. Bioreactors have applications such as animal proteins synthesis, cost benefits, lack of pathogen contamination, and easy scale up that increases the production of plants secondary metabolites. Plants are used as bioreactors and have emerged as an exciting area of research and significant advances, due to low production cost, product safety, and easy scale up. Therefore, plant hairy roots culture systems is a potential method of secondary metabolites production, which is a rapid, simple, and highly efficient *A. rhizogenes*-mediated hairy root transformation system.

References

Abraham, J. and Thomas, T. D. (2017). Hairy root culture for the production of useful secondary metabolites, *Biotechnology and Production of Anti-cancer compounds*. Springer International Publishing, pp: 201–230. doi: https://doi.org/10.1007/978-3-319-53880-8_9.

Acharya, S.K., Hedda, G.V., Kankariya, A.J. and Tamhane, V.A. (2021). Hairy roots of 'dashmula' plant Uraria picta as a promising alternative to its medicinally valued true roots – functional and metabolomic analysis. *Plant Cell, Tissue and Organ Culture (PCTOC)*, Springer Nature, pages 533–544. https://doi.org/10.1007/s11240-021-02024-3.

Alok, A., Kumar, J., and Upadhyay, S.K. (2018). Engineering in hairy roots using CRISPR/Cas9-mediated editing. In: *Hairy Roots* (ed. V. Srivastava, S. Mehrotra and S. Mishra). Singapore: Springer. https://doi.org/10.1007/978-981-13-2562-5_15.

Angulo-Bejarano, P.I., Jimenez, J.L.D.L.F., Paul, S., Donato-Capote, M.D., Castillo-Maldonado, I., Betanzos-Cabrera, G., Valiente-Banuet, J.I. and Sharma, A. (2019). Cell cultures and hairy roots as platform for production of high-value metabolites: current approaches, limitations, and future prospects, *Advances in Plant Transgenics: Methods and Applications*, 23–57. https://doi.org/10.1007/978-981-13-9624-3_2.

Badyal, S., Singh, H., Yadav, A.K. et al. (2020). Plant secondary metabolites and their uses. *Plant Archives* 20 (S2): 3336–3340.

Bhatia, S. and Bera, T. (2015). Classical and nonclassical techniques for secondary metabolite production in plant cell culture. *Modern Applications of Plant Biotechnology in Pharmaceutical Sciences* 231–291.

Cardoso, J.C., de Oliveira, M.E.B., and Cardoso, F.D.C. (2019). Advances and challenges on the *in vitro* production of secondary metabolites from medicinal plants. *Horticultura Brasileira* 37: 124–132. https://doi.org/10.1590/S0102-053620190201.

Chandra, S. and Chandra, R. (2011). Engineering secondary metabolite production in hairy roots. *Phytochemistry Reviews* 10: 371–395.

Chandran, H., Meena, M., Barupal, T., and Sharma, K. (2020). Plant tissue culture as a perpetual source for production of industrially important bioactive compounds. *Biotechnology Reports* 26: e00450. p-1-7.

Cheng, Y., Wang, X., Cao, L. et al. (2021). Highly efcient *Agrobacterium rhizogenes*-mediated hairy root transformation for gene functional and gene editing analysis in soybean. *Plant Methods* 17: 73.

Dixit, S., Shukla, A., Singh, V., and Upadhyay, S.K. (2021). Bioprospecting of natural compounds for industrial and medical applications. In: *Bioprospecting of Plant Biodiversity for Industrial Molecules* (ed. S.K. Upadhyay and S.P. Singh). John Wiley & Sons Ltd. https://doi.org/10.1002/9781119718017.ch3.

Fazili, M.A., Bashir, I., Ahmad, M. et al. (2022). In vitro strategies for the enhancement of secondary metabolite production in plants: a review. *Bulletin of the National Research Centre* 46: 35.

Gandhi, S.G., Mahajan, V., and Bedi, Y.S. (2015). Changing trends in biotechnology of secondary metabolism in medicinal and aromatic plants. *Planta* 241: 303–317.

Gantait, S. and Mukherjee, E. (2020). Hairy root culture technology: applications, constraints and prospect. *Applied Microbiology and Biotechnology*, 105, pages 35–53. https://doi.org/10.1007/s00253-020-11017-9.

Ghorpade, R.P., Chopra, A., and Nikam, T.D. (2011). Infuence of biotic and abiotic ellicitors on four major isomers of boswellic acid in callus culture of "Boswellia serrata" Roxb. *Plant Omics* 4: 169.

Guerriero, G., Berni, R., Muñoz-Sanchez, J.A. et al. (2018). Production of plant secondary metabolites: examples, tips and suggestions for biotechnologists. *Genes* 9: 309. https://doi.org/10.3390/genes9060309.

Guillon, S., Trémouillaux-Guiller, J., Pati, P.K., and Gantet, P. (2008). Chapter 14 Hairy roots: a powerful tool for plant biotechnological advances. In: *Bioactive Molecules and Medicinal Plants* (ed. K. Ramawat and J. Merillon), 271–282.

Gurunani, S.G., Yeole, M.P., Gholse, Y.N., and Chaple, D.R. (2015). Hairy root culture: an experimental system for secondary metabolite production. *Research Journal of Pharmaceuticals and Technology* 8 (6): 728–730.

Gutierrez-Valdes, N., Häkkinen, T.S., Lemasson, C. et al. (2020). Hairy root cultures—a versatile tool with multiple applications. *Frontiers in Plant Science* 11: 33.

Habibi, P., Soccol, C.R., and Grossi-de-Sa, M.F. (2018). Hairy root-mediated biotransformation: recent advances and exciting prospects. In: *Hairy Roots* (ed. V. Srivastava, S. Mehrotra and S. Mishra). Springer Nature Singapore Pte Ltd. https://doi.org/10.1007/978-981-13-2562-5_8.

Halder, M., Majumder, A., Ray, S., and Jha, S. (2021). Medicinal plant research at crossroads: biotechnological approaches for conservation, production and stability in tissue cultures and regenerated plants. In: *Medicinal Plants, Sustainable Development and Biodiversity*, (28) (ed. H.M. Ekiert, K.G. Ramawat and J. Arora), 459–554. Cham: *Springer* https://doi.org/10.1007/978-3-030-74779-4_1.

Häkkinen, S.T. and Oksman-Caldentey, K.M. (2018). Chapter 1 Progress and prospects of hairy root research. In: *Hairy Roots* (ed. V. Srivastava, S. Mehrotra and S. Mishra), 3–19. Springer Nature Singapore Pte Ltd.

Hussain, M.S., Fareed, S., Ansari, S. et al. (2012). Current approaches toward production of secondary plant metabolites. *Journal of Pharmacy & Bioallied Sciences* 4: 10–20.

Karuppusamy, S. (2009). A review on trends in production of secondary metabolites from higher plants by *in vitro* tissue, organ and cell cultures. *Journal of Medicinal Plant Research* 3 (13): 1222–1239.

Khatodia, S. and Biswas, B. (2014). A comparative study of hairy root culture induction efficiency in four medicinally important plants using *Agrobacterium rhizogenes*. *International Journal of Current Microbiology and Applied Sciences* 3 (5): 625–633.

Kolewe, M.E., Gaurav, V., and Roberts, S.C. (2008). Pharmaceutically active natural product synthesis and supply via plant cell culture technology. *Molecular Pharmaceutics* 5 (2): 243–256.

Korde, V.V., Dhas, S.S., and Gurave, N.A. (2016). Hairy root culture: a promising approach in biotransformation. *Asian Journal of Plant Science and Research* 6 (4): 6–11.

Kowalczyk, T., Merecz-Sadowska, A., Picot, L. et al. (2022). Genetic manipulation and bioreactor culture of plants as a tool for industry and its applications. *Molecules* 27 (795): 1–28.

Makhzoum, A.B., Sharma, P., Bernards, M.A., and Trémouillaux-Guiller, J. (2013). Hairy roots: an ideal platform for transgenic plant production and other promising applications. *Recent Advances in Phytochemistry* 42: 95–137.

Mishra, B.N. and Ranjan, R. (2008). Growth of hairy-root cultures in various bioreactors for the production of secondary metabolites. *Biotechnology and Applied Biochemistry* 49 (1): https://doi.org/10.1042/ba20070103.

Mohammed, S. A. H., Tripathi, R. and Sreejith, K. (2021). Plant metabolites: methods for isolation, purification, and characterization. In: *Plant Metabolites: Methods, Applications and Prospects* Editors: Swapna Thacheril Sukumaran, Sabu Abdulhameed, Shiburaj Sugathan Springer Nature Singapore Pte Ltd. 23–56. doi: https://doi.org/10.1007/978-981-15-5136-9_2.

Ochoa-Villarreal, M., Howat, S., Hong, S.M. et al. (2016). Plant cell culture strategies for the production of natural products. *BMB Reports* 49 (3): 149–158.

Rao, S.R. and Ravishankar, G.A. (2002). Plant cell cultures: chemical factories of secondary metabolites. *Biotechnology Advances* 20: 101–153.

Rawat, J.M., Bhandari, A., Raturi, M., and Rawat, B. (2019). *Agrobacterium rhizogenes* mediated hairy root cultures. *New and Future Developments in Microbial, Biotechnology and Bioengineering* 103–118. https://doi.org/10.1016/B978-0-444-63504-4.00008-6.

Ron, M., Kajala, K., Pauluzzi, G. et al. (2014). Hairy root transformation using *Agrobacterium rhizogenes* as a tool for exploring cell type-specific gene expression and function using tomato as a model. *Plant Physiology* 166: 455–469.

Shanks, V. and Morgan, J. (1999). Plant 'hairy root' culture. *Current Opinion in Biotechnology* 10 (2): 151–155.

Sharma, A.K. and Sharma, M.K. (2009). Plants as bioreactors: recent developments and emerging opportunities. *Biotechnology Advances* 27: 811–832.

Siddiqui, Z.H., Mujib, A., Aslam, M.J. et al. (2013). *In Vitro* production of secondary metabolites using elicitor in *Catharanthus roseus*: a case study. *Crop Improvement* 401–418. https://doi.org/10.1007/978-1-4614-7028-1_14.

Singh, R.S., Chattopadhyay, T., Thakur, D. et al. (2018). Hairy root culture for In Vitro production of secondary metabolites: a promising biotechnological approach. In: *Biotechnological Approaches for Medicinal and Aromatic Plants* (ed. N. Kumar), 235–249. Springer Nature.

Srivastava, M., Sharma, S., and Misra, P. (2016). Elicitation based enhancement of secondary metabolites in *Rauwolfia serpentina* and *Solanum khasianum* hairy root cultures. *Pharmacognosy Magazine* 12 (46): S315–S320.

Sujathaa, G., Zdravkovic-Kora, S., Cali, D. et al. (2013). High-efficiency *Agrobacterium rhizogenes*-mediated genetic transformation in Artemisia vulgaris: hairy root production and essential oil analysis. *Industrial Crops and Products* 44: 643–652.

Supriya, R., Kala, R.G., and Thulaseedharan, A. (2020). Hairy root culture: secondary metabolite production in a biotechnological perspective. *Plant Metabolites: Methods, Applications and Prospects* 89–110. https://doi.org/10.1007/978-981-15-5136-9_5.

Sushmita, K.G., Upadhyay, S.K., and Verma, P.C. (2021). An overview of genome-engineering methods. In: *Genome Engineering for Crop Improvement* (ed. S.K. Upadhyay), 1–21. John Wiley & Sons Ltd. https://doi.org/10.1002/9781119672425.ch1.

Thirumurugan, D., Cholarajan, A., Raja, S.S.S. and Vijayakumar, R. (2018). *An Introductory Chapter: Secondary Metabolites*. http://dx.doi.org/10.5772/intechopen.79766.

Upadhyay, S.K. (ed.) (2021). *Genome Engineering for Crop Improvement*. John Wiley & Sons Ltd. https://doi.org/10.1002/9781119672425.

Upadhyay S.K. and Singh S.P. (eds.) (2021). *Bioprospecting of Plant Biodiversity for Industrial Molecules*. John Wiley & Sons Ltd. https://doi.org/10.1002/9781119718017.

Verma, P.C., Trivedi, I., Singh H. et al. (2009). Efficient production of gossypol from hairy root culture of cotton (*Gossypium hirsutum*). *Current Pharmaceutical Biotechnology* 10: 691–700. https://doi.org/10.2174/138920109789542048.

Wang, S., Alseekh, S., Fernie, A.R., and Luo, J. (2019). The structure and function of major plant metabolite modifications. *Molecular Plant* 12: 899–919.

Yang, G.-L., Feng, D., Liu, Y.-T. et al. (2021). Research progress of a potential bioreactor: duckweed. *Biomolecules* 11: 93. https://doi.org/10.3390/.

Yue, W., Ming, Q.-L., Lin, B. et al. (2016). Medicinal plant cell suspension cultures: pharmaceutical applications and high-yielding strategies for the desired secondary metabolites. *Critical Reviews in Biotechnology* 36 (2): 215–232. https://doi.org/10.3109/07388551.2014.923986.

Zhao, J., Zhu, W.H., Hu, Q., and Guo, Y.Q. (2001). Compact callus cluster suspension cultures of *Catharanthus roseus* with enhanced indole alkaloid biosynthesis. *In Vitro Cellular & Developmental Biology. Plant* 37: 68.

Zhou, L.G. and Wu, J.Y. (2006). Development and application of medicinal plant tissue cultures for production of drugs and herbal medicinals in China. *Natural Product Reports* 23: 789–810.

12

Microalgae as Cell Factories for Biofuel and Bioenergetic Precursor Molecules

D. Rodríguez-Zuñiga, A. Méndez-Zavala, O. Solís-Quiroz, J.C. Montañez, L. Morales-Oyervides, and J.R. Benavente-Valdés

Department of Chemical Engineering, Autonomous University of Coahuila, Saltillo, México

12.1 Introduction

Over the last decade, energy demand has increased exponentially due to industrialization, modernization, and population growth. Conventional energy sources, such as coal, oil, and natural gas, cover 85% of the primary energy demand worldwide but are rapidly depleting and causing an increase in greenhouse gas (GHG) emissions. Therefore, the care and preservation of the environment have become a topic of concern and debate, forcing a radical change in the global energy economy to meet the demand for energy supply for the ever-growing population. In this regard, renewable energies are expected to account for nearly 95% of the increase in global energy capacity by 2026 (IEA 2020). Bioenergy refers to renewable electricity, thermal energy, or fuels generated from organic matter; among alternative renewable energies, microalgae-based bioenergy has recently become an even more important alternative.

Microalgae are microscopic (20–200 μm) photosynthetic eukaryotic organisms, considered candidates for renewable energy production and are characterized by high yields, integration with waste streams, and the ability to grow on nonarable land (Yun et al. 2020). Thus, microalgae do not compete with food production; they can capture CO_2 up to 1.7–1.8 times their weight (dry basis) due to their high photosynthetic capacity. Moreover, they mitigate climate change significantly by consuming CO_2 through photosynthesis, given its ability to efficiently convert solar energy and atmospheric carbon dioxide into chemical compounds. Microalgae have the potential to produce up to 60% of their biomass in the form of oil or carbohydrates, from which liquid biofuels such as biodiesel, bioethanol, and gaseous fuels such as biohydrogen (H_2) and biomethane can be obtained (Ahmad et al. 2022).

Microalgae are commercially grown in closed photobioreactors (PBRs) and open raceway ponds. More than 90% of current production comes from open raceways, as they are considered more affordable than closed systems (PBR). However, they are limited by abiotic growth factors and are easily subject to contamination. Raceways are the most widely

Plants as Bioreactors for Industrial Molecules, First Edition.
Edited by Santosh Kumar Upadhyay and Sudhir P. Singh.

used systems among the open cultivation systems, thanks to their relatively low construction cost, ease of maintenance, low energy cost, and easy scalability. PBRs are typically equipped with agitation, aeration, pH control, heat exchange, and addition of medium and CO_2, which effectively control culture conditions and reduce contamination allowing obtaining high-purity biomass. Closed photobioreactors are highly specialized devices, often designed specifically for a particular species, allowing high biomass yields (Sirohi et al. 2022). This chapter examines the main microalgae that produce molecules for bioenergy and biofuels, their biosynthesis, the physical and chemical factors that promote their accumulation, and the main industrial cultivation systems.

12.2 Microalgae that Produce Bioenergy and Biofuel Molecules

Algae are an attractive feedstock to produce bioenergy and biofuel; as autotrophic organisms, algae use solar energy and fix atmospheric CO_2 into primary metabolites, including lipids, carbohydrates, and proteins. The chemical composition of microalgae biomass varies widely between 2 and 37% for lipids and 4 to 65% for carbohydrates. *Scenedesmus* sp., *Botryococcus* spp., *Chlorella* spp., *Chlorococcum* sp., and *Nanochloropsis* sp. are promising sources of lipids as they contain between 26 and 37% lipids. In contrast, *Haematococcus pluvialis, Scenedesmus dimorphus*, and *Chlorella vulgaris* contain up to 56–65% carbohydrates. Microalgae carbohydrates, mainly starch, cellulose, and different polysaccharides, are considered potential feedstocks for biofuel production, i.e., bioethanol, biohydrogen, and biogas. Most of the attention is focused on lipids, such as hydrocarbons, triacylglycerols, free fatty acids, phospholipids, sterols, glycolipids, and waxes; lipid fractions that can be converted into liquid biofuels such as biodiesel by trans etherification processes (de Carvalho Silvello et al. 2022). Table 12.1 lists the main microalgae strains used in lipid and carbohydrate production, production systems, and their bioenergy or biofuels industry applications.

Table 12.1 Summary of the main microalgae strains producing bioenergy molecules and biofuels with their production systems.

Strain	Production system	Obtained metabolites	Application	References
Chlorella sp.	Scale bubble column photobioreactor	Fatty acid	Biodiesel production	Naira et al. (2019)
Scenedesmus obliquus UTEX 393	Airlift, bubble column, and flat panel	Carbohydrate	Hydrogen production	Singh et al. (2019)
Chlorella pyrenoidosa	Closed flat panel photobioreactor (PBR)	Lipid-rich biomass	Biomethane production	Sukačová et al. (2019)
Nannochloropsis oceanica	Closely spaced flat panel	Biomass	Ethanol production	Norsker et al. (2019)

Table 12.1 (Continued)

Strain	Production system	Obtained metabolites	Application	References
Chlorella minutissima	Internal light integrated photobioreactor (ILI-PBR)	Fatty acid	Biodiesel production	Amaral et al. (2020)
Nannochloropsis gaditana	Photobioreactors (PBRs)	Fatty acid		Nogueira et al. (2020)
Scenedesmus incrassulatus	Vertical tubular photobioreactor	Lipids	Biodiesel production	Rattanapoltee et al. (2021)
Chlorella pyrenoidosa	Photobioreactor using saline wastewater	Lipids		Yang et al. (2021)
Dunaliella tertiolecta	Open raceway Pond	Lipids	Diesel fuel	Tizvir et al. (2022)
Chlorella sorokiniana	Bubble column	Biomass	Biohydrogen and biogas production	Yang et al. (2022)
Nannochloropsis oculata	Tubular photobioreactor; helical-tubular photobioreactor	Defatted algal biomass	Ethanol production	Fetyan et al. (2022) and Alami et al. (2021)
Auxenochlorella protothecoides	Tubular photobioreactors	Lipids	Biofuels production	Bolognesi et al. (2022)
Scenedesmus obliquus	Flat panel	Fatty acids	Biofuels production	Trivedi et al. (2022)
Monoraphidium sp.	Flat panel	Carbohydrate	Biofuels production	Kuravi and Mohan (2022)
Tetradesmus sp.	Flat-panel photobioreactor (FP-PBR)	Saturated fatty acid	Biofuels production	Ravi Kiran and Venkata Mohan (2022)
Chlorococcum sp.	Photobioreactor	Lipid (mainly triglycerides)	Biofuels production	Geng et al. (2022)
Phaeodactylum tricornutum	Photobioreactor	Triacylglycerides (TAGs)	Biofuels production	Morais et al. (2021)
Chlorella pyrenoidosa	Membrane photobioreactor (MPBR)	Lipids and biomass	Biofuels production	Gao et al. (2021)

Species such as *Nannochloris* sp. (oil content 56%), *Chlorella* sp. (oil content 53%), *Neochloris oleoabundans* (oil content 65%), *Scenedesmus* sp. (oil content 10–47%), and *Chlamydomonas* sp. (oil content 58.8%) have been primarily reported for the production of biofuels because of their high content of lipids and lipid composition, viz., palmitoleic (16:1), oleic (18:1), linoleic (18:2), linolenic acid (18:3), and the saturated fatty acids of palmitic (16:0) and saturated fatty acids of stearic (18:0). *Spirulina* sp., *Dunaliella* sp., *Porphyridium* sp., and *Scandesmus* sp. are used for ethanol production via fermentation or methane production via anaerobic digestion. The residual defatted (lipid-extracted) algal biomass (DAB) contains a good amount of fermentable carbon that can be used to produce various biofuels such as biohydrogen, biogas, and bioethanol (Rattanapoltee et al. 2021).

12.3 Biosynthesis of Molecules for Bioenergy and Biofuels in Microalgae

Obtaining biofuels and bioenergy derived from microalgal biomass can be defined in terms of its chemical composition, explicitly considering the cellular components that we will call "Biofuel or Bioenergetic Precursor Molecules" (BBPM): lipids, starch, and hydrogen. From these BBPM, it is possible to convert them into bioenergy and biofuels by chemical or biochemical methods. Thus, biomass rich in polysaccharides such as starch can be used for ethanol, acetone, or butanol production by fermentation or methane and hydrogen by anaerobic digestion. On the other hand, biomass with high lipid content, such as triacylglycerols (TAG), is used to produce biodiesel through transesterification.

Thermochemical conversion can convert these same lipids into bio-oils (Gouveia 2011). It is also worth mentioning that the biomass remaining from the lipid or colorant extraction process, or the biomass itself, can generate activated carbon (by pyrolysis), electricity by direct combustion, or biogas (methane) by fermentation. Figure 12.1 shows the production pathways of BBPM and the products derived from them.

It is important to establish the biochemical and physiological bases related to the biosynthesis of these BBPMs to understand the mechanisms that regulate them and establish the potential of microalgae for bioenergetic purposes. However, the simplified description of the metabolic pathways for BBPMs production is not straightforward due to the complex metabolism of microalgae that involves the interaction between several pathways and the accumulation and subsequent consumption of intermediate products and compounds. However, it is possible to estimate and predict the accumulation of such molecules by studying well known and conserved metabolic pathways such as central carbon metabolism (Johnson and Alric 2013). Therefore, in this section, the main mechanisms of BBPM biosynthesis will be briefly reviewed.

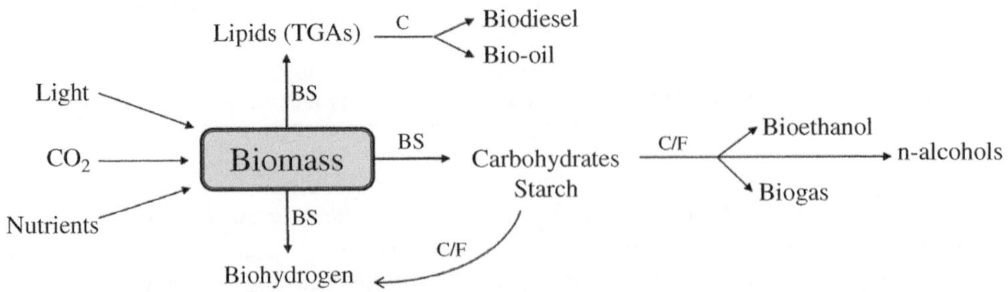

Figure 12.1 Representative scheme of the main molecules produced by microalgae used as precursors of bioenergy and biofuels (BBPM). BS = biosynthesis. C/F = chemical/fermentative process subsequent biomass recovery. C = chemical process.

12.4 Biohydrogen Production

Some microalgae produce molecular hydrogen (H_2) as part of their photosynthetic metabolism, which is described by the following equation that considers oxygenic and anoxic photosynthesis.

$$CO_2 + 2H_2A = (CH_2O) + 2A + H_2O$$

where H_2A can be H_2O, H_2, H_2S, among other compounds more reduced than water.

In some microalgae, H_2 production occurs under light conditions (biophotolysis) after an anaerobic adaptation in the dark, where it uses cellular starch by endogenous catabolism (Tsygankov and Abdullatypov 2016). Under biophotolysis, H_2 production is mediated by hydrogenase linked to ferredoxin in photosystem I, which is inhibited by oxygen concentration that increases with time of illumination. Hydrogenase is located in the thylakoid membrane and is responsible for converting water into H_2 and O_2 (Show and Lee 2014). On the other hand, in some cyanobacteria, such production has been observed under fermentative conditions in the dark mediated by nitrogenase (Benemann 2004).

Currently, the technical-economic feasibility of H_2 production by microorganisms is still under discussion (Wang and Yin 2018). An associated problem is the H_2 per mole of glucose in dark fermentation, which yields only four H_2 and two acetates for each glucose, far from the suggested yields of 12 mol of H_2 for each glucose (Benemann 2004). However, the biomass of microalgae can be used as raw material to generate hydrogen via fermentation (Wang and Yin 2018).

12.5 Starch Biosynthesis

The accumulation and endogenous consumption of starch is a conventional process in microalgae as a mechanism of environmental adaptation; it seems to prepare the microalgae to maintain its cell cycle. Under nonstress conditions, starch production is preferable by the cell compared to lipids as a carbon and energy reserve (Johnson and Alric 2013).

Its biosynthesis has been widely reported in microalgae and can be carried out under different nutritional conditions (Cheng et al. 2017; Debnath et al. 2021; De Carvalho Silvello et al. 2022). In photoautotrophy, starch is accumulated in the chloroplast during the light phase from CO_2 fixation by RuBisCo. The photoreactions allow the generation of ATP used in the Calvin cycle to form glyceraldehyde-3-phosphate (G3P) from 3-phosphoglycerate (3PGA), obtained in the photorespiration cycle. G3P is converted to starch by gluconeogenesis from glucose-6-phosphate (G6P) or fructose-1,6-phosphate (F1,6P) units (Min Tan and Lee 2016). Subsequently, the culture conditions and the presence of other carbon compounds such as acetate regulate its endogenous consumption in the dark phase. The accumulation or consumption of the polysaccharide is regulated by the concentration of ATP that stimulates the carbon flux through enzymatic and molecular control. When there is a high concentration of ATP, the flux is directed toward gluconeogenesis and starch accumulates. On the contrary, in conditions of ATP deficiency, the enzymes α-amylase and starch phosphorylase are activated, which hydrolyze the accumulated starch to obtain glucose that will be metabolized by glycolysis. In the cytosol, ATP is used to produce pyruvate, which is then converted to acetyl-CoA, converted to lipids (Parichehreh et al. 2019). Under heterotrophic conditions, acetate is assimilated, which causes the export of ATP from the mitochondria and is used in the formation of G3P; part of it is imported into the chloroplast and can be used for starch synthesis in the dark (Johnson and Alric 2013).

Accumulated starch in microalgae under nonstress conditions is required for bioenergetic purposes in ethanol production through fermentation or enzymatic hydrolysis in processes that have been documented (Satari and Jaiswal 2021; Abou El-Souod et al. 2021; Banerjee et al. 2016). On the contrary, the biosynthesis of ethanol by microalgae is a little exploited technology for directly converting sugars into ethyl alcohol. The company Algenol Biofuels from Florida, USA (https://www.algenol.com/) has developed patented processes to produce ethanol in genetically modified cyanobacteria based on the plasmids designed for the expression of the enzymes pyruvate decarboxylase (PdC) that converts pyruvate to acetaldehyde, and the Zn^{++}-dependent enzyme alcohol dehydrogenase, to convert acetaldehyde to ethanol (Dühring et al. 2015; Friedrich et al. 2016).

12.6 Lipid Biosynthesis

Like starch, lipids in microalgae are a reserve of carbon and energy that accumulate in cells as oil bodies (Johnson and Alric 2013). These can be produced as free lipids (mainly triacylglycerols used to obtain biodiesel) or as phospholipids and their main precursor is malonyl-CoA (Parichehreh et al. 2019).

The biosynthesis of fatty acids in microalgae such as *Chlamydomonas rendhartii* occurs in the chloroplast from pyruvate. The active enzyme acetyl-CoA carboxylase (ACC) carboxylates acetyl-CoA into malonyl from CO_2. Also, the fatty acid synthase (FAS) elongates the fatty acid chain by two units, resulting in long chains of acyl-CoA or free fatty acids exported to the endoplasmic reticulum to produce TAG by the enzyme acyltransferase (Johnson and Alric 2013; Banerjee et al. 2016). Figure 12.2 presents a simplified scheme of the most important metabolic pathways to produce lipids and starch in microalgae. More details of these pathways can be found in Johnson and Alric (2013), Sun et al. (2018), Banerjee et al. (2017), Min Tan and Lee (2016), Li-Beisson et al. (2015), and Goncalves et al. (2016).

Under conditions of abiotic stress, lipid accumulation seems to be an alternative for microalgae for cell survival and maintenance (Paliwal et al. 2017; Bibi et al. 2022) and a multifactorial process (Johnson and Alric 2013). The variation in the concentration of lipids between the different genera and species of microalgae is a factor to be considered for industrial production as bioenergy. In this sense, physiological studies (especially in gene expression) must be reviewed in a particular way in many cases (Banerjee et al. 2016; LaPanse et al. 2021; Bibi et al. 2022). In this regard, a possible explanation for the greater production of lipids in heterotrophy compared to autotrophy is the positive and increased regulation of the genes that code for the key enzymes in the pentoses phosphate pathway (PPP) during nitrogen deprivation, associated with the use of NADPH in the biosynthesis of fatty acids (Chen and Jiang 2017). In this sense, nutrient limitation, especially nitrogen starvation, is one of the most common strategies applied to increase lipid accumulation in

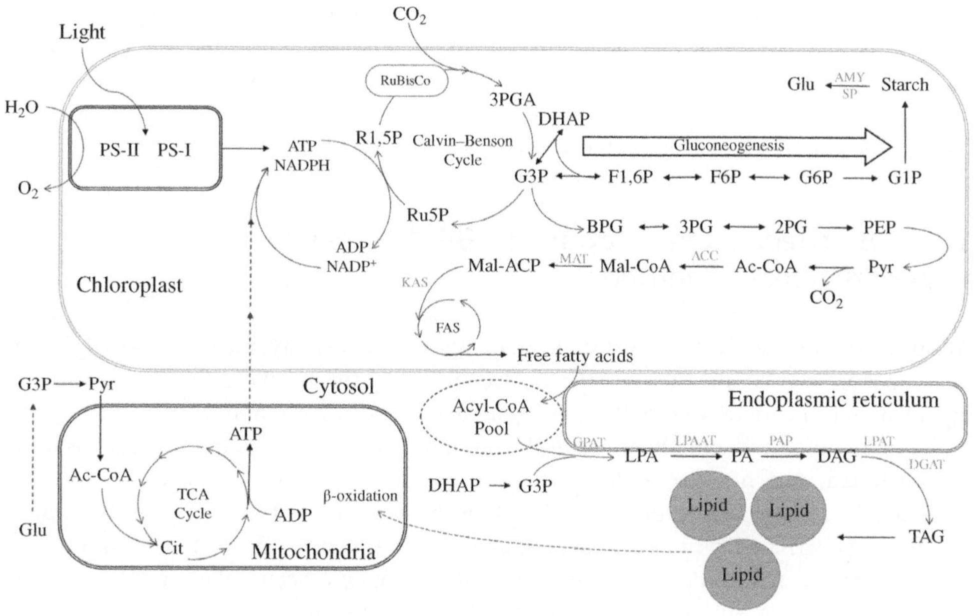

Figure 12.2 Simplified scheme of the metabolic pathways of starch and lipid biosynthesis in microalgae. Photosystem II (PS-II), Photosystem I (PS-I), adenosine triphosphate (ATP), adenosine diphosphate (ADP), nicotinamide adenine dinucleotide phosphate ($NADP^+$), reduced-NADP (NADPH), ribulose 5-phosphate (Ru5P), ribulose 1,5-biphosphate (R1,5P), 3-phosphoglycerate (3PGA), glyceraldehyde-3-phosphate (G3P), dihydroxyacetone phosphate (DHAP), fructose 1,6-biphosphate (F1,6P), fructose 6-phosphate (F6P), glucose-6-phosphate (G6P), glucose-1-phosphate (G1P), glucose (Glu), 1,3-biphospho glycerate (BPG), phospho glycerate (3PG), 2-phospho glycerate (2PG), phophoenolpyruvate (PEP), pyruvate (Pyr), Coenzyme A (Co-A), acyl-carrier protein (ACP), acetyl-CoA (Ac-CoA), malonyl-CoA (Mal-CoA), malonyl-ACP (Mal-ACP), fatty acids synthesis (FAS), lysophosphatidic acids (LPA), phosphatidic acid (PA), diacyl glycerol (DGA), triacylglycerols (TAG), citrate (Cit), a-amylase (AMY), starch phosphorylase (SP), acetyl-CoA carboxylase (ACC), malonyl-CoA: ACP transacylase (MAT), 3-ketoacyl-ACP synthase (KAS), glycerol-3-phosphate acyltransferase (GPAT), lysophosphatidic acid acyltransferase (LPAAT), phosphatidic acid phosphatase (PAP), lysophosphatidylcholine acyltransferase (LPAT), and diacylglycerol acyltransferase (DGAT). The red line represents the export of ATP from the mitochondria to the chloroplast and the metabolism of glucose to produce G3P in heterotrophy.

microalgae (Vitova et al. 2015; Benavente-Valdés et al. 2016). For example, in *C. vulgaris* under N-starvation, the accumulation of up to 43.6% of lipids was possible using simulation of metabolic networks based on flux balance analysis, being CO_2, light, O_2, and the source of nitrogen (as nitrates) the most critical factors (Parichehreh et al. 2019).

Physiological mechanisms associated with the accumulation of lipids under N-starvation are related to cell cycles and the maintenance of metabolic activity and their survival under nonoptimal growth conditions. When an organic carbon such as acetate (heterotrophy) is used without nitrogen, growth stops and the increase in its mass depends only on the reserves accumulated endogenously. Therefore, the control of environmental conditions favors the accumulation of lipids by controlling their photosynthetic capacity; in this way, cell metabolism now depends on the assimilation of an external carbon source, which increases lipid production (Johnson and Alric 2013). Under heterotrophic and mixotrophic conditions, the concentration of chlorophyll tends to decrease, although the ratios of chlorophylls a and b (Chl_a/Chl_b) and chlorophyll-carotenoids (Chl/Car) are maintained (Cecchin et al. 2018). A portion of nitrogen of these molecules in chloroplasts appears to be used to synthesize other nitrogenous molecules; then, lipids accumulate as a mechanism for obtaining energy (electrons and carbon) (Chen and Jiang 2017).

12.7 Biochemical Regulation of BBPM Associated with Nutritional Conditions

Although the metabolic pathways to produce lipids and carbohydrates in microalgae are known, their regulation must be analyzed concerning the biochemical capacity of the organism to grow under different nutritional conditions. A correlation between the accumulation of lipids and carbohydrates in the biomass of many microalgae species has been demonstrated, but not for proteins content which is regulated independently (Bibi et al. 2022). It is also important to consider that the regulation of key genes in its production may be associated with its primary metabolism (Rammers et al. 2018). For example, in *Chlamydomonas reinhardtii*, the direction of starch or lipid accumulation can be calculated in terms of NADPH and ATP content (Johnson and Alric 2013).

The type of nutrition has been shown to affect the growth and production of starch and lipids in *Chlorella sorokiniana*, where acetate assimilation in mixotrophy increases cell growth and product accumulation compared to photoautotrophy. This behavior is associated with carbon flux and energy regulation control mechanisms, particularly in the acetate assimilation that positively regulates the enzyme phosphoenolpyruvate carboxylase in recovering carbons spent in the oxidation of pyruvate and because of increased production of NADH during acetate assimilation. However, the genetic regulation of both metabolisms is not well elucidated, which may imply a complex network of metabolic regulation (Cecchin et al. 2018). On the other hand, in photoautotrophy, cultures of *Nannochloropsis gaditana* have been observed that in the central carbon metabolism, the assimilation of CO_2 in the Calvin cycle is repressed under nitrogen-limiting conditions (tending to accumulate lipids) (Chen and Jiang 2017).

The central metabolism of carbon in microalgae is carried out by two different pathways depending on the nutritional conditions. In heterotrophy, the PPP is the main carbohydrate metabolic pathway, while in photoautotrophy, the glycolytic pathway preferably used is the Embden–Meyerhof–Parnas (EMP). The preference seems to be associated with the metabolic fluxes during the use of carbon and the ATP/ADP ratios in the electron transport in the mitochondria; in addition, in some microalgae, the ability to grow in heterotrophic conditions is associated with the genes that code for hexose transporter proteins in the cell membrane (Chen and Jiang 2017).

In the design and scale-up of PBRs and microalgae cultivation processes for energy and biofuel purposes, different biochemical and molecular aspects related to the metabolism of the strain to be used must be considered. Such aspects include the growth kinetics and accumulation of desired products, the specific growth rate under nutritional conditions (photoautotrophy, mixotrophy, or heterotrophy), the high productivity, and biomass yield, etc. (Bibi et al. 2022). That is, it is necessary to understand the biochemical processes

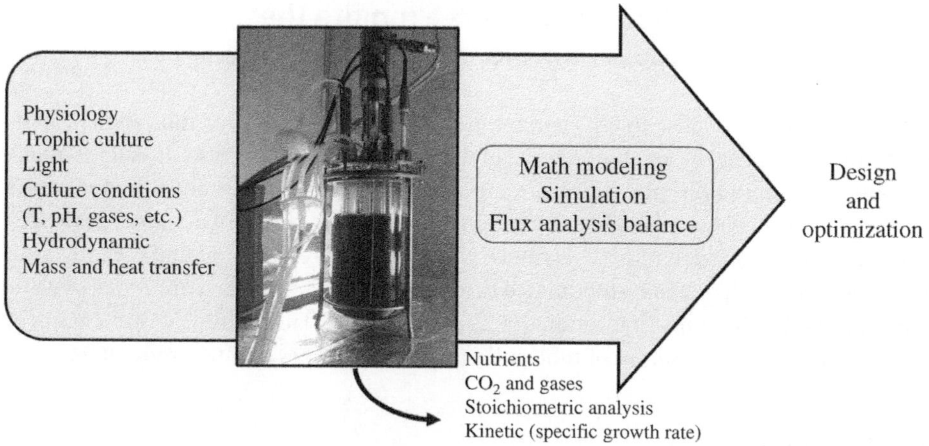

Figure 12.3 Strategies and factors involved in the design and optimization of PBRs and processes for the cultivation of microalgae for bioenergy and biofuel purposes.

involved in the growth of microalgae as a basis for their massive production (Schediwy et al. 2019), which will allow knowing the environmental requirements (such as abiotic stressors), the availability of nutrients, the nutritional and physiological control, gene expression, among others (Banerjee et al. 2016). In addition, the use of mathematical modeling tools, computational fluid dynamics (CFD) simulation, and flux analysis balance allow optimum production (Figure 12.3) (Schirmer and Posten 2016; Bisinella Scheufele et al. 2018; Losa et al. 2020). As the culture requirements differ for each microorganism, it is vital that the bioprocesses development contemplate the production conditions of a given strain, since at industrial scales it seems to be a limiting factor (Gifuni et al. 2018; Li et al. 2022). For example, Rammers et al. (2018) propose that *Nannochloropsis* spp. can be a highly suitable organism to produce lipids under industrial conditions based on its physiological and molecular characteristics.

For the processes scale-up, the stoichiometric factors for converting nutrients and energy sources must meet sufficient yields for their mass production (Figueroa-Torres et al. 2021). For example, during the production of lipids, a yield of 0.5 TAG $mol_{photon}{}^{-1}$ is desirable; however, this scenario is not entirely realistic, assuming that the accumulation of TAG has been associated with a state of N-starvation in batch culture. However, a more realistic scenario is in *Chlorella zofingiensis* of 0.3 g TAG $mol_{photon}{}^{-1}$ (Rammers et al. 2018). In addition, knowing the energy balances is a tool that allows the design and optimization of processes. In this sense, the type of nutrition will determine the accumulation of reserve products, regulated by the cell's energy balances. Starch accumulation is more efficient than lipid synthesis under photoautotrophic conditions where CO_2 is fixed as the sole carbon source. However, the energy return for fatty acids is 6.7 ATP per carbon atom, while for starch synthesis, it is 5.3 ATP per fixed carbon atom. However, more energy is required for lipid production, resulting in 1.5 and 1.37 ATP molecules per NADPH for starch and fatty acid synthesis, respectively. This implies that the regulation of the type of compounds for bioenergy can be carried out in two different stages (Johnson and Alric 2013).

12.8 Physical and Chemical Factors Promote the Accumulation of Molecules for Bioenergy and Biofuels

The growth rate of microalgae in a system intended primarily for the accumulation of molecules to produce bioenergy and biofuels depends on process conditions such as light intensity and duration, agitation, pH, salinity, temperature, turbidity, O_2 removal and nutrient conditions such as nitrates, phosphates, carbohydrates, manganese, cobalt, zinc, molybdenum, and carbon-to-nitrogen ratio (C/N) and CO_2 uptake. Therefore, lipid and carbohydrate composition in microalgae can be modulated through the environmental conditions prevailing during growth (Viswanaathan et al. 2022). As described later, physical and chemical factors promote the accumulation of molecules for bioenergy and biofuel production.

12.9 Light Intensity

Light is one of the key factors controlling the course of physiological processes in microalgae. Low illumination can have a limiting effect on growth, and too high illumination can have an inhibitory effect. Light exposure in microalgae above the optimum level (>300 μmol/m^2/s) leads to photon inhibition that decreases the growth rate of microalgae due to disruption of chloroplast lamellae and inactivation of CO_2-fixing enzymes. Furthermore, energy is diverted mainly to protein synthesis to support cell growth. The illumination period is also of great importance to ensure maximum productivity of microalgae and to minimize energy costs during cultivation. It is a known practice to use continuous lighting or alternating light and dark growing cycles, including pulsed lighting, which is characterized by a rapid change in light and dark periods. Kuravi and Mohan (2022) concluded that 16 hours of continuous light exposure and 8 hours of darkness are the best light–darkness cycle for the development of microalgae. The authors obtained maximum biomass productivity since intermittent light increases photosynthesis of microalgae and, at the same time, improves the quantity and quality of the microalgal biomass.

Chlorella sp. and *Nanochloropsis* sp. were grown under a high light density of 10 000 lx and showed lower lipid productivity, suggesting that the fatty acid content of these two species was positively correlated with light intensity since the higher the light intensity, the lower the lipid productivity (Brindhadevi et al. 2021). *Desmodesmus* sp. and *S. obliquus* were cultured under an irradiance of 300 μmol/m^2/s, reaching a fatty acid content of 6.2% and 5.8%, respectively. When algal cells are cultured with light intensities higher than 300 μmol/m^2/s, the production of lipids and carbohydrates decreases, directing the metabolism to protein synthesis and cell growth (Leonhardt et al. 2020).

12.10 Salts

Salinity is another essential factor that alters the lipid, carbohydrate, and protein composition of microalgae in algal cells (salinity refers primarily to the concentration of sodium chloride). Higher salinity increases the content of neutral lipids (triacylglycerols) to resist unfavorable conditions. A gradual increase in salt concentration has resulted in a significant increase in lipid content of several microalgal species. *Dunaliella tertiolecta* showed a high concentration of total lipids and triacylglycerols (TAG) when stressed with increasing concentration of sodium chloride (NaCl) ranging from 0.5 to 2.0 M. On the other hand, salinity stress is also species specific, and high salt concentration in the culture medium can inhibit microalgal biomass growth and lipid productivity (Teronpi et al. 2021).

12.11 Use of Organic and Inorganic Carbon Sources

In comparison to the inorganic ones, organic carbon sources can generate a higher production of NADPH and ATP. Such response to the carbon sources directly impacts physiological metabolism support. Glucose is the most well-known organic carbon source in mixotrophic and heterotrophic cultures. Glycerol is another broadly used carbon source. It was demonstrated that the algae biomass production was increased by using glycerol as a sole carbon source. Similarly, disaccharides such as fructose and sucrose are also known to have the same effect (Oliveira et al. 2021). Furthermore, it was reported that sodium acetate could promote both growth and lipid content in several microalgae species.

One of the most used inorganic carbon sources for microalgae cultivation is CO_2, bicarbonate (HCO_3^-), and carbonate (CO_3^{2-}). Gao et al. (2021) reported that *C. vulgaris* reached a maximum lipid content as high as 37% using HCO_3^- as an inorganic carbon source and that varying CO_2 concentrations from 4 to 8 and 16% had no significant effect on cell concentration production.

Currently, municipal, aquaculture, and industrial wastewater have been implemented as a nutrient source for microalgae cultivation. Algae use both organic and inorganic carbon present in wastewater as a carbon source and remove elements such as nitrogen and phosphorous. *Scenedesmus quadricauda*, *Chlorella zofingiensis*, and *Monoraphidium* sp. cultured in municipal wastewater reached a lipid content of 5.2 mg l/d, 21.6–25.4 g/l, and 92.3 g/l, respectively, which suggests that microalgae cultivation using wastewater offers a potential strategy to recycle wastewater and obtain high value-added metabolites (Ahmad et al. 2022; Huang et al. 2022).

12.12 Agitation

Agitation enables the correct mixing of the growing system and ensures that nutrients, pH, CO_2, and temperature are evenly distributed throughout the growing system, allowing gas exchange between the growing medium and the environment. Agitation can be applied to the bioreactor in two ways, by mechanical agitation or by supplying air; poor agitation leads to poor mixing and results in inadequate oxygen removal in the culture system affecting the photosynthesis process, which is counterproductive for the microalgae as it decreases the growth rate and accelerates cell death. *C. Vulgaris* exhibited higher specific growth and photosynthetic activity with stirring mixing at a constant top speed of $1.26\,ms^{-1}$. *Scenedesmus obliquus* grown in a PBR magnetically stirred with a speed ratio between 0 and 350 rpm, and an aeration rate from 0 to 1.5 vvm showed slower cell growth (Bautista-Monroy et al. 2022).

12.13 Photobioreactors to Produce Bioenergy and Biofuels

Efficient and cost-effective large-scale microalgal biomass production depends on the cultivation methods, which is a limiting factor for the future of bioenergy and biofuel production. In general, microalgae cultivation has been carried out using large-scale cultivation systems (horizontal tubular PBRs and raceway ponds) and small-scale cultivation systems (flat and vertical PBRs) (Peter et al. 2022), the main cultivation systems used for microalgal biomass cultivation are described later.

12.14 Open Pond Cultivation Systems

The open culture systems (OP) include natural ponds, circular ponds, channel ponds, and tilted systems. They are economical in maintenance and operation; however, they have comparatively lower yields than closed systems due to evaporation problems, temperature variations, light and nutrient limitations, and there is a higher risk of contamination and low control of growing conditions. These systems are characterized by channel depths between 0.2 and 0.4 m since the more profound the water, the less light penetrates the crop. They are generally built-in circular or channel configurations. The largest raceway pond in world is microalgae-based (444 000 m^2 at Calipatria, USA), which is designed to produce *Spirulina* and the products derived from *Spirulina* biomass. A typical OP has biomass productivity varying from 5 to 10 g dw/m^2/d, which is equivalent to 7.4–14.8 tons (dry biomass) $acre^{-1}$ $year^{-1}$. Thus, Thuy Lan Chi et al. (2022) cultivated *Chlorella* sp in an open pond system with a capacity of 500 l, reaching a lipid content of 20.2%, which was used for sustainable biodiesel production. Open systems have become the most commercially used methodology worldwide due to their low construction, maintenance, and operation costs. These systems have been implemented commercially by companies such as algaEnergy for bioenergy production.

12.15 Closed Systems

A closed system, also known as a PBR, is a closed culture vessel provided with light and a CO_2 source that allows the induction of rapid cell propagation. The most widely used system is a tubular and plate PBRs, which are built with parallel transparent plastic tubes (diameter not greater than 0.2 m) that allow the passage of light (Olsen et al. 2021). Faried et al. (2017) indicate that the three main categories are tubular/horizontal, column/vertical, and flat plate/rectangular PBRs.

PBRs are the most efficient systems in biomass generation for biofuel production due to the control of contamination and the control of the main parameters that affect the growth of microalgae. Companies such as algae systems, Algasol, algae international, and Algenol have focused their research on developing new PBR technologies that allow high production of microalgal biomass rich in carbohydrates and lipids for subsequent application in the production of bioethanol, biobutanol, biomethane, biohydrogen, and biofuels. In general, tubular PBRs are widely used for mass algae cultivation. Most tubular PBRs are generally constructed of glass or plastic tubing and can be horizontal/serpentine, vertical, conical, or inclined in shape to maximize sunlight capture. Tubular PBRs are preferred for massive outdoor algae cultivation as they have a large illumination surface (Xiaogang et al. 2020).

12.16 Hybrid Systems

Hybrid systems are an advantageous microalgae cultivation system because they can easily separate the biomass production and biomolecule accumulation phases. The first stage of algal growth occurs in a closed system (PBR), and the second stage occurs in an open system, exposing the cells to nutritional stress, which stimulates the synthesis of desirable metabolites and products. Hybrid systems are represented by a covered open pond, which allows separation of the culture from the outside environment; an example of this type of reactor is the covered raceway pond built by the Chilean biotechnology corporation Atacama Bio Natural Products. It has been reported that the production of lipids and carbohydrates in a hybrid (open raceway pond and a closed PBR system) improves the height/diameter ratio, allowing an increase in the surface/volume ratio, providing higher productivity and lower energy consumption, and decreasing the costs of the process. Hybrid PBRs will enable us to take advantage of both closed and open systems combining characteristics of both systems, overcoming the disadvantages of each one with better control of the cultivation variables, higher productivity, and lower energy consumption (Rehman et al. 2022).

12.17 Conclusions

Microalgal biomass has great potential for bioenergy and biofuel production, given the intrinsic efficiency of microalgae to convert solar energy into chemical energy and their significantly higher potential yield in the production of molecules for bioenergy and biofuel production. The development of industrial-scale cultivation to produce bioenergy biomass faces the challenge of developing integrated technologies that consider all the specific parameters that allow obtaining large quantities of biomass at a low cost.

References

Abou El-Souod, G.W., Morsy, E.M., Hassan, L.H.S., and El-Sheekh, M.M. (2021). Efficient saccharification of the microalga *Chlorella vulgaris* and its conversion into ethanol by fermentation. *Iranian Journal of Science and Technology, Transactions A: Science* 45: 767–774.

Ahmad, A., Banat, F., Alsafar, H., and Hasan, S. (2022). Algae biotechnology for industrial wastewater treatment, bioenergy production, and high-value bioproducts. *Science of the Total Environment* 806: https://doi.org/10.1016/j.scitotenv.2021.150585.

Alami, A.H., Tawalbeh, M., Alasad, S. et al. (2021). Cultivation of Nannochloropsis algae for simultaneous biomass applications and carbon dioxide capture. *Energy Sources, Part A:*

Recovery, Utilization, and Environmental Effects 43 (24): 3227–3433. https://doi.org/10.1080/15567036.2021.1933267.

Amaral, M., Loures, C., Pedro, G. et al. (2020). An unconventional two-stage cultivation strategy to increase the lipid content and enhance the fatty acid profile on Chlorella minutissima biomass cultivated in a novel internal light integrated photobioreactor aiming at biodiesel production. *Renewable Energy* 156: 591–601. https://doi.org/10.1016/j.renene.2020.04.084.

Banerjee, C., Dubey, K.K., and Shukla, P. (2016). Metabolic engineering of microalgal based biofuel production: prospects and challenges. *Frontiers in Microbiology* 7: 432. https://doi.org/10.3389/fmicb.2016.00432.

Banerjee, A., Maiti, S.K., Guria, C., and Banerjee, C. (2017). Metabolic pathways for lipid synthesis under nitrogen stress in Chlamydomonas and Nannochloropsis. *Biotechnology Letters* 39: 1–11.

Bautista-Monroy, S., Chávez-Urbiola, E., Ortega-Palacios, R. et al. (2022). Insights of raceway bioreactor scale-up : effect of agitation on microalgae culture and reduction of the liquid medium speed. *Applied Sciences* 12 (1513): https://doi.org/https://doi.org/10.3390/.

Benavente-Valdés, J.R., Aguilar, C., Contreras-Esquivel, J.C. et al. (2016). Strategies to enhance the production of photosynthetic pigments and lipids in chlorophycae species. *Biotechnology Reports* 10: 117–125.

Benemann, J.R. (2004). Hydrogen and methane production by microalgae. In: *Handbook of Microalgal Culture: Biotechnology and Applied Phycology* (ed. A. Richmond), 566. Oxford, UK: Blackwell Publishing Ltd.

Bibi, F., Jamal, A., Huang, Z. et al. (2022). Advancement and role of abiotic stresses in microalgae biorefinery with a focus on lipid production. *Fuel* 316: 123192. https://doi.org/10.1016/j.fuel.2022.123192.

Bisinella Scheufele, F., Hinterholz, C.L., Zaharieva, M.M. et al. (2018). Complex mathematical analysis of photobioreactor system. *Engineering in Life Sciences* 19: 844–859.

Bolognesi, S., Bañeras, L., Perona-Vico, E. et al. (2022). Carbon dioxide to bio-oil in a bioelectrochemical system-assisted microalgae biorefinery process. *Sustainable Energy and Fuels* 6 (150): 150–161. https://doi.org/10.1039/d1se01701b.

Brindhadevi, K., Mathimani, T., Rene, E. et al. (2021). Impact of cultivation conditions on the biomass and lipid in microalgae with an emphasis on biodiesel. *Fuel* 284: 119058. https://doi.org/10.1016/j.fuel.2020.119058.

de Carvalho Silvello, M.A., Severo Gonçalves, I., Patrícia Held Azambuja, S. et al. (2022). Microalgae-based carbohydrates: a green innovative source of bioenergy. *Bioresource Technology* 344 (September 2021): https://doi.org/10.1016/j.biortech.2021.126304.

Cecchin, M., Benfatto, S., Griggio, F. et al. (2018). Molecular basis of autotrophic vs mixotrophic growth in *Chlorella sorokiniana*. *Scientific Reports* 8: 6465. https://doi.org/10.1038/s41598-018-24979-8.

Chen, H.H. and Jiang, J.G. (2017). Lipid accumulation mechanisms in auto- and heterotrophic microalgae. *Journal of Agricultural and Food Chemistry* 65: 8099–8110.

Cheng, D., Li, D., Yuan, Y. et al. (2017). Improving carbohydrate and starch accumulation in Chlorella sp. AE10 by a novel two-stage process with cell dilution. *Biotechnology for Biofuels* 10: 75. https://doi.org/10.1186/s13068-017-0753-9.

Debnath, C., Bandyopadhyay, T.K., Bhunia, B. et al. (2021). Microalgae: sustainable resource of carbohydrates in third-generation biofuel production. *Renewable and Sustainable Energy Reviews* 150: 111464. https://doi.org/10.1016/j.rser.2021.111464.

Dühring, U., Enke, H., Ziegler, K. and Schwecke, T. (2015). Metabolycally enhanced cyanobacterial cell for the production of ethanol. US Patent US2015/0259708 A1. Filed date: 19 December 2013. Issued date: 25 June 2015.

Faried, M., Samer, M., Abdelsalam, E. et al. (2017). Biodiesel production from microalgae: processes, technologies and recent advancements. *Renewable and Sustainable Energy Reviews* 79: 893–913. https://doi.org/10.1016/j.rser.2017.05.199.

Fetyan, N.A.H., El-Sayed, A.E.-K.B., Ibrahim, F.M. et al. (2022). Bioethanol production from defatted biomass of Nannochloropsis oculata microalgae grown under mixotrophic conditions. *Environmental Science and Pollution Research* 29 (2): 2588–2597. https://doi.org/10.1007/s11356-021-15758-6.

Figueroa-Torres, G.M., Pittman, J.K., and Theodoropoulos, C. (2021). Optimization of microalgal cultivation via nutrient-enhanced strategies: the biorefinery paradigm. *Biotechnology for Biofuels* 14: 64. https://doi.org/10.1186/s13068-021-01912-2.

Friedrich, A., Piven, I., Uliczka, F. and Enke, H. (2016). Chlorogloeopsis sp. host cell for producing ethanol and method for producing ethanol using the same. US Patent: US9399776B2. Filed date: 21 May 2015 Issued date: 17 September 2015.

Gao, F., Yang, Z.-Y., Zhao, Q.-L. et al. (2021). Mixotrophic cultivation of microalgae coupled with anaerobic hydrolysis for sustainable treatment of municipal wastewater in a hybrid system of anaerobic membrane bioreactor and membrane photobioreactor. *Bioresource Technology* 337: https://doi.org/10.1016/j.biortech.2021.125457.

Geng, Y., Cui, D., Yang, L. et al. (2022). Resourceful treatment of harsh high-nitrogen rare earth element tailings (REEs) wastewater by carbonate activated Chlorococcum sp. microalgae. *Journal of Hazardous Materials* 423 (127000): 1–9. https://doi.org/10.1016/j.jhazmat.2021.127000.

Gifuni, I., Olivieri, G., Pollio, A., and Marzocchella, A. (2018). Identification of an industrial microalgal strain for starch production in biorefinery context: the effect of nitrogen and carbon concentration on starch accumulation. *New Biotechnology* 41: 46–54.

Goncalves, E.C., Wilkie, A.C., Krist, M., and Rathinasabapathi, B. (2016). Metabolic regulation of triacylglycerol accumulation in the green algae: identification of potential targets for engineering to improve oil yield. *Plant Biotechnology Journal* 14: 1649–1660.

Gouveia, L. (2011). *Microalgae as a Feedstock for Biofuels*. SpringerBriefs in Microbiology https://doi.org/10.1007/978-3-642-17997-6_1.

IEA (2020). Global energy review 2019. https://doi.org/10.1787/90c8c125-en (accessed 02/03/2022).

Huang, H., Zhong, S., Wen, S. et al. (2022). Improving the efficiency of wastewater treatment and microalgae production for biofuels. *Resources, Conservation and Recycling* 178 (106094): 1–12. https://doi.org/10.1016/j.resconrec.2021.106094.

Johnson, X. and Alric, J. (2013). Central carbon metabolism and electron transport in *Chlamydomonas reinhardtii*: metabolic constraints for carbon partitioning between oil and starch. *Eukaryotic Cell* 12 (6): 776–793.

Kuravi, S.D. and Mohan, V.S. (2022). Mixotrophic cultivation of *Monoraphidium* sp. In dairy wastewater using Flat-Panel photobioreactor and photosynthetic performance. *Bioresource Technology* 348 (126671): https://doi.org/10.1016/j.biortech.2021.126671.

LaPanse, A., Krishnan, A., and Posewitz, M.C. (2021). Adaptive laboratory evolution for algal strain improvement: methodologies and applications. *Algal Research* 53: 102122. https://doi.org/10.1016/j.algal.2020.102122.

Leonhardt, L., Käferböck, A., Smetana, S. et al. (2020). Bio-refinery of Chlorella sorokiniana with pulsed electric field pre-treatment. *Bioresource Technology* 301 (November 2019): https://doi.org/10.1016/j.biortech.2020.122743.

Li, S., Li, X., and Ho, S.H. (2022). Microalgae as a solution of third world energy crisis for biofuels production from wasterwater toward carbon neutrality: an update review. *Chemosphere* 291: 132863. https://doi.org/10.1016/j.chemosphere.2021.132863.

Li-Beisson, Y., Beisson, F., and Riekhof, W. (2015). Metabolism of acyl-lipids in *Chlamydomonas reinhardtii. The Plant Journal* 82: 504–522.

Losa, J.P., Santos, F.M., Alvim-Ferraz, M.C.M. et al. (2020). Dynamic modeling of microalgal growth. In: *Encyclopedia of Marine Biotechnology* (ed. S.K. Kim), 547–467. Wiley.

Min Tan, K.W. and Lee, Y.K. (2016). The dilemma for lipid productivity in green microalgae: importance of substrate provision in improving oil yield without sacrificing growth. *Biotechnology for Biofuels* 9: 255. https://doi.org/10.1186/s13068-016-0671-2.

Morais, K.C.C., Conceição, D., Vargas, J.V.C. et al. (2021). Enhanced microalgae biomass and lipid output for increased biodiesel productivity. *Renewable Energy* 163: 138–145. https://doi.org/10.1016/j.renene.2020.08.082.

Naira, V.R., Das, D., and Maiti, S.K. (2019). Real time light intensity based carbon dioxide feeding for high cell-density microalgae cultivation and biodiesel production in a bubble column photobioreactor under outdoor natural sunlight. *Bioresource Technology* 284 (March): 43–55. https://doi.org/10.1016/j.biortech.2019.03.102.

Nogueira, N., Nascimento, F.J.A., Cunha, C., and Cordeiro, N. (2020). Nannochloropsis gaditana grown outdoors in annular photobioreactors: operation strategies. *Algal Research* 48 (February): 101913. https://doi.org/10.1016/j.algal.2020.101913.

Norsker, N.H., Michiels, M., Slegers, P.M. et al. (2019). Productivity of Nannochloropsis oceanica in an industrial closely spaced flat panel photobioreactor. *Algal Research* 43 (August 2018): 101632. https://doi.org/10.1016/j.algal.2019.101632.

Oliveira, C.Y.B., D'Alessandro, E.B., Antoniosi Filho, N.R. et al. (2021). Synergistic effect of growth conditions and organic carbon sources for improving biomass production and biodiesel quality by the microalga Choricystis minor var. minor. *Science of the Total Environment* 759: 143476. https://doi.org/10.1016/j.scitotenv.2020.143476.

Olsen, M.F.L., Pedersen, J.S., Thomsen, S.T. et al. (2021). Outdoor cultivation of a novel isolate of the microalgae *Scenedesmus* sp. and the evaluation of its potential as a novel protein crop. *Physiologia Plantarum* 173 (2): 483–494. https://doi.org/10.1111/ppl.13532.

Paliwal, C., Mitra, M., Bhayani, K. et al. (2017). Abiotic stresses as tool for metabolites in microalgae. *Bioresource Technology* 244: 1216–1226.

Parichehreh, R., Gheshlaghi, R., Mahdavi, M.A., and Elkamel, A. (2019). Optimization of lipid production in *Chlorella vulgaris* for biodiesel production using flux balance analysis. *Biochemical Engineering Journal* 141: 131–145. https://doi.org/10.1016/j.bej.2018.10.011.

Peter, A.P., Koyande, A.K., Chew, K.W. et al. (2022). Continuous cultivation of microalgae in photobioreactors as a source of renewable energy: current status and future challenges. *Renewable and Sustainable Energy Reviews* 154: https://doi.org/10.1016/j.rser.2021.111852.

Rammers, I.M., Wijffels, R.H., Barbosa, M.J., and Lamers, P.P. (2018). Can we approach theoretical lipid yields in microalgae? *Trends in Biotechnology* 36 (3): 265–276.

Rattanapoltee, P., Dujjanutat, P., Muanruksa, P., and Kaewkannetra, P. (2021). Biocircular platform for third generation biodiesel production: batch/fed batch mixotrophic cultivations of microalgae using glycerol waste as a carbon source. *Biochemical Engineering Journal* 175: https://doi.org/10.1016/j.bej.2021.108128.

Ravi Kiran, B. and Venkata Mohan, S. (2022). Phycoremediation potential of Tetradesmus sp. SVMIICT4 in treating dairy wastewater using Flat-Panel photobioreactor. *Bioresource Technology* 345 (September 2021): 126446. https://doi.org/10.1016/j.biortech.2021.126446.

Rehman, M., Kesharvani, S., Dwivedi, G., and Gidwani Suneja, K. (2022). Impact of cultivation conditions on microalgae biomass productivity and lipid content. *Materials Today: Proceedings* https://doi.org/10.1016/j.matpr.2022.01.152.

Satari, B. and Jaiswal, A.K. (2021). Green fractionation of 2G and 3G feedstocks for ethanol production: advances, incentives and barriers. *Current Opinion in Food Science* 37: 1–9.

Schediwy, K., Trautmann, A., Steinweg, C., and Posten, C. (2019). Microalgal kinetics – a guideline for photobioreactor design and process development. *Engineering in Life Sciences* 19: 830–843. https://doi.org/10.1002/elsc.201900107.

Schirmer, M. and Posten, C. (2016). Modeling of microalgae bioprocesses. In: *Advances in Chemical Engineering. Photobioreaction Enginering*, vol. 48 (ed. J. Legrand), 326. Elsevier.

Show, K.Y. and Lee, D.J. (2014). Production of biohydrogen from microalgae. In: *Biofuels from Algae* (ed. A. Pandey, D.J. Lee, Y. Chisti and C.R. Soccol), 338. USA: Elsevier.

Singh, H., Varanasi, J.L., Banerjee, S., and Das, D. (2019). Production of carbohydrate enrich microalgal biomass as a bioenergy feedstock. *Energy* 188: https://doi.org/10.1016/j.energy.2019.116039.

Sirohi, R., Kumar Pandey, A., Ranganathan, P. et al. (2022). Design and applications of photobioreactors – a review. *Bioresource Technology* 349 (February): 126858. https://doi.org/10.1016/j.biortech.2022.126858.

Sukačová, K., Búzová, D., Trávníček, P. et al. (2019). Optimization of microalgal growth and cultivation parameters for increasing bioenergy potential: case study using the oleaginous microalga Chlorella pyrenoidosa Chick (IPPAS C2). *Algal Research* 40: https://doi.org/10.1016/j.algal.2019.101519.

Sun, H., Zhao, W., Mao, X. et al. (2018). High-value biomass from microalgae production plataforms: strategies and progress based on carbon metabolism and energy conversion. *Biotechnology for Biofuels* 11: 277. https://doi.org/10.1186/s13068-018-1225-6.

Teronpi, H., Baruah, P., and Deka, H. (2021). Salinity stress as a critical factor to trigger lipid accumulation in a freshwater microalga *Lobochlamys* sp. GUEco1006. *Biologia* 76 (12): 3647–3658. https://doi.org/10.1007/s11756-021-00918-3.

Thuy Lan Chi, N., Mathimani, T., Manigandan, S. et al. (2022). Small scale photobioreactor, outdoor open pond cultivation of *Chlorella* sp. and harvesting at log and stationary growth phase towards lipids and methyl ester production. *Fuel* 319. https://doi.org/10.1016/j.fuel.2022.123813.

Tizvir, A., Shojaeefard, M.H., Zahedi, A., and Molaeimanesh, G.R. (2022). Performance and emission characteristics of biodiesel fuel from *Dunaliella tertiolecta* microalgae. *Renewable Energy* 182: 552–561. https://doi.org/10.1016/j.renene.2021.10.018.

Trivedi, J., Agrawal, D., Atray, N., and Ray, A. (2022). Enhanced lipid production in Scenedesmus obliquus via nitrogen starvation in a two-stage cultivation process and evaluation for biodiesel production. *Fuel* 316. https://doi.org/10.1016/j.fuel.2022.123418.

Tsygankov, A. and Abdullatypov, A. (2016). Hydrogen metabolism in microalgae. In: *Photosynthesis: New Approaches to the Molecular, Cellular, and Organismal Levels* (ed. S. Allakhverdiev), 386. Massachusetts, USA: Scribener Publishing.

Viswanaathan, S., Perumal, P.K., and Sundaram, S. (2022). Integrated approach for carbon sequestration and wastewater treatment using algal–bacterial consortia: opportunities and challenges. *Sustainability* 14 (1075): https://doi.org/10.3390/su14031075.

Vitova, M., Bisova, K., Kawano, S., and Zachleder, V. (2015). Accumulation of energy reserves in algae: from cell cycles to biotechnological applications. *Biotechnolgy Advances* 33: 1204–1218.

Wang, J. and Yin, Y. (2018). Fermentative hydrogen production using pretrated microalgal biomass as feedstock. *Microbial Cell Factories* 17: 22. https://doi.org/10.1186/s12934-018-0871-5.

Xiaogang, H., Jalalah, M., Jingyuan, W. et al. (2020). Microalgal growth coupled with wastewater treatment in open and closed systems for advanced biofuel generation. *Biomass Conversion and Biorefinery* 12: 1938–1958.

Yang, Z.-Y., Gao, F., Liu, J.-Z. et al. (2021). Improving sedimentation and lipid production of microalgae in the photobioreactor using saline wastewater. *Bioresource Technology* 347: https://doi.org/10.1016/j.biortech.2021.126392.

Yang, M., Xue, C., Li, L. et al. (2022). Design and performance of a low-cost microalgae culturing system for growing Chlorella sorokiniana on cooking cocoon wastewater. *Algal Research* 62 (January): 102607. https://doi.org/10.1016/j.algal.2021.102607.

Yun, J. H., Cho, D. H., Lee, B., Lee, Y. J., Choi, D. Y., Kim, H. S., & Chang, Y. K. (2020). Utilization of the acid hydrolysate of defatted Chlorella biomass as a sole fermentation substrate for the production of biosurfactant from Bacillus subtilis C9. *Algal Research*, 47(November 2019), 101868. https://doi.org/10.1016/j.algal.2020.101868.

13

Metabolic Engineering for Value Addition in Plant-Based Lipids/Fatty Acids

Himani Thakkar and Vinnyfred Vincent

Department of Biochemistry, All India Institute of Medical Sciences, New Delhi, India

13.1 Introduction

Plant oils represent an important renewable resource from nature. Plant lipids are an indispensable nutritional component of the human diet (Horn and Benning 2016; Zhou et al. 2020) and are being increasingly utilized as biofuel and industrial feedstock for soaps, perfumes, paints, adhesives, and in many chemical industries (Piazza and Foglia 2001; Tao 2007; Carlsson 2009). Worldwide production of plant oils in 2021–2022 accounted for 213.2 million metric tons that included coconut, sunflower, peanut, soybean, cottonseed, palm kernel, and canola seed oil (Statista 2020).

Plants contain a diverse variety of lipids essential for cellular functions. They are the major structural constituent of the cell membrane, serving as permeability barriers for cell and subcellular organelles. Lipids also form protective coatings in the form of cuticles such as cutin and waxes to prevent desiccation and external environmental stresses (Yeats and Rose 2013; Fich et al. 2016; Upadhyay and Singh 2021). In addition, they act as signaling molecules to regulate cell metabolism. Furthermore, lipids stored in the seeds in the form of triacylglycerols (TAG) serve as sources of energy during seed germination and help in seed dispersal of oily fruits like palm (Chapman and Ohlrogge 2012; Theodoulou and Eastmond 2012; Guerin et al. 2020). Lipid composition in plants varies between tissues and cell organelles. Plants accumulate a significant considerable amounts of lipids in seeds or fruits, as a source of energy for the embryo (Theodoulou and Eastmond 2012), while the metabolic process in vegetative tissues is directed mostly toward the synthesis and export of sucrose (Chapman et al. 2013).

Given the wide ranging benefits of plant lipids, strategic value, and growing demand, plant research focused on understanding the major pathways involved in plant lipid metabolism, regulation, and manipulation (Napier and Graham 2010; Bates et al. 2013). With advancement in understanding the lipid biosynthesis pathways, attempts have been made to engineer lipid metabolic pathways for increased production and development of special/novel oils with better agronomical traits (Yoon et al. 2013; Yuan and Grotewold 2015; Haslam et al. 2016; Lin and Eudes 2020).

Plants as Bioreactors for Industrial Molecules, First Edition.
Edited by Santosh Kumar Upadhyay and Sudhir P. Singh.

13.2 Plant Lipids

Plant lipids are normally obtained as liquid vegetable oils either from oilseed crops like rape, soy, or from fruits like avocado, palm, or olive. In majority of the plants, lipids are stored in the form of TAG except for a few plants like jojoba that accumulate wax esters in its seeds. The oil content in plants varies from 20% in soybean (*Glycine max*) to 70% in pecan (*Carya cathayensis*) (Sultan et al. 2015; Huang et al. 2016). Plants accumulate lipids in different tissues as in the embryo for oilseeds such as sunflower, linseed (*Linum usitatissimum*), or rapeseed (*Brassica napus*); endosperm for castor bean, coriander, or carrot; both embryo and endosperm for tobacco; kernel for coconut (*Cocos nucifera*) and oil palm. Oil obtained from plants like soybeans, oil palm, rape, and sunflower constitute about 70% of the world's plant oil production.

A diverse variety of lipid species are produced by plants ranging from simple long linear hydrocarbon chains to glycerolipids with distinct acyl compositions and positioning of fatty acids on glycerol backbones. In addition, plants accumulate unusual fatty acids that are characterized by the difference in the number and arrangement of double bonds, side chains, and type of functional groups (hydroxyls, ketones, epoxyls, etc.). Over 450 different fatty acid structures have been reported to be synthesized by plants. PlantFAdb database (https://plantfadb.org/tree) provides detailed description of the structures of fatty acids and plant species (Ohlrogge et al. 2018; Cahoon and Li-Beisson 2020). Some of the common fatty acids include palmitic (C16:0), stearic (C18:0), oleic (C18:1Δ9), linoleic (C18:2Δ9,12), and α-linolenic (C18:3Δ9,12,15) acids.

13.3 TAG Synthesis in Plants

The biosynthesis of TAG involves two steps in different subcellular compartments (i) biosynthesis of fatty acids (FA) in plastids and (ii) assembly of FA onto the glycerol backbone in the endoplasmic reticulum to form the TAG.

13.3.1 Fatty Acid Synthesis

The fatty acid biosynthesis pathway is the primary metabolic pathway found in the plastidial stroma of every cell of the plant and is essential for cell growth and survival. Photosynthesis is the major source of carbon for de novo fatty acid synthesis, while the glycolytic intermediates (hexose-phosphate, phosphoenolpyruvate, pyruvate, and malate) produced from the breakdown of sucrose in the cytosol or plastids also serve as precursors for FA synthesis. The intermediates formed in the cytosol are transported into the plastid by membrane proteins to synthesize acetyl-CoA, which is the immediate precursor for FA synthesis (Harwood 2005).

The metabolic flux studies have reported four possible pathways for the synthesis of acetyl CoA in plastid. One of the pathways is the activation of free acetate to acetyl-CoA by acetyl-CoA synthetase (ACS). Other route is the conversion of pyruvate formed as a result of glycolytic pathway in the cytosol or plastid to acetyl-CoA via the pyruvate dehydrogenase complex. The third route for synthesis of acetyl-CoA is through a plastid enzyme carnitine acetyltransferase that catalyze the transfer of acetate from acetyl-carnitine to CoA. The fourth pathway that has been seen in leaves of rape, spinach, and tobacco is the ATP-citrate lyase reaction (Fatland et al. 2002; Rawsthorne 2002).

The first step of fatty acid biosynthesis is the conversion of acetyl-CoA to malonyl-CoA by ATP-dependent acetyl-CoA carboxylase (ACCase), where the carboxyl group is added to acetyl-CoA to form malonyl-CoA in the stroma of plastids. Subsequently, fatty acids are synthesized in a stepwise manner by adding two carbons by the enzyme fatty acid synthase. Malonyl-CoA is elongated to produce C16:0-acyl carrier protein (ACP) or C18:0-ACP. In plants like palm and coconut, chain termination occurs earlier than the synthesis of palmitic acid, and as a result, majority of the fatty acids produced and present in TAG of these plants are of carbon length between 8 (caprylic acid; C8:0) and 14 (myristic acid; C14:0).

Following the processes of synthesis is the terminal reaction that includes thioester bond hydrolysis or acyl desaturation. The majority of the fatty acids synthesized are transported to ER for TAG synthesis via the eukaryotic pathway, while the some of the fatty acids undergo prokaryotic pathway in plastid (Li et al. 2016). A series of fatty acid desaturases introduce double bond into fatty acids at different locations, like stearoyl-ACP desaturase, a plastid enzyme that converts stearate (C18:0-ACP) into oleate (C18:1-ACP) (Shanklin and Cahoon 1998; Gagné et al. 2009).

In the eukaryotic pathway, the acyl-CoA thioesterases hydrolysis fatty acids to release C16:0-, C18:0-, or C18:1-acyl carrying protein (ACP) that are transferred to the outer membrane of the plastid. Subsequently, C16:0-, C18:0-, and C18:1- ACP are re-esterified to acyl-CoA by long-chain acyl-CoA synthetase (LACS) that allows them to be exported outside the plastid to undergo modifications on the endoplasmic reticulum (Shockey et al. 2002; Abbadi et al. 2004; Haslam and Kunst 2013; Li et al. 2018). Acyl-CoAs are transported to the ER by the members of acyl-CoA-binding proteins (ACBPs). In plants, ACBPs are expressed during seed development and are classified into different classes based on their size, and they exhibit a different binding affinity for acyl-CoA esters (Raboanatahiry et al. 2018).

The synthesis of very long-chain fatty acids (VLCFAs) also occurs on the ER using the ER-associated acyl-CoA elongases, for instance, in rapeseed (*B. napus*), oleoyl-CoA is converted to form eicosenoyl (C20:1 Δ11-cis)-CoA and erucoyl (C22:1 Δ13-cis)-CoA by elongase enzymes. Both plastid-exported and ER-elongated acyl-CoAs constitute the acyl-CoA pool, which are incorporated into glycerol-3-phosphate to form TAG (Joubès et al. 2008; Bates et al. 2013).

13.3.2 TAG Biosynthesis

The acyl-CoAs synthesized in plastid are assembled on glycerol for TAG synthesis via different pathways. The most common pathway involves sequential acylation of the glycerol backbone of glycerol-3-phosphate, and this pathway is known as the Kennedy pathway. In this pathway, glycerol-3-phosphate is acylated to form lysophosphatidic acid by glycerol-3-phosphate acyltransferase (GPAT), which is further acylated by lysophosphatidate acyltransferase (LPAT) resulting in the production of phosphatidic acid (PA). Thereafter, phosphatidic acid phosphatase (PAP) catalyzes the removal of the phosphate group from phosphatidic acid to generate 1,2-diacylglycerol (DAG). Finally, acylation of DAG catalyzed by acyl-CoA: diacylglycerol acyltransferase (DGAT) generates TAG (Murata and Tasaka 1997; Carman and Han 2006; Liu et al. 2012).

In addition to the Kennedy pathway, more complex pathways are involved in the production of TAG with modified fatty acids like polyunsaturated fatty acids and hydroxylated fatty acids, where phosphatidylcholine (PC) is a central intermediate in the flux of substrates to form TAG. Some of the fatty acids (mostly oleate) produced in the plastid are incorporated into PC through acyl editing in ER. Most de novo synthesized C18:1-CoAs are first incorporated into PC and are then desaturated by fatty acid desaturases. fad2 (ω-6 fatty acid desaturase) convert C18:1-CoA into C18:2-PC, which is then desaturated to C18:3-PC by FAD3 (ω-3 fatty acid desaturase) (Lou et al. 2014).

The modified fatty acids formed on PC are incorporated into glycerol backbone of triacylglycerol via several routes (Bates et al. 2009; Chapman and Ohlrogge 2012): (i) the modified FA formed by acyl editing on PC are incorporated onto DAG after their hydrolysis to form TAG by GPAT, LPAT, and DGAT enzymes, (ii) second route is through enzyme phospholipid:diacylglycerol acyltransferase (PDAT), which catalyzes the synthesis of triacylglycerol utilizing PC as an acyl donor and diacylglycerol as an acyl acceptor. This reaction usually takes place in transferring unusual fatty acids from PC into triacylglycerol (Dahlqvist et al. 2000), (iii) acyl-CoA:lysophosphatidylcholine acyltransferase (LPCAT) or phospholipase A2 (PLA2) generates a pool of polyunsaturated fatty acids or unusual fatty acids from PC using acyl editing reactions catalyzed by. PLA2 catalyzes the hydrolysis of glycerophospholipids to lysophospholipids and free fatty acids, which is then activated to CoA (Land's cycle) for the acyl-CoA-dependent acyltransferases (Mariani and Fidelio 2019). LPCAT plays an important role in regulating the acyl-CoA composition in plant cells by channeling PC-associated fatty acids into TAG. The reversible LPCAT activity contributes to the synthesis of lyso-phosphatidylcholine and acyl-CoA (Lager et al. 2013). An overview of the pathway involved in synthesis of TAG is shown in Figure 13.1.

13.3.3 Lipid Droplets Biogenesis

In seeds, the TAGs once formed accumulate in the cytosol in the form of specialized organelle called lipid droplets (LD), oil bodies, or oleosomes. The lipid droplets consist of a neutral lipid core surrounded by a phospholipid monolayer with structural proteins, such as oleosins, caleosins, and steroleosins. Oleosins are small proteins of molecular weight 15–30 kDa that exhibit three distinct structural domains: an N-terminal domain, a central hydrophobic anchoring domain, and a C-terminal amphipathic α-helical domain. The central large hydrophobic domain interacts with the charged groups of the phospholipid layer on the LD surface. Oleosins also play a major role in regulating the size and stability of lipid droplets. Arabidopsis has 17 genes encoding oleosins of which 5 are expressed in seeds, 3 in both seed and pollen, and 9 (8 in tandem) in the floral tapetum cells. Other minor proteins present in LD fractions are caleosin and steroleosin.

Caleosins, a 27 kDa protein comprises of C- and N-terminal hydrophilic domains and a central hydrophobic region. Arabidopsis express eight caleosin-encoding genes of which AtCLO1 and AtCLO2 are expressed abundantly in developing seeds (Poxleitner et al. 2006; Purkrtova et al. 2007). Steroleosins are sterol-binding dehydrogenases involved in metabolism of brassinosteroids. It possesses an N-terminal hydrophobic LD-binding domain with a proline knob motif and a soluble sterol-binding dehydrogenase/reductase domain

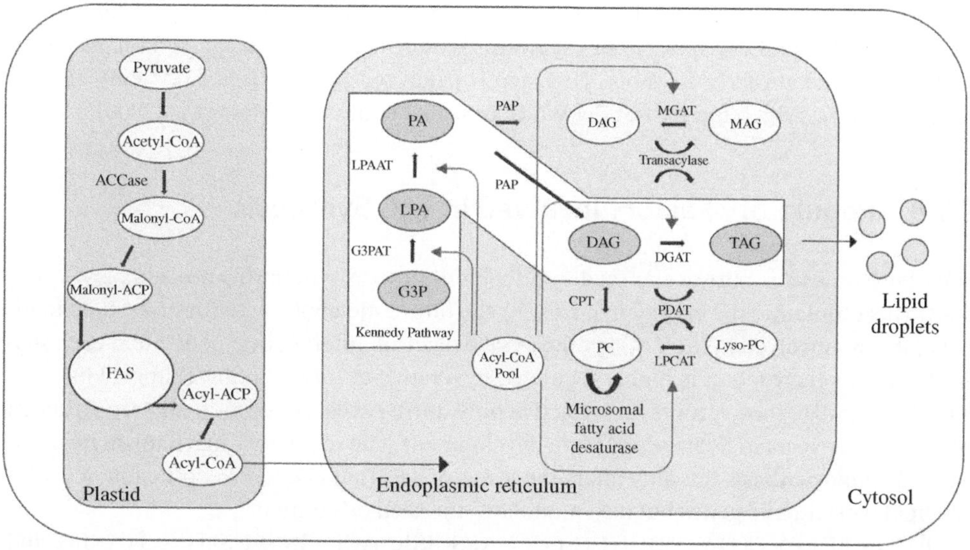

Figure 13.1 Overview of Triacylglycerol synthesis pathway. ACCase, acetyl-CoA carboxylase; ACP, acyl carrier protein; CPT, cytidine-5′ diphosphocholine:diacylglycerol cholinephosphotransferase; DAG, diacylglycerol; DGAT, diacylglycerol acyltransferase; FAS, fatty acid synthase; G3P, glycerol-3-phosphate; LPA, lysophosphatidic acid; MAG, monogalactosyl diacylglycerol; MGT, monoacylglycerol acyltransferase; PA, phosphatidic acid; PAP, phosphatidic acid phosphatase; PC, phosphatidylcholine; PDAT, phospholipid:diacylglycerol acyltransferase; PG, phosphatidylglycerol; TAG, triacylglycerol.

(Shao et al. 2019). Other LD-associated proteins, such as oil body associated protein 1 (OBAP1), and LDAP-interacting protein (LDIP) are also identified in oilseeds (Kretzschmar et al. 2020).

The mechanism of LD biogenesis is still not clear, while the existing evidence suggests that the LD synthesis initiates on the ER membrane. TAGs synthesized in the ER are sequestered in the hydrophobic region of the PL bilayer. Continuous accumulation of TAG at the ER membrane results in budding lipid droplet bordered by a layer of phospholipids. The lipid droplet formed is stabilized by incorporation of oleosins or other proteins on its surface (Galili et al. 1998; Ischebeck et al. 2020). The spherical oil bodies of size 0.5–2.0 μm diameter are formed in such a manner that the aliphatic chains of phospholipids point toward the lumen, where the TAGs accumulate, and the hydrophilic phosphate groups point toward the cytosol. The size and shape of LD are determined by the rate of synthesis of TAG and oleosins (Frandsen et al. 2001; Siloto et al. 2006).

13.3.4 Wax Esters Synthesis

Wax esters (WE) are neutral lipids composed of aliphatic alcohols and fatty acids. Both alcohols and fatty acids in WE are usually long-chain moieties. Wax esters are primarily found in the cuticles, but in some plants, they are storage lipids, like in jojoba (Simmondsia chinensis). WE constitute close to 97% of the seed oil in jojoba. WE are synthesized by a pathway, which involves fatty acid elongase (FAE), fatty acyl reductase (FAR), and wax synthase (WS).

FAE, a complex of four membrane-associated enzymes, elongates C18:1-CoA by two carbons at a time producing C20:1-CoA, C22:1-CoA, and C24:1-CoA. FAR catalyze the synthesis of long-chain fatty alcohols by reduction of the carboxyl group of long-chain acyl-CoAs. The final step of wax ester synthesis is esterification of the long-chain fatty acyl-CoAs to long-chain fatty alcohols. This step is catalyzed by fatty acyl CoA: fatty alcohol acyltransferase, which is a part of the WS (Lardizabal et al. 2000; Metz et al. 2000).

13.4 Regulatory Factors Involved in TAG Synthesis

TAG biosynthesis is part of the seed development process where this process plays a key role in accumulating the energy required for the future metabolic activities. The biochemical pathways involved in the TAG synthesis take place in different organelles, and each step in the metabolic reaction is highly regulated by a range of transcription factors (TFs). The TFs that regulate these processes are active only during seed maturation and are switched off during the vegetative phases of plant development. The regulatory mechanism involved in seed oil biosynthesis not only alters the levels of TFs that affect the expression of multiple lipid biosynthetic genes but also modulates the seed oil content.

The TFs LEAFY COTYLEDON1 (LEC1), ABSCISIC ACID INSENSITIVE3 (ABI3), and FUSCA3 (FUS3) create a transcriptional regulatory network named LAFL that is considered as the main regulator of embryogenesis and seed maturation. LAFL network directly regulates the seed oil content or fatty acid content by directly or indirectly regulating FA or TAG biosynthesis.

These transcription regulators also positively control the expression of genes involved in carbohydrate metabolism like sucrose synthesis and transport and glycolysis (Baud et al. 2007). Increasing the expression of these carbohydrate-metabolism genes results in increased carbon flux into fatty acid synthesis, imperative for higher seed oil content. These TFs also regulate the biosynthesis and signaling of various hormones, such as auxin, abscisic acid, gibberellin, cytokinin, and ethylene, at different stages of seed development.

WRINKLED1 (WRI1), a member of APETALA2 (AP2) family of TFs, is downstream component known to induce glycolytic and fatty acid biosynthesis genes during seed maturation. The expression of WRI1 is controlled by other TF like LEC1, LEC2, and FUS3. The WRI1 gene is so-named because the Arabidopsis plant with a mutation in the specific gene produced wrinkled seeds with 20% less oil content and higher starch levels compared to wild type. In Arabidopsis, WRI1 upregulates the expression of the GLB1 gene that codes for a protein, PII, that inhibit plastidial acyl-CoA carboxylase. Therefore, WRI1 through PII regulates the carbon flux and energy requirements during FA biosynthesis (Baud et al. 2010). The TFs such as DNA BINDING WITH ONE FINGER (DOF), GATA, and MYB TFs containing zinc finger motifs acting downstream of WRI1 have been identified to be regulating the promoters of fatty acid synthesis genes. DOF (GmDof4, GmDof11) regulates lipid synthesis by activating the acetyl-coenzyme A carboxylase (ACCase) gene by direct binding to the cis-DNA elements in the promoter region of the gene, and their constitutive expression in Arabidopsis enhances seed oil content (Wang et al. 2007; Zhang et al. 2014).

GLABRA2 (GL2), a transcription factor belonging to the homeodomain-leucine zipper (HD-ZIP) family, negatively regulates the transcription of oil synthesis genes. GL2 promotes seed coat synthesis via the expression of the gene MUCILAGE MODIFIED 4

(MUM4). MUM4 encodes a rhamnose synthase protein that is required for seed mucilage biosynthesis. The gl2 mutant had elevated seed oil content, which was related to loss of MUM4 function, which could be due to increased carbon flux into oil biosynthesis in the absence of seed coat mucilage formation (Shi et al. 2012). Additionally, the transcription factor AP2 also acts as a negative regulator of seed size. The Arabidopsis AP2 mutants had high hexose to sucrose ratio during seed development, indicating that AP2 modulates oil synthesis though its effect on carbohydrate metabolism (Ohto et al. 2005).

The TFs like TRANSPARENT TESTA 8 (TT8) negatively affect FA synthesis by directly inhibiting the transcription of LEC1, LEC2, and FUS3 (Chen et al. 2014). TT8 forms a ternary complex with other TFs TT2 and TRANSPARENT TESTA GLABRA 1 (TTG1) to control the carbon flow between metabolic pathways including FA, mucilage, and flavonoid synthesis (Li et al. 2018).

13.5 Metabolic Engineering for Lipid/Fatty Acid Synthesis

In order to meet the demands of the growing world's population and to have environment-friendly feedstock for industries, efforts are being made to increase the oil yield by enhancing the accumulation of oil in the seed without reducing the size of the seeds and their number per plant. Plant vegetative tissues such as leaves are source organs that perform metabolic activities like photosynthetic activities and metabolite transport, while tissues like seeds, fruits, and tubers are considered sink organs because of their major role in storing nutrient and energy for future needs. Since the biochemical pathways and their regulatory mechanisms are conserved between seed and vegetative plant tissue, the oil biosynthesis pathway can be upregulated in leaves and seeds in a similar manner for potentially higher oil yields.

Modification of plants for desirable traits has occurred throughout the historical domestication of crops, contributing to the development of improved varieties. The advent of genetic engineering and mutagenesis techniques allowed for the development of a new plant varieties with improved oil quality. The recombinant DNA technology is one of the fundamental techniques used for crop improvement and has replaced conventional plant-breeding methods that involves crossing and selection of new superior genotype combinations.

The genetic engineering technique requires an understanding of key genes and enzymes involved in metabolic pathways and involves the following major steps: (i) gene construction, that is, cloning the desired genes in a vector to overexpress or silence the gene. Insertion/deletion can be done upstream of a suitable promoter to achieve gene expression in the required tissue at the appropriate time. (ii) Transformation or gene insertion. (iii) Selection, regeneration, and analysis of transformants and testing effect of transgene on plant development.

Transgenic plants are the ones whose gene is modified using genetic engineering techniques – it could be the overexpression of a transgene under the transcriptional control of a strong promoter to increase the expression of a target enzyme or gene silencing to reduce the activity of a target enzyme. Several genetic methods that have been employed to engineer plant lipid metabolism include single gene manipulation, multigene stacking, gene silencing, gene knockout, overexpression of individual native genes, or insertion of non-native genes to introduce new trait in host species.

Gene silencing can be performed either by suppressing transcription via blocking a promoter region for the binding of transcriptional machinery in the nucleus or by targeting and degrading mRNA transcripts of specific genes in the cytoplasm (Wesley et al. 2004). Some of the methods of gene silencing are transposon silencing, co-suppression, RNA interference (RNAi)/siRNA, clustered regularly interspaced short palindromic repeats (CRISPR-CAS9), and random disruption by T-DNA insertion (Zebec et al. 2016; Bhalothia et al. 2020; Alok et al. 2021; Shushmita et al. 2021; Upadhyay 2021). T-DNA insertions also known as insertional mutagenesis can result in either complete loss of gene function (knockout) or reduced gene expression or activity (knockdown). Of all the methods, RNAi is the most common method used for silencing. In this method, RNA transcript (hairpin RNA) complementary to the target is transferred into a host cell, the hairpin RNA silences the gene expression by binding to the mRNA of the gene and thereby suppressing protein translation. Transgene silencing also known as co-suppression is an example of post-transcriptional gene silencing.

Overexpression of a single transgene expression depends on the type of promoter upstream of the gene, in plant studies for modulating lipid synthesis pathway, genes are expressed under strong protein promoters like seed specific promoter or oil body protein oleosin promoter. Multiple gene stacking is the expression of more than one gene from a metabolic pathway, and this approach is usually considered when there is a need to modify oil synthesis at different levels or to accumulate non-native fatty acids. Successful development of a transgenic plant expressing multiple transgenes requires transformation with mixtures of independent single transgene binary vectors carrying specific promoters. Spacers and insulators between the genes, orientation and position of a gene within the plasmid, and the regulatory elements affect multiple gene expression (Shockey et al. 2015).

Different methods are employed for the transfer of DNA into plant cells. These methods are divided into indirect methods where foreign gene is inserted into the host plant genome using a vector, and direct methods where DNA is directly inserted into the host genome. Agrobacterium-mediated transformation is an indirect method of genetic transformation that employs bacteria as a vector to insert the gene construct into the target plant cell. The method utilizes a binary vector system that consists of two plasmids in *Agrobacterium*, where one plasmid carries the gene to be transferred to plant cells and the other contains the virulence genes that are necessary for the DNA transfer but are not themselves transferred. The most commonly used binary vectors are pBIN19, pCB301, pPZP200, pCAMBIA-1300, pGreen, pLX, and pLSU (Lee and Gelvin 2008). Among the direct methods, particle bombardment biolistic is the most common method used in plant transformation. In this method, DNA coated with micron-sized gold or tungsten particles are bombarded into the target plant cell under high pressure using a "Gene Gun" (Sanford 1990). Other examples of direct methods are nanoparticle-mediated gene transfer, electroporation, liposome mediated, and polyethylene glycol based.

The aim of plant biotechnology is the development of novel crop varieties with desirable traits without disturbing their normal physiology. The goals of implementation of genetic engineering in plants for crop oil improvement are

- Increasing oil accumulation and yield
- Improving the quality of oil by altering the fatty acid profile

13.5.1 Increasing Oil Accumulation in Plants

Several strategies have been employed for oil content modification in plants, which includes the modulation of carbon flux toward TAG biosynthesis and/or FA synthesis, regulating of oil bodies biogenesis and their associated proteins, and the introduction of genes for alternate pathways for TAG synthesis from mammalian or yeast systems. Combinatorial metabolic strategies involving increased flow of carbon through the glycolysis and fatty acid biosynthesis pathways (Push), optimizing TAG assembly (Pull) and stabilizing lipid droplets or oil bodies (Package), and minimizing lipid turnover (Protect) are termed as "push, pull, package, and protect" strategy (Rogalski and Carrer 2011).

13.5.1.1 Modification of Fatty Acid Synthesis Pathway

One of the examples of metabolic engineering for elevating storage lipid accumulation involved overexpression of the genes involved in the fatty acid synthesis pathway. Overexpression of the acyl-CoA carboxylase from *Arabidopsis thaliana* in potato tubers resulted in a fivefold increase in the amount of TAG (Klaus et al. 2004). High yields of lauric acid (C12:0) were found in seeds of oilseed rape by expressing 12:0-ACP thioesterase from *Umbellularia californica* (Knutzon et al. 1999). Overexpression of *fabH* encoding for 3-ketoacyl-ACP synthase III (KAS III) gene in *B. napus* changed the fatty acid profile of the TAG in seeds without altering the total seed content (Verwoert et al. 1995). Seed-specific RNAi-mediated downregulation of KASII in cottonseed oil increased C16:0 fatty acids predominantly palmitic acid to 51% (Liu et al. 2017).

13.5.1.2 Increasing TAG Synthesis/Assembly Process

Since TAG biosynthesis occurs majorly through the Kennedy pathway, several transgenic studies have altered the enzymes involved in this pathway. One of the examples is the overexpression of safflower plastidial GPAT in Arabidopsis which increased seed oil content by 22%. Further, LPAT and diacylglycerol acyltransferase (DGAT 1) genes expression have been manipulated to elevate oil accumulation in seed as well as vegetative tissue. Furthermore, DGAT is a key enzyme in TAG synthesis regulates the flux of DAG from phospholipids synthesis toward TAG synthesis and hence oil formation in seeds.

Transgenic overexpression of the DGAT 1 gene alone or in combination increased seed oil content and seed weight (Bouvier-Navé et al. 2000; Jako et al. 2001; Taylor et al. 2009). Co-expression of PDAT1 gene with oleosin in Arabidopsis leaf increased TAG content by 6.4% without affecting membrane lipid composition and plant growth (Fan et al. 2013).

13.5.1.3 Increasing Carbon Flux Toward Oil Biosynthesis

The mitochondrial pyruvate dehydrogenase complex (mtPDC) plays a key role in the regulation of lipid biosynthesis by controlling the synthesis and availability of acetyl CoA, the primary substrate of FA synthesis. Pyruvate dehydrogenase kinase (PDHK) downregulates the activity of mtPDC activity through covalent modification (PDHK). Downregulation PDHK using RNAi increased the seed storage lipids and seed weight (Marillia et al. 2003). Furthermore, glucose-6-phosphate dehydrogenase (G6PDH) knock out studies have shown higher seed oil accumulation compared to wild type, which could be because of increased substrate availability for glycolysis and hence more precursors for FA synthesis in absence of pentose phosphate pathway (Andre et al. 2007).

13.5.1.4 Modulating the Expression of Transcription Factors

Another approach to increasing total seed oil is by altering the expression of TFs that are master regulator of metabolic processes in seed and vegetative tissue. A number of TFs like WRI, LEC1, EC2, GLABRA2 (GL2), MYC, and AP2 have been cloned using transgenic engineering; however, there is a need to optimize their expression as they regulate the expression of multiple enzymes in different metabolic pathways (Century et al. 2008). Overexpression of wri1-like gene from *B. napus* in *A. thaliana* observed enlarged seed size with 10–40% increased seed oil content (Liu et al. 2010). LEC2 overexpression in vegetative tissue enhanced oil accumulation by stabilizing TAG containing lipid droplets via upregulation of protein associated with oil bodies (Feeney et al. 2013).

13.5.1.5 Reducing the Hydrolysis of Storage Lipids

Based on the observation that there is a 10% loss in the TAG content of rapeseed (B. napus) during the final stage of seed development, recent genetic engineering studies are aimed at enhancing oil yield by reducing the breakdown of stored lipids by suppressing TAG hydrolysis in seed and vegetative tissues (Chia et al. 2005). Sugar-Dependent 1 (sdp1) is the major lipase observed to be involved in oil body degradation in seeds of Arabidopsis (Eastmond 2006). The Arabidopsis lipase mutants sdp1 had a 50% increase in TAG levels compared to wild type due to disruption of TAG hydrolysis, also the accumulation was more in roots and seeds. The study has also shown that the TAG accumulation increases by double in roots, stem, and leaves in case of transgenic plants constitutively co-expressing WRINKLED1 and DGAT1, sdp1. This is an example of multiple gene stacking. sdp1 gene suppression performed in Arabidopsis, *B. napus*, and *Jatropa curcas* using the RNAi method in different studies showed enhanced seed oil content (Eastmond 2006; Kelly et al. 2011; van Erp et al. 2014).

Additionally, strategies like extending the period of oil synthesis during seed development, introducing new TAG synthesis genes from other species, improving physiological conditions for oil synthesis, and increasing sink size for oil accumulation (endosperm size) have been employed to enhance the seed oil accumulation in plants (Adamski et al. 2009; Shen et al. 2010; Vigeolas et al. 2011; Kanai et al. 2016).

13.5.2 Improving the Quality of Oil by Altering the Fatty Acid Profile

There is also growing interest in modifying the fatty acid composition of seed oils via transgenic metabolic engineering to generate a new quality of oil now termed as "designer oil." The modeling of plant biochemical pathways was employed to generate diverse fatty acids or oils with nutritional and industrial importance like ω-3 long-chain polyunsaturated fatty acids. Different pathways have been introduced into plants to produce DHA in seed oil, one of them is the use of elongases and desaturases from microalgae or fungi to convert plant fatty acids into long-chain polyunsaturated fatty acids. Transgenic *Camelina sativa* and Arabidopsis modified using above method have shown increased DHA content (Usher et al. 2017). Similarly, the synthesis and accumulation of eicosapentaenoic acid (EPA) was increased (up to 29% of total fatty acids) in the Brassica oilseeds (Cheng et al. 2010). Another example of tailoring is the modification of oil composition of soybean by simultaneously reducing the expression of the enzymes fad2 (oleate desaturase) and FatB (palmitoyl thioesterase) in soybean seeds. This manipulation resulted in production of oil with only C18:0 monounsaturated oleic acid instead of palmitic acid (unsaturated C16:0) and other unsaturated fatty acids (Graef et al. 2009). Additionally, other industrial fatty acids or unusual fatty acids have also been produced using the genetic engineering method. Co-expression of fatty acid conjugase (*Vernicia fordii*) and dgat2 increased total Arabidopsis leaf oil content by channeling of eleostearic acid into TAG (Yurchenko et al. 2017). Metabolic engineering using *Ricinus communis* fatty acid hydroxylase 12 (fah12) in Arabidopsis showed an increase in production of hydroxy fatty acids (HFAs) production, while double-transgenic Arabidopsis plants expressing fah12 and dgat2 showed a significant increase (31%) in HFA content, this method of transgene production is an example of gene stacking (Burgal et al. 2008).

A number of studies are attempting combinatorial metabolic engineering by expressing multiple genes in plants. A "Push + Pull + Package" transgenic approach was employed to overexpress PDAT1, oleosin1, and tdg1 gene in Arabidopsis. It resulted in accumulation of up to 8.6% TAG on dry weight basis. A quadruple mutant containing mutations in tgd1, suc2, adg1, and tt4 genes of *A. thaliana* showed 3.8% increase in TAG accumulation in leaf tissue (Zhai et al. 2017). An example of "Push + Pull + Protect" approach was the overexpression of WRI1 and DGAT1 genes in the sdp1 knock out mutant of *A. thaliana* (Kelly et al. 2011). The strategy resulted in a significant increase in the TAG content in the vegetative tissues (leaves, stems, and roots). Other examples of multiple gene stacking and other genetic studies are given in Table 13.1 (Correa et al. 2020).

13.6 Conclusions

Numerous strategies of genetic manipulations in a wide variety of plant species along with multiple flux studies analyzing the key lipid metabolic reactions have opened opportunities to create novel products with desirable characteristics. Multiple challenges still remain in this field. Deep understanding of the complex metabolic pathways responsible for directing beneficial fatty acids into TAG and the roles played by related fatty acid metabolites in plant growth and development will aid in generating transgenics for efficient accumulation of the desired plant oils. Given the benefits of plant oils for both human health and nutrition as well as their role as a sustainable replacement for fast depleting industry feedstock, there is a need to accelerate the process of genome-editing technologies and mutation-based metabolic engineering to create novel plant oil crops.

Table 13.1 List of transgenic plants with the modified fatty acid or lipid.

Engineered fatty acid or lipid	Transgenic plant	Target tissue	Gene	Gene source
Hydroxy fatty acid	*Arabidopsis thaliana*	Seed	*fah12* ↑	*Ricinus communis*
	A. thaliana	Seed	*fah12* ↑ + *dgat2* ↑	*R. communis*
	A. thaliana	Seed	*AtFAE1* ↓ + *rcfah12* ↑ + *RcPDAT1A* ↑	*R. communis; A. thaliana*
	Camelina sativa	Seed	*RcPLCL1* ↑ and rcfah12 ↑	*R. communis*
Lauric acid	*A. thaliana*	Seed	*BTE* ↑	*Umbellularia californica*
Lauric; myristic acid	*A. thaliana*	Seed	*FatB1* ↑ + *FatB2* ↑	*Cuphea wrightii*
Palmitic acid	*A. thaliana*	Seed	*FatB1* ↑	*A. thaliana*
	Brassica napus	Seed	*ChFatB1* ↑	*Cuphea hookeriana*
	Gossypium hirsutum	Seed	*GhKAS2* ↓	*G. hirsutum*
Oleic acid	*A. thaliana*	Seed	*AtROD1* ↓	*A. thaliana*
	A. thaliana	Seed	*AtROD1* ↓, *AtLPCAT1* ↓, *AtLPCAT2* ↓	*A. thaliana*
	A. thaliana	Seed	*fad2* ↓	
	B. napus	Seed	18:1-Δ12-desaturase ↓	*B. napus*
	Brassica juncea	Seed	18:1-Δ12-desaturase ↓	*B. juncea*
Stearic acid	*G. hirsutum*	Seed	*GhSAD1* ↓	*G. hirsutum*
Linoleic acid	*B. napus*	Seed	18:1-Δ12-desaturase ↑	*Myrianthus arboreus*
γ-linolenic acid	*B. napus*	Seed	18:2-Δ6-desaturase ↑, 18:1-Δ12-Desaturase ↑	*Borago pygmea*
Erucic acid	*B. napus*	Seed	*Bn-FAE1.1* ↑, erucoyl-CoA specific-LPAAT ↑	*B. napus*
Eicosapentaenoic acid	*A. thaliana*	Leaves	*IgASE1* ↑ + *EuΔ8* ↑ + *MortΔ5* ↑	*Isochrysis galbana; Euglena gracilis; Mortierella alpina*

	Brassica carinata (zero-erucic line 10H3)	Seed	*CpDesX* ↑ + *Pir-ω3* ↑	*Claviceps purpurea; Pythium irregulare*
	Glycine max	Seed	*PspELO* ↑ + *SaΔ4D* ↑	*Pavlo* sp. *Schizochytrium aggregatum*
Docosahexaenoic acid; Eicosapentaenoic acid	*B. napus*	Seed	*[OrfA, OrfB, OrfC, PFA1, PFA2, PFA3, Hetl]* ↑	*Schizochytrium* sp.
Docosahexaenoic acid	*A. thaliana*	Seed	[LkΔ12 + PpΔ15/ω3 + MpΔ6 + PcΔ6ELO + PcΔ5ELO + PsΔ5 + PsΔ4] ↑	*L. kluyveri; Pichia pastoris; Micromonas pusilla; Pyramimonas cordata; Pavlova salina*
	G. max	Embryo	*PspELO* ↑ + *SaΔ4D* ↑	*Pavlova* sp.; *S. aggregatum*
Very long-chain polyunsaturated fatty acids	*A. thaliana*	Seed	EPA construct [OtΔ6 + PSE1 + TcΔ5] ↑; DHA construct [OtΔ6 + PSE1 + TcΔ5 + TbElo5 + TcΔ4 + OtElo5 and EhΔ4] ↑	*O. tauri; P. patens; Thraustochytrium* sp; *T. brucei; E. huxleyi*
	A. thaliana	Seed	*[McD6DES + AsELOVL5 + PtD5DES]* ↑	*Muraenesox cinereus; Acanthopagrus schlegelii; Phaeodactylum tricornutum*
Polyunsaturated fatty acids	*A. thaliana*	Seed	*Atfad2* ↑ + *AtRDR6* ↓	*A. thaliana*
Total fatty acids	*A. thaliana*	Vegetative tissues	*AtWRI1* ↑ + *AtAPS1* ↓	*A. thaliana*
	B. napus	Seed	*AtACC1* ↑	*A. thaliana*
Total triacylglycerol	*A. thaliana*	Leaves	*CrDGTT2* ↑	*Chlamydomonas reinhardtii*
	A. thaliana	Vegetative tissues	*Atsdp1* ↓ + *AtWRI1* ↑ + *AtDGAT1* ↑	*A. thaliana*
	A. thaliana	Leaves	*PDAT1*↑ + *sdp1* ↓ + *PXA1* ↓	*A. thaliana*
	A. thaliana	Leaves	*AtLEC2 with* pAtNAP promoter ↑	*A. thaliana*
	A. thaliana	Leaves	*AtDGAT1* ↑ + *Cys-oleosin* ↑	*A. thaliana*
	A. thaliana	Leaves	*AtADG1* ↓ + *Atsdp1-4* ↓ or At*PXA1* ↓	*A. thaliana*
	Nicotiana benthamiana	Leaves	*WRI1* ↑	*Avena sativa; Cyperus esculentus; Solanum tuberosum; Populus trichocarpa; A. thaliana*

(*Continued*)

Table 13.1 (Continued)

Engineered fatty acid or lipid	Transgenic plant	Target tissue	Gene	Gene source
	N. benthamiana	Leaves	*CsWRI1* ↑	*C. sativa*
	N. benthamiana	Leaves	*AtWRI1* ↑ + *AtDGAT1* ↑	*A. thaliana*
	N. tabacum	Leaves	*AtWRI1* ↑ + *AtDGAT1* ↑ + *SiOLEOSIN* ↑	*A. thaliana; Sesamum indicum*
	N. tabacum	Leaves	[*AtWRI1* + *AtDGAT1* + *SiOLEOSIN*] ↑ + *AtLEC2* ↑ or *sdp1* (with SAG12 senescence-specific promoter) ↓	*A. thaliana; S. indicum*
	Saccharum s pp.	Leaves	(*AtWRI*1 + *ZmDGAT*1-2 + At OLE1) ↑ + (RNAi dual vector *PXA*1 and *AGP*) ↓	*A. thaliana; Zea mays*
	S. tuberosum	Tubers	*AtACC1* ↑	*A. thaliana*
	S. tuberosum	Tubers	*AtWRI1* ↑ + *AtDGAT1* ↑ + *SiOLEOSIN* ↑	*A. thaliana; S. indicum*
Total lipids	*A. thaliana*	Seed	*WRI1* ↑ + *DGAT1* ↑ + *sdp1* ↓	*A. thaliana*
	B. napus	Seed	*gpd 1* ↑	*Saccharomyces cerevisiae*
	B. napus	Seed	*sdp1* ↓	*B. napus*
Wax esters	*A. thaliana*		*MmWS* ↑ + *MmFAR* ↑	*Mus musculus*

Down arrow represents for downregulation and up arrow represents for up-regulation.

References

Abbadi, A., Domergue, F., Bauer, J. et al. (2004). Biosynthesis of very-long-chain polyunsaturated fatty acids in transgenic oilseeds: constraints on their accumulation. *Plant Cell* 16 (10): 2734–2748.

Adamski, N.M., Anastasiou, E., Eriksson, S. et al. (2009). Local maternal control of seed size by KLUH/CYP78A5-dependent growth signaling. *Proc. Natl Acad. Sci. USA* 106 (47): 20115–20120.

Alok, A., Chauhan, H., Upadhyay, S.K. et al. (2021). Compendium of plant-specific CRISPR vectors and their technical advantages. Life 11: 1021. https://doi.org/10.3390/life11101021.

Andre, C., Froehlich, J.E., Moll, M.R., and Benning, C. (2007). A heteromeric plastidic pyruvate kinase complex involved in seed oil biosynthesis in Arabidopsis. *Plant Cell* 19 (6): 2006–2022.

Bates, P.D., Durrett, T.P., Ohlrogge, J.B., and Pollard, M. (2009). Analysis of acyl fluxes through multiple pathways of triacylglycerol synthesis in developing soybean embryos. *Plant Physiol.* 150 (1): 55–72.

Bates, P.D., Stymne, S., and Ohlrogge, J. (2013). Biochemical pathways in seed oil synthesis. *Curr. Opin. Plant Biol.* 16 (3): 358–364.

Baud, S., Wuillème, S., Dubreucq, B. et al. (2007). Function of plastidial pyruvate kinases in seeds of *Arabidopsis thaliana*. *Plant J.* 52 (3): 405–419.

Baud, S., Feria Bourrellier, A.B., Azzopardi, M. et al. (2010). PII is induced by WRINKLED1 and fine-tunes fatty acid composition in seeds of *Arabidopsis thaliana*. *Plant J.* 64 (2): 291–303.

Bhalothia, P., Yajnika, K., Alok, A., and Upadhayay, S.K. (2020). The current progress of CRISPR/Cas9 development in plants. In: *Genome Engineering via CRISPR-Cas9 Systems* (ed. V. Singh and P.K. Dhar). Elsevier. https://doi.org/10.1016/B978-0-12-818140-9.00011-8.

Bouvier-Navé, P., Benveniste, P., Oelkers, P. et al. (2000). Expression in yeast and tobacco of plant cDNAs encoding acyl CoA:diacylglycerol acyltransferase. *Eur. J. Biochem.* 267 (1): 85–96.

Burgal, J., Shockey, J., Lu, C. et al. (2008). Metabolic engineering of hydroxy fatty acid production in plants: RcDGAT2 drives dramatic increases in ricinoleate levels in seed oil. *Plant Biotechnol. J.* 6 (8): 819–831.

Cahoon, E.B. and Li-Beisson, Y. (2020). Plant unusual fatty acids: learning from the less common. *Curr. Opin. Plant Biol.* 55: 66–73.

Carlsson, A.S. (2009). Plant oils as feedstock alternatives to petroleum – a short survey of potential oil crop platforms. *Biochimie* 91 (6): 665–670.

Carman, G.M. and Han, G.-S. (2006). Roles of phosphatidate phosphatase enzymes in lipid metabolism. *Trends Biochem. Sci.* 31 (12): 694–699.

Century, K., Reuber, T.L., and Ratcliffe, O.J. (2008). Regulating the regulators: the future prospects for transcription-factor-based agricultural biotechnology products. *Plant Physiol.* 147 (1): 20–29.

Chapman, K.D. and Ohlrogge, J.B. (2012). Compartmentation of triacylglycerol accumulation in plants. *J. Biol. Chem.* 287 (4): 2288–2294.

Chapman, K.D., Dyer, J.M., and Mullen, R.T. (2013). Commentary: why don't plant leaves get fat? *Plant Sci.* 207: 128–134.

Chen, M., Xuan, L., Wang, Z. et al. (2014). TRANSPARENT TESTA8 Inhibits seed Fatty acid accumulation by targeting several seed development regulators in Arabidopsis. *Plant Physiol.* 165 (2): 905–916.

Cheng, B., Wu, G., Vrinten, P. et al. (2010). Towards the production of high levels of eicosapentaenoic acid in transgenic plants: the effects of different host species, genes and promoters. *Transgenic Res.* 19 (2): 221–229.

Chia, T.Y.P., Pike, M.J., and Rawsthorne, S. (2005). Storage oil breakdown during embryo development of *Brassica napus* (L.). *J. Exp. Bot.* 56 (415): 1285–1296.

Correa, S.M., Fernie, A.R., Nikoloski, Z., and Brotman, Y. (2020). Towards model-driven characterization and manipulation of plant lipid metabolism. *Prog. Lipid Res.* 80: 101051.

Dahlqvist, A., Stahl, U., Lenman, M. et al. (2000). Phospholipid:diacylglycerol acyltransferase: an enzyme that catalyzes the acyl-CoA-independent formation of triacylglycerol in yeast and plants. *Proc. Natl. Acad. Sci. USA* 97 (12): 6487–6492.

Eastmond, P.J. (2006). Sugar-Dependent 1 encodes a patatin domain triacylglycerol lipase that initiates storage oil breakdown in germinating Arabidopsis seeds. *Plant Cell* 18 (3): 665–675.

van Erp, H., Kelly, A.A., Menard, G., and Eastmond, P.J. (2014). Multigene engineering of triacylglycerol metabolism boosts seed oil content in Arabidopsis 1[W][OPEN]. *Plant Physiol.* 165 (1): 30–36.

Fan, J., Yan, C., Zhang, X., and Xu, C. (2013). Dual role for phospholipid:diacylglycerol acyltransferase: enhancing fatty acid synthesis and diverting fatty acids from membrane lipids to triacylglycerol in Arabidopsis leaves. *Plant Cell* 25 (9): 3506–3518.

Fatland, B.L., Ke, J., Anderson, M.D. et al. (2002). Molecular characterization of a heteromeric ATP-citrate lyase that generates cytosolic acetyl-coenzyme A in Arabidopsis. *Plant Physiol.* 130 (2): 740–756.

Feeney, M., Frigerio, L., Cui, Y., and Menassa, R. (2013). Following vegetative to embryonic cellular changes in leaves of Arabidopsis overexpressing LEAFY COTYLEDON2. *Plant Physiol.* 162 (4): 1881–1896.

Fich, E.A., Segerson, N.A., and Rose, J.K.C. (2016). The plant polyester cutin: biosynthesis, structure, and biological roles. *Annu. Rev. Plant Biol.* 67: 207–233.

Frandsen, G.I., Mundy, J., and Tzen, J.T.C. (2001). Oil bodies and their associated proteins, oleosin and caleosin. *Physiol. Plant.* 112 (3): 301–307.

Gagné, S.J., Reed, D.W., Gray, G.R., and Covello, P.S. (2009). Structural control of chemoselectivity, stereoselectivity, and substrate specificity in membrane-bound fatty acid acetylenases and desaturases. *Biochemistry* 48 (51): 12298–12304.

Galili, G., Sengupta-Gopalan, C., and Ceriotti, A. (1998). The endoplasmic reticulum of plant cells and its role in protein maturation and biogenesis of oil bodies. *Plant Mol. Biol.* 38 (1–2): 1–29.

Graef, G., LaVallee, B.J., Tenopir, P. et al. (2009). A high-oleic-acid and low-palmitic-acid soybean: agronomic performance and evaluation as a feedstock for biodiesel. *Plant Biotechnol. J.* 7 (5): 411–421.

Guerin, C., Serret, J., Montúfar, R. et al. (2020). Palm seed and fruit lipid composition: phylogenetic and ecological perspectives. *Ann. Bot.* 125 (1): 157–172.

Harwood, J.L. (2005). Fatty acid biosynthesis. In: *Plant Lipids* (ed. D.J. Murphy). Blackwell.

Haslam, T.M. and Kunst, L. (2013). Extending the story of very-long-chain fatty acid elongation. *Plant Sci.* 210: 93–107.

Haslam, R.P., Sayanova, O., Kim, H.J. et al. (2016). Synthetic redesign of plant lipid metabolism. *Plant J.* 87 (1): 76–86.

Horn, P.J. and Benning, C. (2016). The plant lipidome in human and environmental health. *Science* 353 (6305): 1228–1232.

Huang, J., Zhang, T., Zhang, Q. et al. (2016). The mechanism of high contents of oil and oleic acid revealed by transcriptomic and lipidomic analysis during embryogenesis in *Carya cathayensis* Sarg. *BMC Genomics* 17 (1): 113.

Ischebeck, T., Krawczyk, H.E., Mullen, R.T. et al. (2020). Lipid droplets in plants and algae: distribution, formation, turnover and function. *Semin. Cell Dev. Biol.* 108: 82–93.

Jako, C., Kumar, A., Wei, Y. et al. (2001). Seed-specific over-expression of an Arabidopsis cDNA encoding a diacylglycerol acyltransferase enhances seed oil content and seed weight. *Plant Physiol.* 126 (2): 861–874.

Joubès, J., Raffaele, S., Bourdenx, B. et al. (2008). The VLCFA elongase gene family in *Arabidopsis thaliana*: phylogenetic analysis, 3D modelling and expression profiling. *Plant Mol. Biol.* 67 (5): 547–566.

Kanai, M., Mano, S., Kondo, M. et al. (2016). Extension of oil biosynthesis during the mid-phase of seed development enhances oil content in Arabidopsis seeds. *Plant Biotechnol. J.* 14 (5): 1241–1250.

Kelly, A.A., Quettier, A.-L., Shaw, E., and Eastmond, P.J. (2011). Seed storage oil mobilization is important but not essential for germination or seedling establishment in Arabidopsis 1[W]. *Plant Physiol.* 157 (2): 866–875.

Klaus, D., Ohlrogge, J.B., Neuhaus, H.E., and Dörmann, P. (2004). Increased fatty acid production in potato by engineering of acetyl-CoA carboxylase. *Planta* 219 (3): 389–396.

Knutzon, D.S., Hayes, T.R., Wyrick, A. et al. (1999). Lysophosphatidic acid acyltransferase from coconut endosperm mediates the insertion of laurate at the sn-2 position of triacylglycerols in lauric rapeseed oil and can increase total laurate levels. *Plant Physiol.* 120 (3): 739–746.

Kretzschmar, F.K., Doner, N.M., Krawczyk, H.E. et al. (2020). Identification of low-abundance lipid droplet proteins in seeds and seedlings. *Plant Physiol.* 182 (3): 1326–1345.

Lager, I., Yilmaz, J.L., Zhou, X.-R. et al. (2013). Plant acyl-CoA:lysophosphatidylcholine acyltransferases (LPCATs) have different specificities in their forward and reverse reactions. *J. Biol. Chem.* 288 (52): 36902–36914.

Lardizabal, K.D., Metz, J.G., Sakamoto, T. et al. (2000). Purification of a jojoba embryo wax synthase, cloning of its cDNA, and production of high levels of wax in seeds of transgenic Arabidopsis. *Plant Physiol.* 122 (3): 645–655.

Lee, L.-Y. and Gelvin, S.B. (2008). T-DNA binary vectors and systems. *Plant Physiol.* 146 (2): 325–332.

Li, N., Xu, C., Li-Beisson, Y., and Philippar, K. (2016). Fatty acid and lipid transport in plant cells. *Trends Plant Sci.* 21 (2): 145–158.

Li, X., Teitgen, A.M., Shirani, A. et al. (2018). Discontinuous fatty acid elongation yields hydroxylated seed oil with improved function. *Nat. Plants* 4 (9): 711–720.

Li, C., Zhang, B., Chen, B. et al. (2018). Site-specific phosphorylation of TRANSPARENT TESTA GLABRA1 mediates carbon partitioning in Arabidopsis seeds. *Nat. Commun.* 9 (1): 571.

Lin, C.-Y. and Eudes, A. (2020). Strategies for the production of biochemicals in bioenergy crops. *Biotechnol. Biofuels* 13: 71.

Liu, J., Hua, W., Zhan, G. et al. (2010). Increasing seed mass and oil content in transgenic Arabidopsis by the overexpression of wri1-like gene from *Brassica napus*. *Plant Physiol. Biochem.* 48 (1): 9–15.

Liu, Q., Siloto, R.M.P., Lehner, R. et al. (2012). Acyl-CoA:diacylglycerol acyltransferase: molecular biology, biochemistry and biotechnology. *Prog. Lipid Res.* 51 (4): 350–377.

Liu, Q., Wu, M., Zhang, B. et al. (2017). Genetic enhancement of palmitic acid accumulation in cotton seed oil through RNAi down-regulation of ghKAS2 encoding β-ketoacyl-ACP synthase II (KASII). *Plant Biotechnol. J.* 15 (1): 132–143.

Lou, Y., Schwender, J., and Shanklin, J. (2014). FAD2 and FAD3 desaturases form heterodimers that facilitate metabolic channeling in vivo. *J. Biol. Chem.* 289 (26): 17996–18007.

Mariani, M.E. and Fidelio, G.D. (2019). Secretory phospholipases A2 in plants. *Front. Plant Sci.* 10: 861.

Marillia, E., Micallef, B.J., Micallef, M. et al. (2003). Biochemical and physiological studies of *Arabidopsis thaliana* transgenic lines with repressed expression of the mitochondrial pyruvate dehydrogenase kinase1. *J. Exp. Bot.* 54 (381): 259–270.

Metz, J.G., Pollard, M.R., Anderson, L. et al. (2000). Purification of a jojoba embryo fatty acyl-coenzyme A reductase and expression of its cDNA in high erucic acid rapeseed. *Plant Physiol.* 122 (3): 635–644.

Murata, N. and Tasaka, Y. (1997). Glycerol-3-phosphate acyltransferase in plants. *Biochim. Biophys. Acta, Lipids Metab.* 1348 (1): 10–16.

Napier, J.A. and Graham, I.A. (2010). Tailoring plant lipid composition: designer oilseeds come of age. *Curr. Opin. Plant Biol.* 13 (3): 330–337.

Ohlrogge, J., Thrower, N., Mhaske, V. et al. (2018). PlantFAdb: a resource for exploring hundreds of plant fatty acid structures synthesized by thousands of plants and their phylogenetic relationships. *Plant J.* 96 (6): 1299–1308.

Ohto, M.-A., Fischer, R.L., Goldberg, R.B. et al. (2005). Control of seed mass by APETALA2. *Proc. Natl. Acad. Sci. USA* 102 (8): 3123–3128.

Piazza, G.J. and Foglia, T.A. (2001). Rapeseed oil for oleochemical usage. *Eur. J. Lipid Sci. Technol.* 103 (7): 450–454.

Poxleitner, M., Rogers, S.W., Lacey Samuels, A. et al. (2006). A role for caleosin in degradation of oil-body storage lipid during seed germination. *Plant J.* 47 (6): 917–933.

Purkrtova, Z., d'Andrea, S., Jolivet, P. et al. (2007). Structural properties of caleosin: a MS and CD study. *Arch. Biochem. Biophys.* 464 (2): 335–343.

Raboanatahiry, N., Wang, B., Yu, L., and Li, M. (2018). Functional and structural diversity of acyl-coA binding proteins in oil crops. *Front. Genet.* 9: 182.

Rawsthorne, S. (2002). Carbon flux and fatty acid synthesis in plants. *Prog. Lipid Res.* 41 (2): 182–196.

Rogalski, M. and Carrer, H. (2011). Engineering plastid fatty acid biosynthesis to improve food quality and biofuel production in higher plants. *Plant Biotechnol. J.* 9 (5): 554–564.

Sanford, J.C. (1990). Biolistic plant transformation. *Physiol. Plant.* 79 (1): 206–209.

Shanklin, J. and Cahoon, E.B. (1998). Desaturation and related modifications of fatty acids1. *Annu. Rev. Plant Physiol. Plant Mol. Biol.* 49: 611–641.

Shao, Q., Liu, X., Su, T. et al. (2019). New insights into the role of seed oil body proteins in metabolism and plant development. *Front. Plant Sci.* 10: 1568.

Shen, B., Allen, W.B., Zheng, P. et al. (2010). Expression of ZmLEC1 and ZmWRI1 increases seed oil production in Maize 1[W][OA]. *Plant Physiol.* 153 (3): 980–987.

Shi, L., Katavic, V., Yu, Y. et al. (2012). Arabidopsis glabra2 mutant seeds deficient in mucilage biosynthesis produce more oil. *Plant J.* 69 (1): 37–46.

Shockey, J.M., Fulda, M.S., and Browse, J.A. (2002). Arabidopsis contains nine long-chain acyl-coenzyme a synthetase genes that participate in fatty acid and glycerolipid metabolism. *Plant Physiol.* 129 (4): 1710–1722.

Shockey, J., Mason, C., Gilbert, M. et al. (2015). Development and analysis of a highly flexible multi-gene expression system for metabolic engineering in Arabidopsis seeds and other plant tissues. *Plant Mol. Biol.* 89 (1): 113–126.

Siloto, R.M.P., Findlay, K., Lopez-Villalobos, A. et al. (2006). The accumulation of oleosins determines the size of seed oilbodies in Arabidopsis. *Plant Cell* 18 (8): 1961–1974.

Statista (2020). World vegetable oil production 2020.

Sultan, S.M., Dikshit, N., and Vaidya, U.J. (2015). Oil content and fatty acid composition of soybean (Glysine max L.) genotypes evaluated under rainfed conditions of Kashmir Himalayas in India. *J. Applied Nat. Sci.* 7 (2): 910–915.

Sushmita, K.G., Upadhyay, S.K., and Verma, P.C. (2021). An overview of genome-engineering methods. In: *Genome Engineering for Crop Improvement* (ed. S.K. Upadhyay), 1–21. John Wiley & Sons Ltd. https://doi.org/10.1002/9781119672425.ch1.

Tao, B.Y. (2007). Chapter 24 - Industrial applications for plant oils and lipids. In: *Bioprocessing for Value-Added Products from Renewable Resources* (ed. S.-T. Yang), 611–627. Amsterdam: Elsevier.

Taylor, D.C., Zhang, Y., Kumar, A. et al. (2009). Molecular modification of triacylglycerol accumulation by over-expression of DGAT1 to produce canola with increased seed oil content under field conditions. This paper is one of a selection of papers published in a Special Issue from the National Research Council of Canada – Plant Biotechnology Institute. *Botany* 87 (6): 533–543.

Theodoulou, F.L. and Eastmond, P.J. (2012). Seed storage oil catabolism: a story of give and take. *Curr. Opin. Plant Biol.* 15 (3): 322–328.

Upadhyay. S.K. (ed.) (2021). *Genome Engineering for Crop Improvement*. John Wiley & Sons Ltd. https://doi.org/10.1002/9781119672425.

Upadhyay S.K. and Singh S.P. (eds.) (2021). *Bioprospecting of Plant Biodiversity for Industrial Molecules*. John Wiley & Sons Ltd. https://doi.org/10.1002/9781119718017.

Usher, S., Han, L., Haslam, R.P. et al. (2017). Tailoring seed oil composition in the real world: optimising omega-3 long chain polyunsaturated fatty acid accumulation in transgenic *Camelina sativa*. *Sci. Rep.* 7 (1): 6570.

Verwoert, I.I.G.S., van der Linden, K.H., Walsh, M.C. et al. (1995). Modification of *Brassica napus* seed oil by expression of the Escherichia coli fabH gene, encoding 3-ketoacyl-acyl carrier protein synthase III. *Plant Mol. Biol.* 27 (5): 875–886.

Vigeolas, H., Hühn, D., and Geigenberger, P. (2011). Nonsymbiotic hemoglobin-2 leads to an elevated energy state and to a combined increase in polyunsaturated fatty acids and total oil content when overexpressed in developing seeds of transgenic Arabidopsis plants 1[OA]. *Plant Physiol.* 155 (3): 1435–1444.

Wang, H.-W., Zhang, B., Hao, Y.-J. et al. (2007). The soybean Dof-type transcription factor genes, GmDof4 and GmDof11, enhance lipid content in the seeds of transgenic Arabidopsis plants. *Plant J.* 52 (4): 716–729.

Wesley, S.V., Helliwell, C., Wang, M.-B., and Waterhouse, P. (2004). Posttranscriptional gene silencing in plants. *Methods Mol. Biol.* 265: 117–129.

Yeats, T.H. and Rose, J.K.C. (2013). The formation and function of plant cuticles. *Plant Physiol.* 163 (1): 5–20.

Yoon, J.M., Zhao, L., and Shanks, J.V. (2013). Metabolic engineering with plants for a sustainable biobased economy. *Annu. Rev. Chem. Biomol. Eng.* 4: 211–237.

Yuan, L. and Grotewold, E. (2015). Metabolic engineering to enhance the value of plants as green factories. *Metab. Eng.* 27: 83–91.

Yurchenko, O., Shockey, J.M., Gidda, S.K. et al. (2017). Engineering the production of conjugated fatty acids in *Arabidopsis thaliana* leaves. *Plant Biotechnol. J.* 15 (8): 1010–1023.

Zebec, Z., Zink, I.A., Kerou, M., and Schleper, C. (2016). Efficient CRISPR-mediated post-transcriptional gene silencing in a hyperthermophilic Archaeon using multiplexed crRNA expression. *G3 (Bethesda)* 6 (10): 3161–3168.

Zhai, Z., Liu, H., Xu, C., and Shanklin, J. (2017). Sugar potentiation of fatty acid and triacylglycerol accumulation. *Plant Physiol.* 175 (2): 696–707.

Zhang, J., Hao, Q., Bai, L. et al. (2014). Overexpression of the soybean transcription factor GmDof4 significantly enhances the lipid content of Chlorella ellipsoidea. *Biotechnol. Biofuels* 7: 128.

Zhou, Y., Zhao, W., Lai, Y. et al. (2020). Edible plant oil: global status, health issues, and perspectives. *Front. Plant Sci.* 11: 1315.

14

Plants as Bioreactors for the Production of Biopesticides

Fernanda Achimón[1,2], *Vanessa A. Areco*[3], *Vanessa D. Brito*[1,2], *María L. Peschiutta*[1,2], *Carolina Merlo*[1,2,4], *Romina P. Pizzolitto*[1,2,4], *Julio A. Zygadlo*[1,2,5], *María P. Zunino*[1,2,5], *and Alejandra B. Omarini*[6]

[1] *Área de Bioplaguicidas, Instituto Multidisciplinario de Biología Vegetal (IMBIV-CONICET), Córdoba, Argentina*
[2] *Área Aromas y Pigmentos, Instituto de Ciencia y Tecnología de los Alimentos (ICTA), Universidad Nacional de Córdoba (UNC), Córdoba, Argentina*
[3] *Instituto Multidisciplinario de Investigación y Transferencia Agroalimentaria y Biotecnológica (IMITAB-CONICET) - Universidad Nacional de Villa María (UNVM), Córdoba, Argentina*
[4] *Departamento de Recursos Naturales, Cátedra de Microbiología Agrícola, Universidad Nacional de Córdoba, Córdoba, Argentina*
[5] *Departamento de Química, Cátedra de Química Orgánica, Universidad Nacional de Córdoba, Córdoba, Argentina*
[6] *Laboratorio de Biotecnología Fúngica y de los Alimentos, Asociación para el Desarrollo de Villa Elisa y Zona (ADVEZ) - Consejo Nacional de Investigaciones Científicas y Técnicas (CONICET), Entre Ríos, Argentina*

14.1 Introduction

Plant bioreactors are attractive production systems of secondary metabolites currently used as medicines for the treatment of different diseases, as additives in food, industrial products and cosmetics, and as pesticides in the agricultural sector. Some of the metabolites produced by plants include flavonoids, alkaloids, quinones, coumarins, tannins, phenolics, saponins, and essential oils (EOs) (Upadhyay and Singh 2021). EOs are hydrophobic substances of complex mixtures of volatile organic compounds (VOCs), *i.e.* compounds that evaporate at normal temperatures and pressures, which are released from specific plant tissues and organs, such as flowers, stems, seeds, and roots. Groups of VOCs as main constituents of EOs are monoterpenes (10-carbon isoprenoids), sesquiterpenes (15-carbon isoprenoids), phenylpropanoids, and benzenoids. In addition, plant tissues can produce volatile aldehydes and their corresponding acids, alcohols, and ketones (Pandey et al. 2017). One of the main functions of these VOCs involve defending plants from insect herbivores and microbial pathogens, such as filamentous fungi and bacteria. Because of these properties, EOs and/or their pure components have been long recognized as effective pesticides, in agreement with studies reporting the potential use of EOs against microbial and insect pests. However, the number of commercial products is low, with only a handful of EOs or pure compounds used as ingredients in manufactured pesticidal formulations. One of the main

Plants as Bioreactors for Industrial Molecules, First Edition.
Edited by Santosh Kumar Upadhyay and Sudhir P. Singh.

obstacles for the massive application of EOs is that they are synthesized by plants in small quantities. In addition, most plant species are annual, yielding in a single harvesting season, a mass whose weight represents a low percentage of the whole plant. The amount of EOs in plant seeds, flowers, stems, and roots is estimated to range from 0.1 to 10% of EO (v/w). Furthermore, a single compound can comprise 40–90% of EO, and it is usually not the most useful one (Gounaris 2010). Accordingly, affording desirable EOs and pure VOCs from plants can be an expensive and difficult process. In this context, biotechnology attempts to facilitate the production of VOCs and EOs, and therefore, reduce the market cost of EO-based products. Most efforts have been devoted to developing transgenic plants through the genetic modification of certain biosynthetic pathways of host plants responsible for the synthesis of the desired EOs and/or VOCs. Other attempts have been made to produce desired VOCs by biotransformation of cheaper precursors into added value products, either by isolated enzyme preparation or use of organisms such as fungi, yeasts, and bacteria.

The present chapter will discuss the use of plants as bioreactors to produce EOs with pesticidal properties for the agricultural area. First, we will focus on the different mechanisms of genetic manipulation of aromatic plants currently used to increase the yield of EOs and their pure components. Second, we will explore the use of different EOs as insecticides, bactericides, and fungicides reported over the past five years. Third, we will address some aspects of the biotransformation of volatile precursors into new bioactive molecules. Finally, we will discuss and compare the production of EOs with transgenic plants and plant cell cultures.

14.2 Plant Metabolic Engineering for the Production of EOs and their Pure Compounds

The use of plants as bioreactors for the synthesis of different valuable products has been recognized as an emerging field of research. The reasons accounting for this significant growth include low production costs, product safety, easy scale-up, and high consumer acceptance (Sharma and Sharma 2009). As high concentrations of desired botanical products are required for commercial purposes, several strategies have been developed to increase their biosynthesis through plant metabolic engineering. Some of these techniques are aimed at raising the level of the existing EOs or VOCs, while others attempt to produce new VOCs in plant species through induced mutations or genomic integration and expression of certain genes in heterologous organisms. In this section, we discuss and illustrate major approaches of plant metabolic engineering.

Some of the transformation approaches commonly used to produce transgenic plants involve *Agrobacterium,* a Gram-negative bacterium responsible for causing tumor in plants through horizontal gene transfer (Matveeva 2018). Aromatic plants are genetically transformed with *Agrobacterium tumefaciens* to produce transformed cells or with *Agrobacterium rhizogenes* to obtain hairy root cultures. In both cases, the transformed products can regenerate a whole viable transgenic plant capable of producing EOs or VOCs absent or present in low amounts in the parent plants. The production level of a certain EO or VOC can be enhanced by increasing the flux of metabolic precursors, blocking a competitive pathway, introducing new genes and metabolic routes, overexpressing or suppressing transcription factors that regulate the pathways of interest, and inducing mutagenesis, among other approaches (Verma et al. 2007; Hendrawati et al. 2010; Alok et al. 2018).

The generation of transgenic plants can lead to increased levels of EOs and/or modified terpene profile of EOs. Terpenes are derived from the 2-methyl-D-erythritol-4-phosphate (MEP; plastid) or from the mevalonate pathway (MVA; cytosol). Both routes produce C5 isoprenoid units, called isopentenyl pyrophosphate (IPP) and the isomer dimethylallyl pyrophosphate (DMAPP). Condensations of one IPP to one DMAPP give rise to geranyl diphosphate (C10), the precursor for monoterpene biosynthesis; a subsequent condensation with another IPP unit generates farnesyl diphosphate (C15), the precursor for sesquiterpene biosynthesis. The committed step in the biosynthesis of monoterpenes is catalyzed by monoterpene synthases, responsible for the biosynthesis of the different parent structural types of monoterpenes from the universal precursor geranyl diphosphate. These parent structures can be further modified by the action of other enzymes, such as hydroxylases, dehydrogenases, oxidoreductases, and transferases to form terpene derivatives (Mahmoud and Croteau 2002). Any of these steps can be engineered through *Agrobacterium*-mediated transformation to increase yield and/or modify the EO composition of targeted plants. Many examples of this transformation can be found among species of Lamiaceae, a plant family that contains numerous plant species with economically important EOs, where monoterpenes are the most common VOCs. Some investigations have shown that the overexpression of genes encoding limiting enzymes of the first steps of the MEP pathway enhances the biosynthesis of isoprene units in plants. This approach showed a significant increase of 50% in the yield of peppermint (*Mentha piperita*) EO (Mahmoud and Croteau 2001) and up to 359.0% in the yield of spike lavender (*Lavandula latifolia*) EO (Muñoz-Bertomeu et al. 2006), without changing the monoterpene profile. Muñoz-Bertomeu et al. (2007) reported that transgenic spike lavender plants expressing the *Arabidopsis thaliana* gene encoding a key enzyme of the MVA pathway that accumulated EOs more than twice than the controls. In the study conducted by Bansal et al. (2018), the overexpression of a key enzyme from the MVA pathway from *Ocimum kilimandscharicum* in different species of *Ocimum* with EOs rich in phenylpropanoids (*Ocimum basilicum*, *Ocimum gratissimum*, and *Ocimum tenuiflorum*) led to terpene accumulation and increase in EO content. A recent study that overexpressed the first enzyme of the MVA pathway reported a significant increase in monoterpenes α-pinene, D-limonene, eucalyptol, β-cyclocitral, and geranial and high concentrations of the sesquiterpenes caryophyllene, β-elemene, aromandendrene, copaene, germacrene D, and α-selinene, which were absent in the control plants (Wu et al. 2020b). These changes in the chemical profile of EOs also evidenced an increased vegetative growth and earlier floriation. Another work showed that the upregulation of other genes of the MEP pathway resulted in increased menthol content, the main constituent of peppermint EO (Taheri 2019).

The monoterpene synthases, responsible for the conversion of geranyl diphosphate (C10) into different classes of monoterpene families, are also good candidates for metabolic engineering. There are many examples of the generation of transgenic plants that overexpress certain monoterpene synthases. The transformation of the limonene synthase gene in peppermint (*M. piperita*) produced plants with EO profiles characterized by a high content of menthone, menthofuran, and pulegone (Krasnyanski et al. 1999). Leaves from transgenic spike lavender plants constitutively expressing the limonene synthase gene from spearmint showed limonene content 450% higher than that in controls (Muñoz-Bertomeu et al. 2008). Similarly, the genetic transformation of *Mentha arvensis* with the limonene synthase gene from *Mentha spicata* led to plants with altered levels of one-step enzymatic products from geranyl diphosphate, 1,8-cineole, and β-ocimene (Diemer et al. 2001). Transgenic plants of spike lavender that overexpressed the linalool synthase gene from *Clarkia breweri* increased their linalool content by 1000%, compared to control plants (Mendoza-Poudereux et al. 2014). To optimize the compositional balance of the main constituents of peppermint EO, Mahmoud and Croteau (2003) reduced the expression of menthofuran synthase gene using an antisense approach. This genetic manipulation resulted in a decrease of up to 75% in the concentration of menthofuran, while increasing the production of menthol, obtaining a high-quality EO. Additionally, the suppression of the gene encoding menthofuran synthase led to the accumulation of two monoterpenes not normally present in this species, (−)-*trans*-piperitone oxide and (+)-piperitenone oxide.

Another metabolic strategy to increase the yield of an EO involves manipulating the expression of transcription factors that regulate different biosynthetic and structural genes of glandular trichomes, the structures responsible for the production of EOs. The transcription factor MsYABBY5 preferentially expressed in glandular trichomes of spearmint was silenced using RNA interference. The reduced expression of MsYABBY5 in transgenic plants led to high levels of terpenes ranging from 20 to 77% (Wang et al. 2016). Similarly, the transgenic lines resulting from the silencing of the MsMYB, a transcription factor specific to glandular trichomes of spearmint, exhibited increased levels of monoterpenes, evidencing their role as repressors (Reddy et al. 2017). Thus, the genetic manipulation seeking to inhibit the expression of transcription factors that negatively impact on the pathway could increase the yield of EOs. Furthermore, the size and density of trichomes also affect the accumulation of EOs. For example, the high EO content in *O. basilicum* was associated with a larger size of glandular trichomes compared to other *Ocimum* species (Maurya et al. 2019). Hence, the higher expression of transcription factors that regulate trichome development was associated with higher EO yields in *O. basilicum* (Chandra et al. 2020).

Induced mutagenesis represents another tool to create genetic variation in aromatic plants since they can affect DNA. Several studies have proved that induced mutagenesis can modify EO quality and quantity with the use of physical or chemical mutagens (Agrawal and Kumar 2021). Different mutant lines of *Chamomilla recutita* were developed by applying gamma irradiation on seeds of the Vallary variety (Lal et al. 2019). These new genotypes showed changes in morphological features such as plant height, flower number, and EO yield and composition. For example, irradiations induced variations in the relative percentage of the VOC chamazulene, bisabolol oxide, bisabolol, bisabolone oxide, matricaria acid methyl ester, and matricaria acid ester. Several mutant genotypes were selected for field trials, yielding encouraging results. Indeed, the most promising mutant line showed a great production potential of 6.00–6.50 kg/ha EO yield, which was released as "CIM-Ujjwala" variety for commercial cultivation. The *M. piperita* varieties "Prankal" and "Kukrail" have also been developed through induced mutation by gamma rays to increase EO yield, both showing a production of up to 90 kg/ha of EO (Kolakar et al. 2018). In addition, another variety of peppermint, also obtained with gamma radiation, showed a high-quality EO, characterized by high concentrations of menthol (48%) and menthone (21%) (Mohamed and El Shimi 2014).

The studies discussed earlier demonstrate that plant biosynthetic pathways can be modified through metabolic engineering to optimize the production of either pure VOCs or EOs.

14.3 Bioactivity of EOs

Research on the biological activity of EOs is extensive since they represent promising alternatives to antibiotics and synthetic pesticides. As opposed to antibiotics and synthetic pesticides, EOs do not show pest resistance or residual toxicity in food commodities (Rosato et al. 2018; da Silva et al. 2019; Siddique et al. 2020). The biological activity of EOs has been widely reported in the literature against insects, bacteria, and filamentous fungi.

14.3.1 Insecticidal Effects of EOs

In tropical and subtropical countries, insects are the main destroyers of stored food products, especially cereals and legumes. Certain storage insect pests, such as legume beetles (*Callosobruchus* spp.), corn weevil (*Sitophilus zeamais*), rice weevil (*Sitophilus oryzae*), and red flour beetles (*Tribolium* spp.), can cause approximately 60% loss in cereals and legumes (Singh and Pandey 2018). Literature on the insecticidal effect of botanical compounds shows that most species studied belong to the family Lamiaceae (Boulogne et al. 2012; Ebadollahi and Sendi 2015). This family includes a great variety of medicinal and aromatic plants, with species such as lavender, thyme, mint, oregano, basil, sage, and rosemary. Most research evaluating the insecticidal effect of Lamiaceae EOs reports their chemical composition; however, 32% of the literature does not include this information or refers to other studies (data not shown).

The EOs of Lamiaceae have been extensively studied against insect pests, with 82.5% of the research focusing on the genera *Mentha*, *Satureja*, *Ocimum*, *Origanum*, *Lavandula*, *Thymus*, *Rosmarinus*, *Salvia*, and *Hyptis* (Figure 14.1a), in decreasing order of frequency.

Insect species from the order Coleoptera are the most commonly used in the studies (61%), followed by those from the order Hemiptera (23%), mostly aphids that attack crops. The remaining percentage comprises insect species from the orders Lepidoptera (10%) and Diptera (6%) (Figure 14.1b). Studies conducted with stored product insects were predominant (56%), with Coleoptera being the most frequent (85%), followed by Lepidoptera (15%) (Figure 14.1c). The genera from Coleoptera more frequently used were *Tribolium* and *Sitophilus*; the genera from Lepidoptera more widely tested were *Plodia* and *Ephestia* (data not shown).

Mentha and *Ociumum* were tested in a greater range of insect species, mainly Coleoptera (*Rhyzopertha dominica, Sitophilus granarius, S. oryzae, S. zeamais,* and *Tribolium castaneum*; data not shown), Hemiptera that attacks stored products, and Diptera that attacks fruits.

14.3.1.1 EO Composition of the Lamiaceae Main Genera with Insecticidal Effect

This study includes the four major components of the EOs from the most widely tested plant species within the main genera. Based on the average percentage of these compounds, we can conclude that *Hyptis* spp. was the only genera whose EOs were characterized by

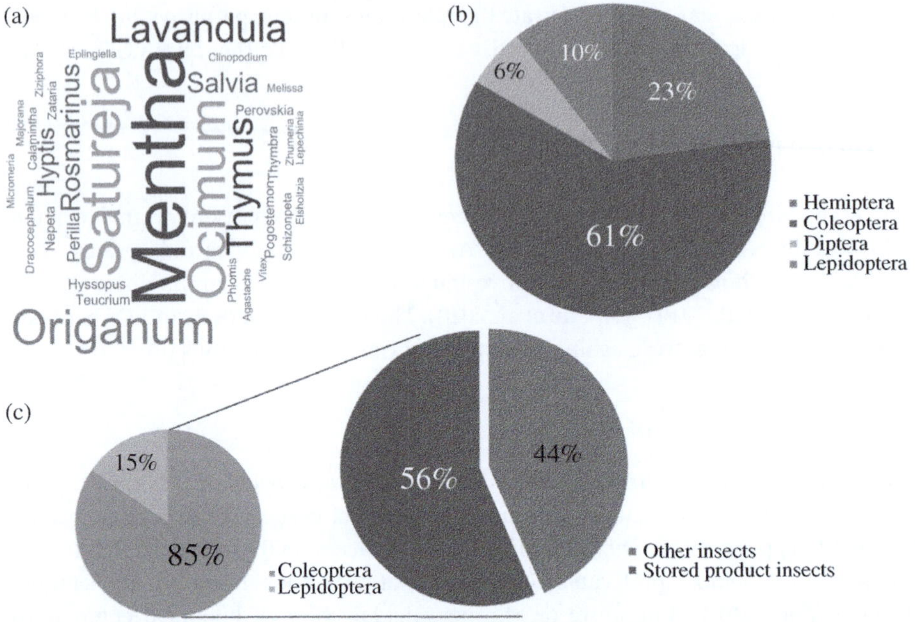

Figure 14.1 (a) Genera of the Lamiaceae family most commonly used against insect pests in agriculture. Larger font size indicates a higher frequency; (b) orders of agronomically important insects (field crops and storage insects) tested with the EOs of the genera belonging to the family Lamiaceae; and (c) orders of stored product insects most regularly used in insecticide assays.

>50% of hydrocarbons, with sabinene, α/β-pinenes, and β-caryophyllene as their main components. *Mentha* spp. and *Salvia* spp. showed about 50% of monoterpene ketones in their EOs, such as menthone, pulegone, piperitenone, and piperitenone oxide in *Mentha* spp. EOs, and camphor and thujone in *Salvia* spp. EOs. On the other hand, *Lavandula* spp. EOs revealed high amounts of terpenes (30–40%) with alcohol (linalool) and epoxy groups (1,8-cineol). Conversely, *Ocimum* spp. EOs showed phenylpropanoids such as eugenol, methyl-eugenol, and estragole as the most common compounds. *Rosmarinus* spp. EOs also presented variability of their major constituents with diverse functional groups, such as terpene hydrocarbons (α-pinene and camphene), alcohols (terpineol and linalool), and ethers (1,8-cineol). Finally, *Origanum* spp., *Satureja,* spp., and *Thymus* spp. exhibited > 50% of phenolic compounds, such as thymol and carvacrol, with the latter compound being, in general, more abundant than the former one.

14.3.1.2 Characteristics of Some Species Within the Main Genera

Mentha spp. is a very important taxon of the Lamiaceae family. The EOs of the species from this genus are enriched with pulegone, menthol, and menthone, with average relative content of 54.21, 36.31, and 9.47%, respectively. These compounds are responsible for the characteristic aromas and flavors of *Mentha* spp. (Singh and Pandey 2018). *Mentha arvensis, Mentha citrata, Mentha haplocalyx, Mentha longifolia, M. piperita, Mentha pulegium, Mirabilis rotundifolia, M. spicata,* and *Mentha suaveolens* are species most largely studied regarding bioactivity against stored product insects. For example, Domingues and Santos

(2019) tested *M. pulegium* EO against *R. dominica* and *S. granarius* in contact toxicity assays, finding LD_{50} =6.95 and 9.11 μl/ml, respectively. Within this group, *M. citrata*, *M. haplocalyx*, and *M. piperita* have alcohols (menthol and linalool) as their major components; *M. arvensis* is rich in both alcohols (menthol) and ketones (menthone and isomenthone), while the remaining species have high proportions of ketones (pulegone and piperitone). The main component (87% of the studies) of the EO from *M. pulegium* is pulegone, with a concentration ranging from 27.3 to 88.2%, or piperitenone (17% of the studies), with concentrations > 90% (Domingues and Santos 2019).

The EOs from the species belonging to *Satureja* spp., *Satureja bachtiarica*, *Satureja hortensis*, *Satureja intermedia*, *Satureja isophylla*, *Satureja khuzestanica*, *Satureja montana*, *Satureja rechingeri*, *Satureja sahendica*, *Satureja spicigera*, *Satureja thymbra*, and *S. wiedemanniana* were the most widely studied against pest insects. *Satureja* spp. was tested in several insect species including Hemiptera, Coleoptera, and Lepidoptera that often attack stored grains and crops. For instance, the species *S. thymbra* showed fumigant toxicity with an LC_{50} =3.43 μl/l against *Plodia interpunctella*, a common flour and dried fruit moth, which is a very low dose for this insect (Ebadollahi and Sendi 2015). The toxicity of this EO could be attributed to carvacrol. Some compounds found in the EOs of this genus include carvacrol, thymol, *y*-terpinene, and *p*-cymene, with average relative content of 49.51, 26.27, 14.85, and 9.24%, respectively. However, a significant variation in the EO composition among specimens of the same species has been reported. For instance, the EO of *S. hortensis* from the Parsabad region (Iran) showed high concentration of phenylpropanoid estragole (Ebadollahi and Setzer 2020), which resulted in LC_{50}<40 μl/l against *T. castaneum*. Other specimens of *S. hortensis* obtained from the Mashhad region (Iran) exhibited high content of carvacrol and γ-terpinene and showed LC_{50}> 100 μl/l against *T. castaneum* (Maede et al. 2011).

Ocimum americanum, O. basilicum, O. gratissimum, Ocimum kenyense, O. kilimandscharicum, and *O. suave* were the main species of the genus *Ocimum* tested as insecticides. In general, the stored grain beetles *R. dominica*, *S. oryzae*, and *S. zeamais* were the insects most commonly used in the assays. Studies on the fumigant toxicity of *O. gratissimum* EO against *R. dominica* and *S. oryzae* showed LC_{50} =0.20–0.50 and LC_{50} =0.50–14.4 μl/l air, respectively (Ebadollahi and Sendi 2015; Campolo et al. 2018). The insecticidal effects of this EO were also evaluated against *S. zeamais* through contact toxicity assays, with LD_{50} =0.20 μl/cm^2 after five hours exposure (Ebadollahi and Sendi 2015). From the EOs of this genus, two chemical compounds were mostly found, linalool and eugenol. The average relative content of linalool was 22.7% in *O. americanum* and 34.4% in *O. basilicum*, while eugenol was highly variable according to the species, with relative percentages ranging from 5.0 to 75.1%. Additionally, *O. americanum* was the species with the major variation in its chemical composition. The main components of this EO may include phenylpropanoids (eugenol and methyl eugenol) or alcohols (terpinen-4-ol and linalool) with relative percentages> 50%. The EO from *O. gratissimum* is generally rich in phenylpropanoid or phenol and hydrocarbon, while the remaining *Ocimum* species have high concentrations of alcoholic compounds such as linalool and terpinen-4-ol. On the other hand, *O. kilimandscharicum* was the only species that presented 70% of the ketone camphor.

The genus *Origanum* includes the plant species most important economically within the Lamiaceae family, having diverse bioactivity. *Origanum acutidens*, *Origanum compactum*,

Origanum glandulosum, *Origanum majorana*, *Origanum minutiflorum*, *Origanum onites*, *Origanum syriacum*, and *Origanum vulgare* are the main species tested for insecticidal activity against storage insect pests. This genus was tested mainly against *Ephestia kuehniella* (Lepidoptera) (LC_{50} = 3.27–9.81 μl/l air) and *S. oryzae* (Coleoptera) (LC_{50} = 1.5–9 μl/l air) (Ebadollahi and Sendi 2015; Tepe et al. 2016; Campolo et al. 2018). The chemical composition of the EOs from most species belonging to the genus *Origanum* showed phenolic compounds higher than 50%. Major compounds in the EOs of this genus include carvacrol and *p*-cymene, with an average relative content of 63.2% and 4.8%, respectively. The average relative content of thymol is 18.8% in the EOs from *O. compactum*, *O. glandulosum*, *O. majorana*, *O. onites*, and *O. vulgare*, whereas that of γ-terpinene is 10.8% in the EOs from *O. compactum*, *O. majorana*, *O. onites*, and *O. vulgare*. The EOs of *O. majorana* and *O. vulgare* were those with the greatest variability in their composition. *O. majorana* presented a totally different profile from the rest of the species of this genus, rich in alcohol (linalool and terpinen-4-ol) and hydrocarbon (γ-terpinene and sabinene) (Karabörklü et al. 2011; Pavela 2016). In the EO of *O. vulgare*, the main component of some plants was carvacrol, while others were rich in ketones (>80%) such as pulegone (Abdelgaleil et al. 2016).

The insecticidal activity of seven species of the genus *Lavandula* were mainly studied on pests of stored products: *Lavandula angustifolia*, *L. latifolia*, *L. intermedia*, *Lysimachia hybrida*, *Lavandula stoechas*, *Lavandula officinalis*, and *Lavandula luisieri*. Dam et al. (2019) studied the insecticidal effects of *Lavandula* EO against the fruit fly, *Drosophila suzukii*. These authors found a half-maximally effective concentration (EC_{50}) that ranged from 0.69 to 7.21% in contact applications and from 3.79 to 8.22 μl/l air in fumigant tests. *Lavandula* EO is characterized by having a high content of oxygenated compounds such as alcohols (linalool), cyclic ethers (1,8-cineol), and esters (linalyl acetate). *Lavandula luisieri* EO has ketone compounds such as camphor and 5-methylene-2,3,4,4-tetramethylcyclopen t-2-enone, as its main components (González-Coloma et al. 2006). Two common oxygenated chemical compounds of the EOs of this genus are 1,8-cineole and camphor, with an average relative percentages of 16.38% and 16.13%, and extreme values ranging from 4.00% to 53.30% and 4.50% to 53.70%, respectively. Linalyl acetate is a main chemical compound of EOs from *L. angustifolia*, *L. intermedia*, and *L. hybrida*, with average relative contents of 33.45%, while linalool is found in the EOs from *L. angustifolia*, *L. latifolia*, *L. intermedia*, *L. hybrida*, and *L. officinalis*, with an average relative content of 29.23%.

The EOs from *Thymus capitatus*, *Thymus carmanicus*, *Thymus daenensis*, *Thymus persicus*, *Thymus satureioides*, *Thymus serpyllum*, *Thymus vulgaris*, and *Thymus zygis* are the ones most frequently studied within the genus *Thymus*. *Thymus* EOs were tested against several insect species, most being pests of grains and stored products, such as beetles *S. oryzae* (LC_{50} = 1.5–9 μl/l air) and *T. castaneum* (LC_{50} = 22.15–236.9 μl/l) (Ebadollahi and Sendi 2015; Campolo et al. 2018; Basaid et al. 2021). Thymol and carvacrol are the main compounds found in these EOs, with average relative content of 37.31% and 30.32%, respectively.

Within the genus *Rosmarinus*, *Rosmarinus officinalis* was the EO most frequently studied as an insecticide, with 1,8-cineole being the main chemical component with an average relative content of 30.88%. The mortality of *Rosmarinus* EO was studied in stored product insects, such as *Bruchus rufimanus*, *T. castaneum*, and *P. interpunctella*, with an LC_{50} = 1.19 μl/l air for *B. rufimanus* (Fierascu et al. 2020), LC_{50} = 1.17 μg/ml for *T. castaneum*, and LC_{50} = 0.93 μl/l for *Plodia interpuntella* (Campolo et al. 2018).

Camphor is one of the main EO components of certain species of the genus *Salvia*, such as *Salvia fruticosa*, *Salvia hydrangea*, *Salvia officinalis*, and *Salvia tomentosa*. This genus offers a vast diversity of major compounds. The genus was mainly studied against *S. oryzae*, with a wide range of LC_{50}, with lower doses from 1.5 to 9 μl/l (Campolo et al. 2018) to higher doses from 15.5 to 40.4 ml/l (Basaid et al. 2021).

Within the genus *Hyptis*, two species have been investigated for insecticidal activity against insect pests, *Hyptis spicigera* and *Mesosphaerum suaveolens*, showing high content of α-pinene (18.7%) and sabinene (37.5%), respectively. These species were evaluated mainly against the stored grain beetles *Callosobruchus maculatus*, *S. oryzae*, *T. castaneum*, and *S. zeamais*. Soujanya et al. (2016) found that *H. spicigera* EO produced fumigant toxicity against *C. maculatus* of LC_{50} = 4.7 mg/l and LD_{50} = 57.0 mg/mg insect in topical applications. Othira et al. (2009) reported that *H. spicigera* EO exhibited high toxicity against adult *S. zeamais*, but low effect on *T. castaneum* in fumigant tests. At a concentration of 2 μl/l air, *H. spicigera* EO caused 68% mortality in *S. zeamais* after two-day exposure, whereas *T. castaneum* showed a low mortality of 3% during the same time frame. *Hyptis* spp. shows hydrocarbon terpenes as main EO components.

14.3.2 Antibacterial Activity of EOs

The indiscriminate use of antibiotic and chemicals has resulted in the emergence of bacteria resistance affecting public health. Bacteria have developed sophisticated means to subvert the action of antibiotics, thus allowing bacterial strains to resist antibiotic packages (multiresistant strains) (Bouyahya et al. 2017; Imane et al. 2020). The use of chemically synthesized additives in food to prevent food spoilage and pathogenic bacteria has also been controversial because of their potential to cause respiratory diseases or other health risks (Tu et al. 2018).

EOs are recognized for their antimicrobial effect that results from complex interactions with their different constituents. These interactions may produce synergistic, additive, or antagonistic effects. In general, the major components of EOs have a central role in bioactivity; yet, those compounds found at lower concentrations can also affect the antimicrobial properties of EOs (Mahdavi et al. 2017; Salah-Fatnassi et al. 2017; Zygadlo et al. 2017; Zhang et al. 2020).

The antibacterial activity of EOs is mainly attributed to the presence phenylpropanoids, the main components of medicinal plant EOs (Salah-Fatnassi et al. 2017; Tu et al. 2018; Desam et al. 2019; Zhang et al. 2020). For example, the high antibacterial activity of cinnamon and clove EOs against four tested foodborne bacteria was attributed to the presence of cinnamaldehyde, eugenol, and anethole. These compounds have a benzene ring and a conjugated double bond structure, which may be responsible for their higher antibacterial activity (Tu et al. 2018). In addition, the EO of *Satureja macrantha*, which demonstrated strong antibacterial activity, showed high percentage of oxygenated monoterpenes, with carvacrol, thymol, *p*-cymene, and γ-terpinene being major compounds (Aghbash et al. 2020). Carvacrol and thymol are considered strong antimicrobial agents and probably the additive/synergistic effects could enhance the effectiveness of EOs. They can align with fatty acid chains of lipid bilayers by interacting with transmembrane proteins and forming channels through the membrane leading to increase of membrane fluidity and alteration of proton motive force and cell permeability (Purkait et al. 2020). In addition, the high antibacterial activity of *M. piperita* EO is ascribed to the presence of menthol and menthone, the main chemical constituents (Desam et al. 2019).

Due to their complexity, EOs could attack multiple molecular targets by blocking DNA replication, altering gene expression, inhibiting cell cycle (S-phase) and protein synthesis, denaturing enzymes, inducing reactive oxygen species, or damaging microbial membranes (Birhan et al. 2011; Dhami 2013). Bacterial cell membrane protects cells from external damage and maintains homeostasis. Due the hydrophobic characteristic and low molecular weight of the constituents of EOs, these compounds accumulate in cell membrane producing structural and functional damages. Cell membrane damage leads to leakage of ions and cytoplasmic molecules (Salah-Fatnassi et al. 2017; Zhang et al. 2020). In addition, EOs produce disturbance in the electron respiratory chain and inhibit the quorum sensing signaling pathways (Bouyahya et al. 2017). Thus, they may serve as a promising antibiotic to overcome the multidrug-resistant bacteria.

In this sense, the EOs of rosemary (*R. officinalis*), ginger (*Zingiber officinale*), tea tree (*Melaleuca alternifolia*), lemon (*Cymbopogon winterianus*), clary sage (*Salvia sclarea*), and clove (*Syzygium aromaticum*) were effective against reference microbial strains and clinical isolates. Moreover, tea tree EO was more efficient than the antibiotic chloramphenicol against multiresistant bacteria (Imane et al. 2020). In another study, Bouyahya et al. (2017) reported that *M. pulegium* EO exhibited a significant antibacterial activity against *Bacillus subtilis* and *Proteus mirabilis* in comparison with commercialized antibiotic (erythromycin and chloramphenicol). The bioactivity of EOs does not seem to depend on the antibiotic susceptibility pattern, even in bacteria with high antibacterial resistance rates, including methicillin-resistant *S. aureus* and extended spectrum beta-lactamase producer *E. coli* (Imane et al. 2020).

Several authors reported that EOs have better effects against Gram-positive than against Gram-negative bacteria, probably due to differences in cell wall composition. Gram-positive cell wall is composed of 90–95% of peptidoglycan, which allows easy diffusion of hydrophobic components like EOs into cells and may, consequently, act on different targets. The peptidoglycan layer of Gram-negative is thinner and the outer membrane of Gram-negative bacteria surrounding the cell wall may restrict the diffusion of hydrophobic compounds like EO constituents through the lipopolysaccharides (Bouyahya et al. 2017). Lipopolysaccharides comprise lipid A and O-side chain, which makes Gram-negative bacteria more resistant to EOs. The hydrophilic transmembrane channels at the surface of outer membrane allow easy penetration of hydrophilic substances through outer membrane, while the hydrophobic molecules such as EOs are slowly traversed into outer membrane through porins due to interaction between EOs and lipopolysaccharide layer (Zhu et al. 2021). Figures 14.2–14.4 show the MIC media of EOs against Gram-positive and Gram-negative bacteria, and in 56.8% of the assays, the EOs acted more strongly against Gram-positive than Gram-negative bacteria (Bączek et al. 2017; Bouyahya et al. 2017; Mahdavi et al. 2017; Salah-Fatnassi et al. 2017; Rosato et al. 2018; Tu et al. 2018; Asadollahi et al. 2019; Desam et al. 2019; Imane et al. 2020; Purkait et al. 2020). By contrast, numerous studies show that some Gram-negative strains are more sensitive than some Gram-positive to certain EOs. Supporting this statement, in 27.3% of the assays, EOs were more effective against Gram-negative than Gram-positive bacteria (Bouyahya et al. 2017; Ács et al. 2018; Tu et al. 2018; Aghbash et al. 2020; Imane et al. 2020), and in 15.9% of the assays, EOs did not show any difference in vulnerability between Gram-negative and Gram-positive bacteria (Figures 14.2–14.4) (Ács et al. 2018; Imane et al. 2020; Siddique et al. 2020). For example, Tu et al. (2018) reported that the EOs of *Pimpinella anisum*, *M. haplocalyx*, *S. aromaticum*, *Zanthoxylum bungeanum*, and *Cymbopogon nardus* were more effective against Gram-positive bacteria, whereas *Cinnamomum camphora* and *Cinnamomum zeylanicum* demonstrated higher bioactivity against Gram-negative bacteria. Additionally, different studies of the antibacterial activity of *C. zeylanicum* EO showed difference in vulnerability between Gram-negative and Gram-positive bacteria (Ács et al. 2018; Tu et al. 2018).

The differences found among the MICs obtained for same EOs may be due to the Gram-positive and Gram-negative strains selected in each study and their sensitivity, cultivation conditions, culture medium, solvents used to dilute the EO, among other factors. Moreover, EO composition might slightly differ according to environmental and geographical factors (Desam et al. 2019; Imane et al. 2020; Siddique et al. 2020).

14.3.3 Antifungal Effect of EOs

Currently, synthetic fungicides are used to protect grains and food commodities from fungal growth. However, the repeated use of these chemicals can lead to the development of

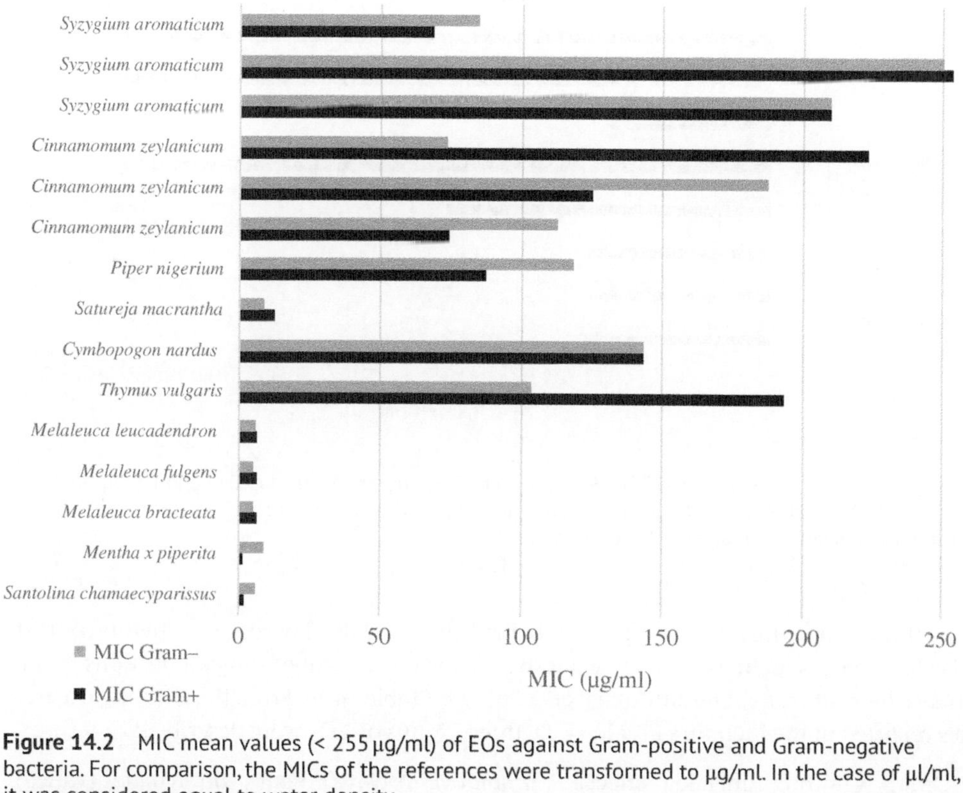

Figure 14.2 MIC mean values (< 255 μg/ml) of EOs against Gram-positive and Gram-negative bacteria. For comparison, the MICs of the references were transformed to μg/ml. In the case of μl/ml, it was considered equal to water density.

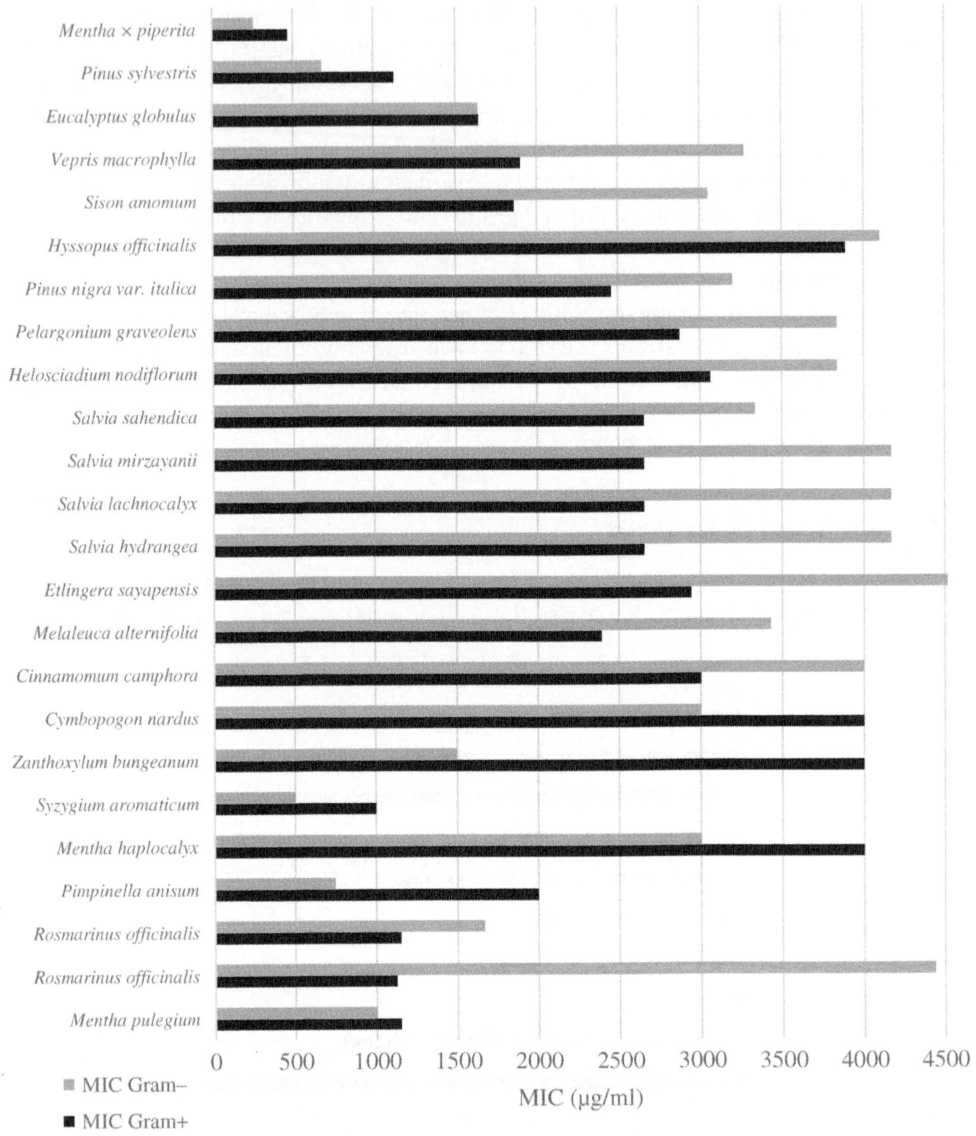

Figure 14.3 MIC mean values (255–4520 μg/ml) of EOs against Gram-positive and Gram-negative bacteria. For comparison, the MICs of the references were transformed into μg/ml. In the case of μl/ml, it was considered equal to water density.

fungal resistance and negative effects on living beings and the environment. In this context, EOs have been sought as a valuable alternative for the control of fungal infections. Many studies have addressed the antifungal effect of EOs (Table 14.1). From the antifungal activities reported in the literature (Table 14.1), three groups of EOs were described:

- Group A (strong fungicidal effect): *S. aromaticum* (clove), *Vatica diospyroides* (vatica), *M. longifolia*, *C. zeylanicum* (cinnamon), *Allium cepa* (onion), *Brassica nigra* (mustard), *T. vulgaris* (thyme), *Lippia berlandieri*, *Zataria multiflora*, *Carum carvi* (caraway),

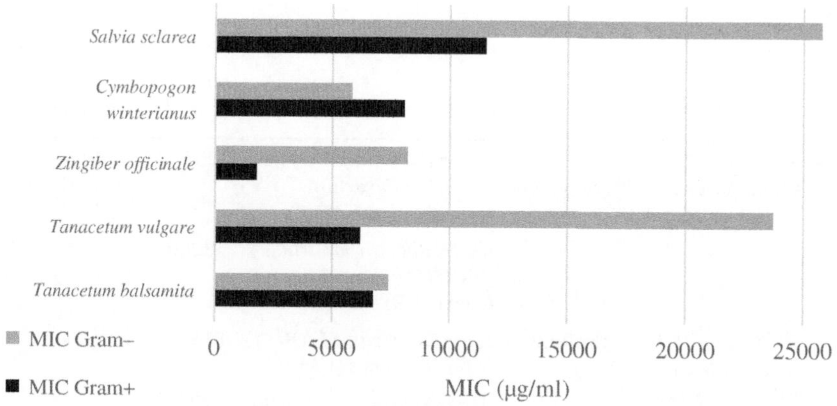

Figure 14.4 MIC mean values (> 1790 μg/ml) of EOs against Gram-positive and Gram-negative bacteria. For comparison, the MICs of the references were transformed into μg/ml. In the case of μl/ml, it was considered equal to water density.

Table 14.1 Antifungal effect of EO against different fungi.

Aromatic plans	Main components	Microorganisms	MIC (μg/ml)	References
Syzygium aromaticum	Eugenol 62.4%, benzenemethanol 21.3%, *trans*-caryophyllene 11.2%	*Aspergillus flavus* PSRDC-2	50	Boukaew et al. (2017)
Vatica diospyroides	Benzyl acetate 48.8%, benzyl benzoate 5.3%, isoeugenol 4.6%		100	
Mentha longifolia	Menthone 39.5%, isopulegone 30,5 %, eucalyptol 10.4%, α-terpineol 3.14 %, linalool 2.4%, β-pinene 2.4%	*A. flavus* (MNHN 994294)	15.8	Desam et al. (2017)
		Trichophyton mentagrophytes	10.5	
		Sclerotinia sclerotiorum	9.8	
		Fusarium solani	9.5	
		Alternaria alternaria (MNHN 843390)	8.5	
		F. solani	8.0	
		Aspergillus fumigatus (MNHN 566)	7.5	
		A. variecolor	5.5	
		Trichophyton rubrum	5.0	
		Fusarium oxyporum (MNHN 963917) *Fusarium acuminatum*	4.5	
		Moliniana fructicola, Rhizoctonia solani, Sarracenia minor	3.5	
		Cladosporium herbarum (MNHN 3369)	2.5	

(*Continued*)

Table 14.1 (Continued)

Aromatic plans	Main components	Microorganisms	MIC (μg/ml)	References
Senecio nutans	Sabinene 27.6%, α-phellandrene 15.7%, o-cymene 9.6%, β-pinene 6.1%	*Fusarium graminearum* (LABI11) *Fusarium verticillioides* (LABI7)	1200	Galvez et al. (2018)
		Aspergillus carbonarius (FRR5690) *Aspergillus niger* (FRR5695)	>1200	
Aloysia gratissima	β-Thujone 36.1%, α-thujone 32.2%, 1,8-cineol 10.7 %, sabinene 6.2%	*F. graminearum* (LABI11), *F. verticillioides* (LABI7)	1200	
		A. carbonarius (FRR5690), *A. niger* (FRR5695)	>1200	
Senecio viridis	9,10-Dehydrofukinone 92.7%	*F. graminearum* (LABI11), *F. verticillioides* (LABI7)	1200	
		A. carbonarius (FRR5690), *A. niger* (FRR5695)	>1200	
Tagetes terniflora	*cis*-Tagetone 33.6%, *cis*-β-ocimene 17.1%, *trans*-tagetone 17.0%, *cis*-ocimenone 8.0%, *trans*-ocimenone 8.2%	*F. graminearum* (LABI11)	1200	
		F. verticillioides (LABI7)	600	
		A. carbonarius (FRR5690), *A. niger* (FRR5695)	>1200	
Pimpinella anisum	ND	*A. niger, Aspergillus oryzae*	1000	Hu et al. (2019)
		Aspergillus ochraceus	500	
Cinnamomum zeylanicum	ND	*A. niger*	60	
		A. oryzae, A. ochraceus	130	
Cymbopogon nardus	ND	*A. niger, A. ochraceus*	2000	
		A. oryzae	1000	
Mentha haplocalyx	ND	*A. niger, A. oryzae*	2000	
		A. ochraceus	1000	
Cinnamomum camphora	ND	*A. niger, A. oryzae, A. ochraceus*	2000	
S. aromaticum	ND	*A. niger, A. oryzae, A. ochraceus*	250	
Zanthoxylum bungeanum	ND	*A. niger, A. oryzae, A. ochraceus*	1000	
Zingiber officinale	2,6-Octadienal,3, 7-dimethyl 76.9%, verbenyl ethyl ether 4.0%, genanic acid 2.6%	*Botrytis cinerea, A. panax, F- oxysporum*	300000	Hussein and Joo (2018)
		Cylindrocarpon destructans, S. sclerotium, S. nivalis	100000	

Table 14.1 (Continued)

Aromatic plans	Main components	Microorganisms	MIC (μg/ml)	References
Allium cepa	Dimethyl-trisulfide 16.6%, methyl-propyl-trisulfide 14.2%, methyl-(1-propenyl)-trisulfide 13.0%	*P. aurantiogriseum, A. niger*	280	Kocic-Tanackov et al. (2016)
		A. carbonarius, A. wentii, A. versicolor, F. oxysporum, F. proliferatum, F. subglutinans, F. verticillioides, P. brevicompactum, P. glabrum, P. chrysogenum	140	
Cymbopogon citratus	Neral 50.0%, geranial 35.0%	*A. tenuissima* (gb/FJ755240.1/), *F. verticillioides* (ATCC 52539)	1000	López-Meneses et al. (2017)
C. zeylanicum	Eugenol 70.0%, caryophyllene 4.5%	*A. tenuissima* (gb/FJ755240.1/), *F. verticillioides* (ATCC 52539)	5000	
S. aromaticum	ND	*A. niger*	200	Jahani et al. (2020)
Brassica nigra	Allyl isothiocyanate 98.4%	*A. fumigatus, A. niger, P. cinnamopurpureum,*	0.02	Reyes-Jurado et al. (2019)
		Eupenicillium hirayamae, A.s nomius	0.06	
		T. rubrum	0.05	
		P. expansum	0.012	
***Thymus vulgaris* (thyme)**	*o*-Cymol 19.8%, linalool 14.6%, thymol 12.1%, γ-terpinene 9.9%, α-pinene 7.55%, carvacrol 7.1%, isoborneol 6.7%, caryophyllene 6.2%	*P. expansum*	0.16	
		A. niger, E. hirayamae, P. cinnamopurpureum	0.2	
		P. viridicatum, T. rubrum, A. fumigatus, A. nomius	0.50	
***Lippia berlandieri* (Mexicano oregano)**	*p*-Cymene 35.5%, carvacrol 26.7%, benzene 6.4%	*A. niger, E. hirayamae, P. cinnamopurpureum, P. expansum*	0.2	
		T. rubrum, A. fumigatus	0.5	
		A. nomius, P. viridicatum	1.0	

(*Continued*)

Table 14.1 (Continued)

Aromatic plans	Main components	Microorganisms	MIC (μg/ml)	References
T. vulgaris	Thymol 41.7%, *p*-cymene 29.0%, γ-terpinene 13.2%	*A. ochraceus* CBS 263.67	300	Moghadam et al. (2019)
Zataria multiflora	Thymol 39.1%, carvacrol 26.6%, *p*-cymene 9.4%, γ-terpinene 6.6%		600	
C. zeylanicum	(E)-Cinnamaldehyde 95.5%		78	
S. aromaticum	Eugenol 93.6%		310	
C. cyminum	*p*-Cymene 37.8%, β-pinene 25.2% cuminal 11.7%, γ-terpinene 10.8%		2500	
C. carvi	*p*-Cymene 34.5%, carene 27.5%, carvacrol 7.4%, estragol 5.9%		62	
C. zeylanicum	ND	*F. graminearum* (KZF1), *F. culmorum* (KZF5)	<800	Perczak et al. (2019)
Origanum vulgare	ND			
Cymbopogon martini	ND	*F. graminearum* (KZF1)	800	
		F. culmorum (KZF5)	310	
Citrus aurantium dulcis	ND	*F. graminearum* (KZF1)	12500	
		F. culmorum (KZF5)	>100000	
Thymus hiemalis	ND	*F. graminearum* (KZF1), *F. culmorum* (KZF5)	<800	
Mentha viridis	ND	*F. graminearum* (KZF1)	12500	
		F. culmorum (KZF5)	50000	
Foeniculum vulgare dulce	ND	*F. graminearum* (KZF1)	<800	
		F. culmorum (KZF5)	>100000	
Aniba rosaeodora	ND	*F. graminearum* (KZF1)	6200	
		F. culmorum (KZF5)	12500	
O. vulgare* spp. *vulgare	Thymol 30%, *cis*-sabinene hydrate 24.7%, γ-terpinene 12.0%, terpeinen-4-ol 7.8%	*F. verticillioides* (M3125)	250	Pizzolitto et al. (2020)
O. applii	Thymol 37.8%, *cis*-sabinene hydrate 31.7%, terpeinen-4-ol 5.9%			
O. vulgare* spp. *virens	Thymol 34.0%, *cis*-sabinene hydrate 19.1%, γ-terpinene 6.8%, terpeinen-4-ol 6.6%			
Chenopodium ambrosioides	Piperitenone 92.3%		500	

Table 14.1 (Continued)

Aromatic plans	Main components	Microorganisms	MIC (μg/ml)	References
S. aromaticum	Eugenol 53.9%, eugenyl acetate 19.3%, β-caryophyllene 10.2%, caryophyllene oxide 4.7%,	*A. niger* (ATCC 16404)	64.6	Purkait et al. (2020)
C. zeylanicum	Cinnamaldehyde 59.2%, cinnamyl acetate 22.7%, β-caryophyllene 8.4%		73.1	
Piper nigrum	ND		94.6	

MICs values were transformed into μg/ml. In the case of μl/ml, it was considered equal to water density.

Cymbopogon martini (palmarosa leaves), *Origanum vulgare* spp. *vulgare*, *O. applii*, *O. vulgare* spp. *virens*, *Chenopodium ambrosioides*, *Piper nigrum* (black pepper).

- Group B (moderate fungicidal effect): *Senecio nutans*, *Aloysia gratissima*, *S. viridis*, *T. terniflora*, *P. anisum* (anise), *C. nardus* (citronella), *M. haplocalyx*, *C. camphora* (camphor), *Z. bungeanum* (pepper), *Cymbopogon citratus* (lemongrass), *Cuminum cyminum* (cumin), *Thymus hiemalis* (verbena).
- Group C (weak fungicidal effect): *Z. officinale* (ginger), *Citrus aurantium dulcis* (orange), *Mentha viridis* (spearmint leaves), *Foeniculum vulgare dulce* (fennel seeds), *Aniba rosaeodora* (rosewood).

In Group A, different species of *Penicillium*, *Aspergillus*, *Eupenicillium*, and *Trichophyton* were highly sensitive to *B. nigra*, *T. vulgaris*, and *L. berlandieri*, showing minimal inhibitory concentration (MIC) values between 0.02 and 1.0 μg/ml (Reyes-Jurado et al. 2019). *M. longifolia* EO also exhibited good antifungal activity at concentrations ranging from 2.5 to 15.8 μg/ml (Desam et al. 2017). By contrast, *M. viridis* (Group C) and *M. haplocalyx* (Group B) EOs showed a low inhibitory effect against *Fusarium graminearum*, *F. culmorum* (Perczak et al. 2019), and *Aspergillus* sp. (Hu et al. 2019). The antifungal effect of *C. zeylanicum* EO differed from species, with *Fusarium* sp. being more resistant than *Aspergillus* sp. The inhibitory effect of *Origanum* sp. EOs against *Fusarium* species reported by Pizzolitto et al. (2020) and Perczak et al. (2019) could be attributed to their main constituents: phenolic compounds thymol, *cis*-sabinene hydrate γ-terpinene, and terpeinen-4-ol. Similarly, the antifungal activities of *S. aromaticum* (clove) EO against *Aspergillus* sp. presented MIC values between 50 and 310 μg/ml (Boukaew et al. 2017, Hu et al. 2019; Moghadam et al. 2019; Jahani et al. 2020). The EOs evaluated by Galvez et al. (2018) from *A. gratissima*, *S. nutans*, *S. viridis*, and *T. terniflora* (group B) showed more bioactivity against *Fusarium* than *Aspergillus* strains. The compounds *cis*-tagetone, *trans*-tagetone, *cis*-β-ocimene, *cis*-ocimenone, *trans*-ocimenone, and 9,10-dehydrofukinone, and several oxygenated sesquiterpenes were responsible for the antifungal activity against *Fusarium verticillioides* and *F. graminearum*, respectively. Moreover, different strains of *Fusarium*, *Aspergillus*, and *Penicillium* were moderately inhibited by *P. anisum*, *C. nardus*, *C. camphora*, *Z. bungeanum*, *C. citratus*, *C. cyminum*, and *T. hiemalis* EOs. On the other hand, EOs from group C have high content of hydrocarbon monoterpenes, ketones, and ethers in their EO chemical profile, which are related to weak antifungal effect (Kurita and Koike 1983; López et al. 2004).

The bioactivity of EOs largely depends on their chemical composition. In this sense, a given EO would probably fail to successfully control both insect and microbial pests. Indeed, according to the evidence presented, certain groups of VOCs are more effectively used as insecticides, while others are more properly used as bactericides or fungicides. For example, EO-containing terpene ketones are usually effective insecticides because of their proven effect on the insect nervous system, inhibiting acetylcholinesterase activity (Herrera et al. 2015). On the other hand, phenolic compounds were reported as effective VOCs against bacteria and filamentous fungi. The presence of a 6-membered aromatic ring and a conjugated double bond structure would interact with ion channel proteins of cell membrane, affecting its permeability and leading to significant consequences in ion homeostasis (López et al. 2004; Achimón et al. 2021).

14.3.4 Bioconversion Process of EOs and Their Components by Microorganisms

Currently, there is an increasing need to develop sustainable methods for industrial biopesticide production, opening up new ways of exploring innovative approaches. New regulatory standards for pesticides tolerance levels in food products have led to the increasing demand for new biopesticides on the market (Kumar et al. 2021). Hence, compared to conventional extraction methods from plant material, biotechnology based on microbial or enzymatic transformations provides more economical and sustainable process for the development of value-added products with improved bioactivity (Omarini et al. 2020; Sodhi et al. 2022). These transformations comprise catalyzing stereo-, regio-, and enantiospecific reactions that are not easily obtained by chemical synthesis, allowing the development of chiral products from racemic mixtures. Thus, microbial biotransformation has considerable potential for the generation of a huge variety of diversified natural compounds, especially with complex structures such as triterpenoids and sesquiterpenoids (Krings and Berger 2010; Sales et al. 2018; Omarini et al. 2020). The main reactions involved in the fungal biotransformation of natural compounds include oxidation, reduction, decarboxylation, hydroxylation, methylation, acetylation glycosylation, and hydrogenation (Hussain et al. 2016; Cano-Flores et al. 2020). In the last years, considerable research has been conducted to form bioactive molecules with diverse structural features to achieve more sustainable industrial processes for biopesticide production (Omarini et al. 2020).

On the other hand, agro-industrial side-streams such as citrus peels, oil cakes, coffee and coconut husks, grape and apple pomace, or forest industry by-products (*Eucalyptus*) were found to be a suitable carbon and nitrogen source to support microbial growth (Laufenberg et al. 2003; Hussain et al. 2016; Omarini et al. 2016; Omarini et al. 2020). Furthermore, these residues represent a rich source of low-cost natural bioactive compounds such as EOs and terpenoids (e.g., mono- and sesquiterpenoids) that can be valorized through biotransformation, creating new opportunities to generate semisynthetic analogues and novel molecules used as biopesticides (Isman 2020; Kumar et al. 2021). Methods like submerged and solid-state fermentation have been extensively applied to recycle different industrial side-streams to form a broad range of bioactive molecules (Hussain et al. 2016; Gounaris 2019; Omarini et al. 2020).

Using favorable biocatalysts (microorganisms or enzymes), substrate, and fermentation conditions, the production yield of bioactive molecules and the functional value of the compounds can be improved through structural modification (Gounaris 2019; Sodhi et al. 2022). For instance, the extract obtained from submerged fermentation of neem leaves and wild garlic mixture using a variety of lactic acid bacteria, yeasts, and phototrophic bacteria was highly effective in reducing the population density of whitefly (*Bemisia tabaci*) and aphids (*Myzus persicae*) in field experiments of tomato crops (Nzanza and Mashela 2012). The solid-state fermentation of *Eucalyptus cinerea* leaves (forestry by-product) by *Pleurotus ostreatus* and *Favolus tenuiculus* resulted in the biotransformation of 1,8-cineole, the main terpene of the EO, into 1,3,3-trimethyl-2-oxabicyclo [2.2.2] octan-6-ol, and 1,3,3-trimethyl-2-oxabicyclo [2.2.2] octan-6-one (Omarini et al. 2016). Similar results were found by bioconversion of 1,8-cineole extracted from *Eucalyptus globulus* using filamentous fungi (Ramos et al. 2015). The repellent and insecticidal properties of 1,8-cineole and its derivatives have been widely reported against several insects (Knight 2009; Tripathi and Mishra 2016; Mączka et al. 2021). Other volatile compounds have been obtained from apple juice fermentation using lactic acid bacteria, such as *trans*-2-hexen-1-ol, 1-octanol, citronellol, geraniol, and (E)-2-hexenal; methylheptenone can also appear during lactic acid bacterial fermentation (Wu et al. 2020a). Several authors have demonstrated that *trans*-2-hexen-1-ol and (E)-2-hexenal have good insecticidal effects against *S. zeamais* (Herrera et al. 2015; Brito et al. 2019). Citrus side streams constitute a good source of natural compounds. For example, numerous compounds with terpenoid structures are seen in EOs, which represent potential precursors for biotransformation. Limonene is the major constituent of citrus EOs (Sales et al. 2018; Achimón et al. 2022). Carveol, carvone, perillyl alcohol, isopiperitenol, α-terpineol, and limonene 1,2-diol were the main compounds originated by biotransformation using different microorganisms (Krings and Berger 2010; Sales et al. 2018). The biotransformation products of limonene obtained from bacteria biotransformation are mainly perillaldehyde, perillic acid, limonene-1,2-epoxide, and limonene-8,9-epoxide, while *cis*/*trans*-mentha-2,8-dienol, limonene-4-ol, and isopiperitenone are commonly formed by fungi and yeasts. For example, limonene dioxide, *cis*-linalool oxide, and limonene-1,2-diol showed neurotoxic effects in insects at different stages of their development (Coats et al. 1991). Kordali et al. (2017) reported that limonene oxide obtained from fermentation caused 100% mortality of *S. granarius* adults compared with limonene leading only to 70% mortality. However, limonene has a broad insecticidal spectrum and is a component of several formulations of commercial products (Tripathi and Mishra 2016). Recently, Cheng et al. (2019) obtained good yields of limonene through fermentation using a genetically modified nonconventional yeast (*Yarrowia lipolytica*) with an optimized limonene and mevalonate synthesis pathway. This offers an alternative to meet the current requirements of limonene. However, an important obstacle that prevents the valorization of the industrial side-streams or by-products such as EOs is the lack of efficient and cost-effective methods for the recovery and purification of targeted bioactive compounds. On the other hand, more efforts are required to scale up production and guarantee the availability of these compounds on the market, in addition to optimizing the efficacy, formulation, and delivery of the formulated products for specific pests under diverse cropping systems. These production problems are essentially critical to biopesticide commercialization.

14.4 *In vitro* Synthesis vs Extraction from Natural Sources: How to Obtain Secondary Metabolites

Obtaining secondary metabolites from plant cell cultures is an invaluable tool for pharmacological, food, agronomic, and cosmetic industries, representing a convenient alternative to their extraction from plant organisms. Steroids, alkaloids, terpenoids, and flavonoids are some of the compounds synthesized by plants that are important for humans and have been produced on a large scale through *in vitro* techniques for commercialization (Chandran et al. 2020).

14.4.1 Factors Affecting the Extraction of Bioactive Compounds from Natural Sources

The amount of secondary metabolites produced by a plant can be strongly affected by seasonality. Gasparetto et al. (2017) evaluated the quantity and composition of EO from *Piper cernuum* leaves collected at the end of each season. According to these authors, the performance of the EO was 1.49%, 1.66%, 2.07%, 2.16% in winter, spring, summer, and autumn, respectively. Seasonal variations in the constitution of the extracted EOs and in their biological activities were also observed. Seasonal variation in the composition and biological activity of the EO of *R. officinalis* has also been tested (Fumiere Lemos et al. 2015), showing a higher yield in autumn, compared to that in the remaining seasons. This difference could be attributed to the nonextreme temperatures at this time of the year, along with the influence of rainfall. In addition, we evaluated possible fluctuations in the antimicrobial activity, observing that both the EO and the ethanolic extract in summer presented a significantly lower MIC than that from the other seasons. The content of camphor and carnosic acid was high, which could partially account for bioactivity.

The fluctuations in the production of secondary metabolites of interest can also be ascribed to the edaphic characteristics of the plantation site. The characteristic microbiota associated with plants can be a conditioning factor. For example, it has been shown that the inoculation of medicinal plants with endophytic bacteria can substantially modify the amount of secondary metabolites synthesized (Maggini et al. 2017). Other factors such as variations in the photoperiod and light intensity may also affect the primary and secondary metabolism of plant species (Thoma et al. 2020). On the other hand, different sources of stress, such as soil contamination with heavy metals (Lajayer et al. 2017), drought (Savoi et al. 2016), thermal stress (Zandalinas et al. 2018), herbivory, and microbial infections (Zaynab et al. 2018) can influence production of secondary metabolites.

Considering that the biosynthesis of bioactive compounds by plant species is commonly scarce, obtaining appreciable quantities implies collecting plant material unsustainably, compromising the persistence of plant species (Gonçalves and Romano 2018). This represents one of the main disadvantages of the traditional extraction methods.

Accordingly, the search for alternatives to traditional extraction methods of secondary metabolites seems necessary. Callus cultures, organ cultures, or cell suspensions are some of the tools available in plant biotechnology aimed at generating bioactive secondary metabolites. Moreover, through these techniques, it is possible to avoid the environmental factors that affect plant growth and secondary metabolite production, since different parameters can be manipulated to maximize yields (Nartop 2018). A case in point is the use of the so-called elicitors. These substances naturally trigger plant defense mechanisms, and their application under laboratory conditions allows artificially inducing the synthesis of secondary metabolites (Nadeem and Ahmad 2019). Other benefits linked with this practice include the high rate at which secondary metabolites are synthesized, their purity (since concomitant extraction with other unwanted secondary metabolites is avoided), the possibility of producing them on large scales, and the predictability of production (Iqbal and Ansari 2019).

However, cultures of plant cells and organs often cause some impairments that can affect the expected outcome, including physiological heterogeneity and genetic instability, especially when cultures are maintained for long periods, and associated production costs (Espinosa-Leal et al. 2018). Yet, the benefits of the synthesis of natural products via *in vitro* methods seem to outweigh its disadvantages.

Arbutin, paclitaxel, shikonin, and coumarins are examples of secondary metabolites industrially obtained from plant tissue cultures (Espinosa-Leal et al. 2018) applied to the field of cosmetics and health. Considering the plethora of secondary metabolites synthesized by plants and useful for human beings, there is still a long way toward the optimization of *in vitro* methods to meet the demands of other industries, such as agriculture. As discussed earlier, there are numerous examples of natural products with full potential for controlling agricultural pests; however, very few are commercially available, like triterpenoid azadirachtin.

14.4.2 Production of Azadirachtin by *Azadirachta indica*. A Case Study

Neem (*Azadirachta indica*) is a tree belonging to the Meliaceae family, from which several metabolites with antitumor, antibacterial, anti-inflammatory, and insecticide activity are obtained (Islas et al. 2020). Azadirachtin is one of its most important secondary metabolites because of its insecticidal properties against a wide range of insects, showing diverse mechanisms of action, such as antifeeding, sterilizing, and growth inhibitory (Kilani-Morakchi et al. 2021). This compound is extracted mainly from mature seeds; yet, its extraction and yield have limitations. Neem blooms only once a year, and from the total seeds obtained, only a third is estimated to be collected. In addition, heterozygosity due to cross-pollination, yield variations due to seasonal changes, and limited geographic distribution of species are the major obstacles encouraging the search for alternatives, thus promoting *in vitro* cultivation methods (Thakore and Srivastava 2017). Several studies have explored mechanisms to achieve a continuous and sustainable production. Srivastava and Srivastava (2014) determined the effect of the addition of biotic and abiotic inducers, precursors, and signal molecules on the biosynthesis of azadirachtin in hairy root cultures. In their work, the incorporation of salicylic acid and methyl jasmonate at low concentrations enhanced the yields of azadirachtin in relation to that of control. The addition of yeast extract and carbohydrate fraction of yeast extract to the medium also increased the production of azadirachtin and biomass of the culture. Similarly, when fungal filtrates of different species (*Phoma herbarium*, *A. alternata*, *Myrothecium* sp., *Fusarium solani*, *Curvularia lunata*, and *Sclerotium rolfsii*) were incorporated into the medium as elicitors, a higher yield of azadirachtin was observed. Since azadirachtin is synthesized in the isoprenoid pathway, the supplementation of certain precursors of this metabolic route to the culture medium was also evaluated by the authors. Among the precursors added (cholesterol, squalene, sodium acetate, isopentenyl pyrophosphate, and mevalonic acid lactone), the highest concentration of azadirachtin was reached with the addition of cholesterol, with no detriment to biomass.

The effect of incorporating elicitors and modifying the carbon/nitrogen source of the culture medium on the synthesis of limonoids by *A. indica* was studied by Vásquez-Rivera et al. (2015). The modification of the Murashige and Skoog medium increased biomass and limonoid, as compared to the control group. The addition of sodium acetate as a precursor also increased biomass and limonoids. The authors attributed this effect to the fact that sodium acetate is transformed into acetyl-CoA inside the cell, which is the precursor of several metabolic pathways. The synergistic effect between elicitors such as jasmonic acid, salicylic acid, and chitosan was also evaluated, showing five times more azadirachtin compared to the control. Regarding biotechnological processes, the two-stage bioreactor process reached three times more limonoids compared to those obtained in the single-stage process.

A further advantage of *in vitro* culture implies the possibility of adding hormones to the culture medium to achieve the dedifferentiation and differentiation of explants, which, in turn, can enhance the production of secondary metabolites. In this context, Ashokhan et al. (2020) evaluated the effect of thidiazuron (TDZ) (cytokine) and 2,4-dichlorophenoxyacetic acid (2,4-D) (auxin) on formation of callus and accumulation of azadirachtin from explants of leaves and petioles of *A. indica*. Their results showed that TDZ produced a better callogenesis compared to 2.4-D, along with an increased synthesis of azadirachtin. Further testing revealed that calluses with the highest azadirachtin were found to be nontoxic to the zebrafish embryos in most of the doses evaluated, evidencing its environmentally friendly nature.

In view of this, it seems evident that plant biotechnology is particularly useful for the development of biopesticides, since it allows optimizing current culture systems (calluses, cell suspensions, hairy root cultures, etc.), increasing the production of secondary metabolites to meet market demands, while preserving natural resources.

14.5 Conclusion

The present chapter discussed the importance of plants as bioreactors for the production of EOs and/or their pure VOCs to be used as effective insecticides, bactericides, and fungicides against many economically important pest organisms. However, there are several obstacles that hinder common use of EOs: the amount of EOs in plants is usually scarce, their composition is subjected to environmental conditions, the extraction of EOs involves collection of the plant, compromising the producing species, among other factors. Additionally, not all aromatic plants provide EOs with good pesticidal properties, with those species belonging to the family Lamiaceae generally being the most active. For these reasons, different techniques of plant biotechnology have emerged as novel alternatives for the production of sufficient amounts of EOs or VOCs, such as induced mutagenesis, transgenic plants, biotransformation of EOs or VOCs into new value-added products, and the synthesis of secondary metabolites through cell culture. Even though some of these techniques have not yet been developed industrially, they have already shown encouraging results. Further research will be required to achieve sustainable sources of EOs capable of meeting current market demands.

References

Abdelgaleil, S.A., Mohamed, M.I., Shawir, M.S., and Abou-Taleb, H.K. (2016). Chemical composition, insecticidal and biochemical effects of essential oils of different plant species from Northern Egypt on the rice weevil, *Sitophilus oryzae* L. *J. Pest. Sci.* 89 (1): 219–229.

Achimón, F., Brito, V.D., Pizzolitto, R.P. et al. (2021). Chemical composition and antifungal properties of commercial essential oils against the maize phytopathogenic fungus *Fusarium verticillioides*. *Rev. Argent. Microbiol.* 53 (4): 292–303.

Achimón, F., Leal, L.E., Pizzolitto, R.P. et al. (2022). Insecticidal and antifungal effects of lemon, orange, and grapefruit peel essential oils from Argentina. *AgriScientia* 39 (1): 71–82.

Ács, K., Balázs, V.L., Kocsis, B. et al. (2018). Antibacterial activity evaluation of selected essential oils in liquid and vapor phase on respiratory tract pathogens. *BMC Complement. Altern. Med.* 18 (1): 1–9.

Aghbash, B.N., Pouresmaeil, M., Dehghan, G. et al. (2020). Chemical composition, antibacterial and radical scavenging activity of essential oils from *Satureja macrantha* C.A. Mey. at different growth stages. *Foods* 9 (4): 494.

Agrawal, L. and Kumar, M. (2021). Improvement in ornamental, medicinal, and aromatic plants through induced mutation. *J. Appl. Biol. Biotechnol.* 9 (4): 1–6.

Asadollahi, M., Firuzi, O., Jamebozorgi, F.H. et al. (2019). Ethnopharmacological studies, chemical composition, antibacterial and cytotoxic activities of essential oils of eleven *Salvia* in Iran. *J. Herb. Med.* 17: 100250.

Ashokhan, S., Othman, R., Abd Rahim, M.H. et al. (2020). Effect of plant growth regulators on coloured callus formation and accumulation of azadirachtin, an essential biopesticide in *Azadirachta indica*. *Plan. Theory* 9 (3): 352.

Alok, A., Kumar, J., and Upadhyay, S.K. (2018). Engineering in hairy roots using CRISPR/Cas9-mediated editing. In: *Hairy Roots* (ed. V. Srivastava, S. Mehrotra, and S. Mishra). Singapore: Springer. https://doi.org/10.1007/978-981-13-2562-5_15

Bączek, K.B., Kosakowska, O., Przybył, J.L. et al. (2017). Antibacterial and antioxidant activity of essential oils and extracts from costmary (*Tanacetum balsamita* L.) and tansy (*Tanacetum vulgare* L.). *Ind. Crop. Prod.* 102: 154–163.

Bansal, S., Narnoliya, L.K., Mishra, B. et al. (2018). HMG-CoA reductase from Camphor Tulsi (*Ocimum kilimandscharicum*) regulated MVA dependent biosynthesis of diverse terpenoids in homologous and heterologous plant systems. *Sci. Rep.* 8 (1): 1–15.

Basaid, K., Chebli, B., Mayad, E.H. et al. (2021). Biological activities of essential oils and lipopeptides applied to control plant pests and diseases: a review. *Int. J. Pest Manag.* 67 (2): 155–177.

Birhan, W., Giday, M., and Teklehaymanot, T. (2011). The contribution of traditional healers' clinics to public health care system in Addis Ababa, Ethiopia: a cross-sectional study. *J. Ethnobiol. Ethnomed.* 7 (1): 1–7.

Boukaew, S., Prasertsan, P., and Sattayasamitsathit, S. (2017). Evaluation of antifungal activity of essential oils against aflatoxigenic *Aspergillus flavus* and their allelopathic activity from fumigation to protect maize seeds during storage. *Ind. Crop. Prod.* 97: 558–566.

Boulogne, I., Petit, P., Ozier-Lafontaine, H. et al. (2012). Insecticidal and antifungal chemicals produced by plants: a review. *Environ. Chem. Lett.* 10 (4): 325–347.

Bouyahya, A., Et-Touys, A., Bakri, Y. et al. (2017). Chemical composition of *Mentha pulegium* and *Rosmarinus officinalis* essential oils and their antileishmanial, antibacterial and antioxidant activities. *Microb. Pathog.* 111: 41–49.

Brito, V.D., Achimón, F., Dambolena, J.S. et al. (2019). Trans-2-hexen-1-ol as a tool for the control of *Fusarium verticillioides* in stored maize grains. *J. Stored Prod. Res.* 82: 123–130.

Campolo, O., Giunti, G., Russo, A. et al. (2018). Essential oils in stored product insect pest control. *J. Food Qual.* 2018: 1–18.

Cano-Flores, A., Gómez, J., Escalona-Torres, I.S., and Velasco-Bejarano, B. (2020). Microorganisms as biocatalysts and enzyme sources. In: *Microorganisms* (ed. M. Blumenberg, M. Shaaban and A. Elgaml), 275–306. London, UK: IntechOpen.

Chandran, H., Meena, M., Barupal, T., and Sharma, K. (2020). Plant tissue culture as a perpetual source for production of industrially important bioactive compounds. *Biotechnol. Rep. 26*: e00450.

Chandra, M., Kushwaha, S., and Sangwan, N.S. (2020). Comparative transcriptome analysis to identify putative genes related to trichome development in *Ocimum* species. *Mol. Biol. Rep.* 47 (9): 6587–6598.

Cheng, B.Q., Wei, L.J., Lv, Y.B. et al. (2019). Elevating limonene production in oleaginous yeast *Yarrowia lipolytica* via genetic engineering of limonene biosynthesis pathway and optimization of medium composition. *Biotechnol. Bioprocess Eng.* 24 (3): 500–506.

Coats, J.R., Karr, L.L., and Drewes, C.D. (1991). Toxicity and neurotoxic effects of monoterpenoids. In: *Insects and Earthworms* (ed. P. Hedin), 305–316. Washington, DC, USA: ACS Symposium Series.

da Silva, R.S., de Oliveira, M.M.G., de Melo, J.O. et al. (2019). Antimicrobial activity of *Lippia gracilis* essential oils on the plant pathogen *Xanthomonas campestris* pv. *campestris* and their effect on membrane integrity. *Pestic. Biochem. Physiol.* 160: 40–48.

Dam, D., Molitor, D., and Beyer, M. (2019). Natural compounds for controlling *Drosophila suzukii*. A review. *Agron. Sustain. Dev.* 39 (6): 1–17.

Desam, N.R., Al-Rajab, A.J., Sharma, M. et al. (2017). Chemical composition, antibacterial and antifungal activities of Saudi Arabian *Mentha longifolia* L. essential oil. *J. Coast. Life Med.* 5: 441–446.

Desam, N.R., Al-Rajab, A.J., Sharma, M. et al. (2019). Chemical constituents, in vitro antibacterial and antifungal activity of *Mentha* × *Piperita* L.(peppermint) essential oils. *J. King Saud Univ.Sci.* 31 (4): 528–533.

Dhami, N. (2013). Trends in pharmacognosy: a modern science of natural medicines. *J. Herb. Med.* 3 (4): 123–131.

Diemer, F., Caissard, J.C., Moja, S. et al. (2001). Altered monoterpene composition in transgenic mint following the introduction of 4S-limonene synthase. *Plant Physiol. Biochem.* 39 (7-8): 603–614.

Domingues, P.M. and Santos, L. (2019). Essential oil of pennyroyal (*Mentha pulegium*): Composition and applications as alternatives to pesticides—New tendencies. *Ind. Crop. Prod.* 139: 111534.

Ebadollahi, A. and Sendi, J. (2015). A review on recent research results on bio-effects of plant essential oils against major Coleopteran insect pests. *Toxin Rev.* 34 (2): 76–91.

Ebadollahi, A. and Setzer, W.N. (2020). Evaluation of the toxicity of *Satureja intermedia* C.A. Mey essential oil to storage and greenhouse insect pests and a predator ladybird. *Foods* 9 (6): 712.

Espinosa-Leal, C.A., Puente-Garza, C.A., and García-Lara, S. (2018). In vitro plant tissue culture: means for production of biological active compounds. *Planta* 248 (1): 1–18.

Gasparetto, A., Cruz, A.B., Wagner, T.M. et al. (2017). Seasonal variation in the chemical composition, antimicrobial and mutagenic potential of essential oils from *Piper cernuum*. *Ind. Crop. Prod.* 95: 256–263.

Fierascu, R.C., Fierascu, I.C., Dinu-Pirvu, C.E. et al. (2020). The application of essential oils as a next-generation of pesticides: recent developments and future perspectives. *Z. Naturforsch. C* 75 (7-8): 183–204.

Galvez, C.E., Jimenez, C.M., Gomez, A.D.L.A. et al. (2018). Chemical composition and antifungal activity of essential oils from *Senecio nutans*, *Senecio viridis*, *Tagetes terniflora* and *Aloysia gratissima* against toxigenic *Aspergillus* and *Fusarium* species. *Nat. Prod. Res.* 34 (10): 1442–1445.

Gonçalves, S. and Romano, A. (2018). Production of plant secondary metabolites by using biotechnological tools. In: *Secondary Metabolites – Sources and Applications* (ed. R. Vijayakumar and S.S. Raja), 81–99. London: IntechOpen https://doi.org/10.5772/intechopen.76414.

González-Coloma, A., Martin-Benito, D., Mohamed, N. et al. (2006). Antifeedant effects and chemical composition of essential oils from different populations of *Lavandula luisieri* L. *Biochem. Syst. Ecol.* 34 (8): 609–616.

Gounaris, Y. (2010). Biotechnology for the production of essential oils, flavours and volatile isolates. A review. *Flavour Fragr. J.* 25 (5): 367–386.

Gounaris, Y. (2019). The role of biotechnology in essential oil production from non-herbaceous plants. In: *Essential Oil Research*, 365–400. Cham: Springer.

Hendrawati, O., Woerdenbag, H.J., Hille, J., and Kayser, O. (2010). Metabolic engineering strategies for the optimization of medicinal and aromatic plants: realities and expectations. *J. Med. Spice Plants* 15: 111–126.

Herrera, J.M., Zunino, M.P., Dambolena, J.S. et al. (2015). Terpene ketones as natural insecticides against *Sitophilus zeamais*. *Ind. Crop. Prod.* 70: 435–442.

Hu, F., Tu, X.F., Thakur, K. et al. (2019). Comparison of antifungal activity of essential oils from different plants against three fungi. *Food Chem. Toxicol.* 134: 110821.

Hussain, A., Bose, S., Wang, J.H. et al. (2016). Fermentation, a feasible strategy for enhancing bioactivity of herbal medicines. *Food Res. Int.* 81: 1–16.

Hussein, K.A. and Joo, J.H. (2018). Antifungal activity and chemical composition of ginger essential oil against ginseng pathogenic fungi. *Curr. Res. Environ. Appl. Mycol.* 8: 194–203.

Imane, N.I., Fouzia, H., Azzahra, L.F. et al. (2020). Chemical composition, antibacterial and antioxidant activities of some essential oils against multidrug resistant bacteria. *Eur. J. Integr. Med.* 35: 101074.

Iqbal, M.S. and Ansari, M.I. (2019). Biotechnological exercises in the production of secondary metabolites and its significance in healthcare practices. In: *Natural Bio-active Compounds* (ed. M. Akhtar and M. Swamy), 183–204. Singapore: Springer.

Islas, J.F., Acosta, E., Zuca, G. et al. (2020). An overview of Neem (*Azadirachta indica*) and its potential impact on health. *J. Funct. Foods* 74: 104171.

Isman, M.B. (2020). Commercial development of plant essential oils and their constituents as active ingredients in bioinsecticides. *Phytochem. Rev.* 19 (2): 235–241.

Jahani, M., Pira, M., and Aminifard, M.H. (2020). Antifungal effects of essential oils against *Aspergillus niger* in vitro and in vivo on pomegranate (*Punica granatum*) fruits. *Sci. Hortic.* 264: 109188.

Karabörklü, S., Ayvaz, A., Yilmaz, S., and Akbulut, M. (2011). Chemical composition and fumigant toxicity of some essential oils against *Ephestia kuehniella*. *J. Econ. Entomol.* 104 (4): 1212–1219.

Kilani-Morakchi, S., Morakchi-Goudjil, H., and Sifi, K. (2021). Azadirachtin-based insecticide: overview, risk assessments, and future directions. *Front. Agron.* 3: 1–13.

Knight A.R. (2009). Preparation and bioactivity of 1,8-cineole derivatives. PhD thesis. Murdoch University, 184 p.

Kocić-Tanackov, S., Dimić, G., Mojović, L. et al. (2016). Antifungal activity of the onion (*Allium cepa* L.) essential oil against *Aspergillus, Fusarium* and *Penicillium* species isolated from food. *J. Food Process. Preserv.* 41 (4): e13050.

Kolakar, S.S., Nadukeri, S., Jakkeral, S.A. et al. (2018). Role of mutation breeding in improvement of medicinal and aromatic crops. *J. Pharmacog. Phytochem.* 425–429.

Kordali, Ş., Usanmaz, A., Bayrak, N., and Çakır, A. (2017). Fumigation of volatile monoterpenes and aromatic compounds against adults of *Sitophilus granarius* (L.) (Coleoptera: Curculionidae). *Rec. Nat. Prod.* 11 (4): 362.

Krasnyanski, S., May, R.A., Loskutov, A. et al. (1999). Transformation of the limonene synthase gene into peppermint (*Mentha piperita* L.) and preliminary studies on the essential oil profiles of single transgenic plants. *Theor. Appl. Genet.* 99 (3): 676–682.

Krings, U. and Berger, R.G. (2010). Terpene bioconversion–how does its future look? *Nat. Prod. Commun.* 5 (9): 1507–1522. 1934578X1000500927.

Kumar, V., Ahluwalia, V., Saran, S. et al. (2021). Recent developments on solid-state fermentation for production of microbial secondary metabolites: challenges and solutions. *Bioresour. Technol.* 323: 124566.

Kurita, N. and Koike, S. (1983). Synergistic antimicrobial effect of ethanol, sodium chloride, acetic acid and essential oil components. *Agric. Biol. Chem.* 47 (1): 67–75.

Lajayer, B.A., Ghorbanpour, M., and Nikabadi, S. (2017). Heavy metals in contaminated environment: destiny of secondary metabolite biosynthesis, oxidative status and phytoextraction in medicinal plants. *Ecotoxicol. Environ. Saf.* 145: 377–390.

Lal, R.K., Chanotiya, C.S., Singh, V.R. et al. (2019). Induced polygenic variations through γ-rays irradiation and selection of novel genotype in chamomile (*Chamomilla recutita* [L.] Rauschert). *Int. J. Radiat. Biol.* 95 (9): 1242–1250.

Laufenberg, G., Kunz, B., and Nystroem, M. (2003). Transformation of vegetable waste into value added products:(A) the upgrading concept;(B) practical implementations. *Bioresour. Technol.* 87 (2): 167–198.

Lemos, M.F., Lemos, M.F., Pacheco, H.P. et al. (2015). Seasonality modifies rosemary's composition and biological activity. *Ind. Crop. Prod.* 70: 41–47.

López, A.G., Theumer, M.G., Zygadlo, J.A., and Rubinstein, H.R. (2004). Aromatic plants essential oils activity on *Fusarium verticillioides* Fumonisin B_1 production in corn grain. *Mycopathologia* 158 (3): 343–349.

López-Meneses, A.K., Sánchez-Mariñez, R.I., Quintana-Obregón, E.A. et al. (2017). In vitro antifungal activity of essential oils and major components against fungi plant pathogens. *J. Phytopathol.* 165 (4): 232–237.

Mączka, W., Duda-Madej, A., Górny, A. et al. (2021). Can eucalyptol replace antibiotics? *Molecules* 26 (16): 4933.

Maede, M., Hamzeh, I., Hossein, D. et al. (2011). Bioactivity of essential oil from *Satureja hortensis* (Lamiaceae) against three stored-product insect species. *Afr. J. Biotechnol.* 10 (34): 6620–6627.

Maggini, V., De Leo, M., Mengoni, A. et al. (2017). Plant-endophytes interaction influences the secondary metabolism in *Echinacea purpurea* (L.) Moench: an in vitro model. *Sci. Rep.* 7 (1): 1–8.

Mahdavi, B., Yaacob, W.A., and Din, L.B. (2017). Chemical composition, antioxidant, and antibacterial activity of essential oils from *Etlingera sayapensis* A.D. Poulsen & Ibrahim. *Asian Pac J Trop Med* 10 (8): 819–826.

Mahmoud, S.S. and Croteau, R.B. (2001). Metabolic engineering of essential oil yield and composition in mint by altering expression of deoxyxylulose phosphate reductoisomerase and menthofuran synthase. *Proc. Natl. Acad. Sci.* 98 (15): 8915–8920.

Mahmoud, S.S. and Croteau, R.B. (2002). Strategies for transgenic manipulation of monoterpene biosynthesis in plants. *Trends Plant Sci.* 7 (8): 366–373.

Mahmoud, S.S. and Croteau, R.B. (2003). Menthofuran regulates essential oil biosynthesis in peppermint by controlling a downstream monoterpene reductase. *Proc. Natl. Acad. Sci.* 100 (24): 14481–14486.

Matveeva, T.V. (2018). Agrobacterium-mediated transformation in the evolution of plants. *Curr. Top. Microbiol. Immunol.* 418: 421–441.

Maurya, S., Chandra, M., Yadav, R.K. et al. (2019). Interspecies comparative features of trichomes in *Ocimum* reveal insights for biosynthesis of specialized essential oil metabolites. *Protoplasma* 256 (4): 893–907.

Mendoza-Poudereux, I., Muñoz-Bertomeu, J., Navarro, A. et al. (2014). Enhanced levels of S-linalool by metabolic engineering of the terpenoid pathway in spike lavender leaves. *Metab. Eng.* 23: 136–144.

Moghadam, Z.A., Hosseini, H., Hadian, Z. et al. (2019). Evaluation of the antifungal activity of cinnamon, clove, thymes, zataria multiflora, cumin and caraway essential oils against Ochratoxigenic *Aspergillus ochraceus*. *Evaluation* 26 (1).

Mohamed, W.M. and El-Shimi, I. (2014). The effect of gamma rays on the pharmaceutical products of essential oils for peppermint (*Mentha piperita* L.). *Int. J. Farm. Sci.* 3 (8): 884–894.

Muñoz-Bertomeu, J., Arrillaga, I., Ros, R., and Segura, J. (2006). Up-regulation of 1-deoxy-D-xylulose-5-phosphate synthase enhances production of essential oils in transgenic spike lavender. *Plant Physiol.* 142 (3): 890–900.

Muñoz-Bertomeu, J., Sales, E., Ros, R. et al. (2007). Up-regulation of an N-terminal truncated 3-hydroxy-3-methylglutaryl CoA reductase enhances production of essential oils and sterols in transgenic *Lavandula latifolia*. *Plant Biotechnol. J.* 5 (6): 746–758.

Muñoz-Bertomeu, J., Ros, R., Arrillaga, I., and Segura, J. (2008). Expression of spearmint limonene synthase in transgenic spike lavender results in an altered monoterpene composition in developing leaves. *Metab. Eng.* 10 (3-4): 166–177.

Nadeem, H. and Ahmad, F. (2019). Prospects for the use of plant cell culture as alternatives to produce secondary metabolites. In: *Natural Bio-active Compounds* (ed. M. Akhtar and M. Swamy), 153–182. Singapore: Springer.

Nartop, P. (2018). Engineering of biomass accumulation and secondary metabolite production in plant cell and tissue cultures. In: *Plant Metabolites and Regulation Under Environmental Stress* (ed. P. Ahmad, M.A. Ahanger, V.P. Singh, et al.), 169–194. London: Academic Press.

Nzanza, B. and Mashela, P.W. (2012). Control of whiteflies and aphids in tomato (*Solanum lycopersicum* L.) by fermented plant extracts of neem leaf and wild garlic. *Afr. J. Biotechnol.* 11 (94): 16077–16082.

Omarini, A., Dambolena, J.S., Lucini, E. et al. (2016). Biotransformation of 1, 8-cineole by solid-state fermentation of *Eucalyptus* waste from the essential oil industry using *Pleurotus ostreatus* and *Favolus tenuiculus*. *Folia Microbiol.* 61 (2): 149–157.

Omarini, A.B., Achimón, F., Brito, V.D., and Zygadlo, J.A. (2020). Fermentation as an alternative process for the development of bioinsecticides. *Fermentation* 6 (4): 120.

Othira, J.O., Onek, L.A., Deng, L.A., and Omolo, E.O. (2009). Insecticidal potency of *Hyptis spicigera* preparations against *Sitophilus zeamais* (l.) and *Tribolium castaneum* (herbst) on stored maize grains. *Afr. J. Agric. Res.* 4 (3): 187–192.

Pandey, A.K., Kumar, P., Singh, P. et al. (2017). Essential oils: Sources of antimicrobials and food preservatives. *Front. Microbiol.* 7: 2161.

Pavela, R. (2016). History, presence and perspective of using plant extracts as commercial botanical insecticides and farm products for protection against insects–a review. *Plant Prot. Sci.* 52 (4): 229–241.

Perczak, A., Gwiazdowska, D., Marchwińska, K. et al. (2019). Antifungal activity of selected essential oils against *Fusarium culmorum* and *F. graminearum* and their secondary metabolites in wheat seeds. *Arch. Microbiol.* 201 (8): 1085–1097.

Pizzolitto, R.P., Jacquat, A.G., Usseglio, V.L. et al. (2020). Quantitative-structure-activity relationship study to predict the antifungal activity of essential oils against *Fusarium verticillioides*. *Food Control* 108: 106836.

Purkait, S., Bhattacharya, A., Bag, A., and Chattopadhyay, R.R. (2020). Synergistic antibacterial, antifungal and antioxidant efficacy of cinnamon and clove essential oils in combination. *Arch. Microbiol.* 202 (6): 1439–1448.

Ramos, A.D.S., Ribeiro, J.B., Teixeira, B.G. et al. (2015). Hydroxylation of 1, 8-cineole by *Mucor ramannianus* and *Aspergillus niger*. *Braz. J. Microbiol.* 46: 261–264.

Reddy, V.A., Wang, Q., Dhar, N. et al. (2017). Spearmint R2R3-MYB transcription factor Ms MYB negatively regulates monoterpene production and suppresses the expression of geranyl diphosphate synthase large subunit (Ms GPPS. LSU). *Plant Biotechnol. J.* 15 (9): 1105–1119.

Reyes-Jurado, F., Cervantes-Rincón, T., Bach, H. et al. (2019). Antimicrobial activity of Mexican oregano (*Lippia berlandieri*), thyme (*Thymus vulgaris*), and mustard (*Brassica nigra*) essential oils in gaseous phase. *Ind. Crop. Prod.* 131: 90–95.

Rosato, A., Maggi, F., Cianfaglione, K. et al. (2018). Chemical composition and antibacterial activity of seven uncommon essential oils. *J. Essent. Oil Res.* 30 (4): 233–243.

Salah-Fatnassi, K.B.H., Hassayoun, F., Cheraif, I. et al. (2017). Chemical composition, antibacterial and antifungal activities of flowerhead and root essential oils of *Santolina chamaecyparissus* L., growing wild in Tunisia. *Saudi J. Biol. Sci.* 24 (4): 875–882.

Sales, A., Afonso, L.F., Americo, J.A. et al. (2018). Monoterpene biotransformation by *Colletotrichum* species. *Biotechnol. Lett.* 40 (3): 561–567.

Savoi, S., Wong, D.C., Arapitsas, P. et al. (2016). Transcriptome and metabolite profiling reveals that prolonged drought modulates the phenylpropanoid and terpenoid pathway in white grapes (*Vitis vinifera* L.). *BMC Plant Biol. 16* (1): 1–17.

Sharma, A.K. and Sharma, M.K. (2009). Plants as bioreactors: recent developments and emerging opportunities. *Biotechnol. Adv.* 27 (6): 811–832.

Siddique, S., Parveen, Z., and Mazhar, S. (2020). Chemical composition, antibacterial and antioxidant activities of essential oils from leaves of three *Melaleuca* species of Pakistani flora. *Arab. J. Chem.* 13 (1): 67–74.

Singh, P. and Pandey, A.K. (2018). Prospective of essential oils of the genus *Mentha* as biopesticides: a review. *Front. Plant Sci.* 9: 1295.

Sodhi, A.S., Sharma, N., Bhatia, S. et al. (2022). Insights on sustainable approaches for production and applications of value added products. *Chemosphere* 286: 131623.

Soujanya, P.L., Sekhar, J.C., Kumar, P. et al. (2016). Potentiality of botanical agents for the management of post harvest insects of maize: a review. *J. Food Sci. Technol.* 53 (5): 2169–2184.

Srivastava, S. and Srivastava, A.K. (2014). Effect of elicitors and precursors on azadirachtin production in hairy root culture of *Azadirachta indica*. *Appl. Biochem. Biotechnol.* 172 (4): 2286–2297.

Taheri, H. (2019). Transcriptional modulation of structural and regulatory genes involved in isoprene biosynthesis and their relevance to oil yield and menthol content in peppermint (*Mentha piperita* L.) upon MeJA and GA3 treatments. *Russ. J. Plant Physiol.* 66 (3): 503–508.

Tepe, B., Cakir, A., and Sihoglu Tepe, A. (2016). Medicinal uses, phytochemistry, and pharmacology of *Origanum onites* (L.): a review. *Chem. Biodivers.* 13 (5): 504–520.

Thakore, D. and Srivastava, A.K. (2017). Production of biopesticide azadirachtin using plant cell and hairy root cultures. *Eng. Life Sci. 17* (9): 997–1005.

Thoma, F., Somborn-Schulz, A., Schlehuber, D. et al. (2020). Effects of light on secondary metabolites in selected leafy greens: a review. *Front. Plant Sci. 11*: 497.

Tripathi, A.K. and Mishra, S. (2016). Plant monoterpenoids (Prospective Pesticides). In: *Ecofriendly Pest Management for Food Security* (ed. O. Omkar), 507–524. Cambridge, MA, USA: Academic Press.

Tu, X.F., Hu, F., Thakur, K. et al. (2018). Comparison of antibacterial effects and fumigant toxicity of essential oils extracted from different plants. *Ind. Crop. Prod.* 124: 192–200.

Upadhyay, S.K. and Singh, S.P. (eds.) (2021). *Bioprospecting of Plant Biodiversity for Industrial Molecules*. John Wiley & Sons Ltd. https://doi.org/10.1002/9781119718017.

Verma, P.C., Trivedi, I., Singh, H. et al. (2009). Efficient production of gossypol from hairy root culture of cotton (*Gossypium hirsutum*). *Current Pharmaceutical Biotechnology* 10: 691–700. https://doi.org/10.2174/138920109789542048.

Vásquez-Rivera, A., Chicaiza-Finley, D., Hoyos, R.A., and Orozco-Sánchez, F. (2015). Production of limonoids with insect antifeedant activity in a two-stage bioreactor process with cell suspension culture of *Azadirachta indica*. *Appl. Biochem. Biotechnol.* 177 (2): 334–345.

Wang, Q., Reddy, V.A., Panicker, D. et al. (2016). Metabolic engineering of terpene biosynthesis in plants using a trichome-specific transcription factor Ms YABBY 5 from spearmint (*Mentha spicata*). *Plant Biotechnol. J.* 14 (7): 1619–1632.

Wu, C., Li, T., Qi, J. et al. (2020a). Effects of lactic acid fermentation-based biotransformation on phenolic profiles, antioxidant capacity and flavor volatiles of apple juice. *LWT* 122: 109064.

Wu, L., Zhao, Y., Zhang, Q. et al. (2020b). Overexpression of the 3-hydroxy-3-methylglutaryl-CoA synthase gene LcHMGS effectively increases the yield of monoterpenes and sesquiterpenes. *Tree Physiol.* 40 (8): 1095–1107.

Zandalinas, S.I., Mittler, R., Balfagón, D. et al. (2018). Plant adaptations to the combination of drought and high temperatures. *Physiol. Plant. 162* (1): 2–12.

Zaynab, M., Fatima, M., Abbas, S. et al. (2018). Role of secondary metabolites in plant defense against pathogens. *Microb. Pathog. 124*: 198–202.

Zhang, Y., Wei, J., Chen, H. et al. (2020). Antibacterial activity of essential oils against *Stenotrophomonas maltophilia* and the effect of citral on cell membrane. *LWT* 117: 108667.

Zhu, Y., Li, C., Cui, H., and Lin, L. (2021). Encapsulation strategies to enhance the antibacterial properties of essential oils in food system. *Food Control* 123: 107856.

Zygadlo, J.A., Zunino, M.P., Pizzolitto, R.P. et al. (2017). Antibacterial and anti-biofilm activities of essential oils and their components including modes of action. In: *Essential Oils and Nanotechnology for Treatment of Microbial Diseases* (ed. M. Rai, S. Zacchino and M.G. Derita), 99–126. Boca Raton, FL: CRC Press.

15

Nutraceuticals Productions from Plants

Isabela Sandy Rosa, Laura Oliveira Pires, and Juliane Karine Ishida

Botany Department, Institute of Biological Sciences, Federal University of Minas Gerais, Belo Horizonte, Brazil

In recent decades, the market for dietary supplements has increased with promises of weight loss, health improvement, and even natural treatment of diseases (Ronis et al. 2018). At the same time, the number of studies on the benefits of specific diets and foods is also increasing (Minuz et al. 2017). In this context, nutraceuticals are mentioned as having great potential. According to DeFelice (1995), nutraceuticals refer to foods or their components that can help human health by treating or preventing disease. The term was coined to raise awareness of the importance of multidisciplinary research in nutrition and shed light on these biochemical substances' mechanisms of action. Dietary supplements are commonly regarded as a safe nutritional supplement (Minuz et al. 2017) or a more natural way to prevent or treat disease. However, many of these substances' actual mechanism of action touted as dietary supplements remains to be elucidated, as does the safe level of consumption and the minimum amount required to achieve their beneficial effects. Therefore, there is an urgent need for more data on the biochemistry of these substances to find the specific benefits, potential risks, and recommended dosages of dietary supplements to allow safe use for the population (Gurley et al. 2012; Ronis et al. 2018). This chapter summarizes currently known nutraceuticals, emphasizing those derived from plants since they are the most critical resource for human nutrition and medicine (Grusak and Dellapenna 2003; Kumar et al. 2020; Valachovičová and Bírošová 2020). We discuss the general mechanism of plant-derived nutraceuticals, their potential risks, and their benefits to human health.

15.1 Plant-Derived Nutraceuticals

Plants have been helping us support and improve our health for thousands of years. In the next century, plant scientists will play a key role in solving one of our biggest problems: the continuing rise of chronic illnesses such as heart disease, cancer, diabetes, and obesity. Scientific evidence gathered over the years links plant-based foods to healthier lives. A classic study from the 1990s showed that daily fat consumption is associated with an increase

Plants as Bioreactors for Industrial Molecules, First Edition.
Edited by Santosh Kumar Upadhyay and Sudhir P. Singh.

in chronic heart disease (Renaud and de Lorgeril 1989; Renaud and de Lorgeril 1992). Thus, cases of cardiovascular disease in places with high-fat consumption, such as Germany, the United Kingdom, and Scandinavian countries, contrast with regions that follow a Mediterranean diet (Portugal, Italy, and Spain). This reasoning did not apply to France, which has a diet more likely in northern Europe, but with a low mortality rate. The phenomenon became known as the French paradox. Later, the scientist attributed this peculiarity to the French custom of drinking red wine with meals (Renaud and de Lorgeril 1989, 1992; Korpela et al. 1992; Artaud-Wild et al. 1993). The high polyphenols content in red grapes is associated with reducing the risk of cardiovascular disease in a similar proportion as pharmacological mediation (Renaud and de Lorgeril 1992; Artaud-Wild et al. 1993; Tousoulis et al. 2008; Lippi et al. 2010; Salazar et al. 2022).

In the 1970s and 1980s, British Richard Doll and Richard Peto pioneered research into the impact of lifestyle and environmental factors as significant causes of cancer. Their review paper, published in 1981, summarized the state of knowledge at the time on how behavior and diet promote cancer and became a worldwide reference, contributing to the introduction of public policies on cancer prevention (Doll and Peto 1981; Blot and Tarone 2015). Queen Elizabeth II later knighted both scientists to recognize their contribution to cancer research. They opened avenues for extensive epidemiological studies suggesting that changing daily diets to include consuming various fruits, legumes, and vegetables can reduce cancer risk (Key et al. 1997; Cohen et al. 2000; Stram et al. 2006; Benetou et al. 2008; Diallo et al. 2016; Kooshki et al. 2016). Biochemical approaches suggest that plant metabolites may alter the development of latent microtumors (Doll and Peto 1981; Key et al. 1997; Cohen et al. 2000; Dorai and Aggarwal 2004; Stram et al. 2006; Béliveau and Gingras 2007; Benetou et al. 2008; Pukkala et al. 2009; Schottenfeld et al. 2013; Blot and Tarone 2015; Rothenberg 2015; Diallo et al. 2016; Kooshki et al. 2016; Kotecha et al. 2016). Other investigations show that a high dietary fiber intake reduces the risk of coronary diseases, diabetes, and obesity (Khaw and Barrett-connor 1987; Humble et al. 1993; Liu 2003; Schottenfeld et al. 2013; Kim and Je 2016; López-González et al. 2022).

Most bioactive compounds with health-promoting properties are secondary metabolites (Aziz et al. 2018; Velu et al. 2018; Albini et al. 2019; Forni et al. 2019; Mahajan et al. 2020; Dixit et al. 2021a,b; Upadhyay and Singh 2021). In plant physiology, the metabolites essential for stable growth are called primary metabolites, while those that occur in a particular lineage, species, or family are called specialized or secondary metabolites (Böttger et al. 2018). The functions of secondary metabolites in plant physiology depend on responses to abiotic and biotic stresses (Erb et al. 2013; Delgoda and Murray 2017; Böttger et al. 2018). During the evolution of land plants, they have evolved to provide tools for coping with variable environmental changes, including the ecological interactions between plants and abiotic stressors (Erb et al. 2013; Delgoda and Murray 2017; Böttger et al. 2018; Dixit et al. 2019). Specialized secondary metabolites often accumulate in specific organs to promote defense against herbivores or pathogen attacks or reward beneficial fungi and insect pollinators (Pichersky and Raguso 2018; Forni et al. 2019). But also, other organisms may use the production of plant metabolites to their advantage. Since ancient times, humanity has been exploring plant metabolisms phytochemicals are primarily responsible for treating various diseases (Velu et al. 2018). Thus, it has been shown that plant chemicals are a richer reservoir of potential nutraceuticals and are, therefore, the focus of this chapter.

15.2 Phytochemicals and their Impacts on Human Health

Nutraceuticals can be classified according to different aspects, including the mechanism of action, chemical nature, food source, or health benefits. Below, the organization follows its functional group or biochemical classification in plant biology, including polyphenols, terpenoids, polysterols, alkaloids, fatty acids, and dietary fiber.

15.2.1 Polyphenols

Polyphenols are a significant herbal supplements class that includes any compounds, including flavonoids, commonly consumed and well studied (Amiot et al. 2016; Andrew and Izzo 2017; Ronis et al. 2018) (see Table 15.1). The production of some polyphenols in plants is associated with a protective function. They are essential for plant survival. Plants are exposed to excessive solar radiation, leading to the overproduction of

Table 15.1 Summary of the polyphenols on human health.

Polyphenolic compounds	Dietary source	Effect on human health	References
Chromones	All plants	Antispasmodic	Andrew and Izzo (2017) and Durazzo et al. (2019)
Coumarins	*Apiaceae, Melilotus officinalis, Culus hippocastanum*	Antioxidant, anti-inflammatory, anticoagulant	Andrew and Izzo (2017) and Durazzo et al. (2019)
Anthocyanins	Berries, red wine	Cancer prevention, antioxidant, anti-inflammatory	Andrew and Izzo (2017), Howes et al. (2020) and Lin et al. (2017)
Flavanols or Catechins	Green tea, black tea, red wine, cocoa	Anti-inflammatory, cancer prevention, cardiovascular protection	Andrew and Izzo (2017), Amiot et al. (2016), Durazzo et al. (2019), Shirakami and Shimizu (2018), Isemura (2019), Chen et al. (2016), Shaterzadeh-Yazdi et al. (2018) and Mechchate et al. (2021)
Flavanones	Citrus fruits, vegetables, spices	Blood pressure improvement, cardiovascular protection, obesity and diabetes prevention	Roleira et al. (2021), Lin and Harnly (2010), Kaparapu et al. (2020), Addi et al. (2021), Najmanová et al. (2019), Fraga et al. (2021), Joshi et al. (2018) and Karim et al. (2018)
Flavones	Spice leaves, hot peppers, and olives	Anti-obesogenic, anti-inflammatory, anticarcinogenic	Dini and Laneri (2021), Bucciantini et al. (2021), Sudhakaran et al. (2020) and Lee and Imm (2018)

(*Continued*)

Table 15.1 (Continued)

Polyphenolic compounds	Dietary source	Effect on human health	References
Flavonols	Fruits, vegetables	Anti-inflammatory, antitumor	Ding et al. (2020), Dabeek and Marra (2019), Sezer et al. (2019) and Gervasi et al. (2022)
Isoflavones	Soy and derivatives	Estrogenic activity, reproductive system, and fertility improvement, estrogen-dependent tumors prevention	Hu et al. (2019), Kim (2021), Zaheer and Humayoun Akhtar (2017) and Rizzo et al. (2022)
Proanthocyanidins	Cinnamon, sorghum, nuts, flowers	Antioxidant, anti-inflammatory, antimicrobial, cardiovascular protection	Rauf et al. (2019), de La Iglesia et al. (2010) and Yang et al. (2018)
Curcumins	*Curcuma longa*	Antioxidant, anti-inflammatory, antiarthritic, cancer prevention, Alzheimer's treatment	Howes et al. (2020), Souyoul et al. (2018) and Kotha and Luthria (2019)
Stillbenes	Red grapes, berries, plums, nuts, wine	Cardiovascular protection, bought weight reduction, cognitive improvement, anti-inflammatory	Ronis et al. (2018), Lippi et al. (2010), Salazar et al. (2022), Durazzo et al. (2019), Howes et al. (2020) and Pecyna et al. (2020)
Xanthones	Vascular plants, mango, mangosteen	Metabolic syndrome regulation, anti-inflammatory	Ansori et al. (2020), Feng et al. (2020) and Klein-Júnior et al. (2020)

photosynthesis-generated reactive oxygen species. Therefore, some groups of polyphenols act by dropping these destructive molecules. Polyphenols have become well known to the general population because they are found in traditional diets, such as the Mediterranean diet, associated with health benefits (Andrew and Izzo 2017). Its general biochemical composition consists of aromatic and hydroxyl rings (Durazzo et al. 2019), antioxidant (Ronis et al. 2018), and anti-inflammatory (Amiot et al. 2016) activities are reported. The most pertinent classes of polyphenols are chromones, coumarins, flavonoids, stilbenes, and xanthones (Andrew and Izzo 2017).

15.2.1.1 Chromones

Chromones are ubiquitous in plants (Andrew and Izzo 2017) (see structure in Figure 15.1). One compound from this group, khellin, was formerly used in health care for its antispasmodic action (Andrew and Izzo 2017; Durazzo et al. 2019).

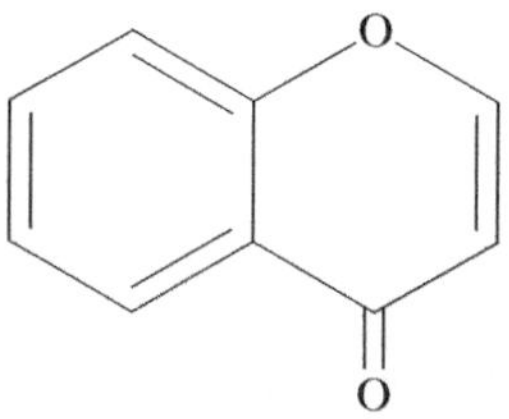

Figure 15.1 Chromones structure.

15.2.1.2 Coumarins

Coumarins are found in the *Apiaceae* family, *Melilotus officinalis*, *Culus hippocastanum* (Andrew and Izzo 2017), and cinnamon (Durazzo et al. 2019) (see structure in Figure 15.2). These compounds are associated with an anticoagulant effect that may reduce the risk of wandering (Andrew and Izzo 2017). In addition, coumarins, like other polyphenols, are associated with antioxidant and anti-inflammatory effects (Durazzo et al. 2019).

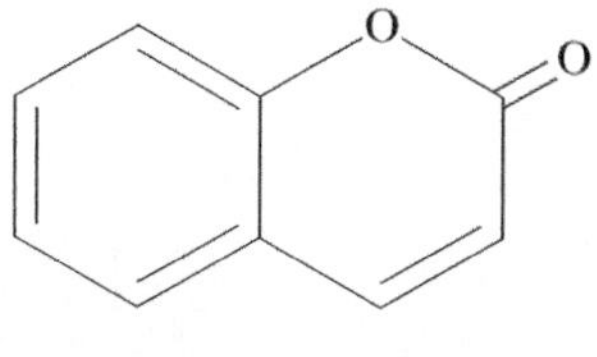

Figure 15.2 Coumarins structure.

15.2.1.3 Flavonoids

Flavonoids are plant pigments found in many plants and are widely consumed (Shen et al. 2022). Their general molecular structure consists of a heterocyclic ring connecting two aromatic rings (Amiot et al. 2016; Ronis et al. 2018; Durazzo et al. 2019; Williamson et al. 2020), dividing into the following subclasses:

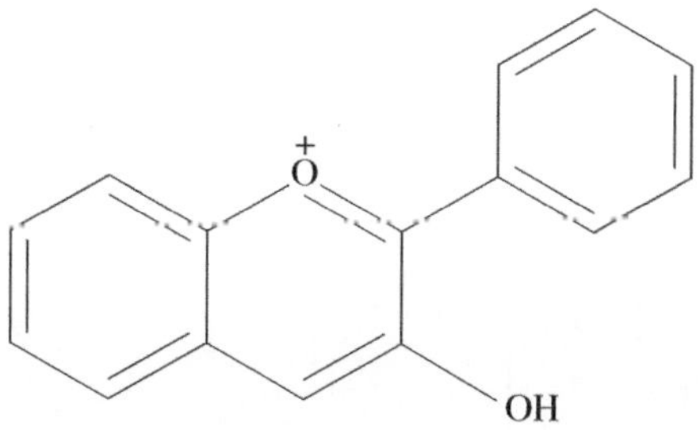

Figure 15.3 Anthocyanidins structure.

Anthocyanins: These flavonoid compounds are found in berries and red wine (Andrew and Izzo, 2017; Howes et al. 2020). They are water-soluble pigments with the chemical structure of flavylium ions (Andrew and Izzo 2017) (see Figure 15.3). Anthocyanins are associated with cancer prevention due to their antioxidant and anti-inflammatory effects (Andrew and Izzo 2017) and induction of apoptosis of tumor cells (Lin et al. 2017).

Flavanols or Catechins: These flavonoids are abundant in green tea, black tea, red wine, and cocoa (Amiot et al. 2016; Andrew and Izzo 2017; Durazzo et al. 2019) (see structure in Figure 15.4). Catechins are generally associated with anti-inflammatory and cancer-preventive effects through cell cycle regulation by inhibiting the tyrosine kinase pathway (Shirakami and Shimizu 2018). Besides, they act in cardiovascular protection (Chen et al. 2016; Shaterzadeh-Yazdi et al. 2018; Isemura 2019). However, the compound alone may not have significant effects, suggesting that some phytochemical activities are enhanced in the presence of other compounds (Mechchate et al. 2021).

Flavanones: These flavonoids are commonly found in citrus fruits, vegetables, and spices (Lin and Harnly 2010; Kaparapu et al. 2020; Addi et al. 2021; Roleira et al. 2021) (see structure in Figure 15.5). Flavanones are associated with a reduced risk of cardiovascular events

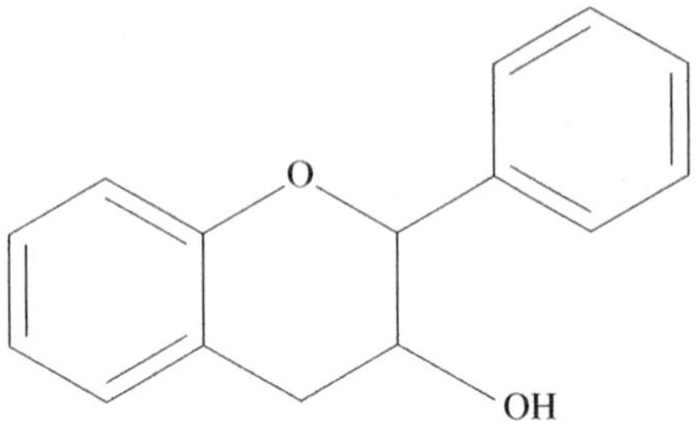

Figure 15.4 Flavanols structure.

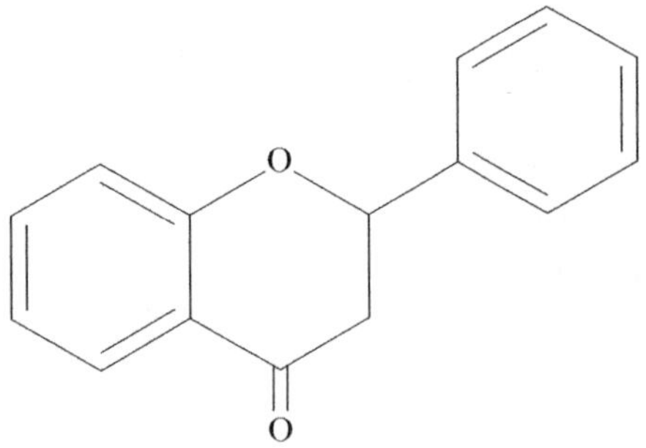

Figure 15.5 Flavanones structure.

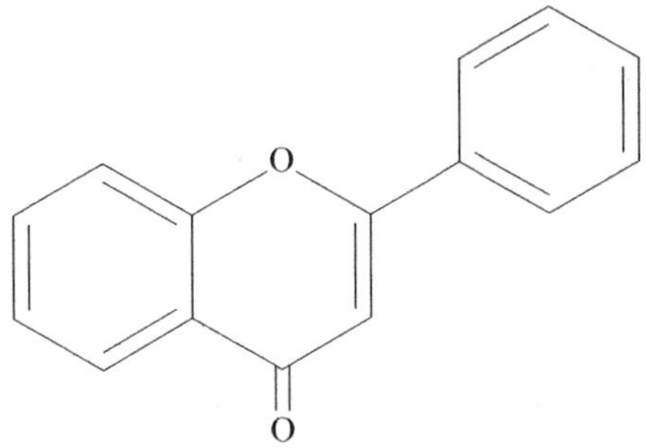

Figure 15.6 Flavones structure.

by improving blood pressure and potential benefits in preventing diabetes and obesity (Najmanová et al. 2019; Macarro et al. 2020; Fraga et al. 2021). Naringenin is a best-studied flavanone. This nutraceutical has an anti-inflammatory activity probably due to its interference in the oxidation of proteins (Karim et al. 2018; Joshi et al. 2018).

Flavones: Flavones are a class of flavonoids commonly found in spice leaves, hot peppers, and olives (Bucciantini et al. 2021; Dini and Laneri 2021) (see structure in Figure 15.6). Flavones are associated with anti-obesogenic, anti-inflammatory, and anticarcinogenic properties (Lee and Imm 2018; Sudhakaran et al. 2020; Bucciantini et al. 2021).

Flavonols: These flavonoids are abundant in fruits and vegetables, with quercetin and kaempferol being the most consumed flavonols (Dabeek and Marra 2019; Sezer et al. 2019; Ding et al. 2020) (see structure in Figure 15.7). These two flavonols are mainly associated with anti-inflammatory and antitumor effects, with quercetin playing a role in regulating some signaling pathways (Ding et al. 2020; Gervasi et al. 2022)

Figure 15.7 Flavonols structure.

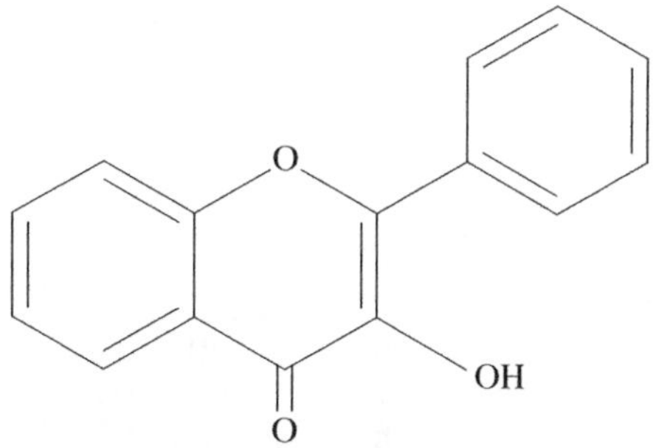

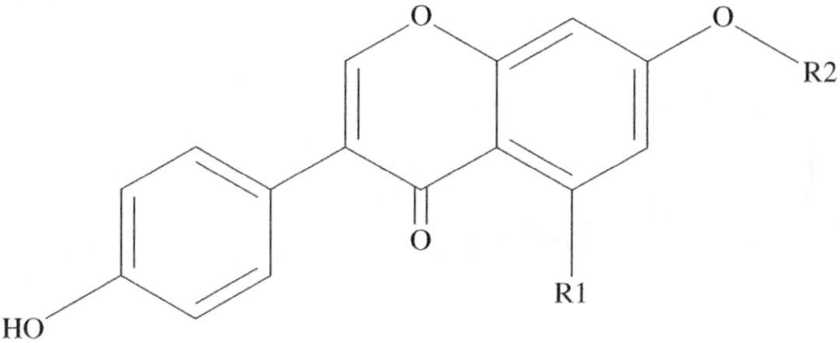

Figure 15.8 Isoflavones structure.

Isoflavones: Isoflavones are a group of flavonoids commonly found in soy and soy derivatives (Hu et al. 2019; Kim 2021) (see structure in Figure 15.8). The well-studied genistein and daidzein are associated with estrogenic activity, known as phytoestrogens (Zaheer and Humayoun Akhtar 2017). Because of their biochemical structure, like estradiol, isoflavones may affect the reproductive system, fertility, and estrogen-dependent tumors (Zaheer and Humayoun Akhtar 2017; Rizzo et al. 2022).

Proanthocyanidins: These flavonoid compounds are highly present in cinnamon, sorghum, nuts, and flowers (Rauf et al. 2019). They have a biochemical structure of flavanol polymers (see Figure 15.9). In addition to the general antioxidant and anti-inflammatory effects of flavonoids, proanthocyanidins are also associated with potential cardiovascular protection and antimicrobial activity (de La Iglesia et al. 2010; Yang et al. 2018; Rauf et al. 2019).

15.2.1.4 Curcumin

Curcumin is a polyphenolic compound in turmeric spice (Souyoul et al. 2018; Kotha and Luthria 2019) (see structure in Figure 15.10). Like other polyphenols, curcumin is also related to anti-inflammatory and antioxidant effects and potential cancer-preventive activity through apoptosis promotion (Souyoul et al. 2018). Furthermore, this polyphenol shows a possible use in arthritis and Alzheimer's therapy (Kotha and Luthria 2019) and cognitive decline due to its antioxidant activity (Howes et al. 2020).

15.2.1.5 Stilbenes

This polyphenol group is commonly found in red grapes, berries, plums, nuts, and wine (Lippi et al. 2010; Ronis et al. 2018; Durazzo et al. 2019; Salazar et al. 2022) (see structure

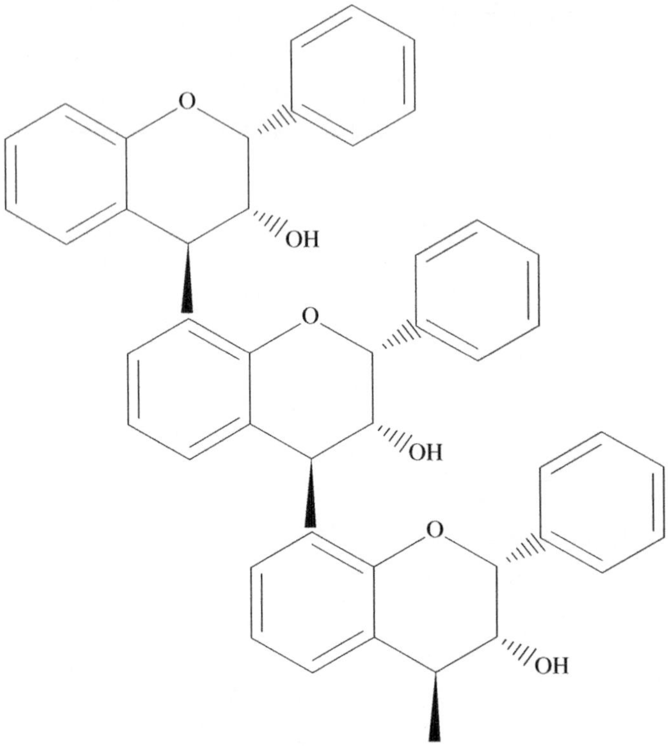

Figure 15.9 Proanthocyanidins structure.

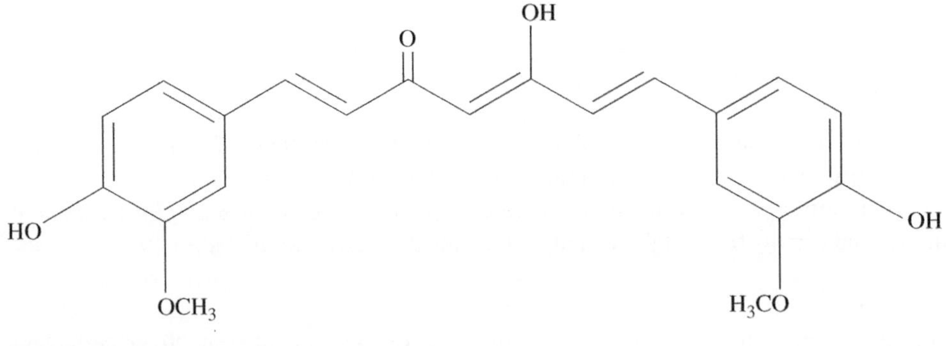

Figure 15.10 Curcumin structure.

in Figure 15.11). Resveratrol is the most studied stilbene and is associated with cardiovascular benefits, body fat loss, anti-inflammatory activity, and cognitive improvement (Howes et al. 2020; Pecyna et al. 2020).

15.2.1.6 Xanthones

Xanthones can be found among vascular plants, mostly in mango and mangosteen (Ansori et al. 2020; Feng et al. 2020) (see structure in Figure 15.12). These compounds are

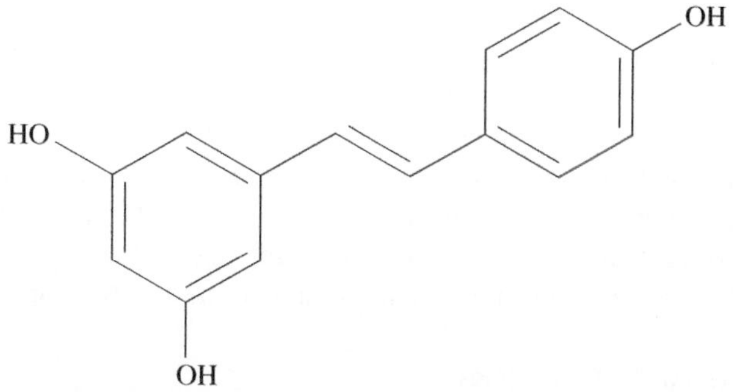

Figure 15.11 Resveratrol structure.

associated with benefits for metabolic syndrome care and anti-inflammatory effects (Feng et al. 2020; Klein-Júnior et al. 2020).

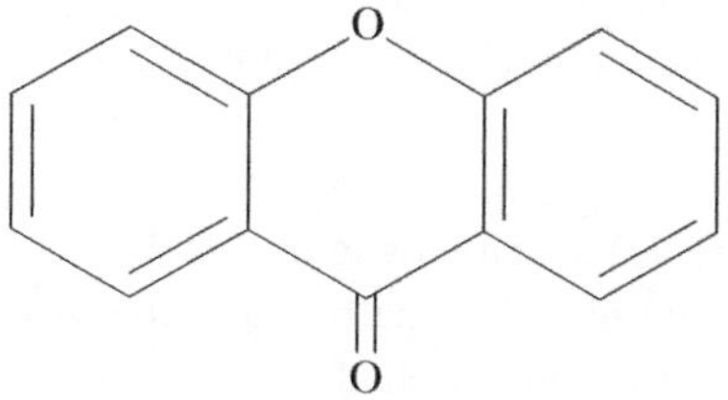

Figure 15.12 Xanthones structure.

15.2.2 Terpenoids

Terpenoids are a group that includes a significant number of plants' secondary metabolites (see Table 15.2). Their general chemical composition consists of isoprene units (Pichersky and Raguso 2018). Regarding their effects on human health, terpenoids are generally associated with anti-inflammatory and antiproliferative activity (Albini et al. 2019).

Table 15.2 Summary of the terpenoids on human health.

Type of terpenoid	Dietary source	Effects on human health	References
Carotenoids	All plants	Antioxidant, cancer prevention, angiogenesis impairment, apoptosis enhancement, free-radical scavenging	Forni et al. (2019), Pichersky and Raguso (2018), Maoka (2020), Ito et al. (2009), Atlante et al. (2020) and Dillard and German (2000)
Ginkgolides	*Ginkgo biloba*	Memory improvement, blood pressure management, cardiac diseases prevention	Fang et al. (2020), Eisvand et al. (2020) and Zhao et al. (2022)
Limonene	Essential oils	Anticancer	Albini et al. (2019)
Oleanolic acid	Citrus fruits peels	Hepatoprotective, anticancer, cirrhosis and fibrosis prevention	Albini et al. (2019)
Phytosterols	Plant oils, seeds, nuts	Lower cholesterol absorption, antiandrogenic	Andrew and Izzo (2017, Li et al. (2022) and Moreau et al. (2018)
Tocopherols/ tocotrienols	Grains	Anticancer, tumor-cell apoptosis induction	Forni et al. (2019), Souyoul et al. (2018) and Dillard and German (2000)

15.2.2.1 Carotenoids

Carotenoids are terpenoids extensively distributed among plants (Pichersky and Raguso 2018). These lipid-soluble compounds can be divided between carotenes and xanthophylls due to the biochemical and structural difference in oxygen presence on xanthophylls (Pichersky and Raguso 2018; Maoka 2020). Carotenoids are components of the provitamin A biosynthesis route (Ito et al. 2009). They are molecules associated with antioxidant effects, cancer prevention, angiogenesis impairment, apoptosis enhancement, and free-radical scavenging (Dillard and German 2000; Forni et al. 2019; Atlante et al. 2020).

15.2.2.2 Ginkgolides

Ginkgolides are diterpenes found on leaves extracts of *Ginkgo biloba* (Fang et al. 2020) (see structure in Figure 15.13). These terpenoid compounds are associated with improvements in memory, blood pressure, and cardiac diseases (Eisvand et al. 2020; Fang et al. 2020; Zhao et al. 2022).

15.2.2.3 Limonene

Limonene is usually found in essential oils. These cyclic terpenes affect some types of cancers (Albini et al. 2019).

15.2.2.4 Oleanolic Acid

A triterpenoid compound is commonly found in citrus fruit peels (Albini et al. 2019). Oleanolic acid may present anticancer potential through activation of the apoptosis cascade. In addition, this triterpenoid is reported to prevent fibrosis and cirrhosis due to hepatoprotective activity (Albini et al. 2019).

15.2.2.5 Phytosterols

Phytosterols are a group of plant-derived triterpenes that resemble cholesterol (Andrew and Izzo 2017; Li et al. 2022). They are 4-desmethyl sterols because they do not have any methyl groups in the fourth positions (Figure 15.14). The most common are 4-desmethylsterols, including stigmasterol (29 carbons), sitosterol (29 carbons) (Figures 15.14 and 15.15), and campesterol (28 carbons). Phytosterols are usually associated with health benefits based on lowering the absorption of cholesterol (Andrew and Izzo 2017). The primary sources of phytosterols in plant food are in oilseeds, especially corn oil (*Zea mays*), rapeseed oil

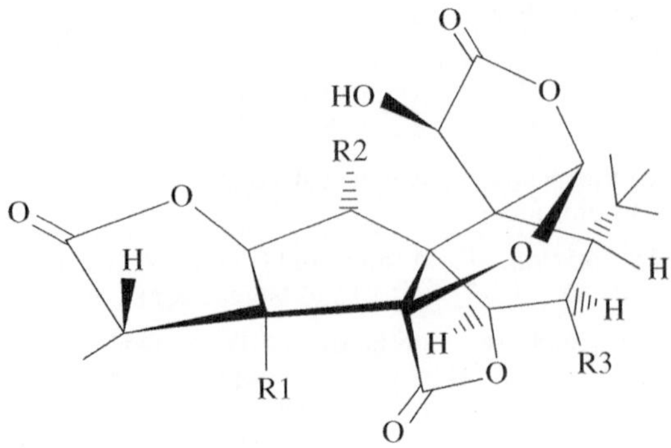

Figure 15.13 Ginkgolide structure.

Figure 15.14 General phytosterols structure.

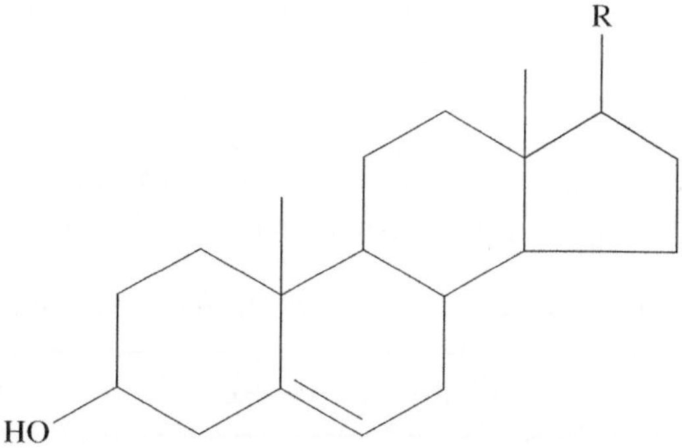

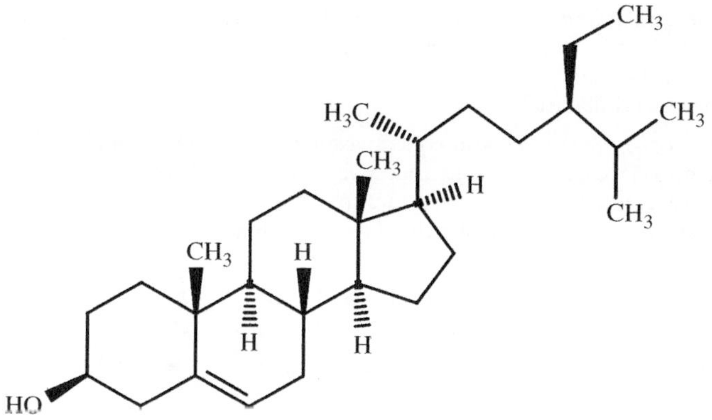

Figure 15.15 Structure of Sitosterol, a type of phytosterol.

(*Brassica napus* var. *napus*), wheat germ oil (*Triticum aestivum*), and nuts, notably in pine nuts (*Pinus sp.*) and pistachio (*Pistacia vera*). The ingestion of phytosterols is associated with the regulation of blood cholesterol and a lower risk of cardiac disease, given their capability of lowering the levels of LDL-cholesterol and triglycerides in the blood (Penugonda and Lindshield 2013; Moreau et al. 2018).

15.2.2.6 Tocopherols and Tocotrienols

Tocopherols and tocotrienols can be found in grains and are associated with anticancer activity as a potential tumor cell apoptosis inductive (Dillard and German 2000). Moreover, alpha-tocopherol has a vital role as vitamin E, highly presenting antioxidant activity (Souyoul et al. 2018; Forni et al. 2019).

15.2.3 Alkaloids

Alkaloids are compounds with one or more nitrogen atoms in a heterocyclic ring. They are classified according to the amino acid carrying the nitrogen atom and the part of its skeleton used to synthesize the alkaloid (Adefegha 2017). Plant alkaloids are very diverse and one of the largest classes of natural products, consisting of approximately 12 000 known substances (Bribi 2018). They can be found mainly in higher plants, especially in

Ranunculaceae, Leguminosae, Papaveraceae, Menispermaceae, and Loganiaceae (Debnath et al. 2018). Plant-derived alkaloids have many applications in human health: anti-malarial action, anticancer properties, blood-circulation promoters, stimulant, diuretic, antioxidant, anticholinergic, antiviral, antihypertensive, antidepressant, and myorelaxant (Adefegha 2017; Debnath et al. 2018; Sachdeva et al. 2020) (see Table 15.3). Caffeine (Figure 15.16) is perhaps one of the most consumed alkaloids worldwide, and it is

Table 15.3 Summary of the alkaloids on human health.

Alkaloid group	Alkaloid	Source	Uses	References
Pyrrolizidine		Boraginaceae, Asteraceae (Senecioneae and Eupatorieae), Orchidaceae, and Fabaceae (Crotalaria)	Anti-inflammatory, demulcent, anesthetic, ganglionic blocking agent, anti-plasmonic	Debnath et al. (2018)
Tropane		Solanaceae, Erythroxylaceae, Convolvulaceae, Brassicaceae, and Euphorbiaceae	Anti-colic, spasmolytic drugs, digestive and urinary tract spasm, ophthalmological eyedrops	Chen and Lin (2020)
Quinolizidine	Sparteine	*Lupinus* spp.	Antibacterial and antifungal	Debnath et al. (2018)
Isoquinoline		Papaveraceae, Berberidaceae, Ranunculaceae, Menispermaceae, Fumariaceae, Rutaceae, and Annonaceae	Antitumor, antibacterial, analgesic, immune regulation, antiplatelet aggregation, antiarrhythmia, antihypertension	Chen and Lin (2020)
	Berberine	*Berberis vulgaris, Berberis aristata, Silybum marianum*	Antidiabetic, antihypertensive, anti-inflammatory, antioxidant, antidepressant, and hepatoprotective	Debnath et al. (2018) and Imenshahidi and Hosseinzadeh (2019)
Quinoline	Quinine	*Cinchona pubescens, Cinchona officinalis*	Antimalarial	Chen and Lin (2020)
	Brucine	*Strychnos nux-vomica*	Anti-inflammatory, analgesic	Debnath et al. (2018)
Glycoalkaloids	Tomatine	*Solanum lycopersicum*	Antioxidant, reduction of cholesterol and triglyceride levels, immune system stimulant, chemopreventive	Tamasi et al. (2019)
Purine	Caffeine	*Theobroma cocoa, Thea sinensis, Coffea spp.*	Central nervous system stimulants, cold medications, analgesics, and anorectal	Debnath et al. (2018)

Table 15.3 (Continued)

Alkaloid group	Alkaloid	Source	Uses	References
Pyridine	Arecoline	*Areca catechu*	Glaucoma, activation of brain function	Debnath et al. (2018) and Chen and Lin (2020)
	Piperine	*Piper nigrum* *Piper longum*	Antioxidant, antitumor, antihypertensive, antiasthmatics, analgesic, antipyretic, antidiarrheal, anti-inflammatory, anxiolytic, antispasmodic, hepatoprotective, antidepressant, antibacterial, immunomodulatory, antifungal, antiapoptotic, antithyroid, antimutagenic, antimetastatic, and antispermatogenic	Haq et al. (2021)
	Ephedrine	*Ephedra* spp.	Vasopressor, ergogenics, weight loss aid, a bronchodilator, central neural system stimulant, mood enhancer	Gad et al. (2021)

present in products like coffee, tea, soft drinks, cocoa beverages, chocolate bars, and maté. Its health benefits include action against cardiovascular, hepatic, pulmonary, and neurodegenerative diseases, type 2 diabetes, obesity, and cancer (Adefegha 2017; Debnath et al. 2018; Sachdeva et al. 2020).

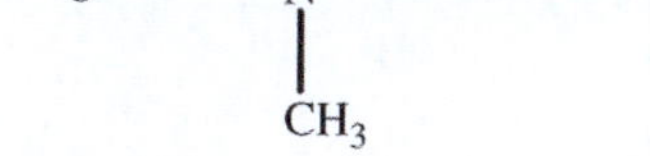

Figure 15.16 Caffeine structure.

15.2.4 Fatty Acids

Fatty acids are long-chain polyunsaturated fatty acids found in oil and fat. In plants, they can be found in seeds like chia (*Salvia hispanica*), perilla (*Perilla frutescens*), flax (*Linum usitatissimum*), nuts, vegetable oils like flaxseed oil, canola oil, and hemp seed oil; some fruits and green leafy vegetables (Alali et al. 2021; Saini and Keum 2018). Fatty acids are divided into omega-3 and omega-6, the former derived from linolenic acid and the latter from linoleic acid (Figure 15.17). Fatty acids have been shown to have anti-inflammatory effects, take part in the growth and normal function of the human brain and retina, maintain cardiac and vascular health, and prevent atherosclerosis (de Carvalho and Caramujo 2018; Saini and Keum 2018; Alali et al. 2021) (see Table 15.4).

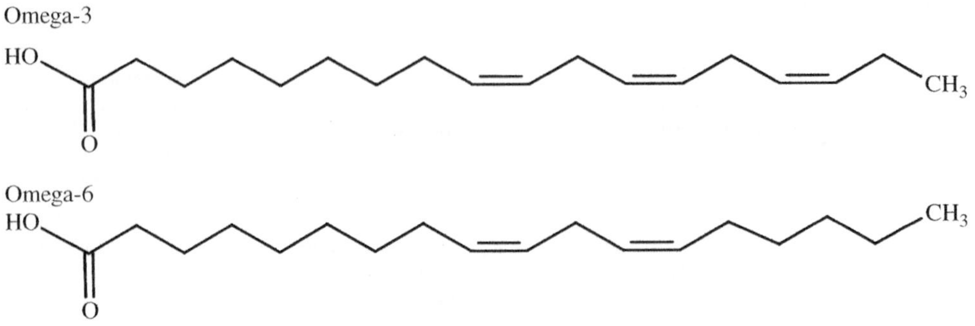

Figure 15.17 Omega-3 (up) and omega 6 (bottom) structures.

Table 15.4 Summary of the fatty acids on human health.

Type	Source	Activity	References
Omega-3	Seed oils from chia, flax, garden cress, camelina, perilla, walnuts	Anti-inflammatory, anticancer, LDL-cholesterol reduction, cardiovascular diseases prevention	Santos et al. (2020) and Saini et al. (2021)
Omega-6	Safflower, sunflower and pumpkin seeds, walnuts, corn, sunflower, safflower, and soybean oils	Cardiovascular diseases prevention, skin function and appearance protection, inflammatory response modulation, total cholesterol, and LDL-cholesterol reduction	Saini and Keum (2018) and Djuricic and Calder (2021)

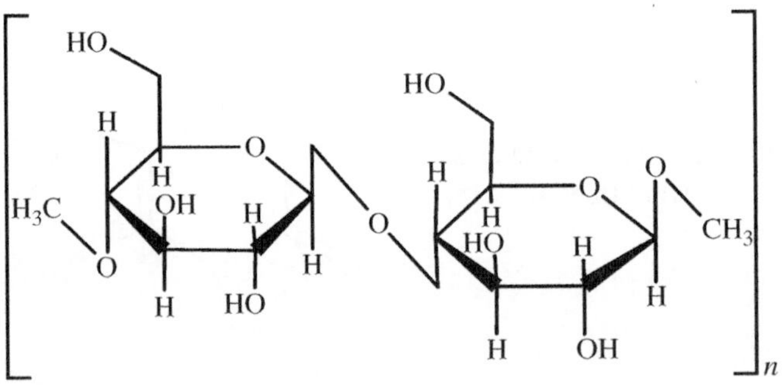

Figure 15.18 Unit of the cellulose, a water-insoluble polysaccharide formed in most plant fiber.

15.2.5 Fiber

Fibers are usually carbohydrate compounds derived from plant products, including cellulose, cutin, pectin, lignin, and suberin (see Figure 15.18). Since the human digestive tract cannot completely break down these substances, they pass through the small intestine undigested. They are partially or entirely processed in the large intestine by the microorganisms of the intestinal microbiome (Cui et al. 2019; Maurya et al. 2021) so that they can

Table 15.5 Summary of the dietary fiber on human health.

Type of fibers	Constituents	Edible parts of plants	Effects on human health	References
Soluble	Pectin, gum, mucilage	Fruit pulps, leafy vegetables	Gastrointestinal transit modulation, blood pressure reduction, gastrointestinal disorders prevention, cancer protection, improved glycemic and insulinemic responses, appetite regulation	Cui et al. (2019) and Müller et al. (2018)
Nonsoluble	Cellulose, lignin, hemicellulose	Fruit peels, whole grains, cereals	Reduced risk of type-2 diabetes, anti-inflammatory, microbial modulation, increased metabolic health	Cui et al. (2019) and Müller et al. (2018)

be absorbed – dietary fiber is classified as soluble or insoluble (Jalili et al. 2015). Soluble fiber mixes with water and reduces the rate at which fecal matter passes through the digestive tract, so the feeling of satiety lasts much longer. Soluble fiber also helps regulate cholesterol and blood sugar levels in people with diabetes (Jalili et al. 2015; Sachdeva et al. 2020). Insoluble fiber does not mix with water. Instead, water is attracted to these fibers and can thus be absorbed, making feces more voluminous and easing its passage through the lower intestine. Both fiber types also function as prebiotics that aid digestion overall (Jalili et al. 2015) (see Table 15.5).

15.3 Engineering Nutraceutical-Enriched Plants

Imagine drinking a juice that removes fat from your body and makes you lose weight. The commonsense already associates eating fruits and vegetables with promoting health, but the nutraceutical research goes beyond this idea that "an apple a day keeps the doctor away." One classic example is golden rice, a genetically engineered rice that produces grains enriched with b-carotene, the precursor of vitamin A (Ye et al. 2000). Roughly, the lack of vitamin A kills millions of children worldwide, mainly those born in low-income regions, turning the genetically modified rice into a powerful strategy to save lives. However, despite the urgency of making b-carotene-enriched rice accessible to these populations, excessive governmental regulation has delayed this innovation's approval for decades (Ye et al. 2000; Wu et al. 2021). This example shows that generating transgenic plants might not be the better strategy to make nutraceutical-enriched plants accessible to the population. Thus, several biofortification programs choose classical breeding to develop plants enriched with healthier phytochemicals (Ma et al. 2005; White and Broadley 2005; Jacob et al. 2008; Thill et al. 2012; Upadhyay 2021). An alternative and sustainable method is the commercialization of a postharvest-stressed plant. As mentioned earlier in this chapter, the plant produces potent antioxidant molecules to protect them against external stresses, including UV-light radiation, wounding, phytohormone exposition, etc. Thus, submitting the crops to severe abiotic stress conditions after harvesting may force them to activate the phenolic biosynthesis pathway, stimulating the accumulation of their nutraceutical content (Jacobo-Velázquez et al. 2011; Jacobo-Velázquez and Cisneros-Zevallos 2012; Villarreal-García et al. 2016; Aviles-Gaxiola et al. 2020; Lone et al. 2020).

Another approach to stimulating innovation and quickly bringing nutraceutical-enriched products to the market is not to target end consumers. Instead, the secondary sector might be the better destination. A promising method is the application of bio-factories using algae or plant cell cultures to rapid turning commercially available healthier alternatives for manufacturing industries. Biofactories platforms may represent a sustainable alternative to delivering more beneficial products. The *in vitro* cultivation of the algae Spirulina (*Arthrospira platensis*) or the tobacco cell plant overexpressing anthocyanin regulatory genes, generates a natural colorant with antioxidative properties for food and cosmetics industries (Sigurdson et al. 2017; Appelhagen et al. 2018).

15.4 Potential Side Effects of Nutraceuticals on Human Health

As mentioned earlier, nutraceuticals present relevant, beneficial properties for human health, such as antioxidant and anti-inflammatory effects. The use of nutraceuticals may present risk depending on the quantity, quality, and timing of intake, especially for those using a controlled medication, as the drug–supplement interaction may cause adverse effects to the user. Therefore, professional advice on the proper use of nutraceuticals is recommended (Alali et al. 2021). The potential negative interaction between drugs and dietary supplements is one of the most well known and common adverse effects next to hepatotoxicity. Some hepatotoxic products to humans are those derived from *Piper nigrum*, *Panax ginseng*, *Panax quinquefolius*, *Garcinia cambogia*, *Valeriana officinalis*, *Viscum coloratum*, and others (Gupta et al. 2018). Long-term consumption of a high concentration of polyphenolic chemicals is associated with metal absorption interference through the chelation of ferric iron (Gurley et al. 2012), therefore possessing destructive potential, especially in cases of a previous iron deficit. However, more clinical studies are still needed to clarify this adverse effect in humans. Some polyphenols can also reduce vitamin C absorption due to their interference with ascorbate transporters in the gastrointestinal tract (Gurley et al. 2012). In addition, polyphenols affect drug absorption through the impairment of intestinal transporter activity, thus presenting the potential for interference with drug therapy (Gurley et al. 2012).

Reports of toxic effects from excessive consumption of fat-soluble vitamins have been more prevalent. High doses of tocopherol as a vitamin E are related to the antiplatelet impacts resulting in bleeding, diarrhea, weakness, blurred vision, and gonadal dysfunction. There is also evidence that vitamin A intake has adverse effects, such as decreasing bone health and bone mineral density and increased risk of fractures. A higher incidence of congenital anomalies has been noted in pregnant women taking vitamin A (Ronis et al. 2018). In animal trials, coumarins over-consumption is presented to cause hepatotoxicity and dermatological phototoxic reactions (Heghes et al. 2022). Cinnamon products have caused allergic reactions and adverse reactions to the skin and oral cavity (Cope 2019). Some ginkgolides may interfere with platelet aggregation, thereby presenting a destructive potential to impair coagulation and cause spontaneous hemorrhage (Gupta et al. 2018; Ronis et al. 2018). Long-term intake of *Ginkgo biloba* extract may cause epileptiform seizures, unconsciousness, leg paralysis, and potential carcinogenic propriety (Gupta et al. 2018).

When taken above 20 mg/kg body weight, caffeine can be toxic to humans, causing several symptoms: dizziness, restlessness, anxiety, irritability, tremors, arrhythmia, tachycardia, hypertension, hypotension, diuresis, gastrointestinal disorders, renal failure, hyperventilation, respiratory failure, and seizures. Overdose can lead to hypokalemia, hyponatremia, defective iron and zinc absorption, rhabdomyolysis, and circulatory collapse (Gupta et al. 2018). Green tea extract (GTE), which is rich in flavanols, has been demonstrated to be hepatotoxic (Gupta et al. 2018; Cerbin-Koczorowska et al. 2021) and cause various side effects such as bloating, nausea, heartburn, abdominal pain, dizziness, headache, and muscle pain when taken in large doses (Gupta et al. 2018).

Soybean (*Glycine max*) is well known to cause allergic reactions (Cope 2019). Women use phytoestrogens derived from soy as fertility enhancers. They show positive effects, but still with uncertain results and suggestive evidence, showing the need for cautious consumption and further research on the safety of these compounds (Rizzo et al. 2022). Nonetheless, nutraceutical manufacturing is at risk of contamination from degradation products, synthetic intermediates, excipients, fillers, and coating materials (Sharif and Khalid 2018). Quality control and regulation of the production of dietary supplements must be observed to ensure the quality and safety of their use. Given the diversity of botanicals that may be considered nutraceuticals, research and appropriate regulation are essential to ensure that the widespread consumption of these dietary supplements brings overall health benefits to the population.

15.5 Final Considerations

Nutraceuticals should not be seen simply as food or medicine. Despite the economic and social importance, there is still no substantial evidence about the actual health benefits or which compounds are responsible for the beneficial results that the plant can present. One of the reasons to explain such a scenario is the multidisciplinary nature of nutraceuticals, which may hinder further advances. It requires an interdisciplinary group to elucidate the phytochemical role in health promotion. A neglected topic is the minimum and maximum recommended dose of a nutraceutical for a given disease, or if there are any interactions among different phytochemical compounds and their potential effect on other synthetic drugs. In this case, there should be a medical concern about the history of consumption of phytochemicals by patients before starting drug therapy. A possible solution to this problem is the advances in research associating phytochemistry with genetic engineering. It aims to generate plants that show an increase in the production of desirable compounds and a decrease in those with adverse effects, providing a plant with a higher dose for consumption or phytochemical extraction. In summary, nutraceuticals represent a future possibility for disease prevention and treatment, with the potential to modify diets precisely according to the needs of each person. Thus, reducing the need for long-term therapy for diseases that phytochemicals can eventually prevent.

References

Addi, M., Elbouzidi, A., Abid, M. et al. (2021). An overview of bioactive flavonoids from citrus fruits. *Applied Sciences* 12 (1): 29.

Adefegha, S.A. (2017). Functional foods and nutraceuticals as dietary intervention in chronic diseases. *Novel Perspectives for Health Promotion and Disease Prevention* 15 (6): 977–1009. https://doi.org/10.1080/19390211.2017.1401573.

Alali, M., Alqubaisy, M., Aljaafari, M.N. et al. (2021). Nutraceuticals: transformation of conventional foods into health promoters/disease preventers and safety considerations. *Molecules* 26 (9): 2540.

Albini, A., Bassani, B., Baci, D. et al. (2019). Nutraceuticals and "Repurposed" drugs of phytochemical origin in prevention and interception of chronic degenerative diseases and cancer. *Current Medicinal Chemistry* 26 (6): 973–987.

Amiot, M.J., Riva, C., and Vinet, A. (2016). Effects of dietary polyphenols on metabolic syndrome features in humans: a systematic review. *Obesity Reviews* 17 (7): 573–586.

Andrew, R. and Izzo, A.A. (2017). Principles of pharmacological research of nutraceuticals. *British Journal of Pharmacology* 174 (11): 1177–1194.

Ansori, A., Fadholly, A., Hayaza, S. et al. (2020). A review on medicinal properties of mangosteen (*Garcinia mangostana* L.). *Research Journal of Pharmacy and Technology* 13: 974–982.

Appelhagen, I., Wulff-Vester, A.K., Wendell, M. et al. (2018). Colour bio-factories: towards scale-up production of anthocyanins in plant cell cultures. *Metabolic Engineering* 48: 218–232.

Artaud-Wild, S.M., Connor, S.L., Sexton, G., and Connor, W.E. (1993). Differences in coronary mortality can be explained by differences in cholesterol and saturated fat intakes in 40 countries but not in France and Finland: a paradox. *Circulation* 88 (6): 2771–2779.

Atlante, A., Amadoro, G., Bobba, A., and Latina, V. (2020). Functional foods: an approach to modulate molecular mechanisms of Alzheimer's disease. *Cell* 9 (11): 1–62.

Aviles-Gaxiola, S., Olivo-Vázquez, G., Cabanillas-Bojórquez, L.A. et al. (2020). Plants as biofactories for phenolic compounds. In: *Plant Phenolics in Sustainable Agriculture* (ed. R. Lone, R. Shuab and A.N. Kamili), 467–500. Singapore: Springer.

Aziz, A., Akram, N.A., and Ashraf, M. (2018). Influence of natural and synthetic vitamin C (ascorbic acid) on primary and secondary metabolites and associated metabolism in quinoa (Chenopodium quinoa Willd.) plants under water deficit regimes. *Plant Physiology and Biochemistry* 123: 192–203.

Béliveau, R. and Gingras, D. (2007). Role of nutrition in preventing cancer. *Canadian Family Physician* 53 (11): 1905.

Benetou, V., Orfanos, P., Lagiou, P. et al. (2008). Vegetables and fruits in relation to cancer risk: evidence from the Greek EPIC cohort study. *Cancer Epidemiology, Biomarkers and Prevention* 17 (2): 387–392.

Blot, W.J. and Tarone, R.E. (2015). Doll and Peto's quantitative estimates of cancer risks: holding generally true for 35 years. *Journal of the National Cancer Institute* 107 (4): djv044.

Böttger, A., Vothknecht, U., Bolle, C., and Wolf, A. (2018). Plant secondary metabolites and their general function in plants. In: *Lessons on Caffeine, Cannabis & Co*, Learning Materials in Biosciences (ed. A. Böttger), 3–17. Springer, Cham.

Bribi, N. (2018). Pharmacological activity of alkaloids: a review. *Asian Journal of Botany* 1: 1–5.

Bucciantini, M., Leri, M., Nardiello, P. et al. (2021). Olive polyphenols: antioxidant and anti-inflammatory properties. *Antioxidants* 10 (7): 1044.

de Carvalho, C.C.C.R. and Caramujo, M.J. (2018). The Various Roles of Fatty Acids. *Molecules* 23 (10): 2583.

Cerbin-Koczorowska, M., Waszyk-Nowaczyk, M., Bakun, P. et al. (2021). Current view on green tea catechins formulations, their interactions with selected drugs, and prospective applications for various health conditions. *Applied Sciences* 11 (11): 4905.

Chen, C. and Lin, L. (2020). Alkaloids in diet. In: *Handbook of Dietary Phytochemicals* (ed. J. Xiao, S.D. Sarker and Y. Asakawa), 1–35. Singapore: Springer.

Chen, X.Q., Hu, T., Han, Y. et al. (2016). Preventive effects of catechins on cardiovascular disease. *Molecules* 21 (12): 1759.

Cohen, J.H., Kristal, A.R., and Stanford, J.L. (2000). Fruit and vegetable intakes and prostate cancer risk. *Journal of the National Cancer Institute* 92 (1): 61–68.

Cope, R.B. (2019). Toxicology and drug interactions of nutraceuticals. In: *Nutraceuticals in Veterinary Medicine* (ed. R.C. Gupta, A. Srivastava and R. Lall), 711–727. Springer.

Cui, J., Lian, Y., Zhao, C. et al. (2019). Dietary fibers from fruits and vegetables and their health benefits via modulation of gut microbiota. *Comprehensive Reviews in Food Science and Food Safety* 18 (5): 1514–1532.

Dabeek, W.M. and Marra, M.V. (2019). Dietary quercetin and kaempferol: bioavailability and potential cardiovascular-related bioactivity in humans. *Nutrients* 11 (10): 2288.

Debnath, B., Singh, W.S., Das, M. et al. (2018). Role of plant alkaloids on human health: a review of biological activities. *Materials Today Chemistry* 9: 56–72.

DeFelice, S.L. (1995). The nutraceutical revolution: its impact on food industry R&D. *Trends in Food Science & Technology* 6 (2): 59–61.

Delgoda, R. and Murray, J.E. (2017). Evolutionary Perspectives on the Role of Plant Secondary Metabolites. In: *Pharmacognosy: Fundamentals, Applications and Strategy* (ed. S. Badal and R. Delgoda), 93–100. Academic Press.

Diallo, A., Deschasaux, M., Galan, P. et al. (2016). Associations between fruit, vegetable and legume intakes and prostate cancer risk: results from the prospective Supplémentation en Vitamines et Minéraux Antioxydants (SU.VI.MAX) cohort. *British Journal of Nutrition* 115 (9): 1579–1585.

Dillard, C.J. and German, J.B. (2000). Phytochemicals: nutraceuticals and human health. *Journal of the Science of Food and Agriculture* 80 (12): 1744–1756.

Ding, T., Cao, K., Fang, W. et al. (2020). Evaluation of phenolic components (anthocyanins, flavanols, phenolic acids, and flavonols) and their antioxidant properties of peach fruits. *Scientia Horticulturae* 268: 109365.

Dini, I. and Laneri, S. (2021). Spices, condiments, extra virgin olive oil and aromas as not only Flavorings, but precious allies for our wellbeing. *Antioxidants* 10 (6): 868.

Dixit, S., Shukla, A., Singh, V., and Upadhyay, S.K. (2021a). Bioprospecting of natural compounds for industrial and medical applications. In: *Bioprospecting of Plant Biodiversity for Industrial Molecules* (ed. S.K. Upadhyay and S.P. Singh), 351–379. John Wiley & Sons Ltd. https://doi.org/10.1002/9781119718017.ch3.

Dixit, S., Shukla, A., Singh, V., and Upadhyay, S.K. (2021b). Engineering of plant metabolic pathway for nutritional improvement. In: *Genome Engineering for Crop Improvement* (ed. S.K. Upadhyay), 351–379. John Wiley & Sons Ltd. https://doi.org/10.1002/9781119672425.ch20.

Dixit, S., Shukla, A., Upadhyay, S.K., and Verma, P.C. (2019). Mode of communication between plants during environmental stress. In: *Molecular Approaches in Plant Biology and Environmental Challenges* (ed. S. Singh, S.K. Upadhyay, A. Pandey and S. Kumar). Energy, Environment, and Sustainability. Singapore: Springer. https://doi.org/10.1007/978-981-15-0690-1_7.

Djuricic, I. and Calder, P.C. (2021). Beneficial outcomes of omega-6 and omega-3 polyunsaturated fatty acids on human health: an update for 2021. *Nutrients* 13 (7): 2421.

Doll, R. and Peto, R. (1981). The causes of cancer: quantitative estimates of avoidable risks of cancer in the United States today. *JNCI: Journal of the National Cancer Institute* 66 (6): 1192–1308.

Dorai, T. and Aggarwal, B.B. (2004). Role of chemopreventive agents in cancer therapy. *Cancer Letters* 215 (2): 129–140.

Durazzo, A., Lucarini, M., Souto, E.B. et al. (2019). Polyphenols: a concise overview on the chemistry, occurrence, and human health. *Phytotherapy Research* 33 (9): 2221–2243.

Eisvand, F., Razavi, B.M., and Hosseinzadeh, H. (2020). The effects of ginkgo biloba on metabolic syndrome: a review. *Phytotherapy Research* 34 (8): 1798–1811.

Erb, M., Huber, M., Robert, C.A.M. et al. (2013). The role of plant primary and secondary metabolites in root-herbivore behaviour, mutrition and physiology. *Advances in Insect Physiology* 45: 53–95.

Fang, J., Wang, Z., Wang, P., and Wang, M. (2020). Extraction, structure and bioactivities of the polysaccharides from ginkgo biloba: a review. *International Journal of Biological Macromolecules* 162: 1897–1905.

Feng, Z., Lu, X., Gan, L. et al. (2020). Xanthones, a promising anti-inflammatory scaffold: structure, activity, and drug likeness analysis. *Molecules* 25 (3): 598.

Forni, C., Facchiano, F., Bartoli, M. et al. (2019). Beneficial role of phytochemicals on oxidative stress and age-related diseases. *BioMed Research International* 2019: 16.

Fraga, L.N., Coutinho, C.P., Rozenbaum, A.C. et al. (2021). Blood pressure and body fat % reduction is mainly related to flavanone phase II conjugates and minor extension by phenolic acid after long-term intake of orange juice. *Food & Function* 12 (22): 11278–11289.

Gad, M.Z., Azab, S.S., Khattab, A.R., and Farag, M.A. (2021). Over a century since ephedrine discovery: an updated revisit to its pharmacological aspects, functionality and toxicity in comparison to its herbal extracts. *Food & Function* 12 (20): 9563–9582.

Gervasi, T., Calderaro, A., Barreca, D. et al. (2022). Biotechnological applications and health-promoting properties of flavonols: an updated view. *International Journal of Molecular Sciences* 23 (3): 1710.

Grusak, M.A. and Dellapenna, D. (2003). Improving the nutrient composition of plants to enhance human nutrition and health. *Annual Review of Plant Physiology and Plant Molecular Biology* 50: 133–161. http://dx.doi.org/10.1146/annurev.arplant.50.1.133.

Gupta, R.C., Srivastava, A., and Lall, R. (2018). Toxicity potential of nutraceuticals. *Methods in Molecular Biology* 1800: 367–394.

Gurley, B.J., Fifer, E.K., and Gardner, Z. (2012). Pharmacokinetic herb-drug interactions (part 2): drug interactions involving popular botanical dietary supplements and their clinical relevance. *Planta Medica* 78 (13): 1490–1514.

Haq, I.U., Imran, M., Nadeem, M. et al. (2021). Piperine: a review of its biological effects. *Phytotherapy Research* 35 (2): 680–700.

Heghes, S.C., Vostinaru, O., Mogosan, C. et al. (2022). Safety profile of nutraceuticals rich in coumarins: an update. *Frontiers in Pharmacology* 13: 48.

Howes, M.J.R., Perry, N.S.L., Vásquez-Londoño, C., and Perry, E.K. (2020). Role of phytochemicals as nutraceuticals for cognitive functions affected in ageing. *British Journal of Pharmacology* 177 (6): 1294–1315.

Hu, C., Wong, W.T., Wu, R., and Lai, W.F. (2019). Biochemistry and use of soybean isoflavones in functional food development. *Critical Reviews in Food Science and Nutrition* 60 (12): 2098–2112. https://doi.org/10.1080/10408398.2019.1630598.

Humble, C.G., Malarcher, A.M., and Tyroler, H.A. (1993). Dietary fiber and coronary heart disease in middle-aged hypercholesterolemic men. *American Journal of Preventive Medicine* 9 (4): 197–202.

Imenshahidi, M. and Hosseinzadeh, H. (2019). Berberine and barberry (Berberis vulgaris): a clinical review. *Phytotherapy Research* 33 (3): 504–523.

Isemura, M. (2019). Catechin in human health and disease. *Molecules* 24 (3): 528.

Ito, M., Yamano, Y., Tode, C., and Wada, A. (2009). Carotenoid synthesis. *Archives of Biochemistry and Biophysics* 483 (2): 224–238.

Jacob, J.K., Hakimuddin, F., Paliyath, G., and Fisher, H. (2008). Antioxidant and antiproliferative activity of polyphenols in novel high-polyphenol grape lines. *Food Research International* 41 (4): 419–428.

Jacobo-Velázquez, D.A. and Cisneros-Zevallos, L. (2012). An alternative use of horticultural crops: stressed plants as biofactories of bioactive phenolic compounds. *Agriculture* 2 (3): 259–271.

Jacobo-Velázquez, D.A., Martínez-Hernández, G.B., del Rodríguez, S.C. et al. (2011). Plants as biofactories: physiological role of reactive oxygen species on the accumulation of phenolic antioxidants in carrot tissue under wounding and hyperoxia stress. *Journal of Agricultural and Food Chemistry* 59 (12): 6583–6593.

Jalili, T., Wildman, R.E.C., and Medeiros, D.M. (2015). Nutraceutical roles of dietary fiber. *Journal of Nutraceuticals, Functional & Medical Foods* 2 (4): 19–34. http://dx.doi.org/10.1300/J133v02n04_03.

Joshi, R., Kulkarni, Y.A., and Wairkar, S. (2018). Pharmacokinetic, pharmacodynamic and formulations aspects of naringenin: an update. *Life Sciences* 215: 43–56.

Kaparapu, J., Pragada, P.M., and Geddada, M.N.R. (2020). Fruits and vegetables and its nutritional benefits. In: *Functional Foods and Nutraceuticals* (ed. C. Egbuna and G.D. Tupas), 241–260. Cham: Springer.

Karim, N., Jia, Z., Zheng, X. et al. (2018). A recent review of citrus flavanone naringenin on metabolic diseases and its potential sources for high yield-production. *Trends in Food Science & Technology* 79: 35–54.

Key, T., Silcocks, P.B., Davey, G.K. et al. (1997). A case-control study of diet and prostate cancer. *British Journal of Cancer* 76 (5): 678–687.

Khaw, K.T. and Barrett-connor, E. (1987). Dietary fiber and reduced ischemic heart disease mortality rates in men and women: a 12-year prospective study. *American Journal of Epidemiology* 126 (6): 1093–1102.

Kim, I.S. (2021). Current perspectives on the beneficial effects of soybean isoflavones and their metabolites for humans. *Antioxidants* 10 (7): 1064.

Kim, Y. and Je, Y. (2016). Dietary fibre intake and mortality from cardiovascular disease and all cancers: a meta-analysis of prospective cohort studies. *Archives of Cardiovascular Diseases* 109 (1): 39–54.

Klein-Júnior, L.C., Campos, A., Niero, R. et al. (2020). Xanthones and cancer: from natural sources to mechanisms of action. *Chemistry & Biodiversity* 17 (2): e1900499.

Kooshki, A., Moghaddam, M.Y., and Akbarzadeh, R. (2016). Study of fruit and vegetable intake in breast cancer patients in the city of Sabzevar. *Electronic Physician* 8 (9): 3011.

Korpela, R., Vesa, T., Renaud, S., and de Lorgeril, M. (1992). Dairy fat intake in Finland. *The Lancet* 340 (8822): 793.

Kotecha, R., Takami, A., and Espinoza, J.L. (2016). Dietary phytochemicals and cancer chemoprevention: a review of the clinical evidence. *Oncotarget* 7 (32): 52517.

Kotha, R.R. and Luthria, D.L. (2019). Curcumin: biological, pharmaceutical, nutraceutical, and analytical aspects. *Molecules* 24 (16): 2930.

Kumar, D., Kumar, S., Gochar, C.S. et al. (2020). Nutritional components in green leafy vegetables: a review. *Journal of Pharmacognosy and Phytochemistry* 9 (5): 2498–2502.

de La Iglesia, R., Milagro, F.I., Campión, J. et al. (2010). Healthy properties of proanthocyanidins. *BioFactors* 36 (3): 159–168.

Lee, D. and Imm, J.Y. (2018). Antiobesity effect of tricin, a methylated cereal flavone, in high-fat-diet-induced obese mice. *Journal of Agricultural and Food Chemistry* 66 (38): 9989–9994.

Li, X., Xin, Y., Mo, Y. et al. (2022). The bioavailability and biological activities of phytosterols as modulators of cholesterol metabolism. *Molecules* 27 (2): 523.

Lin, L.Z. and Harnly, J.M. (2010). Phenolic component profiles of mustard greens, Yu Choy, and 15 other Brassica vegetables. *Journal of Agricultural and Food Chemistry* 58 (11): 6850–6857.

Lin, B.W., Gong, C.C., Song, H.F., and Cui, Y.Y. (2017). Effects of anthocyanins on the prevention and treatment of cancer. *British Journal of Pharmacology* 174 (11): 1226–1243.

Lippi, G., Franchini, M., Favaloro, E.J., and Targher, G. (2010). Moderate red wine consumption and cardiovascular disease risk: beyond the "French paradox.". *Seminars in Thrombosis and Hemostasis* 36 (1): 59–70.

Liu, R.H. (2003). Health benefits of fruit and vegetables are from additive and synergistic combinations of phytochemicals. *The American Journal of Clinical Nutrition* 78 (3): 517S–520S.

Lone, R., Shuab, R., and Kamili, A.N. (ed.) (2020). *Plant Phenolics in Sustainable Agriculture*. Singapore: Springer.

López-González, L., Becerra-Tomás, N., Babio, N. et al. (2022). One-year changes in fruit and vegetable variety intake and cardiometabolic risk factors changes in a middle-aged Mediterranean population at high cardiovascular risk. *European Journal of Clinical Nutrition* 10: 1393–1402.

Ma, J.K.C., Chikwamba, R., Sparrow, P. et al. (2005). Plant-derived pharmaceuticals – the road forward. *Trends in Plant Science* 10 (12): 580–585.

Macarro, M.S., Rodríguez, J.P.M., Morell, E.B. et al. (2020). Effect of a combination of citrus flavones and flavanones and olive polyphenols for the reduction of cardiovascular disease risk: an exploratory randomized, double-blind, placebo-controlled study in healthy subjects. *Nutrients* 12 (5): 1475.

Mahajan, M., Kuiry, R., and Pal, P.K. (2020). Understanding the consequence of environmental stress for accumulation of secondary metabolites in medicinal and aromatic plants. *Journal of Applied Research on Medicinal and Aromatic Plants* 18: 100255.

Maoka, T. (2020). Carotenoids as natural functional pigments. *Journal of Natural Medicines* 74 (1): 1–16.

Maurya, A.P., Chauhan, J., Yadav, D.K. et al. (2021). Nutraceuticals and their impact on human health. In: *Preparation of Phytopharmaceuticals for the Management of Disorders: The Development of Nutraceuticals and Traditional Medicine* (ed. C. Egbuna, A.P. Mishra and M.R. Goyal), 229–254. Academic Press.

Mechchate, H., Es-safi, I., Haddad, H. et al. (2021). Combination of catechin, epicatechin, and rutin: optimization of a novel complete antidiabetic formulation using a mixture design approach. *The Journal of Nutritional Biochemistry* 88: 108520.

Minuz, P., Velo, G., Violi, F., and Ferro, A. (2017). Are nutraceuticals the modern panacea? From myth to science. *British Journal of Clinical Pharmacology* 83 (1): 5–7.

Moreau, R.A., Nyström, L., Whitaker, B.D. et al. (2018). Phytosterols and their derivatives: structural diversity, distribution, metabolism, analysis, and health-promoting uses. *Progress in Lipid Research* 70: 35–61.

Müller, M., Canfora, E.E., and Blaak, E.E. (2018). Gastrointestinal transit time, glucose homeostasis and metabolic health: modulation by dietary fibers. *Nutrients* 10 (3): 275.

Najmanová, I., Vopršalová, M., Saso, L., and Mladěnka, P. (2019). The pharmacokinetics of flavanones. *Critical Reviews in Food Science and Nutrition* 60 (18): 3155–3171. https://doi.org/10.1080/10408398.2019.1679085.

Pecyna, P., Wargula, J., Murias, M., and Kucinska, M. (2020). More than resveratrol: new insights into stilbene-based compounds. *Biomolecules* 10 (8): 1111.

Penugonda, K. and Lindshield, B.L. (2013). Fatty acid and phytosterol content of commercial saw palmetto supplements. *Nutrients* 5 (9): 3617–3633.

Pichersky, E. and Raguso, R.A. (2018). Why do plants produce so many terpenoid compounds? *New Phytologist* 220 (3): 692–702.

Pukkala, E., Martinsen, J.I., Lynge, E. et al. (2009). Occupation and cancer follow-up of 15 million people in five Nordic countries. *Acta Oncologica* 48 (5): 646–790.

Rauf, A., Imran, M., Abu-Izneid, T. et al. (2019). Proanthocyanidins: a comprehensive review. *Biomedicine & Pharmacotherapy* 116: 108999.

Renaud, S. and de Lorgeril, M. (1989). Dietary lipids and their relation to ischaemic heart disease: from epidemiology to prevention. *Journal of Internal Medicine. Supplement* 731 (731 S): 39–46.

Renaud, S. and de Lorgeril, M. (1992). Wine, alcohol, platelets, and the French paradox for coronary heart disease. *Lancet* 339 (8808): 1523–1526.

Rizzo, G., Feraco, A., Storz, M.A., and Lombardo, M. (2022). The role of soy and soy isoflavones on women's fertility and related outcomes: an update. *Journal of Nutritional Science* 11: 1–17.

Roleira, F.M.F., Varela, C.L., Gomes, A.R. et al. (2021). The health components of spices and herbs: the medicinal chemistry point of view. In: *Aromatic Herbs in Food* (ed. C.M. Galanakis), 35–92. Academic Press.

Ronis, M.J.J., Pedersen, K.B., and Watt, J. (2018). Adverse effects of nutraceuticals and dietary supplements. *Annual Review of Pharmacology and Toxicology* 58: 583–601.

Rothenberg, R. (2015). The causes of cancer, revisited. *Annals of Epidemiology* 25 (3): 215–216.

Sachdeva, V., Roy, A., and Bharadvaja, N. (2020). Current prospects of nutraceuticals: a review. *Current Pharmaceutical Biotechnology* 21 (10): 884–896.

Saini, R.K. and Keum, Y.S. (2018). Omega-3 and omega-6 polyunsaturated fatty acids: Dietary sources, metabolism, and significance – a review. *Life Sciences* 203: 255–267.

Saini, R.K., Prasad, P., Sreedhar, R.V. et al. (2021). Omega−3 polyunsaturated fatty acids (PUFAs): emerging plant and microbial sources, oxidative stability, bioavailability, and health benefits – a review. *Antioxidants* 10 (10): 1627.

Salazar, H.M., de Deus Mendonça, R., Laclaustra, M. et al. (2022). The intake of flavonoids, stilbenes, and tyrosols, mainly consumed through red wine and virgin olive oil, is associated with lower carotid and femoral subclinical atherosclerosis and coronary calcium. *European Journal of Nutrition* 61: 2697–2709.

Santos, H.O., Price, J.C., and Bueno, A.A. (2020). Beyond fish oil supplementation: the effects of alternative plant sources of omega-3 polyunsaturated fatty acids upon lipid indexes and cardiometabolic biomarkers – an overview. *Nutrients* 12 (10): 3159.

Schottenfeld, D., Beebe-Dimmer, J.L., Buffler, P.A., and Omenn, G.S. (2013). Current perspective on the global and United States cancer burden attributable to lifestyle and environmental risk factors. *Annual Review of Public Health* 34: 97–117.

Sezer, E.D., Oktay, L.M., Karadadaş, E. et al. (2019). Assessing anticancer potential of blueberry flavonoids, quercetin, kaempferol, and gentisic acid, through oxidative stress and apoptosis parameters on HCT-116 cells. *Journal of Medicinal Food* 22 (11): 1118–1126.

Sharif, M.K. and Khalid, R. (2018). Nutraceuticals: myths versus realities. In: *Therapeutic Foods* (ed. A.L. Holban and A.M. Grumezescu), 3–21. Academic Press.

Shaterzadeh-Yazdi, H., Farkhondeh, T., and Samarghandian, S. (2018). An overview on cardiovascular protective effects of catechins. *Cardiovascular & Hematological Disorders Drug Targets* 17 (3): 154–160.

Shen, N., Wang, T., Gan, Q. et al. (2022). Plant flavonoids: classification, distribution, biosynthesis, and antioxidant activity. *Food Chemistry* 383: 132531.

Shirakami, Y. and Shimizu, M. (2018). Possible mechanisms of green tea and its constituents against cancer. *Molecules* 23 (9): 2284.

Sigurdson, G.T., Tang, P., and Giusti, M.M. (2017). Natural colorants: food colorants from natural sources. *Annual Review of Food Science and Technology* 8: 261–280.

Souyoul, S.A., Saussy, K.P., and Lupo, M.P. (2018). Nutraceuticals: a review. *Dermatol Ther (Heidelb)* 8 (1): 5–16.

Stram, D.O., Hankin, J.H., Wilkens, L.R. et al. (2006). Prostate cancer incidence and intake of fruits, vegetables and related micronutrients: the multiethnic cohort study* (United States). *Cancer Causes and Control* 17 (9): 1193–1207.

Sudhakaran, M., Doseff, A.I., Rupasinghe, H.P.V., and Santos-Buelga, C. (2020). The targeted impact of favones on obesity-induced inflammation and the potential synergistic role in cancer and the gut microbiota. *Molecules* 25 (11): 2477.

Tamasi, G., Pardini, A., Bonechi, C. et al. (2019). Characterization of nutraceutical components in tomato pulp, skin and locular gel. *European Food Research and Technology* 245 (4): 907–918.

Thill, J., Regos, I., Farag, M.A. et al. (2012). Polyphenol metabolism provides a screening tool for beneficial effects of onobrychis viciifolia (sainfoin). *Phytochemistry* 82: 67–80.

Tousoulis, D., Ntarladimas, I., Antoniades, C. et al. (2008). Acute effects of different alcoholic beverages on vascular endothelium, inflammatory markers and thrombosis fibrinolysis system. *Clinical Nutrition* 27 (4): 594–600.

Upadhyay, S.K. (ed.) (2021). *Genome Engineering for Crop Improvement*, 1–394. John Wiley & Sons Ltd. https://doi.org/10.1002/9781119672425.

Upadhyay, S.K. and Singh, S.P. (eds.) (2021). *Bioprospecting of Plant Biodiversity for Industrial Molecules*, 1–431. John Wiley & Sons Ltd. https://doi.org/10.1002/9781119718017.

Valachovičova, M. and Birošova, L. (2020). Health effects of plant foods and the possibility of reducing health risk. In: *Veganism – A Fashion Trend or Food as a Medicine* (ed. M.Z. Jovandaric), 413–521. Belgrade, Serbia: IntechOpen.

Velu, G., Palanichamy, V., and Rajan, A.P. (2018). Phytochemical and pharmacological importance of plant secondary metabolites in modern medicine. In: *Bioorganic Phase in Natural Food: An Overview* (ed. S.M. Roopan and G. Madhumitha), 135–156. Cham: Springer.

Villarreal-García, D., Nair, V., Cisneros-Zevallos, L., and Jacobo-Velázquez, D.A. (2016). Plants as biofactories: postharvest stress-induced accumulation of phenolic compounds and glucosinolates in broccoli subjected to wounding stress and exogenous phytohormones. *Frontiers in Plant Science* 10 (7): 45.

White, P.J. and Broadley, M.R. (2005). Biofortifying crops with essential mineral elements. *Trends in Plant Science* 10 (12): 586–593.

Williamson, E.M., Liu, X., and Izzo, A.A. (2020). Trends in use, pharmacology, and clinical applications of emerging herbal nutraceuticals. *British Journal of Pharmacology* 177 (6): 1227–1240.

Wu, F., Wesseler, J., Zilberman, D. et al. (2021). Allow golden rice to save lives. *Proceedings of the National Academy of Sciences of the United States of America* 118 (51): e2120901118.

Yang, L., Xian, D., Xiong, X. et al. (2018). Proanthocyanidins against oxidative stress: from molecular mechanisms to clinical applications. *BioMed Research International* 2018: 11.

Ye, X., Al-Babili, S., Kloti, A. et al. (2000). Engineering the provitamin A (beta-carotene) biosynthetic pathway into (carotenoid-free) rice endosperm. *Science* 287 (5451): 303–305.

Zaheer, K. and Humayoun Akhtar, M. (2017). An updated review of dietary isoflavones: nutrition, processing, bioavailability and impacts on human health. *Critical Reviews in Food Science and Nutrition* 57 (6): 1280–1293.

Zhao, K., Li, Y., Zhou, Z. et al. (2022). Ginkgolide A alleviates cardiac remodeling in mice with myocardial infarction via binding to matrix metalloproteinase-9 to attenuate inflammation. *European Journal of Pharmacology* 923: 174932.

16

Green Synthesis of Nanoparticles Using Various Plant Parts and Their Antifungal Activity

Chikanshi Sharma[1], *Madhu Kamle*[1], *and Pradeep Kumar*[1,2]

[1] *Applied Microbiology Laboratory, Department of Forestry, North Eastern Regional Institute of Science and Technology, Nirjuli, Arunachal Pradesh, India*
[2] *Department of Botany, University of Lucknow, Lucknow, Uttar Pradesh, India*

16.1 Introduction

In the realm of nanoscience, there has been a rise of interest in the synthesis of antimicrobial nanomaterials using diverse biological, chemical, and physical approaches. Plants have been found to make a variety of nanoparticles, and some of these particles are released into the environment after made outside the organ cells, whereas some of the nanoparticles are released into the environment after produced inside organ by the cells. The study and application of exceedingly small objects are known as nanotechnology with dimensions fewer than 100 nm, this property of nanoparticles has attained a considerable attention. Nanoparticles making nanotechnology a broadly appropriate field in a variety of discipline with the qualities such as selectivity, biocompatibility, and size of these particles can deliver innumerable benefits. Nanoparticles biological and physicochemical features are useful in many ways for instance there are various nanomaterial uses in agriculture, including agricultural product transportation, production, processing, packaging, and storage. In the nanomedicine uses technologies at the nanoscale and nanoenabled techniques to prevent, diagnose, monitor and treat diseases (Farokhzad and Langer 2006). Field of nanotechnology gained an increased interest in biotechnology, especially for organic molecule and agrochemical delivery systems, and more recently, the transfer of DNA molecules or oligonucleotides into plant cells. For skin-penetrating qualities, high solubility and transparent appearance industry of cosmetics use nanotechnology in cosmetic formulations. As far as plant sciences and agriculture are concerned, agriculture predates the industrialized revolution by over 90 centuries, though nearly half a century back nanotechnology research began for industrial applications. Metal nanoparticles, predominantly silver, copper, and zinc, have got a lot of attention lately and as a new class of nonmaterials they have emerged. Nanotechnology has latterly proven to be quite helpful in upgrading the microbiological safety of feed and food matrices. The utilization of plant extracts in nanoparticle synthesis

Plants as Bioreactors for Industrial Molecules, First Edition.
Edited by Santosh Kumar Upadhyay and Sudhir P. Singh.

is commonly referred to as a "green" synthesis approach because it decreases and abolishes the usage of harmful compounds or their manufacturing. Plants are rich source of pigments, antioxidants, enzymes, peptides, products, reducing factors, proteins, antioxidants, triglycerides, polysaccharides, glycoproteins, latex, gums, phytochemical constituents like flavonoids, terpenoids, and vitamins, as well as cofactors (Dixit et al. 2021; Upadhyay and Singh 2021), that are used in the biological method for nanoparticle synthesis (Thombre et al. 2012). The plant-based nanoparticles that are being synthesized are commonly effective alongside a variety of plant diseases, including *Corticium salmonicolor* and *Phytophthora* (Ponmurugan et al. 2016). The utilization of microorganisms in the environmental friendly production of nanomaterials such as silver (Ag) nanoparticles has garnered substantial interest. Because existing chemical synthesis procedures are very expensive and equally use and generate harmful compounds that are harmful to the environment, researchers looked into the biological process that occurs in nature. Plant extract and microbial-based technologies have been resulted as an alternative for nanoparticles synthesis as a survey of research into natural systems (Schlüter et al. 2014). A variety of microorganisms have been indulged including several genera of fungi to biofabricate silver nanoparticles effectively (Dawoud et al. 2021). Furthermore, numerous fungi that are plant pathogenic have been successfully investigated for silver nanoparticle production (Yassin et al. 2016). Cereal infecting fungus like *Curvularia pallescens, Penicillium citrinum, Phoma leveillei,* and *Aspergillus clavatus* have also been employed in green silver nanoparticle manufacturing (Elgorban et al. 2016). The use of biosynthesized silver (Ag) nanoparticles in the control of various phytopathogenic fungi in order to remove or at least reduces their impacts appears propitious (Abdel-Hadi et al. 2014). Copper nanoparticles, on the other hand, have recently synthesized and used in antimicrobial studies due to relatively high activity and lower cost than silver or gold nanoparticles. Pb, Au, Ag, Fe, and ZnO metallic nanoparticles are used in multiple industries, including diagnostics, agriculture, and pharmaceuticals. Silver NPs can be synthesized through biological, physical, and chemical diminution processes, whereas silver and elemental silver nanoparticles have a wide range of medical applications. It is made possible by recent research that silver nanoparticles can now be used in the food, automotive, paints, cosmetics, and packaging industries (Taleb et al. 1997; Sun et al. 2002). Thermal breakdown of organic solvents, silver ions chemical reduction, and reverse micelles photoreduction processes are used to synthesize silver nanoparticles (Esumi et al. 1990; Liz Marzán and Lado-Touriño 1996; Sun et al. 2001). However, because of attributes in optical and magnetic polarizability, catalysis and antimicrobial activity also due to its electrical conductivity all of these procedures are expensive, extremely poisonous, and dangerous chemicals are utilized in synthesis of silver NPs posing environmental risks. Because chemical synthesis has negative effects, biological approaches including microbes, enzymes, and plants were employed for the production of silver nanoparticles(AgNPs), which impart environmental, low cost, and rapid imputes (Nair and Pradeep 2002; Dubey et al. 2009; Pugazhenthiran et al. 2009; Prasad et al. 2016) (Figure 16.1). The metal nanoparticles that are most stable are gold nanoparticles (AuNPs) because of their large surface area, high absorption aspect, bioadjustability, noncytotoxicity, conjugation in a steady covalent linkage among molecules, congregation of various types and purpose to biology gold NPs have a broad range of applications in biochemical, magnetic, physical, electrical, thermal, biological, and optical fields (Paciotti et al. 2004). When opposed to microbial systems, plant-mediated gold nanoparticle synthesis requires a lesser

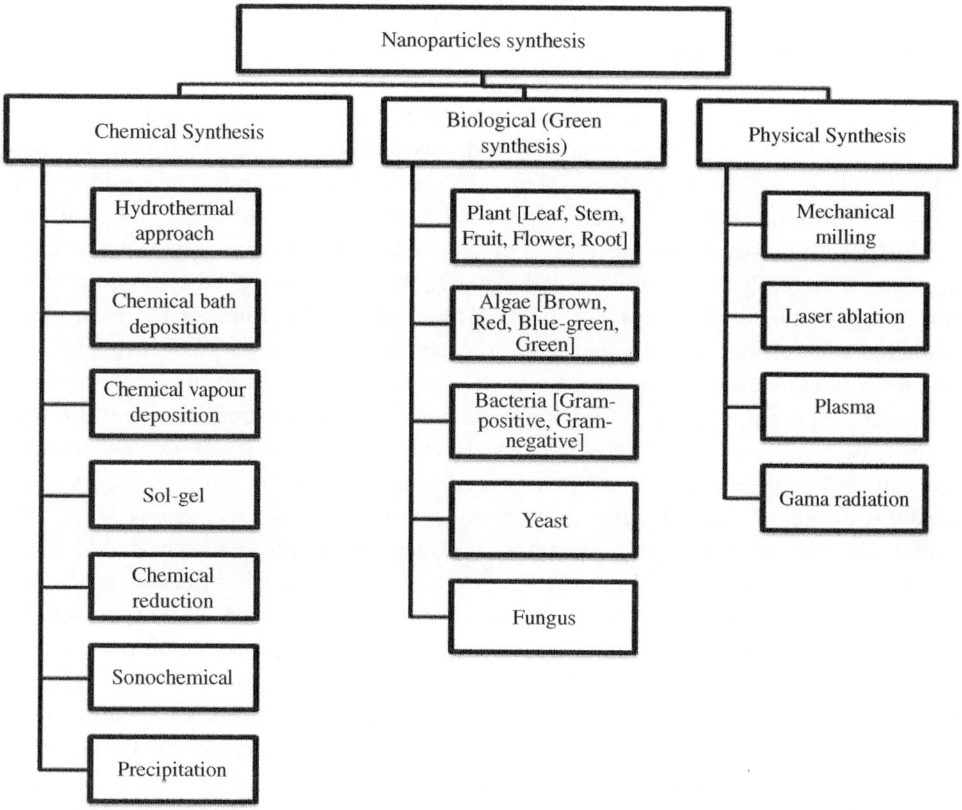

Figure 16.1 Conventional methods used for nanoparticle synthesis.

amount of incubation time can readily be scaled up to meet commercial needs (Murphy et al. 2008; Das et al. 2012). Zinc oxide (ZnO) nanoparticles are stable metal oxide and are main focus as they are considered safe for animals, humans, and environment. Because of its antimicrobial qualities, ZnO has the ability to fight infectious disorders (Ramesh et al. 2014). Gas-sensing, high isoelectric point, magnetic, biocompatibility, piezoelectric, quick electron transfer, outstanding chemically and thermally stability, and various optical features distinguish ZnO from other metal oxides for instance, CaO, MgO, and TiO_2 (Vaseem et al. 2010; Ravichandrika et al. 2012). ZnO possesses high catalytic efficiency and adsorption ability so they can be utilized as sensors, as fungicides, in treatment of sewage, paints, catalysis processing, conversion of solar energy, drug delivery, cosmetics, textiles, and as an antibiotic due to its varied features (Stoimenov et al. 2002; Iravani 2011).

16.2 Gold Nanoparticle Synthesis Using Plant Source

Plant extracts are used in synthesizing Au nanoparicles is advantageous not only because of the lower environmental impact but also because it can create a vast quantity of nanoparicles (Ikram 2015). Gold nanoparticles are made from diverse parts of plants including stems, roots, leaves, and fruits due to their wide application in a range of sectors. Au nanoparicles have qualities that are very different from bulk gold, since the solution of gold nanoparicles is wine-red, whereas gold in bulk is solid and yellow in color. Gold nanoparticles (AuNPs) can be biosynthesized in a variety of morphologies and shapes that includes triangles, hexagons, nanotriangles, rods, decahedrons and icosahedrons, and hierarchical tubes in addition to nodous ribbons. The concentrations and congregation of Au atoms play a significant role in structuring the nanoparticles that have a wide range of uses in industries. Gold (Au) ions have a better chance of colliding with the atom and forming aggregates in the early stages. As a result, large particles can be generated at low concentrations. The reduction or demeaning of the Au III ion into the Au atom causes the Au atom to adhere to the cell surface and other reduced Au binds and aggregates to form gold nanoparticles (Cai et al. 2011). They have been effectively used as delivery molecules to help reduce cancer growth or even induce apoptosis in malignant cells and are also utilized as antiarthritic, antimalarial, and antiangiogenesis drugs having pioneered the way in medicine, cancer therapy, diagnosis, and treatment. The antifungal properties of *Abelmoschus esculentus* examined against *Candida albicans*, *Aspergillus niger*, *Aspergillus lavus*, and *Puccinia graminis tritci* using its aqueous seed extract. As a result, synthesized gold nanoparticles have a lot of potential in drug development for any fungal diseases (Jayaseelan et al. 2013). The fluconazole-coated Au nanoparticles with zone of inhibition showed the best antifungal effectiveness against *Aspergillus flavus – A. niger* and *C. albicans* (Zawrah et al. 2011). Bankar et al. (2010) found that Au NPs mediated by *Musa paradisiaca* (banana peel) extract had effective antifungal activity against the pathogenic fungus *C. albicans* (BH and BX). Extracellular production of stable gold nanoparticles of varying size was achieved utilizing extract from *Pelargonium graveolens* leaves and its endophytic fungus (Shankar et al. 2003). Other important nanoparticles that have been synthesized are given in Table 16.1.

Table 16.1 Various plants and their parts used in the synthesis of AuNPs, AgNPs, and ZnO NPs.

Scientific name of plants–part of plant used	Size in nm	Inference	References
Synthesis of AuNPs			
Lycopersicon esculentum—leaf broth	36.7 nm	Antifungal activity—*Penicillium italicum* had the highest whereas *Fusarium equisetti* had the lowest.	Asmathunisha and Kathiresan (2013)
Mentha piperitta	52 nm	Antifungal activity—*Penicillium italicum* had the highest antifungal activity, while *Candida albicans* had the lowest.	Thanighaiarassu et al. (2014)
Trianthema decandra—root extract	33.7 nm and 99.3 nm	Antifungal activity—lowest antifungal activity against *Candida albicans*	Geethalakshmi and Sarada (2012)

Table 16.1 (Continued)

Scientific name of plants—part of plant used	Size in nm	Inference	References
Phyllanthus emblica—aqueous fruit extract	—	Antifungal activity—*Verticilium dhalea* has the highest followed by *Fusarium oxysporum* and *C. albicans* had the lowest.	Ahmad et al. (2015)
Beta vulgaris-	—	*Fusarium oxysporum* and *Aspergilus niger* showed antifungal activity.	Ahmad et al. (2015)
Helianthus annuus	33.7–99.3 nm	Imipenem-associated AuNPs had a better antifungal impact than Norfloxacin and Vancomycin-associated AuNPs.	Geethalakshmi and Sarada (2012)
Abelmoschus esculentus (Okra)—(crystalline)	62	Antifungal	Jayaseelan et al. (2013)
Artemisia vulgaris (Mugwort)—(spherical, Hexagonal and triangular)	50–100	Antifungal	Sundararajan and Kumari (2017)
Brazilian red propolis-(Pentagonal,hexagonal, tringular and rod)	8–15	Antifungal	Botteon et al.(2021)
Coreopsis lanceolate	—	Aflatoxin detection	Abhijith and Thakur (2012)
Carthamus tinctorius L (spherical and triangular)	40–200 nm	Antifungal	Nagaraj et al. (2012a)
Caesalpinia pulcherrima (spherical)	10–50	Antifungal	Nagaraj et al. (2012b)
Bay cedar (spherical)	20–25 nm	Antifungal	Karthika et al. (2017)
Helianthus annuus (polydispersed)	35 nm	Antifungal	Liny et al. (2012)
Nepenthes khasiana (spherical)	50–80 nm	Antifungal	Bhau et al. (2015)
Punica granatum (spherical)	5–20 nm	Antifungal	Lokina et al. (2014)
Pistacia integerrima (granular)	20–200 nm	Antifungal	Islam et al. (2019)
Rivea hypocrateriformis (spherical)	10–50 nm	Antifungal	Godipurge et al. (2016)
Trianthema decandra L (cuboidal, spherical, and hexadonal)	38–80 nm	Antifungal	Geethalakshmi and Sarada (2013)

(*Continued*)

Table 16.1 (Continued)

Scientific name of plants—part of plant used	**Size in nm**	**Inference**	**References**
Cassia fistula—stem, bark	55.2–98.4 nm	Antifugal activity	Daisy and Saipriya (2012)
Synthesis of AgNPs			
Selaginella bryopteris—Leaf extract	1 mm	Antifungal	Dakshayani et al. (2019)
Citrus limetta—peel extract		Antifungal	Dutta et al. (2020)
Gelidium corneum—Extract	—	Antifungal	Öztürk et al. (2020)
Ligustrum lucidum- leaf extract	3.7 cm	Antifungal	Huang et al.(2020)
Amaranthus retroflexu—Leaf extract	—	Antifungal	Bahrami-Teimoori et al. (2017)
Synthesis of ZnO NPs			
Agathosma betulina (Rutaceae)	Dry leaves	Quaispherical agglomerate	Thema et al. (2015)
Aloe vera (Liliaceae)	Freeze-dried Leaf peel	Spherical, hexagonal	Qian et al. (2015)
A. vera (liliaceae)	Leaf extract	Spherical, oval and hexagonal	Ali et al. (2016)
Anisochilus carnosus (Lamiaceae)	Leaf extract	Hexagonal and Quasispherical	Anbuvannan et al. (2015a)
Azadirachta indica (Meliaceae)	Leaf	Spherical	Bhuyan et al.(2015)
A. indica (Meliaceae)	Fresh leaves	Spherical	Elumalai and Velmurugan (2015)
A. indica (Meliaceae)	Fresh leaves	Hexagonal disk and nonobuds	Madan et al. (2016)
Calatropis gigantea (Apocynaceae)	Fresh leaves	Spherical shape-forming agglomerates	
Cocus nucifera (Arecaceae)	Coconut water	Spherical and predominantly hexagonal without and agglomeration	
Coptidis rhizome (Ranunculaceae)	Dried rhizome	Spherical, rod shaped	
Eichhornia crassipes (Pontederiaceae)	Leaf extract	Spherical without aggregation	
Gossypium (Malvaceae)	Cellulosic fiber	Wurtzite spherical, nanorod	
Moringa oleifera (Moringeaceae)	Leaf	Spherical and granular nanosized shaped with a group of aggregates	Elumalai et al. (2015)

Table 16.1 (Continued)

Scientific name of plants–part of plant used	Size in nm	Inference	References
Nephelium lappaceum (Sapindaceae)	Fruit peels	Spherical and hexagonal	Yuvakkumar et al. (2015)
N. lappaceum L. (Sapindaceae)	Fruit peels	Needle shaped forming agglomerate	Yuvakkumar et al. (2015)
Ocimum basilicum L. var. purpurascens (Limiaceae)	Leaf extract	Hexagonal (Wurtzite)	Salam et al. (2014)
Parthenium hysterophorus L. (Asteraceae)	Leaf extract	Spherical, hexagonal	Rajiv et al. (2013)
Phyllanthus niruri (Phyllanthaceae)	Leaf extract	Hexagonal and quasi spherical	Anbuvannan et al. (2015b)
Plectranthus amboinicus (Lamiaceae)	Leaf extract	Rod-shaped nanoparticle with agglomerates	Fu and Fu (2015)
Pongamia pinnata (Legumes)	Fresh leaves	Spherical, hexagonal, nanorods	
Rosa canina (Rosaceae)	Fruit extract	Spherical	Jafarirad et al. (2016)
Santalum album (Santalaceae)	Leaves	Nanorods	Kavithaa et al. (2016)
Solanum nigrum (Solanaceae)	Leaf extract	Hexagonal wurtzite, quasispherical	Ramesh et al. (2015)
Sphathodea campanulapa (Bignoniaceae)	Leaf extract	Spherical	Ochieng et al. (2015)
Trifolium prapense (Legumes)	Flowers	Spherical	Dobrucka and Długaszewska (2016)
Vitex negundo (Lamiaceae)	Leaf/flowers	Spherical/hexagonal	Ambika and Sundrarajan (2015a, 2015b)

16.3 Silver Nanoparticles Synthesis Using Plants Source

The scientific community is very interested in the synthesis or production of silver nanoparticles (AgNPs) due of their wide variety of uses. Silver nanoparticles are utilized to successfully diagnose and treat cancer (Baruwati et al. 2009; Popescu et al. 2010). Nanoparticles can be made by self-assembling atoms into a new nucleus, which then grew into a tiny particle. In general, there are basically two techniques to create silver nanoparticles: one is "bottom to up" approach and another one is "top to bottom" approach. Plant extracts containing diverse stabilizers and reductants have the potential to be used as greener silver nanoparticle production agents (Narayanan and Sakthivel 2011). An aqueous medium and a biocompatible stabilizing and environmentally benign reducing agent should be used in the green production of AgNPs. Sclerotium-forming fungus was suppressed by silver nanoparticles (AgNPs) in a dose-dependent method according to Min et al. (2009). Kim et al. (2009) investigated the antifungal activity of AgNPs against *Raffaelea* sp., with is an ambrosia fungus and also on other phytopathogenic fungi. AgNPs produced from *Svensonia hyderobadensis* leaf extract showed strong antifungal activity against *Rhizopus* sp., *A. niger, A. flavus, Fusarium* sp., and *Curvularia* sp. indicating that silver nanoparticles synthesized or extracted from any medicinal plant have significant antifungal potential (Rai et al. 2014). To synthesize AgNPs, the use of *Acalypha indica, mangrove, Rumex hymenosepalu, Tribulus terrestris,* and *Arbutus unedo* has already been reported (Krishnaraj et al. 2010; Gnanadesigan et al. 2011; Gopinath et al. 2012; Kouvaris et al. 2012; Rodriguez-León et al. 2013). The key components of these plant extracts are water-soluble heterocyclic compounds, polysaccharides, and polyol components, which are largely responsible for reducing silver ions and stabilizing the nanoparticles generated. Several factors influence the efficiency of the AgNPs fabrication process in green synthesis or production of AgNPs using part of the plant, including temperature, plant source, the concentration and type of pigments excreted from leaf, organic compounds found in the crude extract of leaf from a plants, and concentration of preliminary Ag ions (Leela and Vivekanandan 2008). Some of the important AgNPs that have been synthesized are given in Table 16.1.

16.4 Zinc Oxide Nanoparticles Synthesis Using Plants

Plants and microorganisms have been used in the biological synthesis of zinc oxide nanoparticles, according to a systematic review of the literature. Microbe-mediated synthesis is not an industrially practicable approach of zinc oxide nanoparticle synthesis because of the requirements for extremely aseptic conditions and their upkeep. As a result, plant extracts have been deemed the finest platform for extensive production of nanoparticles due to their low cost and environmentally beneficial characteristics (Bhattacharya and Gupta 2005). The use of plants to synthesize nanoparticles eliminates the need for microbial culture, as well as the costs of culture media and microbe isolation (Kalishwaralal et al. 2010). Milky latex from the plant *Calotropis procera* was used to successfully synthesize zinc-oxide nanoparticles for the first time. TEM and XRD were used to discover spherical zinc oxide nanoparticles in the size range of 5–40 nm (Singh et al. 2011). For the creation of very stable zinc oxide nanoparticles, extract from *Aloe vera* leaf was combined with zinc nitrate. The generated zinc oxide nanoparticles were spherical in form with an average size range of 25–40 nm (Sangeetha et al. 2011). In a variety of research, inorganic metal oxides such as ZnO, TiO_2, and CuO have already been synthesized. ZnO nanoparticles are the most interesting of all these metal oxides since they are less expensive to make, simple to prepare and are safe (Elumalai et al. 2015). Biological systems, optics, and electronics disciplines use zinc oxide nanoparticles widely and have aroused interest in recent years among researches (Azizi et al. 2014). Zinc oxide NPs have been used as an antifungal surface coating agent on materials and textiles, and nanoscale zinc oxide is a good antifungal agent at a significantly lowest concentration (Abramova et al. 2014). Because of their large bandgap ZnO nanoparticles have a wide range of semiconducting properties, including anti-inflammatory, UV filtering, and wound-healing properties (Nagajyothi et al. 2013; Ramesh et al. 2015). They have been widely employed in cosmetics, such as sunscreen creams, due to their UV filtering qualities. Antifungal, anticancer, antibacterial, antidiabetic, and agronomic characteristics are only a few of the biomedical applications (Ramesh et al. 2015). Nanobelts, nanoflakes, nanorods, nanowires, and nanoflowers have all been described as ZnO nanoparticles in various morphologies (Anbuvannan et al. 2015a). ZnO NPs are most often synthesized from the leaves of *Azadirachta indica*, a family of Meliaceae plant. Other important ZnO nanoparticles that have been synthesized from plants are given in Table 16.1.

16.5 Other Nanoparticles Synthesis Using Plant Source

Different Plant sources have been used to produce non-noble metallic nanoparticles such as platinum, lead, copper; indium oxide, selenium, titanium oxide and palladium which are of great significance probably used as antibacterial and antifungal agent in the production industries including cosmetic formulation industries, food and drug industries and also used in medical treatments. In 2010, Song et al. (2010) reported the first platinum nanoparticle synthesis utilising a Persimmon leaf extract (*Diospyros kaki*). The nanoparticles produced ranged in size from 2 to 12 nm and this result demonstrated that 10% concentration of leaf biomass (at 95 °C) converted 90% of the platinum ions in solution. Palladium (Pd) nanoparticles with diameters ranging from 20 to 60 nm and symmetry of cubic crystal in the centre have been synthesized using extracts from coffee (*Coffea arabica*) and tea (*Camellia sinensis*) (Verma et al. 2019). Other important metallic nanoparticles that have been synthesized are given in Table 16.2.

Table 16.2 Other nanoparticles synthesis using plant source.

Plant Source	Nanoparticles	Remarks	References
Aloe vera (Aloe barbadensis Miller)	Indium oxide	5–50 nm, spherical	Maensiri et al. (2008)
Annona squamosa L. peel and larvicidal agent	Palladium	Application as and acaricidal, insecticidal	Roopan et al. (2012)
Aqueous extract of *Hordeum vulgare* and *Rumex acetosa* plants	Iron oxide	—	Makarov et al. (2014)
Capsicum annuum	Selenium	Selenium with protein nonocomposites	
Catharanthus roseus leaf extract	Titanium dioxide	Good effect against *Hippobosca maculapa* and *Bovicola ovis*	Velayutham et al. (2012)
Diospyros kaki	Platinum	15–19 nm	
Dried *Anacardium occidentale* leaf extract	Platinum	Used for thermal and catalytic application	Sheny et al. (2013)
Euphorbiaceae latex	Copper	Very good antimicrobial activity	Valodkar et al. (2011)

(*Continued*)

Table 16.2 (Continued)

Plant Source	Nanoparticles	Remarks	References
Jatropha curcus L. latex (aqueous extract)	Lead	10–12.5 nm	Joglekar et al. (2011)
Ocimum sanctum (electrolysis application)	Platinum	Water	Soundarrajan et al. (2012)
Piper betle leaves broth	Palladium	Used for antifungal studies	Mallikarjuna et al. (2013)
Pulicaria glutinosa extract	Palladium	Catalytic activity toward Suzuki-coupling reaction	Khan et al. (2014a, 2014b)
P. glutinosa extract and supporting of poly (ethylene glycol)	Palladium	Applied for catalytic activity	
Sorghum bran extract	Iron and silver	Synthesized at room temperature	Njagi et al. (2011)
Terminalia chebula aqueous extract	Palladium and iron nonoparticles	—	Kumar et al. (2013a, 2013b)

16.6 Conclusion and Future Perspective

Nature has created the most efficient tiny functional materials in appealing and innovative ways. An emergent awareness of green chemistry and their harmless and eco-friendly processing on green routes for metal nanoparticle synthesis have prompted a quest to develop environmentally acceptable methodologies. For the green production of metallic NPs, microorganisms such as fungus, bacteria, actinomycetes, algae, yeast, and plant sources ranging from angiosperms to gymnosperms are commonly employed. When compared to bacteria, the variety of biomolecules found in various plants is the fundamental rationale for the creation of metal NPs because of phytochemical characteristics. Plant leaves are an excellent source for the synthesis of metallic nanoparticles because they are more stable and produce them at a faster rate. This chapter focused on the green and ecofriendly production of metallic nanoparticles from various plant sources. Plant-based nanoparticles could increase biocompatibility and reduce toxicological in future usage, as different types of nanoparticles have been used in a range of biomedical applications. The plant and their part may be used for the large-scale production of nanoparticles at large scale and application in various field for human health and agriculture.

Acknowledgement

The authors are grateful to head of department forestry and Director, NERIST for the support and co-operation.

Conflicts of Interest

The authors declare no conflicts of interest.

Author Contribution

Pradeep Kumar: conceptualization; Madhu Kamle and Chikanshi Sharma: writing-original draft; Pradeep Kumar: final editing.

References

Abdel-Hadi, A.M., Awad, M.F., Abo-Dahab, N.F., and ElKady, M.F. (2014). Extracellular synthesis of silver nanoparticles by *Aspergillus terreus*: biosynthesis, characterization and biological activity. *Biosciences, Biotechnology Research Asia* 11 (3): 1179–1186.

Abhijith, K.S. and Thakur, M.S. (2012). Application of green synthesis of gold nanoparticles for sensitive detection of aflatoxin B1 based on metal enhanced fluorescence. *Analytical Methods* 4 (12): 4250–4256.

Abramova, A.V., Abramov, V.O., Gedanken, A. et al. (2014). An ultrasonic technology for production of antibacterial nanomaterials and their coating on textiles. *Beilstein Journal of Nanotechnology* 5 (1): 532–536.

Ahmad, B., Hafeez, N., Bashir, S. et al. (2015). Comparative analysis of the biological activities of bio-inspired gold nano-particles of *Phyllantus emblica* fruit and *Beta vulgaris* bagasse with their crude extracts. *Pakistan Journal of Botany* 2 (47): 139–146.

Ali, K., Dwivedi, S., Azam, A. et al. (2016). *Aloe vera* extract functionalized zinc oxide nanoparticles as nanoantibiotics against multi-drug resistant clinical bacterial isolates. *Journal of Colloid and Interface Science* 472: 145–156.

Ambika, S. and Sundrarajan, M. (2015a). Antibacterial behaviour of Vitex negundo extract assisted ZnO nanoparticles against pathogenic bacteria. *Journal of Photochemistry and Photobiology B: Biology* 146: 52–57.

Ambika, S. and Sundrarajan, M. (2015b). Green biosynthesis of ZnO nanoparticles using Vitex negundo L. extract: spectroscopic investigation of interaction between ZnO nanoparticles and human serum albumin. *Journal of Photochemistry and Photobiology B: Biology* 149: 143–148.

Anbuvannan, M., Ramesh, M., Viruthagiri, G. et al. (2015a). Synthesis, characterization and photocatalytic activity of ZnO nanoparticles prepared by biological method. *Spectrochimica Acta Part A: Molecular and Biomolecular Spectroscopy* 143: 304–308.

Anbuvannan, M., Ramesh, M., Viruthagiri, G. et al. (2015b). Anisochilus carnosus leaf extract mediated synthesis of zinc oxide nanoparticles for antibacterial and photocatalytic activities. *Materials Science in Semiconductor Processing* 39: 621–628.

Asmathunisha, N. and Kathiresan, K. (2013). Rapid biosynthesis of antimicrobial silver and gold nanoparticles by in vitro callus and leaf extracts from *Lycopersicon esculentum* Mill. *International Journal of Pharma and Bio Sciences* 4 (1).

Azizi, S., Ahmad, M.B., Namvar, F., and Mohamad, R. (2014). Green biosynthesis and characterization of zinc oxide nanoparticles using brown marine macroalga Sargassum muticum aqueous extract. *Materials Letters* 116: 275–277.

Bahrami-Teimoori, B., Nikparast, Y., Hojatianfar, M. et al. (2017). Characterisation and antifungal activity of silver nanoparticles biologically synthesised by Amaranthus retroflexus leaf extract. *Journal of Experimental Nanoscience* 12 (1): 129–139.

Bankar, A., Joshi, B., Kumar, A.R., and Zinjarde, S. (2010). Banana peel extract mediated novel route for the synthesis of silver nanoparticles. *Colloids and Surfaces A: Physicochemical and Engineering Aspects* 368 (1-3): 58–63.

Baruwati, B., Polshettiwar, V., and Varma, R.S. (2009). Glutathione promoted expeditious green synthesis of silver nanoparticles in water using microwaves. *Green Chemistry* 11 (7): 926–930.

Bhattacharya, D. and Gupta, R.K. (2005). Nanotechnology and potential of microorganisms. *Critical Reviews in Biotechnology* 25 (4): 199–204.

Bhau, B.S., Ghosh, S., Puri, S. et al. (2015). Green synthesis of gold nanoparticles from the leaf extract of *Nepenthes khasiana* and antimicrobial assay. *Advanced Materials Letters* 6 (1): 55–58.

Bhuyan, T., Mishra, K., Khanuja, M. et al. (2015). Biosynthesis of zinc oxide nanoparticles from *Azadirachta indica* for antibacterial and photocatalytic applications. *Materials Science in Semiconductor Processing* 32: 55–61.

Botteon, C.E.A., Silva, L.B., Ccana-Ccapatinta, G.V. et al. (2021). Biosynthesis and characterization of gold nanoparticles using *Brazilian red propolis* and evaluation of its antimicrobial and anticancer activities. *Scientific Reports* 11 (1): 1–16.

Cai, F., Li, J., Sun, J., and Ji, Y. (2011). Biosynthesis of gold nanoparticles by biosorption using Magnetospirillum gryphiswaldense MSR-1. *Chemical Engineering Journal* 175: 70–75.

Daisy, P. and Saipriya, K. (2012). Biochemical analysis of Cassia fistula aqueous extract and phytochemically synthesized gold nanoparticles as hypoglycemic treatment for diabetes mellitus. *International Journal of Nanomedicine* 7: 1189.

Dakshayani, S.S., Marulasiddeshwara, M.B., Kumar, S. et al. (2019). Antimicrobial, anticoagulant and antiplatelet activities of green synthesized silver nanoparticles using Selaginella (Sanjeevini) plant extract. *International Journal of Biological Macromolecules* 131: 787–797.

Das, R.K., Gogoi, N., Babu, P.J. et al. (2012). The synthesis of gold nanoparticles using Amaranthus spinosus leaf extract and study of their optical properties. *Advanced Materials: Physics, Chemistry* 2: 275–281.

Dawoud, T.M., Yassin, M.A., El-Samawaty, A.R.M., and Elgorban, A.M. (2021). Silver nanoparticles synthesized by Nigrospora oryzae showed antifungal activity. *Saudi Journal of Biological Sciences* 28 (3): 1847–1852.

Dixit, S., Shukla, A., Singh, V., and Upadhyay, S.K. (2021). Bioprospecting of natural compounds for industrial and medical applications. In: *Bioprospecting of Plant Biodiversity for Industrial Molecules* (ed. S.K. Upadhyay and S.P. Singh). John Wiley & Sons Ltd. https://doi.org/10.1002/9781119718017.ch3.

Dobrucka, R. and Długaszewska, J. (2016). Biosynthesis and antibacterial activity of ZnO nanoparticles using Trifolium pratense flower extract. *Saudi Journal of Biological Sciences* 23 (4): 517–523.

Dubey, M., Bhadauria, S., and Kushwah, B.S. (2009). Green synthesis of nanosilver particles from extract of Eucalyptus hybrida (safeda) leaf. *Digest Journal of Nanomaterials and Biostructures* 4 (3): 537–543.

Dutta, T., Ghosh, N.N., Das, M. et al. (2020). Green synthesis of antibacterial and antifungal silver nanoparticles using citrus limetta peel extract: experimental and theoretical studies. *Journal of Environmental Chemical Engineering* 8 (4): 104019.

Elgorban, A.M., El-Samawaty, A.E.R.M., Yassin, M.A. et al. (2016). Antifungal silver nanoparticles: synthesis, characterization and biological evaluation. *Biotechnology & Biotechnological Equipment* 30 (1): 56–62.

Elumalai, K. and Velmurugan, S. (2015). Green synthesis, characterization and antimicrobial activities of zinc oxide nanoparticles from the leaf extract of *Azadirachta indica* (L.). *Applied Surface Science* 345: 329–336.

Elumalai, K., Velmurugan, S., Ravi, S. et al. (2015). RETRACTED: green synthesis of zinc oxide nanoparticles using *Moringa oleifera* leaf extract and evaluation of its antimicrobial activity. *Molecular and Biomolecular Spectroscopy* 143: 158–164.

Esumi, K., Tano, T., Torigoe, K., and Meguro, K. (1990). Preparation and characterization of bimetallic palladium-copper colloids by thermal decomposition of their acetate compounds in organic solvents. *Chemistry of Materials* 2 (5): 564–567.

Farokhzad, O.C. and Langer, R. (2006). Nanomedicine: developing smarter therapeutic and diagnostic modalities. *Advanced Drug Delivery Reviews* 58: 1456–1459.

Fu, L. and Fu, Z. (2015). *Plectranthus amboinicus* leaf extract–assisted biosynthesis of ZnO nanoparticles and their photocatalytic activity. *Ceramics International* 41 (2): 2492–2496.

Geethalakshmi, R. and Sarada, D.V.L. (2012). Gold and silver nanoparticles from *Trianthema decandra*: synthesis, characterization, and antimicrobial properties. *International Journal of Nanomedicine* 7: 5375.

Geethalakshmi, R. and Sarada, D.V.L. (2013). Characterization and antimicrobial activity of gold and silver nanoparticles synthesized using saponin isolated from *Trianthema decandra* L. *Industrial Crops and Products* 51: 107–115.

Gnanadesigan, M., Anand, M., Ravikumar, S. et al. (2011). Biosynthesis of silver nanoparticles by using mangrove plant extract and their potential mosquito larvicidal property. *Asian Pacific Journal of Tropical Medicine* 4 (10): 799–803.

Godipurge, S.S., Yallappa, S., Biradar, N.J. et al. (2016). A facile and green strategy for the synthesis of Au, Ag and Au–Ag alloy nanoparticles using aerial parts of *R. hypocrateriformis* extract and their biological evaluation. *Enzyme and Microbial Technology* 95: 174–184.

Gopinath, V., MubarakAli, D., Priyadarshini, S. et al. (2012). Biosynthesis of silver nanoparticles from *Tribulus terrestris* and its antimicrobial activity: a novel biological approach. *Colloids and Surfaces B: Biointerfaces* 96: 69–74.

Huang, W., Yan, M., Duan, H. et al. (2020). Synergistic antifungal activity of green synthesized silver nanoparticles and epoxiconazole against Setosphaeria turcica. *Journal of Nanomaterials* 2020: 9535432.

Ikram, S. (2015). Synthesis of gold nanoparticles using plant extract: an overview. *Nano Research* 1 (1): 5.

Iravani, S. (2011). Green synthesis of metal nanoparticles using plants. *Green Chemistry* 13 (10): 2638–2650.

Islam, N.U., Jalil, K., Shahid, M. et al. (2019). Green synthesis and biological activities of gold nanoparticles functionalized with *Salix alba*. *Arabian Journal of Chemistry* 12 (8): 2914–2925.

Jafarirad, S., Mehrabi, M., Divband, B., and Kosari-Nasab, M. (2016). Biofabrication of zinc oxide nanoparticles using fruit extract of *Rosa canina* and their toxic potential against bacteria: a mechanistic approach. *Materials Science and Engineering: C* 59: 296–302.

Jayaseelan, C., Ramkumar, R., Rahuman, A.A., and Perumal, P. (2013). Green synthesis of gold nanoparticles using seed aqueous extract of *Abelmoschus esculentus* and its antifungal activity. *Industrial Crops and Products* 45: 423–429.

Joglekar, S., Kodam, K., Dhaygude, M., and Hudlikar, M. (2011). Novel route for rapid biosynthesis of lead nanoparticles using aqueous extract of Jatropha curcas L. latex. *Materials Letters* 65 (19-20): 3170–3172.

Kalishwaralal, K., Deepak, V., Pandian, S.R.K. et al. (2010). Biosynthesis of silver and gold nanoparticles using Brevibacterium casei. *Colloids and Surfaces B: Biointerfaces* 77 (2): 257–262.

Karthika, V., Arumugam, A., Gopinath, K. et al. (2017). Guazuma ulmifolia bark-synthesized Ag, Au and Ag/Au alloy nanoparticles: photocatalytic potential, DNA/protein interactions, anticancer activity and toxicity against 14 species of microbial pathogens. *Journal of Photochemistry and Photobiology B: Biology* 167: 189–199.

Kavithaa, K., Paulpandi, M., Ponraj, T. et al. (2016). Induction of intrinsic apoptotic pathway in human breast cancer (MCF-7) cells through facile biosynthesized zinc oxide nanorods. *Karbala International Journal of Modern Science* 2 (1): 46–55.

Khan, A.K., Rashid, R., Murtaza, G., and Zahra, A.J.T.R. (2014a). Gold nanoparticles: synthesis and applications in drug delivery. *Tropical Journal of Pharmaceutical Research* 13 (7): 1169–1177.

Khan, M., Khan, M., Kuniyil, M. et al. (2014b). Biogenic synthesis of palladium nanoparticles using *Pulicaria glutinosa* extract and their catalytic activity towards the Suzuki coupling reaction. *Dalton Transactions* 43 (24): 9026–9031.

Kim, S.W., Kim, K.S., Lamsal, K. et al. (2009). An in vitro study of the antifungal effect of silver nanoparticles on oak wilt pathogen *Raffaelea* sp. *Journal of Microbiology and Biotechnology* 19 (8): 760–764.

Kouvaris, P., Delimitis, A., Zaspalis, V. et al. (2012). Green synthesis and characterization of silver nanoparticles produced using *Arbutus unedo* leaf extract. *Materials Letters* 76: 18–20.

Krishnaraj, C., Jagan, E.G., Rajasekar, S. et al. (2010). Synthesis of silver nanoparticles using *Acalypha indica* leaf extracts and its antibacterial activity against water borne pathogens. *Colloids and Surfaces B: Biointerfaces* 76 (1): 50–56.

Kumar, K.M., Mandal, B.K., Kumar, H.A.K., and Maddinedi, S.B. (2013a). Green synthesis of size controllable gold nanoparticles. *Spectrochimica Acta Part A: Molecular and Biomolecular Spectroscopy* 116: 539–545.

Kumar, K.M., Mandal, B.K., Kumar, K.S. et al. (2013b). Biobased green method to synthesise palladium and iron nanoparticles using *Terminalia chebula* aqueous extract. *Spectrochimica Acta Part A: Molecular and Biomolecular Spectroscopy* 102: 128–133.

Leela, A. and Vivekanandan, M. (2008). Tapping the unexploited plant resources for the synthesis of silver nanoparticles. *African Journal of Biotechnology* 7 (17): 3162–3165.

Liny, P., Divya, T.K., Malakar, B. et al. (2012). Preparation of gold nanoparticles from *Helianthus annuus* (sun flower) flowers and evaluation of their antimicrobial activities. *International Journal of Pharma and Bio Sciences* 3 (1): 439–446.

Liz-Marzán, L.M. and Lado-Touriño, I. (1996). Reduction and stabilization of silver nanoparticles in ethanol by nonionic surfactants. *Langmuir* 12 (15): 3585–3589.

Lokina, S., Suresh, R., Giribabu, K. et al. (2014). Spectroscopic investigations, antimicrobial, and cytotoxic activity of green synthesized gold nanoparticles. *Spectrochimica Acta Part A: Molecular and Biomolecular Spectroscopy* 129: 484–490.

Madan, H.R., Sharma, S.C., Suresh, D. et al. (2016). Facile green fabrication of nanostructure ZnO plates, bullets, flower, prismatic tip, closed pine cone: their antibacterial, antioxidant, photoluminescent and photocatalytic properties. *Spectrochimica Acta Part A: Molecular and Biomolecular Spectroscopy* 152: 404–416.

Maensiri, S., Laokul, P., Klinkaewnarong, J. et al. (2008). Indium oxide (In2O3) nanoparticles using Aloe vera plant extract: synthesis and optical properties. *Journal of Optoelectronics and Advanced Materials* 10 (3): 161–165.

Makarov, V.V., Makarova, S.S., Love, A.J. et al. (2014). Biosynthesis of stable iron oxide nanoparticles in aqueous extracts of *Hordeum vulgare* and *Rumex acetosa* plants. *Langmuir* 30 (20): 5982–5988.

Mallikarjuna, K., Sushma, N.J., Reddy, B.S. et al. (2013). Palladium nanoparticles: single-step plant-mediated green chemical procedure using *Piper betle* leaves broth and their anti-fungal studies. *International Journal of Chemical and Analytical Science* 4 (1): 14–18.

Min, J.S., Kim, K.S., Kim, S.W. et al. (2009). Effects of colloidal silver nanoparticles on sclerotium-forming phytopathogenic fungi. *The Plant Pathology Journal* 25 (4): 376–380.

Murphy, C.J., Gole, A.M., Hunyadi, S.E. et al. (2008). Chemical sensing and imaging with metallic nanorods. *Chemical Communications* 5: 544–557.

Nagajyothi, P.C., An, T.M., Sreekanth, T.V.M. et al. (2013). Green route biosynthesis: characterization and catalytic activity of ZnO nanoparticles. *Materials Letters* 108: 160–163.

Nair, B. and Pradeep, T. (2002). Coalescence of nanoclusters and formation of submicron crystallites assisted by *Lactobacillus strains*. *Crystal Growth & Design* 2 (4): 293–298.

Narayanan, K.B. and Sakthivel, N. (2011). Extracellular synthesis of silver nanoparticles using the leaf extract of *Coleus amboinicus* Lour. *Materials Research Bulletin* 46 (10): 1708–1713.

Nagaraj, B., Malakar, B., Divya, T. et al. (2012a). Synthesis of plant mediated gold nanoparticles using flower extracts of *Carthamus tinctorius* L. (safflower) and evaluation of their biological activities. *Digest Journal of Nanomaterials and Biostructures* 7: 1289–1296.

Nagaraj, B., Divya, T., Malakar, B. et al. (2012b). Phytosynthesis of gold nanoparticles using *Caesalpinia pulcherrima* (Peacock flower) flower extract And evaluation of their antimicrobial activities. *Digest Journal of Nanomaterials and Biostructures* 7: 899–905.

Njagi, E.C., Huang, H., Stafford, L. et al. (2011). Biosynthesis of iron and silver nanoparticles at room temperature using aqueous sorghum bran extracts. *Langmuir* 27 (1): 264–271.

Ochieng, P.E., Iwuoha, E., Michira, I. et al. (2015). Green route synthesis and characterization of ZnO nanoparticles using *Spathodea campanulata*. *International Journal of Biochemistry & Physiology* 23: 53–61.

Öztürk, B.Y., Gürsu, B.Y., and Dağ, İ. (2020). Antibiofilm and antimicrobial activities of green synthesized silver nanoparticles using marine red algae Gelidium corneum. *Process Biochemistry* 89: 208–219.

Paciotti, G.F., Myer, L., Weinreich, D. et al. (2004). Colloidal gold: a novel nanoparticle vector for tumor directed drug delivery. *Drug Delivery* 11 (3): 169–183.

Ponmurugan, P., Manjukarunambika, K., Elango, V., and Gnanamangai, B.M. (2016). Antifungal activity of biosynthesised copper nanoparticles evaluated against red root-rot disease in tea plants. *Journal of Experimental Nanoscience* 11 (13): 1019–1031.

Popescu, M., Velea, A., and Lőrinczi, A. (2010). Biogenic production of nanoparticles. *Digest Journal of Nanomaterials & Biostructures (DJNB)* 5 (4): 1035–1040.

Prasad, R., Pandey, R., and Barman, I. (2016). Engineering tailored nanoparticles with microbes: quo vadis? *Wiley Interdisciplinary Reviews: Nanomedicine and Nanobiotechnology* 8 (2): 316–330.

Pugazhenthiran, N., Anandan, S., Kathiravan, G. et al. (2009). Microbial synthesis of silver nanoparticles by Bacillus sp. *Journal of Nanoparticle Research* 11 (7): 1811–1815.

Qian, Y., Yao, J., Russel, M. et al. (2015). Characterization of green synthesized nano-formulation (ZnO–A. vera) and their antibacterial activity against pathogens. *Environmental Toxicology and Pharmacology* 39 (2): 736–746.

Rai, M., Kon, K., Ingle, A. et al. (2014). Broad-spectrum bioactivities of silver nanoparticles: the emerging trends and future prospects. *Applied Microbiology and Biotechnology* 98 (5): 1951–1961.

Rajiv, P., Rajeshwari, S., and Venckatesh, R. (2013). Bio-Fabrication of zinc oxide nanoparticles using leaf extract of *Parthenium hysterophorus* L. and its size-dependent antifungal activity against plant fungal pathogens. *Spectrochimica Acta Part A: Molecular and Biomolecular Spectroscopy* 112: 384–387.

Ramesh, P., Rajendran, A., and Meenakshisundaram, M. (2014). Green syntheis of zinc oxide nanoparticles using flower extract cassia auriculata. *Journal of Nanoscience and Nanotechnology* 2 (1): 41–45.

Ramesh, M., Anbuvannan, M., and Viruthagiri, G.J.S.A.P.A.M. (2015). Green synthesis of ZnO nanoparticles using *Solanum nigrum* leaf extract and their antibacterial activity. *Spectrochimica Acta Part A: Molecular and Biomolecular Spectroscopy* 136: 864–870.

Ravichandrika, K., Kiranmayi, P., and Ravikumar, R.V.S.S.N. (2012). Synthesis, characterization and antibacterial activity of ZnO nanoparticles. *International Journal of Pharmacy Sciences* 4 (4): 336–338.

Rodríguez-León, E., Iñiguez-Palomares, R., Navarro, R.E. et al. (2013). Synthesis of silver nanoparticles using reducing agents obtained from natural sources (*Rumex hymenosepalus* extracts). *Nanoscale Research Letters* 8: 318.

Roopan, S.M., Bharathi, A., Kumar, R. et al. (2012). Acaricidal, insecticidal, and larvicidal efficacy of aqueous extract of *Annona squamosa* L peel as biomaterial for the reduction of palladium salts into nanoparticles. *Colloids and Surfaces B: Biointerfaces* 92: 209–212.

Salam, H.A., Sivaraj, R., and Venckatesh, R. (2014). Green synthesis and characterization of zinc oxide nanoparticles from *Ocimum basilicum* L. var. purpurascens Benth.-Lamiaceae leaf extract. *Materials Letters* 131: 16–18.

Sangeetha, G., Rajeshwari, S., and Venckatesh, R. (2011). Green synthesis of zinc oxide nanoparticles by aloe barbadensis miller leaf extract: structure and optical properties. *Materials Research Bulletin* 46 (12): 2560–2566.

Schlüter, S., Sheppard, A., Brown, K., and Wildenschild, D. (2014). Image processing of multiphase images obtained via X-ray microtomography: a review. *Water Resources Research* 50 (4): 3615–3639.

Shankar, S.S., Ahmad, A., Pasricha, R., and Sastry, M. (2003). Bioreduction of chloroaurate ions by geranium leaves and its endophytic fungus yields gold nanoparticles of different shapes. *Journal of Materials Chemistry* 13 (7): 1822–1826.

Sheny, D.S., Philip, D., and Mathew, J. (2013). Synthesis of platinum nanoparticles using dried *Anacardium occidentale* leaf and its catalytic and thermal applications. *Spectrochimica Acta Part A: Molecular and Biomolecular Spectroscopy* 114: 267–271.

Singh, R.P., Shukla, V.K., Yadav, R.S. et al. (2011). Biological approach of zinc oxide nanoparticles formation and its characterization. *Advanced Materials Letters* 2 (4): 313–317.

Song, J.Y., Kwon, E.Y., and Kim, B.S. (2010). Biological synthesis of platinum nanoparticles using Diopyros kaki leaf extract. *Bioprocess and Biosystems Engineering* 33 (1): 159–164.

Soundarrajan, C., Sankari, A., Dhandapani, P. et al. (2012). Rapid biological synthesis of platinum nanoparticles using *Ocimum sanctum* for water electrolysis applications. *Bioprocess and Biosystems Engineering* 35 (5): 827–833.

Stoimenov, P.K., Klinger, R.L., Marchin, G.L., and Klabunde, K.J. (2002). Metal oxide nanoparticles as bactericidal agents. *Langmuir* 18 (17): 6679–6686.

Sun, Y.P., Atorngitjawat, P., and Meziani, M.J. (2001). Preparation of silver nanoparticles via rapid expansion of water in carbon dioxide microemulsion into reductant solution. *Langmuir* 17 (19): 5707–5710.

Sun, Y., Yin, Y., Mayers, B.T. et al. (2002). Uniform silver nanowires synthesis by reducing $AgNO_3$ with ethylene glycol in the presence of seeds and poly (vinyl pyrrolidone). *Chemistry of Materials* 14 (11): 4736–4745.

Sundararajan, B. and Kumari, B.R. (2017). Novel synthesis of gold nanoparticles using *Artemisia vulgaris* L. leaf extract and their efficacy of larvicidal activity against dengue fever vector *Aedes aegypti* L. *Journal of Trace Elements in Medicine and Biology* 43: 187–196.

Taleb, A., Petit, C., and Pileni, M.P. (1997). Synthesis of highly monodisperse silver nanoparticles from AOT reverse micelles: a way to 2D and 3D self-organization. *Chemistry of Materials* 9 (4): 950–959.

Thanighaiarassu, R.R., Nambikkairaj, B., Devika, R. et al. (2014). Green synthesis of gold nanoparticles mediated by plant essential oil and its antifungal activity against human pathogenic fungi. *World Journal of Pharmacy and Pharmaceutical Sciences (WJPPS)* 3 (12): 1752–1768.

Thema, F.T., Manikandan, E., Dhlamini, M.S., and Maaza, M.J.M.L. (2015). Green synthesis of ZnO nanoparticles via *Agathosma betulina* natural extract. *Materials Letters* 161: 124–127.

Thombre, R., Parekh, F., Lekshminarayanan, P., and Francis, G. (2012). Studies on antibacterial and antifungal activity of silver nanoparticles synthesized using Artocarpus heterophyllus leaf extract. *Biotechnology, Bioinformatics and Bioengineering* 2 (1): 632–637.

Upadhyay, S.K. and Singh, S.P. (eds.) (2021). *Bioprospecting of Plant Biodiversity for Industrial Molecules*. John Wiley & Sons Ltd. https://doi.org/10.1002/9781119718017.

Valodkar, M., Nagar, P.S., Jadeja, R.N. et al. (2011). *Euphorbiaceae* latex induced green synthesis of non-cytotoxic metallic nanoparticle solutions: a rational approach to antimicrobial applications. *Colloids and Surfaces A: Physicochemical and Engineering Aspects* 384 (1–3): 337–344.

Vaseem, M., Umar, A., and Hahn, Y.B. (2010). ZnO nanoparticles: growth, properties, and applications. *Metal Oxide Nanostructures and their Applications* 5 (1): 10–20.

Velayutham, K., Rahuman, A.A., Rajakumar, G. et al. (2012). Evaluation of *Catharanthus roseus leaf extract*-mediated biosynthesis of titanium dioxide nanoparticles against Hippobosca maculata and *Bovicola ovis*. *Parasitology Research* 111 (6): 2329–2337.

Verma, A., Gautam, S.P., Bansal, K.K. et al. (2019). Green nanotechnology: advancement in phytoformulation research. *Medicine* 6 (1): 39.

Yassin, M.A., Elgorban, A.M., El-Samawaty, A.M.A. et al. (2016). An eco-friendly approach for fabrication of silver nanoparticles using *Aspergillus clavatus*. *Fresenius Environmental Bulletin* 25: 5929–5934.

Yuvakkumar, R., Suresh, J., Saravanakumar, B. et al. (2015). Rambutan peels promoted biomimetic synthesis of bioinspired zinc oxide nanochains for biomedical applications. *Spectrochimica Acta Part A: Molecular and Biomolecular Spectroscopy* 137: 250–258.
Zawrah, M.F., El-Moez, S.A., and Center, D. (2011). Antimicrobial activities of gold nanoparticles against major foodborne pathogens. *Life Sciences Journal* 8 (4): 37–44.

17

Plant-Based/Herbal Nanobiocatalysts and Their Applications

Rajeswaree Gohel[1], Dhara Gandhi[2], and Gaurav Sanghvi[1]

[1] *Department of Microbiology, Marwadi University, Rajkot, Gujarat, India*
[2] *Department of Botany, The Maharaja Sayajirao University of Baroda, Vadodara, Gujarat, India*

17.1 Introduction of Nanobiocatalyst

Biotechnology and nanotechnology are two major technologies that are worked in large-scale productions in the industries, as well as in the field of life science. There is one process that represents the fusion of these two technologies that is immobilization. In this process, two different molecules are belonging to the two different fields of science, for example, enzymes/hormones + nanoparticles. In which, enzymes are biochemicals that work in the living body or living condition as a catalyst. Biocatalysts are enzymes that are used as a catalyst in the living system. The other part is nanoparticles with size ranging between 1 and 100 nm. These types of metal nanoparticles are immobilized with enzymes to increase large-scale production. The fusion products of nanoparticles and enzymes are known as nanobiocatalysts. Nanomaterials are incorporated into the enzyme or protein matrix and form nanobiocatalyst that are being used widely in this century. Biocatalysts are enzymes that are immobilized with the nanoparticles and make a more effective catalysis process, so the rate of reaction will increase and generate more products with high-quality assurance.

The process of biocatalysis was used in ancient Mesopotamia, Japan, and China for the manufacturing of food products and alcoholic beverages.

Around 1950, some groups worked on the immobilization process. Many different plants are producing biocatalysts, but the best plants for biocatalysts production are alkaloid-containing plants, which are more efficient in producing biocatalysts. Generally, many plants are producing many kinds of secondary metabolites; among them, one was alkaloids. Apart from that, it also contains some important enzymes that are used as biocatalysts nowadays. Alkaloid-producing plants are mostly those plants that are widely used in the pharmaceutical industries and herbal medicines.

Herbal alkaloid-producing plants have many different metabolic pathways like

1) Alkaloids through the Shikimate pathway (Figure 17.1)
2) Alkaloid through tyrosine and lysine pathway

Plants as Bioreactors for Industrial Molecules, First Edition.
Edited by Santosh Kumar Upadhyay and Sudhir P. Singh.

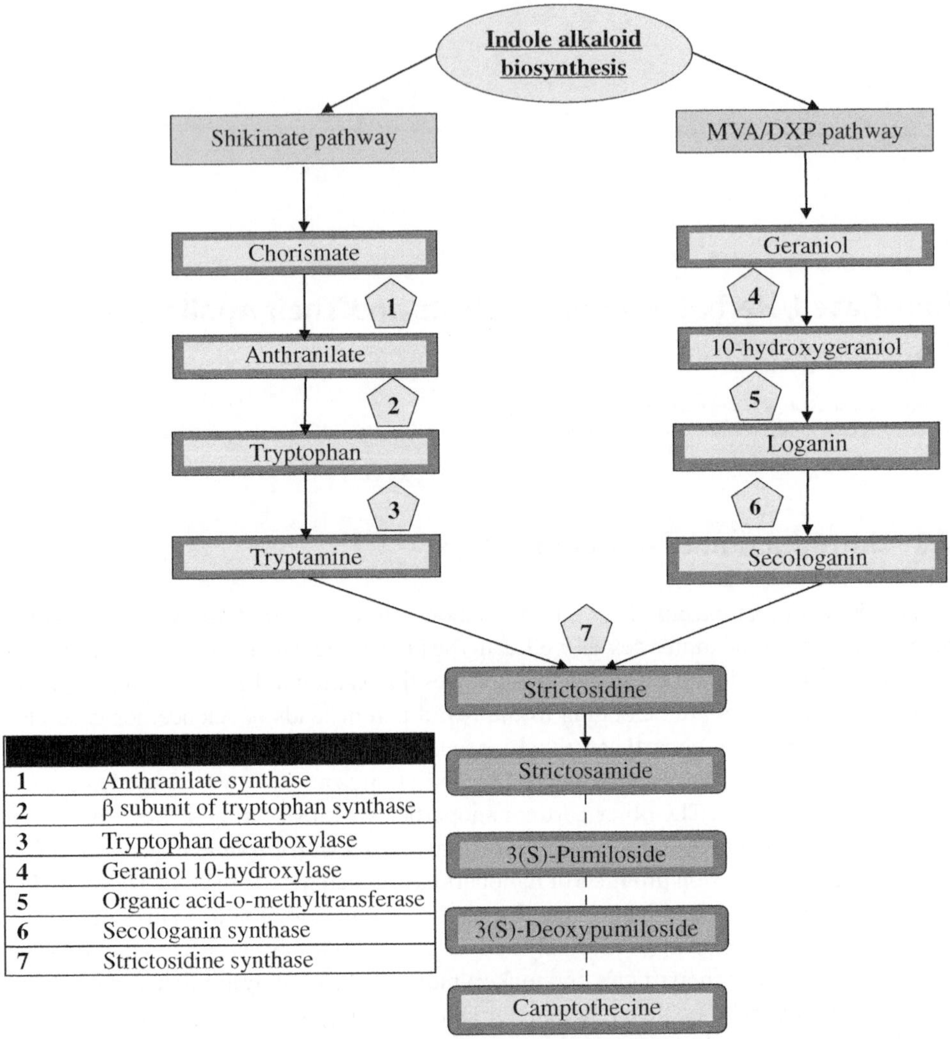

Figure 17.1 Alkaloid biosynthesis pathway.

In this pathway, Shikimate pathway and MVA/DXP pathway are merged. Camptothecin, a quinolone alkaloid, produced in this pathway as an end product, is used as a chemotherapeutic agent in the treatment of leukemia.

17.2 Nanobiocatalysts from Herbal Alkaloid Plants Are Used in Nanotechnology and Bioengineering

- The vanillin synthase from vanilla pods are one of the most important enzymes for the production of vanillin. It is one of the most important flavoring molecules for therapeutic drug discovery purpose. Many nanoparticles and nanocarriers are immobilized with vanillin and produce an important therapeutic drug.

- There are two important plants Madagascar periwinkle (*Catharanthus roseus*) and opium poppy (*Papaver somniferum*), which are mainly used in synthetic biology and nanobiotechnology, their biocatalyst is used in the immobilization process. There is an increase in antioxidants and alkaloids in *C. roseus* using silver nanoparticles on the STR gene (Keshavarzi et al. 2018).
- Berberine bridge enzyme (BBE) produces berberine that is also immobilized with the nanoparticles and binds with nanomatrix.

  There are other plants metabolites that are used as biocatalyst, for example:
  - Bitter gourd peroxidase from bitter gourd
  - Horseradish peroxidase from horseradish plant roots
  - All the dehydrogenases that are present in the herbaceous plants
  - Biocatalyst from Maize plant
- Alcohol dehydrogenase is widely used in nanobiotechnology
- Acetylcholinesterase is one of the most important enzymes for nervous tissue, and it is also immobilized with nanocarriers.

17.3 Why Use Nanobiocatalysts?

There is a biocatalyst that is effective in comparison to the chemical catalyst. Nanobiocatalyst is the advancement of catalysts. There is a limitation in the catalyst and biocatalyst, while nanobiocatalyst is the modified immobilized version of catalysts. It has many different qualities and applications and because of that it will survive in harsh conditions also. Increasing development, increasing pollution, and adverse conditions are the effects of biocatalysts, but it can affect the nanobiocatalyst because of its modifications in the binding site and also on the catalytic site. Some qualities of nanobiocatalysts are

- It has high thermal and chemical stability.
- It has a wide range of pH changing ability and stability at that particular pH.
- It has high catalytic activity.
- The reusability of this catalyst is also high.
- The cost of production is low.
- The application ratio is high compared to original catalysts.
- Survival ratio for a long time is higher than a catalyst.
- Hybrid with single and multienzymes.
- Work very efficiently in any techniques.
- Immobilization methods are applied effectively on nanobiocatalyst.

17.4 Immobilization of Biocatalyst (Enzymes) and Nanoparticles or Nanomatrix

The nanobiocatalyst term itself suggests that biocatalyst is incorporated into the nanoparticles that enhance the activity of biocatalyst and because of that the rate of reaction ultimately increases. Immobilization techniques are generally used to increase enzyme activity and reduce economical costs. There are many different immobilization methods, out of that major methods are adsorption, entrapment, covalent binding, encapsulation, and cross linking explained in Table 17.1.

Table 17.1 Immobilization methods with its advantages and disadvantages.

Name of method	Introduction of method	Advantages	Disadvantages
Adsorption Immobilization	This method is one of the old methods. In this method, the enzyme adsorbs on an insoluble matrix with the help of an adsorbent. Many forces that work in this method are van der Waal forces, ionic, and hydrogen bonding as well as hydrophobic interactions, which are very weak forces, but in large numbers, impart sufficient binding strength.	Its simpler compared to another method with a low cost and reversible approach There is no damage to the biocatalyst Covering or couple with enzyme not required. kcat and km values remain substantially unchanged The higher activity of catalysis because of immobilization	Feedback inhibition occurred and reversible interaction. Bioleaching and desorption are high.Enzyme activity is lost after some time.Control lost over packing density of enzyme.
Covalent-bonding	In this method, the enzyme binds with the support matrix by covalent linkage. The active site of the enzyme is unaffected by this bonding and able to perform its catalytic activity. It is one of the strongest interactions of enzymes and nanomatrix.	It is stable binding with strong bondage activity.Bioleaching of enzyme prevention by this method.Increase the stability of thermal activity.	Deactivation of enzymes occurred frequently. The affinity of substrate and enzyme decrease with time. Conformational change of catalytic activity restricted.
Entrapment	It is the method in which enzyme movement is restricted by the porus gel or nanofibers. It is only convenient when low molecular substrate and products are there.	Enzyme protected by the effect of mechanical sheer effect, hydrophobic solvents, and gas bubbles.Continuous suitability of downstream processing Retain protein integrity and efficacy.	Enzyme loading activity is low Transfer of mass limitation occurred.
Cross-linking	In this method, biocatalysts are bound to each other with bi- or multifunctional reagent ligands. And it also helps in the adsorption process.	Matrix of support is not required and stability will be also increased and stable Decrease in desorption Ease of recycling and reuse.	Conformational change happened because of that loss of enzyme catalytic activity. The diffusion rate is also decreased.

Source: Ahmad et al. (2013), Nath et al. (2014), Ahmad and Sardar (2015).

17.5 Application of the Nanobiocatalyst

Biocatalyst can work in a biological system as a catalyst. Many factors that affect the activity of biocatalyst are such as pH, temperature, osmotic pressure, and other physical and chemical conditions. So, after many years of invention researchers were find that if nanoparticles bind to the biocatalyst, then the activity of biocatalyst will be increased. There are many types of nanomaterials like polymers, chitosan, magnetic nanoparticles, silica, zirconia, graphene, hybrid of organic-inorganic nanoflower, and zinc oxide.

17.5.1 Application of Enzyme Immobilized on Graphene-Based Nanomaterial

Graphene is a two-dimensional carbon-based nanomaterial that provides a platform for enzymes to bind with it. It contains high stability, high electronic conductivity, and high biocompatibility. It provides a matrix to the enzymes for binding.

There are many applications and qualities present in the two-dimensional structural matrix. Functionalized graphene-based nanomaterial (GBNs) are changed and manipulate the microenvironment and activity of enzymes. Facilitating the regulation of its activity or specifying the activity, the improvement of transfer in electrons in proteins, the increase of enzymes' operational, chemical and thermal stability, and the protection from proteolytic cleavage.

The development of conjugates between enzymes and GBNs (Garg et al. 2015; Mandotra et al. 2018; Fatima et al. 2020) can be applied on preparation of biosensor, enzyme-based biofuel cell, biocatalytic transformation, degradation of pollution, degradation of agroindustrial waste, wastewater treatment, proteomic analysis, and development of microchip bioreactor.

17.5.2 Enzyme-Based Biosensor

Biosensors have measured the change in electrochemical signaling and detected the target molecules. The graphene-based matrix contains detector molecular that detect its target by binding with it (Figure 17.2).

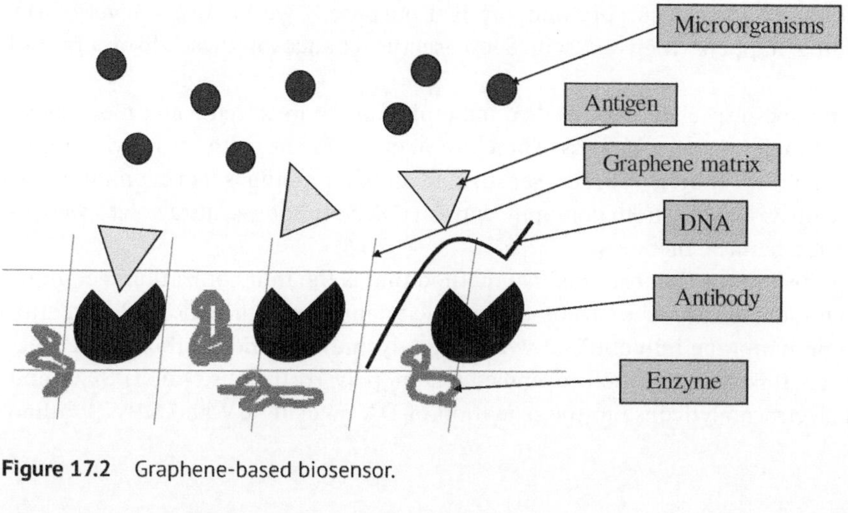

Figure 17.2 Graphene-based biosensor.

Some antibodies, enzyme-conjugated matrix, and some nucleotide-like DNA are embedded in the graphene. Antibodies detected the antigens present in the medium with the help of the matrix and detected the result. Many enzymes are embedded in the graphene matrix that worked as a biosensor. The three major plant-based enzymes that are isolated from herbal plants are horseradish peroxidase (found in horseradish plant roots), alcohol dehydrogenase (found in herbaceous and woody plants, alkaloid plants), and acetylcholinesterase (found in maize or zea maize).

17.5.2.1 Horseradish Peroxidase Immobilized with the Graphene Oxide (GO)

Horseradish peroxidase when binds with GO its catalytic activity will ultimately increase. HRP and GO immobilized form that shows more thermal stability, a wide range of pH, and removal of phenolic compounds from aqueous solutions and industrial wastewater. There are shreds of evidence and experimental proof that this shows great activity under the immobilized condition. GO-immobilized HRP has great efficiency of phenolic compound removal than soluble HRP especially 2-chlorophenol and 2,4-dimethoxyphenol.

The immobilization was carried out between GO and HRP in phosphate buffer under the pH of 7. GO contains various oxide groups on its surface so the HRP loading efficiency is high and direct. It is the electrostatic interaction between the GO and HRP.

Electrostatic interaction between GO and HRP is stated and that GO has an overall negative charge, while it binds with positively charged residues that are present on the surface of the HRP. HRP did not contain more than 10 positively charged residues, but the biggest advantage is that the substrate-binding site of HRP was free so because of that it is easy for GO to binds with HRP.

17.5.2.2 HRP Biosensor Towards the Detection of Dopamine

Dopamine (DA) is one of the most important hormones and catecholamine neurotransmitters in mammalian brain tissue. The level of DA is low then it leads to many neurodegenerative disorders like Parkinson's and Alzheimer's. DA is also widely applied in the treatment of circulatory collapse syndrome caused by myocardial infarction trauma renal failure, cardiac surgery, or congestive cardiac failure so the detection of the DA will increase the chance of surgery and survival purpose if we knew the level of DA before anything happens then we will decrease the chance of these disorders and conditions.

Carbon nanotubes especially multiwalled nanotubes are more specific and more sensitive to work in the biosensing process. There are many different methods in which they detect the level of DA, but it is the most sensitive in all other methods like chemiluminescence, HPLC, ion exchange chromatography, capillary electrophoresis, ultraviolet spectroscopy, and electrochemical methods.

Other than this, there is a most sensitive method that is the immobilization technique of silica gel entrapment with the carbon multiwalled nanotube with HRP. HRP and multiwalled carbon nanotube immobilized with poly (glycine) modified carbon paste electrode and then this whole structure embedded in poly (orthosilisic acid)/SiSG and generate an electrocatalyticonse to the oxidation of DA by using CV and DPV (Prathna et al. 2011).

17.5.2.3 HRP – Inorganic Hybrid Nanoflower

HRP and copper phosphate immobilized with each other in an aqueous solution. It is a flower-like spherical structure that has hundreds of petals in a structure (Figure 17.3). The catalytic activity of HRP shows that o-phenylenediamine (OPD) was oxidized in the presence of hydrogen peroxide (H_2O_2), and this reaction evaluated the whole enzymatic activity of HRP hybrid nanoflower.

The hybrid nanoflower of HRP is 506% more efficient than free HRP. Because of visual detection of phenols and H_2O_2 in the range of 0.5 and 1 μM minimum, we can observe the color change and applied in the medical treatment and natural water. This proposed method detects the low amount of H_2O_2 in spiked human serum and phenols. The recoveries of this method are great and because of that there is 93% of excellent reusability and reproducibility in hybrid nanoflower of HRP in cyclic analysis (Lin et al. 2014; Li et al. 2016).

17.5.3 Bitter Gourd Peroxidase Immobilized with TiO_2 Nanoparticles

Bitter gourd peroxidase and TiO_2 particles are immobilized by covalent bonding, and this bioactive monoconjugate increases the catalytic efficiency; this monoconjugate will add into the aqueous solution that contains dyes and phenol. It is useful in wastewater treatment where phenol and dyes are pollutants.

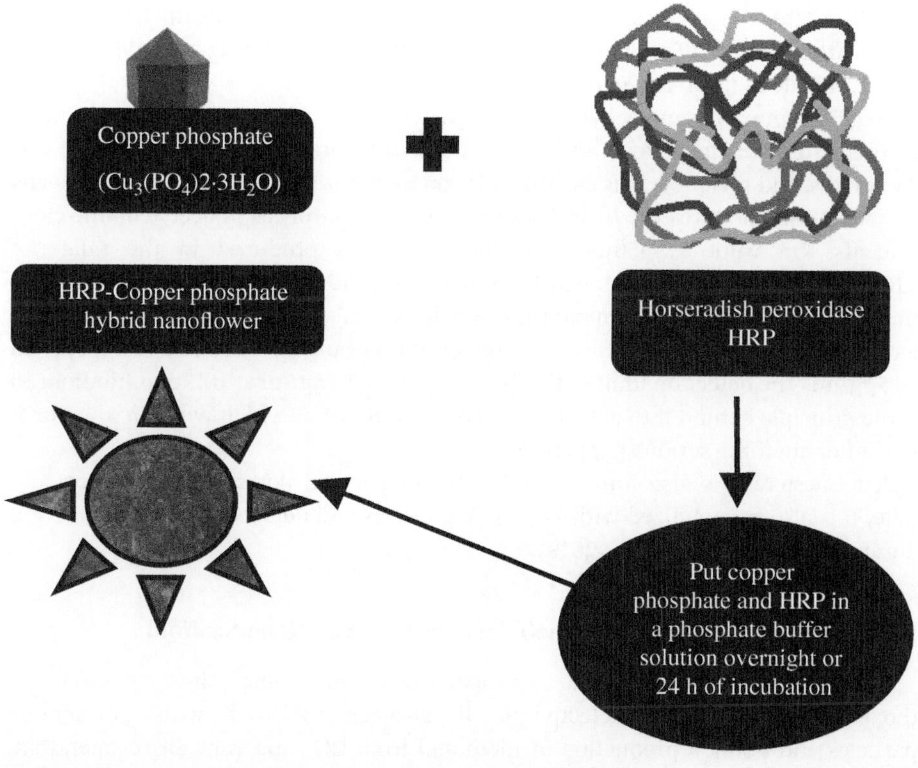

Figure 17.3 Diagramatic representation of the hybrid nanoflower.

Bitter gourd peroxidase with TiO_2 particle modified on the APTS. TiO_2 particles first bind with aminopropyletriethoxysilan (APTS). TiO_2 particles work as the photocatalyst; these particles absorb UV rays from sunlight or illuminated light source. Modified TiO_2 is widely used for environmental purposes. Modified TiO_2 has high efficiency compared to the original form. It has high photoactivity, low cost, and high stability.

When both the enzymes are compared, then it is observed that the free peroxidase enzyme shows less activity than the conjugate enzyme, which is 82–84%. The free enzyme has less Vmax/Km (100), while the bioconjugate has a higher value range of Vmax/Km (165–167). It has higher thermostability at 60–65 °C in comparison with its soluble counterpart. Removal of dyes and phenols from aqueous solution is also high in bioconjugate compared to the free enzyme, its reusability is also high, and enzyme activity will not be lost.

This TiO_2 and peroxidase conjugate greatly work in the removal of the dyes and phenolic compounds. The residual activity of this conjugate is 100%, and the catalytic activity is very high in comparison to the free enzyme. The other applications of peroxidase in the biofuel cell and in the oxidization of cyclic or aromatic compounds that are hazardous for the environment.

17.5.4 Immobilization of Acetylcholinesterase on Gold Nanoparticles Embedded in Sol–Gel Nanomatrix

Organophosphorus (OP) is an important pest control or insecticide, but it produces toxic compounds. It inhibits the activity of acetylcholinesterase, which is the most important enzyme of the central nervous system and causes nerve paralysis and death. So to control and measure the level of OP, enzyme acetylcholinesterase immobilized with gold nanoparticles was embedded on sol–gel matrix.

The sol–gel structure has many hydroxyl groups, which provide stability and biocompatible microenvironment to perform the catalytic activity. Catalyzation of thiocholine was done by AuNPs, and this process is electrooxidation that increases the activity as well. The immobilized AchE catalyzes the hydrolysis process. The hydrolysis of acetylcholine chloride and its Km value is 450 μM and then signal is produced in the range of 10.05–1000.0 μM. This signal generated by thiocholine-oxidized process. Monocrotophos was an ideal compound for optimizing the whole condition. Its inhibition is based on monocrotophos and is proportional to the concentration range in between 0.001–1 μg/ml and 2–15 μg/ml. The detection limit of this biosensor was 0.6 ng/ml at 10% of inhibition. So this whole principle behind it to detect the good amount of OP, which will cure the previous effect after anything serious happened.

Acetylcholinesterase is also immobilized with sol–gel and detect the Malathion and Acephate. It is also immobilized with ZnO nanoparticles and detects the chlorpyrifos and carbofuran compounds (Dobrucka 2019).

17.5.5 Alcohol Dehydrogenase Immobilized with Carbon Nano Scaffold

ADH immobilized with magnetic nanoparticles and doing the reduction of 7-methoxy-2-tetralone (enantioselectivity). It also immobilized with polystyrene nanoparticles and doing a production of methanol from CO_2 and cofactor regeneration.

Other dehydrogenases are also immobilized with polystyrene nanoparticles like glutamate dehydrogenase, formate dehydrogenase, and formaldehyde dehydrogenase.

ADH is immobilized with a carbon nanoscaffold rather than more amount of GO because this nanoscaffold increases the ability of electron transfer in ADH. It enhances the ability of electron transfer. For that purpose, it has been using poly (2-(trimethylamine) ethyl methacrylate) (MADQUAT) cationic polymer and carbon nanoscaffold. The carbon nanoscaffold was composed of a single-walled nanotube with a reduced amount of GO. ADH was entrapped in MADQUAT that was present on the nanoscaffold. It generated a high electron exchangeability with the cofactor of NADH. This nicotinamide dinucleotide reduces in disodium salt hydrate redox reaction. This method is demonstrated by the CV (cyclic volumetry) method and EIS (electrochemical impedance spectroscopy). It generates best bioanode, which is very useful for the transfer of an electron. It is very fruitful for ethanol powder biofuel cells (Touqeer et al. 2020).

17.5.6 Vanillin or Vanillin Synthase is Used as a Therapeutic Drug by Immobilizing with Nanoparticles

Vanillin or vanillin synthase encapsulated with the nanoparticles or capping of nanoparticles on it. It is used to target the particular site of action and that there are many uses of vanillin in which it sustains prolonged time because of nanoparticles. Here is the bioactive profile of vanillin.

Vanillin is used previously just for fragrance or flavored purposes. It is derived from pods of *Vanilla planifolia* orchid. This vanillin is produced by the catalytic activity of vanillin synthase enzyme that acts as a nanobiocatalyst. Vanillin is used in many desired conditions. Here are some conditions in which vanillin as a bioactive molecule used in a wide range of therapeutic products like neuroprotective, nephroprotective, cardioprotective, antiviral, sickle cell anemia, antioxidant, prevent insulin glycation, wound healing, antimicrobial, antibiotic potentiation, anticancerous, and anti-inflammatory. These are all beneficial uses of vanillin and its immobilization with nanocarriers and nanoparticles.

17.5.7 STR Gene Regulation with the Help of Silver Nanoparticles

STR is one of the most important enzymes to produce indole alkaloids. As we could observe in the biosynthesis pathway of alkaloids strictosidine synthase is important to produce secondary metabolites. This enzyme is regulated by the STR gene which is targeted by our nanoparticles, and it will increase the production of all alkaloids. These enzyme researchers are derived from *C. roseus*.

According to the global market, agriculture is one of the most important things nowadays and with the time crop yielding capacity, we must increase and because of that, we have to control the genes and regulate it. There is also another purpose that is to reduce total waste. Alkaloids have numerous numbers of secondary metabolites, which are important in cellular functions and biochemical activity for essential physiological conditions. And for that particular purpose silver nanoparticles are boosting the alkaloids accumulation and collection in the *C. roseus* plant. Silver nanoparticles are boosting the *C. roseus* mitogen-activated protein 3 (CrMPK3) and STR gene. It is triggered by the map kinase signaling pathway. There are five different concentrations of AgNPs that are utilized and double distilled water is used as a control. Results show dual positive relation in different concentrations between AgNPs. The stress of superoxide dismutase and sulfur oxidative compounds and ascorbate peroxidase indicated that increased amount of malondialdehyde contents and hydrogen peroxide, and it is also indicated that regulation in the expression of CrMPK3 gene and biosynthesis of alkaloid gene which is STR and collection of alkaloid or accumulation of alkaloids. So it will be indicated that silver nanoparticles trigger the signaling pathway and increase production (Fatima 2021).

17.5.8 Effect of Titanium Dioxide Nanoparticles and Different Enzymes of Alkaloid Plants Conjugate on the Bioengineering Pathway

Enzymes like BBE, cheilanthifoline synthase (CFS), stylopine synthase (STS), tetrahydro protoberberine cis-N-methyltransferase (TNMT), protopine 6-hydroxylase (P6H), S-cis-N-methylstylopine 14-hydroxylase (MSH), and dihydro benzophenanthridine oxidase (DBOX) play a major role in different bioengineering pathways.

The sanguinarine plant is one of the most used herbal remedies. Sanguinarine is also extracted from the bloodroot plant and used as an herbal remedy. Sanguinaria is majorly used in the conditions like cardiovascular disease, asthma, Type 2 diabetes, blood pressure, cancer, and hypertension. These are conditions in which it increase its production. There are many methods, but the most effective method is to titanium dioxide nanoparticle immobilized with the 7 enzymes (which are mentioned earlier), which ultimately increase the production of sanguinarine. Titanium dioxide affects the celandine sanguinarine biosynthetic pathway genes that are included in three important factors: concentration of titanium dioxide, spraying time, and post induction time.

Analysis of variation in the expression of *STS*, *P6H*, *BBE*, and *MSH* shows significant differences in titanium dioxide concentrations, post-induction times, spraying period, different organs, and their interactions at $P \leq 0.05$, 0.01 levels of probability. The accumulation of *P6H* and *STS* transcripts in the roots was 2–3 times higher than the leaves and stems. Transcripts of the *BBE* and *MSH* genes were four times higher in the root than the stem. The concentration of 500 mg/l titanium dioxide increased the expression of *STS*, *P6H*, *BBE*, and *MSH* genes by 63, 66, 500, and 91%, respectively, relative to control treatment. 24 and 48 hours post-induction time is the maximum expression in all four studied genes in comparison to 72 hours post-induction time, and 500 mg/l titanium dioxide showed the highest expression for *STS*, *P6H*, and *BBE* genes, and 1000 mg/l for *MSH* gene. Result indicates that the titanium dioxide nanoparticles are effective in increasing the production of celandine sanguinarine through the regulation of its biosynthetic pathway genes.

17.5.9 Application of Plant Extract Biocatalyst Which is Useful to Make Different Nanoparticles and Used as a Remedy

See Table 17.2.

Table 17.2 Plant extracts biocatalyst applied to make nanoparticles and its application.

Plant extract	Nanoparticles	Application
Blackberry, blueberry, pomegranate, and turmeric extracts	Gold and silver NPs	Green synthesis technology
Wormwood aqueous extract	Silver NPs	Green synthesis technology
Tree oyster mushroom Pleurotusostreat4us	Silver NPs	It shows the inhibition against pathogens.
Sweet marjorum and sweet orange	Silver NPs	Antibacterial activity
Yellow fragrant bush rose petal extract	Gold NPs	Green synthesis technology
Flame lily leaf extract	Cerium oxide NPs	Antipathogenic bacterial properties
Indian chrysanthemum leaf extract	Silver NPs	Antibacterial and antifungal cytotoxic effects: an in outer study
Citrus limon (lemon) aqueous extract	Silver NPs	Biomimetic synthesis
Thymbra spicata extract	Silver NPs	Heterogeneous and recyclable nanocatalyst for catalytic reduction of a variety of dyes in water
Trigonellafoenum-graecum extract	TiO_2 NPs	For the medicinal purpose
Geranium leaf aqueous extract	Silver NPs	Oil-in-water microemulsion and nanoemulsion using geranium leaf aqueous extract as a reducing agent
Wild mussaenda or dhobi tree leaf extract	Gold and silver NPs	Environmental application and dye degradation
Joseph's coat leaf extract	Silver NPs	Antimicrobial activity
Hairy Spurge, Ara Tanah, Asthma Weed. leaf extract	Gold NPs	Green synthesis technology, antimicrobial activity
Trigonella foenum-graecum	Gold NPs	Green synthesis technology
Chinese Arborvitae twig and leaf extract	Gold NPs	Simulated solution of the plant-mediated biochemical mechanism
Fumariae herba extract	Platinum NPs	Biofabrication
Green tea (*Camellia sinensis*)	ZnO NPs	Antimicrobial activity
Chinese hibiscus	ZnO NPs	Green synthesis technology
Asian ginseng leaves	Gold and silver NPs	Green synthesis technology

(*Continued*)

Table 17.2 (Continued)

Plant extract	Nanoparticles	Application
Indian gooseberry or amla fruit extract	Gold and silver NPs	Transmetallation in an organic solution
Sesbania grandiflora	Gold NPs	Catalytic reduction of methylene blue
Triticum aestivum	Gold NPs	Green synthesis technology
Tamarindus indica L leaves	Gold NPs	Green synthesis technology
Lemongrass	Gold NPs	Boost the predation efficiency of copepod *Mesocyclops aspericornis* against malaria and dengue mosquitoes
Alfalfa plants (*Medicago Sativa*)	Gold NPs	Insight in phytomining

Source: Ankamwar et al. (2005), Chandran et al. (2006), Noruzi et al. (2011), Zhan et al. (2011), Aromal and Philip (2012), Annamalai et al. (2013), Park et al. (2013), Arokiyaraj et al. (2014), Das and Velusamy (2014), Devi and Gayathri (2014), Nadagouda et al. (2014), Pardha-Saradhi et al. (2014), Senthilkumar and Sivakumar (2014), Arumugam et al. (2015), Murugan et al. (2015), Ali et al. (2016), Correa et al. (2016), Singh et al. (2016), Singh and Rawat (2016), Al-Bahrani et al. (2017), Francis et al. (2017), Rivera-Rangel et al. (2018), Veisi et al. (2018), Reddy et al. (2019), Nisha Singh and Dhanya Madan (2020).

17.6 Conclusion

Many herbal plants are used everywhere either in green technology agriculture or in the health sector. In every system, biological processes are most important and in that biocatalyst play a major role. Biocatalyst has some limitations. To minimize the limitations, when biocatalyst reacts with nanoparticles, nanotubes, and other nanocarriers, then they increase the stability, reactivity, and catalytic activity. There are huge applications of nanobiocatalyst compared to chemical and biological catalysts because of that they are useful in every field like textile industries, wastewater treatment, detection of the toxic compound and another level of chemicals, biofuel production, biosensor mechanism, bioremediation, bioleaching, phytoremediation, antimicrobial activity, therapeutic drug production, and bioengineering of pathways. There are vast numbers of applications of nanobiocatalysts and many more.

References

Ahmad, R. and Sardar, M. (2015). Enzyme immobilization: an overview on nanoparticles as immobilization matrix. *Biochemistry and Analytical Biochemistry* 4 (2): 1.

Ahmad, R., Mishra, A., and Sardar, M. (2013). Peroxidase-TiO_2 nanobioconjugates for the removal of phenols and dyes from aqueous solutions. *Advanced Science, Engineering and Medicine* 5 (10): 1020–1025.

Al-Bahrani, R., Raman, J., Lakshmanan, H. et al. (2017). Green synthesis of silver nanoparticles using tree oyster mushroom *Pleurotus ostreatus* and its inhibitory activity against pathogenic bacteria. *Materials Letters* 186: 21–25.

Ali, M., Kim, B., Belfield, K.D. et al. (2016). Green synthesis and characterization of silver nanoparticles using *Artemisia absinthium* aqueous extract – a comprehensive study. *Materials Science and Engineering: C* 58: 359–365.

Ankamwar, B., Damle, C., Ahmad, A., and Sastry, M. (2005). Biosynthesis of gold and silver nanoparticles using *Emblica officinalis* fruit extract, their phase transfer, and transmetallation in an organic solution. *Journal of Nanoscience and Nanotechnology* 5 (10): 1665–1671.

Annamalai, A., Christina, V.L.P., Sudha, D. et al. (2013). Green synthesis, characterization, and antimicrobial activity of Au NPs using *Euphorbia hirta* L. leaf extract. *Colloids and Surfaces. B, Biointerfaces* 108: 60–65.

Arokiyaraj, S., Arasu, M.V., Vincent, S. et al. (2014). Rapid green synthesis of silver nanoparticles from *Chrysanthemum indicum* L and its antibacterial and cytotoxic effects: an in vitro study. *International Journal of Nanomedicine* 9: 379.

Aromal, S.A. and Philip, D. (2012). Green synthesis of gold nanoparticles using *Trigonella foenum*-graecum and its size-dependent catalytic activity. *Spectrochimica Acta Part A: Molecular and Biomolecular Spectroscopy* 97: 1–5.

Arumugam, A., Karthikeyan, C., Hameed, A.S.H. et al. (2015). Synthesis of cerium oxide nanoparticles using *Gloriosa superba* L. leaf extract and their structural, optical, and antibacterial properties. *Materials Science and Engineering: C* 49: 408–415.

Chandran, S.P., Chaudhary, M., Pasricha, R. et al. (2006). Synthesis of gold nano triangles and silver nanoparticles using *Aloe vera* plant extract. *Biotechnology Progress* 22 (2): 577–583.

Correa, S.N., Naranjo, A.M., and Herrera, A.P. (2016). Biosynthesis and characterization of gold nanoparticles using extracts of *Tamarindus indica* L leaves. *In Journal of Physics* 687 (1): 012082.

Das, J. and Velusamy, P. (2014). Catalytic reduction of methylene blue using biogenic gold nanoparticles from *Sesbania grandiflora* L. *Journal of the Taiwan Institute of Chemical Engineers* 45 (5): 2280–2285.

Devi, R.S. and Gayathri, R. (2014). Green synthesis of zinc oxide nanoparticles by using Hibiscus rosa-Sinensis. *International Journal of Current Engineering and Technology* 4 (4): 2444–2446.

Dobrucka, R. (2019). Biofabrication of platinum nanoparticles using Fumariae herbal extract and their catalytic properties. *Saudi Journal of Biological Sciences* 26 (1): 31–37.

Fatima Z (2021) Use of nanomaterials for the immobilization of industrially important enzymes. *Preprints,* 2021050329 doi: https://doi.org/10.20944/preprints202105.0329.v1.

Fatima, A., Mumtaz, M.W., Mukhtar, H. et al. (2020). Synthesis of lipase-immobilized CeO_2 nanorods as a heterogeneous nano-bio catalyst for optimized biodiesel production from *Eruca sativa* seed oil. *Catalysts* 10 (2): 231.

Francis, S., Joseph, S., Koshy, E.P., and Mathew, B. (2017). Green synthesis and characterization of gold and silver nanoparticles using Mussaenda glabrate leaf extract and their environmental applications to dye degradation. *Environmental Science and Pollution Research* 24 (21): 17347–17357.

Garg, B., Bisht, T., and Ling, Y.C. (2015). Graphene-based nanomaterials as efficient peroxidase mimetic catalysts for biosensing applications: an overview. *Molecules* 20 (8): 14155–14190.

Keshavarzi, M., Davoodi, D., Pourseyedi, S., and Taghizadeh, S. (2018). The effects of three types of alfalfa plants (*Medicago sativa*) on the biosynthesis of gold nanoparticles: an insight into phytomining. *Gold Bulletin* 51 (3): 99–110.

Li, P.W., Wang, G., Yang, Z.M. et al. (2016). Development of drug-loaded chitosan-vanillin nanoparticles and their cytotoxicity against HT-29 cells. *Drug Delivery* 23 (1): 30–35. https://doi.org/10.3109/10717544.2014.900590.

Lin, Z., Xiao, Y., Yin, Y. et al. (2014). Facile synthesis of enzyme-inorganic hybrid nanoflowers and its application as a colorimetric platform for visual detection of hydrogen peroxide and phenol. *ACS Applied Materials & Interfaces* 6 (13): 10775–10782.

Mandotra, S.K., Kumar, R., Upadhyay, S.K., and Ramteke, P.W. (2018). Nanotechnology: a new tool for biofuel production. In: *Green Nanotechnology for Biofuel Production* (ed. N. Srivastava, M. Srivastava, H. Pandey, P.K. Mishra, P.W. Ramteke). Biofuel and Biorefinery Technologies. Springer International. https://doi.org/10.1007/978-3-319-75052-1_2.

Murugan, K., Benelli, G., Panneerselvam, C. et al. (2015). Cymbopogon citratus-synthesized gold nanoparticles boost the predation efficiency of copepod *Mesocyclops aspericornis* against malaria and dengue mosquitoes. *Experimental Parasitology* 153: 129–138.

Nadagouda, M.N., Iyanna, N., Lalley, J. et al. (2014). Synthesis of silver and gold nanoparticles using antioxidants from blackberry, blueberry, pomegranate, and turmeric extracts. *ACS Sustainable Chemistry & Engineering* 2 (7): 1717–1723.

Nath, N., Mitchell, J.B.O., and Caetano-Anollés, G. (2014). The natural history of biocatalytic mechanisms. *PLoS Computational Biology* 10 (5): e1003642.

Nisha Singh, B.S. and Dhanya Madan, L.V. (2020). Nano-immobilized biocatalysts and their potential biotechnological applications in bioenergy production. *Materials Science for Energy Technologies* 3: 808–824, ISSN 2589-2991. doi: 2020.09.006.

Noruzi, M., Zare, D., Khoshnevisan, K., and Davoodi, D. (2011). Rapid green synthesis of gold nanoparticles using *Rosa hybrida* petal extract at room temperature. *Spectrochimica Acta Part A: Molecular and Biomolecular Spectroscopy* 79 (5): 1461–1465.

Pardha-Saradhi, P., Yamal, G., Peddisetty, T. et al. (2014). The root system of live plants is a powerful resource for the green synthesis of Au-nanoparticles. *RSC Advances* 4 (14): 7361–7367.

Park, J.M., Kim, M., Park, H.S. et al. (2013). Immobilization of lysozyme-CLEA onto electrospun chitosan nanofiber for effective antibacterial applications. *International Journal of Biological Macromolecules* 54: 37–43.

Prathna, T.C., Chandrasekaran, N., Raichur, A.M., and Mukherjee, A. (2011). Biomimetic synthesis of silver nanoparticles by Citrus limon (lemon) aqueous extract and theoretical prediction of particle size. *Colloids and Surfaces. B, Biointerfaces* 82 (1): 152–159.

Reddy, P.N.K., Shaik, D.P., Ganesh, V. et al. (2019). Structural, optical and electrochemical properties of TiO_2 nanoparticles synthesized using medicinal plant leaf extract. *Ceramics International* 45 (13): 16251–16260.

Rivera-Rangel, R.D., González-Muñoz, M.P., Avila-Rodriguez, M. et al. (2018). Green synthesis of silver nanoparticles in oil-in-water microemulsion and nano-emulsion using geranium leaf aqueous extract as a reducing agent. *Colloids and Surfaces A: Physicochemical and Engineering Aspects* 536: 60–67.

Senthilkumar, S.R. and Sivakumar, T. (2014). Green tea (*Camellia sinensis*) mediated the synthesis of zinc oxide (ZnO) nanoparticles and studies their antimicrobial activities. *International Journal of Pharmacy and Pharmaceutical Sciences* 6 (6): 461–465.

Singh, D. and Rawat, D. (2016). Microwave-assisted synthesis of silver nanoparticles from *Origanum majorana* and *Citrus sinensis* leaf and their antibacterial activity: a green chemistry approach. *Bioresources and Bioprocessing* 3 (1): 1–7.

Singh, P., Kim, Y.J., and Yang, D.C. (2016). A strategic approach for rapid synthesis of gold and silver nanoparticles by Panax ginseng leaves. *Artificial Cells, Nanomedicine, and Biotechnology* 44 (8): 1949–1957.

Touqeer, T., Mumtaz, M.W., Mukhtar, H. et al. (2020). Fe_3O_4-PDA-Lipase as surface functionalized nano biocatalyst for the production of biodiesel using waste cooking oil as feedstock: characterization and process optimization. *Energies* 13 (1): 177.

Veisi, H., Azizi, S., and Mohammadi, P. (2018). Green synthesis of the silver nanoparticles mediated by *Thymbra spicata* extract and its application as a heterogeneous and recyclable nanocatalyst for catalytic reduction of a variety of dyes in water. *Journal of Cleaner Production* 170: 1536–1543.

Zhan, G., Huang, J., Lin, L. et al. (2011). Synthesis of gold nanoparticles by Cacumen Platycladi leaf extract and its simulated solution: toward the plant-mediated biosynthetic mechanism. *Journal of Nanoparticle Research* 13 (10): 4957–4968.

18

Potential Plant Bioreactors

Karishma Seem and Simardeeep Kaur

Division of Biochemistry, ICAR-Indian Agricultural Research Institute, New Delhi, India

18.1 Introduction

Over the past few decades, there have been intensive efforts to produce secondary metabolites, under sustainable and controlled conditions for the pharmaceutical, food, and cosmetic industries. Expression hosts that are available for recombinant proteins include bacteria, yeast, filamentous fungi, unicellular algae, mammalian cells, and transgenic plants (Demain and Vaishnav 2009; McKenzie and Abbott 2018; Owczarek et al. 2019; Puetz and Wurm 2019; O'Flaherty et al. 2020; Shanmugaraj et al. 2020; Sharma and Upadhyay 2020; Shanmugaraj and Phoolcharoen 2021). Different expression systems have their own unique strengths and constraints, such as scalability, difficulty to express mammalian proteins, requirement of successful downstream technique, risks, regulatory approval, applicable and cost-effective platform. Hence, the need to overcome these hurdles has prompted research into alternatives. Plants have emerged as promising platform offering several potential benefits compared to more established platforms. This has led to the advancements in the establishment of transgenic plants as "BIOREACTORS," which is relatively new bioscience and is gaining momentum. Plant molecular farming involves the genetic modification of the host plant through the insertion and expression of new genes for production of useful recombinant proteins. The products that are currently being produced in plants include bioactive peptides, vaccine antigens, antibodies, diagnostic proteins, nutritional supplements, enzymes, industrial proteins/enzymes, and biopolymer (Upadhyay and Singh 2021). Plant production systems include microalgae, cells, hairy roots, moss, *in vitro* plant cell culture, and whole plants via both stable and transgenic expression (Figure 18.1). Plants allow the capital-efficient design of upstream *manufacturing* capacity enabling significant reduction in overall manufacturing costs on an agricultural scale. In addition, low risk of contamination with animal pathogens opens the possibility of expressing vaccine candidate in edible plant organ, allowing the production of orally administered antigens. Although recombinant proteins can be functionally expressed in different plant systems, it is imperative to determine the platform that offers

Plants as Bioreactors for Industrial Molecules, First Edition.
Edited by Santosh Kumar Upadhyay and Sudhir P. Singh.

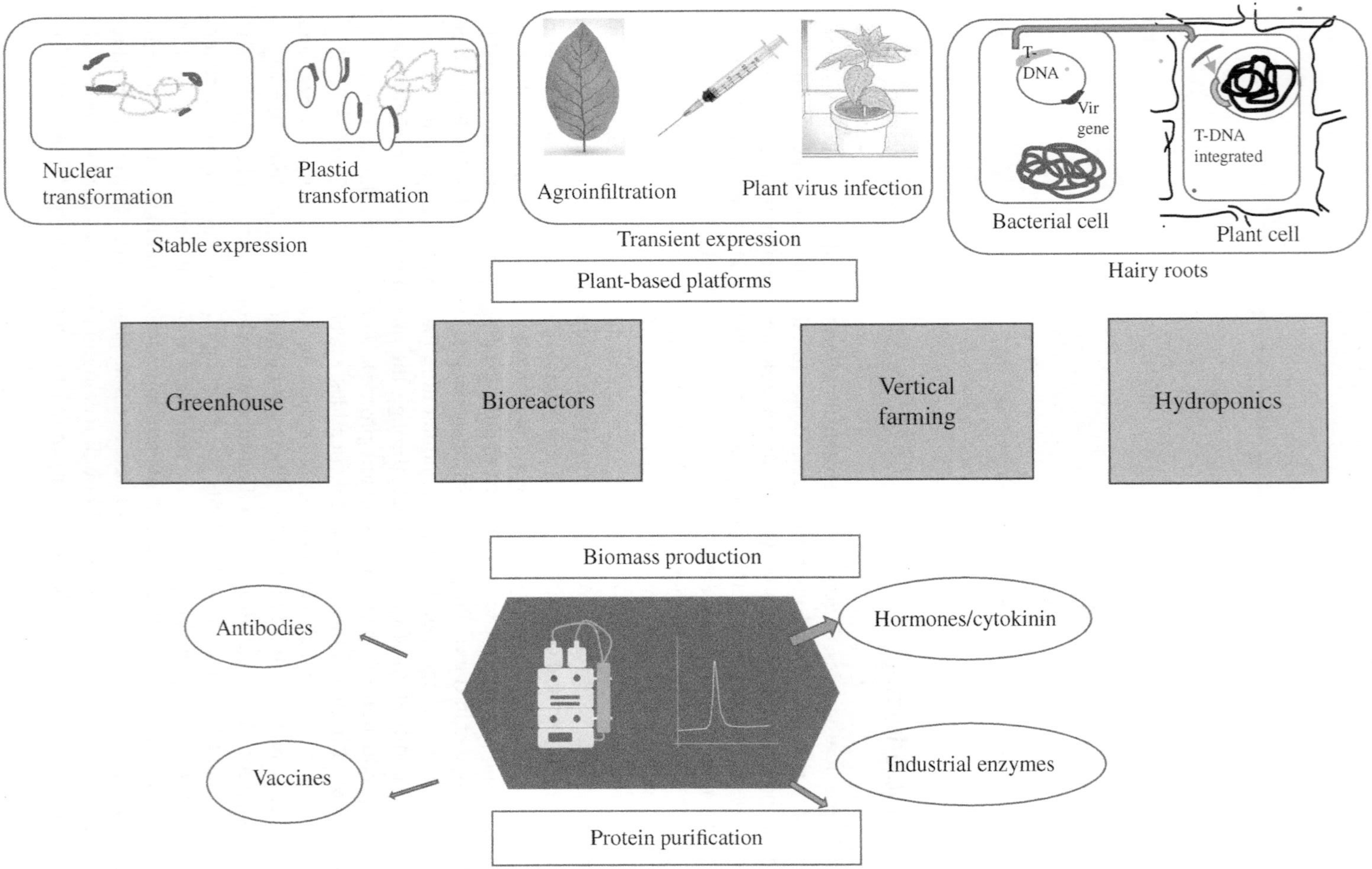

Figure 18.1 Overview of production system of plant-based recombinant proteins and plant biomass.

the most advantageous conditions for the expression and recovery of a particular protein (Ma et al. 2003). Basically, there are multiple options for plant expression systems, there are three strategies for recombinant protein production in plant-based systems: (i) use of cell-culture-based systems that are equivalent to mammalian, microbial, and insect cell systems; (ii) transient expression of foreign genes in plant tissues that are transformed by either agro-injection or by viral infection; and (iii) development of transgenic plants carrying stably integrated transgenes (Rech et al. 2008; Rech 2012). This chapter is concerned with existing platforms as bioreactors for different plant cell and tissue culture types producing bioactive compounds. We have also summarized future prospects and current challenges associated with each type of plant production platform and production strategy.

18.2 Whole Plants: Stable and Transient Expression Systems

The expression of recombinant proteins in plants in whole plants or plant cell cultures is based on either transient expression system via viral or nonviral vectors or stably transformed plants with expression of transgenes targeted either into the nuclear or chloroplast genome. Stable transformation offers several advantages including, a very high expression level, heritable transgene, high scale-up capacity, low risk from animal-borne contaminants, and inexpensive storage costs. However, there are still limitations mainly related to economic aspects. However, there are problems associated in achieving transgene expression such as position effects, multicopy gene silencing, and GMO environmental concerns (Yao et al. 2015; Ganapathy 2016). Stable transgenic are often produced by the integration of recombinant DNA (nuclear vs. plastid), recombinant protein accumulation in different subcellular compartments (e.g. cytosol, apoplast, endoplasmic reticulum, and vacuole), and the specific plant tissue gene expression target (leaves vs. seeds).

18.2.1 Stable Expression (Whole Plant Based)

18.2.1.1 Leaf Based

Leaf tissues of nonfood crops have been adopted regularly as a viable production platform for either research or production purposes. One of the earliest clinical trial of plant-produced recombinant protein, an immunoglobulin, was expressed in field grown transgenic tobacco leaves (Kaiser 2008; Ma et al. 1998). Tobacco can provide sustainable and cheap platform to produce high-value protein for many reasons such as s high biomass yield, easy genetic manipulation, year-round growth and harvesting, and the existence of large-scale infrastructure for processing. Tobacco is often used for producing recombinant proteins such as pharmaceuticals, vaccines, hormones, cytokines, growth regulators, and industrial products henceforth is considered as a molecular biology workhouse of the plant world. Based on conventional breeding, development of low nicotine and low alkaloid tobacco cultivars has been a preferred route to overcome potential toxicity carryover into economically valuable recombinant protein products and making it a suitable candidate for direct oral delivery (Menassa et al. 2001). Most importantly, tobacco being a nonfood, nonfeed crop minimizes regulatory hurdles by eliminating risk of plant-made recombinant proteins inadvertent entering the human food chain. Besides tobacco, many other leafy crops such as lettuce, alfalfa, soybean, and clover have been investigated as suitable hosts for stable expression of proteins. The use of crops like alfalfa and soybean produces lower amounts of leaf biomass than tobacco, but they have additional benefits of being perennial that fixes nitrogen resulting in reduced fertilizer inputs. Alfalfa is particularly useful because it can be grown all the year round with high leaf protein content (up to 30 mg total protein per gram in fresh weight) and offers advantage of notable homogeneity of N-glycosylated recombinant. Experiments focusing on lettuce (Lactuca sativa) plants evaluated their applicability as a production host for potential edible recombinant vaccines (Huy et al. 2011).

Stable leaf nuclear and plastid expression systems have been exploited for integration of target gene for the production of recombinant proteins (Mett et al. 2008). Plastid transformation offers several advantages, especially high expression levels that the nuclear transformation lacks, transgenic containment via maternal inheritance, and multigene expression in a single transformation event. Nuclear integration for expression of functional glycoproteins allows protein processing in the endomembrane system, thus successfully expressing over 100 functional glycosylated proteins including antibodies (Fiedler et al. 1997; Johansen et al. 1999), vaccines (Pogrebnyak et al. 2005; Golovkin et al. 2007), cytokines (Menassa et al. 2007, Wang et al. 2008), growth hormones, and industrial enzymes in plant leaves. However, one of the limitation of the chloroplast transgenic system for the production of biopharmaceutical proteins is the absence of glycosylation in the plastids required for the function of many human proteins. However, since the proteins in chloroplast are not glycosylated plastid-based expression technology currently are not suitable for the production of nonglycosylated products (Maliga 2003; Daniell 2006). As a result of its polyploidy (up to 10 000 copies of the chloroplast genome in each cell), chloroplast permits thousands of copies of a given transgene per plant cell (Gao et al. 2010; Bock 2014; Waheed et al. 2015). Chloroplast transformation expresses recombinant proteins at extremely high levels up to 70% of the total leaf protein. Several recombinant proteins including human insulin-like monoclonal antibody (Triguero et al. 2011; Ma et al. 2015; Hehle et al. 2016) cytokines (Zhang et al. 2003; Turchinovich et al. 2004), and industrial enzymes (Zhang 2013; Harrison et al. 2014), and vaccines against *Bacillus anthracis*, *Yersina pestis*, *Entamoeba histolytica*, and HIV (Pogrebnyak et al. 2005) have been successfully expressed in plant chloroplasts. In lyophilized plant cells, therapeutic proteins except for exceptionally high-level expression are stable at ambient temperature for several years, maintaining their folding and efficacy thereby eliminating costs of cold storage and transportation (Kwon et al. 2013). Another disadvantage of chloroplast transformation is the lack of efficient protocols for the transformation of cereals (Egelkrout et al. 2012). Chloroplast transformation *has* emerged as a promissory platform to provide a unique bioproduction system for oral administration of medicines (Kwon and Daniell 2016). When a pharmaceutical protein is accumulated in plant leaves, protein drugs in plant cells are protected by cell wall from stomach acids and enzymes via bioencapsulation when administered orally. However, target proteins are subsequently released into the gut lumen by microbes that digest the plant cell wall (Kwon and Daniell 2016). The large *intestinal mucosa* can be considered to act as an ideal system for oral drug delivery. *Attached* to a *CPP*, therapeutic cargo efficiently crosses the intestinal epithelium and enter systemic circulation or immune system.

18.2.1.2 Seed Based

Recombinant protein expression targeted to plant seeds can overcome the major limitations of protein stability and storage associated with PMF-based recombinant products (Ma et al. 2003; Lau and Sun 2009). Currently, most pharmaceutical proteins are synthesized in leafy crops for optimum biomass. One of the major challenges associated with the leafy crops is their long-term storage, protein degradation rate is high in leaf proteins after they are harvested hence immediate processing of the harvested material must be done (Burnett and Burnett 2020). Also, further, a purification strategy purification may be optimized for the readily separation and purification of the protein of interest from leaf constituents like pigments, alkaloids, and other secondary metabolites. Seeds have high rate of protein synthesis and accumulate high protein content (7–10%), during developmental stage, hence expression via seed-specific promoters is preferred in many cases. Most soluble proteins in seeds are stored within membrane-bound organelles *called* storage vacuole (PSV) or protein body to be used upon seed germination. This stable accumulation in a small volume has the advantage that purification would be easier. Also, proteins can be stored for a longer time in storage avoiding degradation at ambient temperature, owing to low protease activity, and low water content seeds representsan attractive alternative as a *platform* for *production* of recombinant proteins (Ma et al. 2003; Stoger et al. 2005a). The only disadvantage is that it takes a long time for seed set depending on the lifecycle of the plant, hence transgene expression can be assessed only then. Edible seeds such as maize and rice *can* be *orally administered* as flour or *flakes* suitable for *oral vaccine* production (Sabalza et al. 2013). In the past decade, significant advances using seeds have created new opportunities including the commercialization of first plant-produced bio-molecules. Cereals, e.g. rice, wheat, barley, soybean, and maize, are the most suitable platform for *seed* recombinant protein production (Ramessar et al. 2008a). Other seed crops having the potential to provide platform for recombinant protein production include legumes such as pea, cowpea, and soybean and oil crops such as canola and safflower, *Arabidopsis*, and tobacco. Recombinant *proteins* targeted to *seeds have been* reported to accumulate stable expression yields up to 10% of total seed proteins (Stoger et al. 2005b; Lau and Sun 2009). Products include therapeutic proteins such as antibodies (Rademacher et al. 2008), vaccines (Van Droogenbroeck et al. 2007), and cytokines (Sardana et al. 2007) further industrial enzymes such as trypsin (Woodard et al. 2003), phytase (Chen et al. 2008), and cellulase (Hood et al. 2007) and biopolymers such as spider silk protein (Weichert et al. 2016). Maize seeds are exemplary system for the production for the commercial therapeutic proteins and industrial enzymes (Sabalza et al. 2013). In contrast to other cereals, maize possesses relatively higher proportion of endosperm owing to large size of grain (82% of the seed) and a higher yield per hectare at lower production costs (Giddings et al. 2000; Naqvi et al. 2011; Sabalza et al. 2013). The *maize seed* system *has been successfully* used to market *several industrial* proteins including GUS, cellulase, laccase, and trypsin (Basaran and Rodriguez-Cerezo 2008; Naqvi et al. 2011). Downstream processing techniques have also been used for therapeutic proteins such as the HIV neutralizing antibody 2G12 (Ramessar et al. 2008b), influenza virus H3N2 nucleoprotein (Nahampun et al. 2015), and α-galactosidase (Yang et al. 2015) in maize seeds to minimize the production cost. Other *seed-based* recombinant protein production *platforms* include rice seed-produced human transferrin (Zhang 2013) by US-based Ventria Biosciences and endotoxin-free *growth factors* in *barley* seeds (Orfeus™ unique expression system) (Magnusdottir et al. 2013) by ORF Genetics Ltd. Via the *oleosin fusion* technology, endogenous oil body bound recombinant proteins with specific applications are expressed in rapeseed (McLean et al. 2012; Vargo et al. 2012). Fusion proteins accumulated in *oil bodies* are free from contaminating *proteins* and can be *easily isolated* from other cellular compartment in the lipid fraction from the bulk seed.

Although a seed-based platform is considered as superior choice, major hurdles still exist. Some of the challenges include relatively lower biomass and high possibility of spread of transgene into the environment through the seed or from pollen dispersal (Vitale and Pedrazzini 2005; Conley et al. 2009). General public and regulatory agencies are reluctant to use seed-based systems (e.g. maize, rice, and wheat) for recombinant protein production due to inevitable contamination of nontransgenic crop and entry in the food chain (Vitale and Pedrazzini 2005). The creation of a transgenic plant line requires considerable time and input (Saberianfar et al. 2015). Stability of the recombinant protein, reduced production cost, and postharvest processing make them excellent source for large-scale production of recombinant (Schillberg et al. 2005).

18.2.2 Transient Expression

Among the new and emerging enabling technologies that have turned plants into viable hosts, transient expression systems mediated via *Agrobacterium tumefaciens or* recombinant viral vector has been the most transformative. For the transient expression of transgene in plant, the target gene is transferred into plant cells through a process called agroinfiltration system and scaled-up. In simple words, it is achieved by the vacuum *infiltration* of *plant cell or tissue* with recombinant *A. tumefaciens* carrying the target gene within a binary vector. It allows gene in the episomal form without inserting into the genome of the target cells and can have transcriptionally competent during certain time (Leuzinger et al. 2013). Depending on the vector and protein, the *in-planta* incubation time required for the production of maximum yield of recombinant protein production are no more than two weeks. Transgenic plants generated through stable transformation, when used as production systems are grown only in sterile and contained greenhouse conditions, or in areas where no weed or food crop relatives are grown. Plants are often produced under fully-*contained cultivation* systems prior to infiltration, which is considered to be a highly automated process at production scale. These systems have potential for rapid transgene assessment and offer *rapid production with high product yield.* These systems have the ability to easily co-express multiple genes simultaneously (e.g. Medrano et al. 2009; Huang et al. 2010). Transient expression can optimise downstream processing costs, however, there is immediate need of processing of harvested leaf biomass and an additional step is required to purify the protein by removing endotoxins derived from the infiltrated *Agrobacterium.* Transient expression has been demonstrated in other plants including *Arabidopsis,* lettuce, potato, green pea, and other *Nicotiana* species such as *N. maritime, N. debneyi, N. exigua, N. excelsior* and *N. simulans* (Negrouk et al. 2005; Sheludko et al. 2007; Sindarovska et al. 2008; Bhaskar et al. 2009; Green et al. 2009; Kim et al. 2009). There are essentially two approaches for transient expression based on the method of transgene delivery into the plant cell: autonomously replicating viral vectors or plant binary vector-based (Matoba et al. 2011; Rivera et al. 2012). In a transient expression system using "conventional" *binary vectors* carrying a gene of interest within a *plant viral vector* is a rapid and efficient method with protein yields typically ranging from ~0.1 to 180 μg/g fresh leaf (Liu et al. 2008; Condori et al. 2009; Medrano et al. 2009). Recently, *transient-expressed xylanase localised in* the *endomembrane* system was reported with considerable high product recoveries (1 mg/g fresh leaf) (Llop-Tous et al. 2011). The plasticity of transient expression in plants, speed of scalability and relatively low capital costs, highlight the great potential of this technology in the future of human and animal health.

Expression of biopharmaceuticals in plant is a widely accepted strategy that generates a large number of target genes in relatively short time, high protein yields. Transient expression has obvious advantages compared to develop stable transgenic lines, which are expensive and *time-consuming* methodologies (Xia et al. 2020). Transient expression in plants involves short timeframe to achieve high-level expression of the recombinant proteins produced. Particularly, when new pathogens and diseases are prone to sudden outbreaks, many researchers and commercial applications have found transient expression for the production of medicinal proteins or antibodies in plants to be a powerful tool for social security and disease treatment (Peyret and Lomonossoff 2015). Plant culture for transient expression is usually divided into two stages, the preinoculation and postinoculation processes, which require different optimal environments. Optimum cultivation conditions, such as the nutrient composition of the culture medium before agro-infiltration (Fujiuchi et al. 2014) and dehydration of plant material after agro-infiltration (Fujiuchi et al. 2014), are key factors enable plants to produce a wide range of recombinant proteins on a rapid timescale. Preinoculation environmental *factors* associated with a high-nitrate nutrient solution are known to accelerate recombinant protein content (Fujiuchi et al. 2014). Very recently, interest in the surrounding environment has extended beyond research into the plant itself (Moon et al. 2012; Twyman et al. 2013; Fujiuchi et al. 2014). Nitrate-enriched fertilizers that enable plant growth (Fujiuchi et al. 2017), immediate desiccation after agro-infiltration (Fujiuchi et al. 2016), and slightly lower plant density are critical factors for improving plant biomass thus enhancing the efficiency of *transient systems* of recombinant protein production. Key factors influencing the gene expression are optimized, including the plant density, the number of plants per unit growth area, and the effect of leaf position on harvested biomass (Fujiuchi et al. 2017). In addition, removal of residual water from a bacterial suspension on detached leaves after agroinfiltration had significant impact on the yield of recombinant hemagglutinin (HA) protein, and yield from detached leaves in a transient over-expression system commensurate with intact leaves (Fujiuchi et al. 2016). Merlin et al. (2016) demonstrated the highest yields of an optimized variant protein in a transient expression system using the MagnICON vector. They hypothesized that a significant increase in the yield of the TOI could be achieved using transient expression. The relevance of alternative plant sources for integrated protein production in green biorefineries depends on many biological, environmental, economic, and social factors, such as nutritional profile, environmental implications, consumer acceptance, digestibility, and bioavailability. The plasticity of transient expression in plants, speed of scalability and relatively low capital costs, highlight the great potential of this technology in the future of human and animal health.

18.2.3 *In vitro* Culture Systems

Plant biomass (e.g. suspension cells, hairy roots, and moss) can be excised and cultivated indefinitely in confined bioreactors *under sterile conditions* for large-scale production of recombinant proteins. The status of plant cells as suitable alternative to the current production systems for recombinant pharmaceutical proteins began to change after the first bubble of commercial interest in molecular farming collapsed due to concern surrounding *GM foods* and crops commonly focussing on human and environmental safety and the lack of support from an industry already heavily invested in fermenters. *in vitro* culture provides precise control over cell growth and protein production, batch-to-batch product consistency, and a production process that ensures good manufacturing practices (cGMP) (Wilson and Roberts 2012). However, recovery and purification of recombinant proteins from plant biomass is an expensive and technically challenging business that has limited its use for commercial high-value protein therapeutics. Among the plant-based platforms that have been explored, *in vitro* cultures offer a promising system for production of biopharmaceuticals with lower regulatory burden and environmental impact (Xu and Zhang 2014). Like other bioreactor-based culture systems, *in vitro* cultures have significant limitations in terms of *scalability*. However, because the product can be recovered from the culture medium allowing continuous production, the process of separation and purification of expressed proteins becomes less expensive than from whole plants (Schillberg et al. 2013). Indeed, the first licensed plant made recombinant pharmaceutical protein, taliglucerase alfa (Elelyso™), produced in genetically engineered carrot cells, as an orphan drug for Gaucher's disease, and Gaucher's disease has finally approved for human use by US Food and Drug Administration (FDA)

18.2.3.1 Plant Suspension Cultures

Plant suspension culture cells provide a fast system for producing secondary metabolites, biologically active recombinant proteins, and antibodies Plant cell lines most widely used to provide a fast system for producing secondary metabolites, biologically active recombinant proteins, and antibodies are most widely derived from tobacco (*Nicotiana tabacum*), particularly cultivar *Bright Yellow-2 cell* (*N. tabacum* cv. BY2) cells. BY-2 cells have appealing features, including high growth rate, being robust and can multiply their numbers up to 100-fold in a week. Moreover, they are easily transformed by *Agrobacterium* transformation and can be highly synchronized (Hellwig et al. 2004; Ozawa and Takaiwa 2010; Xu et al. 2011). Other commonly used cell lines generally include common edible crop species such a rice (*Oryza sativa*), alfalfa (*Medicago sativa*), tomato (*Lycopersicon esculentu*), and carrot (*Daucus carota*). However, these cell lines derived from common edible crop species may be more favorable than tobacco cells in terms of by-product levels and regulatory approval (Hellwig et al. 2004). Rice cell suspension cultures are also extensively used as tobacco BY-2 cells due to availability of the carbohydrate-sensitive α-amylase expression (Trexler et al. 2005). This sugar starvation induced highly active promoter has enabled successful competitive yield of many recombinant proteins in rice cells, e.g. α1-antitrypsin (rAAT) (McDonald et al. 2005; Trexler et al. 2005), hGM-CSF (Shin et al. 2003), interleukin-12 (Shin et al. 2010), and human serum albumin (Liu et al. 2015), with highest product yield of recombinant protein reaching 247 mg/l for rAAT (McDonald et al. 2005). Rice cell lines are considered inferior to those of tobacco BY-2 cell lines. Sucrose starvation has a negative impact on the viability of rice cell lines. Plant suspension cultured cells provide a fast system for producing secondary metabolites, biologically active recombinant proteins, and antibodies (Francisco et al. 1997; Terashima et al. 1999; Torres et al. 1999). The recombinant proteins expressed can either be transported to subcellular organelles or exported out of the cell *into* the *extracellular* space via the default pathway in plant suspension cultured cells (Denecke et al. 1990; Liu et al. 1997). The latter secretion method provides a cost-effective strategy for stable accumulation and purification of the secreted recombinant proteins in the absence of large quantities of contaminating proteins. In addition, suspension cultures also provide a rapid and high-throughput screening of various recombinant proteins before they are further expressed in transgenic plants in seed bioreactors. The plant cell wall was considered as barrier for *large*-scale manufacturing of large lysosomal protein. However, 80-kDa human protein, under the control of the 35S CaMV constitutive promoter and a signal peptide from a plant protein, has been successfully expressed and secreted into the cultured media of transgenic tobacco BY-2 cells. These secreted proteins have been shown to acquire N-linked glycosylation and had very high enzymatic activity, a result demonstrating the correct folding of this BY-2 cells-derived human lysosomal enzyme (Fu et al. 1999). There was a concern that during transit through the Golgi apparatus, recombinant protein expressed in plant suspension cultures may acquire complex-glycan modifications that could be highly immunogenic in humans. Much effort was expended to ensure that plant glycans were avoided, including inhibition of the plant Golgi enzyme glycosyltransferases, expression of the mammalian glycosyltransferases, and the proteins to be retained in the ER resulting in generic high-mannose glycans, (Gomord and Faye 2004; Wang et al. 2006).

18.2.3.2 Hairy Root System

Hairy roots are induced from the infected sites of *Rhizobium rhizogenes* (previously referred to as *Agrobacterium rhizogenes*) in a wide variety of higher plants and allow the production of several secondary metabolites (Chilton et al. 1982; Shanks and Morgan 1999). Similar to suspension cells, hairy roots can be cultured under aseptic *and well-controlled environment systems* suitable for developing robust cost-effective GMP production of pharmaceutical proteins. However, as a more organized culture, hairy roots offer additional benefits, including high genetic and biochemical stability and hormone autotrophy. Hairy roots are capable for the production of complex active glycoproteins and high scalability from a large spectrum of organisms (Häkkinen and Ritala 2010; Stoger et al. 2014; Dixit et al. 2021a,b). It can synthesize recombinant proteins by roots grown in bioreactor and their secretion in the culture medium under controlled and confined conditions. The fast growth property, long-term genetic, and biosynthetic stability have made "hairy root" an excellent bioreactor system. Hairy roots can successfully produce metabolites or secondary metabolites, which correspond to complex molecules naturally present in the plants and displaying interesting features at a pharmacological, cosmetic, and nutraceutical level. Hairy root can synthesize the same components as those of the intact roots of transgenic plants (Kim et al. 2002). Several recombinant proteins have been successfully produced and secreted by the hairy root system including reporter proteins (example the green fluorescent protein (GFP)) (Medina-Bolívar and Cramer 2004), human acetyaQlcholinesterase (Woods et al. 2008), murine interleukin (Liu et al. 2009), thaumatin sweetener (Pham et al. 2012), human interferon alpha-2b (Luchakivskaia et al. 2012), recombinant human EPO (rhEPO) (Gurusamy et al. 2017), and recombinant alpha-L-iduronidase (Cardon et al. 2019) (Table 18.1). In addition, hairy roots have capability to produce even *complex glycosylated proteins* with highly homogeneous post-translational profile. Hairy roots secrete expressed proteins from

Table 18.1 Comparison of different plafftorms and representative proteins produced *in vitro* cultures.

Platform	Viable species	Time for production	Recombinant proteins
Whole plants (leaf-based and seed-based system)			
Stable transgenic plants	Corn, soybean, safflower, rice, tobacco	3–6 months	Leaf based; antibodies (Ma et al. 1995; Fiedler et al. 1997; Johansen et al. 1999), vaccines (Pogrebnyak et al. 2005), cytokines (Wang et al. 2008), growth hormones, and industrial enzymes Seed-based: antibodies (Rademacher et al. 2008), vaccines (Van Droogenbroeck et al. 2007), and cytokines (Sardana et al. 2007) further industrial enzymes such as trypsin, (Woodard et al. 2003) phytase (Chen et al. 2008), and cellulase (Hood et al. 2007)
Transient plants	*N. benthamiana*, lettuce	2–7 days	Therapeutic proteins, xylanase (Llop-Tous et al. 2011).
In vitro culture system			
Plant cell suspension	Tobacco (*Nicotiana tabacum* cv. BY-2), carrot, rice	7–14 days	α1-antitrypsin (rAAT) (McDonald et al. 2005; Trexler et al. 2005), hGM-CSF (Shin et al. 2003), interleukin-12 (Shin et al. 2010), and human serum albumin (Liu et al. 2015)
Hairy roots	*N. tabacum*	14–30 days	Green fluorescent protein (GFP) (Medina-Bolívar and Cramer 2004), human acetyaQlcholinesterase (Woods et al. 2008), murine interleukin (Liu et al. 2009), thaumatin sweetener (Pham et al. 2012), human interferon alpha-2b (Luchakivskaia et al. 2012), recombinant human EPO (rhEPO) (Gurusamy et al. 2017), and recombinant alpha-L-iduronidase (Cardon et al. 2019)
Moss	*Physcomitrella patens*	14–30 days	Tumor-directed monoclonal antibody with enhanced antibody-dependent cell-mediated cytotoxicity (ADCC) (Schuster et al. 2007; Kircheis et al. 2012), keratinocyte growth factor (FGF7/KGF) (Niederkruger et al. 2014), asialo-erythropoietin (asialo-EPO) (Weise et al. 2007; Parsons et al. 2012), α-galactosidase, and β-glucocerebrosidase (Niederkruger et al. 2014).

Table 18.1 (Continued)

Platform	Viable species	Time for production	Recombinant proteins
Aquatic plants			
Microalgae	*Lemna* sp., *Spirodela* sp	20–40 days	Cosmetics, pharmaceuticals, biofuels, and health supplements (Vickers et al. 2017).
Duckweed	*Chlamydomonas reinhardtii, Dunaliella salina*	14–30 days	Monoclonal antibody (mAb) (Cox et al. 2006), plasminogen (Spencer et al. 2010), anti-interferon α2 (De Leede et al. 2008), vaccine antigen avian influenza including H5N1 hemagglutinin (Guo et al. 2009; Bertran et al. 2015), and M2e peptide (Firsov et al. 2015) human growth hormone, hemagglutinin, human granulocyte colony-stimulating factor, and porcine epidemic diarrhea virus (PEDV) protective antigen.

cultured tissues, termed rhizosecretion, which offers a simplified method for the isolation of target *protein* in the *hydroponic thus simplifying* the downstream process (Borisjuk et al. 1999; Sivakumar 2006), where recombinant proteins can be collected continuously over the life span of a transgenic plant (Sivakumar 2006). Since, root biomass is not destroyed for extracting secondary metabolites, a given culture can be used for several cycles of bioproduction. Using design of experiments approach to develop an *optimized induction protocol* for the cultivation of tobacco HR secreting the mAb M12, extra KNO3, α-naphthaleneacetic acid, and polyvinylpyrrolidone in the culture medium enhanced antibody yield by 30-fold, yielding 5.9 mg/l. Silencing of transgenes is a common problem in the hairy root system and thus a major commercial obstacle to its application (Dunoyer et al. 2007). These hairy roots are characterized by large number of highly branched phenotype vigorous growth without added phytohormones. Furthermore, they often produce secondary metabolites for a long period of time, unlike intact roots (Verma et al. 2009; Alok et al. 2018). For these reasons, switching from undifferentiated cell culture to hairy root culture is considered an attractive alternative for the production of many valuable secondary metabolites that originally accumulated in root tissues.

The bottleneck to exploiting hairy root technology as a robust bioproduction platform for *applications* like large-scale production of secondary metabolites and recombinant protein has been low protein productivity (Georgiev et al. 2012). Expression systems have been strategically created to identify a strong promoter such as a double-enhanced CaMV35S promoter (Woods et al. 2008) for transcriptional control, a chimeric super-promoter (Aocs, 3AmasPmas) (Peebles et al. 2009), and inducible promoters (Peebles et al. 2009). In addition, the special morphological characteristics of hairy roots including nonhomogeneous cell growth and multiple branches present major constraints to culture scale-up in bioreactors for the commercial exploitation.

18.2.3.3 Moss

Currently, Moss protonema, a nonseed plant, is being investigated as another promising platform for the production of recombinant proteins in bioreactors. Moss is an attractive photoautotrophic organism requiring minimal medium containing only water and inorganic salts and has a fully sequenced genome that can be easily engineered through homologous recombination. This ability of moss cells to photosynthesize in culture leads to reduced production cost and facilitates recovery of recombinant protein from the medium (Schillberg et al. 2013). While plant cell cultures show a high degree of genetic instability, the so-called somaclonal variation, moss cells rely on differentiated instead of undifferentiated plant cell cultures. Moss cultures are characterized by their genetic stability over long periods of time (von Stackelberg et al. 2006). Cultures of the moss, *Physcomitrella patens* with its genome fully sequenced, are the main species within bioreactors to enable protein production (Zimmer et al. 2013). Several biopharmaceutical human proteins have been produced in this system. Among the products are tumor-directed monoclonal antibody (Schuster et al. 2007; Kircheis et al. 2012), keratinocyte growth factor (FGF7/KGF) (Niederkruger et al. 2014), asialo-erythropoietin (asialo-EPO) (Weise et al. 2007; Parsons et al. 2012), α-galactosidase, and β-glucocerebrosidase, etc. (Niederkruger et al. 2014). (www.greenovation.com). Some of the recombinant biopharmaceuticals produced from moss showed superior characteristics compared to the *mammalian cell*-based products *and* are catalogued as **biobetters.** For example, *mannose* receptor-mediated delivery of *moss-made alpha-galactosidase* yields improved pharmacokinetics in Fabry mice (Reski et al. 2015). Moreover, moss N-glycans are free from the α-1,6-fucose present in the core region, typical of higher eukaryotes. The absence of this sugar moiety in moss-made IgG leads to more efficient antibody-dependent cell-mediated cytotoxicity (ADCC) *than* their fucosylated *counterparts in* mammals (Schuster et al. 2007; Kircheis et al. 2012; Reski et al. 2015). *P. patens* receives increased scientific interest since its fully sequenced genome can be readily engineered through gene targeting with homologous recombination (Schillberg et al. 2013) The *glyco-engineering approaches* have been efficiently used for knocking out nonhuman N- and O-glycosylation on target *proteins produced* in *moss* bioreactors thereby allowing for production of glycoproteins with authentic *human*-like glycosylation. (Schuster et al. 2007). For example, moss mutants were engineered with genes encoding plant typical glycosyl transferases identified and deleted from the *moss genome* (Koprivova et al. 2004; Parsons et al. 2012) and further knocking out the gene encoding β-1,4-galactosyltransferase into the xylosyltransferase or fucosyltransferase locus, respectively (Huether et al. 2005). Genes responsible for prolyl hydroxylation were identified and deleted to avoid unwanted potential O-glycosylation of human (Parsons et al. 2013).

18.2.4 Aquatic Plants

Some aquatic plants also are best suited as plant-based bioproduction platforms, including duckweed and microalgae. Several functional industrial enzymes and human therapeutic proteins have been expressed in duckweeds or microalgae has been cost effective.

18.2.4.1 Duckweed

Duckweed, with scientific name Lemnaceae, a monocot plant family is divided into four major genera: *Lemna, Spirodela, Wolffia,* and *Wolffiella*. Duckweed is the smallest and fastest-growing aquatic plant, which is easy to survive and widely distributed. The exponential growth of duckweed without the need for pollen or seeds plays a vital role in the rapid production of proteins, which simplifies line management, a shorter production cycle and propagation, and the process feed stream (Ziegler et al. 2015). Duckweed has a fast growth rate, usually with a doubling time of one to two days in a suitable environment (Stomp 2005). Other desirable characteristics of large-scale production are simple and easy culture conditions, culture medium capable of making complex proteins, fast clonal growth, ease of harvest and has a high protein content (up to 45% dry weight) (Appenroth et al. 2017). Cultivation only requires scalable inexpensive upstream facilities. The absence of pollen or seeds in Duckweed make it environmentally safer than other transgenic flowering plants. Duckweed is an edible aquatic plant, which would be a promising new alternative to oral delivery (Popov et al. 2006; Rival et al. 2008). During the production of recombinant proteins for medicinal purposes, the animal expression may get contamination of agents and oncogenic sequences, which can cause diseases. However, duckweed can be a sustainable practice for providing easy access to sterile pure culture without microbial and chemical pollutants. Two main stable genetic transformation systems used in duckweed include the callus transformation system (CTS) and the frond transformation system (FTS) (Yang et al. 2018). Duckweed can be transformed using either *A. tumefaciens* or biolistics genetic transformation. Fast and efficient nuclear transformation protocols using FTS for two species of duckweed *L. gibba and L. minor* (Yamamoto et al. 2001) almost 20 recombinant proteins were produced (Vunsh et al. 2007). Using a genetic transformation system and expression system, a variety of exogenous proteins including industrial enzymes, e.g. E1 endoglucanase (Sun et al. 2007), and therapeutic products have been successfully expressed in duckweed (Table 18.1). Until now, a few vital medicinal proteins have been successfully expressed in duckweeds, such as monoclonal antibody (mAb) (Cox et al. 2006), plasminogen (Spencer et al. 2010), anti-interferon α2 (De Leede et al. 2008), vaccine antigen avian influenza including H5N1 hemagglutinin (Bertran et al. 2015), M2e peptide (Firsov et al. 2015) human growth hormone (Hawkins and Nakamura 1999), hemagglutinin, human granulocyte colony-stimulating factor, and porcine epidemic diarrhoea virus (PEDV) protective antigen. US-based Biolex, Inc. developed therapeutics produced bioactive recombinant polypeptides in duckweed (Lemna)-based expression (LEX) system (Paul and Ma 2011).

18.2.4.2 Microalgae

Over the past few years, there has been a surge of interest in natural source of valuable compounds and as bioreactors for recombinant protein production. Microalgae are a diverse photosynthetic group, consisting of eukaryotic organisms and prokaryotic cyanobacteria that can use sunlight to synthesize chemical energy carriers such as carbohydrates, lipids, and proteins. Microalgae represent a promising and sustainable emerging biotechnology platform for the light-driven synthesis of therapeutic proteins such as vaccines, antibodies, and antimicrobials (Vavitsas et al. 2018). Moreover, numerous microalgal species are valuable source of novel biologically active products making them an economical and effective bioreactor for obtaining high added-value compounds for healthy food. In addition to naturally produced compounds, microalgae with rapid growth and ease of culture offer another promising platform for cost-effective production of recombinant proteins other valuable natural products, such as cosmetics, pharmaceuticals, biofuels, and health supplements (Vickers et al. 2017). Microalgae have unique advantages, including a very simple structure, high growth rate in simple media, production of large amounts of biomass with short life cycles, low growth costs. Metabolic pathways to produce large amounts of biomass with short life cycles are similar to those of higher plants, leading to the same post-transcriptional and post-translational modifications that occur in higher plants. Recombinant proteins are typically expressed by the genome of nucleus or chloroplast, Downstream purification of proteins from microalgae is similar to yeast and bacterial systems and thus is generally cost effective than from whole plants (Kagermann et al. 2013). Many species are generally recognized as safe (GRAS) fit for human consumption and are promising to produce oral vaccines. Recombinant proteins are typically expressed by the genome of nucleus or chloroplast of microalgae. However, due to inefficient nuclear transgenic expression in microalgae possibly due to gene silencing occurring in nuclear transformants which generally accumulate less recombinant protein than chloroplast transformants (Jinkerson and Jonikas 2015). The proteins expressed in chloroplasts are currently regarded as more feasible for commercial production owing to higher levels of proteins expressed than those expressed in the nucleus production. The disadvantage of plastid transformation is the absence of machinery to perform post-translational modification, e. g. glycosylation. However, chloroplast-expressed nonglycosylated antibodies are superior as they do not activate the immune system in humans (Mayfield et al. 2007). *Chlamydomonas reinhardtii*, a single cell green alga, is a classical model organism for studying genetic engineering under laboratory conditions to a vast extend. The green microalga *C. reinhardtii* is under development as a production host for various recombinants including endolysins, subunit vaccines, and bioactive compounds from either the nuclear or plastid transformation (Scaife et al. 2015). Most of these were produced in the *C. reinhardtii* chloroplast, but other microalgae species such as *Dunaliella salina* and *Phaeodactylum tricornutum* could accumulate upto 60% of protein.

Chlymadomonas displays rapid growth rate with doubling time of ~10 hours, supports easy nuclear and chloroplast transformation, and can be cultivated under photoautotrophic or heterotrophic mode with acetate as a carbon source (Xu et al. 2011). Development of economically viable bioproduction is, however, hindered due to lack of efficient and reproducible genetic transformation systems for a wider range of species, as nuclear expression or inconsistent plastid expression for recombinant protein yields (Bai et al. 2013). However, recombinant proteins produced from algae do not undergo certain post translational modifications, and as a result, may not be suitable for the production of some glycoproteins. For example, algae may not be able to produce human forms of glycosylated proteins due to a lack of the proper enzymatic machinery. However, a variety of therapeutic and diagnostic recombinant proteins, including vaccines, enzymes, and antibodies, have been produced in algal systems. In some cases, however, foreign genes are only expressed transiently in algae. Efficient and stable expression of foreign genes in algae is achieved under control of strong promoters, proper codon usage, intron integration, and stable transformation methods. An optimized system to commercialize microalgal production with indoor photobioreactors has been developed by a USA-based algae bioreactor company, PhycoBiologics. Photobioreactors yielding axenic algae with recombinant protein yields up to 20% of total soluble protein in the chloroplast have been obtained, which makes the algal platform a promising approach for the production of commercial pharmaceuticals (Yao et al. 2015).

Till now, we have studied that recombinant protein production in plants is one of the best and cost-effective strategies and various plant models can be used for production in transient manner; however, if we look into the production processes, microbial and mammalian systems still dominate the market as they offer higher production capacities at lower costs and possess ability to produce complex biopharmaceutical products. However, plant-based proteins in various industries such as research, pharmaceutical have reached the market and proves that plant-based systems can excel in a better manner in some defined niches. Here, in the following sections, we address the unique features and opportunities of plant-based production systems over microbial and mammalian systems and also strategies to enhance/develop the plant-based production systems for commercial uses in near future to produce the valuable proteins.

18.3 Unique Features of Using Plant-based Production Over Microbial and Mammalian Systems

Plant cell culture provides a promising alternative bioproduction platform. It became apparent that current *plant* expression *platforms offer* several potential benefits over conventional eukaryotic and prokaryotic expression platforms (Table 18.2). Following features provide insights to the potential of plant-based production systems as a substitute to the other methods:

Table 18.2 Advantages of using plant molecular pharming over mammalian and prokaryotes-based cell culture production systems.

S. No.	Advantages over bacterial and mammalian cell culture systems	Explanation
1.	Safety	No risk of replication of human pathogens in plants, low pathogen load, and lesser risks of process-related contamination.
2.	Simplicity	No need of sterile environment because plants rely on their indigenous immunity to defend themselves against pathogens.
3.	Cost effective	Inexpensive as predefined organic/fertilizer solutions are sufficient enough to cultivate the plants and no expensive media is required.
4.	Scalability	Can be expanded to the agricultural scale, facilitation of recombinant proteins production at the multitonne scale even in a greenhouse or in a vertical farm can also be done.
5.	Post-translational modifications	Plants possess the ability to perform PTMs same as that of mammalian cells; thus, they are far superior than prokaryotes for the expression of many complex proteins.
6.	Systemic proteins/ secondary products	Plants possesses potential to produce intrinsically disordered proteins that are present naturally in plants; for instance, production of viscumin (a lectin which possesses anticancerous activity). However, it is difficult to them in prokaryotes and mammals due to their toxic and complex nature.
7.	Rapid production	Protein expression can be attained in nearly 8 weeks after inculcation of specific DNA sequence; therefore, this kind of fast production allows rapid responses to various threats (epidemic or pandemic)

18.3.1 Better Protein Functionality

Most of the biopharmaceuticals possess N-linked glycans, and plant glycans are somewhat different than that of the human glycans, mainly the core "β (1,2)-xylose and α (1,3)-fucose" residues, which are present in plants but not in mammalian glycans. Thus, many studies have worked on knocking out the plant glycosyltransferases enzymes and introducing the human glycan chains so as to eliminate any kind of potential risks while injecting plant-derived pharmaceuticals to humans (Montero-Morales and Steinkellner 2018). On the other hand, various vaccines and few biopharmaceuticals required for cancer treatment may get benefits form plant glycans because the immunogenicity stimulates the activity of antigen-presenting cells by dendritic cells (Rosales-Mendoza et al. 2016). Several studies also reported that some pharmaceutical proteins that carry plant-based glycans work in a manner superior to their counterparts. For instance, Elelyso, the recombinant form of human glucocerebrosidase is produced in carrot cells and is targeted to the vacuole because the vacuole-specific glycans improve the uptake of the protein by human macrophages (Rup et al. 2017).

18.3.2 Plant Matrix

Plant matrix has the potential to increase the vaccines and therapeutics efficacy as they can encapsulate and protect them from digestion. For example, a poultry vaccine generated in suspension cells of tobacco was injected as a crude extract into chickens to protect them against the Newcastle disease virus (Schillberg et al. 2013). Therefore, plants provide an affordable and cost-effective approach to produce therapeutics and animal vaccines, mainly if they need to administer orally or topically as it prevents requirement of costlier downstream processing and reduces the costs of manufacturing (Zimmermann et al. 2009).

18.3.3 Speed and Scalability of Production

Due to the transient expression in plant-based production systems, recombinant proteins can be produced within just few days. Therefore, providing a greater opportunity to produce emergency vaccines such as influenza vaccines that are required within just few weeks after confirmation of their gene sequence. Furthermore, it is always easy to scale up the transient expression to generate the primary products in bulk manner (Table 18.2) (Spiegel et al. 2015).

18.3.4 Consumer Acceptance

Since plants are considered as more natural, organic, safe and environmentally stable, and sustainable as compared to that of microbial and mammalian cells, therefore acceptance of the plant-derived products is higher in consumers as compared to the counterparts (microbial and mammalian cell-based products).

18.3.5 Animal-free Production thus Lower Risks of Pathogen Invasion

In case of plants recombinant proteins can be produced without any need of animal derived agents, thereby risks of pathogen invasion and various allergies and chronic diseases are minimized. Additionally, it is always preferable to produce therapeutic proteins without any contamination of animal materials or other toxins so that there will not be any adverse impacts or interference while testing on mammalian cells, tissues, or organs (Schillberg et al. 2019).

18.4 Strategies to Enhance the Potential of Plant-based Production Systems

Various strategies (Figure 18.2) for developing the potential of plant-based production systems are discussed below

18.4.1 To Minimize Ecological Footprint via Inherent Carbon dioxide Fixation and Improved and Sustainable Fertilizer Use

If plant-based production systems are operated in close proximity with industries such as steel, coal power plants, i.e. industries that are emission intensive then can act as a rapid and cost-effective CO_2 refixation tool. In these cases, all the emissions including CO_2 from air can be conditioned (to maintain proper purity and quality) and can then be shifted to cultivation regions like vertical farms thus, CO_2 can be exploited as a gaseous fertilizer to cultivate the plants. Therefore, a symbiosis can be created between conventional plan production processes and sustainable biotechnology (Osborne 2016; Zhu et al. 2016).

18.4.2 Use of Pant Bioreactors to Harvest Multiple Products from a Single Process

Various side stream/additional products generated during plant-based production system can be used to harvest other additional products, thus leading to the enhanced process profitability (Octave and Thomas 2009). As it is evident from Figure 18.2 that when plants are used as a model to produce a primary product/recombinant protein, the product accounts only less than 10% of the total protein content. Exploring options for the rest of the 90% can improve both potential of plant-based systems as well as the economy.

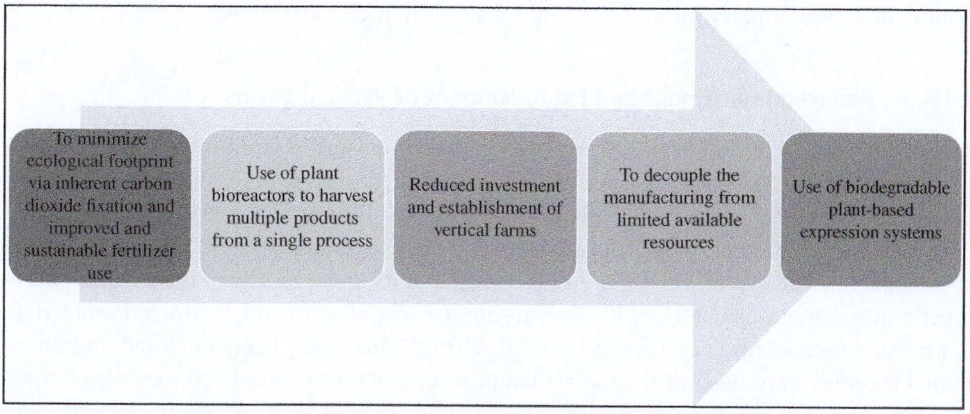

Figure 18.2 Strategies to enhance the potential of plant-based production systems.

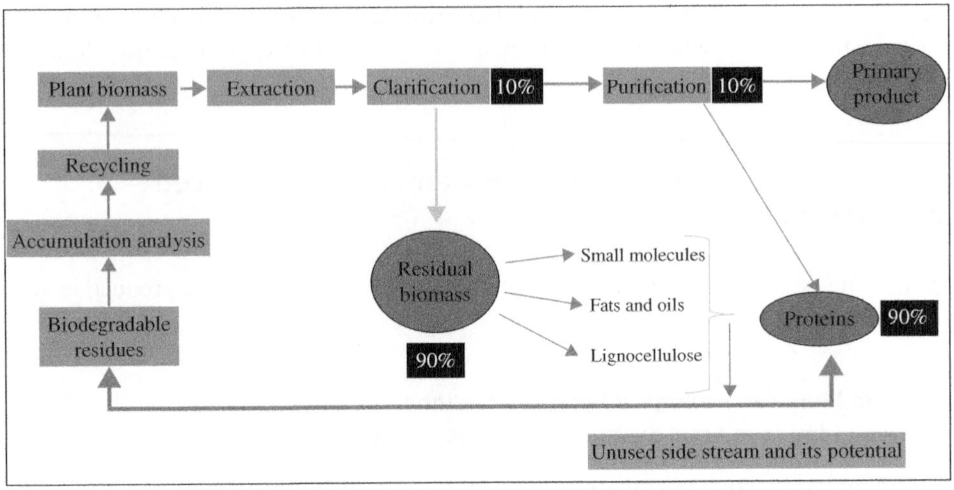

Figure 18.3 Potential of side stream products in plant molecular farming. Generally, primary products comprise less than 1% of the total biomass produced during the process, therefore, additional revenue can be generated by using small molecules, lipids, and other secondary products that further broadens the diversity of the value chain.

Furthermore, secondary metabolites that are produced during the process are generally discarded; however, these metabolites can be used as precursors of various pharmaceuticals and other chemicals of industrial importance (Baranska et al. 2013; Isikgor and Becer 2015). Although extraction environment of secondary metabolites generated during the process is different than that of the proteins, therefore orthogonal processing by sequential extraction can be exploited as a starting purification step (Santos-Buelga et al. 2012; Routray and Orsat 2013; Buyel 2015). Residual biomass mainly composed of small molecules, fats, and oils and other lignocellulosic material can be processed further to produce biofuels and other chemical building blocks (Li et al. 2014; Klose et al. 2015; Lambertz et al. 2016). There is no harm in utilizing these side stream products as biomass from plants do not contain any human pathogens, thus associated risks are low as compared to that of mammalian and prokaryotic systems (Figure 18.3) (Bethencourt 2009).

18.4.3 Reduced Investment and Establishment of Vertical Farms

During the growth period from seeds, plants require preinstalled equipment's that are very costly as prices for necessary commodities for infrastructure set up are subjected to considerable fluctuations, thereby directly affecting the cost of multiuse machinery, equipment, and conventional bioreactors. Furthermore, taxes and trade regulations may increase the costs, thus impacting the overall volatility. Additionally, quality and quantity of the products from greenhouse conditions change with the varying seasons, thereby only restricted protection is provided by the infrastructure (Sack et al. 2015). Thus, completely enclosed and automated facilities were need of the hour, therefore vertical farms have been recently coming into force and they possess partial automated facility (Holtz et al. 2015). Studies suggest that these kinds of farms are cost effective and are cheaper than conventional processes.

18.4.4 Use of Biodegradable Plant-based Expression Systems

After the extraction of products in plant-based production processes, remaining plant materials form various organic/inorganic substances due to the action of bacteria, fungi, and other nematodes. Therefore, sustainability is promoted in these systems, and it also assists in minimizing the amount of waste which requires costly disposal facilities unlike in mammalian systems that require expensive and dedicated equipment's for the waste removal. Furthermore, amount of liquid waste generated is also lesser in plants as plant growth promoting fertilizer solutions can easily be used for several production batches as component concentrations can be individually re-adjusted and a sterile handling is not required. Additionally, there is no requirement of heat inactivation of solid and liquid wastes, if transgenic and noninfiltrated plants are being exploited for production process (Zhu et al. 2013; Ebeling et al. 2014; Baldrian 2017). A brief comparison of side stream generated in plants and mammalian systems is depicted in Figure 18.4.

18.5 Concluding Remarks and Future Perspectives

Since we have seen that plants provide multiple benefits in producing recombinant proteins and other important chemicals. However, due to better establishment and characterization of mammalian and prokaryote-based cell production systems, adoption of plant molecular pharming is still hindered. Thus, numerous additional strategies based on plants system need to be explored to produce sustainable products derived from mainly from plants. We have discussed that how integration of various side stream products has potential to double the revenue. Furthermore, using plants as model systems to produce recombinant proteins is more advantageous, cost-effective strategy, and reduces the risks of pathogen invasion and deadly infections as occur in mammals and bacterial systems. Moreover, use of plants as a sustainable production platform also prevents penalties by regulatory authorities as plants can better match with the mandates and goals of

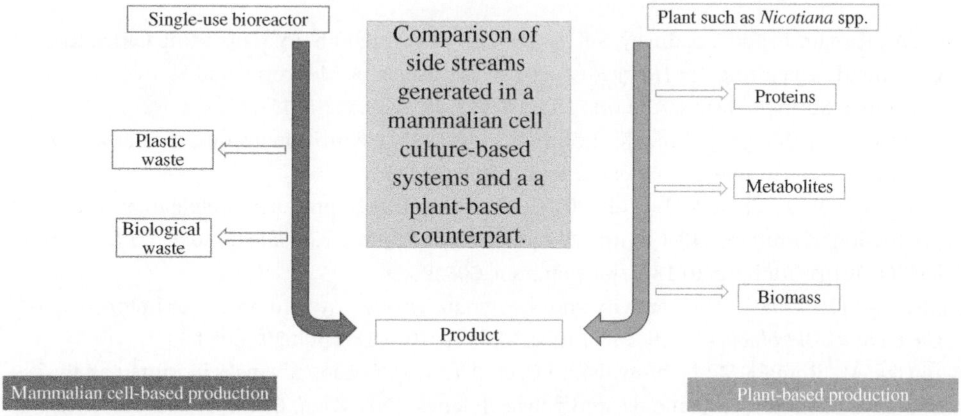

Figure 18.4 Comparative overview of waste/side streams produced during mammalian and plant cell-based systems.

environmental protection legislation as compared to that of mammals and prokaryotes. Identification of novel side stream products from residual biomass would be a great approach to increase revenues, therefore elaborated research on integrating and implementing plant-based production systems such a vertical farm is a need of the hour. Furthermore, progression toward value chain needs comprehensive studies such as toxicity analysis, biomarker identification, therapeutic interventions, diverse clinical trials, and consent of health and other regulatory agencies.

Despite all strengths of *plant* expression systems for specific applications there are major bottlenecks that must be addressed before the plant cell culture platform can compete with conventional bacterial and mammalian cell systems. Limitations include low productivity, *nonmammalian* glycan pattern, genetic instability, proteolytic degradation, scale up of cell in bioreactors, cell aggregation, and culture heterogeneity, as well as difficulty with cryopreservation. Large cell size and complex morphology represent two distinctive aspects of plant cell culture that determine bioreactor process development. Moreover, considerable investment and cooperation with various industrial sectors such as pharmaceuticals is utmost important to launch more biopharmaceuticals that are mainly derived from plants. Critical analysis after comparing major types of plant cultivation methods such as open field, greenhouse, and vertical farm should be done to find out the impact of fertilizer use, energy, socioeconomic acceptance in order to find best method and to mitigate all the limitations. For the ultimate adoption and success of plant bioreactor technologies, it is crucial to overcome the scalability issue in downstream processing involving recovery and the purification of biosynthetic protein hence vastly expand the approval pipeline of plant-made proteins.

Conflict of Interest

The authors declare no conflict of interest.

References

Alok, A., Kumar, J., and Upadhyay, S.K. (2018). Engineering in hairy roots using CRISPR/Cas9-mediated editing. In: *Hairy Roots* (ed. V. Srivastava, S. Mehrotra, and S. Mishra). Singapore: Springer. https://doi.org/10.1007/978-981-13-2562-5_15.

Appenroth, K.-J., Sowjanya Sree, K., Böhm, V. et al. (2017). Nutritional value of duckweeds (Lemnaceae) as human food. *Food Chem.* 217: 266–273.

Bai, L.L., Yin, W.B., Chen, Y.H. et al. (2013). A new strategy to produce a defensin: stable production of mutated NP-1 in nitrate reductase-deficient *Chlorella ellipsoidea*. *PLoS One* 8: e54966. https://doi.org/10.1371/journal.pone.0054966.

Baldrian, P. (2017). Microbial activity and the dynamics of ecosystem processes in forest soils. *Curr. Opin. Microbiol.* 37: 128–134. https://doi.org/10.1016/j.mib.2017.06.008.

Baranska, M., Roman, M., Dobrowolski, J.C. et al. (2013). Recent advances in raman analysis of plants: alkaloids, carotenoids, and polyacetylenes. *Curr. Anal. Chem.* 9: 108–127. https://doi.org/10.2174/157341113804486455.

Basaran, P. and Rodriguez-Cerezo, E. (2008). Plant molecular farming: opportunities and challenges. *Crit. Rev. Biotechnol.* 28: 153–172.

Bertran, K., Thomas, C., Guo, X. et al. (2015). Expression of H5 hemagglutinin vaccine antigen in common duckweed (Lemna minor) protects against H5N1 high pathogenicity avian influenza virus challenge in immunized chickens. *Vaccine* 33: 3456–3462. https://doi.org/10.1016/j.vaccine.2015.05.076.

Bethencourt, V. (2009). Virus stalls genzyme plant. *Nat. Biotechnol.* 27: 681–681. https://doi.org/10.1038/nbt0809-681a.

Bhaskar, P.B., Venkateshwaran, M., Wu, L. et al. (2009). *Agrobacterium* – mediated tran-sient gene expression and silencing: a rapid tool for functional gene assay in pota-to. *PLoS One* 4: e5812.

Bock, R. (2014). Genetic engineering of the chloroplast: novel tools and new applications. *Curr. Opin. Biotechnol.* 26 (7–13): 158.

Borisjuk, N.V., Borisjuk, L.G., Logendra, S. et al. (1999). Production of recombinant proteins in plant root exudates. *Nat. Biotechnol.* 17: 466–469.

Burnett, M.J.B. and Burnett, A.C. (2020). Therapeutic recombinant protein production in plants: challenges and opportunities. *Plants People Planet* 2: 121–132.

Buyel, J.F. (2015). Process development strategies in plant molecular farming. *Curr. Pharm. Biotechnol.* 16: 966–982. https://doi.org/10.2174/1389201016111509021154l3.

Cardon, F., Pallisse, R., Bardor, M. et al. (2019). Brassica rapa hairy root based expression system leads to the production of highly homogenous and reproducible profiles of recombinant human alpha-L-iduronidase. *Plant Biotechnol. J.* 17: 505–516. https://doi.org/10.1111/pbi.12994.

Chen, R., Xue, G., Chen, P. et al. (2008). Transgenic maize plants expressing a fungal phytase gene. *Transgenic Res.* 17: 633–643.

Chilton, M.D., Tepfer, D., Petit, A. et al. (1982). *Agrobacterium rhizogenes* inserts T-DNA into the genomes of host plant root cells. *Nature* 295: 432–434.

Condori, J., Medrano, G., Sivakumar, G. et al. (2009). Functional characterization of a stilbene synthase gene using a transient expression systemin planta. *Plant Cell Rep.* 28: 589–599.

Conley, A.J., Joensuu, J.J., Menassa, R., and Brandle, J.E. (2009). Induction of protein body formation in plant leaves by elastin-like polypeptide fusions. *BMC Biol.* 7: 48.

Cox, K.M., Sterling, J.D., Regan, J.T. et al. (2006). Glycan optimization of a human monoclonal antibody in the aquatic plant Lemna minor. *Nat. Biotechnol.* 24: 1591–1597. https://doi.org/10.1038/nbt1260.

Daniell, H. (2006). Production of biopharmaceuticals and vaccines in plants via the chloroplast genome. *Biotechnol. J.* 10: 1071–1079.

De Leede, L.G., Humphries, J.E., Bechet, A.C. et al. (2008). Novel controlled-release lemna-derived IFN-α 2b (locteron): pharmacokinetics, pharmacodynamics, and tolerability in a phase I clinical trial. *J. Interf. Cytokine Res.* 28: 113–122.

Demain, A.L. and Vaishnav, P. (2009). Production of recombinant proteins by microbes and higher organisms. *Biotechnol. Adv.* 27: 297–306. https://doi.org/10.1016/j.biotechadv.2009.01.008.

Denecke, J., Botterman, J., and Deblaere, R. (1990). Protein secretion in plant cells can occur via a default pathway. *Plant Cell* 2: 51–59.

Dixit, S., Shukla, A., Singh, V., and Upadhyay, S.K. (2021a). Bioprospecting of natural compounds for industrial and medical applications. In: *Bioprospecting of Plant Biodiversity for Industrial Molecules* (ed. S.K. Upadhyay and S.P. Singh). John Wiley & Sons Ltd. https://doi.org/10.1002/9781119718017.ch3.

Dixit, S., Shukla, A., Singh, V., and Upadhyay, S.K. (2021b). Engineering of plant metabolic pathway for nutritional improvement. In: *Genome Engineering for Crop Improvement* (ed. S.K. Upadhyay). John Wiley & Sons Ltd. https://doi.org/10.1002/9781119672425.ch20.

Dunoyer, P., Himber, C., Ruiz-Ferrer, V. et al. (2007). Intra- and intercellular RNA interference in *Arabidopsis* thaliana requires components of the micro RNA and heterochromatic silencing pathways. *Nat. Genet.* 39: 848–856.

Ebeling, A., Meyer, S.T., Abbas, M. et al. (2014). Plant diversity impacts decomposition and herbivory via changes in aboveground arthropods. *PLoS One* 9: e106529. https://doi.org/10.1371/journal.pone.0106529.

Egelkrout, E., Rajan, V., and Howard, J.A. (2012). Overproduction of recombinant proteins in plants. *Plant Sci.* 184: 83–101. 164.

Fiedler, U., Phillips, J., Artsaenko, O., and Conrad, U. (1997). Optimization of scFv antibody production in transgenic plants. *Immunotechnology* **3**: 205–216.

Firsov, A., Tarasenko, I., Mitiouchkina, T. et al. (2015). High-yield expression of M2e peptide of avian influenza virus H5N1 in transgenic duckweed plants. *Mol. Biotechnol.* 57: 653–661. https://doi.org/10.1007/s12033-015-9855-4.

Francisco, J.A., Gawlak, S.L., Miller, M. et al. (1997). Expression and characterization of bryodin 1 and a bryodin 1-based single-chain immunotoxin from tobacco cell culture. *Bioconjug. Chem.* 8: 708–713.

Fu, T.J., Sinh, G., and Curtis, W.R. (1999). *Plant Cell and Tissue Culture for the Production of Food Ingredients*. US, Boston, MA: Springer.

Fujiuchi, N., Matsuda, R., Matoba, N., and Fujiwara, K. (2014). Effect of nitrate concentration in nutrient solution on hemagglutinin content of Nicotiana benthamiana leaves in a viral vector-mediated transient gene expression system. *Plant Biotechnol.* 31: 207–211.

Fujiuchi, N., Matoba, N., and Matsuda, R. (2016). Environment control to improve recombinant protein yields in plants based on *Agrobacterium* – mediated transient gene expression. *Front. Bioeng. Biotechnol.* 4: 23.

Fujiuchi, N., Matsuda, R., Matoba, N., and Fujiwara, K. (2017). Effects of plant density on recombinant hemagglutinin yields in an *Agrobacterium* – mediated transient gene expression system using Nicotiana benthamiana plants. *Biotechnol. Bioeng.* 114: 1762–1770.

Ganapathy, M. (2016). Plants as bioreactors-a review. *Adv. Tech. Biol. Med.* 4: 161.

Gao, L., Su, Y.J., and Wang, T. (2010). Plastid genome sequencing, comparative genomics, and phylogenomics: current status and prospects. *J. Syst. Evol.* 48: 77–93.

Georgiev, M.I., Agostini, E., Ludwig-Müller, J., and Xu, J. (2012). Genetically transformed roots: from plant disease to biotechnological resource. *Trends Biotechnol.* 30: 528–537. https://doi.org/10.1016/j.tibtech.2012.07.001.

Giddings, G., Allison, G., Brooks, D., and Carter, A. (2000). Transgenic plants as factories for biopharmaceuticals. *Nat. Biotechnol.* 18: 1151–1155.

Golovkin, M., Spitsin, S., Andrianov, V. et al. (2007). Smallpox subunit vaccine produced in planta confers protection in mice. *Proc. Natl. Acad. Sci. USA.* 104: 6864–6869.

Gomord, V. and Faye, L. (2004). Posttranslational modification of therapeutic proteins in plants. *Curr. Opin. Plant Biol.* 7: 171–181.

Green, B.J., Fujiki, M., Mett, V. et al. (2009). Transient protein expression in three Pisum sativum (green pea) varieties. *Biotechnol. J.* 4: 230–237.

Guo Y, Rumschlag-Booms E, Wang J, Xiao H, Yu J, Wang J, Guo L, Gao GF, Cao Y, Caffrey M, Rong L. Analysis of hemagglutinin-mediated entry tropism of H5N1 avian influenza. *Virol. J.* 2009;6:39. doi: https://doi.org/10.1186/1743-422X-6-39.

Gurusamy, P.D., Schäfer, H., Ramamoorthy, S., and Wink, M. (2017). Biologically active recombinant human erythropoietin expressed in hairy root cultures and regenerated plantlets of *Nicotiana tabacum* L. *PLoS One* 12 (8): e0182367. https://doi.org/10.1371/journal.pone.0182367.

Häkkinen, S.T. and Ritala, A. (2010). Medicinal compounds produced in plant cell factories. In: *Medicinal Plant Biotechnology* (ed. R. Arora), 13–35. Oxfordshire, U. K.: CABI Publishing https://doi.org/10.1079/9781845936785.001.

Harrison, M.D., Zhang, Z., Shand, K. et al. (2014). The combination of plant-expressed cellobiohydrolase and low dosages of cellulases for the hydrolysis of sugar cane bagasse. *Biotechnol. Biofuels* 7 (131): 153.

Hawkins, R.L. and Nakamura, M. (1999). Expression of human growth hormone by the eukaryotic alga. *Chlorella. Curr. Microbiol.* 38: 335–341.

Hehle, V.K., Paul, M.J., Roberts, V.A. et al. (2016). Site-targeted mutagenesis for stabilization of recombinant monoclonal antibody expressed in tobacco (*Nicotiana tabacum*) plants. *FASEB J.* 30: 1590–1598.

Hellwig, S., Drossard, J., Twyman, R.M., and Fischer, R. (2004). Plant cell cultures for the production of recombinant proteins. *Nat. Biotechnol.* 22: 1415–1422. 82.

Holtz, B.R., Berquist, B.R., Bennett, L.D. et al. (2015). Commercial-scale biotherapeutics manufacturing facility for plant-made pharmaceuticals. *Plant Biotechnol. J.* 13: 1180–1190. https://doi.org/10.1111/pbi.12469.

Hood, E.E., Love, R., Lane, J. et al. (2007). Subcellular targeting is a key condition for high-level accumulation of cellulase protein in transgenic maize seed. *Plant Biotechnol. J.* 5: 709–719.

Huang, Z., Phoolcharoen, W., Lai, H. et al. (2010). High-level rapid pro-duction of full-size monoclonal antibodies in plants by a single-vector DNA repli-con system. *Biotechnol. Bioeng.* 106: 9–17.

Huether, C.M., Lienhart, O., Baur, A. et al. (2005). Glyco-engineering of moss lacking plant-specific sugar residues. *Plant Biol.* 7: 292–299. https://doi.org/10.1055/s-2005-837653.

Huy, N.X., Yang, M.S., and Kim, T.G. (2011). Expression of a cholera toxin B subunit-neutralizing epitope of the porcine epidemic diarrhea virus fusion gene in transgenic lettuce (*Lactuca sativa* L.). *Mol. Biotechnol.* 48 (3): 201–209.

Isikgor, F.H. and Becer, C.R. (2015). Lignocellulosic biomass: a sustainable platform for the production of bio-based chemicals and polymers. *Polym. Chem.* 6: 4497–4559. https://doi.org/10.1039/C5PY00263J.

Jinkerson, R.E. and Jonikas, M.C. (2015). Molecular techniques to interrogate and edit the *Chlamydomonas* nuclear genome. *Plant J.* 82: 393–412.

Johansen, F.E., Pekna, M., Norderhaug, I.N. et al. (1999). Absence of epithelial immunoglobulin A transport, with increased mucosal leakiness, in polymeric immunoglobulin receptor/secretory component-deficient mice. *J. Exp. Med.* 190: 915–922.

Kagermann, H., Wahlster, W., and Helbig, J. (2013). *Recommendations for Implementing the Strategic Initiative INDUSTRIE 4.0: Securing the Future of German Manufacturing Industry; Final Report of the Industrie 4.0 Working Group*. Frankfurt: acatech – National Academy of Science and Engineering.

Kaiser, J. (2008). Is the drought over for pharming? *Science* 320 (5875): 473–475. doi: 10.1126/Science.320.5875.473.

Kim, Y., Wyslouzil, B.E., and Weathers, P.J. (2002). Secondary metabolism of hairy root cultures in bioreactors. *In vitro Cell Dev. Biol. Plant.* 38: 1–10. https://doi.org/10.1079/IVP2001243.

Kim, M.J., Baek, K., and Park, C.M. (2009). Optimization of conditions for transient *Agrobacterium* – mediated gene expression assays in *Arabidopsis*. *Plant Cell Rep.* 28: 1159–1167.

Kircheis, R., Halanek, N., Koller, I. et al. (2012). Correlation of ADCC activity with cytokine release induced by the stably expressed, glyco-engineered humanized Lewis Yspecific monoclonal antibody MB314. *MAbs* 4: 532–541.

Klose, H., Gunl, M., Usadel, B. et al. (2015). Cell wall modification in tobacco by differential targeting of recombinant endoglucanase from Trichoderma reesei. *BMC Plant Biol.* 15: 54. https://doi.org/10.1186/s12870-015-0443-3.

Koprivova, A., Stemmer, C., Altmann, F. et al. (2004). Targeted knockouts of *Physcomitrella* lacking plant-specific immunogenic N-glycans. *Plant Biotechnol. J.* 2: 517–523.

Kwon, K.C. and Daniell, H. (2016). Oral delivery of protein drugs bioencapsulated in plant cells. *Mol. Ther.* 24: 1342–1350.

Kwon, K.C., Verma, D., Singh, N.D. et al. (2013). Oral delivery of human biopharmaceuticals, autoantigens and vaccine antigens bioencapsulated in plant cells. *Adv. Drug Deliv. Rev.* 65: 782–799.

Lambertz, C., Ece, S., Fischer, R., and Commandeur, U. (2016). Progress and obstacles in the production and application of recombinant lignin-degrading peroxidases. *Bioengineered* 7: 145–154. https://doi.org/10.1080/21655979.2016.119 1705.

Lau, O.S. and Sun, S.S. (2009). Plant seeds as bioreactors for recombinant protein production. *Biotechnol. Adv.* 27: 1015–1022.

Leuzinger, K., Dent, M., Hurtado, J. et al. (2013). Efficient agroinfiltration of plants for high-level transient expression of recombinant proteins. *J. Vis. Exp.* 77: 50521.

Li, Q., Song, J., Peng, S. et al. (2014). Plant biotechnology for lignocellulosic biofuel production. *Plant Biotechnol. J.* 12: 1174–1192. https://doi.org/10.1111/pbi.12273.

Liu, J.H., Selinger, L.B., Cheng, K.J. et al. (1997). Plant seed oil-bodies as an immobilization matrix for a recombinant xylanase from the rumen fungus Neocallimastix patriciarum. *Mol. Breed.* 3: 463–470.

Liu, J., Dolan, M., Reidy, M., and Cramer, C. (2008). Expression of bioactive single-chain murine IL-12 intransgenic plants. *J. Interf. Cytokine Res.* 28: 381–392.

Liu, C.Z., Towler, M.J., Medrano, G. et al. (2009). Production of mouse interleukin-12 is greater in tobacco hairy roots grown in a mist reactor than in an airlift reactor. *Biotechnol. Bioeng.* 102: 1074–1086.

Liu, Y.K., Li, Y.T., Lu, C.F., and Huang, L.F. (2015). Enhancement of recombinant human serum albumin in transgenic rice cell culture system by cultivation strategy. *New Biotechnol.* 32: 328–334.

Llop-Tous, I., Ortiz, M., Torrent, M., and Ludevid, M.D. (2011). The expression of a xylanase targeted to ER-protein bodies provides a simple strategy to produce active insoluble enzyme polymers in tobacco plants. *PLoS One* 6: e19474.

Luchakivskaia I. S., Olevinskaia Z. M., Kishchenko E. M., Spivak N. I., Kuchuk N. V (2012). Obtaining of hairy-root, callus and suspension carrot culture (*Daucus carota* L.) able to accumulate human interferon alpha-2b. *Tsitol. Genet.* 46, 18–26. https://doi.org/10.3103/S0095452712010057

Ma, J.K., Hikmat, B.Y., Wycoff, K. et al. (1998). Characterization of a recombinant plant monoclonal secretory antibody and preventive immunotherapy in humans. *Nat. Med.* 4: 601–606.

Ma, J.K., Drake, P.M., and Christou, P. (2003). The production of recombinant pharmaceutical proteins in plants. *Nat. Rev. Genet.* 2003 (4): 794–805.

Ma, J.K., Drossard, J., Lewis, D. et al. (2015). Regulatory approval and a first-in-human phase I clinical trial of a monoclonal antibody produced in transgenic tobacco plants. *Plant Biotechnol. J.* 13: 1106–1120.

Magnusdottir, A., Vidarsson, H., Bjornsson, J.M., and Orvar, B.L. (2013). Barley grains for the production of endotoxin-free growth factors. *Trends Biotechnol.* 31: 572–580.

Maliga, P. (2003). Progress towards commercialisation of plastid transformation technology. *Trends Biotechnol.* 21: 20–28.

Matoba, N., Davis, K.R., and Palmer, K.E. (2011). Recombinant protein expression in Nicotiana. *Methods Mol. Biol.* 701: 199–219.

Mayfield, S.P., Manuell, A.L., Chen, S. et al. (2007). *Chlamydomonas reinhardtii* chloroplasts as protein factories. *Curr. Opin. Biotechnol.* 18: 126–133. https://doi.org/10.1016/j.copbio.2007.02.001.

McDonald, K.A., Hong, L.M., Trombly, D.M. et al. (2005). Production of human alpha-1-antitrypsin from transgenic rice cell culture in a membrane bioreactor. *Biotechnol. Prog.* 21: 728–734.

McKenzie, E.A. and Abbott, W.M. (2018). Expression of recombinant proteins in insect and mammalian cells. *Methods* 147: 40–49. https://doi.org/10.1016/j.ymeth.2018.05.013.

McLean, M.D., Chen, R., Yu, D. et al. (2012). Purification of the therapeutic antibody trastuzumab from genetically modified plants using safflower Protein A-oleosin oilbody technology. *Transgenic Res.* 21: 1291–1301.

Medina-Bolívar F, Cramer C. Production of recombinant proteins by hairy roots cultured in plastic sleeve bioreactors. *Methods Mol. Biol.* 2004;267:351–63. doi: https://doi.org/10.1385/1-59259-774-2:351.

Medrano, G., Reidy, M.J., Liu, J. et al. (2009). Rapid system for evaluatingbioproduction capacity of complex pharmaceutical proteins in plants. *Methods Mol. Biol.* 483: 51–56.

Menassa, R., Nguyen, V., Jevnikar, A., and Brandle, J. (2001). A self-contained system for the field production of plant recombinant interleukin-10. *Mol. Breed.* 8: 177–185.

Menassa, R., Du, C., Yin, Z.Q. et al. (2007). Therapeutic effectiveness of orally administered transgenic low-alkaloid tobacco expressing human interleukin-10 in a mouse model of colitis. *Plant Biotechnol. J.* 5: 50–59.

Merlin, M., Gecchele, E., Arcalis, E. et al. (2016). Enhanced GAD65 production in plants using the MagnICON transient expression system: optimization of upstream production and downstream processing. *Biotechnol. J.* 11: 542–553.

Mett, V., Farrance, C.E., Green, B.J., and Yusibov, V. (2008). Plants as biofactories. *Biologicals* 36 (6): 354–358.

Montero-Morales, L. and Steinkellner, H. (2018). Advanced plant-based glycan engineering. *Front. Bioeng. Biotechnol.* 6: 81. https://doi.org/10.3389/fbioe.2018.00081.

Moon, K.-B., Jeon, J.-H., Lee, W.-S., and Kim, H. (2012). Transient erythropoietin overexpression with cucumber mosaic virus suppressor 2b in Nicotiana benthamiana. *Hortic. Environ. Biotechnol.* 53: 434–440.

Nahampun, H.N., Bosworth, B., Cunnick, J. et al. (2015). Expression of H3N2 nucleoprotein in maize seeds and immunogenicity in mice. *Plant Cell Rep.* 34: 969–980.

Naqvi, S., Ramessar, K., Farre, G. et al. (2011). High-value products from transgenic maize. *Biotechnol. Adv.* 29: 40–53.

Negrouk, V., Eisner, G., Lee, H. et al. (2005). Highly efficient transient expression of functional recombinant antibodies in lettuce. *Plant Sci.* 169: 433–438.

Niederkruger, H., Dabrowska-Schlepp, P., and Schaaf, A. (2014). Suspension € culture of plant cells under phototrophic conditions. In: *Industrial Scale Suspension Culture of Living Cells* (ed. H.-P. Meyer and D.R. Schmidhalter), 259–292. Weinheim: Wiley-VCH Verlag GmbH & Co. KGaA.

O'Flaherty, R., Bergin, A., Flampouri, E. et al. (2020). Mammalian cell culture for production of recombinant proteins: a review of the critical steps in their biomanufacturing. *Biotechnol. Adv.* 43: 107552. https://doi.org/10.1016/j.biotechadv.2020.107552.

Octave, S. and Thomas, D. (2009). Biorefinery: toward an industrial metabolism. *Biochimie* 91: 659–664. https://doi.org/10.1016/j.biochi.2009.03.015.

Osborne, C.P. (2016). Crop yields: CO_2 fertilization dries up. *Nat. Plants* 2: 16138. https://doi.org/10.1038/nplants.2016.138.

Owczarek, B., Gerszberg, A., and Hnatuszko-Konka, K. (2019). A brief reminder of systems of production and chromatography-based recovery of recombinant protein biopharmaceuticals. *Biomed. Res. Int.* 2019: 1–13. https://doi.org/10.1155/2019/4216060.

Ozawa, K. and Takaiwa, F. (2010). Highly efficient *Agrobacterium* – mediated transformation of suspension-cultured cell clusters of rice (*Oryza sativa* L.). *Plant Sci.* 179: 333–337.

Parsons, J., Altmann, F., Arrenberg, C.K. et al. (2012). Moss-based production of asialo-erythropoietin devoid of Lewis A and other plant-typical carbohydrate determinants. *Plant Biotechnol. J.* 10: 851–861.

Parsons, J., Altmann, F., Graf, M. et al. (2013). A gene responsible for prolyl-hydroxylation of moss-produced recombinant human erythropoietin. *Sci. Rep.* 3: 3019.

Paul, M. and Ma, J.K. (2011). Plant-made pharmaceuticals: leading products and production platforms. *Biotechnol. Appl. Biochem.* 58: 58–67. https://doi.org/10.1002/bab.6.

Peebles, C.A.M., Sander, G.W., Li, M. et al. (2009). Five year maintenance of the inducible expression of anthranilate synthase in catharanthus roseus hairy roots. *Biotechnol. Bioeng.* 102: 1521–1525. https://doi.org/10.1002/bit.22173.

Peyret, H. and Lomonossoff, G.P. (2015). When plant virology met *Agrobacterium*: the rise of the deconstructed clones. *Plant Biotechnol. J.* 13 (8): 1121–1135.

Pham, N.B., Schäfer, H., and Wink, M. (2012). Production and secretion of recombinant thaumatin in tobacco hairy root cultures. *Biotechnol. J.* 7: 537–545. https://doi.org/10.1002/biot.201100430.

Pogrebnyak, N., Golovkin, M., Andrianov, V. et al. (2005). Severe acute respiratory syndrome (SARS) S protein production in plants: development of recombinant vaccine. *Proc. Natl. Acad. Sci. USA.* 102: 9062–9067.

Popov, S.V., Golovchenko, V.V., Ovodova, R.G. et al. (2006). Characterisation of the oral adjuvant effect of lemnan, a pectic polysaccharide of Lemna minor L. *Vaccine* 24: 5413–5419.

Puetz, J. and Wurm, F.M. (2019). Recombinant proteins for industrial versus pharmaceutical purposes: a review of process and pricing. *PRO* 7: 476. https://doi.org/10.3390/pr7080476.

Rademacher, T., Sack, M., Arcalis, E. et al. (2008). Recombinant antibody 2G12 produced in maize endosperm efficiently neutralizes HIV-1 and contains predominantly single-GlcNAc N-glycans. *Plant Biotechnol. J.* 6: 189–201.

Ramessar, K., Capell, T., and Christou, P. (2008a). Molecular pharming in cereal crops. *Phytochem. Rev.* 7: 579–592.

Ramessar, K., Rademacher, T., Sack, M. et al. (2008b). Cost-effective production of a vaginal protein microbicide to prevent HIV transmission. *Proc. Natl. Acad. Sci. USA.* 105: 3727–3732.

Rech, E.L. (2012). Seeds recombinant DNA and biodiversity. *Seed Sci. Res.* 22: 36–44.

Rech, E.L., Vianna, G.R., and Aragão, F.J.L. (2008). High-efficiency transformation by biolistics of soybean common bean and cotton transgenic plants. *Nat. Protoc.* 3 (410-418): 5.

Reski, R., Parsons, J., and Decker, E.L. (2015). Moss-made pharmaceuticals: from bench to bedside. *Plant Biotechnol. J.* 13: 1191–1198.

Rival, S., Wisniewski, J.-P., Langlais, A. et al. (2008). *Spirodela* (duckweed) as an alternative production system for pharmaceuticals: a case study, aprotinin. *Transgenic Res.* 17: 503–513. https://doi.org/10.1007/s11248-007-9123-x.

Rivera, A.L., Gómez-Lim, M., Fernández, F., and Loske, A.M. (2012). Physical methods for genetic plant transformation. *Phys Life Rev* 9 (3): 308–345.

Rosales-Mendoza, S., Salazar-Gonzalez, J.A., Decker, E.L., and Reski, R. (2016). Implications of plant glycans in the development of innovative vaccines. *Expert Rev. Vaccines* 15: 915–925. https://doi.org/10.1586/14760584.2016.1155987.

Routray, W. and Orsat, V. (2013). Preparative extraction and separation of phenolic compounds. In: *Natural Products: Phytochemistry, Botany and Metabolism of Alkaloids, Phenolics and Terpenes* (ed. K.G. Ramawat and J.-M. Mérillon), 2013–2045. Berlin: Springer.

Rup, B., Alon, S., Amit-Cohen, B.C. et al. (2017). Immunogenicity of glycans on biotherapeutic drugs produced in plant expression systems – the taliglucerase alfa story. *PLoS One* 12: e0186211. https://doi.org/10.1371/journal.pone.0186211.

Sabalza, M., Vamvaka, E., Christou, P., and Capell, T. (2013). Seeds as a production system for molecular pharming applications: status and prospects. *Curr. Pharm. Des.* 19: 5543–5552.

Saberianfar, R., Joensuu, J.J., Conley, A.J., and Menassa, R. (2015). Protein body formation in leaves of Nicotiana benthamiana: a concentration-dependent mechanism influenced by the presence of fusion tags. *Plant Biotechnol. J.* 13: 927–937.

Sack, M., Rademacher, T., Spiegel, H. et al. (2015). From gene to harvest: insights into upstream process development for the GMP production of a monoclonal antibody in transgenic tobacco plants. *Plant Biotechnol. J.* 13: 1094–1105. https://doi.org/10.1111/pbi.12438.

Santos-Buelga, C., Gonzalez-Manzano, S., Duenas, M., and Gonzalez-Paramas, A.M. (2012). Extraction and isolation of phenolic compounds. *Methods Mol. Biol.* 864: 427–464. https://doi.org/10.1007/978-1-61779-624-1_17.

Sardana, R., Dudani, A.K., Tackaberry, E. et al. (2007). Biologically active human GM-CSF produced in the seeds of transgenic rice plants. *Transgenic Res.* 16: 713–721.

Scaife, M.A., Nguyen, G.T., Rico, J. et al. (2015). Establishing *Chlamydomonas reinhardtii* as an industrial biotechnology host. *Plant J.* 82 (3): 532–546. https://doi.org/10.1111/tpj.12781.

Schillberg, S., Twyman, R.M., and Fischer, R. (2005). Opportunities for recombinant antigen and antibody expression in transgenic plants–technology assessment. *Vaccine* 23: 1764–1769.

Schillberg, S., Raven, N., Fischer, R. et al. (2013). Molecular farming of pharmaceutical proteins using plant suspension cell and tissue cultures. *Curr. Pharm. Des.* 19: 5531–5542. https://doi.org/10.2174/1381612811319310008.

Schillberg, S., Raven, N., Spiegel, H. et al. (2019). Critical analysis of the commercial potential of plants for the production of recombinant proteins. *Front. Plant Sci.* 10: 720. https://doi.org/10.3389/fpls.2019.00720.

Schuster, M., Jost, W., Mudde, G.C. et al. (2007). *In vivo* glyco-engineered antibody with improved lytic potential produced by an innovative non-mammalian expression system. *Biotechnol. J.* 2: 700–708.

Shanks, J.V. and Morgan, J. (1999). Plant 'hairy root' culture. *Curr. Opin. Biotechnol.* 10: 151–155.

Shanmugaraj, B. and Phoolcharoen, W. (2021). Addressing demand for recombinant biopharmaceuticals in the COVID-19 era. *Asian Pac J Trop Med* 14: 49–51. https://doi.org/10.4103/1995-7645.306736.

Shanmugaraj, B., Bulaon, C.J.I., and Phoolcharoen, W. (2020). Plant molecular farming: a viable platform for recombinant biopharmaceutical production. *Plan. Theory* 9: 842. https://doi.org/10.3390/plants9070842.

Sharma, H. and Upadhyay, S.K. (2020). Enzymes and their production strategies. In: *Advances in Enzyme Catalysis and Technologies* (ed. S.P. Singh, R.R. Singhania, Z. Li et al.). Elsevier. https://doi.org/10.1016/B978-0-12-819820-9.00003-X.

Sheludko, Y.V., Sindarovska, Y.R., Gerasymenko, I.M. et al. (2007). Compari-son of several Nicotiana species as hosts for high-scale *Agrobacterium* – mediatedtransient expression. *Biotechnol. Bioeng.* 96: 608–614.

Shin, Y.J., Hong, S.Y., Kwon, T.H. et al. (2003). High level of expression of recombinant human granulocyte-macrophage colony stimulating factor in transgenic rice cell suspension culture. *Biotechnol. Bioeng.* 82: 778–783.

Shin, Y.J., Lee, N.J., Kim, J. et al. (2010). High-level production of bioactive heterodimeric protein human interleukin-12 in rice. *Enzym. Microb. Technol.* 46: 347–351.

Sindarovska, Y.R., Sheludko, Y.V., Gerasymenko, I.M. et al. (2008). Purifca-tion of recombinant GFP produced by *Agrobacterium* – mediated transient expres-sion in Nicotiana excelsior. *Tsitol. Genet.* 42: 16–20.

Sivakumar, G. (2006). Bioreactor technology: a novel industrial tool for high-tech production of bioactive molecules and bioparmaceutical from plant roots. *J. Biotechnol.* 1: 1419–1427.

Spencer, D., Dickey, L.F., Gasdaska, J.R. et al. (2010). Expression of Plasminogen and Microplasminogen in Duckweed. U.S. Patent AU2005212431(B2), 15 July 2010.

Spiegel, H., Boes, A., Voepel, N. et al. (2015). Application of a scalable plant transient gene expression platform for malaria vaccine development. *Front. Plant Sci.* 6: 1169. https://doi.org/10.3389/fpls.2015.01169.

von Stackelberg, M., Rensing, S.A., and Reski, R. (2006). Identification of genic moss SSR markers and a comparative analysis of twenty-four algal and plant gene indices reveal species-specific rather than group-specific characteristics of microsatellites. *BMC Plant Biol.* 6: 9.

Stoger, E., Sack, M., Nicholson, L. et al. (2005a). Recent progress in plantibody technology. *Curr. Pharm. Des.* 11: 2439–2457.

Stoger, E., Ma, J.K., Fischer, R., and Christou, P. (2005b). Sowing the seeds of success: pharmaceutical proteins from plants. *Curr. Opin. Biotechnol.* 16: 167–173.

Stoger, E., Fischer, R., Moloney, M., and Ma, J.K.-C. (2014). Plant molecular pharming for the treatment of chronic and infectious diseases. *Annu. Rev. Plant Biol.* 65: 743–768. https://doi.org/10.1146/annurev-arplant-050213-035850.

Stomp, A.M. (2005). The duckweeds: a valuable plant for biomanufacturing. *Biotechnol. Annu. Rev.* 11: 69.

Sun, Y., Cheng, J.J., Himmel, M.E. et al. (2007). Expression and characterization of *Acidothermus cellulolyticus* E1 endoglucanase in transgenic duckweed Lemna minor 8627. *Bioresour. Technol.* 98: 2866–2872. https://doi.org/10.1016/j.biortech.2006.09.055.

Terashima, M., Murai, Y., Kawamura, M. et al. (1999). Production of functional human alpha 1-antitrypsin by plant cell culture. *Appl. Microbiol. Biotechnol.* 52: 516–523.

Torres, E., Vaquero, C., Nicholson, L. et al. (1999). Rice cell culture as an alternative production system for functional diagnostic and therapeutic antibodies. *Transgenic Res.* 8: 441–449.

Trexler, M.M., McDonald, K.A., and Jackman, A.P. (2005). A cyclical semicontinuous process for production of human alpha 1-antitrypsin using metabolically induced plant cell suspension cultures. *Biotechnol. Prog.* 21: 321–328.

Triguero, A., Cabrera, G., Rodriguez, M. et al. (2011). Differential N-glycosylation of a monoclonal antibody expressed in tobacco leaves with and without endoplasmic reticulum retention signal apparently induces similar in vivo stability in mice. *Plant Biotechnol. J.* 9: 1120–1130.

Turchinovich, A.A., Deineko, E.V., Filipenko, M.L. et al. (2004). Transgenic tobacco plants producing human interleukin-18. *Dokl. Biochem. Biophys.* 395: 104–107. 151.

Twyman, R.M., Schillberg, S., and Fischer, R. (2013). Optimizing the yield of recombinant pharmaceutical proteins in plants. *Curr. Pharm. Des.* 19: 5486–5494.

Upadhyay, S.K. and Singh, S.P. (eds.) (2021). *Bioprospecting of Plant Biodiversity for Industrial Molecules*. John Wiley & Sons Ltd. https://doi.org/10.1002/9781119718017.

Van Droogenbroeck, B., Cao, J., Stadlmann, J. et al. (2007). Aberrant localization and underglycosylation of highly accumulating single-chain Fv-Fc antibodies in transgenic *Arabidopsis* seeds. *Proc. Natl. Acad. Sci. USA.* 104: 1430–1435.

Vargo, K.B., Parthasarathy, R., and Hammer, D.A. (2012). Self-assembly of tunable protein suprastructures from recombinant oleosin. *Proc. Natl. Acad. Sci. USA.* 109: 11657–11662.

Vavitsas, K., Fabris, M., and Vickers, C.E. (2018). Terpenoid metabolic engineering in photosynthetic microorganisms. *Gene* 9: E520.

Verma, P.C., Trivedi, I., Singh, H. et al. (2009). Efficient production of gossypol from hairy root culture of cotton (*Gossypium hirsutum*). *Current Pharmaceutical Biotechnology* 10: 691–700. https://doi.org/10.2174/138920109789542048.

Vickers, C.E., Williams, T.C., Peng, B., and Cherry, J. (2017). Recent advances in synthetic biology for engineering isoprenoid production in yeast. *Curr. Opin. Chem. Biol.* 40: 47–56. https://doi.org/10.1016/j.cbpa.2017.05.017.

Vitale, A. and Pedrazzini, E. (2005). Recombinant pharmaceuticals from plants: the plant endomembrane system as bioreactor. *Mol. Interv.* 5: 216–225.

Vunsh, R., Li, J., Hanania, U. et al. (2007). High expression of transgene protein in Spirodela. *Plant Cell Rep.* 26 (9): 1511–1519.

Waheed, M.T., Ismail, H., Gottschamel, J. et al. (2015). Plastids: the green frontiers for vaccine production. *Front. Plant Sci.* 6 (1005): 159.

Wang, J., Miao, Y., and Jiang, l. (2006). Response to Gomord et al.: Golgi-bypassing: delivery of biopharmaceutical proteins to protein storage vacuoles in plant bioreactors. *Trends Biotechnol.* 24: 147–149.

Wang, D.J., Brandsma, M., Yin, Z.Q. et al. (2008). A novel platform for biologically active recombinant human interleukin-13 production. *Plant Biotechnol. J.* 6: 504–515.

Weichert, N., Hauptmann, V., Helmold, C., and Conrad, U. (2016). Seed-specific expression of spider silk protein multimers causes long-term stability. *Front. Plant Sci.* 7: 6.

Weise, A., Altmann, F., Rodriguez-Franco, M. et al. (2007). High-level expression of secreted complex glycosylated recombinant human erythropoietin in the Physcomitrella D-fuc-t D-xyl-t mutant. *Plant Biotechnol. J.* 5: 389–401.

Wilson, S.A. and Roberts, S.C. (2012). Recent advances towards development and commercialization of plant cell culture processes for the synthesis of biomolecules. *Plant Biotechnol. J.* 10 (3): 249–268.

Woodard, S.L., Mayor, J.M., Bailey, M.R. et al. (2003). Maize (Zea mays)-derived bovine trypsin: characterization of the first large-scale, commercial protein product from transgenic plants. *Biotechnol. Appl. Biochem.* 38: 123–130.

Woods, R.R., Geyer, B.C., and Mor, T.S. (2008). Hairy-root organ cultures for the production of human acetylcholinesterase. *BMC Biotechnol.* 8: 95.

Xia, P., Hu, W., Liang, T. et al. (2020). An attempt to establish an *Agrobacterium* – mediated transient expression system in medicinal plants. *Protoplasma* 257 (6): 1497–1505.

Xu, J. and Zhang, N. (2014). On the way to commercializing plant cell culture platform for biopharmaceuticals: present status and prospect. *Pharm. Bioprocess* 2: 499–518.

Xu, L., Brilman, D.W.W., Withag, J.A. et al. (2011). Assessment of a dry and a wet route for the production of biofuels from microalgae: energy balance analysis. *Bioresour. Technol.* 102: 5113–5122. https://doi.org/10.1016/j.biortech.2011.01.066.

Yamamoto, Y.-T., Rajbhandari, N., Lin, X. et al. (2001). Genetic transformation of duckweed Lemna gibba and Lemna minor. *In vitro Cell. Dev. Biol. Plant.* 37: 349–353.

Yang, W., Zhang, Y., Zhou, X. et al. (2015). Production of a highly protease-resistant fungal alpha-galactosidase in transgenic maize seeds for simplified feed processing. *PLoS One* 10: e0129294.

Yang, J., Li, G., Hu, S. et al. (2018). A protocol for efficient callus induction and stable transformation of Spirodela polyrhiza (L.) Schleiden using *Agrobacterium tumefaciens. Aquat. Bot.* 151: 80–86. https://doi.org/10.1016/j.aquabot.2018.08.004.

Yao, J., Weng, Y., Dickey, A., and Wang, K.Y. (2015). Plants as factories for human pharmaceuticals: applications and challenges. *Int. J. Mol. Sci.* 16: 28549–28565.

Zhang, D. (2013). Plant seed-derived human transferrin: expression, characterization, and applications. *Biotechnol.* 2: 17.

Zhang, B., Yang, Y.H., Lin, Y.M. et al. (2003). Expression and production of bioactive human interleukin-18 in transgenic tobacco plants. *Biotechnol. Lett.* 25: 1629–1635. 152.

Zhu, J., Yang, W., and He, X. (2013). Temporal dynamics of abiotic and biotic factors on leaf litter of three plant species in relation to decomposition rate along a subalpine elevation gradient. *PLoS One* 8: e62073. https://doi.org/10.1371/journal.pone.0062073.

Zhu, Z., Piao, S., Myneni, R.B. et al. (2016). Greening of the earth and its drivers. *Nat. Clim. Chang.* 6: 791–795. https://doi.org/10.1038/nclimate3004.

Ziegler, P., Adelmann, K., Zimmer, S. et al. (2015). Relative *in vitro* growth rates of duckweeds (Lemnaceae)—the most rapidly growing higher plants. *Plant Biol.* 17: 33–41.

Zimmer, A.D., Lang, D., Buchta, K. et al. (2013). Reannotation and extended community resources for the genome of the non-seed plant *Physcomitrella patens* provide insights into the evolution of plant gene structures and functions. *BMC Genomics* 14: 498.

Zimmermann, J., Saalbach, I., Jahn, D. et al. (2009). Antibody expressing pea seeds as fodder for prevention of gastrointestinal parasitic infections in chickens. *BMC Biotechnol.* 9: 79. https://doi.org/10.1186/1472-6750-9-79.

19

Production of Nutraceuticals Using Plant Cell and Tissue Culture

Elif Karlik and Elif Aylin Ozudogru

Department of Molecular Biology and Genetics, University of Istinye, Istanbul, Turkey

19.1 Introduction

Nutraceuticals is the term combined by "nutrition" and "pharmaceutical" words and was first used by De Felice in 1989 (Kalra 2003). Nutraceuticals constitute a mosaic product group including herbal/natural products, dietary supplements, and functional foods such as cereals, soups, and beverages (Dureja et al. 2003; Upadhyay and Singh 2021). Although they are mostly isolated from the plants, they can also be isolated and purified from animals, microorganisms, and marine sources. Accordingly, they are categorized depending on the different parameters, such as their sources, chemical properties, or the established stages (Kokate et al. 2002; Kalia 2005; Singh and Upadhyay 2021). Among some popular nutraceuticals are ginseng, coenzyme Q10, green tea, omega-3, folic acid, lutein, prebiotics, yoghurt as probiotics, flavonoids, dietary fibers, carotenoids, etc. Both *in vitro* and *in vivo* studies evaluate the beneficial effects of such nutraceuticals (Liu et al. 2019; Sayed et al. 2020) and indeed the usage and applications of these bioactive compounds, which are mostly flagged as a "possible beneficial" to health, cover a wide range of industrial sectors (Santini and Novellino 2014; Santini 2018; Cheng et al. 2020). Recent studies also demonstrated that nutraceuticals had promising results in improving health by the treatment and prevention of diseases, such as diabetes, atherosclerosis, cardiovascular diseases, cancer, and neurological disorders (Anwar et al. 2021; Dixit et al. 2021a,b; Cox et al. 2022; Le et al. 2022; Sottero et al. 2022; Ullah et al. 2022), as well as delaying aging process, increasing life expectancy, and supporting structure and body functions (Nasri et al. 2014). Although some of these nutraceuticals consist of compounds known as potential nutraceuticals which are accepted for their therapeutical benefits but are lacking compliance with large-scale clinical research, for the others, nutraceuticals' benefits are well established by clinical data (Pandey et al. 2010; Khalaf et al. 2021).

Plant-derived nutraceuticals, also known as secondary metabolites, are compounds that are not directly related to the growth and development of the plant; however, they are involved in the generation of the plant stress response to its biotic and/or abiotic environment

Plants as Bioreactors for Industrial Molecules, First Edition.
Edited by Santosh Kumar Upadhyay and Sudhir P. Singh.

with the aim to provide protection (Pagare et al. 2015; Dixit et al. 2019). When supplemented with the primary metabolites (i.e. basic nutrients required for growth and development), secondary metabolites provide health benefits (Holst and Williamson 2008). Accordingly, some of these nutrients also act as antioxidants and help the body remove toxic substances known as free radicals (also known as reactive oxygen species, ROS), which may result in cell damage and diseases if accumulated within the cell. They also serve as drugs, artificial flavors, pigments, cosmetic additives, and food supplements and that increase their economic value (Jaiwal 2006). Although many plants contain such compounds, it is not easy to cultivate them at a large scale throughout the year via traditional propagation techniques, as their synthesis within the plant is highly affected by seasonal and ecological conditions as well as the developmental stage of the plant. *In vitro* cultures, however, not only permit the production and extraction of useful secondary metabolites from propagated plantlets independently from the seasonal and ecological conditions but may also provide elevated levels of secondary metabolites through *in vitro* manipulations such as the use of elicitors. An elicitor can be defined as a low-molecular-weight chemical or biochemical compound that is introduced to a plant to trigger plant's immune response by activating signal cascade and hence to promote the biosynthesis of the target secondary metabolite (Patel et al. 2020). The treatment of plants with elicitors causes an array of defense reactions, including the over production and accumulation of a range of plant defensive secondary metabolites (Zhao et al. 2005).

As indicated earlier, plants are the major sources of nutraceuticals and hence they have been used for their medicinal effects through the worldwide since ancient times. Phytochemical components of plants are valuable sources for bioactive compounds that can be used in several sectors, including pharmaceuticals, cosmetics, and nutraceuticals, etc. (Nasri et al. 2014; Krishnan et al. 2021). These natural compounds are named as secondary metabolites that are not essential for the growth and development that are also divergent in their structure and metabolic pathways. Secondary metabolites mainly function as signaling molecules or defense agents that are classified as steroids, saponin, terpenes, lipids, alkaloids, terpenoids, and enzyme cofactors (Ncube and Van Staden 2015; Hussein and El-Anssary 2018). Especially today, plants as nutraceuticals are gaining more attention because of their accessibility, low-cost comparable to high-cost synthetic drugs, affordability, and eco-friendly properties. Therefore, increasing demands of nutraceuticals are required new strategies to enhance their production without disturbing their natural sources such as plants (Atanasov et al. 2015; Chandran et al. 2020).

Plants are conventionally propagated either sexually through seed germination or asexually by cutting and grafting. However, both approaches harbor disadvantages alongside their advantages. In some instances, propagation through seeds is not possible as some species either do not produce seeds or the seeds are recalcitrant to germination. In some other instances, seed propagation may not be the method of preference as the germinated seedlings are not true-to-type plants due to cross-pollination. Seed-derived plants are also known to bear inferior fruit quality (Mishra et al. 2011). Propagation by cutting or grafting harbors difficulties related to rooting and tissue competence, respectively, and require a manipulation by an expert agronomist. *In vitro* propagation through plant tissue culture techniques may overcome these issues. The method is of great importance especially when the attainment of seeds or crops from wild plants is extremely difficult in particular seasons. Indeed, *in vitro* propagation enables the maintenance of the explant source at all times of the year.

In vitro propagation of plants dates back to 1900s and includes rapid multiplication of plantlets from various plant parts (i.e. cells, tissues, or organs) in aseptic conditions, on an artificial nutrient medium, under controlled temperature and light conditions. True-to-type and disease-free plants can be produced in a large scale via micropropagation, while mass propagation of nonclonal plants can be achieved by direct and indirect organogenesis or somatic embryogenesis. Selection of appropriate explant type (i.e. plant fragments such as apical or axillary shoots, leaf, node, internode, or root segments, zygotic and somatic embryos that are used for the initiation of *in vitro* cultures) is a factor markedly significant in the redirection of plant morphogenesis. Explants composed of meristematic tissues, such as meristems, apical and axillary buds, possess genetic stability and hence are suitable for clonal propagation via micropropagation, while nonmeristematic tissues, such as internodes, leaf and root segments, seeds, zygotic, and somatic embryos, are ideal for organogenesis and somatic embryogenesis. As all these so-called explant types have high potential to possess totipotency and competence (i.e. two unique characteristics of plant tissues that make possible the production of whole plants from a given plant tissue), one is only to determine the aim of propagation and then to select the proper explant type accordingly.

In vitro culture techniques are useful tools also to improve the production of plant-derived nutraceuticals, which allows priming a sustainable and nature-friendly plants (Verma et al. 2009; Lucchesini and Mensuali-Sodi 2010; Alok et al. 2018). Plant tissue culture allows plants to grow and reproduce continuously for secondary metabolites as nutraceuticals under aseptic conditions. By using the controlled growth chambers, researchers can eliminate the variable climatic conditions that effects the production of these bioactive compounds. Moreover, controlled conditions reduce the toxic residues and biotic contaminants depending on the sterility of plant tissue culture techniques. *In vitro* propagation also leads to make clones facilitating to manipulate phenotypic variation in the concentration of bioactive compounds. Therefore, plant tissue culture aims to increase concentration, uniformity, and consistency of extracts and reduce the toxicity. Additionally, collection, storage, and transportation problems can be eliminated based on the extraction times, which can be carried out at regular intervals all over the year. The protection of wild, endangered, and/or endemic plant species is also guaranteed by plants tissue culture techniques leading to maintenance of biodiversity of germplasms in nature (Colin 2001; Lucchesini and Mensuali-Sodi 2010; Sang et al. 2018).

Survey in the literature indicates numerous studies that pointed out the presence of secondary metabolites in the *in vitro*-derived plantlets, quantified the number of certain compounds in those plantlets, and provided comparisons to the plants in nature (e.g. Sulaiman and Balachandran 2012; Yasir et al. 2017; Kruczek et al. 2020; Mamdouh et al. 2021). However, only scanty of reports are available on the production and/or extraction of those compounds from *in vitro*-derived plantlets, although plant tissue culture-based approaches possess a huge potential to enhance the production of desired secondary metabolites. Present report focuses on the latter; production and/or extraction of secondary metabolites, at the elevated levels, when possible, in *in vitro* cultures.

19.2 Production of Secondary Metabolites as Nutraceuticals in *In vitro* Cultures

19.2.1 Nutraceuticals Used in Pharmaceuticals Industry

Isoprenoids represent the most abundant and structurally diverse group of nutraceuticals, known to be involved in biochemical processes. Among these, coenzyme Q (CoQ, 2,3-dim ethoxy-5-methyl-6-polyprenyl-1,4-benzoquinone), which is naturally occurring, fat soluble, vitamin-like, membrane-bound isoprenoid quinone (Bonakdar and Guarneri 2005; Nowicka and Kruk 2010), acts as an effective natural antioxidant and thus gained a special attention in medicine and industry (Bentinger et al. 2007). It is also known as ubiquinone due to its ubiquitous distribution (Yuan et al. 2012). CoQ is naturally present in the plasma membranes of prokaryotic organisms, while in eukaryotes, all the endomembrane, (i.e. not only the plasma membrane but also the membrane of all the cellular organelles) contain it (Crane 2007).

Living organisms may contain several different forms of CoQ, but usually one of them is dominantly used, and the form that is used differs according to the organisms. For instance, *Saccharomyces cerevisiae* uses CoQ_6, *Escherichia coli* uses CoQ_8, *Schizosaccharomyces pombe* uses CoQ_{10}, and mammals like humans use mainly CoQ_{10} and small amounts of CoQ_9 as an electron and proton transporter of their aerobic respiratory electron transport chain for the production of cellular ATP energy (Lenaz et al. 2007). In the case of plants, mitochondria of higher plants contain $CoQ_{10,}$ and cereals contain CoQ_9 (Takahashi et al. 2006). The difference between these structurally different forms derives from the length of the isoprenoid side chain (Okada et al. 1998; Nowicka and Kruk 2010). CoQ_{10}, a mitochondrial isoprenoid, is one of the most potent antioxidants, and in humans, it serves as a dietary supplement and cosmetic ingredient, boosts energy, and enhances immune system (Parmar et al. 2015). Its use in therapeutic applications for several diseases (such as Alzheimer, cancer, and Parkinson) have been reported (Portakal et al. 2000; Beal 2004). Notwithstanding, although CoQ_{10} is found in wide variety of plant-based foods, their CoQ_{10} content is very limited, and it further decreases during food processing, and thus its supplement from food is not sufficient. CoQ_{10} is synthesized endogenously in humans (as well as in most other aerobic organisms) (Ernster and Dallner 1995), however, also this is not sufficient as it is significantly affected by the physiological, genetic, and pathological conditions (Bentinger et al. 2010). Therefore, an extensive attempt has been made to increase its production. Among these, its extraction from biological tissues or its chemical synthesis uses solvents and chemicals that are not environmentally friendly and thus are not preferred (Yuan et al. 2012). Biotechnological alternatives for the production of CoQ_{10} in microbial systems exist, *Agrobacterium tumefaciens* being the most promising one among all investigated strains, but even with this most attractive strain which is relatively easier to control and has relatively low production cost (Choi et al. 2005; Gu et al. 2006; Ha et al. 2007, 2008), still the struggle against low yield and limited scalability continues due to manufacturing constraints (Cluis et al. 2007; Jeya et al. 2010) and its low specific CoQ_{10} content (Yuan et al. 2012).

Recently, thanks to biotechnological applications, such as plant cell and tissue culture techniques that can enhance their CoQ_{10} content, and plants are considered as very promising production hosts. Indeed, plants are capable of providing high biomass and seed yield at relatively low cost, as well as relatively easier scalability. One additional advantage is that no specific processing of the product is required when the nutraceutical is accumulated in an edible plant tissue (Kawakatsu and Takaiwa 2010). Initial plant cell and tissue culture studies focused their attention on the selection of plants with high CoQ content. For instance, Ikeda and his co-workers managed to develop tobacco cell lines which produced 360 μg/g CoQ_{10} dry wt. (Ikeda et al. 1976) and rice cell lines that produced < 600 μg/g CoQ_9 dry wt. in *in vitro* conditions (Ikeda and Kagei 1979). Matsumoto and his co-workers isolated strains producing high levels of CoQ_{10} from *Nicotiana tabaccum* cv. By-2 cell suspension cultures by a cell cloning technique. For this, they evaluated three strains with high CoQ yields in suspension cultures, which were initially investigated for the optimization of appropriate culture conditions. Among the tested parameters, sucrose and 2,4-dichlorophenoxyacetic acid (2,4-D) concentration and culture temperature did not affect markedly the CoQ production. Similarly, although addition of yeast extract to the medium promoted cell growth remarkably, it did not enhance CoQ production. Therefore, the selection of CoQ-producing strains was carried out by a cell cloning technique. Selected CoQ producing strains were cultured in liquid medium, and their CoQ content was determined. Some strains producing large quantities of CoQ in suspension culture were selected, and those strains were subjected to further cell cloning. CoQ content was remarkably increased by the repeated cell cloning. After six re-cloning, three strains producing high levels of CoQ were selected from more than 2000 cell clones that were tested. The highest levels of production in suspension culture were 1848 μg/g dry wt. These results showed that the By-2 cells contained much more CoQ than their parent plants and that the CoQ productivities of the cultured cells were similar to those of microorganisms which have been reported to contain large amounts of CoQ. Thus, the authors concluded that the By-2 cells might provide a suitable source for the large-scale production of CoQ and for the investigation of some of the cultural conditions necessary for CoQ formation (Matsumoto et al. 1981). These reports show that, as also commented by Takahashi and his co-workers, utilization of plants as a production host of CoQs is encouraging with successful attainment of CoQ_{10} levels sufficient for supplying a daily dose (Takahashi et al. 2009).

Alkoloids represent another important group of plant secondary metabolites, involved in adaptation and defense against different types of stress. For instance, oxidative stress accompanied with the generation of ROS is triggered when plants are subjected to phytotoxic concentrations of heavy metals. This provokes ascorbate glutathione cycle enzymes, together with catalase, peroxidases and superoxide dismutase, and metabolic activities are directed towards pathways which eventually promotes the production of alkoloids (Lajayer et al. 2017; Tyagi et al. 2019; Madhu et al. 2022a,b). This phenomenon was demonstrated by the application of several heavy metals to *Fagonia indica* and *Catharanthus roseus* with the aim to increase its alkaloid production.

Fagonia indica is a rich source of pharmacologically active compounds and hence is one of the most important medicinal plants of the Zygophyllaceae family, with multiple therapeutic properties such as anti-inflammatory, hepato-protective, anticancer, antidiabetic, antimicrobial, antioxidant, antihemorrhagic, anthelminthic, and thrombolytic features (Shehab et al. 2011; Bagban et al. 2012; Waheed et al. 2012). *F. indica* shows a restricted distribution in several parts of the world (Rathore et al. 2012). Besides, the variation in the metabolites of interest is one of the major issues in wild plants due to different environmental factors. The addition of chemical elicitors can be considered as one of the effective strategies to trigger the biosynthetic pathways for the release of a higher quantity of bioactive compounds. Accordingly, a study was designed by Khan et al. (2021) to investigate the effects of chemical elicitors, aluminum chloride ($AlCl_3$), and cadmium chloride ($CdCl_2$), on the biosynthesis of secondary metabolites, biomass, and the antioxidant system in callus cultures. Among various treatments applied, 0.1 mM $AlCl_3$ provided the highest biomass accumulation in comparison to the control (404.72 g/l FW vs. 269.85 g/l). The exposure of cultures to 0.01 mM $AlCl_3$ enhanced the accumulation of secondary metabolites, total phenolic contents (7.74 mg/g DW), and total flavonoid contents (1.07 mg/g DW) in comparison to the cultures that were exposed to 0.01 mM $CdCl_2$. Likewise, $AlCl_3$ and $CdCl_2$ also promoted the free radical scavenging activity (89.4% and 90%, respectively) at a concentration of 0.01 mM, as compared to the control (65.48%). Cultures grown in the presence of $AlCl_3$ triggered higher quantities of secondary metabolites than those grown in the presence of $CdCl_2$. Moreover, $AlCl_3$ at 0.1 mM also enhanced the biosynthesis of superoxide dismutase and peroxidase enzymes. These results suggested that $AlCl_3$ was a better elicitor in terms of a higher and uniform productivity of biomass, secondary cell products, and antioxidant enzymes in *F. indica* callus cultures, compared to $CdCl_2$.

Catharanthus roseus is tropical evergreen plant and is a natural source of a wide range of pharmaceutically valuable alkoloids, among which are antihypertensive compound ajmalicine and antileukemic bis-indole alkaloids vincristine and vinblastine. Especially for the latter two, *C. roseus* is the only plant source in nature (Moreno et al. 1995). However, their isolation from *C. roseus* is still very expensive as the natural concentration of these pharmaceuticals within the plant are very low, and their artificial synthesis is not effective and economic either due to the complex chemical structures of these alkoloids (Heijden et al. 2004; Carqueijeiro et al. 2013). The pharmacological significance of alkoloids and their low amounts in the plant which is the unique source of them (around 0.0005% of dry mass) have motivated broad research on the manipulation of the plant metabolism in order to enhance alkaloid amounts. Accordingly, that made the *C. roseus* a valuable candidate for researchers that focus on the stimulation of secondary metabolites by the application of elicitors.

Among these, Fathalla et al. (2011) investigated the effects of mercuric chloride ($HgCl_2$) on the accumulation rate of indole alkaloids in callus cultures of Egyptian *C. roseus* (L.) G. (Don). $HgCl_2$ was used at 0.0, 0.2, 0.4, and 0.8 μm/l. Authors reported that all tested concentrations of Hg^{2+} significantly decreased the different measured growth parameters of shoot- and root-derived calli; however, the percentage of dry matter content significantly increased in comparison to the control treatment. Moreover, the percentage of total indole-alkaloids was significantly increased.

As previous studies have demonstrated the existence of *C. roseus* alkoloids in aerial parts of the plant, Paeizi et al. (2018) investigated the simultaneous effects of methyl jasmonate (100 μM) and silver ions (Ag^+ as silver nitrate, $AgNO_3$, 50 and 100 μM) individually and simultaneously on the production of important medicinal alkaloids (vincristine, vinblastine, ajmalicine, vindoline, and catharanthine) and the expression profile of related regulatory and biosynthetic genes in micropropagated shoots of *C. roseus*. *In vitro* propagation was selected as the mode of proliferation, as it was more controllable than the other culture conditions, eliminated unwanted factors and made the results more accurate. The effects of the treatments were also investigated on nonenzymatic defensive metabolites (total phenolics, flavonoids, and carotenoids) and antioxidant enzymes activities (peroxidase, EC 1.11.1.7, catalase, EC 1.11.1.6, and superoxide dismutase, EC 1.15.1.1). Changes of dry weight, quantity of lipid peroxidation, and photosynthetic pigments contents have been measured as well. The maximum yields of the alkaloids were observed under 100 μM methyl jasmonate in combination with 100 μM of $AgNO_3$ after seven days. The employed treatments induced increased lipid peroxidation, higher levels of enzymatic antioxidants activities, and more production of nonenzymatic defensive metabolites which shows activity of plant defensive system. The results suggested that silver nitrate and methyl jasmonate signaling pathways might have cross talks, and their simultaneous application make an effective combination for elicitation of medicinal alkaloids biosynthesis in *C. roseus* micropropagated shoots. Therefore, authors suggested the use of this combination as an effective elicitor for the alkoloid production and accumulation in *C. roseus* tissues.

The capabilities of cobalt ions (Co^{2+}) in enhancing alkaloids accumulation in *C. roseus* suspension cultures were evaluated at four concentrations (5, 10, 15, and 20 mg/l). Positive correlations were detected between both cobalt concentration and expression of CrMPK3 gene on one hand and the activities of antioxidant enzymes (SOD and APX) and alkaloids content on the other hand. Such correlations suggested CrMPK3 gene as a common player in cobalt-induced stress signaling (Fouad and Hafez 2018).

Khataee et al. (2019) aimed to determine the effects of methyl jasmonate (Mj) combined with chromium (Cr^{2+}) as elicitor on production of medicinal alkaloids, its antioxidant potential, and its effects on the expression of signaling and biosynthetic enzymes. Combined treatment had positive effects on secondary metabolite production and changed gene expression levels of mitogen-activated protein kinase 3 (MAPK3) of the *in vitro* shoot cultures of. *C. roseus*. The maximum total yield of vincristine was 1.52-fold more than control in response to 100 μM Mj after one week. This increase was 2.16, 4.01, 2.39, and 1.97-fold for ajmalicine, vinblastine, vindoline, and catharanthine, respectively, in response to 100 μM Mj + 50 μM Cr. Authors concluded that such combined abiotic treatments could elevate alkaloid production by induction of MAPK3 and ORCA3 (a transcription factor known as octadecanoid-responsive Catharanthus AP2-domain 3) signaling pathway, which induces expression of downstream terpenoid indole alkaloids biosynthetic enzymes. As suggested by them, this can be a promising approach to enhance commercially the overproduction of such medicinal chemicals with high pharmaceutical values.

However, one should not underestimate the fact that the triggered stress may, in addition to enhance alkoloid production, negatively affect the growth of the plant (Fouad and Hafez 2018). Hence, it is clear that developing strategies for the maximum enhancement in alkaloids production with minimum growth retardation is of great importance.

Nanoparticles are small-sized (1–100 nm) substances of silver, gold, zinc, copper, titanium, silicon, or magnesium, each with unique physiochemical properties and diverse applications in biology, agriculture, and medicine (Ruttkay-Nedecky et al. 2017). Among these, silver nanoparticles (AgNPs) have received particular attraction and hence became one of the most commonly used ones (Oukarroum et al. 2012), due to their tremendous physiological properties (such as their significant antimicrobial potential) and their potential in influencing plant cell growth, biomass production, and production of secondary metabolites in plant cell cultures (Elechiguerra et al. 2005). Their application on intact plants or cell cultures is still growing. Indeed, AgNPs may act as a stimulator of plant growth, with a significant potential to trigger germination, improve induction and proliferation of shoots, and provide higher pigment content (Yan and Chen 2019; Salem et al. 2020). Potential use of AgNPs to enhance secondary metabolite production in *in vitro* cultures was demonstrated in *Corylus avellana* (Jamshidi and Ghanati 2017), *Cucumis anguria* (Chung et al. 2018), *Caralluma tuberculata* (Ali et al. 2019), and *C. roseus* (Fouad et al. 2021).

Taxol is an outstanding secondary metabolite that functions as an anticancer agent against breast, ovarian, and nonsmall cell lung cancer. It is mainly produced by one of the slowest growing plants, Taxus, thus its attainment from natural sources is not efficient. Chemical synthesis and partial synthesis of Taxol from its precursors are very expensive and time consuming, as well. Sustainable production of Taxol and related taxanes by hazel cell culture is a very promising method and can be further improved by elicitation. With this aim, Jamshidi and Ghanati (2017) evaluated the taxanes production in suspension-cultured hazel (*C. avellana* L.) cells after treatment with AgNPs at different concentrations (0, 2.5, 5, and 10 ppm). The growth of cells and their membrane integrity decreased; however, treatment of cells with AgNPs (in particular of 5 ppm) rapidly and remarkably increased the yields of two major taxanes, i.e. Taxol and Baccatin III; their contents reaching up to 378 and 163% of the control, respectively, after 24 hours of the treatment. Increase of Taxanes was accompanied by the increase of total soluble phenols. The results suggested the elicitation of suspension-cultured hazel cells with AgNPs as a procedure for rapid enhancement of anticancer taxanes biosynthesis by the cells. In addition, as the extract of hazel cells is very effective in killing cancer cells, especially shortly after the treatment with AgNPs, authors suggested that this might open new perspectives in cancer therapy.

The study that was conducted by Chung et al. (2018) described the elicitor effect of silver ion (Ag^+) and biologically synthesized AgNPs to enhance the biomass accumulation and phenolic compound production as well as biological activities (antioxidant, antimicrobial, and anticancer) in genetically transformed hairy root cultures of *C. anguria*. *C. anguria* L. is a highly nutritious cucurbit vegetable that also contains many potential phytochemicals that can be used in traditional medicine (Yoon et al. 2015). In the conducted study, the biomass of hairy root cultures was significantly increased by AgNPs, whereas decreased in Ag^+ elicitation at 1 and 2 mg/l. AgNPs-elicited hairy roots produced a significantly higher amount of individual phenolic compounds (flavonols, hydroxycinnamic, and hydroxybenzoic acids), total phenolic, and flavonoid contents than Ag^+-elicited hairy roots. Moreover, antioxidant, antimicrobial, and anticancer activities were significantly higher following AgNPs-elicitation compared to that in Ag^+-elicited hairy roots. These findings suggested that AgNPs could be an efficient elicitor in hairy root cultures to increase the phytochemical production.

Ali et al. (2019) supplemented the *in vitro* callus cultures of *C. tuberculata* with different ratios of AgNPs and plant growth regulators (PGRs) for the sustainable production of biomass and antioxidant secondary metabolites. *C. tuberculata* of the Asclepediaceae family is a leafless, fleshy, edible medicinal plant that contains a variety of bioactive metabolites with a high antioxidant potential. The plant is mostly used in treatment against diabetes, inflammation, asthma, paralysis, obesity, joints pains, and fever. Nonetheless, the plant extracts have shown positive results through *in vitro* assays, against human carcinoma cell lines (Ansari et al. 2005; Rauf et al. 2013). Results of the study conducted by Ali et al. (2019) indicated that various concentrations of AgNPs significantly affected the callus proliferation and substantially increased the callus biomass, when combined with PGRs in the MS (Murashige and Skoog 1962) media. The highest fresh (0.78 g/l) and dry (0.051 g/l) biomass accumulation of callus was observed at 60 μg/l AgNPs in combination with 0.5 mg/l 2,4-D plus 3.0 mg/l benzyl adenine (BA). Phytochemical analysis of the callus cultures showed higher production of phenolics (3.0 mg), flavonoids (1.8 mg), phenylalanine ammonialyase activity (5.8 U/mg), and antioxidant activity (90%), respectively, in the callus cultures established on MS media in the presence of 90 μg/l AgNPs. Moreover, enhanced activities of antioxidant enzymes such as superoxide dismutase (4.8 U/mg), peroxidase (3.3 U/mg), catalase (2.5 U/mg), and ascorbate peroxidase (1.9 U/mg) were detected at higher level (90 μg/l) of AgNPs tested alone for callus proliferation in the MS media. Accordingly, authors concluded that the AgNPs could be effectively utilized for the enhancement of bioactive antioxidants in the callus cultures of *C. tuberculata*.

Fouad et al. (2021) conducted a study to demonstrate potential effects of using AgNPs in the stimulation of alkaloids biosynthesis in *C. roseus* suspension cultures. AgNPs that were used in the study were roughly spherical 20–30 nm particles dispersed in deionized water (DW) in a concentration of 1 mg/ml. *In vitro* cultures were established by the introduction of hypocotyl explants, obtained from *in vitro*-germinated seedling, to Murashige and Skoog medium (MS Medium, Murashige and Skoog 1962). Callus proliferation was observed on MS medium enriched with 0.5 mg/l kinetin, 1.0 mg/l 2,4-D, and 1.0 mg/l indole acetic acid (IAA). Filter sterilized aqueous AgNPs solution was aseptically introduced into cultures to reach a final concentration of 5, 10, 15, 20, and 25 mg/l. The results showed that the inclusion of 5 mg/l AgNPs to the medium provided about 15% increase in both fresh and dry weights. Dry weight remained unaffected at 10 mg/l then decreased reaching about 73% of control at 15 mg/l that remained insignificantly changed with further increases in AgNPs concentration. Higher concentrations of AgNPs were associated with decrease in fresh weight reaching about 55% of the corresponding control at 25 mg/l. Regarding the alkaloid contents, control cultures accumulate about 3 mg alkaloids in each gram of dried cells and about 35 mg of the secondary metabolites in each liter liquid growth medium after filtering the cells. Although alkaloid accumulation remained insignificant at AgNPs concentration at 5 mg/l, higher AgNPs concentrations provided significant enhancements in alkaloids accumulation in cells to reach 1.42 folds of the corresponding control at 25 mg/l. The study also demonstrated that AgNPs can trigger the accumulation of alkaloids in plant cells through a signaling pathway that involves hydrogen peroxide and MAPKs, leading to upregulation of the biosynthetic genes, including STR gene (Fouad et al. 2021). MAPKs (mitogen-activated protein kinases) play a crucial role for alkoloid biosynthesis during stress conditions (Khataee et al. 2019).

19.2.2 Nutraceuticals Used in Food and/or Cosmetic Industry

Anthocyanins are the color pigments that are highly used in food, nutraceuticals, pharmaceutical, and cosmetic industry as natural colorants and/or for their biological functions. Although these anthocyanins are a color agent, they are also utilized as flavor scavengers and natural preservatives to protect food ingredients against environmental stresses during storage and transportation (Shipp and Abdel-Aal 2010; Liu et al. 2018). Colored fruits (acai, blackcurrant, Aronia, blackberry, black raspberry, blueberry, cherry, plum, blood orange, redcurrant, and Pyrus, grape) and some vegetables (eggplant, purple corn, red cabbage, Okinawan sweet potato, and purple yam) are the main natural sources that synthesize anthocyanins depending on genetic, climatic, and edaphic factors (Pourcel et al. 2010;

Belwal et al. 2019; Iorizzo et al. 2019). The higher demand of anthocyanins is estimated to be USD 489.18 million by the end of the 2025 at a compound annual growth rate (CAGR) of 4.5% from 2022 to 2027 (Global Anthocyanin Market Report 2022). The high demand of these precious compounds exhibits enormous pressure on the availability of raw materials (plants) and its supply chain along with climate change and other environmental problems. Therefore, the production of plants containing anthocyanins requires alternative solutions. *In vitro* plant cell or tissue culture is one of the alternative ways to produce anthocyanins from *Vitis vinifera*, *Daucus carota*, and *Rosa hybrida*, which are the most studied plant species (Belwal et al. 2020). Under laboratory conditions, Cabernet Sauvignon cell suspension cultures do not spontaneously produce anthocyanins. However, application of exogenous abscisic acid (ABA) induced both structural and regulatory genes associated with anthocyanin production up to 400 μg/g FW (fresh weight) (Gagné et al. 2010). In another study, effects of polysaccharide elicitors have been investigated in cell suspension cultures of *V. vinifera*. Especially, anthocyanin production was increased by chitosan, pectin, and alginate with levels of 2.5-, 2.5-, and 2.6-fold, respectively, compared to control (Cai et al. 2011). Influence of salicylic acid and methyl jasmonate on anthocyanin production was investigated in callus cultures of *R. hybrica* by Ram et al. (2013). Different concentrations of salicylic acid and methyl jasmonate included individually or in combination with *Euphorbia millii* medium supplemented with 204.5 mM sucrose, 2.45 μM indole butyric acid (IBA), and 2.33 μM kinetin resulted in positive effects on both callus biomass and anthocyanin accumulation. Treatment with 0.5 μM methyl jasmonate resulted in higher anthocyanin induction. Interestingly, combination of salicylic acid (400 μM) and methyl jasmonate (5.0 μM) decreased the frequency of color response (Ram et al. 2013).

Plant-based bioreactors are used in production of nutraceuticals as cosmetic ingredients. The pigmented carrot *D. carota* have been widely used to produce polyphenols (Ceoldo et al. 2009), including eight specific cyanidin derivatives, some acylated with coumaric, caffeic, ferulic, and sinapic acids (Edgar Gläßgen and Ulrich Seitz 1992; Ceoldo et al. 2009). Especially, anthocyanin-producing suspension cultures (Rosa1 and Rosa3) of *D. carota* cv. Flakkese were selected by independent visual selection of rare pigmented cells from the established K1 cell line (Ceoldo et al. 2005). In further study, pigmented cell lines (R4G) were derived from one of these R3G based on its unusual ability to accumulate anthocyanins in the dark (Ceoldo et al. 2009). Recently, the activities of an extract of the anthocyanin-accumulating R4G were evaluated to reveal its antioxidant and anti-inflammatory activities when applied to mouse J774A.1 monocyte/macrophage cells for its ability to promote the nuclear translocation of NF-kB, and the ability to upregulate vascular endothelial growth factor (VEGF-A) expression in two *in vitro* human skin models. The findings suggested that R4G extracts have potential to protect mammalian cells from the oxidative stress triggered by exposure to bacterial lipopolysaccharides and H_2O_2. Additionally, R4G extracts prevented the nuclear translocation of NF-kB in an epidermal skin model and stimulated the expression of VEGF-A to promote the microcirculation in a dermal microtissue model (Bianconi et al. 2020). In 2018, Appelhagen et al. (2018) used tobacco lines to generate cell suspension cultures as customizable and sustainable alternatives to conventional anthocyanin production platforms. They have engineered the expression of regulatory genes together with expression of genes encoding side chain decorating enzymes. The results showed that the cell lines were stable for the production rate. Additionally, this approach is transferrable to other species, including *Arabidopsis thaliana* (Appelhagen et al. 2018). Anthocyanin producing callus cultures of *Angelica archangelica* were established by Siatka (2018). Effects of different mediums were tested for callus proliferation and accumulation of anthocyanins. Anthocyanin contents were found to higher with 114%, 41%, 33%, 31%, 25%, and 15% on Linsmaier and Skoog, Gamborg B5, Schenk and Hildebrandt, Woody plant, Nitsch and Nitsch, and Heller medium, respectively, in comparison with that of MS medium. Also, anthocyanin production was 2% higher on MS medium supplemented with 2 mg/l 2,4-D acid, and 0.4 mg/l benzylaminopurine (BAP) compared to control (Siatka 2018). Medium compositions were optimized to enhance the anthocyanin production on cell suspension cultures of *D. carota*. Change in ammonium to potassium nitrate ratio and phosphate concentration in MS medium had an extensive effect on the cell biomass and anthocyanin accumulation. The lower concentration of KH_2PO_4 (0.45 mM) in the suspension culture resulted in 1.63-fold increase, while the addition of NH_4NO_3 to KNO_3 ratio (20.0 mM: 37.6 mM) in MS medium resulted in a 2.85-fold increase in anthocyanin content (Saad et al. 2018). In another study, Saad et al. (2021) investigated the effects of salt stress to scale-up the production of anthocyanins on cell culture of *D. carota*. The findings exhibited that salt stress enhanced the accumulation of anthocyanins by stimulating the changes in the expression of key genes involved in anthocyanin production, including chalcone synthase, cinnamate-4-hydroxylase, eucoanthocyanidin dioxygenase, and flavonoid 3-O-glucosyltransferase and sequestration such as MATE (multidrug and toxic compound extrusion) transporter (Saad et al. 2021). Some *in vitro* approaches exhibiting anthocyanin production in different plant species have been summarized in Table 19.1.

The berry crops in genus *Vaccinium*, which are the richest sources of antioxidant metabolites, consist of about 450 species, including *Cyanococcus* (cluster-fruited blueberries), *Oxycoccus* (cranberries), *Vitis-idea* (lingonberries), and *Myrtillus* (bilberries) that are used for commercial production (Hancock et al. 2003; Song 2014). Several *Vaccinium* species are used as medicinal plant due to having high phenolic compounds with strong antioxidative properties (Debnath and Sion 2009; De Alvarenga et al. 2019). Especially, blueberries are the most popular ones known as a "super fruit" due to their nutritional value and increased level of bioactive phenolic molecules (Kalt et al. 2019). Many varieties of blueberries also include a wide range of organic acids, antioxidants such as flavanols, phenolic acids, anthocyanins, and ellagitannins; and non-nutritive phytosterols, including stigmasterol and sitosterol (Nile and Park 2014; Skrovankova et al. 2015). Pharmaceutical studies exhibited that bioactive compounds of blueberry are effective against carcinogenesis and several degenerative diseases, thus it is considered that blueberries could help preventing human body from these diseases (Bomser et al. 1996; Skupień et al. 2006; Skrovankova et al. 2015). Additionally, blueberries demonstrated the ability to reduce the cardiovascular risk factors in obese women and men with metabolic syndrome (Basu et al. 2010). The market demand of *Vaccinium* species has been dramatically increased according to their health-promoting properties. Propagation of *Vaccinium* by stem or rhizome cutting method is easy; however, it is insufficient and time consuming for large-scale production. Tissue culture is becoming more attractive method due to its fast propagation capacity (Goyali et al. 2013; Vyas et al. 2013). The early efforts of micropropagation of *Vaccinium* spp. were carried out with the three-stage method due to subculture of shoot explants on different medium supplemented with (0–4 mg/l) and 2-isopentenyladenine (2iP, 5–10 mg/l) and rooting shoots

Table 19.1 *In vitro* anthocyanin production by plant tissue cultures.

Species	Explants used	Culture conditions	Total anthocyanin content	References
Arabidopsis thaliana	*In vitro* germinated seed-derived leaves	MS supplemented with 1/2 strength of NH_4NO_3 and $KN0_3$	1.5 mg/g FW	Shi and Xie (2010)
Azadirachta indica	Leaves	MS supplemented with 0.2 mg/l TDZ	262.54 mg/g DW	Ashokhan et al. (2020)
Cleome rosea	Leaves	MS supplemented with 11.41 μM IAA and 0.93 μM Kin + 0.45 mM KH_2PO_4	20.90 mg/g FW	Saad et al. (2018)
Daucus carota	Cell suspension culture	B5 supplemented with 0.5 mg/l 2.4-D	120 μg/g FW	Ceoldo et al. (2005)
		MS supplemented with 11.4 μM IAA and 0.93 μM Kin + 20.0 mM NH_4NO_3 and 37.6 mM KNO_3	166 μg/g FW	Saad et al. (2021)
Melissa officinalis	Apical part	MS supplemented with 0.5 mg/l BA, O_3 fumigation (200 ppb, 3 h)	0.32 mg/g FW	Tonelli et al. (2014)
***Nicotiana tabacum* L.**	Cell suspension culture	LS supplemented with 1 mg/l 2.4-D	20–25 mg/g DW.	Appelhagen et al. (2018)
Rosa hybrida	Leaves	Modified EM supplemented with 2.45 μM IBA and 2.33 μM Kin +0.5 μM MeJA	3.48 ± 0.07 CV/g FW	Ram et al. (2013)
Rosa gallica	Leaves	MS supplemented with 2 mg/l 2,4 D and 1 mg/l BAP	138 μmol/g FW	Tarrahi and Rezanejad (2013)
Raphanus sativus	Root tip	MS supplemented with 0.1 mg/l IBA and 0.5 mg/l NAA	0.15% DW	Betsui et al. (2004)
Vitis vinifera	Cell suspension culture	B5 supplemented with 0.1 mg/l NAA and 0.2 mg/l Kin + 30 g/l sucrose, 250 mg/l casein hydrolysate + 20 μ MeJA	22.62 CV/g FCW	Zhang et al. (2002)
Vitis vinifera* L. cv. *Gamay Freaux	Berries	B5 supplemented with 0.5 mg/l NAA, and 0.12 mg/l BA 20 g/l sucrose, 250 mg/l casein hydrolysate + 10–4 M ABA	400 μg/g FW	Gagné et al. (2010)

B5 Medium-Gamborg B5 medium (1968); LS medium-Linsmaier and Skoog (1965); MS medium-Murashige and Skoog (1962) medium; EM -Euphorbia Millii medium; DW-dry weight; FW-fresh weight; FCW-fresh cell concentration; TDZ-thidiazuron; IAA-indole acetic acid; Kin- Kinetin; BA-6 Benzylaminopurine; ABA-abscisic acid; MeJA-methyl jasmonate; NAA-naphthaleneacetic acid; IBA-Indole-3-butyric acid; BAP-benzyl adenine.
Source: Adapted from Belwal et al. (2020).

in vivo (Litwińczuk 2013). Ghosh et al. (2018) used TDZ to induce somatic embryogenesis for changing the antioxidant properties in blueberries. In that study, they used the four half-high blueberry cultivars "St. Cloud," "Patriot," "Northblue," and "Chippewa." TDZ, a plant growth regulator, was effective for embryo formation at high concentrations (9 μM), although low concentration (2.3 μM) was enough for germination. Combinations of plant growth regulators were assessed to promote embryo maturation, and the highest percentage was observed at 2.3 μM TDZ followed by 4.6 μM zeatin. Comparisons of antioxidant properties between donor plants and somatic embryogenesis regenerated plantlets exhibited that propagation method and genotype have an impact on half-high blueberries under certain condition to synthesize phenols and flavonoids (Ghosh et al. 2018). Some studies exhibiting *in vitro* propagation of bioactive compounds in *Vaccinium* berries, and their biological properties have been summarized in Table 19.2.

Cappai et al. (2020) highlighted some points associated with tissue culture of *Vaccinium* spp. that (i) zeatin should be used in the 1–2 mg/l range to promote plant architecture, and (ii) meristem and biomass production is mainly induced with red-blue LED lights than white LED or fluorescent lights. In another study, cloudberry, lingonberry, stone berry, arctic bramble, and strawberry were cultivated *in vitro*, and their polyphenol and carotenoid composition, antioxidant activity, antihemolytic activity, and cytotoxicity effects were assessed on cancerous cells. Cell suspension cultures were established to produce high-value plant-derived secondary metabolites, i.e. phenolics, anthocyanins, lignans, and carotenoids, which can be an alternative source of bioactive ingredients for food/beverage models with antioxidant potential. Although *in vitro* cultivated strawberry exhibited the lowest antioxidant activity, stone berry and cloudberry showed the highest antiproliferative activity. Additionally, lingonberry demonstrated the highest antioxidant capacity, especially anthocyanins content (7.07 ± 0.1 mg/g DM) was higher compared to the other plants (Rischer et al. 2022).

Strawberry is one of the most popular and important sources of phenolic compounds, including p-coumaric, syringic, ferulic, caffeic, sinapic acids, vanillic, and flavonoids contributing to health as anti-inflammatory, anticancer, antioxidant, and antidiabetic agents (White et al. 2010; Fernández-Ochoa et al. 2017; Gasparrini et al. 2017; López de las Hazas et al. 2016). Recently, possible use of extracts of strawberry has been investigated in ameliorating skin conditions. Skin is continuously exposed to environmental stress factors, such as excessive UV radiation, smoke, and microorganisms, which lead to oxidative stress, inflammation, and cell death via the production of ROS (Sirerol et al. 2015). Different strawberry-based formulations have been assessed for their potential photoprotective capacity and these formulations were also enriched with nanoparticles of Coenzyme Q10 and with sun protection factor 10 (SPF10). Strawberry extracts (25 μg/ml–1 mg/ml) exhibited a remarkable photoprotection in human dermal fibroblasts (HuDe), enhancing cell viability and potentiated by the presence of CoQ10red (100 μg/ml). The study confirmed that strawberry-based formulations included a high antioxidant capacity with polyphenols, and vitamin content resulted in a protective effect on skin cells against UVA-induced damage (Gasparrini et al. 2015). Żebrowska and colleagues (2019) compared the antioxidant properties between the conventionally and *in vitro* propagated strawberry (*Fragaria × ananassa* Duch.). Their study exhibited antioxidant properties have been affected by *in vitro* propagation, thus microplants showed higher increase for

Table 19.2 Examples of *in vitro* propagation of bioactive compounds in *Vaccinium* berries and their biological properties.

Berry types	Bioactive compounds	Biological properties	Micropropagation via	Explants used	References
Highbush blueberry	Anthocyanins, polyphenol, carotene, tannins, lutein, and zeaxanthin	Anticancer, anti-inflammatory, antimicrobial activities; decrease of cardiovascular risks, retardation of type II diabetes, juvenile idiopathic arthritis and osteoarthritis	Somatic embryogenesis and shoot proliferation	Leaf segments, buds, micro shoots, axillary shoots, nodal segments, single node, shoot tip, and segments	Skupień et al. (2006), Basu et al. (2010), Vescan et al. (2012), Pervin et al. (2013), Tsuda et al. (2014), Skrovankova et al. (2015), Kang et al. (2015) and Debnath (2016)
Lowbush blueberries	Anthocyanins, phenolics, flavonoids, and proanthocyanin fractions	Retardation of liver and prostate cancer, Inhibition of urinary tract infections, reverse signs of aging; protection from brain against ischemia-damage, strengthen blood vessels, and arteries, neuroprotective	Somatic Embryogenesis, shoot proliferation, Callus formation	Leaf segments, single nodes, internodes, axillary buds, shoot tip, and segments, hypocotyl and cotyledons	Debnath (2004), Matchett et al. (2005), Schmidt et al. (2006), Leduc et al. (2006), Kaldmäe et al. (2006), Papandreou et al. (2009) and Ghosh et al. (2018)
Rabbiteye blueberries	Anthocyanins, polyphenol, and tannins	Inhibition of colon and liver cancer	Shoot multiplication	Nodal segments	Yi et al. (2005), Yi et al. (2006) and Tetsumura et al. (2012)
Lingonberries	Anthocyanins, polyphenol, and proanthocyanin fractions	Anti-inflammatory, anticarcinogenic, antimicrobial, antiadhesion activities, Prevention of the detrimental metabolic e_ects induced by high-fat diet, protection from kidney against ischemia–reperfusion induced kidney injury, prevention of leukemia and colon cancer.	Shoot proliferation	Nodal segments; leaf segments	Debnath (2005), Kylli et al. (2011), Debnath et al. (2012), Vyas et al. (2013), Grace et al. (2013) and Brown et al. (2014)

Cranberries	Anthocyanins, polyphenol, and proanthocyanin fractions	Antibacterial, anticarcinogenic activities; reduction of cardiovascular risk in patients with metabolic syndrome; protection from diet-induced obesity and insulin resistance; prevention of intestinal oxidative stress, inflammation and urinary tract infection.	Shoot proliferation	Nodal segments, shoot tips	Debnath (2005), Debnath (2008), Kylli et al. (2011), Simão et al. (2013), Grace et al. (2013), Skrovankova et al. (2015), Fu et al. (2017) and Caldas et al. (2018)
Bilberries	Anthocyanins, carotenoid, flavanols, lutein, and zeaxanthin	Anticarcinogenic; reduction of inflammation and progression of chronic hypertension; prevention of development of glaucoma, cataract, and macular degeneration.	Shoot proliferation	Auxiliary buds	Faria et al. (2005), Faria et al. (2010), Shim et al. (2012), Mykkänen et al. (2014) and Georgieva et al. (2016)

Source: Adapted from Debnath and Goyali (2020).

antioxidant activity. Microplants were cultivated in MS medium supplemented with IAA (1 mg/dm^3), BAP (1 mg/dm^3), and gibberellic acid (GA_3) (0.01 mg/dm^3). In another study, combination of berry leaves, including *Rubus fruticosus* L., *Vaccinum myrtillus* L., *Ribes nigrum* L., and *Fragaria vesca* L., were assessed for the antioxidant potential for scavenging of intracellular free radicals. *R. fruticosus* and *V. myrtillus* ferments demonstrated the higher radical scavenging capacity to increase the viability of fibroblasts and keratinocytes. Additionally, the study showed the antiaging potential by inhibiting the activity of metalloproteinases (Ziemlewska et al. 2022).

19.3 Conclusions

Nutraceuticals contain physiologically and biologically active ingredients that can improve health and prevent the diseases. Plants are an important source of nutraceuticals and have been used for centuries for medicinal purposes. Increasing demand of cost-effective and sustainable bioactive compounds from plants results in an increasing necessity to mass propagate them through plant tissue culture. *In vitro* approaches are shortening the duration of time for the production of valuable phytochemicals compared to conventional techniques. In this chapter, the tissue culture methods applied to obtain the commercially valuable plant-derived nutraceuticals are summarized. In the future, development of optimal conditions of *in vitro* approaches may meet the increasing demands of nutraceuticals.

References

Ali, A., Mohammad, S., Khan, M.A. et al. (2019). Silver nanoparticles elicited *in vitro* callus cultures for accumulation of biomass and secondary metabolites in *Caralluma tuberculata*. *Artificial Cells, Nanomedicine, and Biotechnology* 47 (1): 715–724. https://doi.org/10.1080/21691401.2019.1577884.

Alok, A., Kumar, J., and Upadhyay, S.K. (2018). Engineering in hairy roots using CRISPR/Cas9-mediated editing. In: *Hairy Roots* (ed. V. Srivastava, S. Mehrotra and S. Mishra). Singapore: Springer. https://doi.org/10.1007/978-981-13-2562-5_15.

de Alvarenga, L.A., Borges, N.A., de Moreira, L.S.G. et al. (2019). Cranberries – potential benefits in patients with chronic kidney disease. *Food & Function* 10 (6): 3103–3112. https://doi.org/10.1039/C9FO00375D.

Ansari, N.M., Houlihan, L., Hussain, B., and Pieroni, A. (2005). Antioxidant activity of five vegetables traditionally consumed by south-Asian migrants in Bradford, Yorkshire, UK. *Phytotherapy Research* 19 (10): 907–911. https://doi.org/10.1002/ptr.1756.

Anwar, H., Rasul, A., Iqbal, J. et al. (2021). Dietary biomolecules as promising regenerative agents for peripheral nerve injury: an emerging nutraceutical-based therapeutic approach. *Journal of Food Biochemistry* 45 (12): https://doi.org/10.1111/jfbc.13989.

Appelhagen, I., Wulff-Vester, A.K., Wendell, M. et al. (2018). Colour bio-factories: towards scale-up production of anthocyanins in plant cell cultures. *Metabolic Engineering* 48: 218–232. https://doi.org/10.1016/j.ymben.2018.06.004.

Ashokhan, S., Othman, R., Abd Rahim, M.H. et al. (2020). Effect of plant growth regulators on coloured callus formation and accumulation of Azadirachtin, an essential Biopesticide in *Azadirachta indica*. *Plants* 9 (3): 352. https://doi.org/10.3390/plants9030352.

Atanasov, A.G., Waltenberger, B., Pferschy-Wenzig, E.-M. et al. (2015). Discovery and resupply of pharmacologically active plant-derived natural products: a review. *Biotechnology Advances* 33 (8): 1582–1614. https://doi.org/10.1016/j.biotechadv.2015.08.001.

Bagban, I.M., Roy, S.P., Chaudhary, A. et al. (2012). Hepatoprotective activity of the methanolic extract of *Fagonia indica* Burm in carbon tetra chloride induced hepatotoxicity in albino rats. *Asian Pacific Journal of Tropical Biomedicine* 2 (3): S1457–S1460. https://doi.org/10.1016/s2221-1691(12)60437-7.

Basu, A., Du, M., Leyva, M.J. et al. (2010). Blueberries decrease cardiovascular risk factors in obese men and women with metabolic syndrome. *The Journal of Nutrition* 140 (9): 1582–1587. https://doi.org/10.3945/jn.110.124701.

Beal, M.F. (2004). Mitochondrial dysfunction and oxidative damage in Alzheimer's and Parkinson's diseases and coenzyme Q10 as a potential treatment. *Journal of Bioenergetics and Biomembranes* 36 (4): 381–386. https://doi.org/10.1023/b:jobb.0000041772.74810.92.

Belwal, T., Huang, H., Li, L. et al. (2019). Optimization model for ultrasonic-assisted and scale-up extraction of anthocyanins from Pyrus communis "Starkrimson" fruit peel. *Food Chemistry* 297: 124993. https://doi.org/10.1016/j.foodchem.2019.124993.

Belwal, T., Singh, G., Jeandet, P. et al. (2020). Anthocyanins, multi-functional natural products of industrial relevance: recent biotechnological advances. *Biotechnology Advances* 43: 107600. https://doi.org/10.1016/j.biotechadv.2020.107600.

Bentinger, M., Brismar, K., and Dallner, G. (2007). The antioxidant role of coenzyme Q. *Mitochondrion* 7: S41–S50. https://doi.org/10.1016/j.mito.2007.02.006.

Bentinger, M., Tekle, M., and Dallner, G. (2010). Coenzyme Q – biosynthesis and functions. *Biochemical and Biophysical Research Communications* 396 (1): 74–79. https://doi.org/10.1016/j.bbrc.2010.02.147.

Betsui, F., Tanaka-Nishikawa, N., and Shimomura, K. (2004). Anthocyanin production in adventitious root cultures of *Raphanus sativus* L. cv. Peking Koushin. *Plant Biotechnology* 21 (5): 387–391. https://doi.org/10.5511/plantbiotechnology.21.387.

Bianconi, M., Ceriotti, L., Cuzzocrea, S. et al. (2020). Red carrot cells cultured *in vitro* are effective, stable, and safe ingredients for skin care, nutraceutical, and food applications. *Frontiers in Bioengineering and Biotechnology* 8: 575079. https://doi.org/10.3389/fbioe.2020.575079.

Bomser, J., Madhavi, D., Singletary, K., and Smith, M. (1996). *In vitro* anticancer activity of fruit extracts from *Vaccinium* species. *Planta Medica* 62 (3): 212–216. https://doi.org/10.1055/s-2006-957862.

Bonakdar, R.A. and Guarneri, E. (2005). Coenzyme Q10. *American Family Physician* 72 (6): 1065–1070.

Brown, E.M., Nitecki, S., Pereira-Caro, G. et al. (2014). Comparison of *in vivo* and *in vitro* digestion on polyphenol composition in lingonberries: potential impact on colonic health. *BioFactors* 40 (6): 611–623. https://doi.org/10.1002/biof.1173.

Cai, Z., Kastell, A., Mewis, I. et al. (2011). Polysaccharide elicitors enhance anthocyanin and phenolic acid accumulation in cell suspension cultures of *Vitis vinifera*. *Plant Cell, Tissue and Organ Culture (PCTOC)* 108 (3): 401–409. https://doi.org/10.1007/s11240-011-0051-3.

Caldas, A.P.S., Coelho, O.G.L., and Bressan, J. (2018). Cranberry antioxidant power on oxidative stress, inflammation and mitochondrial damage. *International Journal of Food Properties* 21 (1): 582–592. https://doi.org/10.1080/10942912.2017.1409758.

Cappai, F., Garcia, A., Cullen, R. et al. (2020). Advancements in low-chill blueberry *Vaccinium corymbosum* L. Tissue culture practices. *Plants* 9 (11): 1624. https://doi.org/10.3390/plants9111624.

Carqueijeiro, I., Noronha, H., Duarte, P. et al. (2013). Vacuolar transport of the medicinal alkaloids from *Catharanthus roseus* is mediated by a proton-driven antiport. *Plant Physiology* 162 (3): 1486–1496. https://doi.org/10.1104/pp.113.220558.

Ceoldo, S., Levi, M., Marconi, A.M. et al. (2005). Image analysis and *in vivo* imaging as tools for investigation of productivity dynamics in anthocyanin-producing cell cultures of *Daucus carota*. *New Phytologist* 166 (1): 339–352. https://doi.org/10.1111/j.1469-8137.2004.01308.x.

Ceoldo, S., Toffali, K., Mantovani, S. et al. (2009). Metabolomics of *Daucus carota* cultured cell lines under stressing conditions reveals interactions between phenolic compounds. *Plant Science* 176 (4): 553–565. https://doi.org/10.1016/j.plantsci.2009.01.011.

Chandran, H., Meena, M., Barupal, T., and Sharma, K. (2020). Plant tissue culture as a perpetual source for production of industrially important bioactive compounds. *Biotechnology Reports* 26: e00450. https://doi.org/10.1016/j.btre.2020.e00450.

Cheng, L., Khalaf, A.T., Lin, T. et al. (2020). Exercise promotes the Osteo induction of HA/β-TCP biomaterials via the Wnt signaling pathway. *Metabolites* 10 (3): 90. https://doi.org/10.3390/metabo10030090.

Choi, G.-S., Kim, Y.-S., Seo, J.-H., and Ryu, Y.-W. (2005). Restricted electron flux increases coenzyme Q10 production in *Agrobacterium tumefaciens* ATCC4452. *Process Biochemistry* 40 (10): 3225–3229. https://doi.org/10.1016/j.procbio.2005.03.038.

Chung, I.-M., Rajakumar, G., and Thiruvengadam, M. (2018). Effect of silver nanoparticles on phenolic compounds production and biological activities in hairy root cultures of *Cucumis anguria*. *Acta Biologica Hungarica* 69 (1): 97–109. https://doi.org/10.1556/018.68.2018.1.8.

Cluis, C.P., Burja, A.M., and Martin, V.J.J. (2007). Current prospects for the production of coenzyme Q10 in microbes. *Trends in Biotechnology* 25 (11): 514–521. https://doi.org/10.1016/j.tibtech.2007.08.008.

Colin, H.A. (2001). Secondary product formation in plant tissue cultures. *Plant Growth Regulation* 34: 119–134.

Cox, F.F., Misiou, A., Vierkant, A. et al. (2022). Protective effects of curcumin in cardiovascular diseases – impact on oxidative stress and mitochondria. *Cell* 11 (3): 342. https://doi.org/10.3390/cells11030342.

Crane, F.L. (2007). Discovery of ubiquinone (coenzyme Q) and an overview of function. *Mitochondrion* 7: S2–S7. https://doi.org/10.1016/j.mito.2007.02.011.

Debnath, S.C. (2004). *In vitro* culture of lowbush blueberry (*Vaccinium angustifolium* Ait.). *Small Fruits Review* 3 (3–4): 393–408. https://doi.org/10.1300/j301v03n03_16.

Debnath, S.C. (2005). Effects of carbon source and concentration on development of lingonberry (*Vaccinium vitis-idaea* L.) shoots cultivated *in vitro* from nodal explants. *In vitro Cellular & Developmental Biology. Plant* 41 (2): 145–150. https://doi.org/10.1079/ivp2004590.

Debnath, S.C. (2008). Zeatin-induced one-step *in vitro* cloning affects the vegetative growth of cranberry (*Vaccinium macrocarpon* Ait.) micropropagules over stem cuttings. *Plant Cell, Tissue and Organ Culture* 93 (2): 231–240. https://doi.org/10.1007/s11240-008-9366-0.

Debnath, S.C. (2016). Temporary immersion and stationary bioreactors for mass propagation of true-to-type highbush, half-high, and hybrid blueberries (*Vaccinium* spp.). *The Journal of Horticultural Science and Biotechnology* 92 (1): 72–80. https://doi.org/10.1080/14620316.2016.1224606.

Debnath, S.C. and Goyali, J.C. (2020). *In vitro* propagation and variation of antioxidant properties in micropropagated *Vaccinium* berry plants – a review. *Molecules* 25 (4): 788. https://doi.org/10.3390/molecules25040788.

Debnath, S.C. and Sion, M. (2009). Genetic diversity, antioxidant activities, and anthocyanin contents in lingonberry. *International Journal of Fruit Science* 9 (2): 185–199. https://doi.org/10.1080/15538360903005061.

Debnath, S.C., Vyas, P., and Igamberdiev, A.U. (2012). Characteristics of lingonberry plants propagated by *in vitro* and *ex vitro* propagation methods. *Acta Horticulturae* 926: 259–264. https://doi.org/10.17660/actahortic.2012.926.35.

Dixit, S., Shukla, A., Upadhyay, S.K., and Verma, P.C. (2019). Mode of communication between plants during environmental stress. In: *Molecular Approaches in Plant Biology and Environmental Challenges* (ed. S. Singh, S.K. Upadhyay, A. Pandey and S. Kumar). Energy, Environment, and Sustainability. Singapore: Springer.

Dixit, S., Shukla, A., Singh, V., and Upadhyay, S.K. (2021a). Bioprospecting of natural compounds for industrial and medical applications. In: *Bioprospecting of Plant Biodiversity for Industrial Molecules* (ed. S.K. Upadhyay and S.P. Singh). John Wiley & Sons Ltd. https://doi.org/10.1002/9781119718017.ch3.

Dixit, S., Shukla, A., Singh, V., and Upadhyay, S.K. (2021b). Engineering of plant metabolic pathway for nutritional improvement. In: *Genome Engineering for Crop Improvement* (ed. S.K. Upadhyay). John Wiley & Sons Ltd. https://doi.org/10.1002/9781119672425.ch20.

Dureja, H., Kaushik, D., and Kumar, V. (2003). Developments in nutraceuticals. *Indian Journal of Pharmacology* 35: 363–372.

Edgar Gläßgen, W. and Ulrich Seitz, H. (1992). Acylation of anthocyanins with hydroxycinnamic acids via 1-O-acylglucosides by protein preparations from cell cultures of *Daucus carota* L. *Planta* 186 (4): 582–585. https://doi.org/10.1007/bf00198039.

Elechiguerra, J., Burt, J.L., Morones, J.R. et al. (2005). Interaction of silver nanoparticles with HIV-1. *Journal of Nanobiotechnology* 3 (1): 6. https://doi.org/10.1186/1477-3155-3-6.

Ernster, L. and Dallner, G. (1995). Biochemical, physiological and medical aspects of ubiquinone function. *Biochimica et Biophysica Acta (BBA). Molecular Basis of Disease* 1271 (1): 195–204. https://doi.org/10.1016/0925-4439(95)00028-3.

Faria, A., Oliveira, J., Neves, P. et al. (2005). Antioxidant properties of prepared blueberry (*Vaccinium myrtillus*) extracts. *Journal of Agricultural and Food Chemistry* 53 (17): 6896–6902. https://doi.org/10.1021/jf0511300.

Faria, A., Pestana, D., Teixeira, D. et al. (2010). Blueberry anthocyanins and pyruvic acid adducts: anticancer properties in breast cancer cell lines. *Phytotherapy Research* 24 (12): 1862–1869. https://doi.org/10.1002/ptr.3213.

Fathalla, M.A.E.K., Abd-El Kawy, A.M., and Taha, H.S. (2011). Effect of heavy metal ($HgCl_2$) on accumulation and production of total indole alkaloids, vinblastine and/or vincristine from Egyptian *Catharanthus roseus* (L.) G. Don. Calli cultures. *Journal of Applied Sciences Research* 7 (4): 542–549.

Fernández-Ochoa, Á., Borrás-Linares, I., Pérez-Sánchez, A. et al. (2017). Phenolic compounds in rosemary as potential source of bioactive compounds against colorectal cancer: in situ absorption and metabolism study. *Journal of Functional Foods* 33: 202–210. https://doi.org/10.1016/j.jff.2017.03.046.

Fouad, A.S. and Hafez, R.M. (2018). Effect of cobalt nanoparticles and cobalt ions on alkaloids production and expression of CrMPK3 gene in *Catharanthus roseus* suspension cultures. *Cellular and Molecular Biology* 64 (12): 62–69. https://doi.org/10.14715/cmb/2018.64.12.13.

Fouad, A., Hegazy, A.E., Azab, E. et al. (2021). Boosting of antioxidants and alkaloids in *Catharanthus roseus* suspension cultures using silver nanoparticles with expression of CrMPK3 and STR genes. *Plants* 10 (10): 2202. https://doi.org/10.3390/plants10102202.

Fu, Z., Liska, D., Talan, D., and Chung, M. (2017). Cranberry reduces the risk of urinary tract infection recurrence in otherwise healthy women: a systematic review and meta-analysis. *The Journal of Nutrition* 147 (12): 2282–2288. https://doi.org/10.3945/jn.117.254961.

Gagné, S., Cluzet, S., Mérillon, J.-M., and Gény, L. (2010). ABA initiates anthocyanin production in grape cell cultures. *Journal of Plant Growth Regulation* 30 (1): 1–10. https://doi.org/10.1007/s00344-010-9165-9.

Gasparrini, M., Forbes-Hernandez, T., Afrin, S. et al. (2015). A pilot study of the photoprotective effects of strawberry-based cosmetic formulations on human dermal fibroblasts. *International Journal of Molecular Sciences* 16 (8): 17870–17884. https://doi.org/10.3390/ijms160817870.

Gasparrini, M., Forbes-Hernandez, T.Y., Giampieri, F. et al. (2017). Anti-inflammatory effect of strawberry extract against LPS-induced stress in RAW 264.7 macrophages. *Food and Chemical Toxicology* 102: 1–10. https://doi.org/10.1016/j.fct.2017.01.018.

Georgieva, M., Badjakov, I., Dincheva, I. et al. (2016). *In vitro* propagation of wild Bulgarian small berry fruits (bilberry, lingonberry, raspberry and strawberry). *Bulgarian Journal of Agricultural Science* 22 (1): 46–51.

Ghosh, A., Igamberdiev, A.U., and Debnath, S.C. (2018). Thidiazuron-induced somatic embryogenesis and changes of antioxidant properties in tissue cultures of half-high blueberry plants. *Scientific Reports* 8 (1): 16978. https://doi.org/10.1038/s41598-018-35233-6.

Global Anthocyanin Market Report (2022). *Mordor intelligence*. Mordorintelligence.com. https://www.mordorintelligence.com/ (accessed 12 April 2022).

Goyali, J.C., Igamberdiev, A.U., and Debnath, S.C. (2013). Morphology, phenolic content and antioxidant capacity of lowbush blueberry (*Vaccinium angustifolium* Ait.) plants as affected by in vitro and ex vitro propagation methods. *Canadian Journal of Plant Science* 93 (6): 1001–1008. https://doi.org/10.4141/cjps2012-307.

Grace, M.H., Esposito, D., Dunlap, K.L., and Lila, M.A. (2013). Comparative analysis of phenolic content and profile, antioxidant capacity, and anti-inflammatory bioactivity in wild alaskan and commercial *Vaccinium* berries. *Journal of Agricultural and Food Chemistry* 62 (18): 4007–4017. https://doi.org/10.1021/jf403810y.

Gu, S.-B., Yao, J.-M., Yuan, Q.-P. et al. (2006). Kinetics of *Agrobacterium tumefaciens* ubiquinone-10 batch production. *Process Biochemistry* 41 (8): 1908–1912. https://doi.org/10.1016/j.procbio.2006.04.002.

Ha, S.-J., Kim, S.-Y., Seo, J.-H. et al. (2007). Optimization of culture conditions and scale-up to pilot and plant scales for coenzyme Q10 production by *Agrobacterium tumefaciens*. *Applied Microbiology and Biotechnology* 74 (5): 974–980. https://doi.org/10.1007/s00253-006-0744-4.

Ha, S.-J., Kim, S.-Y., Seo, J.-H. et al. (2008). Lactate increases coenzyme Q10 production by *Agrobacterium tumefaciens. World Journal of Microbiology and Biotechnology* 24 (6): 887–890. https://doi.org/10.1007/s11274-007-9547-8.

Hancock, J.F., Luby, J.J., and Beaudry, R. (2003). Fruits of temperate climates/fruits of the ericaceae. In: *Encyclopaedia of Food Science, Food Technology, and Nutrition* (ed. L. Trugo, P.M. Finglas and B. Caballero), 2762–2768. Academic Press.

Heijden, R.v.d., Jacobs, D., Snoeijer, W. et al. (2004). The *Catharanthus* alkaloids: pharmacognosy and biotechnology. *Current Medicinal Chemistry* 11 (5): 607–628. https://doi.org/10.2174/0929867043455846.

Holst, B. and Williamson, G. (2008). Nutrients and phytochemicals: from bioavailability to bioefficacy beyond antioxidants. *Current Opinion in Biotechnology* 19 (2): 73–82. https://doi.org/10.1016/j.copbio.2008.03.003.

Hussein, R.A. and El-Anssary, A.A. (2018). Plants secondary metabolites: the key drivers of the pharmacological actions of medicinal plants. *Herbal Medicine* https://doi.org/10.5772/intechopen.76139.

Ikeda, M. and Kagei, K. (1979). Ubiquinone content of eight plant species in cell culture. *Phytochemistry* 18 (9): 1577–1578. https://doi.org/10.1016/s0031-9422(00)98506-6.

Ikeda, T., Matsumoto, T., and Noguchi, M. (1976). Formation of ubiquinone by tobacco plant cells in suspension culture. *Phytochemistry* 15 (4): 568–569. https://doi.org/10.1016/s0031-9422(00)88982-7.

Iorizzo, M., Cavagnaro, P.F., Bostan, H. et al. (2019). A cluster of MYB transcription factors regulates anthocyanin biosynthesis in carrot (*Daucus carota* L.) root and petiole. *Frontiers in Plant Science* 9: 1927. https://doi.org/10.3389/fpls.2018.01927.

Jaiwal, P.K. (2006). *Plant Genetic Engineering*. Studium Press.

Jamshidi, M. and Ghanati, F. (2017). Taxanes content and cytotoxicity of hazel cells extract after elicitation with silver nanoparticles. *Plant Physiology and Biochemistry* 110: 178–184. https://doi.org/10.1016/j.plaphy.2016.04.026.

Jeya, M., Moon, H.-J., Lee, J.-L. et al. (2010). Current state of coenzyme Q10 production and its applications. *Applied Microbiology and Biotechnology* 85 (6): 1653–1663. https://doi.org/10.1007/s00253-009-2380-2.

Kaldmäe, H., Starast, M., Karp, K., and Paal, T. (2006). Effect of donor plant physiological condition on *in vitro* establishment ff *Vaccinium Angustifolium* shoot explants. *Acta Horticulturae* 715: 433–438. https://doi.org/10.17660/actahortic.2006.715.65.

Kalia, A. (2005). *Textbook of Industrial Pharmacognocy*, 204–208. CBS Publisher and Distributor.

Kalra, E.K. (2003). Nutraceutical-definition and introduction. *American Association of Pharmaceutical Scientists* 5 (3): 27–28. https://doi.org/10.1208/ps050325.

Kalt, W., Cassidy, A., Howard, L.R. et al. (2019). Recent research on the health benefits of blueberries and their anthocyanins. *Advances in Nutrition* 11 (2): https://doi.org/10.1093/advances/nmz065.

Kang, B.-H., Racicot, K., Pilkenton, S.J. et al. (2015). Blueberry extract inhibits carbohydrate-hydrolyzing enzymes and these inhibitory activities are not proanthocyanidin dependent. *Journal of Korean Society for Applied Biological Chemistry* 58 (1): 127–136. https://doi.org/10.1007/s13765-015-0001-6.

Kawakatsu, T. and Takaiwa, F. (2010). Cereal seed storage protein synthesis: fundamental processes for recombinant protein production in cereal grains. *Plant Biotechnology Journal* 8 (9): 939–953. https://doi.org/10.1111/j.1467-7652.2010.00559.x.

Khalaf, A.T., Wei, Y., Alneamah, S.J.A. et al. (2021). What is new in the preventive and therapeutic role of dairy products as nutraceuticals and functional foods? *BioMed Research International* 2021: 8823222. https://doi.org/10.1155/2021/8823222.

Khan, H., Khan, T., Ahmad, N. et al. (2021). Chemical elicitors-induced variation in cellular biomass, biosynthesis of secondary cell products, and antioxidant system in callus cultures of *Fagonia indica*. *Molecules* 26 (21): 6340. https://doi.org/10.3390/molecules26216340.

Khataee, E., Karimi, F., and Ravazi, K. (2019). Chromium-induced alkaloid production in *Catharanthus roseus* (L.) G.Don *in vitro* cultured shoots and related gene expression patterns particularly for the novel gene GS. *Acta Agriculturae Slovenica* 113 (1): 95. https://doi.org/10.14720/aas.2019.113.1.09.

Kokate, C.K., Purohit, A., and Gokhale, S.B. (2002). Nutraceutical and cosmaceutical. In: *Pharmacognosy*, 21e (ed. C.K. Kokate, A. Purohit and S.B. Gokhale), 542–549. Pune: Nirali Prakashan.

Krishnan, R., Singh, S.P., and Upadhyay, S.K. (2021). An introduction to plant biodiversity and bioprospecting. In: *Bioprospecting of Plant Biodiversity for Industrial Molecules* (ed. S.K. Upadhyay and S.P. Singh), 1–13. John Wiley & Sons Ltd. https://doi.org/10.1002/9781119718017.ch1.

Kruczek, A., Krupa-Małkiewicz, M., Lachowicz, S. et al. (2020). Health-promoting capacities of *in vitro* and cultivated goji (*Lycium chinense* Mill.) fruit and leaves; polyphenols, antimicrobial activity, macro- and microelements and heavy metals. *Molecules* 25 (22): 5314. https://doi.org/10.3390/molecules25225314.

Kylli, P., Nohynek, L., Puupponen-Pimiä, R. et al. (2011). Lingonberry (*Vaccinium vitis-idaea*) and european cranberry (*Vaccinium microcarpon*) proanthocyanidins: isolation, identification, and bioactivities. *Journal of Agricultural and Food Chemistry* 59 (7): 3373–3384. https://doi.org/10.1021/jf104621e.

Lajayer, B.A., Ghorbanpour, M., and Nikabadi, S. (2017). Heavy metals in contaminated environment: destiny of secondary metabolite biosynthesis, oxidative status and phytoextraction in medicinal plants. *Ecotoxicology and Environmental Safety* 145: 377–390. https://doi.org/10.1016/j.ecoenv.2017.07.035.

Le, Y., Wang, B., and Xue, M. (2022). Nutraceuticals use and type 2 diabetes mellitus. *Current Opinion in Pharmacology* 62: 168–176. https://doi.org/10.1016/j.coph.2021.12.004.

Leduc, C., Coonishish, J., Haddad, P., and Cuerrier, A. (2006). Plants used by the Cree Nation of Eeyou Istchee (Quebec, Canada) for the treatment of diabetes: a novel approach in quantitative ethnobotany. *Journal of Ethnopharmacology* 105 (1–2): 55–63. https://doi.org/10.1016/j.jep.2005.09.038.

Lenaz, G., Fato, R., Formiggini, G., and Genova, M.L. (2007). The role of Coenzyme Q in mitochondrial electron transport. *Mitochondrion* 7: S8–S33. https://doi.org/10.1016/j.mito.2007.03.009.

Linsmaier, E.M. and Skoog, F. (1965). Organic growth factor requirements of tobacco tissue cultures. *Physiologia Plantarum* 18 (1): 100–127. https://doi.org/10.1111/j.1399-3054.1965.tb06874.x.

Litwińczuk, W. (2013). Micropropagation of *Vaccinium* sp. by *in vitro* axillary shoot proliferation. In: *Protocols for Micropropagation of Selected Economically-Important Horticultural Plants* (ed. E.A. Ozudogru and S.M. Jain). Humana Press.

Liu, Y., Tikunov, Y., Schouten, R.E. et al. (2018). Anthocyanin biosynthesis and degradation mechanisms in *Solanaceous* vegetables: a review. *Frontiers in Chemistry* 6: 52. https://doi.org/10.3389/fchem.2018.00052.

Liu, W.-N., Shi, J., Fu, Y., and Zhao, X.-H. (2019). The stability and activity changes of apigenin and luteolin in human cervical cancer hela cells in response to heat treatment and Fe^{2+}/Cu^{2+} addition. *Food* 8 (8): 346. https://doi.org/10.3390/foods8080346.

López de las Hazas, M.C., Mosele, J.I., Macià, A. et al. (2016). Exploring the colonic metabolism of grape and strawberry anthocyanins and their *in vitro* apoptotic effects in HT-29 colon cancer cells. *Journal of Agricultural and Food Chemistry* 65 (31): 6477–6487. https://doi.org/10.1021/acs.jafc.6b04096.

Lucchesini, M. and Mensuali-Sodi, A. (2010). Plant tissue culture – an opportunity for the production of nutraceuticals. *Advances in Experimental Medicine and Biology* 698: 185–202. https://doi.org/10.1007/978-1-4419-7347-4_14.

Madhu, Sharma, A., Kaur, A. et al. (2022a). Glutathione peroxidases in plants: innumerable role in abiotic stress tolerance and plant development. *Journal of Plant Growth Regulation* https://doi.org/10.1007/s00344-022-10601-9.

Madhu, Kaur, A., Tyagi, S. et al. (2022b). Exploration of glutathione reductase for abiotic stress response in bread wheat (*Triticum aestivum* L.). *Plant Cell Reports* 41: 639–654. https://doi.org/10.1007/s00299-021-02717-1.

Mamdouh, D., Mahgoub, H.A.M., Gabr, A.M.M. et al. (2021). Genetic stability, phenolic, flavonoid, ferulic acid contents, and antioxidant activity of Micropropagated Lycium schweinfurthii plants. *Plants* 10 (10): 2089. https://doi.org/10.3390/plants10102089.

Matchett, M.D., MacKinnon, S.L., Sweeney, M.I. et al. (2005). Blueberry flavonoids inhibit matrix metalloproteinase activity in DU145 human prostate cancer cells. *Biochemistry and Cell Biology* 83 (5): 637–643. https://doi.org/10.1139/o05-063.

Matsumoto, T., Kanno, N., Ikeda, T. et al. (1981). Selection of cultured tobacco cell strains producing high levels of ubiquinone 10 by a cell cloning technique. *Agricultural and Biological Chemistry* 45 (7): 1627–1633. https://doi.org/10.1080/00021369.1981.10864769.

Mishra, M., Shukla, N., and Chandra, R. (2011). Aonla. In: *Advances in Horticulture Biotechnology (Vol. 1): Regeneration Systems of Fruit Crops, Plantation Crops and Spices* (ed. H.P. Singh, V.A. Parthasarathy and K. Nirmal Babu), 185–190. Westville Publishing.

Moreno, P.R.H., van der Heijden, R., and Verpoorte, R. (1995). Cell and tissue cultures of *Catharanthus roseus*: a literature survey. *Plant Cell, Tissue and Organ Culture* 42 (1): 1–25. https://doi.org/10.1007/bf00037677.

Murashige, T. and Skoog, F. (1962). A revised medium for rapid growth and bio assays with tobacco tissue cultures. *Physiologia Plantarum* 15 (3): 473–497. https://doi.org/10.1111/j.1399-3054.1962.tb08052.x.

Mykkänen, O.T., Huotari, A., Herzig, K.-H. et al. (2014). Wild blueberries (*Vaccinium myrtillus*) alleviate inflammation and hypertension associated with developing obesity in mice fed with a high-fat diet. *PLoS One* 9 (12): e114790. https://doi.org/10.1371/journal.pone.0114790.

Nasri, H., Baradaran, A., Shirzad, H., and Rafieian-Kopaei, M. (2014). New concepts in nutraceuticals as alternative for pharmaceuticals. *International Journal of Preventive Medicine* 5 (12): 1487–1499.

Ncube, B. and Van Staden, J. (2015). Tilting plant metabolism for improved metabolite biosynthesis and enhanced human benefit. *Molecules* 20 (7): 12698–12731. https://doi.org/10.3390/molecules200712698.

Nile, S.H. and Park, S.W. (2014). Edible berries: bioactive components and their effect on human health. *Nutrition* 30 (2): 134–144. https://doi.org/10.1016/j.nut.2013.04.007.

Nowicka, B. and Kruk, J. (2010). Occurrence, biosynthesis and function of isoprenoid quinones. *Biochimica et Biophysica Acta* 1797 (9): 1587–1605. https://doi.org/10.1016/j.bbabio.2010.06.007.

Okada, K., Kainou, T., Matsuda, H., and Kawamukai, M. (1998). Biological significance of the side chain length of ubiquinone in *Saccharomyces cerevisiae*. *FEBS Letters* 431 (2): 241–244. https://doi.org/10.1016/s0014-5793(98)00753-4.

Oukarroum, A., Bras, S., Perreault, F., and Popovic, R. (2012). Inhibitory effects of silver nanoparticles in two green algae, *Chlorella vulgaris* and *Dunaliella tertiolecta*. *Ecotoxicology and Environmental Safety* 78: 80–85. https://doi.org/10.1016/j.ecoenv.2011.11.012.

Paeizi, M., Karimi, F., and Razavi, K. (2018). Changes in medicinal alkaloids production and expression of related regulatory and biosynthetic genes in response to silver nitrate combined with methyl jasmonate in *Catharanthus roseus* in vitro propagated shoots. *Plant Physiology and Biochemistry* 132: 623–632. https://doi.org/10.1016/j.plaphy.2018.10.015.

Pagare, S., Bhatia, M., Tripathi, N. et al. (2015). Secondary metabolites of plants and their role: overview. *Current Trends in Biotechnology and Pharmacy* 9 (3): 293–304.

Pandey, M., Verma, R., and Saraf, S. (2010). Nutraceuticals: new era of medicine and health. *Asian Journal of Pharmaceutical and Clinical Research* 3 (1): 11–15.

Papandreou, M.A., Dimakopoulou, A., Linardaki, Z.I. et al. (2009). Effect of a polyphenol-rich wild blueberry extract on cognitive performance of mice, brain antioxidant markers and acetylcholinesterase activity. *Behavioural Brain Research* 198 (2): 352–358. https://doi.org/10.1016/j.bbr.2008.11.013.

Parmar, S.S., Jaiwal, A., Dhankher, O.P., and Jaiwal, P.K. (2015). Coenzyme Q10 production in plants: current status and future prospects. *Critical Reviews in Biotechnology* 35 (2): 152–164. https://doi.org/10.3109/07388551.2013.823594.

Patel, Z.M., Mahapatra, R., and Jampala, S.S.M. (2020). Role of fungal elicitors in plant defense mechanism. In: *Molecular Aspects of Plant Beneficial Microbes in Agriculture* (ed. V. Sharma, R. Salwan and L.K. Tawfeeq Al-Ani), 143–158. Academic Press https://doi.org/10.1016/b978-0-12-818469-1.00012-2.

Pervin, M., Hasnat, M.A., and Lim, B.O. (2013). Antibacterial and antioxidant activities of *Vaccinium corymbosum* L. leaf extract. *Asian Pacific Journal of Tropical Disease* 3 (6): 444–453. https://doi.org/10.1016/s2222-1808(13)60099-7.

Portakal, O., Özkaya, Ö., Erden Inal, M. et al. (2000). Coenzyme Q10 concentrations and antioxidant status in tissues of breast cancer patients. *Clinical Biochemistry* 33 (4): 279–284. https://doi.org/10.1016/s0009-9120(00)00067-9.

Pourcel, L., Irani, N.G., Lu, Y. et al. (2010). The formation of anthocyanic vacuolar inclusions in *Arabidopsis thaliana* and implications for the sequestration of anthocyanin pigments. *Molecular Plant* 3 (1): 78–90. https://doi.org/10.1093/mp/ssp071.

Ram, M., Prasad, K.V., Singh, S.K. et al. (2013). Influence of salicylic acid and methyl jasmonate elicitation on anthocyanin production in callus cultures of *Rosa hybrida*

L. *Plant Cell, Tissue and Organ Culture (PCTOC)* 113 (3): 459–467. https://doi.org/10.1007/s11240-013-0287-1.

Rathore, A.S., Lohar, V., Kumar, R. et al. (2012). Chemical composition and anti-inflammatory activity of various extracts of *Fagonia schweinfurthii* Hadidi. *Medicinal Chemistry and Drug Discovery* 3 (1): 30–36.

Rauf, A., Rehman, W., Jan, M., and Muhammad, N. (2013). Phytochemical, phytotoxic and antioxidant profile of *Caralluma tuberculata* N. E. Brown. *Wudpecker Journal of Pharmacy and Pharmacology* 2 (2): 21–25.

Rischer, H., Nohynek, L., Puupponen-Pimiä, R. et al. (2022). Plant cell cultures of Nordic berry species: phenolic and carotenoid profiling and biological assessments. *Food Chemistry* 366: 130571. https://doi.org/10.1016/j.foodchem.2021.130571.

Ruttkay-Nedecky, B., Krystofova, O., Nejdl, L., and Adam, V. (2017). Nanoparticles based on essential metals and their phytotoxicity. *Journal of Nanobiotechnology* 15 (1): https://doi.org/10.1186/s12951-017-0268-3.

Saad, K.R., Parvatam, G., and Shetty, N.P. (2018). Medium composition potentially regulates the anthocyanin production from suspension culture of *Daucus carota*. *3 Biotech* 8 (3): https://doi.org/10.1007/s13205-018-1146-x.

Saad, K.R., Kumar, G., Mudliar, S.N. et al. (2021). Salt stress-induced anthocyanin biosynthesis genes and MATE transporter involved in anthocyanin accumulation in *Daucus carota* cell culture. *ACS Omega* 6 (38): 24502–24514. https://doi.org/10.1021/acsomega.1c02941.

Salem, S.S., EL-Belely, E.F., Niedbała, G. et al. (2020). Bactericidal and *in-vitro* cytotoxic efficacy of silver nanoparticles (Ag-NPs) fabricated by endophytic actinomycetes and their use as coating for the textile fabrics. *Nanomaterials* 10 (10): 2082. https://doi.org/10.3390/nano10102082.

Sang, Y.L., Cheng, Z.J., and Zhang, X.S. (2018). Plant stem cells and de novo organogenesis. *New Phytologist* 218 (4): 1334–1339. https://doi.org/10.1111/nph.15106.

Santini, A. (2018). Nutraceuticals: redefining a concept. *Annals of Pharmacology and Pharmaceutics* 3 (2): 1147.

Santini, A. and Novellino, E. (2014). Nutraceuticals: beyond the diet before the drugs. *Current Bioactive Compounds* 10 (1): 1–12. https://doi.org/10.2174/157340721001140724145924.

Sayed, S., Ahmed, M., El-Shehawi, A. et al. (2020). Ginger water reduces body weight gain and improves energy expenditure in rats. *Food* 9 (1): 38. https://doi.org/10.3390/foods9010038.

Schmidt, B.M., Erdman, J.W., and Lila, M.A. (2006). Differential effects of blueberry proanthocyanidins on androgen sensitive and insensitive human prostate cancer cell lines. *Cancer Letters* 231 (2): 240–246. https://doi.org/10.1016/j.canlet.2005.02.003.

Shehab, N.G., Mahdy, A., Khan, S.A., and Noureddin, S.M. (2011). Chemical constituents and biological activities of *Fagonia indica* Burm F. *Research Journal of Medicinal Plant* 5 (5): 531–546. https://doi.org/10.3923/rjmp.2011.531.546.

Shi, M.-Z. and Xie, D.-Y. (2010). Features of anthocyanin biosynthesis in pap1-D and wild-type *Arabidopsis thaliana* plants grown in different light intensity and culture media conditions. *Planta* 231 (6): 1385–1400. https://doi.org/10.1007/s00425-010-1142-9.

Shim, S.H., Kim, J.M., Choi, C.Y. et al. (2012). *Ginkgo biloba* extract and bilberry anthocyanins improve visual function in patients with normal tension glaucoma. *Journal of Medicinal Food* 15 (9): 818–823. https://doi.org/10.1089/jmf.2012.2241.

Shipp, J. and Abdel-Aal, E.-S.M. (2010). Food applications and physiological effects of anthocyanins as functional food ingredients. *The Open Food Science Journal* 4 (1): 7–22. https://doi.org/10.2174/1874256401004010007.

Siatka, T. (2018). Production of anthocyanins in callus cultures of *Angelica archangelica*. *Natural Product Communications* 13 (12): 1934578X1801301. https://doi.org/10.1177/1934578x1801301219.

Simão, T.N.C., Lozovoy, M.A.B., Simão, A.N.C. et al. (2013). Reduced-energy cranberry juice increases folic acid and adiponectin and reduces homocysteine and oxidative stress in patients with the metabolic syndrome. *British Journal of Nutrition* 110 (10): 1885–1894. https://doi.org/10.1017/s0007114513001207.

Singh, S.P. and Upadhyay, S.K. (eds.) (2021). *Bioprospecting of Microorganism Based Industrial Molecules*, 1–424. John Wiley & Sons Ltd. https://doi.org/10.1002/9781119717317.

Sirerol, J.A., Feddi, F., Mena, S. et al. (2015). Topical treatment with pterostilbene, a natural phytoalexin, effectively protects hairless mice against UVB radiation-induced skin damage and carcinogenesis. *Free Radical Biology and Medicine* 85: 1–11. https://doi.org/10.1016/j.freeradbiomed.2015.03.027.

Skrovankova, S., Sumczynski, D., Mlcek, J. et al. (2015). Bioactive compounds and antioxidant activity in different types of berries. *International Journal of Molecular Sciences* 16 (10): 24673–24706. https://doi.org/10.3390/ijms161024673.

Skupień, K., Oszmiański, J., Kostrzewa-Nowak, D., and Tarasiuk, J. (2006). *In vitro* antileukaemic activity of extracts from berry plant leaves against sensitive and multidrug resistant HL60 cells. *Cancer Letters* 236 (2): 282–291. https://doi.org/10.1016/j.canlet.2005.05.018.

Song, G.-Q. (2014). Blueberry (*Vaccinium corymbosum* L.). *Methods in Molecular Biology* 1224: 121–131. https://doi.org/10.1007/978-1-4939-1658-0_11.

Sottero, B., Testa, G., Gamba, P. et al. (2022). Macrophage polarization by potential nutraceutical compounds: A strategic approach to counteract inflammation in atherosclerosis. *Free Radical Biology and Medicine* 181: 251–269. https://doi.org/10.1016/j.freeradbiomed.2022.02.010.

Sulaiman, C. and Balachandran, I. (2012). Total phenolics and total flavonoids in selected Indian medicinal plants. *Indian Journal of Pharmaceutical Sciences* 74 (3): 258. https://doi.org/10.4103/0250-474x.106069.

Takahashi, S., Ogiyama, Y., Kusano, H. et al. (2006). Metabolic engineering of coenzyme Q by modification of isoprenoid side chain in plant. *FEBS Letters* 580 (3): 955–959. https://doi.org/10.1016/j.febslet.2006.01.023.

Takahashi, S., Ohtani, T., Iida, S. et al. (2009). Development of CoQ10-enriched rice from giant embryo lines. *Breeding Science* 59 (3): 321–326. https://doi.org/10.1270/jsbbs.59.321.

Tarrahi, R. and Rezanejad, F. (2013). Callogenesis and production of anthocyanin and chlorophyll in callus cultures of vegetative and floral explants in *Rosa gallica* and *Rosa hybrida* (Rosaceae). *Turkish Journal of Botany* 37: 1145–1154. https://doi.org/10.3906/bot-1205-42.

Tetsumura, T., Kajiwara, Y., Honsho, C. et al. (2012). Effective micropropagation of rabbiteye blueberries for leaf tea production. *Environmental Control in Biology* 50 (3): 289–296. https://doi.org/10.2525/ecb.50.289.

Tonelli, M., Pellegrini, E., D'Angiolillo, F. et al. (2014). Ozone-elicited secondary metabolites in shoot cultures of *Melissa officinalis* L. *Plant Cell, Tissue and Organ Culture (PCTOC)* 120 (2): 617–629. https://doi.org/10.1007/s11240-014-0628-8.

Tsuda, H., Kunitake, H., Aoki, Y. et al. (2014). Efficient in vitro screening for higher soil pH adaptability of intersectional hybrids in blueberry. *HortScience* 49 (2): 141–144. https://doi.org/10.21273/hortsci.49.2.141.

Tyagi, S., Shumayla, Singh, S.P., and Upadhyay, S.K. (2019). Role of superoxide dismutases (SODs) in stress tolerance in plants. In: *Molecular Approaches in Plant Biology and Environmental Challenges* (ed. S. Singh, S.K. Upadhyay, A. Pandey and S. Kumar). Energy, Environment, and Sustainability. Singapore: Springer. https://doi.org/10.1007/978-981-15-0690-1_3.

Ullah, M.F., Ahmad, A., Bhat, S.H. et al. (2022). Diet-derived small molecules (nutraceuticals) inhibit cellular proliferation by interfering with key oncogenic pathways: an overview of experimental evidence in cancer chemoprevention. *Biologia Futura* 73 (1): 55–69. https://doi.org/10.1007/s42977-022-00110-x.

Upadhyay, S.K. and Singh, S.P. (eds.) (2021). *Bioprospecting of Plant Biodiversity for Industrial Molecules*. John Wiley & Sons Ltd. https://doi.org/10.1002/9781119718017.

Verma, P.C., Trivedi, I., Singh, H. et al. (2009). Efficient production of gossypol from hairy rootculture of cotton (*Gossypium hirsutum*). *Current Pharmaceutical Biotechnology* 10: 691–700. https://doi.org/10.2174/138920109789542048.

Vescan, L.A., Pamfil, D., Clapa, D. et al. (2012). Efficient micropropagation protocol for highbush blueberry (*Vaccinium corymbosum* L.) cv. "Elliot". *Romanian Biotechnology Letters* 17: 6893–6902.

Vyas, P., Debnath, S.C., and Igamberdiev, A.U. (2013). Metabolism of glutathione and ascorbate in lingonberry cultivars during *in vitro* and *ex vitro* propagation. *Biologia Plantarum* 57 (4): 603–612. https://doi.org/10.1007/s10535-013-0339-8.

Waheed, A., Barker, J., Barton, S.J. et al. (2012). A novel steroidal saponin glycoside from *Fagonia indica* induces cell-selective apoptosis or necrosis in cancer cells. *European Journal of Pharmaceutical Sciences* 47 (2): 464–473. https://doi.org/10.1016/j.ejps.2012.07.004.

White, B.L., Howard, L.R., and Prior, R.L. (2010). Release of bound procyanidins from cranberry pomace by alkaline hydrolysis. *Journal of Agricultural and Food Chemistry* 58 (13): 7572–7579. https://doi.org/10.1021/jf100700p.

Yan, A. and Chen, Z. (2019). Impacts of silver nanoparticles on plants: a focus on the phytotoxicity and underlying mechanism. *International Journal of Molecular Sciences* 20 (5): 1003. https://doi.org/10.3390/ijms20051003.

Yasir, M., Sultana, B., and Anwar, F. (2017). LC–ESI–MS/MS based characterization of phenolic components in fruits of two species of Solanaceae. *Journal of Food Science and Technology* 55 (7): 2370–2376. https://doi.org/10.1007/s13197-017-2702-9.

Yi, W., Fischer, J., Krewer, G., and Akoh, C.C. (2005). Phenolic compounds from blueberries can inhibit colon cancer cell proliferation and induce apoptosis. *Journal of Agricultural and Food Chemistry* 53 (18): 7320–7329. https://doi.org/10.1021/jf051333o.

Yi, W., Akoh, C.C., Fischer, J., and Krewer, G. (2006). Effects of phenolic compounds in blueberries and muscadine grapes on HepG2 cell viability and apoptosis. *Food Research International* 39 (5): 628–638. https://doi.org/10.1016/j.foodres.2006.01.001.

Yoon, J.-Y., Chung, I.-M., and Thiruvengadam, M. (2015). Evaluation of phenolic compounds, antioxidant and antimicrobial activities from transgenic hairy root cultures of gherkin (*Cucumis anguria* L.). *South African Journal of Botany* 100: 80–86. https://doi.org/10.1016/j.sajb.2015.05.008.

Yuan, Y., Tian, Y., and Yue, T. (2012). Improvement of coenzyme Q10 production: mutagenesis induced by high hydrostatic pressure treatment and optimization of fermentation conditions. *Journal of Biomedicine and Biotechnology* 2012: e607329. https://doi.org/10.1155/2012/607329.

Zhang, W., Curtin, C., Kikuchi, M., and Franco, C. (2002). Integration of jasmonic acid and light irradiation for enhancement of anthocyanin biosynthesis in *Vitis vinifera* suspension cultures. *Plant Science* 162 (3): 459–468. https://doi.org/10.1016/s0168-9452(01)00586-6.

Zhao, J., Davis, L.C., and Verpoorte, R. (2005). Elicitor signal transduction leading to production of plant secondary metabolites. *Biotechnology Advances* 23 (4): 283–333. https://doi.org/10.1016/j.biotechadv.2005.01.003.

Ziemlewska, A., Nizioł-Łukaszewska, Z., Zagórska-Dziok, M. et al. (2022). Evaluation of cosmetic and dermatological properties of kombucha-fermented berry leaf extracts considered to be by-products. *Molecules* 27 (7): 2345. https://doi.org/10.3390/molecules27072345.

20

Algal Bioreactors for Polysaccharides Production

Michele Greque de Morais[1], Priscilla Quenia Muniz Bezerra[2], Kricelle Mosquera Deamici[2], Suelen Goettems Kuntzler[1], Juliana Botelho Moreira[1], Céline Laroche[3], and Jorge Alberto Vieira Costa[2]

[1] *Laboratory of Microbiology and Biochemistry, College of Chemistry and Food Engineering, Federal University of Rio Grande, Rio Grande, Rio Grande do Sul, Brazil*
[2] *Laboratory of Biochemical Engineering, College of Chemistry and Food Engineering, Federal University of Rio Grande, Rio Grande, Rio Grande do Sul, Brazil*
[3] *Institut Pascal, Université Clermont Auvergne, Clermont-Ferrand, France*

20.1 Introduction

Microalgae and cyanobacteria are photosynthetic microorganisms able to produce high-added value compounds (Moreira et al. 2022). Polysaccharides from these microorganisms have been studied, characterized, and biologically evaluated, focusing on their biologically active activities to promote human and animal health (Colusse et al. 2022). Thus, there is industrial interest in these polysaccharides to produce pharmaceuticals, nutraceuticals, cosmetics, and food/feed products (Costa et al. 2021; Moreira et al. 2022).

Exopolysaccharides (EPS) have gained prominence among microalgae polysaccharides. Microalgae EPS can be found solubilized in the medium or bound to cells (Costa et al. 2021). EPS are excreted by microalgae through a regular physiological process or under stress conditions. EPS protects cells from microalgae by acting as a physical barrier to biotic and abiotic stresses (Delattre et al. 2016; Morais et al. 2022). The nitrogen limitation during microalgae cultivation is a strategy to accumulate these compounds. Other parameters such as salinity and luminosity also interfere with the synthesis of polysaccharides (Costa et al. 2021; Morais et al. 2022). Structural versatility, low nutrient requirements, and the possibility of manipulating the cultivation conditions to improve the yields of the compound of interest are some of the advantages of microalgae and cyanobacteria for industrial production of polysaccharides (Parwani et al. 2021).

The industrial production of EPS must be based on the proper design of photobioreactors (PBRs) for photosynthetic efficiency, through the adequate supply of light, temperature, and agitation of the culture medium. The suitable PBRs configurations will contribute to carbon dioxide (CO_2) capture, and biomass production (Zhou et al. 2020; Morais et al. 2022).

Plants as Bioreactors for Industrial Molecules, First Edition.
Edited by Santosh Kumar Upadhyay and Sudhir P. Singh.

Microalgae cultivation can be conducted in open or closed PBRs. Open PBRs are easier to build/operate than closed systems. The high surface area to volume ratio allows greater use of CO_2 by open reactors. Open systems also have cost advantages when using sunlight for photosynthesis and biomass production. Thus, open bioreactors are generally used for the production of microalgae biomass, corresponding to approximately 98% of commercially produced biomass (Morais et al. 2020). On the other hand, open systems are more prone to contamination and water loss through evaporation. Therefore, they can be covered with transparent material to minimize water loss in these systems. Besides, closed PBRs emerges in this context, presenting advantages such as resistance to contamination and better control of cultivation conditions (water evaporation, pH, temperature, and light intensity) (Morais et al. 2020; Colusse et al. 2022). Although PBRs are widely used in microalgae production, research is still optimizing the processes and increasing the EPS accumulation (Ekelhof and Melkonian 2017; Deamici et al. 2021).

In this way, this chapter provides an overview of the main advances in the production of polysaccharides from microalgae, highlighting the influence of different types of cultivation systems, presenting the chemical structure and biological activity of these macromolecules, as well as future perspectives.

20.2 Algae

Algae contribute approximately to 50% of photosynthetic productivity on Earth and range from microscopic cells to macroscopic aggregates (Barsanti and Gualtieri 2014). Microalgae are found as part of phytoplankton and on aquatic surfaces, such as freshwater, seawater, brackish, and wastewater (Patel et al. 2022). Prokaryotic photosynthetic microorganisms are cyanobacteria known as blue-green and filamentous algae. Eukaryotic cells also belong to the microalgae group, such as Chlorophyta, Chromophyta, Euglenophyta, and Dinophyta (Alvarez et al. 2021a; Tomaselli 2004; Barsanti and Gualtieri 2014). Microalgae and cyanobacteria adapt to adverse conditions with rapid growth and high efficiency for carbon biofixation, which contributes to the reduction of greenhouse gases (Choi et al. 2019; Hong et al. 2019). Besides, these photosynthetic microorganisms have a high potential to accumulate proteins, polysaccharides, lipids, and pigments. Therefore, microalgae biomass application is explored in several areas of industry (Patel et al. 2020a, 2020b).

20.2.1 Algae Producers of Polysaccharides

Microalgae and cyanobacteria present different metabolic pathways to synthesize intracellular monosaccharides, structural polysaccharides, and complex EPS (Rossi and Philippis 2016). The polysaccharides' content and composition can be affected by cultivation conditions and different strains (Patel et al. 2022). The advantages of producing polysaccharides by microalgae and cyanobacteria are the ability to optimize cultivation parameters and use different carbon sources that allow low-cost production (Pereira et al. 2019). Cyanobacteria produce EPS composed of heteropolysaccharides with a strong anionic nature, presence of sulfate groups, structural variability, and amphiphilic behavior (Pereira et al. 2009; Rossi and Philippis 2016).

Some species of cyanobacteria have been studied as producers of complex polysaccharides, such as *Anabaena laxa*, *Oscillatoria limosa*, and *Phormidesmis molle* (Georgiev et al. 2021). Georgiev et al. (2021) observed a high total carbohydrate content in cyanobacteria that varied between 53.0 and 63.5% (w/w), with the main monomer in the composition being uronic acids. In addition, neutral exopolysaccharide of the (hetero)xylan type was found in the strain of *P. molle*. *Spirulina* (Brezeştean et al. 2021; Deamici et al. 2021; Winahyu and Primadiamanti 2021), *Synechococcus* (Hussein et al. 2019), and *Nostoc* (Gonzales et al. 2020; Alvarez et al. 2021b; Uhliariková et al. 2022) also produce EPS.

Sulfated EPS is also produced by different species of *Chlorella* (Mousavian et al. 2022) and diatoms such as *Phaeodactylum* (Nur et al. 2019). Mousavian et al. (2022) found that the sulfated EPS content varied according to the microalgae species. *Chlorella sorokiniana*, *Chlorella* sp., and *Picochlorum* sp. presented 93.73 ± 5.95, 76.95 ± 2.35, and 35.79 ± 2.29 mg/g of EPS, respectively. Other green microalgae are EPS producers such as *Botryococcus* (Gouveia et al. 2019; Wang et al. 2021) and *Dunaliella* (Goyal et al. 2019).

Red microalgae producing EPS that have attracted much interest from researchers are the class Porphyridiophyceae and Rhodellophyceae (Borjas Esqueda et al. 2022). *Porphyridium* EPS has some characteristics such as solubility, nontoxicity, the approximate molecular weight of $1\text{–}7 \times 10^6$ Da, and can produce a viscous solution at relatively low concentration (0.25%) (Liberman et al. 2020; Singh et al. 2000; Risjani et al. 2021). Borjas Esqueda et al. (2022) observed red microalgae *Timspurckia oligopyrenoides*, *Erythrolobus coxiae*, *Erythrolobus madagascarensis*, *Porphyridium sordidum*, *Neorhodella cyanea*, and *Corynoplastis japonica* had 1.12, 1.43, 1.63, 1.91, 2.32, and 1.59 g/l of EPS production.

20.2.2 Types of Algae Polysaccharides

Polysaccharides constitute an important fraction of the biomass of microalgae and cyanobacteria being classified into storage, structural, and EPS (Bernaerts et al. 2019; Patel et al. 2022). All types of microalgal polysaccharides have functionality as texturizing, stabilizing, emulsifying, bioflocculating, biosurfactant, and biosorbent (Bernaerts et al. 2019; Moreira et al. 2022). The storage polysaccharides include three types of polyglucans with α-1,4 and α-1,6 bonds in different proportions and branching levels (Aikawa et al. 2015). Compounds of glucose residues linked in β-1,3 and β-1,6 make up another polysaccharide called chrysolaminarin found in diatoms (Yang et al. 2020). Moreover, paramylon is constituted of glucose residues linked to β-1,3 belonging to the division of Euglenophyta (Wu et al. 2021). These polysaccharides can be stored in chloroplasts, vacuoles, or granules in the cytosol of microalgal cells (Suzuki and Suzuki 2013).

Structural polysaccharides are generally composed of multiple monosaccharide residues and are more sensitive to growing conditions (González-Fernández and Ballesteros 2012; Markou et al. 2012). Sulfated polysaccharides are found in the cell wall of diatoms due to the increase in uronic acid, sulfate, and fucose during stress conditions (Gügi et al. 2015; Tiwari et al. 2021). In cyanobacteria, such as *Arthrospira platensis*, *Scenedesmus* sp., and *Nostoc* sp., EPS are formed by rhamnose, galacturonic acid, glucuronic acid, glucose, mannose, arabinose, and glucosamine (Rossi and Philippis 2016).

Cyanobacteria such as *Spirulina* can produce high EPS concentrations. Jesus et al. (2019) cultivated *Spirulina* sp. LEB 18 in an outdoor raceway tank containing 250 l of Zarrouk medium for 30 days. The total EPS production of this microalga during 30 days of cultivation was 9.5 g/l, and maximum productivity of 0.6 g/l/d occurred on the 10th day. Chentir et al. (2017) determined the EPS concentration and total polysaccharides produced by *Spirulina* sp. after the desalination and deproteinization processes. The EPS concentration found after desalting was 60.56 ± 1.16%, and total polysaccharides was 67.3 ± 1.1% after deproteinization.

Zampieri et al. (2020) cultivated *Phormidium* sp. and analyzed the monosaccharides constituting the EPS of this red microalgae. The composition and concentration of each EPS monosaccharide were xylose (28.2%), rhamnose (18.4%), glucose (18.0%), galactose (8.7%), arabinose (4.7%), fucose (7.2%), glucosamine (3.6%), mannose (6.1%), glucuronic acid (2.4%), galacturonic acid (1.4%), traces of galactosamine, fructose, and ribose. In another study by Medina-Cabrera et al. (2021), the microalgae *P. sordidum* and *Porphyridium purpureum* were cultivated in 4.5 l bioreactors. At the end of the cultivation, soluble EPS was collected from the cell-free supernatant, resulting in a concentration of 0.21 and 0.17 g/l for *P. sordidum* and *P. purpureum* cultures, respectively.

20.3 Biological Activity of Algal Polysaccharides

Microalgae polysaccharides are alternative biocompounds for application in pharmaceutical, cosmetic, food, and biotechnological fields. This potential is mainly due to their antioxidant, antitumor, anti-inflammatory, antiviral, anticoagulant, and antimicrobial activities (Costa et al. 2021; Morais et al. 2022). These properties depend on several factors, such as molecular weight and the chemical structure of the polysaccharide (Colusse et al. 2022; Parwani et al. 2021).

Antioxidant activity is one of the most explored biological activities of microalgae polysaccharides. In addition to molecular weight and chemical structure, salt ions are also influential factors in antioxidant activity since they can change the three-dimensional structure of the compound and expose antioxidant sites (Burg and Oshrat 2015; Colusse et al. 2022). Sun et al. (2009) observed that the low molecular weight (6.55 kDa) of *Porphyridium cruentum* EPS fragments contributes to higher antioxidant activity compared to high molecular weight fractions (61–256 kDa), in addition to having a higher content of sulfate (17.3%). *Scenedesmus*, *Chlorococcum*, and *Chlorella pyrenoidosa* are also capable of producing EPS with antioxidant properties, inhibiting 1,1-diphenyl-2-picrylhydrazyl and hydroxyl radicals (Zhang et al. 2019a, 2019b).

The interest in natural compounds with antitumor activity has increased the demand for research in this area. Microalgae polysaccharides are highlighted by showing that these compounds inhibit cell lines of different cancers. These biocompounds, for example, inhibit the proliferation of human leukemic monocyte lymphoma cell lines (U937) (Sadovskaya et al. 2014). EPS from *Chlorella vulgaris*, *Chlorella zofingiensis* (Zhang et al. 2019a), *C. pyrenoidosa*, *Scenedesmus* sp., and *Chlorococcum* sp. (Zhang et al. 2019b) exhibited an antitumor effect on human colon cancer cell lines. In addition to the potential to inhibit the proliferation of HeLa cells from cervical cancer, Zhong et al. (2021) suggested that *Chlorella* EPS may regulate the genes that cause cancer cells to resist apoptosis (BCL-2 gene), thus allowing cell death. Parra-Riofrío et al. (2020) demonstrated the cytotoxic effect of sulfated EPS from *Tetraselmis suecica* on breast adenocarcinoma, lung cancer, and leukemia cell lines. Besides, polysaccharides isolated from *Spirulina*, *Aphanothece*, *Nostoc*, *Aphanizomenon*, and *Synechocystis* also have antitumor properties (Parwani et al. 2021). EPS produced by *C. vulgaris* pointed out its pharmacological potential for asthma as it was effective against airway inflammation (Barboríková et al. 2019; Capek et al. 2020). Decreased proinflammatory cytokine IL-4 levels (Barboríková et al. 2019) and increased production of interleukin-12 and interferon-γ (Capek et al. 2020) explained the observed anti-inflammatory property for *Chlorella* EPS. Zampieri et al. (2020) evaluated the *in vivo* anti-inflammatory potential of EPS isolated from *Phormidium*. These authors associated recovery from chemically and mechanically induced inflammation with the regulation of NF-κB transcription factor signaling pathways. In another study, polysaccharides from the *Aphanothece sacrum* also presented anti-inflammatory activity (Motoyama et al. 2016).

Sulfated polysaccharides are known for their antiviral property, inhibiting the activities of retroviral reverse transcriptases. As a result, these compounds interfere with the absorption and penetration of virus into host cells (Parwani et al. 2021; Talyshinsky et al. 2002). The anionic charge and the presence of sulfate groups contribute to the antiviral property of microalgae polysaccharides (Ahmadi et al. 2015). According to Raposo et al. (2015), sulfated polysaccharides can interact with the virus due to their chemical similarity to the cell receptors to which the virus binds. In this context, EPS produced by *P. purpureum* and *Spirulina platensis* showed activity against vaccinia and ectromelia viruses (Radonić et al. 2010). The sulfated polymer isolated from *S. platensis* is known as calcium spirulan. This polysaccharide has antiviral (Parwani et al. 2021) and anticoagulant properties (Mousavian et al. 2022).

Spirulina EPS aqueous extract reduced the growth of *Staphylococcus epidermis* and *Salmonella enterica* ser. Typhimurium. On the other hand, the methanolic extract of EPS showed activity against *Pseudomonas aeruginosa*, *S. enterica* ser. Typhimurium, and *Micrococcus luteus* (Challouf et al. 2011). Bhatnagar et al. (2014) reported the antibacterial activity of EPS from *Tolypothrix tenuis* and *Anabaena* against *P. aeruginosa*, *Escherichia coli*, *Staphylococcus aureus*, and *Bacillus licheniformis*. EPS from *Porphyridium marinum* was also shown to have antibacterial effect against *E. coli*, *Salmonella enteritidis*, and to a lesser extent on *S. aureus* Methicillin Resistant. If the EPS from *P. marinum* was found unable to affect *Candida albicans* growth, in contrast it can prevent biofilm formation (Gargouch et al. 2021). In another study, the EPS of *Gloecapsa* sp. *Gacheva* 2007/R-06/1 and *Synechocystis* sp. showed potential against foodborne pathogens (Najdenski et al. 2013). Thus, the different biological properties of microalgae polysaccharides actuate new opportunities for their use in several areas. Furthermore, these statements make microalgae a promising and relevant source of bioactive compounds.

20.4 Parameters that Iinfluence the Polysaccharides Production by Microalgae

The polysaccharides content is variable depending on the microalgae strain and culture conditions. Besides, depending on the photosynthetic microorganisms, this compound can be excreted in the medium or remain linked to the cells (Phélippé et al. 2019). The studies demonstrated that the parameters chemical and physical have high importance on final

Table 20.1 Polysaccharide content for different strains of microalgae, cultivation conditions, and bioreactors.

Microalga/ Cyanobacteria	Cultivation conditions	Bioreactor type	Total polysaccharide	References
Mixed microalgae	Two-stage cultivation: complete nutrient medium followed by nitrogen free, sulfur free, or phosphorus free medium; 175 mmol/m^2/s; 24 h light: 0 h dark; 24 °C; 10 d	Scott bottle (500 ml)	Polysaccharide intracellular (starch) content variation 28–50% (w/w)	Danesh et al. (2017)
Chlamydomonas biconvexa	Residue addition-vinasse: diluted (50%) and clarified (100%); 400 mmol/m^2/s; 12 h light:12 h dark; 37 °C; air supplemented with 5% CO_2 was provided; 8 d	Airlift flat plate (13 l)	Polysaccharide intracellular (total carbohydrates) content variation 11–31% (w/w)	Santana et al. (2017)
Amphora sp.	Natural seawater þ sterile nutrient solution with a salinity range of 35–233 ppt; 150 mmol/m^2/s; 12 h light:12 h dark; 25 °C.	Flasks (250 ml)	Polysaccharide intracellular (total carbohydrates) content variation 15–43% (w/w)	Ishika et al. (2018)
Chlorella zofingiensis; Chlorella vulgaris	Mixotrophic cultivation (BG-11; 5.0 g/l glucose); 40 μE/m^2/s; 25 °C; 5 d	Flask	EPS - 208.4 mg/l (*C. zofingiensis*) and 364.3 mg/l (*C. vulgaris*)	Zhang et al. (2019a)

polysaccharides content (Table 20.1). According to Seviour et al. (2011), the strategy applied in cultivation mode depends on the moment the cell produces more polysaccharides and how long it takes to achieve.

Regarding the stress conditions applied to microalgae and cyanobacteria, the most significant inducers of polysaccharides/EPS production are nitrogen source, salinity, and luminosity (Costa et al. 2021). Besides, these parameters can be divided into chemical and physical, such as culture medium, salinity, micronutrients, and culture conditions, such as temperature, luminosity, and cultivation's modes.

20.4.1 Chemical Parameters

The nutritional optimization is one of the most important parameters that influence the production of polysaccharides by microalgae, especially the limitation of nutrient supply, as nitrogen and phosphorus (Delattre et al. 2016). On the other hand, biomass production can be reduced. Thus, it is necessary to combine conditions that stimulate biomass production with adverse growth parameters, as stress conditions imposed to promote de polysaccharides production (González-Fernández and Ballesteros 2012; Deamici et al. 2021).

The most common nitrogen source used in the culture medium is sodium nitrate ($NaNO_3$). Otherwise, the studies in the literature also tested ammonium, urea, potassium nitrate, ammonium chloride (NH_4Cl), and arginine, combined or not. Mota et al. (2013) cultivated *Cyanothece* sp. CCY 0110 with $NaNO_3$ as nitrogen source and observed the improvement of the released polysaccharide production, achieving 1.8 g/l. For *A. platensis*, the nitrate limitation induces specifically the glycogen fraction increase with a decrease of the EPS fraction (Phélippé et al. 2019). *Nostoc flagelliforme* was cultured with four different nitrogen sources by Han et al. (2017), using urea, $NaNO_3$, NH_4Cl, and arginine. The best result was in the cultivations with urea (increase in 217.3% of the EPS content). *Botryococcus braunii* UC58 was cultured with three different nitrogen sources (nitrate, ammonium, and urea), and the nitrate (2.5 g/l) improved the EPS concentration (Lupi et al. 1994).

Other components that may cause a significant effect in polysaccharides production are $NaHCO_3$ and glucose. Vergnes et al. (2019) cultivated *Arthropira platensis* with $NaHCO_3$ and observed a positive influence on EPS production and flocculation with almost 90% harvest efficiency. On the other hand, the use of glucose in the mixotrophic cultivation of *N. flagelliforme* did not interfere with EPS production (Lama et al. 1996; Shen et al. 2018).

A high salinity medium is also reported as an improver of polysaccharides content since high salt concentrations may cause stress on microalgal cells due to excess sodium and chloride ions (Vo et al. 2020). Some studies show that this stress can increase the production of this compound since the culture in a medium with high salinity could be a protective cellular response. Chen et al. (2006) demonstrated that *Microcoleus vagiantus* produced 63% more EPS with high amount of salt, the sodium chloride (NaCl 0.3–0.7 mol/l), while the microalga *Cyanothece* sp. also showed an increase in the EPS production when cultivated with high salinity (70 g/l) (Chi et al. 2007). Vo et al. (2020) cultivated *Chlorella* sp. with seawater (3.5% more salt than freshwater) that induced the EPS production 30 times. Many authors reported the increased EPS productivity through NaCl stress (Borah et al. 2018; Jindal et al. 2011, Shen et al. 2018), while other studies reported no significant effect on production (Khattar et al. 2010; Mota et al. 2013).

20.4.2 Physical Parameters

Each microalga strain demands a specific cultivation with optimum values of the parameters, such as temperature, luminosity, aeration, and photoperiod. This strategy is evaluated depending on the purpose of the cultivation and the downstream process, always aiming to increase the variables of interest. Changing the cultivation mode is used as a strategy to increase biomass and biocompounds production as the EPS. Many studies reporting the batch mode as the best one to obtain EPS from cyanobacteria instead the fed-batch (Tan et al. 2012; Di Pippo et al. 2013). Di Pippo et al. (2013) cultivated cyanobacterial strains in different cultivation modes. The batch mode for *Trichormus variabilis* VRUC162 and *Anabaena augstumalis* VRUC163 produced more capsular polysaccharides, while *Calothrix* sp. VRUC 166 and *Nostoc* sp. 167 synthesized more released polysaccharides. As well known, nutrient limitation tends to increase the EPS production (Pereira et al. 2009) and, in batch cultures, the medium was not renewed, that led to potential decrease in the nutrient concentration.

The light intensity and color also influence the polysaccharides/EPS production and composition. The effects of light are confirmed by (Aikawa et al. 2012; Ge et al. 2014; Han et al. 2014). The microalga *Nostoc* sp. cultivated under 80 μE/m^2/s enhanced the EPS and capsular polysaccharides in its biomass (25% and 20%, respectively) compared to 40 μE/m^2/s (Ge et al. 2014). A further increase in the light intensity was evaluated by Aikawa et al. (2012) that cultivated *A. platensis* under 700 μmol/m^2/s with deprivation of nitrate source and observed the maximum production of glycogen (1.03 g/l and glycogen productivity of 0.29 g/l/d). These results were associated with the influence of both parameters (higher light intensity and low nitrate concentration). Han et al. (2014) evaluated wavelengths of light (red 660 nm, yellow 590 nm, green 520 nm, blue 460 nm, purple 400 nm) on *N. flagelliforme* cells. This study concluded that blue and red lights application in the cultivations improved EPS and capsular polysaccharides production compared to the white light. On the other hand, yellow light decreased its production.

Temperature has also been reported to induce growth of microalga and EPS production (Yu et al. 2010; Mota et al. 2013). *Scytonema tolypothrichoides* and *Tolypothrix bouteillei* were cultivated in a petri dish under different crossed gradients of temperature (8–40 °C) and light intensity (3–21 W/m^2) to obtain the optimal range for growth and EPS production. The temperature was the most important controlling factor (Kvíderova et al. 2018). Phélippé et al. (2019) realized meta-bibliographic research to determine main cultivation factors for the EPS and glycogen production in *A. platensis*. The results demonstrated that it is not easy to define one condition to increase the polysaccharides/EPS production. There are many different operating conditions and metholodologies applied. Moreover, each microalga presents a different behavior faced to stress conditions. However, the authors concluded that light intensity is the main factor, and the higher ratio EPS/glycogen was found with lower intensities (Phélippé et al. 2019).

As exposed, stress on microalga culture increases polysaccharides/EPS production; however, reduce biomass concentration/yield, resulting in a problem if the aim is to obtain both. Therefore, it is important to evaluate the engineering design and know the optimum conditions for each microorganism.

20.5 Algal Bioreactors

The structure of PBRs affects microalgae cultivation (Chanquia et al. 2021). The configuration of these PBRs needs efficient devices for the adequate supply of light, temperature, and agitation of the culture medium and contribute to the capture and reduction of CO_2 dissipation (Zhou et al. 2020). These PBRs can be classified as open or closed (Figure 20.1) and enable the cultivation of microalgae in the production of biomass, remediation of wastewater, and obtaining compounds with high-added value such as polysaccharides (Gaignard et al. 2019; Zhou et al. 2020; Deamici et al. 2021). Although PBR are widely used in microalgal biotechnology, they are still being studied and improved to optimize processes in the production of polysaccharides (Ekelhof and Melkonian 2017; Deamici et al. 2021).

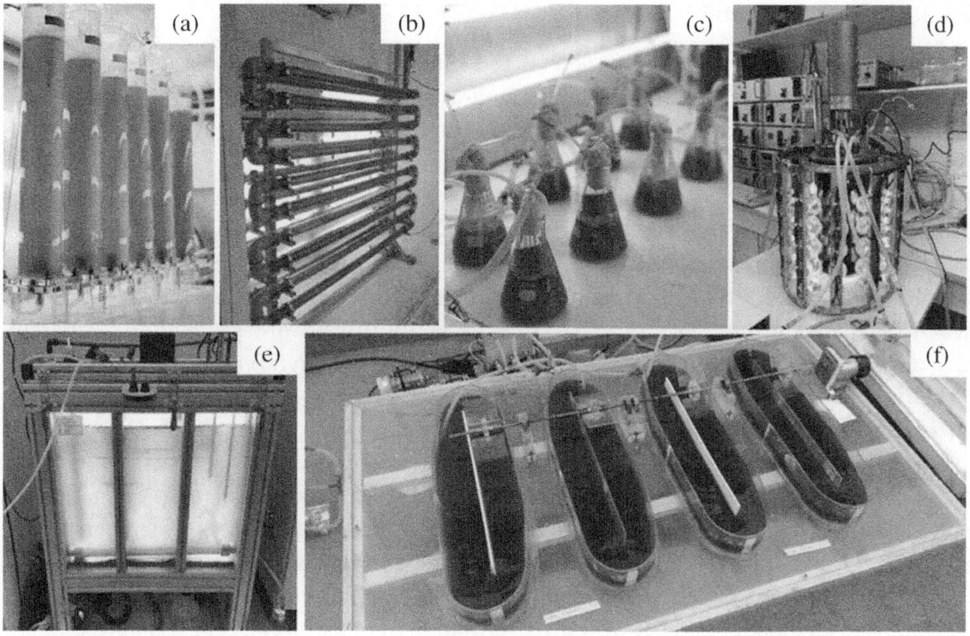

Figure 20.1 Types of bioreactors for microalgae cultivation: (a) vertical tubular, (b) horizontal tubular, (c) erlenmeyer, (d) cylindrical, (e) plate, and (f) raceway.

20.5.1 Open System

Open systems, such as raceway or circular tanks, when compared to closed systems, are more used in large-scale microalgae cultivation, since they entail lower costs of biomass production, in terms of construction, operation, and maintenance (Chanquia et al. 2021). Deamici et al. (2021) cultivated *Chlorella fusca* LEB 111 in a raceway (5 l) with static magnetic fields and obtained an increase in starch content of up to 23% in the microalgae biomass. The growth of microalgae (2.30 g/l; 24% higher than the control condition) and the increase of the polysaccharide were possible under the monitoring of temperature (34 °C), light intensity (300 μmol photons/m^2/s), and optimization of the emission of the magnetic intensity (25 mT; 24 hours).

Spirulina sp. LEB 18 was grown in open tanks (250l) by Andrade et al. (2019) to obtain biomass with a carbohydrate content of up to 58% (w/w). According to these researchers, these metabolites were promising to produce bioethanol. In this segment, de Jesus et al. (2019) monitored EPS production during the growth of *Spirulina* sp. LEB 18 in a raceway (250 l) evaluated the productivity and composition of the exopolymers obtained. EPS production occurred in all phases of microalgae growth (10 times higher than the final concentration of microalgae biomass in the culture). The researchers highlighted the bioflocculant potential of EPS and its high thermal stability. It should be noted that strategies for producing polysaccharides from microalgae grown in these open PBRs are not widely carried out. Most studies on this topic have performed microalgae cultures in closed PBRs, as they allow greater control of the parameters that potentiate the increase of this metabolite (Ekelhof and Melkonian 2017).

20.5.2 Closed System

Closed PBRs, including tubular, flat plate, and column bioreactors, allow better control of temperature, pH, and reduction of contamination by other microorganisms. Besides, these PBRs enable higher biomass productivity and optimization of microalgal cell recovery processes due to the smaller volume of culture medium required (Zhou et al. 2020). In addition, these systems allow greater control of gas exchange between the crop and the atmosphere, occupy less space, and can be installed indoors or outdoors, under natural and/or artificial light (Johansen 2012; Morais et al. 2015). According to Bezerra et al. (2012), tubular-type closed PBRs allow better light distribution and photosynthetic efficiency when compared to raceways. The researchers justify this advantage to the transparent structure of the materials of these PBRs (glass, plastic, or acrylic), which enable greater penetration of light throughout the reactor. These PBRs can also be classified according to their arrangement: vertical, horizontal, or inclined (Wang et al. 2012).

Some closed PBR designs have been improved to promote microalgae growth and increase polysaccharides synthesis (Ekelhof and Melkonian 2017; Deamici et al. 2021). The quality and quantity of light, for example, is another parameter that is studied by researchers in the closed PBR. This PBR can be adapted to emit controlled artificial light in order to stimulate and/or modify the polysaccharide composition in microalgae cultures (Phélippé et al. 2019). According to Phélippé et al. (2019), the glycogen and EPS contents of *A. platensis* were strongly affected by the light intensity of the closed PBR, with the EPS/glycogen ratios decreasing from 3.1 at 100 $\mu mol_{h\nu}/m^2/s$ to 0.6 at 1200 $\mu mol_{h\nu}/m^2/s$. The authors also observed a significant decrease in the EPS uronic acids for high light incidences. In addition to controlling this parameter, closed PBRs can be programmed to provide controlled temperature and pH, CO_2 emission, and magnetic fields at different concentrations and intensities (Deamici et al. 2021).

Deamici et al. (2021) proposed a new concept of using static magnetic fields to induce polysaccharide synthesis in *A. platensis* SAG 21, cultivated in closed PBR (glass bottles (1 l)) and magnets. The authors evaluated which polysaccharide fraction could be potentiated by static magnetic fields and the possible changes in the EPS composition. From this, it was possible to observe that the production of EPS by the microalgae did not increase significantly, and the global composition was similar to the control (80–86% of neutral sugars and 13–19% of uronic acids). However, a significant change in the composition of EPS was reported with an increase in the content of fucose, arabinose, rhamnose, galactose, and glucuronic acid and a decrease in glucose.

20.6 Conclusions and Future Perspectives

The knowledge about microalgae and cyanobacteria producing EPS has strongly increased during the last decades. The diversity of structures and biological activities of these macromolecules have made them attractive for potential applications in several industrial sectors such as pharmaceutical, cosmetical, food, or agricultural sectors. However, commercial products are still to develop, and lot of work remains to perform. If systems production and culture conditions are partially mastered for some strains, the production cost remains too high to be sustainable. Indeed, open ponds allow a low-cost production, whereas closed

PBR are required for control and constant composition, prerequisite of these commercial applications. Some improvements are thus still to develop to increase production yields, probably by a better comprehension of the metabolic pathways involved. Additionally, even if of the number of producing strains described has strongly increased, the phylogenetic diversity of microalgae and cyanobacteria is much more important, letting possibility to discover a lot of new and original structures. If some biological activities are described at laboratory scale, only a few commercial products are available, mainly for the cosmetical field, due to some regulatory issues such as "novel foods" list and a lack of clinical trials for pharmaceutical applications. Besides the strong potential of these polymers, the route for effective use in industrial sectors will then still be long.

Acknowledgments

This research was developed within the scope of the Capes-PrInt Program (Process # 88887.310848/2018-00). The authors also are grateful to the Coordenação de Aperfeiçoamento de Pessoal de Nível Superior – Brasil (CAPES) – Finance Code 001 and Program CAPES-COFECUB (Sv 945/19).

References

Ahmadi, A., Moghadamtousi, S.Z., Abubakar, S. et al. (2015). Antiviral potential of algae polysaccharides isolated from marine sources: a review. *BioMed Research International* 2015: 1–10.

Aikawa, S., Izumi, Y., Matsuda, F. et al. (2012). Synergistic enhancement of glycogen production in *Arthrospira platensis* by optimization of light intensity and nitrate supply. *Bioresource Technology* 108: 211–215.

Aikawa, S., Ho, S.H., Nakanishi, A. et al. (2015). Improving polyglucan production in cyanobacteria and microalgae via cultivation design and metabolic engineering. *Biotechnology Journal* 10 (6): 886–898.

Alvarez, A.L., Weyers, S.L., Goemann, H.M. et al. (2021a). Microalgae, soil and plants: a critical review of microalgae as renewable resources for agriculture. *Algal Research* 54: 102200.

Alvarez, X., Alves, A., Ribeiro, M.P. et al. (2021b). Biochemical characterization of *Nostoc* sp. exopolysaccharides and evaluation of potential use in wound healing. *Carbohydrate Polymers* 254: 117303.

Andrade, B.B., Cardoso, L.G., Assis, D.J. et al. (2019). Production and characterization of *Spirulina* sp. LEB 18 cultured in reused Zarrouk's medium in a raceway-type bioreactor. *Bioresource Technology* 284: 340–348.

Barboríková, J., Šutovská, M., Kazimierová, I. et al. (2019). Extracellular polysaccharide produced by *Chlorella vulgaris* – chemical characterization and anti-asthmatic profile. *International Journal of Biological Macromolecules* 135: 1–11.

Barsanti, L. and Gualtieri, P. (2014). *Algae: Anatomy, Biochemistry, and Biotechnology*, 1–48. Boca Raton: CRC Press, Taylor & Francis Group.

Bernaerts, T.M., Gheysen, L., Foubert, I. et al. (2019). The potential of microalgae and their biopolymers as structuring ingredients in food: a review. *Biotechnology Advances* 37 (8): 107419.

Bezerra, R.P., Matsudo, M.C., Sato, S. et al. (2012). Effects of photobioreactor configuration, nitrogen source and light intensity on the fed-batch cultivation of *Arthrospira* (*Spirulina*) *platensis*. Bioenergetic aspects. *Biomass and Bioenergy* 37: 309–317.

Bhatnagar, M., Parwani, L., Sharma, V. et al. (2014). Exopolymers from *Tolypothrix tenuis* and three *Anabaena* sp. (Cyanobacteriaceae) as novel blood clotting agents for wound management. *Carbohydrate Polymers* 99: 692–699.

Borah, D., Nainamalai, S., Gopalakrishnan, S. et al. (2018). Biolubricant potential of exopolysaccharides from the cyanobacterium *Cyanothece epiphytica*. *Applied Microbiology and Biotechnology* 102: 3635–3647.

Borjas Esqueda, A., Gardarin, C., and Laroche, C. (2022). Exploring the diversity of red microalgae for exopolysaccharide production. *Marine Drugs* 20 (4): 246.

Brezeştean, I., Bocăneală, M., Gherman, A.M.R. et al. (2021). Spectroscopic investigation of exopolysaccharides purified from *Arthrospira platensis* cultures as potential bioresources. *Journal of Molecular Structure* 1246: 131228.

Burg, A. and Oshrat, L.O. (2015). Salt effect on the antioxidant activity of red microalgal sulfated polysaccharides in soy-bean formula. *Marine Drugs* 13 (10): 6425–6439.

Capek, P., Matulová, M., Ŝutovská, M. et al. (2020). *Chlorella vulgaris* α-L-arabino-α-L-rhamno-α, β-D-galactan structure and mechanisms of its anti-inflammatory and anti-remodelling effects. *International Journal of Biological Macromolecules* 162: 188–198.

Challouf, R., Trabelsi, L., Ben Dhieb, R. et al. (2011). Evaluation of cytotoxicity and biological activities in extracellular polysaccharides released by cyanobacterium *Arthrospira platensis*. *Brazilian Archives of Biology and Technology* 54: 831–838.

Chanquia, S.N., Vernet, G., and Kara, S. (2021). Photobioreactors for cultivation and synthesis: Specifications, challenges, and perspectives. *Engineering in Life Sciences* 13: 1618–0240.

Chen, L.-Z., Li, D.-H., Song, L.R. et al. (2006). Effects of salt stress on carbohydrate metabolism in desert soil alga *Microcoleus vaginatus* Gom. *Journal of Integrative Plant Biology* 48: 914–919.

Chentir, I., Hamdi, M., Doumandji, A. et al. (2017). Enhancement of extracellular polymeric substances (EPS) production in *Spirulina* (*Arthrospira* sp.) by two-step cultivation process and partial characterization of their polysaccharidic moiety. *International Journal of Biological Macromolecules* 105: 1412–1420.

Chi, Z., Su, C.D., and Lu, W.D. (2007). A new exopolysaccharide produced by marine *Cyanothece* sp. 113. *Bioresource Technology* 98: 1329–1332.

Choi, Y.Y., Patel, A.K., Hong, M.E. et al. (2019). Microalgae bioenergy carbon capture utilization and storage (BECCS) technology: an emerging sustainable bioprocess for reduced CO_2 emission and biofuel production. *Bioresource Technology Reports* 7: 100270.

Colusse, G.A., Carneiro, J., Duarte, M.E.R. et al. (2022). Advances in microalgal cell wall polysaccharides: a review focused on structure, production, and biological application. *Critical Reviews in Biotechnology* 42 (4): 562–577.

Costa, J.A.V., Lucas, B.F., Alvarenga, A.G.P. et al. (2021). Microalgae polysaccharides: an overview of production, characterization, and potential applications. *Polysaccharides* 2: 759–772.

Danesh, A.F., Ebrahimi, S., Salehi, A. et al. (2017). Impact of nutrient starvation on intracellular biochemicals and calorific value of mixed microalgae. *Biochemical Engineering Journal* 125: 56–64.

Deamici, K.M., de Morais, M.G., Santos, L. et al. (2021). Static magnetic fields effects on polysaccharides production by different microalgae strains. *Applied Sciences* 11: 5299.

Delattre, C., Pierre, G., Laroche, C. et al. (2016). Production, extraction and characterization of microalgal and cyanobacterial exopolysaccharides. *Biotechnology Advances* 34: 1159–1179.

Di Pippo, F., Ellwood, N.T.W., Gismondi, A. et al. (2013). Characterization of exopolysaccharides produced by seven biofilm-forming cyanobacterial strains for biotechnological applications. *Journal of Applied Phycology* 25: 1697–1708.

Ekelhof, A. and Melkonian, M. (2017). Microalgal cultivation in porous substrate bioreactor for extracellular polysaccharide production. *Journal of Applied Phycology* 29: 1115–1122.

Gaignard, C., Laroche, C., Pierre, G. et al. (2019). Screening of marine microalgae: Investigation of new exopolysaccharide producers. *Algal Research* 44: 101711.

Gargouch, N., Elleuch, F., Karkouch, I. et al. (2021). Potential of exopolysaccharide from *Porphyridium marinum* to contend with bacterial proliferation, biofilm formation, and breast cancer. *Marine Drugs* 19: 66–84.

Ge, H., Xia, L., Zhou, X. et al. (2014). Effects of light intensity on components and topographical structures of extracellular polysaccharides from the cyanobacteria *Nostoc* sp. *Journal of Micrbiology* 52: 179–183.

Georgiev, Y.N., Batsalova, T.G., Dzhambazov, B.M. et al. (2021). Immunomodulating polysaccharide complexes and antioxidant metabolites from *Anabaena laxa, Oscillatoria limosa* and *Phormidesmis molle. Algal Research* 60: 102538.

Gonzales, K.N., Troncoso, O.P., Torres, F.G. et al. (2020). Molecular α-relaxation process of exopolysaccharides extracted from *Nostoc commune* cyanobacteria. *International Journal of Biological Macromolecules* 161: 1516–1525.

González-Fernández, C. and Ballesteros, M. (2012). Linking microalgae and cyanobacteria culture conditions and key-enzymes for carbohydrate accumulation. *Biotechnology Advances* 30: 1655–1661.

Gouveia, J.D., Moers, A., Griekspoor, Y. et al. (2019). Effect of removal of bacteria on the biomass and extracellular carbohydrate productivity of *Botryococcus braunii. Journal of Applied Phycology* 31 (6): 3453–3463.

Goyal, M., Baranwal, M., Pandey, S.K. et al. (2019). Hetero-polysaccharides secreted from *Dunaliella salina* exhibit immunomodulatory activity against peripheral blood mononuclear cells and RAW 264.7 macrophages. *Indian Journal of Microbiology* 59 (4): 428–435.

Gügi, B., Costaouec, T., Burel, C. et al. (2015). Diatom-specific oligosaccharide and polysaccharide structures help to unravel biosynthetic capabilities in diatoms. *Marine Drugs* 13: 5993–6018.

Han, P.-P., Sun, Y., Jia, S.-R. et al. (2014). Effects of light wavelengths on extracellular and capsular polysaccharide production by *Nostoc flagelliforme. Carbohydrate Polymers* 105: 145–151.

Han, P., Yao, S., Guo, R. et al. (2017). Influence of culture conditions on extracellular polysaccharide production and the activities of enzymes involved in the polysaccharide sythesis of *Nostoc flagelliforme. RSC Advances* 7: 45075–45084.

Hong, M.E., Chang, W.S., Patel, A.K. et al. (2019). Microalgae based carbon sequestration by converting LNG-fired waste CO_2 into red gold astaxanthin: the potential applicability. *Energies* 12: 1718.

Hussein, M.H., Hamouda, R.A., Elhadary, A.M.A. et al. (2019). Characterization and chromium biosorption potential of extruded polymeric substances from *Synechococcus mundulus* induced by acute dose of gamma irradiation. *Environmental Science and Pollution Research* 26: 31998–32012.

Ishika, T., Bahri, P.A., Laird, D.W. et al. (2018). The effect of gradual increase in salinity on the biomass productivity and biochemical composition of several marine, halotolerant, and halophilic microalgae. *Journal of Applied Phycology* 30: 1453–1464.

Jesus, C.S., Jesus Assis, D., Rodriguez, M.B. et al. (2019). Pilot-scale isolation and characterization of extracellular polymeric substances (EPS) from cell-free medium of *Spirulina* sp. LEB-18 cultures under outdoor conditions. *International Journal of Biological Macromolecules* 124: 11061114.

Jindal, N., Singh, D.P., and Khattar, J.I.S. (2011). Kinetics and physico-chemical characterization of exopolysaccharides produced by the cyanobacterium *Oscillatoria formosa*. *World Journal of Microbiology and Biotechnology* 27: 2139–2146.

Johansen, M.N. (2012). *Microalgae: Biotechnology, Microbiology and Energy*, 2012. New York: Nova Science Publishers, Inc.

Khattar, J.I.S., Singh, D.P., Jindal, N. et al. (2010). Isolation and characterization of exopolysaccharides produced by the cyanobacterium *Limnothrix redekei* PUPCCC 116. *Applied Biochemistry and Biotechnology* 162: 1327–1338.

Kvíderova, J., Kumar, D., Lukavský, J. et al. (2018). Estimation of growth and exopolysaccharide production by two soil cyanobacteria, *Scytonema tolypothrichoides* and *Tolypothrix bouteillei* as determined by cultivation in irradiance and temperature crossed gradients. *Engineering in Life Sciences* 19: 184–195.

Lama, L., Nicolaus, B., Calandrelli, V. et al. (1996). Effect of growth conditions on endo-and e polymer biosynthesis in *Anabaena cylindrica* 10 C. *Phytochemistry* 42: 655–659.

Liberman, G.N., Ochbaum, G., Mejubovsky-Mikhelis, R. et al. (2020). Physico-chemical characteristics of the sulfated polysaccharides of the red microalgae *Dixoniella grisea* and *Porphyridium aerugineum*. *International Journal of Biological Macromolecules* 145: 1171–1179.

Lupi, F.M., Fernandes, H.M.L., Tomé, M.M. et al. (1994). Influence of nitrogen source and photoperiod on exopolysaccharide synthesis by the microalga *Botryococcus braunii* UC 58. *Enzyme and Microbial Technology* 16: 546–550.

Markou, G., Angelidaki, I., and Georgakakis, D. (2012). Microalgal carbohydrates: an overview of the factors influencing carbohydrates production, and of main bioconversion technologies for production of biofuels. *Applied Microbiology and Biotechnology* 96: 631–645.

Medina-Cabrera, E.V., Gansbiller, M., Rühmann, B. et al. (2021). Rheological characterization of *Porphyridium sordidum* and *Porphyridium purpureum* exopolysaccharides. *Carbohydrate Polymers* 253: 117237.

Morais, M.G., Vaz, B.S., Morais, E.G. et al. (2015). Biologically active metabolites synthesized by microalgae. *BioMed Research International* 2015: 1–15.

Morais, M.G., Morais, E.G., Cardias, B.B. et al. (2020). Microalgae as a source of sustainable Biofuels. In: *Recent Developments in Bioenergy Research* (ed. V.J. Gupta, R.C. Kuhad, H. Treichel and S. Rodriguez-Cout), 253–271. Elsevier.

Morais, M.G., Santos, T.D., Moraes, L. et al. (2022). Exopolysaccharides from microalgae: production in a biorefinery framework and potential applications. *Bioresource Technology Reports* 18: 101006.

Moreira, J.B., Kuntzler, S.G., Bezerra, P.Q.M. et al. (2022). Recent advances of microalgae exopolysaccharides for application as bioflocculants. *Polysaccharides* 3 (1): 264–276.

Mota, R., Guimarães, R., Büttel, Z. et al. (2013). Production and characterization of extracellular carbohydrate polymer from *Cyanothece* sp. CCY 0110. *Carbohydrate Polymers* 92: 1408–1415.

Motoyama, K., Tanida, Y., Hata, K. et al. (2016). Anti-inflammatory effects of novel polysaccharide sacran extracted from cyanobacterium *Aphanothece sacrum* in various inflammatory animal models. *Biological and Pharmaceutical Bulletin* 39: 1172–1178.

Mousavian, Z., Safavi, M., Azizmohseni, F. et al. (2022). Characterization, antioxidant and anticoagulant properties of exopolysaccharide from marine microalgae. *AMB Express* 12 (1): 1–16.

Najdenski, H.M., Gigova, L.G., Iliev, I.I. et al. (2013). Antibacterial and antifungal activities of selected microalgae and cyanobacteria. *International Journal of Food Science & Technology* 48: 1533–1540.

Nur, M.M.A., Swaminathan, M.K., Boelen, P. et al. (2019). Sulfated exopolysaccharide production and nutrient removal by the marine diatom *Phaeodactylum tricornutum* growing on palm oil mill effluent. *Journal of Applied Phycology* 31: 2335–2348.

Parra-Riofrío, G., García-Márquez, J., Casas-Arrojo, V. et al. (2020). Antioxidant and cytotoxic effects on tumor cells of exopolysaccharides from *Tetraselmis suecica* (Kylin) butcher grown under autotrophic and heterotrophic conditions. *Marine Drugs* 18: 1–23.

Parwani, L., Bhatt, M., and Singh, J. (2021). Potential biotechnological applications of cyanobacterial exopolysaccharides. *Brazilian Archives of Biology and Technology* 64: e21200401.

Patel, A.K., Choi, Y.Y., and Sim, S.J. (2020a). Emerging prospects of mixotrophic microalgae: way forward to bioprocess sustainability, environmental remediation and cost-effective biofuels. *Bioresource Technology* 300: 122741.

Patel, A.K., John, J., Hong, M.E. et al. (2020b). A sustainable mixotrophic microalgae cultivation from dairy wastes for carbon credit, bioremediation and lucrative biofuels. *Bioresource Technology* 313: 123681.

Patel, A.K., Vadrale, A.P., Singhania, R.R. et al. (2022). Algal polysaccharides: current status and future prospects. *Phytochemistry Reviews* 48: 451–463.

Pereira, S., Zille, A., Micheletti, E. et al. (2009). Complexity of cyanobacterial exopolysaccharides: composition, structures, inducing factors and putative genes involved in their biosynthesis and assembly. *FEMS Microbiology Reviews* 33: 917–941.

Pereira, S.B., Sousa, A., Santos, M. et al. (2019). Strategies to obtain designer polymers based on cyanobacterial extracellular polymeric substances (EPS). *International Journal of Molecular Sciences* 20 (22): 5693.

Phélippé, M., Gonçalves, O., Thouand, G. et al. (2019). Characterization of the polysaccharids chemical diversity of the cyanobacteria *Arthrospira platensis*. *Algal Research* 38: 101426.

Radonić, A., Thulke, S., Achenbach, J. et al. (2010). Anionic polysaccharides from phototrophic microorganisms exhibit antiviral activities to Vaccinia virus. *Journal of Antivirals & Antiretrovirals* 2: 51–55.

Raposo, M.F.J., De Morais, A.M.B., and De Morais, R.M.S.C. (2015). Marine polysaccharides from algae with potential biomedical applications. *Marine Drugs* 13: 2967–3028.

Risjani, Y., Mutmainnah, N., Manurung, P. et al. (2021). Exopolysaccharide from *Porphyridium cruentum* (*purpureum*) is not toxic and stimulates immune response against vibriosis: the assessment using zebrafish and white shrimp Litopenaeus vannamei. *Marine Drugs* 19: 133.

Rossi, F. and Philippis, F. (2016). Exocellular polysaccharides in microalgae and cyanobacteria: chemical features, role and enzymes and genes involved in their biosynthesis. In: *The Physiology of Microalgae. Developments in Applied Phycology* (ed. M. Borowitzka, J. Beardall and J. Raven), 565–590. Cham: Springer.

Sadovskaya, I., Souissi, A., Souissi, S. et al. (2014). Chemical structure and biological activity of a highly branched (1→3,1→6)-β-D-glucan from *Isochrysis galbana*. *Carbohydrate Polymers* 111: 139–148.

Santana, H., Cereijo, C.R., Teles, V.C. et al. (2017). Microalgae cultivation in sugarcane vinasse: selection, growth and biochemical characterization. *Bioresource Technology* 228: 133–140.

Seviour, R.J., McNeil, B., Fazenda, M.L. et al. (2011). Operating bioreactors for microbial exopolysaccharide production. *Critical Reviews in Biotechnology* 31: 170–185.

Shen, S., Jia, S., Wu, Y. et al. (2018). Effect of culture conditions on the physicochemical properties and antioxidant activities of polysaccharides from *Nostoc flagelliforme*. *Carbohydrate Polymers* 198: 426–433.

Singh, S., Arad, S.M., and Richmond, A. (2000). Extracellular polysaccharide production in outdoor mass cultures of *Porphyridium* sp. in flat plate glass reactors. *Journal of Applied Phycology* 12: 269–275.

Sun, L., Wang, C., Shi, Q. et al. (2009). Preparation of different molecular weight polysaccharides from *Porphyridium cruentum* and their antioxidant activities. *International Journal of Biological Macromolecules* 45: 42–47.

Suzuki, E. and Suzuki, R. (2013). Variation of storage polysaccharides in phototrophic microorganisms. *Journal of Applied Glycoscience* 60: 21–27.

Talyshinsky, M.M., Souprun, Y.Y., and Huleihel, M.M. (2002). Anti-viral activity of red microalgal polysaccharides against retroviruses. *Cancer Cell International* 2: 1–7.

Tan, N., Jia, S.R., Han, P.P. et al. (2012). The open culture of *Nostoc flagelliforme* with a 25 l open pond. In: *Advanced Materials Research*, vol. 554 (ed. S. Chen, Z.-T. Liu and Q. Zeng), 1009–1012. Bäch, Switzerland: Trans Tech Publications Ltd.

Tiwari, A., Melchor-Martínez, E.M., Saxena, A. et al. (2021). Therapeutic attributes and applied aspects of biological macromolecules (polypeptides, fucoxanthin, sterols, fatty acids, polysaccharides, and polyphenols) from diatoms—a review. *International Journal of Biological Macromolecules* 171: 398–413.

Tomaselli, L. (2004). The microalgal cell. In: *Handbook of Microalgal Culture: Biotechnology and Applied Phycology* (ed. A. Richmond), 3–19. Oxford: Blackwell Science Ltd.

Uhliariková, I., Matulová, M., Košťálová, Z. et al. (2022). Lactylated acidic exopolysaccharide produced by the cyanobacterium *Nostoc cf. linckia*. *Carbohydrate Polymers 276*: 118801.

Vergnes, J.-B., Gernigon, V., Guiraud, P. et al. (2019). Bicarbonate concentration induces production of exopolysaccharides by *Arthrospira platensis* that mediate bioflocculation and enhance flotation harvesting efficiency. *ACS Sustainable Chemistry & Engineering* 7: 13796–13804.

Vo, H.N.P., Ngo, H.H., Guo, W. et al. (2020). Selective carbon sources and salinities enhance enzymes and extracellular polymeric substances extrusion of *Chlorella* sp. for potential co-metabolism. *Bioresource Technology* 303: 122877.

Wang, B., Lan, C.Q., and Horsman, M. (2012). Closed photobioreactors for production of microalgal biomasses. *Biotechnology Advances* 30: 904–912.

Wang, W.N., Li, T., Li, Y. et al. (2021). Exopolysaccharides from the energy microalga strain *Botryococcus braunii*: purification, characterization, and antioxidant activity. *Food* 11 (01): 110.

Winahyu, D.A. and Primadiamanti, A. (2021). Formulation and evaluation of the exopolysaccharide compound extract lotion from the microalgae *Spirulina* sp. *Journal of Physics: Conference Series* 1882 (1): 012107.

Wu, M., Du, M., Wu, G. et al. (2021). Water reuse and growth inhibition mechanisms for cultivation of microalga *Euglena gracilis*. *Biotechnology for Biofuels* 14 (1): 1–15.

Yang, R., Wei, D., and Xie, J. (2020). Diatoms as cell factories for high-value products: chrysolaminarin, eicosapentaenoic acid, and fucoxanthin. *Critical Reviews in Biotechnology* 40 (7): 993–1009.

Yu, H., Jia, S., and Dai, Y. (2010). Accumulation of exopolysaccharides in liquid suspension culture of *Nostoc flagelliforme* cells. *Applied Biochemistry and Biotechnology* 160: 552–560.

Zampieri, R.M., Adessi, A., Caldara, F. et al. (2020). Anti-inflammatory activity of exopolysaccharides from *Phormidium* sp. ETS05, the most abundant cyanobacterium of the therapeutic euganean thermal muds, using the zebrafish model. *Biomolecules* 10: 582.

Zhang, J., Liu, L., and Chen, F. (2019a). Production and characterization of exopolysaccharides from *Chlorella zofingiensis* and *Chlorella vulgaris* with anti-colorectal cancer activity. *International Journal of Biological Macromolecules* 134: 976–983.

Zhang, J., Liu, L., Ren, Y. et al. (2019b). Characterization of exopolysaccharides produced by microalgae with antitumor activity on human colon cancer cells. *International Journal of Biological Macromolecules* 128: 761–767.

Zhong, R., Li, J.Q., Wu, S.W. et al. (2021). Transcriptome analysis reveals possible antitumor mechanism of *Chlorella* exopolysaccharide. *Gene* 779: 145494.

Zhou, W., Lu, Q., Han, P. et al. (2020). Microalgae cultivation and photobioreactor design. In: *Microalgae Cultivation for Biofuels Production* (ed. A. Yousuf), 31–50. Academic Press.

Index

Plants as Bioreactors for Industrial Molecules, First Edition.
Edited by Santosh Kumar Upadhyay and Sudhir P. Singh.

f

k

l

m

Printed and bound by CPI Group (UK) Ltd, Croydon, CR0 4YY

26/01/2023

03184193-0001